ECONOMIC DEPOSITS AND THEIR TECTONIC SETTING

'On a high ridge between the two countries grows a tree that bears red blossoms on the Siamese side and white blossoms on the Perak side. When the red blossoms fall they turn to gold, the white blossoms on the Perak side fall and turn into tin. Could anything be more satisfactory? This theory is worth a whole book on the pneumatolytic genesis of tin, and on the question whether the gold is of magmatic origin or laterally secreted.'

(J. B. Scrivenor, *A Sketch of Malayan Mining,* 1928, p. 9; quoted by Hosking, 1973a)

Economic Deposits and their Tectonic Setting

Charles S. Hutchison, BSc, PhD, FGS, FIMM

Professor of Applied Geology
Department of Geology
University of Malaya
Kuala Lumpur
Malaysia

A WILEY-INTERSCIENCE PUBLICATION

JOHN WILEY & SONS
NEW YORK

First published 1983 by
The Macmillan Press Ltd
London and Basingstoke

Published in the USA by
Wiley-Interscience, a Division of
John Wiley & Sons, Inc., New York

Filmsetting by Vantage Photosetting Co. Ltd., Eastleigh and London

Printed in Hong Kong

ISBN 0 471-87281-4

With eternal gratitude to my father Charles McDonald Hutchison (1905–1981) for giving me educational and other opportunities denied to him.

Contents

Preface xi

Acknowledgements xiii

CHAPTER 1: INTRODUCTION 1
Some basic definitions 1
Classification and genetic terms 4
Hydrothermal solutions 5
The principal ore-forming processes 7
Magmatic processes 7
Hydrothermal processes 8
Metamorphic processes 8
Surface processes 9

CHAPTER 2: OCEANIC LITHOSPHERE AND MINERALIZATION 11
Mineralization on the ocean floor 14
Manganese nodules 15
Ophiolite 15
Podiform chromitite 19
Nickel laterite 23
Bedded ferromanganese deposits 26
Lahn–Dill iron-ore deposits 30

CHAPTER 3: A SPECTRUM OF SEA-FLOOR SULPHIDES 31
Cyprus-type 31
Cyprus 32
Newfoundland 34
Philippines 34
Turkey 34
Kuroko deposits 34
Spanish–Portuguese massive sulphide deposits 37
Some other examples of Kuroko deposits 40
Besshi-type massive sulphide deposits 40
Relationship between Kuroko and porphyry-copper deposits 42

CHAPTER 4: MINERALIZATION IN INTRACRATONIC BASINS 44
The Red Sea 44
Normal Red Sea sediments 44
Oxide facies 44
Sulphide facies 46
Sulphate facies 46
Carbonate facies 46
Silicate facies 46

Mississippi Valley-type ore deposits 47
Fluid-inclusion studies 47
Lead-isotope studies 47
Sulphur isotopes 49
Oxygen isotopes 49
Model for ore deposition 49
The Viburnum Trend, southeast Missouri 50
The Pennines of England 53
Pine Point, Northwest Territories, Canada 56
Appalachian zinc–lead deposits, eastern USA 58
Fluorite deposits 59
Mexico 59
Southern Illinois–Kentucky, USA 60
Barite 62
Bedded deposits 62
Replacement deposits 63
Vein and cavity-filling deposits 64
Residual deposits 66
Sedimentary ironstone deposits of shallow epicratonic seas 68
Middle Silurian Clinton Group ironstones 69
Europe 72
Fresh-water ferromanganese deposits 74
Uranium mineralization in continental basins 75
Sandstone uranium fields of the western USA 75

CHAPTER 5: OTHER EPICONTINENTAL SEA DEPOSITS 82
Phosphorites 82
Distribution of phosphorites 82
Petrology and classification 83
Types of phosphate deposit 85
Genesis of phosphorites 88
Potash–salt–anhydrite–sulphur evaporites 89
The Middle Devonian Prairie evaporite 89
Permian evaporites of the North Sea basin 91
The question of water depth 93
Sulphur 94
Uraniferous marine black shales 95
Swedish black shales 96
USA 96
Mercury deposits 96
The Almaden mercury deposit, Spain 97
The Idria mine, Yugoslavia 99
New Almaden mercury district, California 100

CHAPTER 6: INTRUSIVE BODIES EMPLACED IN STABLE CRATONIC TERRAIN 102
The Sudbury Basin 102
Creighton Embayment 104
Strathcona Mine 104
Levack Embayment 104
Bushveld igneous complex 104
Tectonic setting 105
The Transvaal System 107
The layered sequence 108

Epicrustal rocks 110
Economic deposits of the Bushveld complex 110
Great dyke of Zimbabwe 113
Alkaline igneous rocks 114
Apatite deposits of igneous origin 114
Mineralization in carbonatites 116
Uraniferous peralkaline plutons 119
Kimberlite and diamonds 119
Cratonic rift-related tin– (niobium–tantalum–tungsten) mineralization 123
Pegmatite mineralization 125

CHAPTER 7: BATHOLITH-ASSOCIATED MINERALIZATION 130
General features of batholiths 130
Batholith-type and orogenic association 131
Alpinotype orogenies 131
Andinotype orogenies 131
Hercynotype orogenies 131
Relationships between granitoids and hydrothermal deposits 133
Temperature of hydrothermal formation and alteration 134
The Malay Peninsula 135
The eastern volcano–plutonic arc 136
Main-range belt (Malaysia) 139
Phuket, Thailand 143
The tungsten–(tin) deposits of China 144
Quartz vein deposits (southern Kiangsi-type) 145
Skarns of scheelite (southeast Hunan-type) 145
Wolframite–scheelite reticulated veins (east Kwantung-type) 145
Vein-type zoning 147
Cornwall, southwest England 148
Types of deposits 148
The economic minerals 150
Pasto Buena, north Peru 150
Panasqueira, Portugal 152
Hamme tungsten district, North Carolina 154
Strata-bound tungsten and tin deposits 154
Namaqualand (Southwest Africa) 154
Eastern Alps 154
Norway 154
Zimbabwe 154
Sangdong mine, South Korea 154
King Island, Bass Strait, Australia 155
Renison–Bell tin field 155
Cleveland mine 156
Mount Bischoff tin ore body 156
Iron contact–hydrothermal deposits 156
Philippines 157
Bukit Ibam Iron Mine, Peninsular Malaya 157
China 159
Japan 159
Canada 159
Cornwall, Pennsylvania 159
Uranium deposits associated with granitoids 160
The Hercynian granitic deposits 161
Spokane Mountain region, Washington State 165
Rössing, Namibia 167

CHAPTER 8: EPIGENETIC DEPOSITS OF VOLCANIC AND EPIZONAL PLUTONIC ASSOCIATION 168
Porphyry-copper deposits 168
Economic significance 168
Classification 168
Host-rock characteristics 169
Stable isotope studies as an indicator of derivation 171
Comparison between island-arc and continental deposits 171
Hydrothermal alteration–mineralization 171
The Panguna deposit of Bougainville 173
Copper-bearing skarns 175
Porphyry-molybdenum deposits 177
Urad and Henderson mines 177
The Endako deposit, British Columbia 177
The Climax mine, Colorado 178
Discussion 178
Epithermal vein deposits 180
Characteristic features 180
Mineralogy 181
Textures 181
Depth of formation 181
Age relations 181
Nature of the mineralizing solutions 181
Origin of the metals 181
Distribution of the deposits 182
North-American cordillera 182
Western Pacific 185
Gold–antimony association 186
Antimony 188
Volcanogenic tin deposits 188

CHAPTER 9: ARCHAEAN-STYLE MINERALIZATION 196
Archaean tectonics 196
Geotectonic setting 196
Intrusive assemblages 196
Volcanic assemblages 196
Sedimentary assemblages 196
Successions 198
Typical structures and metamorphism 198
Mineralization associated with greenstone belts 198
Iron formations 198
Synvolcanic nickel sulphide deposits 202
Gold in greenstone belts 204
Massive base-metal sulphides 207
Antimony in greenstone belts 210
Model of Archaean mineralization 212

CHAPTER 10: PROTEROZOIC-STYLE MINERALIZATION 214
Iron formations 214
Biwabik–Gunflint Iron Formation 214
Supergene-enriched iron formations 217
Kiruna-type iron ore bodies 219
The Swedish ores 219
Pea Ridge, Missouri 222
Bedded manganese ore 222

Kalahari manganese field 223
Madhya Pradesh and Maharashtra, India 223
West Africa 225
Massive sulphide deposits 225
Mineralization in the Belt–Purcell aulacogen 225
Coeur d'Alene mining district 228
The great Australian stratiform ore bodies 229
The Central African Copper Belt 234
Early Proterozoic hydrothermal uranium deposits 238
Alligator Rivers–Rum Jungle, Northern Australia 238
Mary Kathleen deposit, Queensland 242
Athabasca basin uranium deposits 243

CHAPTER 11: SURFICIAL DEPOSITS OF CONTINENTAL AREAS 249
Bauxite 249
Classification 249
Geographical distribution 250
Distribution in time 250
Composition 250
Origin of bauxite 251
Jamaica 251
Australia 253
Bedded deposits of China 253
The Naurzum sedimentary deposit 254
Uranium deposits in calcrete 254
Placer deposits 256
Tin placers of Southeast Asia 257
Gold placers 261
Diamond placers 262
Basal Proterozoic uranium placers 264
Ilmenite, rutile and zircon placers 272

CHAPTER 12: FUELS 276
Petroleum and natural gas 276
Source beds 276
Primary migration 281
Secondary migration 284
Some examples of oil fields 285
Classification of oil traps 292
Classification of oil basins 293
Worldwide oil generation and preservation 299
Coal 300
Origin 300
Tectonic setting of coal basins 303
Climate and coal formation 303
Sapropelic coals 303
Petrology of coal 305
Coalification 306
The Carboniferous coal fields of the Rhine–Ruhr, Germany 308
Pennsylvanian coal basins of the eastern USA 310
Sydney Basin, New South Wales, Australia 310
Uranium resources 310
Strata-controlled deposits 313
Structure or fracture-controlled (vein-type and similar deposits) 315
Intrusion controlled 315

APPENDIX: THE ECONOMIC COMMODITIES 317
Alphabetically listed, giving the uses, the world production and major producers, with cross-references to geological descriptions

REFERENCES 328

Author and Subject Index 357

Preface

The great majority of university teachers continually express their dissatisfaction with the available textbooks, and economic geology is not unusual in this respect. The present choice is rather limited — the best available texts are either somewhat outdated or not sufficiently comprehensive. This volume is an attempt to rectify this unsatisfactory situation.

The collections of papers edited by Strong (1976) and Tarling (1981), and the outstanding work of my friend Andrew Mitchell (for example, Mitchell, A.H.G. and Garson, 1976) gave the impetus for me to write this text. It was felt that more modern tectonic terminology needs to be used in the interpretation of ore deposits. This book describes the spectrum of ore deposits, coal, oil and gas, and attempts to say something meaningful about their tectonic setting. There is no separate discussion of the theory of plate tectonics, for it is felt that by now most readers will be familiar with the broad concepts. For those who are not, an introduction may be obtained from Bird (1980). However, at appropriate places, certain aspects of plate tectonics are amplified, wherever they are thought to be relevant to the understanding of the particular deposits.

The book is not aimed at students beginning the study of geology, and it is assumed that the reader will already have a broad geological background. The book is comprehensive enough to form the basis for either a one or two semester undergraduate course in economic geology. Primarily it is aimed at senior undergraduates; there is no discussion of ore mineralogy and ore petrology, which are well described by Ramdohr (1980) and Uytenbogaart and Burke (1971), but emphasis is placed on the geology and details of the deposits themselves. It is hoped that the text will provide a valuable reference for graduate students and practising geologists employed by surveying and commercial companies. With this readership in mind, I have attempted to include a comprehensive up-to-date list of nearly 1000 references to guide the reader to deposits other than those selected for amplification in the text.

One thing became obvious during my writing – that it is possible to keep abreast of the rapid developments in the field of economic geology through the pages of two outstanding journals: *Economic Geology* (abbreviated to *Econ. Geol.* in the text) and the *Bulletin of the American Association of Petroleum Geologists* (abbreviated to *Amer. Assoc. Petrol. Geol. Bull.*), both of which are highly recommended.

In conclusion, the general textbooks detailed in the following references provide useful additional reading on mineral deposits: Lindgren (1933), Bateman (1950), Routhier (1963), Park and MacDiarmid (1970), Stanton (1972), Smirnov (1976), Dixon (1979), Jensen and Bateman (1979) and Evans (1980). Additional reading on fossil fuels is detailed in: Stach *et al.* (1975), Tissot and Welte (1978), Tiratsoo (1979), Crelling and Dutcher (1980) and Hobson and Tiratsoo (1981).

Ithaca, New York, 1981

C.S.H.

Acknowledgements

I gratefully acknowledge the large amount of typing assistance given by Ms Fauzian Hanim Mohad. Saidi and Ms Zaimah bt. Ahanad Saleh in Kuala Lumpur and Ms Gemma Tod in Ithaca. The figures were drafted by Mr. Mohammad bin Haji Majid (who also suggested the design for the cover), Mr. S. Sriniwass, Mr. Ching Yu Hay and Mr. Roslin bin Ismail. Mr. Jaafar bin Haji Abdullah produced the photo-reductions.

I am grateful to both the University of Malaya and Cornell University for facilities to carry out my research and writing. I have benefited greatly from discussions with colleagues both in Kuala Lumpur and Ithaca, but would particularly like to mention the help given by Dr. Teh Guan Hoe and Dr. Allan Gibbs.

I wish to thank the editors and staff of the Macmillan Press for their help in producing this book.

1

Introduction

What is economic geology? It may be defined as the study of geological bodies and materials that may be useful to mankind. The materials included are fuels, metallic and non-metallic minerals, rocks and water. This book will exclude considerations of water and building materials, and will concentrate on ores and fuels.

Some people may say that economic geology is a subdiscipline within the geological sciences, but this is not strictly correct, for it is unlike other subdisciplines, such as palaeontology, igneous petrology, structural geology, and so on, which all go to make up the science of geology, for each of these can exist independently. Economic geology cannot exist as a separate discipline, for it has to be based on the broad spectrum of the geological sciences. A good economic geologist is also a good geologist, that is a person with a wide background experience of the geological sciences. It must therefore be stressed that no book can serve as a self-contained economic geology text, for to understand economic deposits it is necessary to understand a wide spectrum of geological processes, and this can only be achieved by having a wide background experience in most branches of the geological sciences. Petroleum, coal, evaporite and phosphorite geology require a strong background in sedimentology and structural geology. Ore geology is so extremely varied as to encompass deposits that have been laid down or concentrated by igneous, metamorphic and sedimentary processes, and by circulating magmatic and ground-water hydrothermal fluids, sometimes further concentrated by weathering processes. These cover almost the whole spectrum of the geological sciences. Fundamental also to the study of economic geology is a good background in structural geology, for the foregoing processes are all likely to have been controlled or modified by geological structures. Therefore a suitable background for a practising economic geologist would include igneous petrology, sedimentology, structural geology and, of lesser necessity, metamorphic petrology. Geochemistry is also important because one is dealing with mobilization and concentration of metals and other materials. This would include both organic and inorganic geochemistry to cover the spectrum from fuels to ores. It is clear, therefore, that economic geology is not a branch of, but a culmination of a broad training in, the geological sciences. This book cannot hope to do more than outline the various processes that may have been active in the formation of economic deposits. The necessary background in the various branches of the geological sciences may be obtained from the textbooks detailed in the following references. *Structural geology*: Dennis (1972), Hobbs, B.E. *et al.* (1976), Spencer, E.W. (1977); *sedimentology*: Friedman and Sanders (1978), Reading (1978); *igneous petrology*: Cox, K.G. *et al.* (1979), Yoder (1979); *metamorphic petrology*: Miyashiro (1973), Mason, R. (1978); *geochemistry*: Krauskopf (1979).

Although not necessary for the understanding of the deposits, their discovery and assessment will depend on a variety of techniques, including geophysical methods, exploration geochemistry and eventual drilling. The books detailed in the following references summarize these useful methods: Parasnis (1973), Levinson (1974), Peters (1978), Reedman (1979), Sittig (1980).

Some Basic Definitions

Geological and mining terms may be checked against a good glossary, such as the *Glossary of Geology* (Bates and Jackson, 1980) or *A Dictionary of Mining, Mineral, and Related Terms* (Thrush, 1968).

Deposit is a rather unsatisfactory term that is widely used in geology. It may mean material laid down by water, or clastic material coming out of suspension, or it may refer to precipitation from solution in response to different chemical conditions. Such deposits are sedimentary. The term also refers to ore material concentrated by igneous processes, either as cumulates of heavy minerals sinking to the base of a magma chamber, or to concentrations of minerals in the final phases of crystallization towards the top of a magma chamber. The term is also applied to mineral concentrations precipitated from hydrothermal solutions that pass through the pores or fissures in a rock. *Economic deposits* therefore comprise any materials of economic interest that have been concentrated by any process to an extent and degree of concentration that invite exploitation.

Ore may be defined as a naturally occurring material from which a mineral or minerals of economic importance may be extracted. In most cases the term applies to metallic deposits and is then qualified, for example, iron ore or zinc ore. Ore is made up of *ore minerals*, which are the part of the ore, usually metallic, that is desirable for extraction. Ore is usually composed of both ore minerals and *gangue*, the unwanted matrix that dilutes the grade of the economic commodity, and needs to be extracted and rejected. It is separated in ore-dressing plants, and discarded in the form of tailings. Frequently an ore concentrate may contain minor amounts of a valuable commodity, which are recovered only at the smelting stage.

There is a tendency for definitions of ore to include the phrase 'which may be mined at a profit'. Obviously this need not apply to state-run mines that remain open for political reasons or to provide employment.

The major non-renewable fuels are coal, petroleum, natural gas and nuclear materials. They provide the energy necessary for the smelting of the ores into metals, their further fabrication, electricity generation, transportation and all other energy-using activities upon which civilization is dependent.

Protore is the mineral material in which initial but uneconomic concentrations of the metals have occurred. The protore may be further concentrated into ore by processes such as near-surface reaction with ground water to give an enriched subsurface concentration above the protore. The process is known as *supergene enrichment*. Tropical weathering of some rock-types may be sufficient to concentrate economic deposits of some metals in the residual blanket laterite or bauxite.

Economically mineable concentrations of ore minerals are known as ore deposits, ore bodies, or ore shoots. *Ore shoot* is a large elongate, pipe-like or chimney-like mass of ore within a deposit. It may be part of a vein or part of a disseminated deposit. It represents the most valuable part of the deposit, that which is economically capable of being mined. The size and dimensions of an ore shoot vary with the cost of the metal commodity at any particular time (*Figure 1.1*). *Lode* is defined as an ore deposit consisting of a zone of veins. It therefore has the same meaning as ore shoot when applied to a veined deposit. However, ore shoot is a more generally applicable term that encompasses both veined, massive and disseminated mineralization.

Both lode and ore shoot are outlined by the *cut-off grade*, which is the concentration of the ore mineral or minerals below which the mining operation is unprofitable. Obviously if there is an increase in demand for a commodity, with an attendant increase in marketable price, the cut-off grade may be decreased and the outline of the ore shoot expanded (*Figure 1.1*). Many mining operations, initially set up with a particular cut-off grade, continue mining lower-grade material as the price of the commodity increases and mining methods become more efficient.

Ore reserves mean the amount of ore that has been defined by geological investigations, including drilling of the deposit before mining. As mining proceeds, considerably more geological information is obtained, so that improved estimates of the reserves become available. Hence reserves of a mine are dependent on the up-to-date known geological information, and they will be modified as this information becomes more accurate through development. Some reserves may not be realistic in the sense that they include unmineable material. For example, the ore shoots may extend downwards to depths at which mining operations are impossible because of high temperatures. These are potential reserves, which are uneconomic at present, but may be mined in future if the demand is great enough for the company to air-condition the mine workings. Ore reserves of any mining region may be further classified as proved, probable and possible. Proved ore has been so thoroughly sampled that there can be no doubt of its outline. Elsewhere sampling may not have been so thorough, so that the other categories become proved only after further information and exploitation. The concepts of ore shoot, cut-off and reserves are illustrated in *Figure 1.1*, based on a hypothetical example of a hydrothermal ore deposit.

Hydrothermal simply means heated water. Many different mineral deposits are hydrothermally deposited, meaning that they have been deposited from hot aqueous solutions that have circulated through pores or channelway within the country rocks. The ore deposition may be dissemi-

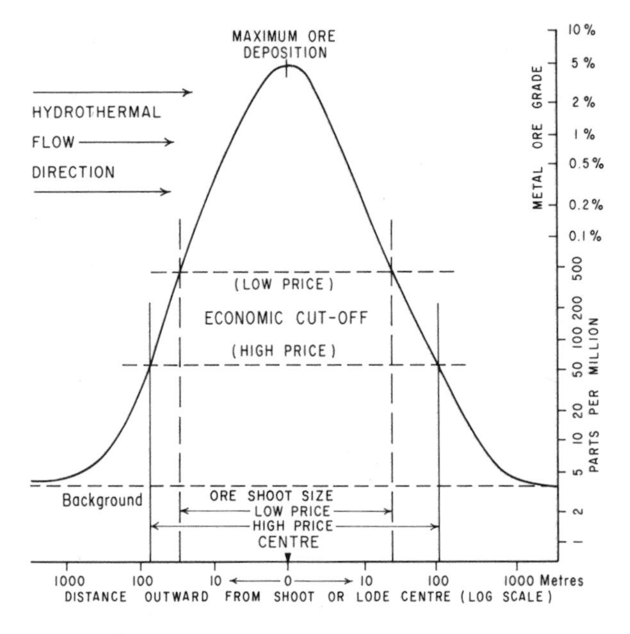

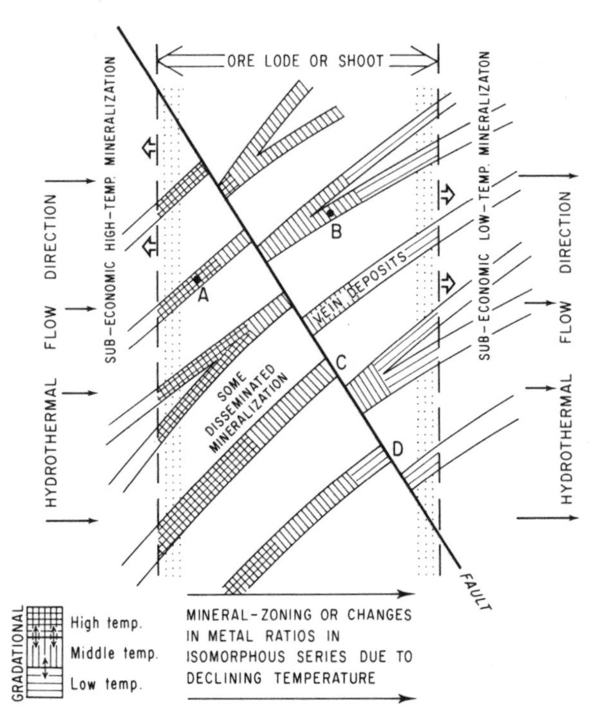

Figure 1.1. Top: schematic representation of the log normal distribution of mineralization deposited from a moving hydrothermal solution. The ore zone (ore shoot) is defined on the basis of economics (cost of extraction and commodity price). Bottom: minerals and mineral chemistry show a zonation down-flow, which can be used to decide whether the mining is approaching the centre of the shoot or its edge. A faulted vein at D may therefore not be worth the effort of tracing farther to the right, but that at C will be because it is still in high-grade ore.

nated within the pores or, in the case of fractured rocks, deposited within fissures in the rocks. The hydrothermal solutions may be demonstrably related to igneous intrusions or extrusions. The water may have been magmatic, or meteoric ground water, or water released from dehydration reactions resulting from metamorphic reactions. In many cases an igneous body, intruded high within the crust, has provided a localized heat source that has acted as a centre for hydrothermal circulation through the country rocks. In many other cases the hydrothermal solutions are entirely ground water or formation water that already occupied the pore spaces before the igneous heat centre was emplaced. The only contribution from the magmatic body may have been its heat. In other cases, late-stage hydrothermal–magmatic solutions may have evolved from the magma itself, but with declining temperature there is usually mixing with meteoric water. In the case of hydrothermal circulation, there will be a definite, directed flow through the rocks, away from the heat source, and therefore the solutions will cool in that direction (*Figure 1.1*). The ore elements will be deposited within a temperature range characteristic of the particular elements present in solution, and the pH and E_h conditions of the environment. In hydrothermal deposits, the thermal gradient will be demonstrated by a change in mineralogy or mineral composition down the thermal gradient. Ideally, the grade of the ore deposition defines a log normal distribution parallel to the flow direction. The actual boundary between a shoot and barren ground is arbitrary, as shown in *Figure 1.1*. Therefore the reserves of any particular deposit need to be defined against a quoted cut-off grade.

Figure 1.1 also illustrates that details of the already mined vein within a lode or shoot can be used to predict the direction in which the ore grade is likely to fall. Thus, a comparison of the detailed mineralogy or mineral chemistry at points A and B will show that the ore content is likely to decline in grade towards the left at A, and mining towards the right of B will still be profitable but of a different character from A. Veins are commonly cut off by faults, and a major problem in mining geology is whether the fault represents the end of mineralization, or whether it is profitable to look for the vein continuation beyond the fault. The example at C shows that the vein mined from the left and terminating at the fault should be sought towards the right, because the mineralogical details and metal ratios in the ore minerals clearly indicate that the mineralization still continues significantly towards the right. It is of commercial interest to try to find the vein continuation to the right of the fault. On the other hand, if the fault occurs near the boundaries of the shoot or lode (D) there is no reason to look for the vein continuation.

Classification and Genetic Terms

The purpose of a classification should be to provide a framework for thought and discussion and to define a useful terminology. The classification should undoubtedly aid the exploration geologist to find new deposits by clearly defining the environments of various types of deposit. Unfortunately, established classification systems have often hindered the understanding of deposits because mine geologists have commonly forced their deposits into rigid established pigeonholes, in a way similar to the universal application of Alpine models to all orogens. It is now known that this is a futile exercise. Perhaps the most attractive way to classify deposits at the present time of imperfect knowledge is by comparison with well-known deposits. Thus it may be said that the deposit is a Mount Isa-type, or an Almaden-type, and so on.

A good illustration of misclassification is the Witwatersrand gold field of South Africa (Chapter 11), which traditionally was regarded as *epigenetic*, meaning that the mineralization was of later origin than the enclosing rocks and imposed on the rocks by infiltrating hydrothermal solutions. It was envisioned that these hydrothermal fluids rose from an undefined mysterious source at depth and deposited gold and uranium minerals hydrothermally and epigenetically. This serious misclassification (Pretorius, 1976) undoubtedly resulted from a worldwide obsession with epigenetic theories of ore genesis, because of the great influence of Lindgren (1933), whose classification was so logically argued that most practising geologists uncritically followed it. Misclassifications also resulted from the fact that few sedimentologists ever became ore geologists. In recent years it has been clearly demonstrated that the gold and uranium of the Witwatersrand field were detritally sedimented and that the deposit accordingly is a *palaeo-placer*. A placer is a surficial mineral deposit formed by mechanical concentration of mineral particles transported from an eroding source area. The transporting agent is usually water in an alluvial or lacustrine environment. Had sedimentologists worked in the Witwatersrand gold field, its sedimentary character would long have been recognized and this would undoubtedly have aided the mining strategy. Traditionally, mining geologists had a background of hard-rock geology with little knowledge of sedimentology. I can quote my Southeast Asian experience to further illustrate the problem.

The geologists of the Quaternary alluvial tin fields had always traditionally trained in hard-rock schools of mines where they had little exposure to sedimentology. Their professional employment later consisted of evaluating placer cassiterite deposits for which they were little trained, and for which their hard-rock background was of little value. Accordingly, the extensive placer deposits of Southeast Asia have not been properly studied from a sedimentological point of view, and the mining strategy has suffered severely.

Some economic deposits, such as coal and evaporites, are predictable, because it is possible to see present-day actualistic examples forming in swamps and land-locked seas, respectively. However, most economic deposits are not predictable; unless it is known how they occur in existing active environments, and models are made for their genesis, such deposits cannot be predicted because of the lack of actualistic models. Very few deposits can be seen in the process of formation. It is known that fumaroles in volcanic regions often contain sulphur concentrations, and accordingly one mentally links volcanism with ore-forming processes. It is also expected that magma differentiation in a plutonic environment must concentrate ore minerals in the early stages by cumulate settling of the high-temperature heavy minerals. It is further known that certain elements have a low affinity for silicates, and therefore concentrate in the late residual melts, which then evolve into hydrothermal solutions that must accumulate in the cupolas of plutons. Although these things cannot be seen in the process of formation, one can imagine them to be a logical development of igneous behaviour, and therefore batholith-related deposits are largely predictable.

Only in recent years, with the increase in oceanographic research, did it become apparent that sediments deposited in certain parts of the ocean floors may become enriched in metals while they are accumulating. The agent responsible is the hot brine that circulates through the uncompacted sediments and underlying basalt. The circulation is driven by localized high heat-flow fluxes, associated with spreading axes. Examples are in the Red Sea, the Salton Sea and many areas on and adjacent to the mid-ocean ridges. Hence the possibility of syngenetic metal enrichment during the deposition or diagenesis of sediments can now be appreciated. *Syngenetic* means formed contemporaneously with the enclosing rocks. Such mineral deposits are likely to be conformable with the sedimentary bedding or other primary structures. However the deeper parts of a syngenetic ore deposit may have epigenetic characteristics, resulting from the deeper circulation of hydrothermal fluids through the underlying older rocks. *Syndiagenetic* means formed during compaction and diagenesis of the sediments. The term *strata-bound* applies to a mineral deposit confined to a single stratigraphic unit. The term can apply to a stratiform deposit or to a randomly oriented ore body contained within a single stratigraphic unit. *Stratiform* applies to a layered mineral deposit of either sedimentary or igneous origin. The layered cumulate deposits of chromite in the Bushveld complex, which resulted from igneous 'sedimentation' from a cooling magma, are said to be stratiform. Truly syngenetic, synsedimentary ore deposits, in which the mineralization is controlled by the sedimentation processes, are stratiform. Strata-bound deposits may be synsedimentary or may have been introduced epigenetically by later hydrothermal fluids flowing parallel to the bedding, or through the whole rock mass, depositing the metals in favourable strata-controlled porous horizons or horizons whose chemistry was favourable to reaction with the ore solutions to cause precipitation. The growing importance of strata-bound and stratiform deposits is shown by the increasing number of books specialized to such deposits. In particular I can recommend the huge seven-volume work edited by Wolf (1976b).

Hydrothermal Solutions

Hydrothermal fluids play an extremely important role in the concentration and modification of ore deposits. A good general review may be obtained in Barnes (1979). Hydrothermal fluids may have many origins. When a mineral deposit is found, the hydrothermal solutions from which it was precipitated have long departed. Fortunately, however, there are remnants of these solutions, trapped as small *fluid inclusions* in such minerals as quartz, sphalerite, cassiterite, calcite and fluorite. A study of these fluids in the inclusions can be made under a microscope on a heating–cooling stage. This gives valuable information about the nature of the fluids and the temperature at which the minerals were precipitated (Roedder, 1979). Analyses of the fluids shows that hydrothermal solutions varied widely in composition and salt concentration, but generally tended to be brines whose salinities waned with falling temperature. It is also known that the deposit-forming fluids are similar chemically to fluids encountered in present-day geothermal systems. Therefore a study of present-day systems gives an understanding of the spectrum of hydrothermal deposits because actualistic models are available. Such fluids are widespread throughout the crust and deposits form only where the fluids encounter suitable fissures or suitably reactive rocks. There are four types of hydrothermal waters (Skinner, B.J., 1979): surface or *meteoric* water, including rain water, lake and river water, sea water and ground water — 'meteoric' implies that the water recently had a connection with the atmosphere; *connate* water, which has been trapped in the pores of sedimentary rocks since the time of deposition — this category also includes old waters trapped in joints within igneous rocks, and deeply penetrating meteoric water that has long been out of touch with the atmosphere; *metamorphic* water released from regions undergoing dehydrating metamorphic reactions; and *magmatic* water, released from the late stages of crystallizing igneous plutons.

Measurement of the relative abundance of the isotopes of hydrogen and oxygen (H/D and $^{16}O/^{18}O$ ratios) in hydrothermal solutions can provide useful information on the sources of the water in any particular system (Faure, 1977). However, uncertainties in interpretation arise from rock–water reactions, and from the mixing of waters of different origins. An isotopic continuum therefore results, and it is still often impossible to prove that a single kind of water produced any given deposit (White, D.E., 1974). Similar deposits can be formed from quite different kinds of waters. The actual source of the water is apparently not the controlling factor in the formation of hydrothermal ore deposits (Skinner, B.J., 1979). Much more important is the geotectonic setting in which the hydrothermal solutions circulated. For example, volcanic activity in island arcs is likely to drive circulation cells of sea water that has filled the pore spaces of the submarine volcanic pile. The hydrothermal circulation cells are likely to emanate on the sea floor to form submarine hydrothermal mineral deposits within the sediments, which are likely to be tuffaceous because of the proximity of the volcanic arc. Thus *Kuroko*-type deposits are formed. The ore deposits may be concentrated close to the volcanic centres to give proximal deposits, or carried far from the centres to give distal deposits. Much submarine hydrothermal ore formation appears to be closely and genetically associated with felsic phases of bimodal volcanic activity, and this is a feature of ore genesis that has occurred throughout geological time. It is found in Archaean greenstone belts, as well as in the Miocene green-tuffs of Japan. *Porphyry*-type ore deposition occurs when the volcanoes occur on continental margins or on island arcs that are mature enough to be subaerial. Thus, as an island arc grows from submarine to subaerial, Kuroko-type deposits may be followed in time by porphyry-type deposits, but cordilleran plate margins are devoid of kuroko-type deposits because the volcanic activity was always subaerial. Thus it is the tectonic setting that determines the style of mineralization, rather than the source of the water in the hydrothermal fluids. In both kuroko-type and porphyry-type deposits, the hydrothermal fluids may start off as predominantly magmatic, but the fluids become quickly diluted by connate and meteoric water, and eventually the role of magmatic water disappears. The source of the metals that are concentrated into the ore deposit is therefore not to be regarded as only the plutonic stock that first supplied the magmatic water. As the circulation cells enlarge, the hydrothermal fluids are capable of leaching metals from the older volcanic and volcaniclastic pile that makes up the whole of the country rocks.

High geothermal gradients associated with constructive rift margins are also known to drive hydrothermal circulation through the upper crust. The water in such cases is sea water and connate or formation water, which become highly concentrated brines. In the case of intracratonic rifts such as the Red Sea, the spreading rift is covered with thick sediments, which become highly enriched in sulphide mineralization. The source of the metals is assumed to be the underlying basaltic crust over and around the spreading axis, leached by large-scale circulation cells centred on the active heat sources. Thus, some of the mineralization is syngenetic with the sedimentation, but beneath the surface the mineralization is epigenetic. One should therefore not try to pursue the distinction between epigenetic and syngenetic to its logical conclusion, for this is unrealistic in the light of an example like the Red Sea. The mineralization in the near-surface sediments is predominantly syngenetic, but strictly epigenetic in that it has been brought into the rocks soon after their deposition. Many ores are formed during sediment diagenesis, when the pore water is expelled and the sediments lithify. If these rocks originally contained metal concentrations, the textures of the ore-mineral assemblages will be modified during diagenesis, and may more resemble epigenetic rather than syngenetic textures.

Magmatic water evolved from the roof zones of granitic batholiths may permeate through the already solid outer zones of the batholiths, depositing high-temperature minerals such as cassiterite and wolframite and of necessity reacting with the country rock and causing extensive hydrothermal alteration. Extensive alteration is also characteristic of the surrounding dioritic and granodioritic stocks with which porphyry deposits are associated. If the plutons are at a high level, emplaced into an epizonal environment, the magmatic fluids may become diluted by meteoric water and circulate widely throughout the overlying country rocks, depositing their mineral load in suitable fracture systems, and batholith-initiated hydrothermal systems may therefore even break the land surface and emanate into sediments that are in the process of being deposited. In this way stratiform syngenetic shoots may be formed, which may grade downwards into epigenetic lodes.

It is known from the studies of large sedimentary basins that the migration of oil, water and gas has been effected by large-scale migration of the formation waters from the basin centre, predominantly up-dip towards the shallower sediments along the basin margins. This is a familiar mechanism for the filling of oil reservoirs. The source rocks were in the deeper parts of the basin, and the reservoirs in the younger and shallower parts. The source of the metals can be the sedimentary strata or intercalated volcanic rocks of the deeper parts of the basin. The large-scale migration of formation waters can be envisioned as carrying and concentrating metals in addition to oil and gas. The connections are shown by the petroleum contents of fluorite deposits and the uranium contents of natural petroleum. The *Mississippi Valley*-type deposits generally are located along the margins of major basins, which also contain oil and gas. In this way a similarity can be seen between the migration and trapping of petroleum and the accumulation of ore minerals such as galena, sphalerite and fluorite. The hydrothermal solutions that carried the metals were connate or meteoric waters that were heated in the deeper parts of the basins, simply on account of the higher geothermal gradients, and deeper burial, and migrated up-dip outwards towards the basin margins. The metals were deposited in favourable country rocks such as reef limestones, caves, fissures or along unconformities around the basin margins. It is clear that igneous processes played no role in the formation of such deposits. However, there is no reason to exclude igneous processes as having an influence, for an igneous intrusion into the basin sediments would have provided a suitable heat source to drive a local circulation superimposed on the larger-scale basin-wide circulation. The regional pattern would be intensified and modified. Uranium is an interesting metal that can show a complete spectrum from transportation by magmatic hydrothermal solutions to be deposited within vein systems in the outer zones of granitic batholiths, to epigenetic low-temperature deposition resulting from large-scale fluid migrations through porous sandstone horizons on the edges of major basins. In many cases the igneous connection may not exist. In others the igneous body may be limited to the role of providing the heat source to drive a circulation system of connate or meteoric waters. This is so for many gold–quartz vein deposits that are far removed from the nearest batholith. The batholith provided the heat source, and the gold was leached from large volumes of country rocks through which the solutions flowed.

In the past, it was conventional to regard all hydrothermal solutions as rising upwards from an igneous source, *per ascensum*. However, just as some oil deposits may have migrated downwards from the overpressurized source mudstones into suitable underlying reservoir rocks, so hydrothermal circulations may be capable of leaching and concentrating metals in a downwards direction, *per decensum*. Many of the traditional views need now to be applied with caution.

A generally acceptable classification of economic deposits is therefore largely impossible. Some classifications, instead of aiding thought, tend rather to provoke controversy. I wish to avoid this aspect of economic geology. In applying a comparison-type of classification, the geotectonic environment of occurrence appears to be fundamentally important. Thus, there are some deposits that are characteristically associated with granitoid batholiths, others with oceanic lithosphere. Many types of deposit have formed in environments for which no actualistic models exist. Therefore I have tried to avoid a rigid classification system, and the chapter headings represent only a broad grouping of deposit-types into geotectonic settings.

A classification by commodity is geologically useless; for example, uranium deposits may occur within sandstones and also within granites, and their origins are radically different. However, many readers may wish to refer to commodities. Therefore I have included an appendix, which gives page cross-references to descriptions of types of deposits in which particular commodities occur.

The Principal Ore-forming Processes

A brief review of the principal theories of ore genesis may serve as a useful introduction to the chapters that follow. The reader should, however, be aware that very often several processes may contribute to the formation of any particular ore body. For this reason, there may be disagreement about the interpretation of some deposits and the significance of some of their features. An epigenetic deposit may gradually pass upwards into one that is predominantly syngenetic, so the interpretation depends on the location of the features studied. A deposit-type such as the Kiruna magnetite ore, which traditionally has been regarded as intrusive magmatic, has subsequently been interpreted as volcanic exhalative, and with time the theory of formation may again revert to magmatic intrusive. Classification of ore deposits is not a static subject; like every healthy science (or art) it is in a continual state of flux, and the interpretation of any particular deposit needs to be reviewed in the light of new evidence as mining develops, and as new data become available. Deposits do not change in the human span of time, but their interpretation may change as geologists fall under the influence of contemporary thought.

Magmatic Processes

Certain minerals, which usually constitute only accessory amounts in most igneous bodies, may occur within some varieties in sufficient concentration to constitute ores.

DISSEMINATIONS

Certain categories of igneous rocks contain accessory minerals of sufficiently high economic value for their selective mining. Such minerals have not undergone any concentration, so that the deposits are referred to as *disseminated*. The alkaline igneous rocks provide good examples. Kimberlites are mined for diamonds, and other alkaline bodies yield apatite, niobium, rare earths and uranium (*see* Chapter 6).

CUMULATES

Gravity separation is the most powerful mechanism for igneous rock differentiation (Cox, K.G. *et al.*, 1979). High-temperature crystal phases may settle to the solidified floor of a magma chamber to form cumulate layers of economic importance. Less commonly, some phases may float upwards in the magma, though this may apply only to feldspar-rich and feldspathoid-rich rocks. Chromite, magnetite and platinum minerals were concentrated by gravity settling in, for example, the spectacular Bushveld igneous complex (*see* Chapter 6). Crystal concentration by cumulate settling may also have played an important part in magma chambers that formed in the gabbro layer of ophiolite complexes (Chapter 2).

Liquid Immiscibility

In the process of magma fractionation, the supernatant magma may become progressively enriched in a particular group of elements, for example, iron, copper and sulphur, leading to the immiscible separation of droplets of an ore magma within the silicate magma. Indeed the glassy matrix of many oceanic tholeiitic basalts contain spherules of rhyolitic glass that must have separated immiscibly at liquidus temperatures. Immiscible separation is therefore well known geologically. The dispersed droplets of one liquid phase may coalesce to form discrete magma bodies quite distinctly different from the parent magma. Thus it may be possible that a gabbroic magma, fractionating in a plutonic chamber, may evolve to iron overenrichment, by which stage an iron-rich magma may immiscibly separate from the daughter diorite. The separated metal-rich magma may segregate in place beneath the silicate magma, or may be independently injected into the country rocks. Possible examples of immiscibility-derived ore magmas are the copper–nickel ores of the Archaean greenstone belts (Chapter 9) and the magnetite ores of Proterozoic age (Chapter 10). The latter may have analogues within Phanerozoic volcano–plutonic arcs (Chapter 7).

Pegmatites

Late-stage igneous phenomena include pegmatite bodies containing mineral phases that have incorporated elements that have been strongly partitioned towards the late-stage liquid. Pegmatite phases of felsic magmatism may be enriched in lithium, caesium, tin and uranium (*see* Chapter 6).

Astrobleme or Cryptovolcanic Activity

It is possible that certain igneous bodies may have resulted from melting of the crust following the impact of a cosmic body. The Sudbury Complex (Chapter 6) may have such an origin.

Hydrothermal Processes

Igneous-Related

As a felsic magma crystallizes, an aqueous fluid is liberated from the magma by second boiling (Burnham and Ohmoto, 1980). Chlorine, sulphur, alkalis and chalcophile elements are strongly partitioned towards the aqueous fluid, which commonly mixes with meteoric water contained in the pore spaces, fissures and channelways of the country rocks into which the magma was intruded. The heat of the pluton drives circulation cells of this hydrothermal liquid, which has the ability to leach further metals from the rocks through which it passes, and deposit them in a process of alteration–mineralization centred around the plutonic heat source; porphyry-copper and porphyry-molybdenum deposits (Chapter 8) provide excellent examples of this process. Tin, tungsten and copper vein and skarn deposits associated with granitoid batholiths also result from a similar process (Chapter 7).

No Igneous Relation

Low-temperature hydrothermal solutions may have no genetic relationship with any igneous body but an intrusion may have acted as a heat source to drive hydrothermal circulation of meteoric water through the rocks. The process is often referred to as *lateral secretion.* Epigenetic deposits of gold and antimony may have formed by this process (*see* Chapter 8, for example). Ore and gangue minerals may have been transported long distances by brines migrating from the hotter and deeper parts of sedimentary basins, outwards and upwards towards the basin margins, to be deposited epigenetically in porous rocks, palaeo-karst features, faults and fissures. The Mississippi Valley-type deposits of lead–zinc–fluorite–barite offer excellent examples, and the brines that transported the materials apparently also transported the petroleum from the source rocks in the deeper parts of the basins to the reservoirs in the basin edges (*see* Chapter 4). Igneous activity played no role in the large-scale migration of these low-temperature hydrothermal solutions.

Metamorphic Processes

Pyrometasomatic Contact Deposits

These commonly occur where the country rock was calcareous. Mineralization in skarns may be equated with magmatic hydrothermal mineralization, because the process is largely a result of the escape of the separated aqueous solution from the magma into the country rocks. Copper, tungsten and iron deposits may form in this way (*see* Chapters 7 and 8).

PARTIAL MELTING OF CONTINENTAL CRUSTAL ROCKS

This is a powerful mechanism for recycling and concentrating certain elements by polycyclic anatectic events in volcano–plutonic arcs. Tin, tungsten, antimony, uranium and mercury may all be strongly partitioned in granitic melts relative to the restite material left behind in the lower crust (*see* Chapters 7 and 8).

NICKEL, CHROMIUM AND PLATINUM METALS

These may become concentrated in the refractory residuum resulting from withdrawal of basaltic magma formed at oceanic spreading centres by partial melting of the peridotitic upper mantle (*see* Chapter 2).

Surface Processes

MECHANICAL ACCUMULATION

Durable heavy minerals, released from outcrops by weathering and erosion, may be transported by fluvial systems and deposited in placer concentrations. Gold, cassiterite, rutile, ilmenite and diamond are important examples (*see* Chapter 11). Uranium was transported detritally only before the atmosphere became oxygenated in the early Proterozoic era.

SEDIMENTARY PRECIPITATES

Precipitation of particular elements in suitable sedimentary environments, with or without the intervention of bacteria, may lead to important sedimentary economic deposits. The precipitation is predominantly of material carried by the water in solution, but may also include material carried in suspension. Evaporite deposits form when the water becomes saturated in salt (*see* Chapter 5); phosphorites form when the waters are enriched in phosphate (*see* Chapter 5); banded-iron formations (Chapters 9 and 10), and certain iron-bedded and manganese-bedded deposits (Chapter 4) are formed by sedimentary precipitation.

RESIDUAL PROCESSES

Exposure to the atmosphere under a tropical climate may result in intense leaching of some elements and hence the relative enrichment of others in the regolith. Bauxite (Chapter 11) and nickel laterite (Chapter 2) are outstanding examples of ores formed by surface weathering of suitable primary materials that were not ores themselves.

SECONDARY OR SUPERGENE ENRICHMENT

This is a process that may affect most ore bodies to some degree. After the deposit and its country rocks have been uplifted near the land surface, circulating ground water may leach some of the metals out of the section of the ore body that lies above the ground-water table. These dissolved elements are deposited in that part of the ore body that lies beneath the ground-water table, leading to a considerable supergene enrichment in metal values. The zone of supergene enrichment usually overlies primary mineralization, which frequently may not be of ore grade (protore). Supergene-enriched iron ores are described in Chapter 10.

Gossan is the oxidized outcropping cellular mass of limonitic material and gangue that overlies sulphide ore shoots, formed by supergene leaching of the sulphide ore. Gossans may extend some distance beneath the surface, as far down as the ground-water table, and in some cases the gossan may constitute ore.

VOLCANIC-EXHALATIVE

Exhalations of hydrothermal solutions directly on to the floor of a sea or lake cause syngenetic mineral enrichments of the rocks that floor the basin. This is more common in marine basins, and leads to the formation of syngenetic stratiform sulphide ore bodies and, if sedimentation is proceeding at or near the exhalative centre, the ore bodies so formed are referred to as synsedimentary. The exhalations may come out into the water around an igneous centre, or submarine caldera, in which case the mineralized rocks are volcanic and the deposit is called proximal. Alternatively the exhaled hydrothermal solutions may mix with the sea water, to be carried far from the exhalative centre, leading to mineral enrichment of sedimentary or tuffaceous sediments distal from the volcanic centres. A large number of stratiform ore bodies are considered to have resulted from exhalations. Present-day examples of such mineralization occur along the spreading axes of major oceans (Chapter 2) and within the deeps of the Red Sea (Chapter 4). Submarine exhalations lead to a wide spectrum of stratiform ore bodies, including Kuroko-type, Cyprus-type and Besshi-type deposits (Chapter 3), and many of the great Precambrian stratiform ore bodies are considered to have resulted from submarine exhalations of hydrothermal solutions connected with a phase of rhyolitic magmatism (Chapters 9 and 10). However, the volcanic-exhalative source of some deposits is frequently inferred, rather then proven, for massive sulphide deposits may occur entirely within sedimentary sequences. Bacteria frequently play an important role in the fixing of mineralization within exhalative deposits.

2

Oceanic Lithosphere and Mineralization

The floor of the ocean has many major features such as continental rises, oceanic basins, mid-ocean ridges, submerged ridges, island ridges, deep-sea trenches, volcanic peaks and sea mounts, all with different crustal and mantle structures beneath them. Despite this variability, it is possible to propose an average or standard ocean structure (Worzel, 1974). This structure varies quite dramatically from that of the continental areas (*Table 2.1*). The sea water is an important layer of the oceanic crust. It contains trace elements and obviously has an important role in the formation of the layer of sediments and metal concentrations on the sea floor.

Since the oceans and continents are in close isostatic equilibrium, it is important to determine whether their deduced standard structural sections should be in equilibrium. If they are, the mass per unit area for each column should be essentially the same; they can be calculated from *Table 2.1.* The total mass per unit area beneath the main continental crust would be 95.71×10^5 g cm^{-2} and beneath the normal ocean crust 95.83×10^5 g cm^{-2}. This difference would be equal to only about 5 mgal of gravity anomaly (1 mgal=0.001 cm s^{-2}), which is a reasonable agreement and shows that the average crustal models of *Table 2.1* are essentially correct.

Table 2.1 Comparison of standard oceanic and continental lithosphere

Oceanic layers					*Continental layers*				
Name	*Composition*	*Average thickness (km)*	*Seismic velocity (km s^{-1})*	*Density (g cm^{-3})*	*Name*	*Composition*	*Average thickness (km)*	*Seismic velocity (km s^{-1})*	*Density (g cm^{-3})*
Water	H_2O + dissolved elements	4.8	1.5	1.03	Water	Layer absent except on shelf seas			
Crust, layer 1	Sediments such as chert, micrite and mud	0.8	2.0	1.9	Continental crust	Tectonic complex of sedimentary, igneous and metamorphic rocks	33.7	6.36	2.84
Crust, layer 2	Ocean-floor basaltic volcanic rocks commonly pillowed	1.68	5.12	2.55					
Crust, layer 3	Plutonic basic igneous rocks with ultrabasic cumulates	4.77	6.70	2.86					
Upper Mantle	Ultrabasic rocks and serpentinite	—	8.08	3.30	Upper Mantle	Ultrabasic rocks and serpentinite	—	8.11	3.35

The tectonic elements of the oceans that are considered to be important associates of economic metal concentration will be briefly reviewed by reference to *Figure 2.1*, a diagrammatic segment through the earth's crust and upper mantle (after Hutchison, 1978a). It is assumed that most readers are familiar with plate tectonic theory, which is the most widely accepted for how the earth works. Readers unfamiliar with the theory are referred to Wyllie (1976) for an elementary introduction, to more detailed papers in Bird and Isacks (1972) and especially to the classic paper by Isacks *et al.* (1968).

New oceanic crust and upper mantle (together called lithosphere) is forming continuously along the active spreading axes, which form the oceanic rises of the major oceans (*Figure 2.1*). The mechanism is partial melting of the upper mantle directly beneath the spreading axes caused by elevated temperatures resulting from rising convection currents in the upper mantle. Newly formed crust along the active spreading axes continuously pushes the older crust away from the axes, towards the continents, so that the age of the oceanic crust increases directly away from the axes, as shown by the magnetic stripes on the ocean floor (indicated by 1 in *Figure 2.1*). This movement away from the axes (sea-floor spreading) is unlikely to be simply a push from the magma injected into the fracture system of the ocean rise. It is more likely to be a result of mantle convection beneath the oceanic plate, resulting from buoyancy forces. The question of what causes an oceanic plate, for example, the 7000 km wide Pacific Plate, to move intact clear across the ocean towards Asia has puzzled geophysicists. The question has been discussed by Bostrom (1978), who concludes that the passage of the earth's tidal bulge assists the mantle buoyancy forces that cause the convection.

Local rising currents from deep within the mantle are thought to be the cause of relatively fixed hot-spots on the earth's crust (2). Since they are small in radius and relatively fixed with respect to the deeper mantle, the effect of a hot-spot mantle current will be shown on the crust by a line as the crust moves over the relatively fixed rising current. The present position of a hot-spot is recognized as a locus of igneous activity. The background heat flux from the mantle is thought to be concentrated into the hot-spots (Crough, 1979), and over a hot-spot the lithosphere is thinned, causing isostatic uplift. The lithosphere again thickens and subsides when it passes away from the hot-spot; by that time any igneous activity has ceased. The trace of the movement of oceanic lithosphere over a hot-spot may be referred to as a hot-spot track, for example, the Emperor seamounts—the Hawaiian Ridge of the Pacific (Dickinson, 1978). Likewise the Ninetyeast Ridge of the Indian Ocean is considered to be a hot-spot track as India and the Indian Plate passed northwards (Curray *et al.*, 1982).

Oceanic lithosphere is thought to be continuously consumed at deep-sea trenches, where the spreading lithosphere inclines downwards and descends below dipping seismic zones called Benioff Zones. This consumption may be of oceanic lithosphere descending beneath oceanic lithosphere, giving rise to an islands arc (3), or beneath continental lithosphere giving a cordilleran margin as in South America (4). There are also conservative plate margins (5), or Atlantic-type margins, in which continental plates, such as the Americas, move ahead of the spreading oceanic plate with no tectonic action at the contact between continental and oceanic lithosphere.

The intersection of an inclined Benioff-Zone surface with the spheroid earth gives an arcuate line, whose radius of curvature varies with the inclination of the Benioff Zone. A vertically dipping inclined surface would intersect the sphere in a great circle. With increasing shallowness of dip, the intersecting line becomes a smaller small circle. The curvature of the trench and parallel volcanic arc is, as a rule, convex towards the spreading axis and concave in the direction of the dip of the Benioff Zone. An arc–trench system is said to face towards the up-dip direction of the Benioff Zone, towards its associated spreading axis. An interesting but rare departure from the rule of convexity is the descent of the Nazca Plate beneath Chile–Peru where the reversed curvature of the trench requires that the descending oceanic plate must split up into several segments on its descent (Rodriguez *et al.*, 1976). Because of this exception, the curvature of an arc–trench system cannot by itself indicate the direction it faces. The relative position of volcanic arc and trench reliably indicate the direction of facing since the volcanic arc normally overlies a Benioff-Zone depth of 150–100 km depth and the trench is the locus of outcrop of the Benioff Zone.

In the back-arc region, two opposite extremes of behaviour may occur (Dickinson, 1978). (1) Behind some intraoceanic arcs, sea-floor spreading within an interarc basin widens the marginal sea that separates the island arc from the nearby continental margin; this process involves separation of lithosphere and up-welling of asthenosphere to form new oceanic lithosphere in the back-arc region. Such marginal basins have spreading axes and sea-floor spreading and they are analogous to

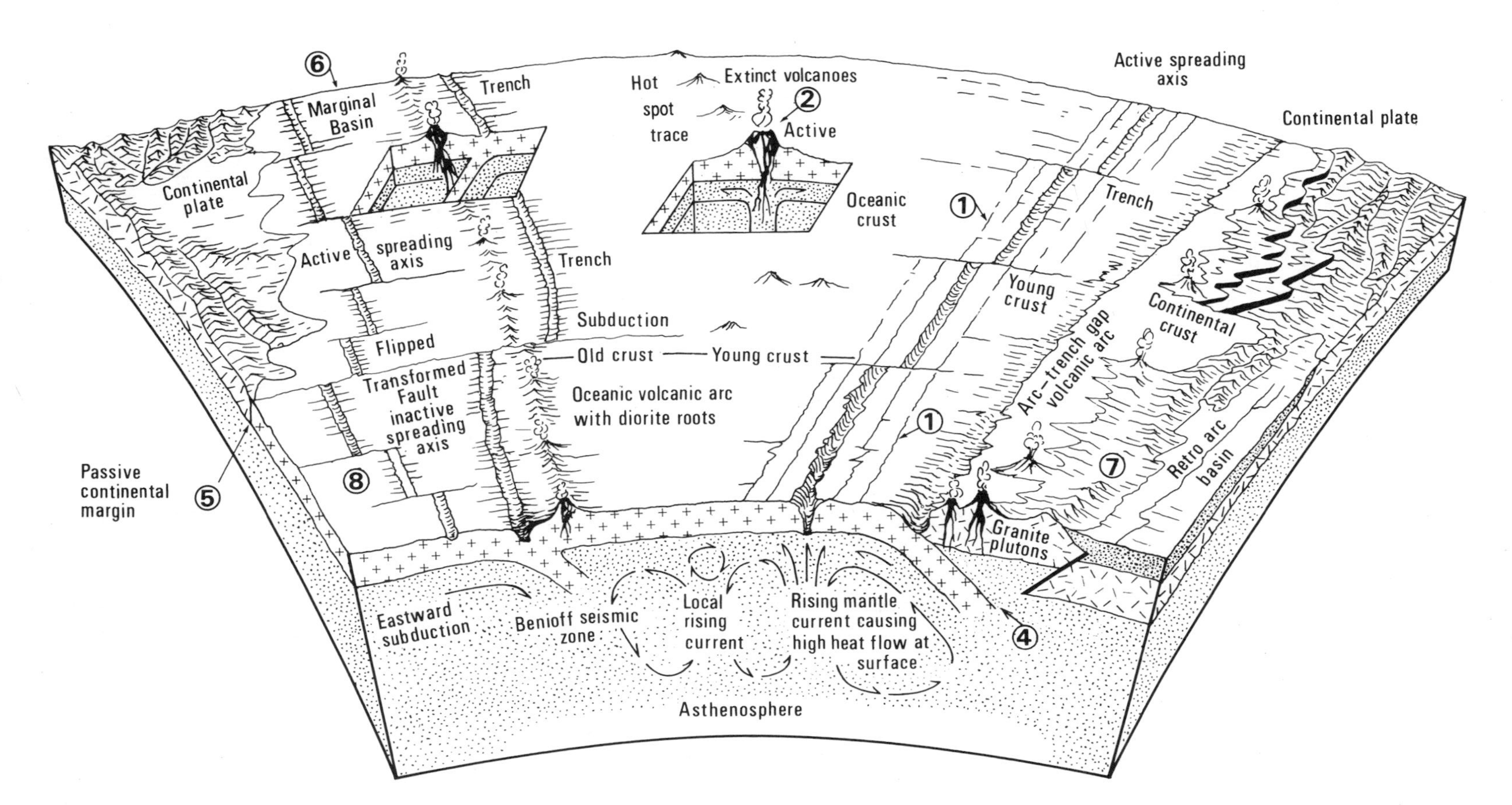

Figure 2.1. Diagrammatic segment through the crust and upper mantle of the earth (after Hutchison, 1978a). The overlying sea water has been omitted to show features of the ocean floor. The two rectangular diagrams in the upper centre and upper left centre represent vertical cross-sections as seen through windows in the surface. The numbered localities are referred to in the text.

the larger oceans except for their smaller scale (6); they are more ephemeral and liable to extinction as the major plate motions reorganize. Examples of active marginal basins are the Sea of Japan and the Andaman Sea. There is a larger number of extinct examples such as the South China Sea and Sulu Sea. Behind some continental-margin cordilleran arcs, fold-thrust belts form a subsidiary mountain chain that rises between the volcanic arc and the adjacent continental interior (7): this process involves contraction of lithosphere, crustal thickening within the mountain belt and the development of a retro-arc foreland basin on a depressed part of the continent in the back-arc region. Examples occur to the east of the cordilleran volcanic arc of North America. However, for some cordilleran plate margins neither marginal-basin nor retro-arc basin formation occurs in the back-arc region.

Modern arc–trench system patterns show that back-arc spreading (marginal basin) is characteristic of those margins that face eastwards (subduction downwards towards the west), whereas back-arc thrusting is most widespread in those that face west. Neutral systems without spreading or thrusting typically face southwards.

Bostrom (1978) has accounted for these differences by suggesting that whereas the passage of the tidal bulge reinforces the sea-floor spreading to account for marginal basins in the western Pacific, any marginal basins that may have formed on the Andean orogenic front are overriden and destroyed by the westward-advancing American Plates.

The extinction of volcanic arcs and marginal basins gives rise to inactive or remnant arcs, which are numerous in the western Pacific (Karig, 1972). These remnant arcs separate small extinct marginal seas or basins (8).

Mineralization on the Ocean Floor

Only recently have present-day subsea-floor geochemical systems been recognized in the ocean basins in the vicinity of sites of sea-floor spreading. They are difficult to locate compared with terrestrial counterparts because the heat-flow anomaly is rapidly dispersed into the overlying water. Perhaps the best studied is the mid-Atlantic Ridge (Rona, 1980). Abundant hydrothermal manganese oxide encrustations coat the basalt talus on the rift-valley wall (Scott *et al.*, 1976). The hydrothermal activity of the mid-Atlantic rift valley is replaced by hydrogenous ferromanganese oxide precipitation on the ocean crust older than 0.7 Ma (1 Ma = one million years) on the ridge crest highlands.

The Bauer Deep in the southeast Pacific is the only area of extensive metalliferous sediments so far discovered away from a mid-ocean crest (Anderson, R.N. and Halunen, 1974). Metalliferous sediments have been reported throughout the geological succession. Concentrations of iron up to 18% and manganese up to 6.5% and also anomalously high copper, nickel and zinc contents have been reported (Sayles and Bischoff, 1973). Precipitation of the metals from hydrothermal saline fluids or brines emanating from the hot oceanic crust may be the metallogenetic mechanism.

Basal sediments of the western flank of the East Pacific Rise are enriched in iron, manganese, nickel, copper, lead and zinc (Cronan, 1976). Iron contents may be as high as 30% and manganese as high as 9%. Most of the iron is probably of direct volcanic origin, and most of the manganese and minor elements are derived from the sea water but ultimately are of volcanic origin. The bulk composition of the deposits varies with age. This is thought to be due to variation in the incidence of volcanic activity at the East Pacific Rise crest, where the deposits were formed and carried westwards by sea-floor spreading. Following the model of Spooner and Fyfe (1973) for subsea-floor metamorphism at spreading centres, the hydrothermal brine circulating through the basaltic crust, driven by the hot basalt at depth, is assumed to be the cause of chemical exchange with the overlying basaltic crust and sediments. Inorganic precipitation from the hot brine can account for heavy-metal enrichment in the overlying sediments. The isotopic composition of the metal-enriched sediments is in keeping with formation of the metals by sea water. Iron–manganese, manganese and iron sulphide enrichments at ridge crests have been ascribed to submarine hydrothermal activity. The iron–manganese-rich sediments are associated with fast spreading ridges. The suite of metal-rich deposits can be explained as the manifestation of the degree of subsurface mixing between high-temperature acid, reducing hydrothermal fluids and the ground water in the sediments (Edmond *et al.*, 1979).

Massive sulphide ore deposits have not been directly observed yet on the existing floors of the oceanic areas. However massive sulphide deposits are known from the land masses associated closely with deep-sea sediments and pillow basalts and they have to be presumed to have originated on the ocean floor (Fleet and Robertson, 1980). They may be interpreted as the submarine expression of subsea-floor geothermal activity (Henley and Thornley, 1979). The individual characteristics of massive sulphide deposits may depend on the mode of discharge from the underlying hydrothermal reservoir. High-power discharge of the metallogenic fluid to the sea floor may occur through breccia pipes and vents generated by hydrothermal explosions on the sea floor. This allows hypotheses to be made about the formation of the brecciated footwall of ancient stratiform base-metal deposits.

Manganese Nodules

Ferromanganese nodules are common on the deep-ocean floor, especially in the areas with little sedimentation and in the absence of bottom currents. Their abundance and composition varies from place to place. Reviews of the extensive literature are given by Glasby (1976), Glasby and Read (1976) and Cronan (1977). The nodules reach their greatest concentration in deep water below the carbonate compensation depth, and range in diameter from 1 mm to 20 cm. Usually each nodule is of a layered concentric structure deposited round a nucleus, such as a shark's tooth or other solid object. Such concentrations are not confined to the deep sea, but also occur in shallow seas, and periglacial lakes and bogs. The manganese/iron ratios vary with degree of oxidation: in the deep sea the ratio is generally about 1. The total manganese content is variable, with an average of 14% for the Pacific Ocean. The nodules also contain cobalt and nickel. The world average is 16% manganese and 15.6% iron, with 0.30% cobalt and 0.49% nickel. The variation is often diagnostic of environment. Nodules found on seamounts are high in cobalt, with an average content of 1.15%. Continental borderland nodules are very rich in manganese, with average values of 38.69%, so that their manganese/iron ratio is around 28.8 in contrast to the usual ratio of 1. Abyssal nodules are highest of all in copper with average values of 0.37%. The factors that determine the compositional variations include sea-water composition, biological productivity, the nature of the substrate and the degree of oxidation of the environment of deposition.

Fossil nodules associated with deep-sea sediments have been found in the Cretaceous red clays of Timor, and the Eocene chert–spilite association of Borneo and Late Oligocene deposits of Sicily (Jenkyns, 1977).

The source of the submarine nodules is an open question. Some authors suggest a derivation from decomposition of the submarine volcanic rocks, but others suggest a precipitation directly from the sea water. Whatever their origin, they are widespread both in time and space and their presence must be incorporated into any metallogenic theory for explaining deposits associated with oceanic sediments and oceanic lithosphere.

From time to time, programmes for mining the nodules from the ocean floor have been proposed (Mero, 1977), so that the nodules themselves can be regarded as the economic deposit. A minimum mining grade of nodules would be a content of at least 2.8% of nickel/copper/cobalt on a dry-weight basis. The minimum concentration of nodules would have to be about 5 kg m^{-2} of sea floor and a deposit would have to cover an area of about 10 000 km^2 to be economic to mine by deep-sea dredges. The deposits of greatest economic interest are in the Pacific Ocean between the equator and 20° N, and between 110° and 180° W.

Ophiolite

Ophiolite is a distinctive assemblage of mafic-to-ultramafic rocks with minor sodium-rich acid igneous rocks and a characteristic association of pelagic oceanic sedimentary rocks. It is not a rock name, but an assemblage. Because the assemblage is characteristically tectonically emplaced at plate junctures or in old suture zones between major tectonic terrains, ophiolite complexes are generally folded, faulted and frequently metamorphosed. The variability of the tectonic complexity and size of intact ophiolite bodies results in a range of completeness and fragmentation of the ophiolite assemblage. Hence some ophiolites may be considered as complete, others dismembered, others incomplete. The voluminous literature has been summarized by Coleman (1971, 1977), Coleman and Irwin (1974) and Gass (1977).

The ideal complete ophiolite is composed of a characteristic succession of rocks, derived from studies of classic localities such as the Troodos complex of Cyprus (Gass and Masson-Smith, 1963), Papua New Guinea (Davies, H.L., 1971), Vourinos in Greece (Moores, 1969), east Borneo (Hutchison, 1975) and New Caledonia (Brothers and Blake, 1973). In a completely developed ophiolite (*Figure 2.2*) the rock-types occur in the following sequence from the bottom upwards: ultramafic complex, consisting of variable proportions of harzburgite, iherzolite and dunite, usually with a metamorphic tectonite fabric, and variably serpentinized; a gabbroic complex, ordinarily commonly layered with cumulus textures containing cumulus peridotite and pyroxenite and usually much less deformed than the ultramafic complex;

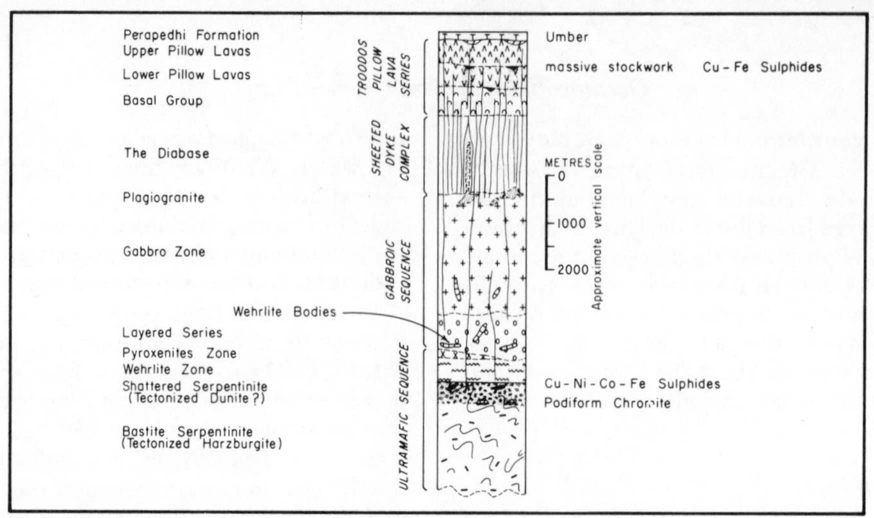

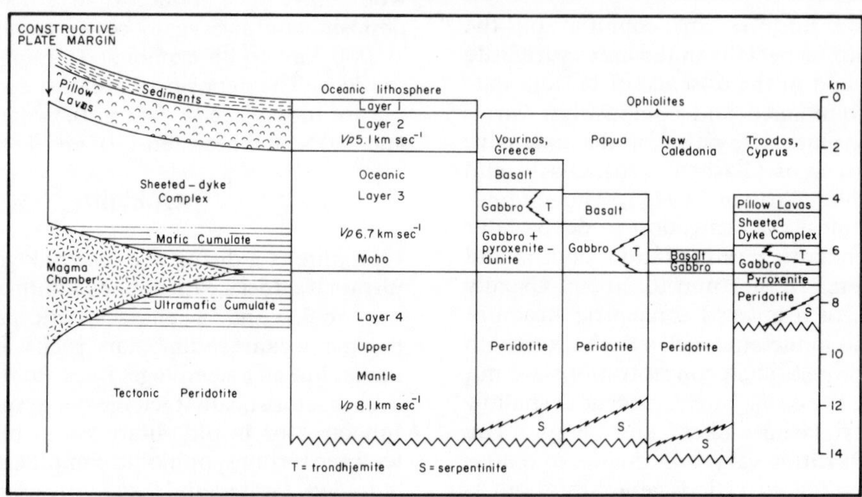

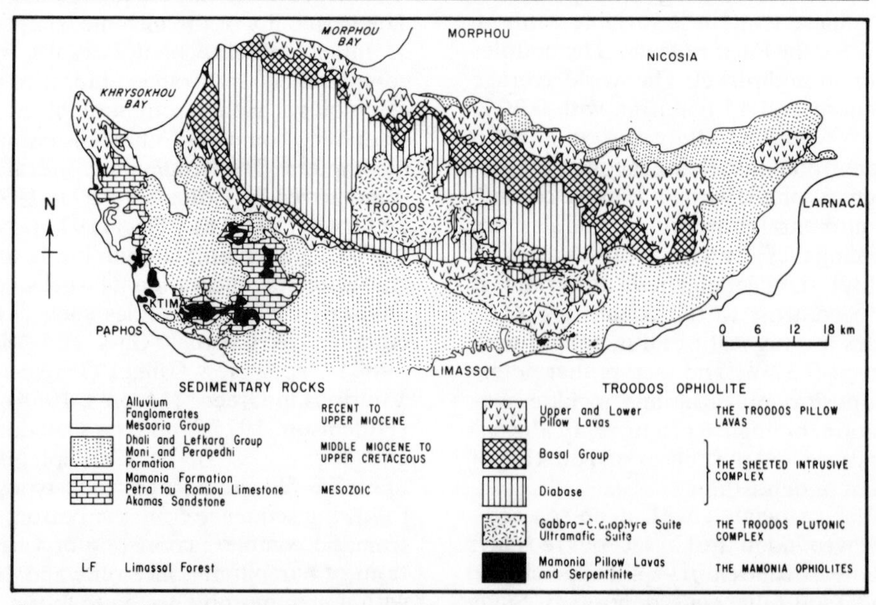

Figure 2.2. Top: idealized stratigraphic column of the ophiolite sequence at Limassol Forest (from Panayiotou, 1978). Centre: comparison of other ophiolite sequences and the standard oceanic lithosphere (from Gass, 1977). Bottom: simplified geological map of western Cyprus (from Constantinou, 1976), showing the location of Limassol Forest.

mafic sheeted-dyke complex composed of swarms of diabase dykes forming a zone separating the underlying gabbro–plagiogranite from the overlying extrusive pillow lavas (sheeted-dyke complexes are not always present); and mafic volcanic complex, commonly composed of pillow spilite.

Rocks characteristically associated with the above assemblage include: an overlying sedimentary section typically including deep-sea facies such as chert, shale and micritic limestone; podiform chromitites contained within the dunite; and sodic felsic intrusive rocks (trondhjemite and plagiogranite) and extrusive keratophyre associated with the gabbro layer and the basalt layer, respectively.

The ophiolite suite is commonly and characteristically impoverished in the large-radii incompatible elements of potassium, rubidium, strontium, uranium, thorium and the light rare earths. There is frequently a horizontally disposed low-pressure/high-temperature metamorphism imposed on the ophiolite rocks, ranging downwards from zeolite through greenschist to amphibolite facies (increasing grade with stratigraphic depth). The metamorphism is often sporadically incomplete with poor equilibration between the mafic minerals and the plagioclase (Hutchison, 1978b). Stable-isotope studies on the metamorphosed basalts indicate oxygen ratios compatible with reactions between the basalt and sea water at elevated temperatures.

The idealized sequence shown in *Figure 2.2* is rarely present in most ophiolite occurrences and the thicknesses of the lava, dyke and layered plutonic units vary from ophiolite to ophiolite. The sequence and rock assemblage, however, are common to all, implying a common genesis. The field relationships, and the chemical and mineral composition of the tectonized peridotite suggest that it is a formation from which a liquid basaltic fraction has been melted and removed to higher levels. The peridotite is therefore a refractory residue. The basaltic magma body, or bodies, some 2–3 km deep and perhaps 25 km long, can be assumed to have collected at shallow depths of 1–2 km below the sea floor. In an extensional environment, magma is tapped off, and injected as dykes and extruded as pillow lavas (*Figure 2.2*). At the same time, the magma remaining within the chamber fractionates on cooling to give the layered complex, ranging from dunite with chromite, through peridotite and olivine gabbro to gabbro with minor differentiated sodium-rich granite (trondhjemite). These are the events that are envisioned to have produced an ophiolite sequence. The submarine nature is clear because of the submarine metamorphism and the associated overlying pelagic sedimentary rocks. Ophiolite is not defined as ocean-floor lithosphere. However, the ophiolite sequence strongly resembles the oceanic crust and upper mantle in that the ophiolite lavas are very similar to lavas dredged from the present-day ocean ridges—both are low-potassium tholeiites. The low-grade metamorphic lavas of ophiolites also have direct equivalents that have been dredged from oceanic ridges (Hutchison, 1978b). The geophysical characteristics of the ocean floor (*Table 2.1*) can allow an interpretation of it being similar to the ophiolite sequence (*Figure 2.2*). The inferred sequence of magmatic events at an ocean or marginal-basin spreading axis can be imagined as the one that would give rise to an ophiolite sequence. The only interpretative step necessary is the correlation of the peridotitic substratum of an ophiolite with the upper mantle of the oceanic lithosphere. The present view is that they were formed at the spreading axes of an ocean basin. The size of the ocean basin is irrelevant. It could have been a major ocean, or something like the Red Sea or a back-arc marginal basin. It is likely that most known ophiolites, especially those of the Tethys, were produced in small ocean basins. After formation at an oceanic spreading centre, the rock sequence migrates away from the axis by sea-floor spreading. The principal change in the rocks is metamorphism caused by interaction with hot sea-water brine.

The precise mechanism whereby fragments of this oceanic lithosphere are emplaced on continental margins or in island arcs is unknown. Normally the oceanic lithosphere should descend beneath a Benioff Zone. However, it is easy to speculate what unusual processes could be involved to cause remnants of the oceanic lithosphere to be uplifted along the edges of the ocean basin. The special term ‘obduction’, the converse of subduction, has been coined to describe the overall process (Coleman, 1971). The least deformed ophiolite sequences appear to be vertically uplifted pieces of ocean floor that have not yet undergone any significant horizontal deformation. Two such sequences are the Macquarie Island south of New Zealand (Varne and Rubenach, 1973) and the Troodos massif of Cyprus (Moores and Vine, 1971) (*Figure 2.2*). Most other ophiolites are much more deformed. The possible mechanism for uplift of an intact ophiolite is shown in *Figure 2.3*; extensive serpentinization of the mantle wedge overlying a Benioff Zone is presumed, followed by uplift, a direct result of volume expansion during serpentinization. The domal uplift of the ocean lithosphere of the mantle wedge can result in gravity

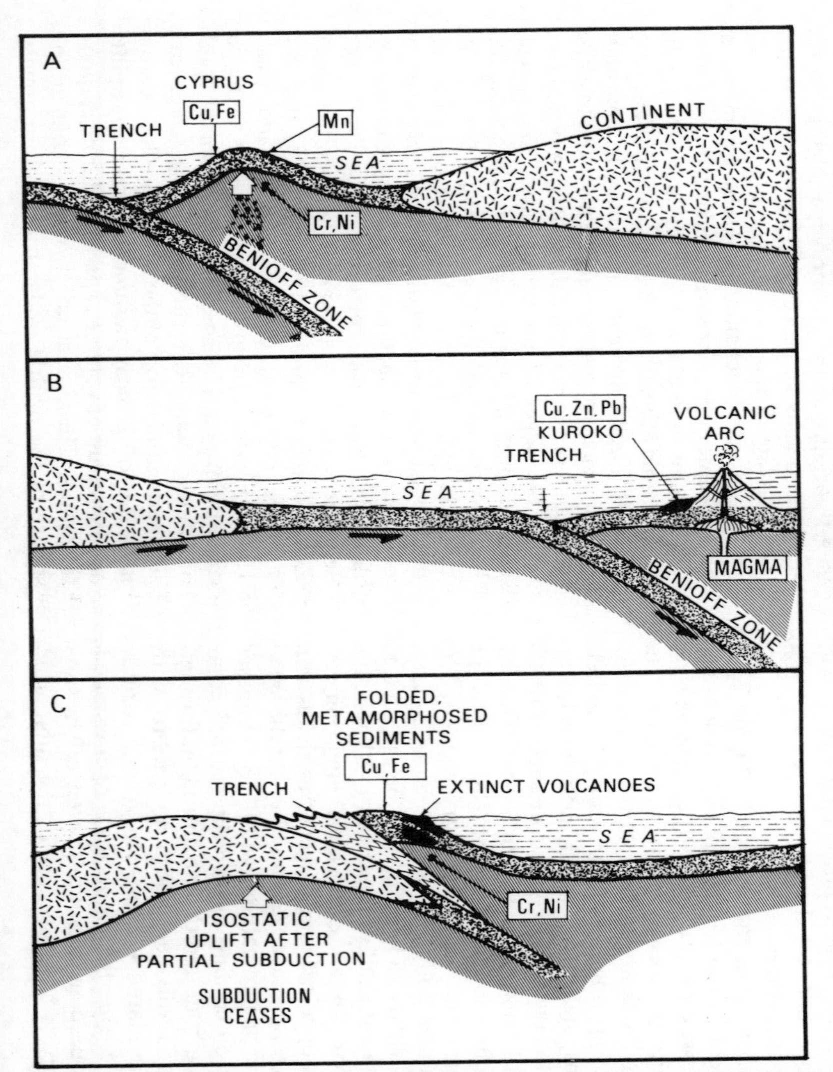

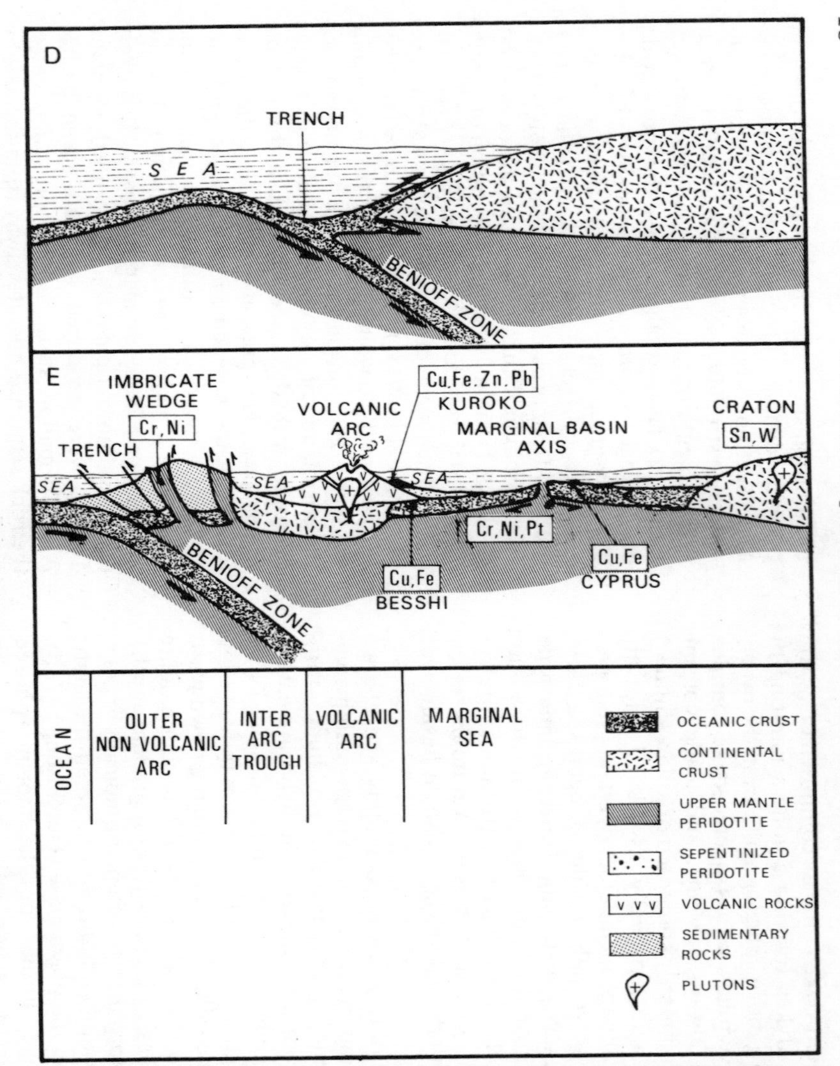

Figure 2.3. Diagrammatic plate tectonic cross-sections. (A) Uplift of oceanic lithosphere by serpentinite diapirs as proposed for the Cyprus ophiolite. (D) Ophiolite slivers removed from the descending slab and pushed over a continental margin as proposed for the Franciscan of California (after Gass, 1977). (B) and (C) Sequence of events to explain the Papua New Guinea ophiolite obduction as Australia converges on the Indonesian island-arc region (after Davies, H.L., 1981). (E) A convergent margin with back-arc spreading, showing the possible location of mineralization (from Mitchell, A.H.G. and Garson, 1976).

sliding and olistostrome formation, and this model can be used to explain serpentinite conglomerates that frequently occur in association with ophiolite complexes. These ophiolites above serpentinite diapirs in the mantle wedge may be referred to as autochthonous, but many deformed ophiolites are clearly allochthonous, and have been thrust upwards or obducted either in island arcs, such as New Caledonia, or over continental margins, as in Papua New Guinea. A reasonable model for obduction of large ophiolite slabs is the underthrusting of the leading edge of a continental plate, followed by isostatic uplift when subduction ceases, as in the case of Papua New Guinea. *Figure 2.3* gives diagrammatic sketches of the possible mechanisms for allochthonous obduction. They are purely models, but it is important to have some idea of the way in which an ophiolite sheet may be obducted, for it is only by this process that the ophiolite and its ore deposits become accessible for exploitation. In older orogenic belts, the mechanism of obduction may be impossible to resolve, obscured through tectonic complexity, erosion and the cover of younger sedimentary formations. However, by analogy with younger ophiolites, it has to be concluded that the obduction mechanisms can only be envisioned at or near plate margins, so that older ophiolite belts in orogenic zones must represent former plate margins or suture zones between major plates. During allochthonous obduction, folding and deformation of the oceanic lithosphere may superimpose on the sea-floor metamorphism a subsequent metamorphism related to the tectonic activity.

Podiform Chromitite

Podiform chromites, or Alpine-type chromite deposits, are known from many ophiolite dunite or harzburgite bodies. Late Mesozoic and Cenozoic examples occur in the Philippines, Cuba, Oman and Cyprus. The harzburgite or dunite country rock in which the chromite occurs is considered as a refractory residuum resulting from partial melting of the upper mantle at oceanic or marginal-basin ridges or related to hot-spot activity in oceanic lithosphere. Chromite layers may also have settled as cumulates from large gabbroic magma chambers emplaced in the lower crust at the spreading ridges. The possible origins of the chromite include incongruent partial melting of the primitive pyrolite mantle, or crystal settling within basic magma chambers resulting in ultrabasic cumulates (*Figure 2.2*).

The features of podiform chromitites have been summarized by Thayer (1964). The ore bodies are fundamentally tabular, pencil-shaped, or irregular. The chromite is characteristically anhedral and commonly shows the effects of granulation and magmatic corrosion. Flow-layering, foliation and lineation are parallel in most chromite deposits and peridotite host rocks, and normally pass through major rock units, including the chromite. Locally the foliation and lineation may cross the layering. Most podiform chromitites are oriented with their longer dimensions essentially parallel to the layering or foliation of the host peridotite, but some are crosswise. In all podiform chromitites, whether parallel or crosswise, internal flow structures are harmonious with the country rock, except where disturbed by post-magmatic faulting. The great difference between podiform and stratiform chromitites may be considered analogous to the similarities and differences between metamorphosed and unmetamorphosed sedimentary rocks. Relict structures in the form of crystal cumulates are sometimes preserved in podiform chromites, but most features are due to extensive flowage during replacement under magmatic conditions and to recrystallization during metamorphism and emplacement of the ophiolite. The chromite crystals commonly exhibit pull-apart textures. Just as sedimentary principles apply to the description of stratiform chromite deposits, so metamorphic principles apply to the podiform deposits.

Chromite has the general formula $(Mg, Fe^{2+})(Cr, Al, Fe^{3+})_2O_4$. Its composition strongly reflects the tectonic setting in which it was formed (Thayer, 1970). In the stratiform chromitites, total Fe content increases substantially or rapidly as Cr_2O_3 decreases below about 55%, but in the podiform chromitites Al_2O_3 increases reciprocally with the Cr_2O_3 decline, whilst Fe stays approximately constant. This shows that the fractionation trends in large mafic plutons emplaced in stable cratonic crust are different from those at ocean or marginal-basin spreading centres, where the ophiolite complexes are thought to have formed. The podiform chromitites are thought to show that crystal settling on a large scale is one of the important processes in the upper 25–30 km of the mantle. All economically important stratiform chromitites (Chapter 6) are Precambrian and occur in stable cratons. Podiform chromitites occur along island arcs and mobile mountain belts of Palaeozoic or younger age. The disparity in ages between these two major groups of chromite deposits may simply reflect the slow denudation rates of cratonic terrains.

ZAMBALES OPHIOLITE, PHILIPPINES

The Zambales Range of western Luzon includes all the lithological rock-types considered typical of an ophiolite (*Figure 2.4*). The ophiolite has been uplifted to the level of the island-arc volcanic series. Sedimentological data suggest a Late Eocene age of uplift. The ultramafic rocks include chromium-rich dunite containing Fo_{92-93} olivine, harzburgite and diopsidic pyroxenite, and are widely serpentinized. Layered igneous rocks include anorthosite (An_{95-98}), cumulate textured and massive gabbro, troctolite and norite. They all show recrystallized cataclastic textures formed at near-solidus temperatures. Basalt, diabase and amphibolite dykes cut the layered and the ultramafic rocks. Altered pillow basalts contain clinopyroxene, olivine and plagioclase An_{75}. They have low titanium, zirconium and strontium contents and resemble island-arc tholeiites (Hawkins, J.W. and Batiza, 1977). The ultramafic complex, the layered igneous rocks and the pillow basalt are separated by shear zones, and the original ophiolite stratigraphy has been largely lost. Although Hawkins, J.W. and Batiza (1977) favour a source of the ophiolite from a marginal basin that formerly lay to the east of the Zambales Range, Hutchison and Taylor (1978) and Schweller and Karig (1979) concluded that the Zambales ophiolite is South China Sea marginal basin upper lithosphere uplifted in the Late Eocene epoch.

The chromite deposits have been described by Fernandez, N.S. (1960), Stoll (1962), and summarized by Bryner (1969) and Gervasio (1973). They are both tabular and pod-like. Present production is confined to the Acoje and Masinloc chromite belts, which are separated by a north-northeast-trending belt of layered gabbro, 40 km long and 5 km wide.

The Acoje metallurgical chromitite belt extends a distance of 20 km. The present mining area appears to be part of the eastern limb of an anticline extending southwest to nearly south. The western limb is barely explored. The mining zone occupies a belt of dunite and olivine-rich peridotite that forms the basal sequence of the layered folded ultrabasic rocks. The layered sequence includes basal dunite, lherzolite, harzburgite, and an upper troctolitic and noritic gabbro intercalated with peridotite or harzburgite. The ore bodies occur as sporadic lenses, bands, layers and as shoe-string deposits in dunite with clinopyroxenite, which are more or less serpentinized. The ore deposits lie hundreds of metres west of the olivine-deficient norite. The ore lenses are offset from one another by en-echelon faults. In the northern half of the mining area, the layers strike northeast and dip east, while in the southern half they strike northnorthwest and dip west. A major fault is inferred between these domains. The lenses coincide with the major-layering of the host rocks. There are at least 27 ore bodies distributed over a north–south interval of about 1 km and over a vertical interval of 350 m. The chromite is fine grained and euhedral with local cataclastic structures. The reserves of the Acoje Mine amount to 2 500 000 t (t= metric ton or million grams) of ore containing, on average, 35% chromium oxide (Cr_2O_3) and 0.8% nickel. The ore has average molecular ratios magnesium oxide/iron oxide (MgO/FeO) = 1.8:1, chromium/iron = 2.7:1 and chromium/aluminium = 2.6:1. Disseminated pentlandite and other nickel sulphides have been mined from the lower mine levels. They occur as irregular blebs in serpentinized dunite, containing, on average, 1.5% nickel, within the range 0.25–8%.

The Masinloc refractory chromite belt lies parallel to and about 12 km east of the Acoje belt. It is the world's largest producer of refractory chromite. The ore bodies are enclosed in serpentinized harzburgite close to its contact with olivine gabbro (*Figure 2.4*). There is one big lens and twelve smaller ones, all strung out in a belt 2.5 km long. A rough parallelism exists between the ultramafic layering and the string of chromite lenses, which all have a definite layered structure. Of the ore bodies shown in *Figure 2.4* the Coto is remarkable for its size. According to Stoll (1962) it is an oval flat-lying lens about 600 m × 300 m and estimated to have been 80 m thick before erosion. The original ore body contained 15 000 000 t of ore of average composition chromium oxide (Cr_2O_3) 36% and aluminium oxide (AL_2O_3) 30%. The chromite of the Masinloc, as with other refractory chromite occurrences, lies close to olivine-rich gabbro. The average molecular ratios are magnesium oxide/iron oxide (MgO/FeO) = 2.8:1, chromium/iron = 2.2:1 and chromium/-aluminium = 0.8:1. The belt of chromite ore bodies coincides with parts of a zone of intense serpentinization and fracturing of the harzburgite. Many of the fractures are occupied by swarms of diabase dykes, which cut the chromite bodies.

SABAH (NORTH BORNEO)

The Philippine island arc continues westwards along the Sulu Archipelago island arc into Sabah. Chromite occurs as layers and pods in dunite and serpentinized lenses within most of the peridotite bodies of the non-volcanic arc. However, the volume of chromite is considered to be too small for

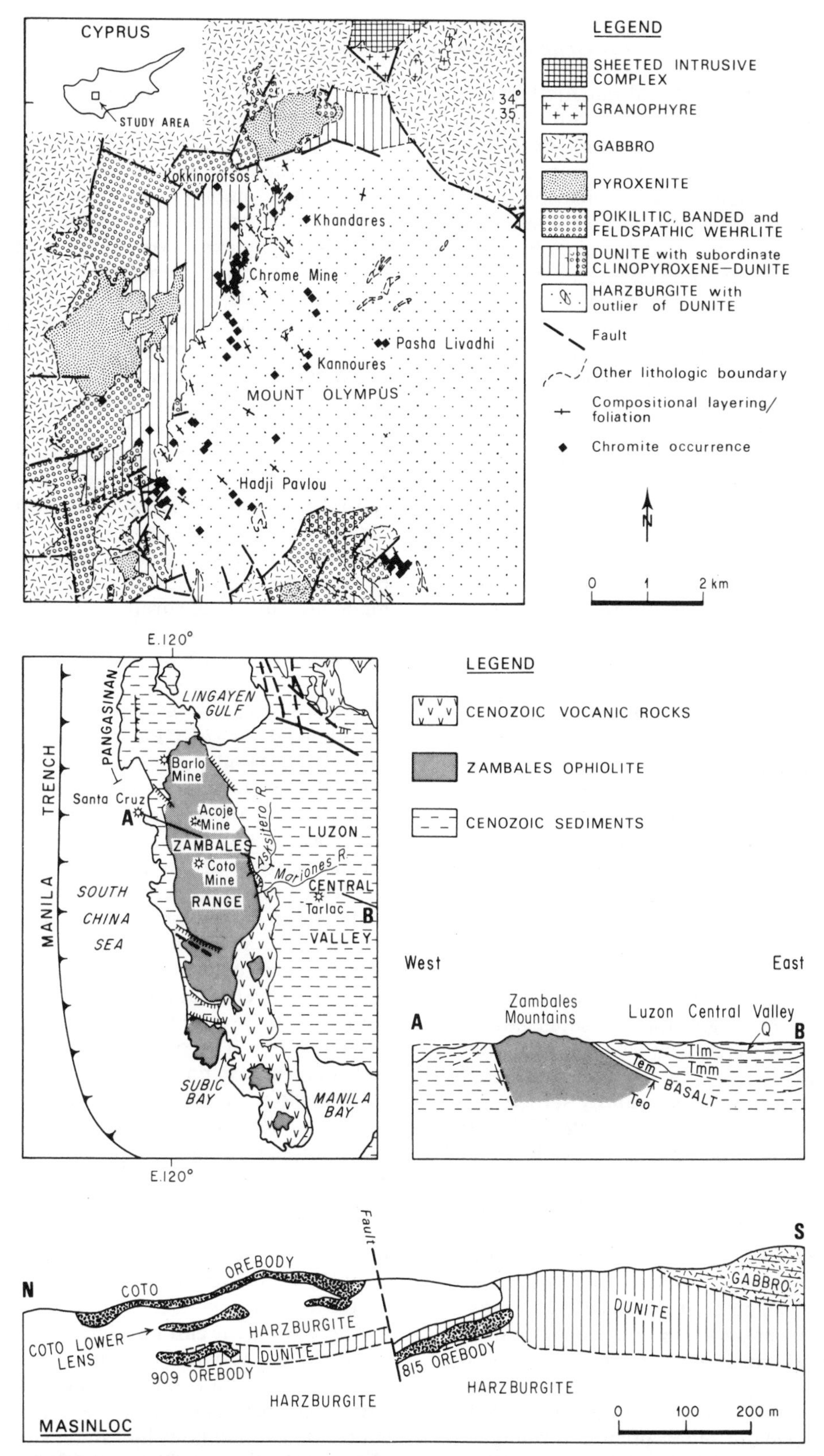

Figure 2.4. The podiform chromitites of the Zambales of the Philippines and of the Troodos of Cyprus and their geological setting (after Bryner, 1969; Greenbaum, 1977; CCOP, 1980).

economic exploitation. The ultramafic bodies of the Sabah ophiolite contain, on average, 2770 ppm chromium and 1530 ppm nickel, whilst the layered gabbros contain, on average, 230 ppm chromium and 96 ppm nickel (Hutchison, 1972). The distribution coefficients ultramafic/mafic are therefore 15.9 for nickel and 11.9 for chromium, which are in close agreement with values given by Naldrett and Cabri (1976), showing how nickel and chromium may be regarded as refractory elements enriched in the ultramafic residuum. Hence magmatic events at ocean or marginal-basin spreading centres are suitable processes for chromium-enrichment and nickel-enrichment. It can be shown that platinum metals are also enriched.

CYPRUS

The very complete ophiolite complex of Cyprus is subdivided into two complexes known as the Troodos Plutonic complex, centred on Mount Olympus (Greenbaum, 1977), and the Limassol Forest Complex (Panayiotou, 1978). Both comprise a sequence of basaltic volcanic rocks overlying a plutonic complex of upper gabbros and lower peridotites containing podiform chromitites. The chromitiferous rocks range from disseminations and minor segregations of chromite in dunite to ore bodies of economic size and grade composed of nearly pure chromitite (Greenbaum, 1977). They display both primary textures, such as chromite net and occluded silicate texture of cumulate magmatic origin, and schlieren structures produced by deformation. Dendritic texture in nodular chromite may be interpreted as a product of crystallization of a magma that was supersaturated or supercooled with respect to chromium. Orbicular structures are thought to have formed by mechanical accretion of previously settled chromite grains around a nucleus of dunite. Unlike the Zambales, there are no tabular ore bodies and all are podiform.

The chromite deposits occur mainly as isolated pods and layers either within dunite near the harzburgite contact (*Figure 2.4*), in the harzburgite within dunite lenses, or enclosed in a narrow dunitic envelope. The chromite shows a wide range of reciprocal chromium–aluminium variation and low iron oxidation values typical of ophiolitic chromitites.

Chromite in the Limassol Forest is restricted to the early ultramafic differentiates, forming discrete deposits in the basal parts of dunite (Panayiotou, 1978). It is concentrated as small pods, lenses and schlieren in serpentinites (*Figure 2.2*). The massive chromite is highly cataclastic and extremely altered to ferrit-chromite. Pervasive deformation, tectonic and metamorphic, caused serpentinization of the host rock and brecciation and remobilization of the segregations.

It may be generally concluded that the Cyprus plutonic complexes evolved as a differentiated body of magmatic cumulates laid down upon a basement of depleted mantle harzburgite, from which the basic magma was withdrawn by partial melting. The chromitites originated as isolated magmatic segregation ore deposits near the base of the cumulate dunite during episodic crystallization at localized centres within the magma. Postcumulus mobilization and accompanying penetrative deformation of the lower ultramafic rocks caused the tectonic overprint of schlieren structures on earlier magmatic textures, and associated large-scale rock flowage resulted in deep infolding between the lithological units, leading to the formation of outliers of chromite-bearing dunite.

TURKEY

There are two east–west-trending ophiolite zones (Brinkmann, 1976; Pinar-Erdem and Ilhan, 1977). The Middle Anatolian zone extends from the Aegean coast to Armenia. A southern zone extends from the western Taurus via Cyprus to southeast Anatolia and Iran. The ophiolites are presumed to be the original floor of the Tethys Sea, or its marginal basins, tectonically obducted in the Mesozoic and especially in the Late Cretaceous period. Sedimentary rocks of Liassic and younger age contain detrital debris from the ophiolite sheets, indicating that the ophiolites were uplifted several times and eroded during the Mesozoic era. The ophiolites were also tectonically deformed in post-Mesozoic times. As is typical of ophiolite masses, there is a sporadic association of unmetamorphosed closely associated with metamorphosed ophiolite or greenstone. The assemblages include spilite, gabbro and harzburgite, associated with pelgic sediments such as cherts, and also closely associated with high-pressure lawsonite–albite glaucophane metamorphic assemblages, characteristic of metamorphic regimes in trench areas.

There are many chromite localities, especially in eastern Turkey. The Guleman–Soridag district is the most important source of metallurgical-grade chromite after South Africa, and it has been the source of a voluminous literature, summarized by Helke (1962) and Thayer (1964). The most famous ore body is the Golalan, which contained more than 1 200 000 t of ore yielding 51–52%

chromium oxide (Cr_2O_3). The chromite is contained in serpentinite within a gabbro–peridotite–serpentinite complex. The area is the least tectonically disturbed of all the Turkish deposits. Some tabular ore bodies are as much as 1.5 km in length, 100 m or more in dip dimension, and only 0.5–4 m thick. Helke (1962) argued that they were relicts of a truly stratiform-type chromite deposit. The ore textures, however, are not of the stratiform-type. Most of the deposits are massive. Nodular ore occurs and pull-apart texture is well shown in many deposits. The chromite crystals are not euhedral, but strongly rounded, as is common in ophiolite chromites.

CUBA

The Domingo Belt of Cuba (Pardo, 1975) consists of an ophiolite association of banded gabbroic and ultramafic rocks, which outcrop widely over the island; serpentinites are also common. The ophiolites are overlain by Cretaceous sedimentary rocks, but have themselves been thrust over young sedimentary formations that are obviously oil-bearing. Oil seepages through the overlying ophiolite have resulted in asphalt dykes cutting through the basic igneous rocks, and oil seeps occur in association with the ophiolite. The ophiolite assemblage includes peridotite, dunite, pyroxenite, gabbro, troctolite and anorthosite. The peridotite is frequently serpentinized and weathered to a deep residual blanket of laterite. Podiform chromitite occurs within the serpentinite and peridotite. The bodies are tabular to lenticular and are usually surrounded by a jacket of dunite of width variable from a thickness of 30 cm to several metres. The deposits have been summarized by Thayer (1942), and detailed accounts of the better-known regions given by Guild (1947) and Flint *et al.* (1948). The Moa area of eastern Cuba is the best known, and has been a well-known producer of refractory-grade chromite ore of average composition 33% chromium oxide (Cr_2O_3). The chromite–dunite pods occur within peridotite. The deposits are tabular to lenticular and range upwards in size to 220 m × 100 m × 28 m.

CALIFORNIA

Podiform chromite ore bodies occur frequently in the ophiolite bodies within the Franciscan melange terrain of California. They are not usually mined except in times of difficulty of supply from overseas sources. The deposits have been summarized by Thayer (1966). The main occurrences are in the Klamath Mountains, the northern Sierra Nevada and the Southern Coastal Ranges. They are in difficult mountainous country. Most are of metallurgical-grade. The larger deposits are of disseminated chromite in peridotite or serpentinite.

Nickel Laterite

Deep tropical weathering of the ultramafic parts of ophiolite bodies can result in valuable nickel ore in the form of blanket residual deposits of nickel-bearing laterite, which is becoming an increasingly important source of nickel. Just as with bauxitic laterite, a blanket deposit of sufficient size for strip mining develops only on a relatively flat surface such as a peneplain or plateau. On the steeper slopes, erosion prevents the build up of economic deposits. The typical nickel oxide content of fresh peridotite is 0.44% or less. Tropical weathering may result in a sevenfold enrichment to give typical mining values of 2.7%.

The processes of nickel laterite formation have been elucidated by Zeissink (1969), Reynolds, C.D. *et al.* (1973), Trescases (1973) and Golightly (1979a). The major units of a laterite profile are, from top to bottom: the limonite zone, which is dominated by goethite; the smectite-quartz zone consisting of soft nontronite clay and quartz; and the saprolite zone consisting of altered bedrock in which bedrock textures are preserved (Golightly, 1979a). Serpentinized peridotite alters to montmorillonite, aluminous goethite and quartz. Stable chromitiferous chlorite becomes concentrated in the weathering profile. There is no kaolinite and the alumina occurs within the goethite. In the upper weathering levels, serpentinite and montmorillonite disappear completely and the aluminium content of the goethite diminishes. With increasing depth, iron oxide (Fe_2O_3) and aluminium oxide (AL_2O_3) contents fall, while silica (SiO_2) and magnesium oxide contents increase. The trace elements nickel, cobalt, manganese, copper and chromium are all concentrated during weathering. There is strong evidence that much of the nickel is contained in the goethite.

There are two types of laterite horizons. One is high in silica and magnesia and has low iron oxide (Fe_2O_3) contents. It occurs low in the weathering profile, is green in colour and distinct nickel-bearing minerals are visible. The second group is low in magnesium oxide and silica and high in iron oxide (Fe_2O_3). It occurs high in the weathering profile, is red or reddish-brown and no nickel-bearing minerals are visible. Both types occur within any deposit. The three nickel-bearing minerals of common occurrence are pimelite (a montmorillonite-like mineral), schuchardite (a nickel-containing chlorite) and nickel-containing serpentine. The name garnierite is used loosely for these hydrous nickel minerals (Faust, 1966). The main bulk of the nickel is associated with goethite, and high nickel values are often associated with the Mn–Ni–Co nodules that occur in some laterite profiles. The concentration of nickel can be seen as a result of percolating ground water, causing release of the nickel that was fixed in the newly formed minerals.

In many nickel-containing laterites, the nickel is richest just above the fresh bedrock. For example Reynolds, C.D. *et al.* (1973) presented analyses for fresh peridotite core boulders in a laterite. The core of the boulders contains 0.48% nickel, while the weathered rim contains 1.76% nickel. Likewise cobalt increases from 0.013% to 0.02%. Core boulders are tumbled to remove the nickel-rich rims before being discarded. If the nickel is fixed in the aluminium-containing goethite high in the weathering profile, then downward migration of nickel to concentrate just above the fresh bedrock will be severely restricted. How this concentration takes place in the rich commercial deposits is not yet understood. The analyses of chromuim oxide (Cr_2O_3) and nickel oxide (NiO) in a completely analysed weathering profile are shown in *Figure 2.5* and the strong enrichment in nickel just above the fresh bedrock is obvious.

Nickel-containing laterites are now being mined in the southwest Pacific at Soroako and Pomala in southeast Sulawesi, Indonesia, on Nonoc island in the Philippines, and in New Caledonia (Hutchison and Taylor, 1978). In the Caribbean region, they are mined in Cuba (Kesler, S.E., 1978) at Nicaro and Moa. The geology of the deposits in the Dominican Republic is described by Haldeman *et al.* (1979), and that of the Greenvale deposit of Queensland, Australia, by Burger (1979).

NEW CALEDONIA

New Caledonia is of interest because of its huge ultrabasic masses, glaucophane schists, and extensive basalts, and it is always quoted in summaries of ophiolite emplacement (Coleman, 1977). Of prime economic interest are the widespread deposits of nickel and subsidiary chromite found in the laterite surfaces developed on the ultramafic bodies. The geology of the island has been summarized by Lillie and Brothers (1970) and Guillon (1974).

The ultramafic rocks comprise well-layered harzburgite, dunite and pyroxenite, with sheets of serpentinite occupying a total area of about 7000 km^2. The main ultramafic mass is 3000 m thick. The contact with the underlying formations is either horizontal or dips northnortheast at from 10° to 50° (Guillon and Lawrence, 1973). The ultramafic mass rests upon Upper Eocene tholeiitic basalts and upon folded sedimentary rocks of Cretaceous and Eocene age with a strong unconformity (*Figure 2.5*). The tectonic overthrusting of the ultramafic rocks as a flat-lying sheet is thought to have taken place in the Oligocene epoch.

The concentration of nickel at the base of the laterite surface belongs to the Cycle 1, which is a high erosion peneplain forming a dissected plateau. The nickel laterites are thought to have formed in the Miocene epoch. There is no evidence to suggest that much active nickel concentration is taking place at present.

The formation of garnierite (hydrous nickel silicate) is influenced by the composition of the underlying rocks, the climate and the physiography (Troly *et al.*, 1979). This involves the chemical alteration of olivine-rich nickel-bearing rocks at near-surface conditions under a tropical climate. The nickel in the garnierite ores was derived from nickel in the silicate minerals such as olivine. However, it is also evident that a small proportion was derived from nickel sulphides occurring as accessory minerals in the ultramafic rocks (Guillon and Lawrence, 1973). The laterization has in many places been facilitated by brecciation of the ultramafic rocks along major fault zones (*Figure 2.5*). The most common type of ore contains fresh peridotite core boulders surrounded by yellow onion-skinned nickel-rich layers (Troly *et al.*, 1979).

SULAWESI, INDONESIA

Sulawesi consists of a western Tertiary volcanic belt, a central schist belt and an eastern belt characterized by ultramafic rocks and melange (Hamilton, 1979). On the southeast arm of the island, the ultramafic rocks are mostly massive peridotite, largely harzburgite, with little associated gabbro and basalt (Bemmelen, 1970a). Local zones of serpentinization occur. The ultramafic rocks probably represent the lower parts of an ophiolite that are in fault contact with Cretaceous and Tertiary limestones. Melange south of Lake Matano has a serpentinite matrix (Silver *et al.*, 1976).

Geophysical studies indicate a westwards thickening of the ultramafics. Their eastwards thinning, their apparent dip beneath the schist belt and their close association with the melange belt all fit the

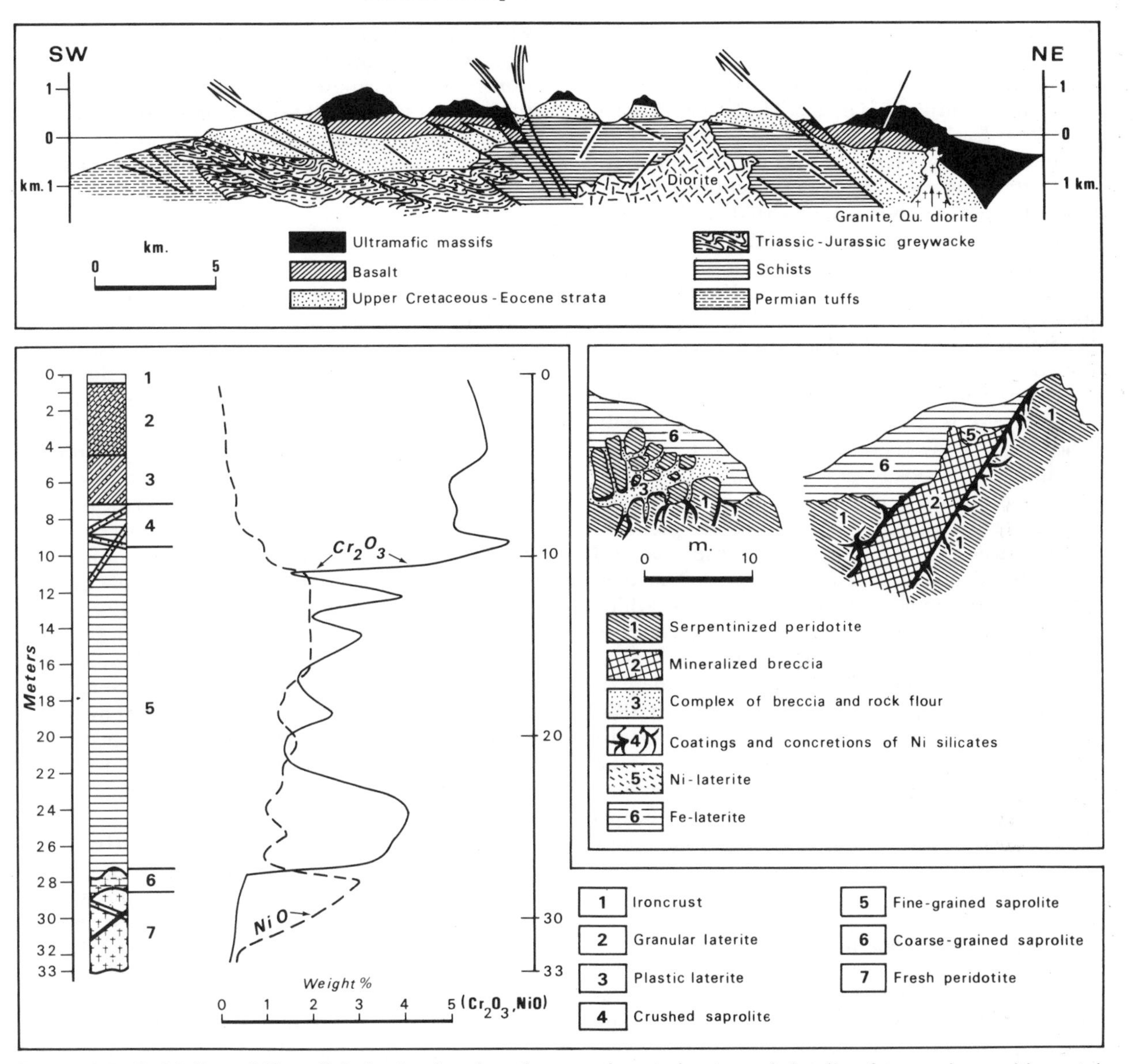

Figure 2.5. Ophiolite of New Caledonia showing the overthrust sheets and details of two mines with metal contents of a nickel-containing laterite profile (after Raguin, 1961; Trescases, 1973; Guillon, 1974).

model that the ophiolite is composed of fault slices concordant with the direction of subduction (Silver *et al.*, 1978). Uplift has been aided by the tectonic inclusion of large masses in the melange, and finally aided by the collision of Sulawesi with the Banggai–Sula islands.

Virtually all the ultramafic rocks are overlain by residual nickel-containing laterite (Golightly, 1979b). The weathered profiles commonly range from 5 m to 30 m thickness. The lower saprolitic profiles are strongly enriched in nickel. The saprolite contains core boulders of unserpentinized harzburgite surrounded by yellow-to-orange rims and fracture fillings of garnierite and manganese oxides. The garnierite has been determined to be $(Ni,Mg)_3Si_4O_{10}(OH)_2.H_2O$, with nickel in the range 10–25% by weight. Nickel is predominantly contained in silicates in the saprolite zone and in goethite in the overlying limonite zone. A mass balance calculated from selected weathering profiles indicates that *in situ* leaching of a column of peridotite 20–50 m thick has taken place with only minor lateral migration of nickel in solution (Golightly, 1979b).

GREECE

Ophiolite outcrops over about 3300 km^2 or about 2.5% of the surface of Greece. The bodies range in age from Early Tertiary to pre-Mesozoic, but the main complexes, such as the Mount Vourinos, are of Cretaceous age. The mineralization in the ultramafic rocks has been summarized by Mossoulos (1964) and Zachos (1964). The nickel-containing laterites of central Greece are the most important, and contain an average of 2–2.5% chromium oxide (Cr_2O_3) and variable nickel oxide contents. There are some tens of millions of tonnes of lateritic ore.

Two types of deposit exist. The autochthonous deposits are residual-formed on the serpentinite and have a thickness of about 10 m. The allochthonous deposits are the most interesting. These secondary deposits were derived from the autochthonous deposits. As a result of the Cretaceous transgression, they were transported and redeposited as sedimentary formations. The most interesting example is at Agios Ioannis, in which more than 10 m of redeposited stratified laterite has been deposited over dolomitic limestone with a basal unconformity. The nickel oxide content of the secondary deposits varies from 0.5% to 6.0%, and their chromium oxide (Cr_2O_3) content is uniform at 2–2.5%. As with the autochthonous, the allochthonous deposits have the richest nickel concentrations in the lower sections, which indicates leaching of nickel downwards from the upper layers.

Bedded Ferromanganese Deposits

Since manganese and iron are often enriched in the surface sediments and coat the basaltic outcrops above ocean rises, it should be expected that older manganese and iron deposits may show a close association with pillow basalt and pelagic oceanic sediments, implying that their formation was a result of a hydrothermal system at oceanic or marginal-basin spreading axes (*Figure 2.6*). The clearest examples are in Cyprus and Oman, but others occur in the Philippines, Cuba, the Japanese islands, and in central Europe and Turkey. These deposits are all associated with obducted ophiolite and eugeoclinal sequences.

The stratified manganese deposits have been reviewed by Shatskiy (1964) and Hewett (1966). They recognize a common association of the deposits with basic volcanic rocks and cherts, implying their formation upon oceanic-type crust under submarine conditions. There is also an association with intermediate volcanic rocks. Their reviews indicate that most of the large accumulations of iron-free manganese minerals in sedimentary rocks, and especially those of Upper Mesozoic and Cenozoic age, are derived from hydrothermal solutions emanating from submarine centres of volcanism.

CYPRUS

The *umber* is a brown extremely fine-grained manganiferous iron-rich sedimentary formation that lies unconformably over the Troodos pillow lavas (Constantinou and Govett, 1972). It is the first sediment deposited over the lava surface and does not occur again higher in the succession (*Figure 2.2*). Mines are situated at Skouriotissa, Mathiati and Mousoulos; the deposit is associated with red jasper and chert and fills irregularities in the pillow lava surface; although poorly crystalline it contains identified geothite. The origin may be considered as analogous to that of occurrences on areas of the present-day ocean floor where basalts are coated with iron/manganese oxides at crest rises (*see* page 14). The metal content is undoubtedly derived from the underlying lavas. The presence of radiolaria in the umber suggests a marine environment, but algal and plant fragments suggest that the environment of formation was not too far removed from a land mass, hence a marginal basin is more likely than a major ocean.

The *ochre* is a manganese-poor iron-rich bedded deposit containing varying proportions of interbedded chert and tuff. It occurs at the same localities as the umber and conformably overlies the Cyprus sulphide ore, contains bands and fragments of sulphides and is itself enriched in copper and zinc; it is orange-yellow to brown. Its fine banding is due to graded bedding. The interlayered massive sulphide consists of blocks of pyrite in a soft friable matrix. The commonest component is goethite. At Skouriotisa and Mathiati it is interbedded with siliceous limestone. It is thought that the umber is a subaqueous oxidation product of the Cyprus sulphide ore body, resulting from material leached from the ore exposed on the sea floor (Constantinou and Govett, 1972).

OMAN

The lavas of the Semail ophiolitic nappe of Oman, which formed at or very near a Late Cretaceous oceanic spreading axis, are intercalated with, and overlain by, a range of metalliferous sediments, including cupriferous sulphides, ferruginous ochres and ferromanganese umbers (Fleet and

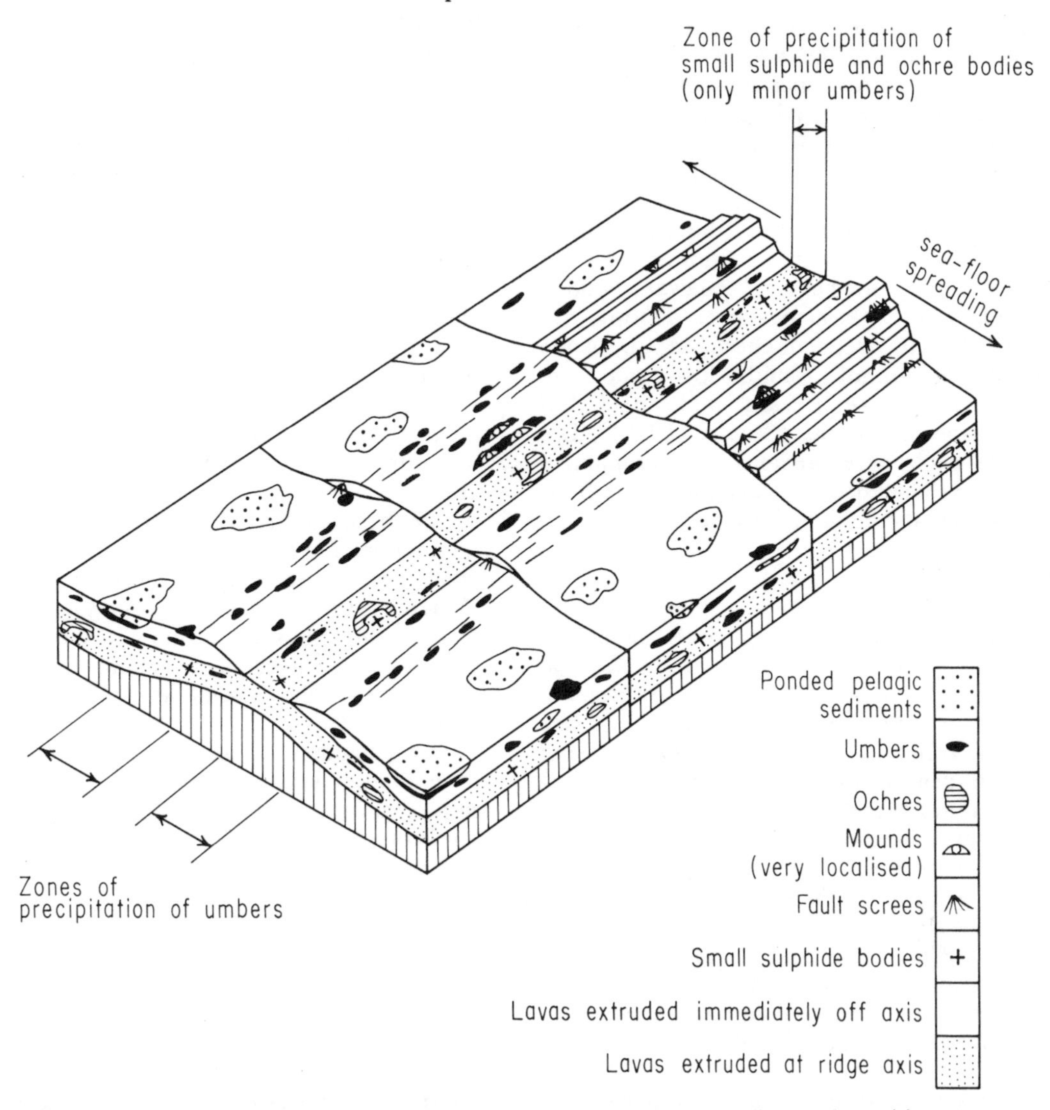

Figure 2.6. Schematic block diagram for a model of metalliferous sediment deposition at or near an oceanic spreading axis, showing the relationship between umbers, ochres and sulphide deposits as seen in the Semail Nappe, Oman (after Fleet and Robertson, 1980).

Robertson, 1980). Small massive cupriferous sulphides, gossans, ochres and minor umbers occur within the lower lavas. They also contain mounds of ferruginous oxide sediments associated with umbers. The mounds may have resulted from bacterial precipitation of sulphides and/or iron oxyhydroxides. Localized umbers, deposited from hydrothermal solutions, occur throughout and immediately above the upper lavas, including areas where subsea-floor faulting has produced thick volcaniclastic breccias. The uppermost umbers are overlain by radiolarian chert and foraminiferal chalk.

Most of the metalliferous deposits and pelagic sediments can be related to a spreading oceanic ridge affected by subsea-floor fault zones, leading to a generalized schematic model for subsea-floor mineralization (*Figure 2.6*).

PHILIPPINES

Manganese ore lenses and fracture fillings in pillow joints and interstices and in reddish volcanic shale, sometimes limy, are found in limited quantity within the chert–spilite association and more commonly the spilitic pillow lavas (Gervasio, 1973). The ore is commonly of braunite, cryptomelane and psilomelane, with secondary supergene pyrolusite. The grade averages 30–40% manganese. The intense deformation of the uplifted spilite sequence and lenticularity of the ore bodies hamper exploration. Occurrences are known from Vintar, Iliocos Norte, Luar and Malinao, Quezon Sierra Madre in Isabela, and Catanduanes. Low-grade ore (20–35% manganese) has been mined from bedded deposits interstratified with Miocene tuffs on Siquijor island. On Busuanga island, northeast of Palawan, bedded high-grade (48–55% manganese) deposits are conformable with bedded chert. In the Anda Peninsula of Bohol island, low-grade ore (25–30% manganese) is associated with basalt and Miocene limestone (Sorem, 1956). The Philippine manganese ore may be presumed to have been exhalated onto the basaltic sea floor before uplift of the ophiolite sequence.

CUBA

Manganese deposits associated with submarine volcanic rocks are found in southern Central America and throughout the Greater Antilles, but are of economic importance because of their size only in Cuba (Simons and Straczek, 1958). Most of the layered deposits occur in the upper few hundreds of metres of the Cobre Volcanics, which are associated with limestone. The oldest deposits are associated with the San Cayetano sedimentary sequence, which contains the Matahambre massive sulphide deposit. The most important deposits are of Late Cretaceous to Eocene age, situated in the Vincent–Cobre sequence of eastern Cuba, which contains the Cauto–Wagwater massive sulphide deposit. The largest manganese deposits are in the upper parts of the sequence; they form layers that follow and partly replace the pyroclastic sediments. Smaller, but important, deposits have formed by replacement of an extensive limestone layer in the upper parts of the Cobre formation.

Two of the largest deposits are the Quinto mine near Cristo and the Chario Kedondo in the foothills of the Sierra Maestra range, Oriente Province. The larger deposits are bedded, and contain 15–55% manganese, composed mainly of psilomelane and pyrolusite. The metals have been transported some distance from their submarine exhalative centre and deposited within limestone and tuff.

JAPAN

Strata-bound manganese deposits occur generally above massive chert and below laminated chert in close association with basalts (Ishihara, 1978). The geology of the deposits has been summarized by Watanabe *et al.* (1970a). There are many bedded deposits in Palaeozoic, Mesozoic and Miocene formations (*Figure 2.7*). There are metamorphosed equivalents, an example of which is described by Watanabe *et al.* (1970b). They occur in the Sanbagawa and Ryoke metamorphic belts and some occur in the metamorphic aureoles of Mesozoic granites.

The deposits are intimately associated with cherty sediments and associated with basic tuffs or lava flows. Hence they are assumed to have formed on the ocean floor. Pisolitic textures are known. The principal minerals are rhodochrosite, hausmannite and braunite. In the metamorphosed deposits there is a much wider diversity of minerals.

The best studied deposits are in the Ashio mountains (*Figure 2.7*). Bedded manganese deposits occur in at least four chert horizons and the Palaeozoic sequence contains abundant basic tuff. In the contact aureole of the Cretaceous granite, the ore is converted to manganese silicate–carbonate–oxide assemblages. The beds are strongly folded.

The manganese deposits can be subdivided into three types. Type A occurs in a geoclinal succession composed predominantly of chert, with slate, sandstone and a small amount of basic volcanic and pyroclastic material. These deposits contain rhodochrosite, hausmannite and manganese silicates. Type B is similar to A except for a closer association with basic volcanic and pyroclastic material including pillow lavas. Braunite is the characteristic mineral and the wall rocks often contain haematite. Type C occurs in the Neogene shallow-water rocks, associated with basic-to-intermediate volcanic materials in the marginal areas of the Green Tuff sedimentary basin. Manganese oxide (MnO_2) minerals are the major components, and they are commonly associated with jasper.

The absence of wall-rock alteration indicates that the deposits are syngenetic with sedimentation and that they are distal from the submarine hot spring or fumarole source.

The Noda–Tamagawa mine is one of Japan's major manganese producers, located 16 km southeast of Kuji City (*Figure 2.7*). The geology has been summarized by Watanabe *et al.* (1970b). The occurrence of sillimanite, andalusite and cordierite in pelitic hornfels near the manganese ore bodies indicates a high grade of contact metamorphism in the Cretaceous granite aureole. The pre-Cretaceous rock succession is of metachert and biotite hornfels. The manganese ore is conform-

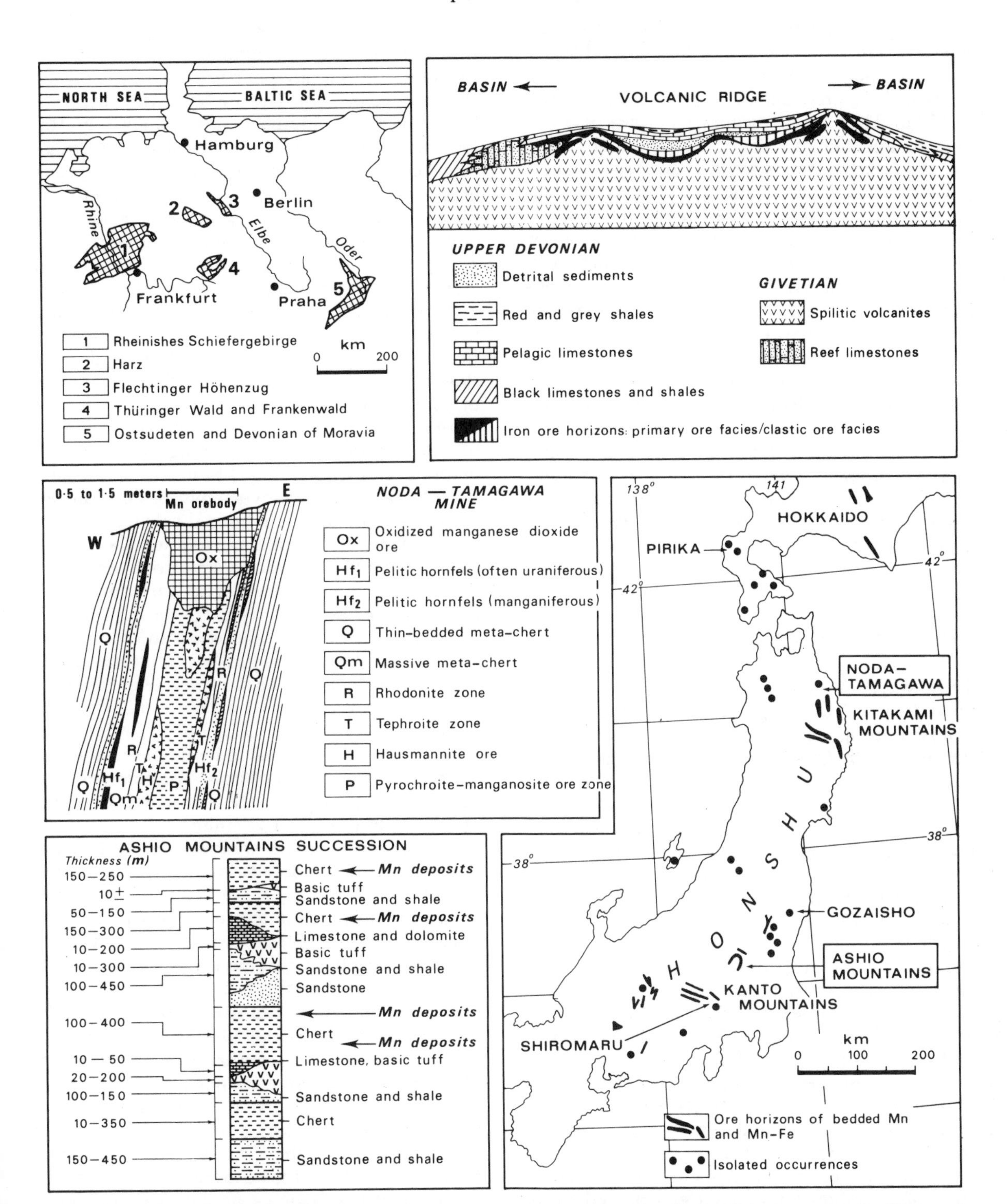

Figure 2.7. Bedded Fe-Mn deposits associated with pillow basalt and pelagic sedimentary rocks. (Bottom) the manganese deposits of Japan (after Watanabe *et al.*, 1970a, 1970b) and (top) the Lahn–Dill iron deposits of Central Europe (after Quade, 1976).

able with the chert and strikes 10–20° with a steep dip westwards. The manganese ore is found exclusively in the metachert. The ore zone extends 1200 m along strike and has been traced to a depth of 450 m. The ore is lenticular and occurs between bedded and massive chert. It is of high grade (30–55% manganese) and variable type, including pyrochroite ore, manganese carbonate ore, hausmannite ore, braunite ore, tephroite, rhodonite and albandite ore. There is a low-grade ore of the manganese hornfels-type containing only 10–20% manganese. There is also a supergene-enriched zone containing 60–85% manganese dioxide. Most of the ores are banded.

Associated with the manganese ore is a uraniferous pelitic hornfels of lenticular form, which is itself associated with massive chert. The uranium is considered to have been deposited on the sea floor under submarine conditions from hot geothermal systems, probably related to an active spreading axis. The deposits have been modified by uplift and metamorphism and subsequently enriched by supergene processes.

Lahn–Dill Iron-ore Deposits

These volcano–sedimentary iron ores are characterized by Devonian and Lower Carboniferous deposits of the Central European Variscan, well summarized by Quade (1976), and similar Devonian deposits of the Taurus mountains of Turkey (Vaché, 1966).

The iron ores occur as syngenetic conformable layers and lenticular bodies within sedimentary, volcanic and mixed sequences in the Variscan eugeocline. The most important mining districts have been (*Figure 2.7*): the eastern Sauerland anticline (northeastern part of area 1); the Dill and Lahn synclines (southern part of area 1); the Kellerwald (southeastern part of area 2); the Harz and Elbingerede complex and the Oberharz diabase belt (area 2); the Thüringer Wald and the Permitz fold belt (northern part of area 4); the Frankenwald (southern part of area 4); and the Ostsudente (northern part of area 5), Altvatergebirge, Hohes Gesenk and Devonian deposits of the Upper Moravia Valley (central area 5).

The ore bodies are found in areas with volcanic rocks of the spilite–keratophyre group, predominantly with spilitic tuffs, belonging to the pre-flysch period of the geoclinal development. The ore deposits of economic importance occur only in the zones of maximum accumulation of spilitic tuffs within a volcanic arc near to the active volcanic centres (*Figure 2.7*). The ore bodies are underlain by spilitic tuffs, sometimes by pillow lavas, and are overlain by pelagic limestones or shales with intercalations of clastic material. There are also transgressions of ore sediments on nearby reef complexes. The ore sequences are well bedded and internally laminated, showing features of rapid precipitation, erosion, transport and redeposition. Generally silica-rich deposits occur at the bottom and calcareous ones at the top, with horizontal transitions from the upper part of a volcanic eruptive centre to calcareous accumulations along its slopes. The ore sediments are rhythmic sequences of extremely varied iron contents. The iron is incorporated as primary and secondary mineral phases including haematite, magnetite, siderite, limonite, iron-containing chlorite, stilpnomelane, specularite and melnikovite–pyrite. The main accessory components are quartz, calcite, dolomite and chlorite. Common impurities are clastic fragments, pyroclastic particles, limestone lenses and argillaceous bands. Small-scale variation of ore facies and ore quality is typical of the deposits.

Because of their formation on the top of volcanic rises, the ore bodies occur only in or on anticlinal structures. The ore reserves of any single deposit are low and rarely exceed 5 million t. The largest is in the still-operating Fortuna mine in the Lahn area, with about 10 million t of ore containing about 3.5 million t of iron. The combined ore deposits of any one volcanic centre may, however, amount to as much as 100 million t.

The water-enriched spilitized basaltic ocean floor has to be considered as the source for the iron and silica. The primary ore sediment seems to have been a rapidly precipitated gelatinous mud of high viscosity that was able to resist strong waves and currents along the top of the volcanic rise.

The similarity of the Lahn–Dill iron ore deposits and the Archaean iron ores has been demonstrated by Kimberley (1978b) who classified them both together as shallow-water volcanic platform-type deposits. In that the Archaean deposits are probably related to rifting of the Archaean cratons with formation of oceanic-type or marginal basins, they must have analogues through the Phanerozoic period. However, the higher heatflow regimes of the Archaean crust provided a more fruitful supply of the economic elements.

3

A Spectrum of Sea-floor Sulphides

Conformable massive sulphide deposits represent an important world resource of base and precious metals. They are often of high grade and their associated rock assemblage allows their tectonic setting of formation to be deduced fairly accurately. It is generally accepted that they formed at or close to the ocean floor by volcanic-exhalative processes. Summaries of the deposits are given by Hutchinson (1973, 1980) and Sawkins (1976). Their site of formation may range from ocean rises to volcanic island arcs, and some may have eventually been incorporated in deep-sea trenches or imbricate sedimentary wedges (*Figure 3.1*). Some may have formed adjacent to the centres of submarine volcanic–geothermal activity, known as exhalative centres, and they are classified as proximal. Others may have been deposited by sea water far from the centres, and they are called distal (*Figure 3.1*). There therefore exists a spectrum of deposit-types, some more associated with volcanic rocks, others more with sedimentary formations. In many cases, the Phanerozoic examples have many features in common with Precambrian deposits, described in Chapters 9 and 10. Submarine volcanic-exhalative processes of ore formation seem to have existed on earth since the Archaean era. However, the plate-tectonic interpretations for the Phanerozoic deposits differ somewhat from those of the Precambrian. The general association and tectonic positions of the various types of deposits are shown diagrammatically in *Figure 3.1*.

Cyprus-type

This type is of cupriferous pyrite. It has virtually no lead, only minor zinc and carries significantly higher proportions of gold relative to silver than other massive sulphides. Copper contents are several percent,and silver/gold ratios normally range from 2 to 10. The associated igneous rocks are all of the ophiolite suite, the closest association being with pillow basalt. Lenses of massive pyrite–chalcopyrite ore are contained in spilitic basaltic pillow lava, which resembles modern oceanic tholeiite. Felsic volcanic rocks are usually absent but, if present, are of keratophyre. The associated sedimentary rocks are insignificant but include pelagic facies – radiolarian chert especially – and tectonic formations such as mudstone melange. Very few clastic or pyroclastic sediments are associated. The examples of such deposits are all Phanerozoic, with the best known in the Mesozoic, including the numerous deposits of Cyprus (Constantinou and Govett, 1972, 1973; Constantinou, 1976). Tertiary deposits are represented by those of Ergani–Maden in Turkey (Griffiths *et al.*, 1972) and of the Philippines (John, J.U., 1963; Bryner, 1969). Ordovician examples are represented by occurrences in Newfoundland (Upadhyay and Strong, 1973).

A model of hydrothermal ore deposition has been presented by Heaton and Sheppard (1977) based on hydrogen-isotope and oxygen-isotope studies of the Cyprus ores. In the stockwork zones, which underlie the sea-floor sites of massive sulphide ore bodies (*Figure 3.2*), isotopic temperatures declined from initial values greater than 300°C (with water/rock ratios greater than 1 by weight) to less than 220°C during post-sulphide, later-stage vein quartz formation; these temperatures were higher than in the country rocks, which were undergoing zeolite facies subsea-floor metamorphism. The stockwork zones are interpreted as the discharge zones of the sea water–hydrothermal fluids responsible for the regional mass transfer and the mineralization. At least some of the sulphur in the ore bodies and stockworks is of sea-water origin; but most may be magmatic. The formation of the massive ore deposits requires the combination of several factors, including the penetration of brine to deep crustal levels (*Figure 3.2*), where temperatures were greater than 300°C in an active spreading centre, leaching of the ore metals and discharge of the hot

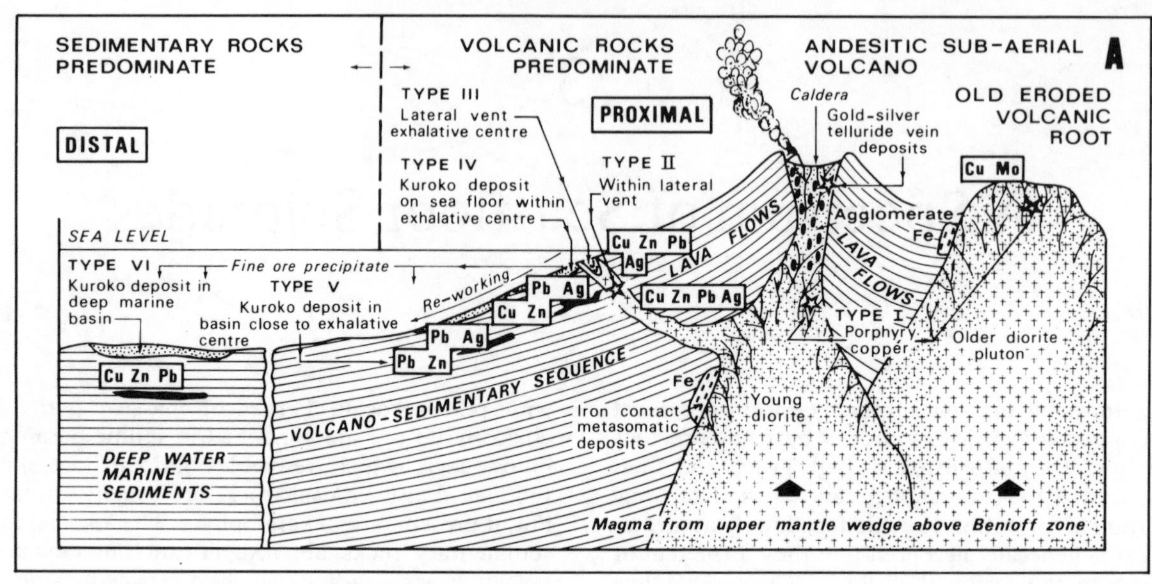

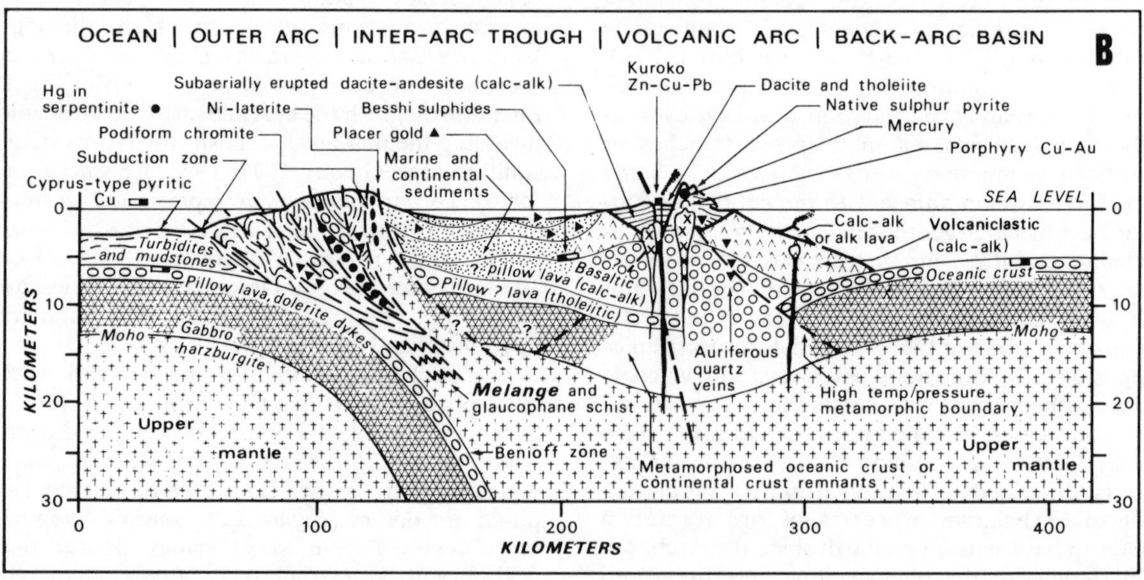

Figure 3.1. (A) Diagrammatic sketch cross-section of a volcanic island arc showing the relationship between the spectrum of Kuroko deposits and porphyry-copper deposits (after Colley, 1976). (B) Schematic cross-section of an island-arc consuming plate margin showing possible locations of mineral deposition (after Garson and Mitchell, 1977).

springs into basins on the sea floor where sulphide deposition could both occur and accumulate.

Cyprus

There are about 15 ore bodies that outcrop in the Troodos pillow lavas that flank the Troodos igneous complex. Most of them are centred around Skouriotissa and Mathiati, in a belt trending northwest–southeast, lying on the northeastern side of the Troodos Range. The geology has been summarized by Constantinou (1976) and Constantinou and Govett (1972, 1973). Each deposit had an initial saucer-shape and lies on a hydrothermally altered and brecciated zone of the pillow lavas containing disseminated sulphides. There is a distinct vertical zoning comprising massive ore containing more than 40% sulphur, now absent in some deposits, underlain by a pyrite–quartz zone (30–40% sulphur) and then a stockwork zone containing less than 30% sulphur. Sulphur continues to decrease downwards.

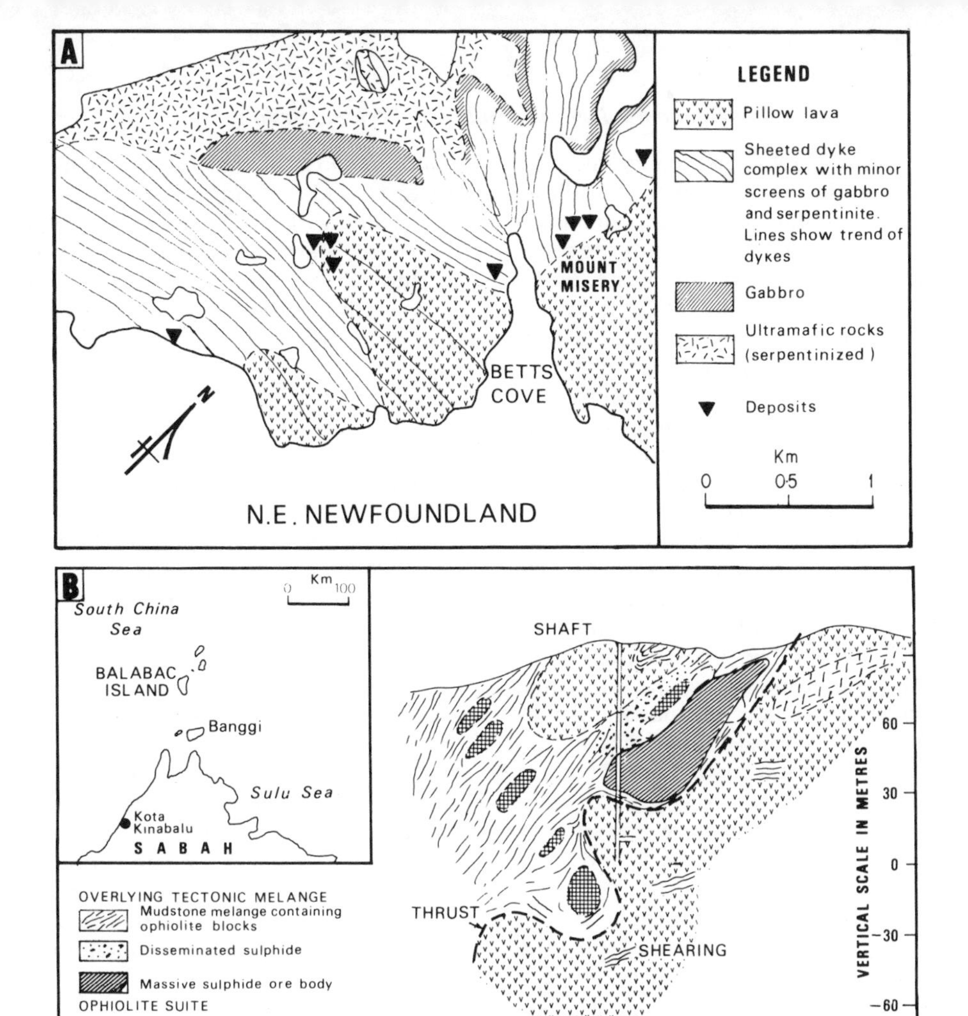

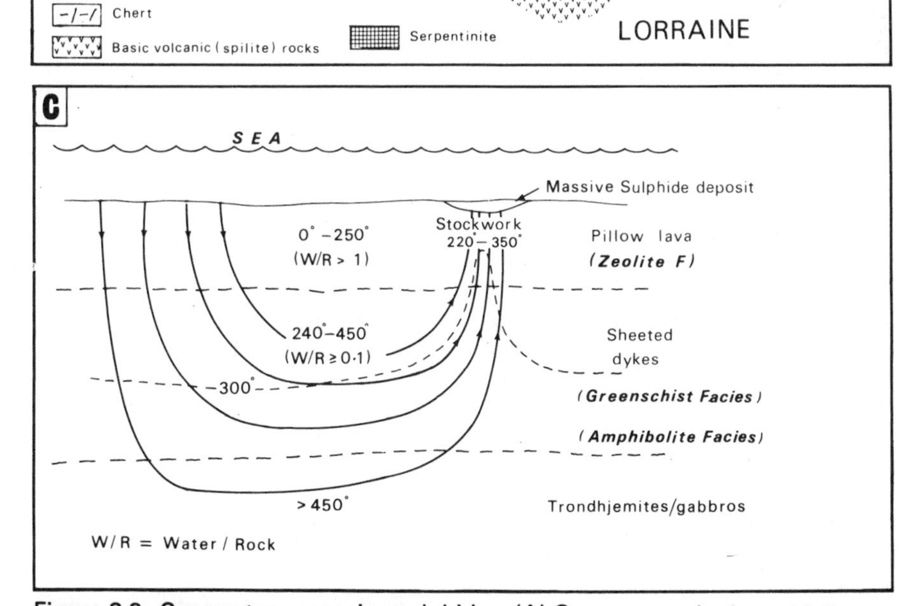

Figure 3.2. Cyprus-type massive sulphides. (A) Occurrences in the ophiolite sequence of Betts Cove, Newfoundland, Canada (after Upadhyay and Strong, 1973). (B) Lorraine ore body of Balabac island, north of Borneo (after John, J.U., 1963). (C) Diagrammatic model of sea-water brine circulation through Cyprus-type ocean crust to give rise to ore deposits on the pillow basalt ocean floor (from Heaton and Sheppard, 1977).

The massive ore is of two types: conglomeratic, composed of sulphide blocks commonly pillow-shaped in a sugary sulphide matrix; and an underlying compact ore with large blocks of pyrite separated by fractures containing sugary pyrite. In the conglomerate ore, the size of the blocks increases downwards. The conglomeratic structure is considered secondary, produced by electrochemical leaching as the ore body lay on the sea floor.

There is also a characteristic collomorphic banding of some ore. The collomorphic pyrite texture is thought to be genetically related to fragmentation of the ore and to have been acquired as a result of differential leaching of the euhedral zoned pyrite. Subsequent covering of the ore body by its own decomposition products and by later basalt flows isolated it from the sea water and thus protected it from complete oxidation. The ore is predominantly of pyrite with variable marcasite, and interstitial chalcopyrite, covellite, sphalerite, quartz and gypsum. The compact ore consists of pyrite with thin coatings of chalcopyrite and covellite on fractures and pyrite crystal faces.

Sulphides in the stockwork zone beneath the ore bodies occur chiefly between the pillows and in the fractured pillows.

The ore bodies are clearly not a replacement-type and owe their origin to fumarole-type solutions in caldera-like depressions on the pillow-lava sea floor. Submarine leaching led to secondary enrichment. Obduction of the ophiolite led to faulting and other tectonic structures in the originally simple saucer-shaped ore bodies.

Newfoundland

The Betts Cove copper deposits occur in one of the eight Lower Ordovician ophiolite complexes of Newfoundland (*Figure 3.2*). The deposits consist predominantly of pyrite, chalcopyrite, with minor sphalerite, displaying such features as sedimentary slump folds, fracture-filling, and replacement features (Upadhyay and Strong, 1973). The ore body localities are strictly controlled by the ophiolite stratigraphy, localized near the base of the volcanic sequence. They make excellent examples of Cyprus-type ore bodies that have subsequently undergone remobilization and re-emplacement along fault zones imposed on the ophiolite during tectonic obduction.

Philippines

The Barlo mine of West Luzon is of Cyprus-type low-grade fine-grained pyrite surrounded by small random zones of chalcopyrite-rich ore containing up to 8% copper and/or 6% zinc. The ore is closely associated with the Zambales ophiolite pillow lava but has been remobilized and reconcentrated within breccia (Bryner, 1969). An alternative interpretation is that the volcanic breccia was the exit of the hydrothermal ore fluids on to the basaltic ocean floor.

The Lorraine ore body of Balabac island (*Figure 3.2*) is a massive cupriferous pyrite deposit averaging over 4% copper (John, J.U., 1963). The massive sulphide occurs within a mudstone melange lying immediately above spilitic pillow basalt closely associated with pelagic chert. The melange contains blocks of serpentinite, basalt and chert. There is no hydrothermal wall-rock alteration around the ore body, showing that it was not emplaced directly into its present setting. It is likely that the tectonic mudstone melange, which contains huge ophiolite bodies, has also incorporated the ore body with its associated spilite. The ore body is 122 m long with a maximum down-dip dimension of 95 m and an average width of 15 m. At its widest it is 25 m thick. The contact with the country rock is sharp. Pyrite is predominant, followed by chalcopyrite, with subordinate sphalerite, and there are local haematite concentrations. It is capped by a gossan containing azurite and malachite.

Turkey

The Ergani–Maden area of the Taurus mountains of southeastern Turkey contains important massive sulphides (Griffiths *et al.*, 1972). The largest ore body is at Ana Yatak. It occurs in mudstone melange lying on top of serpentinized ophiolite. The mudstone is chloritized and contains disseminated sulphides. The main ore body is about $550 \times 300 \times 50$ m in maximum dimensions. It consists of pyrite and chalcopyrite and locally magnetite and pyrrhotite. The gangue is of chlorite, quartz, minor calcite and barite. A zone rich in supergene chalcocite, covellite and bornite lies between the sulphide ore body and the well-developed gossan.

Kuroko Deposits

It is likely that the Kuroko deposits of Japan are the best studied of all mineral deposits. Interest is growing because similar-type deposits are being discovered and described from many other parts of the world, ranging in age back to the Palaeozoic era. The large number of mines in Japan has attracted eminent Japanese as well as non-Japanese geologists largely because the Japanese Miocene occurrences are so young that they have suffered little tectonic deformation so that their study has allowed the formulation of a genetic model for submarine exhalative mineralization associated with a calc-alkaline volcanic arc. The vast literature has been summarized in the following outstanding publications on which the subsequent account is based: Matsukuma and Horikoshi (1970), Ishihara (1974), Lambert and Sato (1974), Sato (1974), Sato (1977) and Ohmoto (1978). Interpretations of isotope studies are based on Ohmoto and Rye (1974) and Ishihara and Sasaki (1978).

The Kuroko deposits of Japan are zoned massive base-metal deposits that are largely stratiform in volcano–sedimentary sequences dominated by felsic tuffs, rhyolite lavas and shallow intrusives of Miocene age. Base-metal veins and stockwork deposits also occur in these sequences, which are referred to as green-tuffs. The total Japanese reserves of Kuroko are around 50 million t of ore. The overall average composition for the Hokuroko district, which is the pre-eminent ore region, is 2% copper, 5% zinc, 1.5% lead, 21% iron, 12% barium, 1.5 g t^{-1} gold and 95 g t^{-1} silver. Some of the black ores contain uranium concentrations. The highest content of uranium oxide (U_3O_8) recorded is 2300 ppm with an average around 100–1000 ppm. The Hokuroko district contains the three largest mines: Hanaoka (35 million t), Kosaka (25 million t) and Shakanai (10 million t) (*Figure 3.3*).

Individual deposits commonly consist of two or three closely spaced ore bodies of variable size. The largest at Kosaka was originally 700 × 300 × 500 m. The common features of the ore bodies are shown in *Figure 3.3*. Zoned massive stratiform ore, typically oval-shaped in plan, grades downwards into less economically important stockwork ore, which is referred to as siliceous ore, composed of disseminated and network or stockwork mineralization distributed in an irregular funnel-shape in silicified felsic lava (called white rhyolite) and in pyroclastic rocks. Thin beds or small lenses of ferruginous chert are commonly present either directly overlying the stratiform ore body or within the overlying tuffs. Lenticular or irregular masses of gypsum and/or anhydrite are usually present. The boundary between the stratiform ore body and the overlying tuff or mudstone is usually sharp. The stratiform mineralogy usually lenses out rather sharply laterally, and the stockwork ore grades into unmineralized volcanic rocks and tuffs.

The ore bodies have a distinct vertical zoning in the following ascending order. (1) Siliceous ore (keiko): pyrite–chalcopyrite–quartz stockwork ore. The volcanic rock textures are frequently obscured by an extensive development of cryptocrystalline quartz. (2) Gypsum ore (sekko-ko): gypsum–anhydrite–(pyrite–chalcopyrite–sphalerite–galena–quartz–clay) strata-bound ore. (3) Pyrite ore (ryukako): pyrite–(chalcopyrite–quartz) ore usually stratiform. (4) Yellow ore (oko): pyrite–chalcopyrite–(sphalerite–barite–quartz) stratiform ore. (5) Black ore (Kuroko): sphalerite–galena–chalcopyrite–pyrite–barite stratiform ore. Towards the top there are significant amounts of tetrahedrite–tennantite. Bornite may be abundant in some deposits. (6) Barite ore: thin well-stratified bedded ore of almost pure barite. (7) Ferruginous chert: a thin bed of cryptocrystalline quartz and haematite. The relative abundance of these ore facies varies from deposit to deposit. The minerals are usually very fine grained (200–200 μm) in the black Kuroko ore and the texture is massive.

All the Kuroko deposits are restricted to the green-tuff region where differential subsidence and intense submarine volcanic activity occurred in the Miocene period. The economic deposits were formed only at a time of large-scale mixing of magmatic fluids and sea-water hydrothermal systems. This happened in Japan 13 Ma ago (1 Ma = 1 million years). Present Japanese hot springs are mostly of 100% meteoric or sea-water origin and hence it is extremely doubtful if there is any present-day active mineralization going on. The deposits lie above a present-day Benioff-Zone depth of about 150 km and the Kuroko deposits lie within the arc tholeiite zone (*Figure 3.3*). They were deposited during a very limited period when the green-tuff zone experienced a high geothermal heat flow.

The currently popular model for the ore formation, based largely on isotopic studies substantiated by water–rock experiments, implies that the ores were formed near the sea floor from hydrothermal solutions. Fluid-inclusion studies on quartz in the stockwork ore suggest a temperature range of deposition of 200–270°C. There appears to have been a 50°C drop in temperature from the stockwork to the top of the massive ore. A water depth of 300–500 m has been estimated and it is likely that the ore solutions boiled before they reached the sea floor.

Isotope studies indicate that the mineralizing solutions were coeval with sea water. Lead-isotope and sulphur-isotope data suggest that magmatic water also played a dominant role. The sulphur-

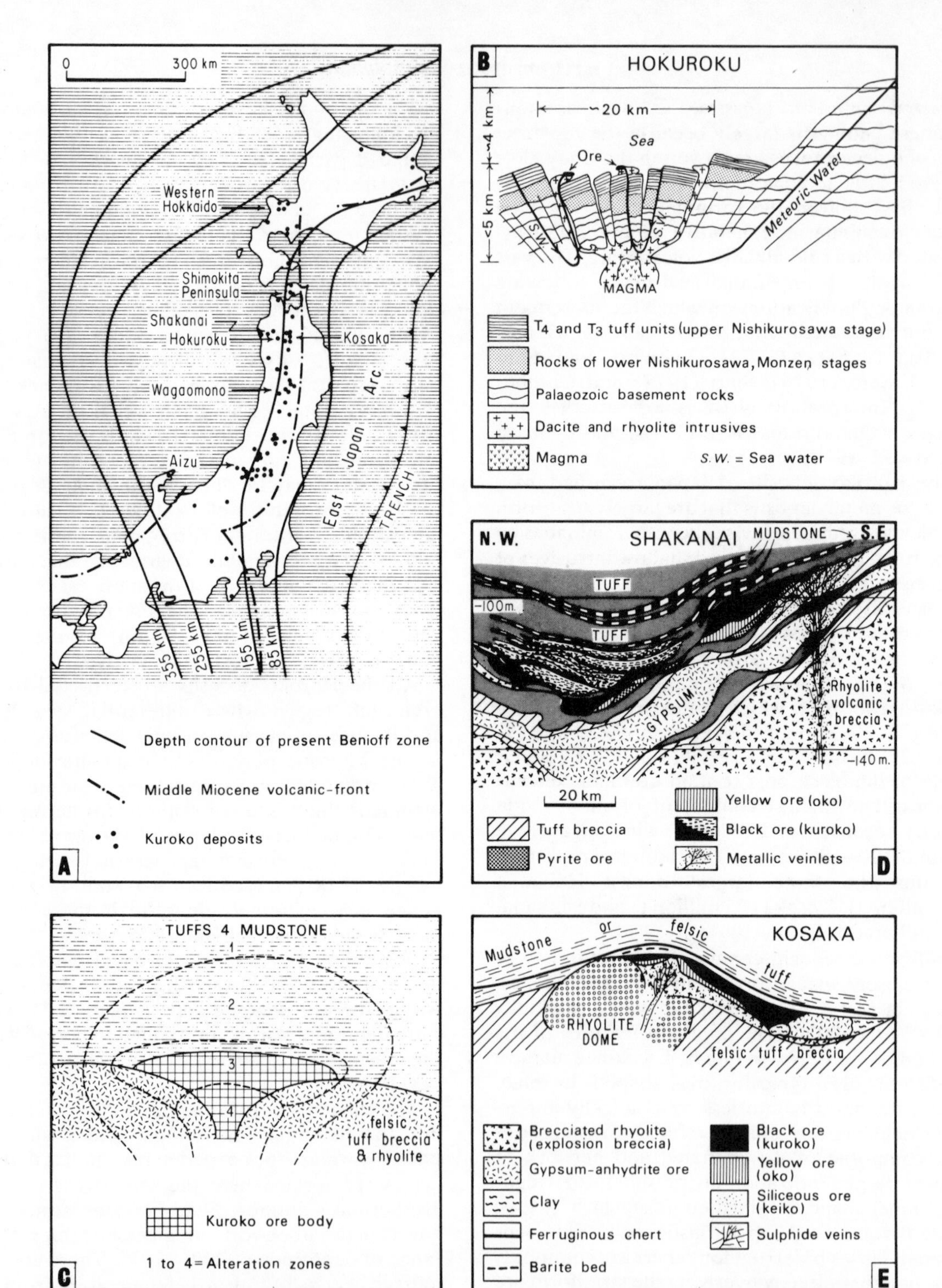

Figure 3.3. (A) Kuroko ore deposits—their main distribution in relation to Benioff-Zone depth and volcanic front (after Sato, 1977). (B) Concept of ore-body formation in submarine caldera based on the Hokuroku district (Ohmoto, 1978). (C) Clay alteration zones around an ore body (Lambert and Sato, 1974); (1) montmorillonite, zeolite, cristobalite, (2) sericite, sericite – montmorillonite, Fe/Mg-chlorite, (albite, K-feldspar), (3) sericite, sericite – montmorillonite, Mg-chlorite, kaolinite, (4) quartz, sericite, Mg-chlorite. (D) Cross-section of ore deposit (1), Shakanai mine (after Kajiwara, 1970; Ohtagaki *et al.*, 1974). (E) Schematic section of a typical kuroko ore body based on Kosaka Mine (after Lambert and Sato, 1974).

isotope data suggest that the ores, the plutonic rocks and the volcanic rocks all had a common magmatic sulphur source, which was likely to be deep seated. Thus, Ishihara and Sasaki (1978) conclude that the Miocene green-tuff mineralization of volcanic affinity is closely associated with coeval tonalite plutonic activity.

It is concluded that the hydrothermal process occurred as a result of the creation of submarine calderas (Ohmoto, 1978). The Hokuroku Basin, which has the largest concentration of ores, was a submarine resurgent caldera (*Figure 3.3*). The initiation and termination of the mineralization was controlled by subsidence, and then by resurgency of the caldera.

It must be emphasized that Kuroko formation is not always to be expected when there is submarine volcanism. The Japanese deposits formed in a short time span of about 0.3 Ma, occurring at the last stage of a large-scale magmatic cycle that continued for 10 Ma. Although there was large-scale hydrothermal activity on the sea floor throughout the green-tuff formation, only the Middle-Miocene activity gave economic deposits.

The rocks of the green-tuff basins have undergone regional zeolite facies metamorphism. They have never been deeply buried and the geothermal gradient has been estimated to be of the order of 100°C km^{-1} for the mineralization episode. The clay alteration zones are shown in *Figure 3.3*. The most significant changes are: silica is enriched in zone 4; potassium is slightly enriched in zones 2–4 and sodium leached from all zones, while there is a progressive increase in potassium/sodium ratio towards ore bodies over a distance of hundreds of metres; magnesium is enriched in all zones; iron is strongly enriched in zones 3 and 4; and calcium is depleted in the tuffs of zones 3 and 4 but gypsum–anhydrite ores can occur there, while calcium is concentrated in zones 1 and 2.

Spanish–Portuguese Massive Sulphide Deposits

The Iberian massive sulphide bodies of Upper Devonian–Lower Carboniferous age have characteristics of the Kuroko deposits and some characteristics of the Cyprus-type. They are therefore intermediate between ocean or marginal basin-spreading ridge and submarine volcanic-arc formation, and serve to display the spectrum of massive sulphide deposits.

The well-known Iberian Pyrite Belt, 230 km long, 40 km wide, extends westwards from Seville and swings northwards into Portugal (*Figure 3.4*). It contains many important massive sulphide ore bodies, including the world's largest at Rio Tinto. The geology of the pyrite belt and of the mines has been summarized by Williams, D. (1962), Strauss and Madel (1974), Williams, D. *et al.* (1975) and Soloman *et al.* (1980). The region has reserves of approximately 1000 million t of sulphide ores with an average content of 46% sulphur, 42% iron and 2–4% copper + lead + zinc. There is a close association with many small manganese deposits.

The Devonian–Carboniferous eugeocline succession begins with a basal slate–quartzite group of Upper Devonian age. It contains rare conglomerates and limestones; the thickness exceeds 500 m. This is followed by the volcanic-sedimentary complex of Lower Carboniferous age, composed of slates, cherts, jaspers, greywackes, felsic and mafic volcanic rocks, together with manganese and pyrite base-metal ores; the thickness is variable up to 800 m. Overlying this is the Culm group of slates and greywackes of thickness variable from 500 m to 1000 m. During the Hercynian orogeny these formations were intensely folded and locally overturned and a chlorite-grade of metamorphism was imposed.

The volcanic–sedimentary complex, which houses the mineralization, is characterized by quartz keratophyre volcanic rocks in which albite is the predominant feldspar; there are also dacites. The acid eruptive centres are grouped in lineaments parallel to the regional strike, east–west in Spain and northwest in Portugal. Acid volcanic rocks always outnumber the basic. The explosive volcanic activity gave predominantly pyroclastic formations; basic volcanic rocks include spilitic pillow lavas.

The stratiform pyrite ores are always linked to the submarine explosive acid volcanism, and were formed mostly at its final stages. The deposits are confined to the vicinity of the extrusive centres, lying directly on pyroclastic rocks, or sometimes interbedded with black shales. The manganese ores are associated with jaspers and with the pyrite ore deposits.

As with the Kuroko deposits, the Rio Tinto stratiform deposit and almost all the other pyrite ore bodies are underlain by stockwork pipes, which are mineralized and mined at San Dionisio and Cerro Colorado.

RIO TINTO (CERRO COLORADO)

It is estimated that prior to their long history of weathering and subsequent mining, the Rio Tinto deposits consisted of at least 500 000 000 t of pyrite. From 1875 to 1976 the pyrite production was 128 000 000 t. The pyritic ores extend over 5 km^2;

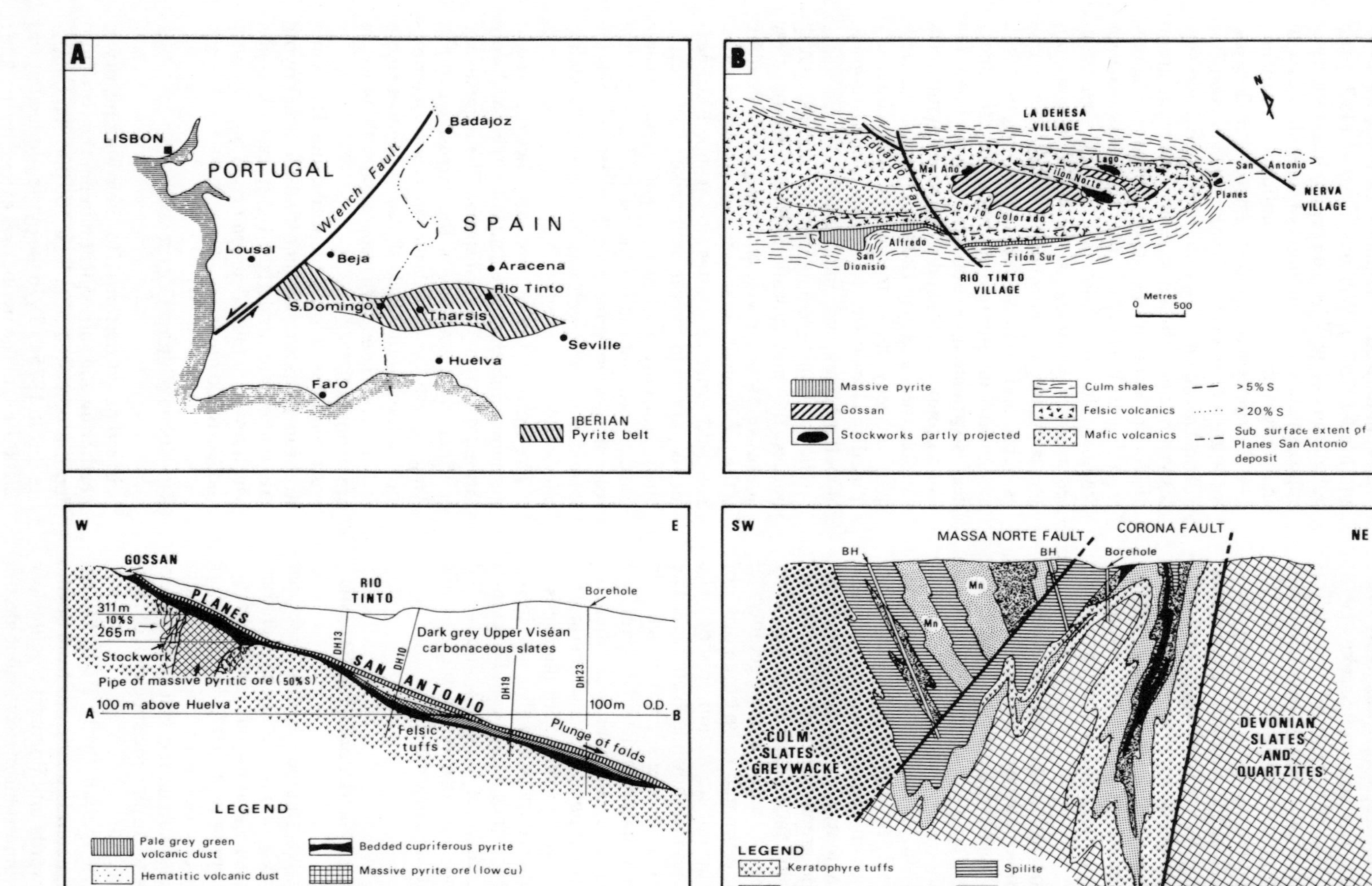

Figure 3.4. (A) Kuroko-type massive sulphide stratiform deposits of the southwest Iberian Early Carboniferous pyrite belt. (B) Simplified geological map of the Rio Tinto–Cerro Colorado area (after Solomon *et al.*, 1980). (C) Cross-section through the Planes–San Antonio deposit (after Williams, D. *et al.*, 1975). (D) Cross-section through the Lousal deposit (after Strauss and Madel, 1974).

they are overlain by carbonaceous slates of Upper Viséan age and underlain by Lower Viséan to Tournaisian volcanic rocks, all folded into the Rio Tinto anticline. Most of the mining production has come from the northern and southern flanks of this fold. A general description of the geology of the mining region is given by Solomon *et al.* (1980). The steeply dipping and crumpled stratiform pyrite sheets overlie altered and mineralized volcanic rocks (*Figure 3.4*). A typical mine succession, from the San Dionisio area, from top downwards is: more than 500 m of dark-grey carbonaceous Culm shales; 0–100 m of transition beds, including tuffs, haematitic shales and jaspers; 50 m of fine-grained massive pyrite; 0.3 m of banded granular pyrite; 0.2 m of pyritic sericitic tuff, with minor lenses of sphalerite and galena; 100 m of stockwork of pyrite and chalcopyrite veinlets cutting chloritized felsic volcanic rocks; about 30 m of breccias containing fragments of shale and tuff with local stockworks; and more than 200 m of mafic lavas and tuffs, with local chloritic stockwork.

The transition beds beneath the massive sulphide support the view that the pyritite formed in a marine environment. The massive geothite gossans that originally covered much of the Cerro Colorado ore body are thought to have been derived from the massive pyritite that originally extended over the whole anticlinal hinge area. The stockworks beneath the pyritite and gossan are now being worked in the San Dionisio and Cerro Colorado areas (*Figure 3.4*). It is clear that chlorite–quartz–pyrite–chalcopyrite stockwork (called the *clorita*) passes to sericite–quartz–pyrite stockwork near the limits of the massive pyritite. The clorita contains zones of intense pyrite–chalcopyrite veining of chloritized volcanic rocks with grades of 1.7% copper and 31% sulphur in the San Dionisio mine. The mineralization is generally confined to the felsic volcanics, but high values also extend into the mafic volcanics. At Cerro Colorado, grades of 0.8% copper are mined.

The formation of the mineralization is assumed to be a result of submarine hydrothermal plumes emanating from fumaroles on the sea floor. Several plumes may have interfered with each other. Modelling can account for the massive pyritite and the distribution of copper, lead and zinc (Solomon *et al.*, 1980).

The Planes–San Antonio Deposit

The Planes pipe consists of sulphur-rich copper-poor sulphide ore formed by an almost complete replacement of a stockwork conduit that traverses the felsic pyroclastics (*Figure 3.4*). Overlying the pipe, and extending far beyond its margin, is the genetically related Planes–San Antonio cupriferous pyrite ore body, a sedimentary–volcanogenic deposit precipitated as a chemical sediment derived from sea-floor hot springs associated with the Lower Carboniferous volcanic island arc of southwest Iberia (Williams, D. *et al.*, 1975). Most of the pyrite ore bodies of Rio Tinto were deposited directly above or close to their underlying stockwork feeder channels. The Planes–San Antonio pyrite ore body displays well-preserved stratification, slump structures and interdigitating tuff bands, and it was mainly precipitated comparatively distant from the source Planes stockwork.

Across-layer zoning in the sheet base-metal ore body is characterized by an increase in copper, and to a lesser extent in zinc and lead, towards the base of the deposit. Much of this enrichment may have been due to late hypogene processes involving the ascent of juvenile-meteoric solutions that leached base metals from the underlying volcanic pile and deposited them as chalcopyrite, sphalerite and galena by partial replacement of the lower part of the pre-existing, pyrite-rich, stratiform sulphide sheet. A notable feature of the pyritite is its abundance in colloform and framboidal pyrite.

The Lousal Mine

This is in the westernmost extension of the pyrite belt. The ore-bearing facies of predominantly fine-grained acid pyroclastic rocks grading into tuffites and black shales, interbedded with abundant siliceous shales, occupies the overturned and intensely folded southern flank of an anticline (*Figure 3.4*). The ore bodies are lined up along one horizon of 1.5 km length in strike and their downward extension is mined to 460 m depth (Strauss and Madel, 1974). They occur on either limb of the narrow mine syncline. The ore-bearing horizon is made up of black and grey shales with occasional thin lenses of crinoidal dolomite. The first basic volcanism produced pillow spilite lavas, which were extruded after the ore formation. The manganese formation is composed of siliceous shales, radiolarian chert and tuffite shales. The manganese minerals are principally of rhodochrosite and rhodonite, occasionally with spessartite and accessory pyrite. The gangue is of quartz, chalcedony, ankerite, sericite and chlorite. The sedimentary structures are the same as in the massive pyritite, except that the manganese may be oolitic; the ooliths formed around radiolaria or debris. The mines work the oxides in the upper parts of the beds but the carbonate-rich layers only at deeper horizons.

The 18 ore lenses being worked range in size from 100 000 t to 3 million t each. All are composed of pyrite, chalcopyrite, sphalerite, galena and gangue. The average ore contains 45% sulphur, 39% iron, 0.7% copper, 0.8% lead and 1.4% zinc. The following sedimentary structures characterize the ore: interbedded layers of shales, tuffs and cherts (sterile); fine-grained ore layers alternating with coarser-grained layers, rhythmic bedded on a millimetre scale; graded, oblique and cross bedding; and synsedimentary and ore breccias, slump structures and scour fill channels. Apart from the massive ore, there is a disseminated copper ore body related to chloritized footwall keratophye tuffs, ash tuffs and tuffite shales, spatially distributed on both limbs of the anticline; it is of pyrite, pyrrhotite, chalcopyrite, sphalerite and galena with a gangue of quartz and approximate grade 25% sulphur, 0.7% copper, 0.6% zinc and 0.1% lead. This disseminated ore is probably epigenetic and represents a feeder to the exhalative stratiform ore body.

The sulphur-isotope data support an interpretation that the sulphur in the feeder stockworks is isotopically heavier than that in the overlying stratiform ore bodies. Mantle-derived sulphur on its upward passage became mixed with sea-water sulphate entrapped in the buried volcanic rocks and then ascended along the stockwork systems on to the sea floor. It is visualized that the principal part of the pyrite ore bodies were deposited as volcanic sediments, and partly resedimented penecontemporaneously. The similarity to the manganese sediments suggests a similar origin, hydrothermal submarine emanations of mixed magmatic–meteoric origin. The close association with felsic explosive volcanic activity, and the underlying stockwork systems, indicate a close affinity with Kuroko-type deposits. The association with a keratophye rather than a rhyolite felsic assemblage, and with spilitic pillow basalts, show some affinity with Cyprus-type deposits. However there can be no doubt that the deposits are submarine-exhalative and that sedimentary processes played a most important role in their deposition.

Some Other Examples of Kuroko Deposits

The Palaeozoic rocks in southeast Ireland consist of a Cambrian–Lower Ordovician sequence of clastic and basic volcanic rocks overlain unconformably by Upper Ordovician felsic volcanic rocks and argillites and then by Silurian argillites. The Avoca sulphide-ore deposits are contained within a belt of felsic volcanics preserved in the synclinal cores of Caledonian folds; the geology and details of the deposits have been described by Platt (1977) and Badham (1978). The massive ore is associated with extensive zones of stockwork ore and has been interpreted as Kuroko-type with synsedimentary slump structures.

The Buchans polymetallic stratiform massive volcanogenic sulphide deposits of Newfoundland are intimately associated with volcaniclastic siltstone and breccias within dacitic pyroclastic flows of Ordovician–Silurian age; they have been described by Thurlow *et al.* (1975). Most of the massive sulphide bodies are underlain by stockwork mineralization. The average grade is 14.88% zinc, 7.7% lead and 1.36% copper. The deposits have a spatial relationship to depressions in the footwall rock, and in this and many other characteristics can be likened to Kuroko deposits.

The Kuroko model is currently being applied to deposits widely spread throughout the world and of different ages. For example, Vokes (1976) reviews the Scandinavian occurrences, and Scheibner and Markham (1976) review those of New South Wales, Australia. Certain Precambrian deposits have some similarities to the Kuroko deposits, but they are described separately in Chapters 9 and 10 because it is not clear to what extent the Precambrian processes resemble those of the Miocene of Japan. The plate tectonic elements obviously differed, although the genetic relationship to felsic volcanic and pyroclastic rocks is clear.

Besshi-type Massive Sulphide Deposits

The strata-bound base-metal deposits in the Japanese island-arc system include an important lead-free set of chalcopyrite–pyrite deposits known as the Besshi-type, named from the largest deposit of this type at Besshi in Shikoku island (*Figure 3.5*). This type of deposit differs from the Kuroko in the quantity of gelena and calcium sulphate. The cobalt/nickel ratios of 1:1 to 1:30 are high, making them similar to the Cyprus-type and distinct from the Kuroko deposits.

The Besshi deposits are low in zinc compared with the Kuroko. They occur mainly in the glaucophane-bearing metamorphic rocks of the Sanbagawa metamorphic belt but some are recorded in the weakly metamorphosed to nonmetamorphosed rocks elsewhere. In the San-

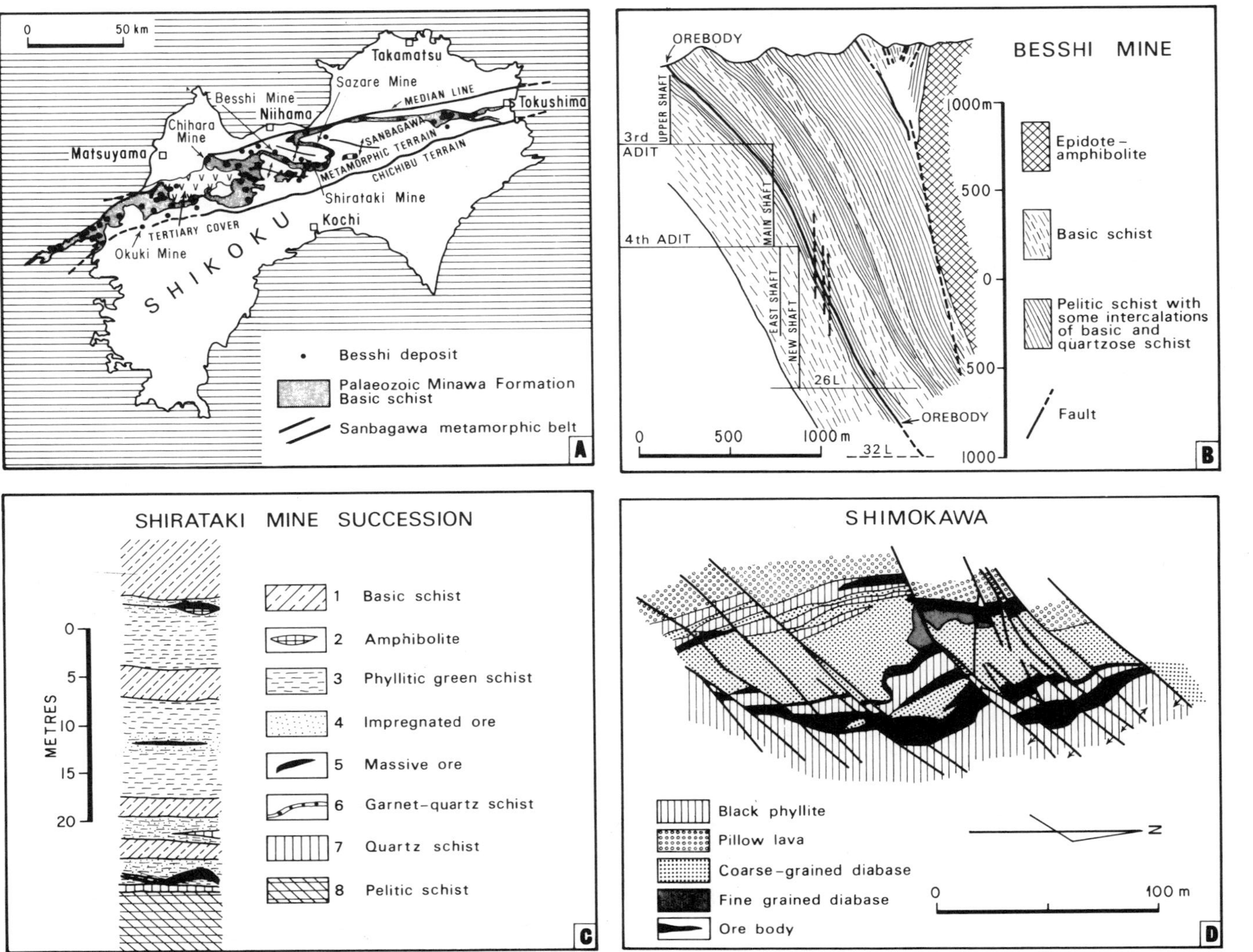

Figure 3.5. Besshi-type stratiform massive sulphide deposits. (A) Their distribution in the Sanbagawa metamorphic belt of Shikoku, Japan. (B) Cross-section through the Besshi mine. (C) Geological column of the ore section of the Palaeozoic Minawa formation at Shirataki mine. (D) Spatial relationships of the country rocks and the ore at Shimokawa mine, Hokkaido (after Kanehira and Tatsumi, 1970; Takeda, 1970; Suzuki and Kubota, 1980).

bagawa belt, hundreds of ore deposits and prospects occur at a definite stratigraphic horizon—the middle member of the Carbo–Permian Minawa Formation (Kanehira and Tatsumi, 1970), which was metamorphosed during the Mesozoic Sanbagawa orogeny. There is a close positive correlation between the number of Besshi-type deposits and the frequency of occurrence of basic schists in each stratigraphic unit, so that a genetic relationship with basic volcanism is implied. The basic schists contain such minerals as albite, chlorite, actinolite, epidote, glaucophane, hornblende, haematite and most are greenschists. Besshi deposits also occur in the Jurassic and Cretaceous but there are no Cenozoic examples, hence their exact setting of formation is difficult to elucidate.

The Shikoku ore bodies have been interpreted in terms of early island-arc tholeiitic volcanism adjacent to a continent, but other settings are possible. A tensional tectonic setting is favoured by Sawkins (1976), and the occurrence of glaucophane facies assemblages within the enclosing rocks must imply that they were, subsequent to their formation, incorporated in imbricate wedges and trench zones in a subduction system.

Most of the Besshi deposits occur in mafic schist alternating with pelitic and psammitic schists. The basic schist can be interpreted as originally of basaltic pyroclastic material, associated with a shale–sandstone sequence. Pelitic and psammitic schists occur even within the ore horizons at several mines (Ishihara, 1978). However, at Okuki Mine (Watanabe *et al.*, 1970c) and at Shimokawa mine in Hokkaido (Suzuki and Kubota, 1980) the immediate host rocks are basaltic lavas and diabase intrusives; these basic rocks are chemically similar to mid-ocean ridge abyssal tholeiites. At Shimokawa, the footwall basalts are more fractionated than the hangingwall basalts. The glaucophane schists of the Sanbagawa Belt are composed of pyroclastic alkaline basalts, which are dominant in the upper horizons of the Minawa Formation; the latter contains most of the Besshi deposits. However the Besshi deposits show no preference for any particular basaltic-type. This fact may explain their variation in lead-isotope data.

The deposit at Besshi is entirely conformable within the enclosing crystalline schists of basic and pelitic origin and is accompanied by quartz schist in the hangingwalls and footwalls (*Figure 3.5*). These quartz schists may originally have been cherty formations. The ore body is tabular and averages 3 m thickness along a strike of 1600 m. It dips steeply northwards and is known to extend down-dip at least 2000 m. The ore consists of pyrite, chalcopyrite and sphalerite with a small gangue content. Pyrrhotite occurs at deeper levels. The copper-rich ore is chiefly of chalcopyrite. Schistosity is conspicuous in the banded ore and the ore body is tightly isoclinally folded together with the country rocks (Kanehira and Tatsumi, 1970).

A pyrite–chalcopyrite–sphalerite–(bornite) assemblage is ubiquitous in most Besshi deposits in the Sanbagawa metamorphic belt. The Shirataki and Sazare deposits of Shikoku, like Besshi itself, are enclosed in the Minawa Formation and are closely associated with basic schists (Takeda, 1970). Regional metamorphism has obscured the relationships between the ore bodies and the country rocks but they are clearly conformable.

The ore bodies of the Iimori district, 50 km south of Osaka, have been described by Kanehira (1970). Association of the copper pyrite deposits with the basic schists suggests that the deposits were formed as a result of submarine basaltic volcanism during the Late Palaeozoic era. During the Mesozoic Sanbagawa orogeny they were deformed and metamorphosed either to glaucophane greenschist or epidote amphibolite facies together with the enclosing rocks.

The Besshi deposits are distinguished from Cyprus-type deposits by their association with a thick sedimentary sequence. However, the tectonic setting, unlike the Cyprus-type, is by no means clear. The Shimokawa deposit of Hokkaido has similarities with Besshi and it is associated with pillow basalt. Suzuki and Kubota (1980) have described amphibolite alteration of the diabase country rock around the ore body. They regard a green schistose rock as a metasomatic zone implying some epigenetic processes during formation. Clearly discussion will continue regarding the mode and settling of the formation of Besshi-type deposits.

Some of the small Late Palaeozoic to Early Mesozoic strata-bound massive sulphide deposits of central Taiwan correlate with the Besshi deposits (Biq, 1974; Tan, 1976). They are associated with schist, greenschist and sandstone.

Sawkins (1976) has suggested that several Norwegian deposits are of Besshi-type, but the value of classifying deposits in other countries in this obscure category is lost when their origin is uncertain.

Relationship between Kuroko and Porphyry-Copper Deposits

The apparent lack of porphyry-copper deposits in Japan has led to early speculations that Kuroko ores are genetically related to porphyry ores and form instead of them when the sulphides and magma reach the surface (Garson and Mitchell, 1977). Kuroko ores, however, are of submarine origin, whereas porphyry ores have formed at depths of less than 5 km in elevated volcanic settings that vent to the atmosphere. The concept of a genetic relationship between the two types has been formalized by Colley (1976) in his classification of Kuroko deposits based on occurrences in Fiji (*Figure 3.1*). Five types of deposit have been proposed. Since they form a spectrum with porphyry copper, it is more logical to consider porphyry as one end of the spectrum (Type I) with five types of Kuroko deposit, whose settings of formation are farther and farther from the volcanic-exhalative centre. Type II forms within a submarine volcanic conduit emanating from a diorite pluton with porphyry-style mineralization, and consists of breccia-pipe and stockwork mineralization (quartz, pyrite, chalcopyrite). Type III is produced where a volcanic conduit reaches the sea bottom on the slopes of a volcanic pile. It consists of siliceous Keiko ore and massive pods of Kuroko zinc–copper–lead ore and copper-rich Oko ore within pipe-like exhalative centres. Types IV and V are near-horizontal and stratiform, composed of chemical precipitates and partial replacement ore bodies in unconsolidated sediment, with some reworking down-slope progressively farther from the volcanic centre.

Type VI (*Figure 3.1*) has not yet been found on Fiji, and is believed to form distant from the volcanic centre by fine ore precipitation, perhaps partly from a brine–ore solution lighter than sea water. This distal-type of deposit would show sedimentary structures and only minor evidence of volcanic activity. The Devonian Rammelsberg deposit of Germany (Kraume, 1960) is perhaps of this type. It is a stratiform barite-rich zinc, lead and copper deposit of modest dimensions, containing only about 6×10^6 t of zinc, lead and copper. The presence of a nearby volcanic centre may be suggested by the occurrence, stratigraphically below parts of the basal pyrite–chalcopyrite-rich portions of the deposit, of siliceous stringer or stockwork ore known as Kneist, which comprises siliceous and chloritized mudstone ramified by cracks carrying quartz, carbonate, pyrite and chalcopyrite. Sulphur-isotope studies of the Rammelsberg deposit (Anger *et al.*, 1966) have led to the conclusion that the galena, sphalerite and chalcopyrite of the ore body contain sulphur of magmatic–hydrothermal origin, whereas the sulphur of pyrite is bacteriogenic in origin. The sulphur in the barite is identical to that established for Devonian sea water, indicating the submarine nature of the deposit.

Taking the Cyprus occurrences as typical of the ophiolite association, the Kuroko of Japan as an island-arc-type associated with caldera formed during felsic phases of explosive submarine volcanism, and the Besshi of Japan as an unusual stratiform-type of deposit associated apparently with a trench or imbricate wedge, metamorphism and deformation, one therefore has displayed a whole spectrum of types of massive stratiform sulphide deposits that can all be brought together at a consuming plate margin, as displayed diagrammatically by Garson and Mitchell (1977) and shown in *Figure 3.1*. With the present incomplete knowledge of the actualistic formation of ore deposits, it is not to be expected that each deposit can be fully understood. However, it should be clear that island arcs are beyond doubt the ideal settings for this spectrum of base-metal stratiform deposits. They are seen in the Japanese and Philippine islands and Taiwan (Taylor, D. and Hutchison, 1979). Therefore they can be expected to, and indeed do, exist in older island-arc settings now forming part of continental orogenic belts. There are some similarities with Precambrian deposits, but it is not clear to what extent Phanerozoic plate tectonic models apply to the Precambrian. One obvious similarity is that the Kuroko base-metal deposits and the Precambrian stratiform sulphide deposits have a common association with felsic volcanic rocks. This shows that similar economic metal fractionation patterns have characterized the earth since its earliest history.

4

Mineralization in intracratonic basins

An oceanic spreading axis like the Mid-Atlantic Ridge, began life as an intracratonic rift, when the Americas were a continuous part of Laurasia or Gondwanaland. Sea-floor spreading along the major rift system led to the progressive westwards drift of the Americas. The Benue Trough of Nigeria–Cameroons, however, failed to achieve complete separation and ended up as a failed arm or aulacogen (*Figure 12.5*). Intracratonic spreading axes are closely similar to mid-ocean ridges but there are obvious differences because of the proximity of continental land masses. For one thing, they are easier to locate and study than the mid-ocean ridges because the geothermal field associated with the spreading is not dissipated by great volumes of oceanic water. Not all intracratonic basins develop truly oceanic characteristics. They develop by thinning of the underlying continental crust, but this process may not always continue right to the stage of developing a floor of oceanic crust. In all cases, however, they represent regions of high geothermal gradients, which are responsible for the continental crust thinning and the possible eventual oceanic character.

Ore bodies composed of hydrothermally impregnated metal-rich sediments are in the process of formation in the Red Sea, which is one of three known remarkable systems. The others are the Salton Sea–Imperial Valley of western California and the Cheleken geothermal system of the USSR. Deep geothermal wells in the Imperial Valley near the Salton Sea of California have a geothermal gradient of 36°C/100 m and have tapped metal-rich brines containing up to 2000 ppm iron, 1400 ppm manganese, 500 ppm zinc, 400 ppm strontium, 390 ppm boron, 750 ppm barium, 90 ppm lead, 970 ppm zinc, 10 ppm copper and 15 ppm silver of the original brine (Doe *et al.*, 1966; White, D.E., 1968). The isotopic data suggest that the metals were leached from the sediments by hydrothermal meteoric water. The sulphur and some of the water may be from volcanic sources at depth, the heat of which drives the hydrothermal system. The area is transected by a large number of active faults and a down-warping of the Salton Sea floor at a rate of 2–3 cm year^{-1} indicates the possibility that continental crustal thinning or sea-floor spreading may be taking place beneath the Sea (Morton, 1977).

The Red Sea

The Red Sea floor is characterized by a number of deeps within which ore bodies appear to be in the process of formation. The metalliferous deposits have been reviewed by Bischoff (1969), Hackett and Bischoff (1973) and Bignell *et al.* (1976). The widespread presence of metal-rich sediments within the median valley of the Red Sea shows the direct relationship between sea-floor spreading and the ore genesis. Not all depressions on the sea floor contain metal-rich sediments, but all are filled with brine (*Figure 4.1*). The sediments of the Red Sea have been classified into six groups. Considerable intermixing of the facies occurs where several ore-forming processes have been operating at the same time within one deep.

Normal Red Sea Sediments

These are the most common and are interbedded with the metal-rich sediments. They are light-brown and biogenic, and calcite is the main component, with minor amounts of detrital quartz, feldspars and clay. The rate of deposition is about 10 cm/1000 years, which is very low compared with the metal-rich sediments, estimated to be in excess of 40 cm/1000 years.

Oxide Facies

These consist of oxides and hydroxides of iron and manganese. Goethite is the most common mineral, being present as layers of orange-yellow clay-sized sediments. The goethite beds are rich in

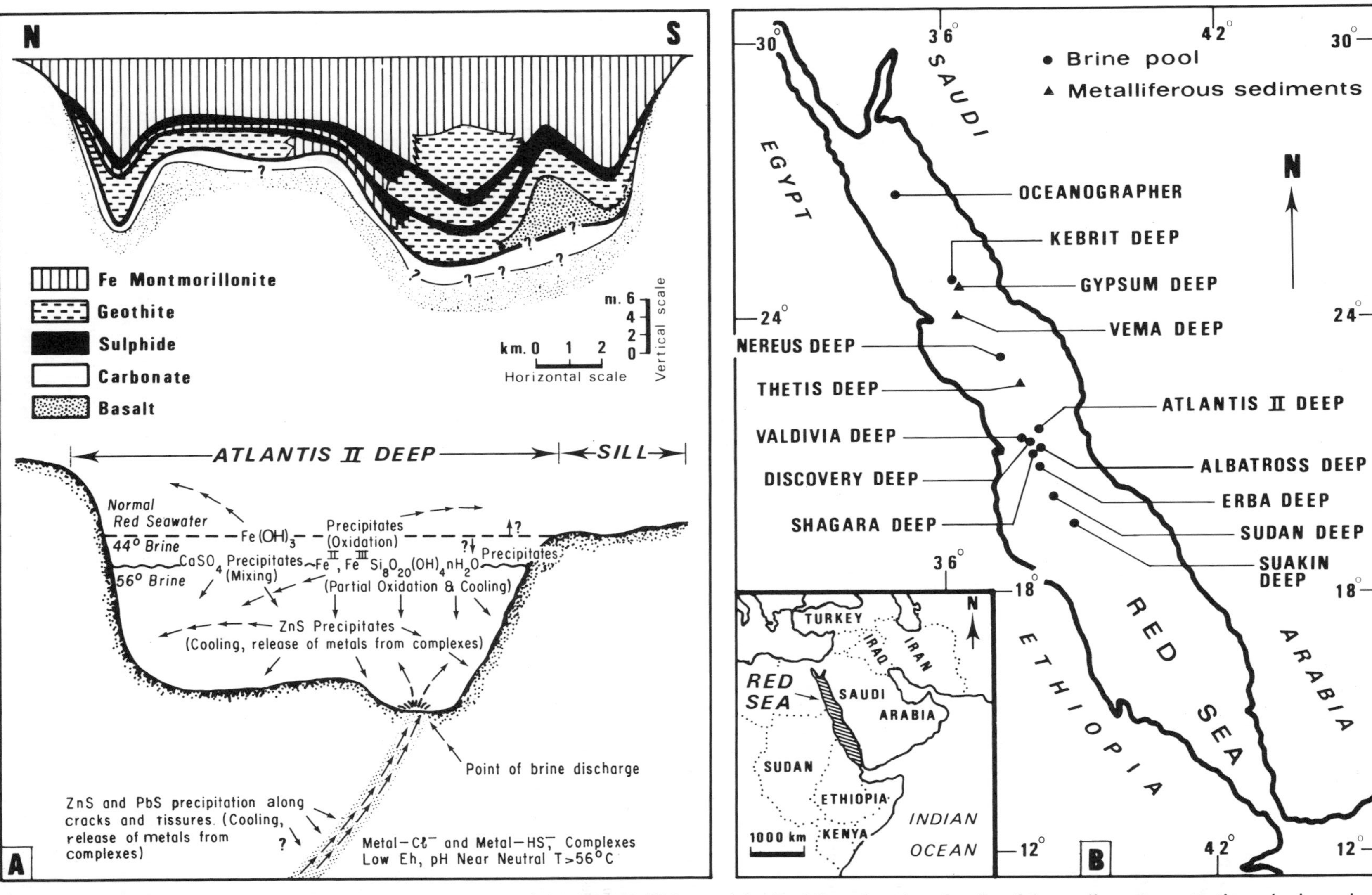

Figure 4.1. (A) Cross-section through the hot brine area of the Atlantis II deep of the Red Sea showing details of the sedimentary stratigraphy based on core information, and model of metal deposition (after Bischoff, 1969; Hackett and Bischoff, 1973). (B) Location of deeps from Bignell *et al.* (1976).

iron, zinc and sometimes copper. Values are: 30–64% iron, 0.03–0.45% manganese, 700–6650 ppm zinc, 9–340 ppm copper and 15–670 ppm lead. Black magnetite layers have been found in the Atlantis II and Thetis Deeps (*Figure 4.1*).

Sulphide Facies

These may be divided into two subgroups: mixed sulphides of iron, copper and zinc; and pyrite, which may be separate or mixed. The sulphide facies occur as black massive beds containing a large amount of material of less than 2 μm grain size. The major minerals present are pyrite, sphalerite, chalcopyrite and poorly crystallized iron sulphide. The composition is: 15–29% iron, 0.09–2.81% manganese, 0.05–10.39% zinc and 0.06–2.22% copper. Lead may range up to 1850 ppm and mercury up to 3100 ppm. Small amounts of sulphides are precipitated from the incoming brine in the Atlantis II Deep. Brines rich in hydrogen sulphide have been found in the Kebrit and Oceanographer Deeps. Injection of such brines into deeps already containing heavy metal-rich brines could account for the widespread occurrence of sulphide layers.

Sulphate Facies

These occur as anhydrite and gypsum. The anhydrite results from the mixing of brine with sea water. It is a relatively minor facies.

Carbonate Facies

These are represented by rhodochrosite and manganosiderite in the Atlantis II Deep and by siderite in the Gypsum Deep

Silicate Facies

These are typically rich in iron and silica and low in aluminium. In the Atlantis II Deep, they consist of dark-brown amorphous silicate with a high brine content.

The distribution, unconsolidated nature and age relations of the facies indicate that the solids were precipitated out of the overlying brine, that the area of brine discharge is very local within the Atlantis II Deep and that the chemistry of the brine has changed considerably with time. Average metal assays in the Atlantis II Deep deposits are 29% iron, 3.4% zinc, 1.3% copper, 0.1% lead and 54 ppm silver. The deposits contain, on average, 85% interstitial brine. The total tonnage of economic minerals in the Atlantis II Deep has been realistically estimated from isopach maps by Hackett and Bischoff (1973) as: zinc 32.2×10^5, copper 8.05×10^5, lead 0.8×10^5, silver 45×10^2 and gold 45 t. However, these deposits may never be mined, for political reasons. Dark-brown, finely-bedded montmorillonite mud comprises the uppermost facies throughout the Atlantis II Deep (*Figure 4.1*). Orange-to-yellow beds of goethite–amorphous facies underlie the montmorillonite beds. Beneath these lie the sulphide facies, occurring as continuous black beds throughout the deep bottom.

A possible idealized sequence of events for formation of the metalliferous sediments in the Red Sea is as follows. Fossil Red Sea water, long in contact with the evaporite strata, has formed brines; these should contain metals originally present at the time of deposition, modified by exchange with the evaporites, and enriched by metals leached from sediments and the underlying basalts, carried largely as chlorides and sulphate complexes. The discharge of these brines within the deeps of the sea is possibly a result of hydrodynamic continuity between the margins of the Red Sea and its centre, perhaps enhanced by local, high-geothermal gradients. The hydraulic head would have changed in the past as the level of the Red Sea altered, accounting for the changes in deposition at different places with time. The cooling and mixing of the metalliferous brines with normal sea water would result in precipitation of the various metalliferous facies described. White, D.E. (1968) has suggested a similar process for sulphide deposition in the Salton Sea region.

The absence of metalliferous sediments from deeps such as the Kerbit and Oceanographer suggests that the effectiveness of brines in transporting metals is reduced by low temperatures and by the presence of sulphide. Isotopic data of sulphur, strontium and carbon suggest that these elements could have been partially derived from young basic volcanic rocks that underlie the sediments of the Red Sea. This is in agreement with oceanic mineralization in the Atlantic and Pacific Oceans as described in Chapter 2. The ore deposition is clearly syngenetic with the sedimentation. However the brine must pass through fissures, faults and brecciated zones and along bedding planes of the sedimentary formations that underlie the present Red Sea floor. The fluid-inclusion data indicate that chloride brines are the mineralizing solutions and that the temperatures of deposition were around 150°C.

The connection between the Red Sea metalliferous sediments and the following groups of lead–zinc deposits of intracratonic basins is inferred, but cannot be directly demonstrated. One could well imagine that as the Red Sea widens, the metal enrichments in the deeper parts of the basin might be transported outwards to the basin edges up-dip into shallow-water miogeoclinal sediments as the basin sediments diagenetically compact. There would be no direct igneous source for these metals, and indeed by the time they are deposited in basin-margin sediments, a considerable amount of time may have elapsed since their igneous-related initial introduction into the basin deeps. With this possibility in mind, the Mississippi Valley-type deposits will now be considered.

Mississippi Valley-Type Ore Deposits

This major class of ore deposits represents stratiform concentrations predominantly of galena and sphalerite, most commonly, but not exclusively, in carbonate host rocks. Fluorite and/or barite, pyrite and chalcopyrite are usually also present. The concentrations are characteristically within a single or a few sedimentary formations, but minor amounts of similar sulphides occur in the overlying and underlying major host (Snyder, 1968). The deposits tend to occur along zones of sedimentary facies changes such as reef to off-reef, shelf dolomite to basin limestone, or carbonate to shale changes. The host rocks are exclusively shallow-water, marine carbonate-dominated sequences, and they occur always on the rim of a large sedimentary basin or over basement highs at or near limestone/dolomite interfaces and at any depth (*Figure 4.2*).

The deposits are truly epigenetic, that is, not formed contemporaneously with the host rocks. The low-temperature saline hydrothermal solutions clearly had the ability first to dolomitize the reef rocks, and then subsequently to deposit the sulphides and fluorite. The evidence points to the solutions moving laterally out of the major adjacent sedimentary basin. The best terminology therefore is that they are laterally secreted epigenetic deposits. In some cases the solutions led to silicification of the host rocks. The ore bodies are contained in stratigraphic traps in organic reefs and associated sediments, and in breccias of submarine slides, solution cavities, or fault voids. All these structures may have resulted from diagenesis or karstification of the reef environment during periods of uplift.

Ore bodies are controlled by some sedimentary feature or structure. Mineralization may be disseminated, concentrated along a particular bedding plane, or form the matrix of a conglomerate or breccia, in which case the whole formation is selectively mined. The ore may be massive or colliform and stratified. In some districts the ore has become concentrated into opened faults that were active during the regional flow of the mineralizing fluids, in which case the fault zones are mined. The ore minerals are simple sulphides of zinc, lead and copper with or without barium, fluorine, nickel, cadmium, germanium and silver (Snyder, 1968). Direct magmatism plays no role in the deposition of Mississippi Valley-type deposits. They are products of sedimentary basins and diagenesis (Dozy, 1970).

Fluid-inclusion Studies

The nature of the fluids that were responsible for the deposits can be deduced from the results of numerous studies of the fluid inclusions, well summarized by Roedder (1976). With rare exceptions, the overall fluid-inclusion densities (at the time of trapping) are found to be greater than 1.0, and frequently higher than 1.1 g cm^{-3}. The ore fluids are believed to have moved very slowly at the site of deposition. There is no evidence for boiling of the fluids during deposition. The ore-forming fluids were generally in the temperature range 100–150°C, rarely over 200°C. Temperatures of late-stage calcite deposition were less than 100°C. The fluid salinity usually exceeds 15% salts by weight and frequently exceeds 20% by weight. Mississippi Valley deposits have formed from fluids that were essentially sodium–calcium chloride brines, generally containing appreciable methane in solution and a yellow-brown fluorescent oil phase that smells of petroleum. They are therefore different from hydrothermal deposits of magmatic affinity (Roedder, 1976).

Lead-isotope Studies

A great number of lead-isotope studies have been carried out, in particular by Doe and Delevaux (1972), and recent summaries are available in Köppel and Saager (1976) and Sangster (1976). A characteristic feature of many deposits is the abnormally radiogenic lead, which cannot be used for dating because it yields a negative or future age. The age of the deposition is accordingly difficult to pin down. The enrichment in radiogenic isotopes has generally been interpreted as indicating a shal-

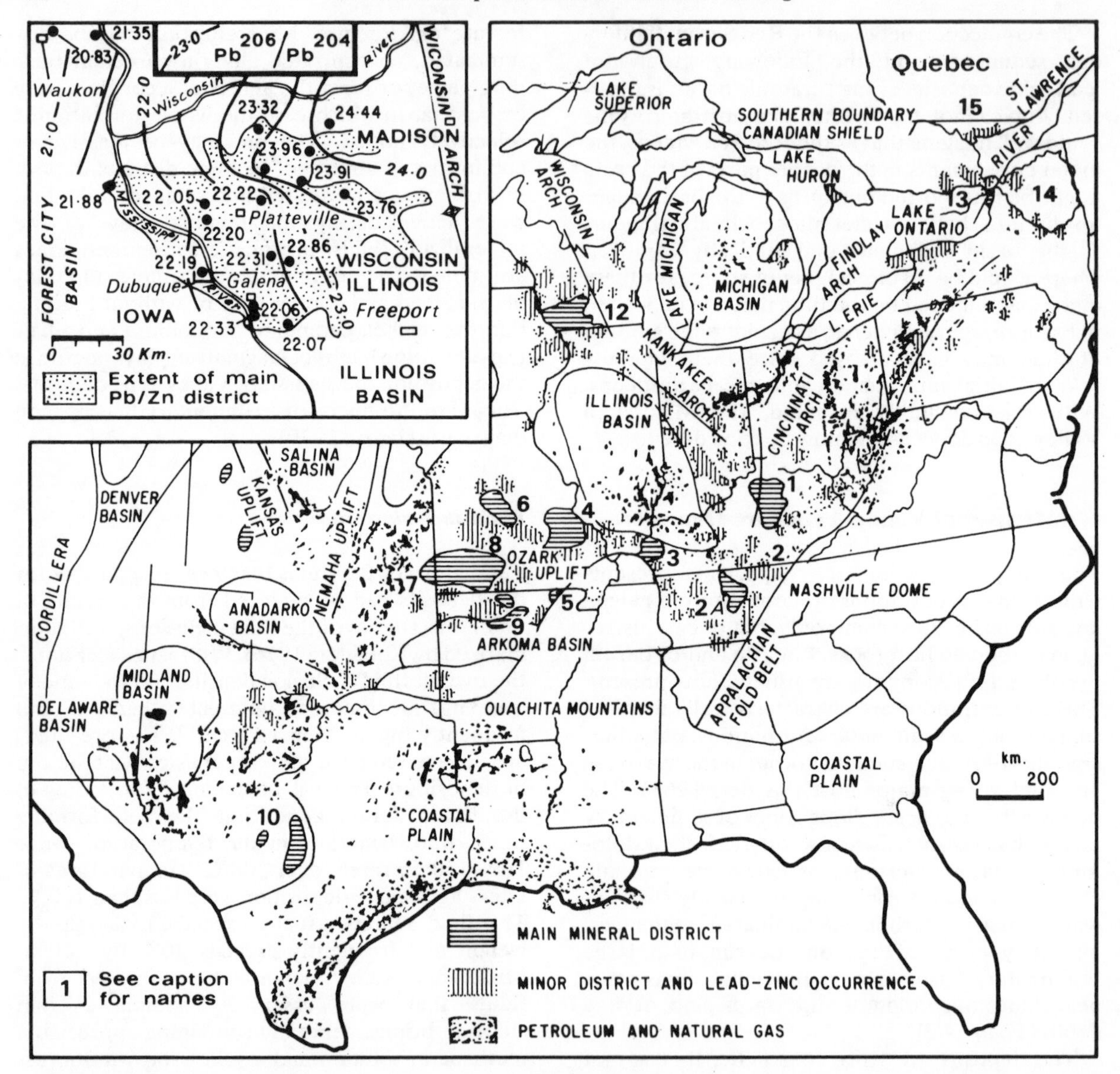

Figure 4.2. Mississippi Valley-type lead–zinc districts of North America. The metals are concentrated in the uplifts and arches between the basins, which contain the petroleum deposits (after Anderson, G.M., 1978). Main mineral occurrences are: (1) central Kentucky; (2) Cumberland River; (2A) central Tennessee; (3) Illinois–Kentucky; (4) southeast Missouri; (5) northeast Arkansas; (6) central Missouri; (7) Tri-state; (8) Seymour; (9) north Arkansas; (10) central Texas; (12) Upper Mississippi Valley; (13) Madoc and Kingston; (14) Rossie; (15) Ottawa. The smaller map shows the regional zonation as shown by $^{206}Pb/^{204}Pb$ isotopic ratios for the Upper Mississippi Valley district (12) (after Heyl *et al.*, 1974).

low crustal source for the lead, from underlying Precambrian basement rocks and/or Palaeozoic sandstone and carbonates. Each district has a distinctive regional lead-isotope pattern that reflects the direction of flow of the metal-bearing solutions, possible buried heat sources and areas of localization of major deposits. The deposits have frequently been compared with the metalliferous sediments of the Red Sea, which exhibit only slight radiogenic enrichment with $^{206}Pb/^{204}Pb$ ratios averaging 18.92. Average ratios for the Mississippi Valley-type deposits are: Illinois–Kentucky 20.36, southeast Missouri 20.81, Tri-state 22.11 and Wisconsin–Illinois 22.54. These ratios are greater than those initially present in any igneous rock of any kind anywhere in the world. The Appalachian zinc districts have ratios somewhat lower than the Mississippi Valley areas, between

18.5 and 19.5. The Pine Point deposits of northwest Canada have ratios around 18.3 and those of England 18.4, much closer to the Red Sea deposits than those of the Mississippi Valley.

Each studied district has its own distinctive regional zonation of $^{206}Pb/^{204}Pb$ and $^{208}Pb/^{206}Pb$ ratios, suggesting a mixing of two isotopic varieties of lead, one more and the other less radiogenic. The isotope ratios may even vary within a single crystal. The most remarkable example is given in the Upper Mississippi Valley (*Figure 4.2*). The regional zonation is consistent with the following interpretation (Heyl *et al.*, 1974). An ordinary lead may have originated in heated brines in the Forest City and Illinois Basins to the west and south. These solutions migrated up-dip towards the regional arch. The ordinary lead would have become progressively more contaminated with the indigenous lead in sandstone and basement rocks as the solution flow proceeded. The anomalously radiogenic lead could have resided in the sandstone aquifer (Doe and Delevaux, 1972) and in the underlying granite.

Ores of the Mississippi Valley-type show a documented relationship to volcanic and plutonic rocks in the Illinois–Kentucky fluorspar district *only*. However this district shows evidence of district-wide ore-solution flow not related to the carbonatite-like mafic–alkalic intrusions. The zonation suggests that the ore-solutions flowed along faults and aquifers in the underlying aquifer and basement, and a large buried pluton may have acted as a heat source for the circulation. The zonation is centred on Hicks Dome, with the least radiogenic lead in the centre, and the more radiogenic at its periphery (Heyl *et al.*, 1974).

Sulphur Isotopes

The bulk of the sulphur is of crustal origin. There is a wide range of $\delta^{34}S$ values, which argues against a deep-seated magmatic major source. The range exhibited by the deposits exactly coincides with that of evaporites and petroleum. The source for most of the sulphur may be crude oil, associated hydrogen sulphide gas, oil shales, basin brine sulphate or evaporite or connate sea water and coals. Temperatures of up to 227°C have been calculated from sulphur-isotope fractionation. They are high compared with the fluid-inclusion filling temperatures of 50–121°C, but the fluid inclusions have been measured on the later ore mineral phases, and it may well be that the earliest phases of mineral deposition were at these higher temperatures. Where sulphur-isotope temperature estimation has been available for ore bands that were also measured by fluid inclusions, there has been a good agreement, giving a value in the region of 100°C. Sulphur-isotope composition of the Bonneterre deposits of southeast Missouri is nearly identical with sea-water sulphur, and closest to algal reef rock in the Bonneterre Dolomite. Brown, J.S. (1967) proposed that the source of the sulphur may be pre-ore diagenetic pyrite in black shale beds that were replaced during the mineralization of the dolomite.

The isotopic evidence supports the view that the ore-fluids were heated oil-field brines having a largely crustal source for lead and sulphur.

Oxygen Isotopes

Oxygen-isotope studies of the Pine Point deposit of northwest Canada, and the Tri-state district of the USA, indicate the mixing of two waters (Fritz, 1969, 1976; Hall, W.E. and Friedman, 1969). The initial fluid has similarities to oil-field brines. During the late stages of mineralization, these hydrothermal fluids were diluted by surface water, which resulted in considerable decrease of salinity and ^{18}O contents. This influx of cool dilute water into a hydrothermal system during the late stages of mineralization is not an unusual phenomenon. The chemical and isotopic similarity of oil-field brines and hydrothermal fluids has been noticed not only for Pine Point but also for other Mississippi Valley deposits (Roedder, 1976).

Model for Ore Deposition

The voluminous literature on this subject illustrates the remarkable concensus of opinions on the genesis. A major review has been given by Wolf (1976a). Some of the remaining problems have been reviewed by Ohle (1980). There are two main theories of ore genesis: precipitation from a single fluid system, which seems to be negated by the isotope data (Fritz, 1979); and a mixing of at least two subsurface waters (Beales and Onasick, 1970), supported by the isotope data. There are, however, strong stratigraphic restraints to the deposition. The stratiform ores were deposited in voids without metamorphism and commonly with little concurrent dissolution of the host rock. The distribution of lead–zinc Mississippi Valley-type deposits in highs on the margins of petroleum and anhydrite basins of the USA (*Figure 4.2*) implies that the movement of the basinal brines outwards and up-dip must have been diagenetically produced and the transportation of petroleum and metals must have been effected by the same salt connate waters (Anderson, G.M., 1978). The subject of ore genesis influenced by compacting sediments has been exhaustively reviewed by Wolf (1976a) and he does justice to the various models that have been proposed for the metal transportation and deposition. The subject has also been discussed by Amstutz and Bubenicek (1967), who showed that clear carbonate and quartz were mobilized and deposited in the early diagenetic stages of the southern Illinois fluorite district, and bituminous oolite, sphalerite and calcite in the main burial stage, while calcite, pyrite, galena and fluorite were deposited at a later stage of compaction. There is compelling evidence that the hot brines deposited their metal content around the margins of the basins where they met and mingled with cooler, fresher water. The following are the important salient features of Mississippi Valley deposits that any theory needs to explain: the deposits generally occur towards the edges of large deep sedimentary basins; carbonates usually form the host rocks for the stratigraphically controlled ores; the presence of evaporates somewhere in the sedimentary basin must be regarded as a favourable sign, because the ore fluids are invariably very saline; there were abundant available cavities for the ore to be deposited in; unconformities in the limestone succession provides karst features, breccias and other features, which may be mineralized; dolomitization is frequently associated with the deposits (dolomitization is known to lead to cavity formation, while silicification is not a good guide to mineralization); the presence of petroleum in the basin appears to be related, and it may be that the same large-scale saline connate-water flow that carried the petroleum along to its reservoirs somehow also carried the minerals farther out of the basin; and carbonate fronts (shale–carbonate facies changes) were of extreme importance for the deposition of the ores. It is envisioned that when the fluids moved out of the shales into the carbonates, they soon found open spaces and hydrogen sulphide to promote precipitation (Anderson, G.M., 1978). However the persistence of uniform banding of sphalerite, which can be correlated many kilometres across the Mid-Ordovician southwest Wisconsin zinc–lead deposits (McLimans *et al.*, 1980), combined with the ore textures, suggest that the hydrothermal fluids carried both the metals and the sulphur and that a persistent and extensive plumbing system must have existed. The colour banding in the sphalerite is related to the iron sulphide content.

Three-quarters of the world's present sulphur supply comes from a bacteriogenic origin (elemental sulphur in evaporites, hydrogen sulphide in sour natural gas and sulphur compounds in petroleum). Even greater amounts of sulphur are contained in related coal, oil shale and tar sands (Sangster, 1976). The bacterial production of hydrogen sulphide from connate brines or evaporites and its consequent migration to the site of ore deposition around the basin margins, where it precipitates base metals from presumed chloride-rich sulphur-poor brines, is an attractive hypothesis for Mississippi Valley-type deposits.

It is obvious that local magmatism played no part in the formation of these deposits. Nevertheless it is possible, as happened in the northern Pennines of England and in southeast Missouri, that the migrating hydrothermal fluids may have leached the metals from a variety of underlying basement igneous rocks. Magmatic activity may have played a distant and earlier role towards the centre of the intracratonic sedimentary basin. It is clear that the hydrothermal solutions migrated up-dip away from any possible magmatic activity towards the basin margins (Dozy, 1970). The Mississippi Valley-type deposits may have a genetic relation to the Red Sea-type metalliferous brines, but the hydrothermal fluids have travelled so far from the deep-basin interior that no genetic relationship to a rifting centre can be demonstrated in any example. Some outstanding problems remain to be solved by a comprehensive theory of ore genesis; in particular, the identification of the source bed, the timing of mineralization and the temperature (Ohle, 1980). The temperatures shown by fluid-inclusion studies would not have been possible unless there had been very high geothermal gradients because some of the basins were shallow. It may therefore be necessary to invoke circulation of the connate water deep into the basement rocks to acquire temperatures in excess of 100°C.

The Viburnum Trend, Southeast Missouri

This mining region is the world's foremost producer of lead, and a major producer of zinc. The ores also contain significant amounts of copper, silver and cadmium. At present it accounts for about 85% of the production of the USA and 15% of world production. The Viburnum mineralized area lies near the southern edge of the central stable region of the North American craton (Thacker and Anderson, 1977). The eroded Precambrian igneous rocks of the St. Francois Mountains are unconformably overlain by the time-transgressive marine orthoquartzite, known as the Lamotte Sandstone (*Figure 4.3*). Shelf sedimentation continued with the Upper Cambrian Bonneterre Formation, in which the mineral deposits occur. The St. Francois Mountains formed a positive feature on the Cambrian shelf, dominating the depositional framework of the Bonneterre Formation (Larsen, 1977). The highland area became the nucleus for a carbonate platform, and a stromatolite reef developed in part of the platform margin. All the mines are located in close proximity to the stromatolite reef—the most characteristic feature of this mining region (*Figure 4.3*).

Ore bodies are controlled by some sedimentary feature or structure, so that it is the sedimentary structure that is mined (Snyder and Gerdemann, 1968; Gerdemann and Myers, 1972). Where the Precambrian knob stood high enough to cut out sedimentation of the Bonneterre Formation, the ore bodies are referred to as pinch-out type. They are exemplified by the Mine La Motte body. The ore body is narrow, from 15 m to 45 m wide. Mineralization in the form of disseminated and bedding plane ore may be only a few metres thick away from the Precambrian knob and up to 12 m thick close to the knob. At the Hayden Creek mine, the Precambrian knob is overlain by a conglomerate formed of granite boulders, which has been mineralized together with the contiguous Bonneterre dolomite. In the western part of the district, ridges of coarse-grained sediments form anticlinal structures trending northeast. Ore may occur on the flanks of the ridges, related to shoaling of certain formations against the ridge. Mineralization may occur in shale overlying the ridges. The northeast-trending bar reef complexes are gigantic compound ridges around and in which the main ore deposits have been formed. The structures range in size up to 4500 m long and 300 m wide. They include the lower 60 m of the Bonneterre Formation. Epigenetic mineralization may occur at any of the major bedding plane contacts or disseminated throughout any of the Bonneterre units. Reef mineralization lies directly above the bedding plane and disseminated ore of the lower units. The entire 60 m, the lower half of the formation, may be mined in this case. Ore within the reef itself is variable in grade because of the influence of internal reef structures. The major ore-controlling structure within an algal reef is the contact between the superimposed colonial algal structures and clastic carbonates. The contact zone of organic and clastic sediments, containing abundant black shales, is invariably more strongly mineralized than either the organic rocks or the clastic carbonate away from the contact. The ore minerals occur along bedding-plane and growth-line contacts, in fracture zones and disseminated within the organic carbonates. However, vast amounts of reef rock are unmineralized. Major disconformities in the Bonneterre Formation are the loci of blanket-type mineralization, which become ore grade when slight relief on the disconformity simulates a ridge or dome structure. Submarine slide breccia bodies along the flanks of calcarenite bars are often strongly mineralized.

Although sedimentary structures were important, the fault pattern played a major role in localization of mineralization (Snyder and Gerdemann, 1968). Although much of the faulting is post-ore, most occurred before the end of mineralization. Some faults are filled with seams of galena up to 8 cm thick. The seams may also be deformed, indicating continued fault movement. Where faulting is pre-ore, the base of mineralization may be at different stratigraphic levels on either side of the fault, and the intensity of mineralization may be greatly different.

MINERALOGY

The primary sulphides include galena, sphalerite, chalcopyrite, siegenite, bravoite, pyrite, marcasite and possibly bornite and millerite. Galena occurs as bedded deposits along disconformities and bedding planes, in part as replacements, but also as open-space fillings. It also occurs as disseminated crystals and aggregates in black shales and as fracture fillings. Massive bedding seams of solid galena, up to 48 cm thick, frequently have a layered appearance due to alternations of fine and coarse sulphides. Curved cleavage planes are common. Sphalerite is minor throughout the district, but may be important locally. It is fine grained and occurs as disseminations in dolomite. Cadmium and silver are important minor elements, but no cadmium or silver minerals have been identified. The paragenetic sequence is generally pyrite, followed by marcasite, galena, siegenite, chalcopyrite and spahlerite, though there is a great overlap. The only gangue mineral of any importance is the host dolomite, but there are minor introductions of calcite, dolomite, dickite, quartz and jasperoid. Wall-rock alteration consists of minor silicification. The dolomitization is regarded as diagenetic and pre-ore.

Fluid inclusions in sphalerite from ores of the Viburnum Trend indicate a deposition from very saline brines at temperatures of 94–120°C (Roedder, 1977).

Lead-isotope studies of galena indicate that the bulk of the lead in the Palaeozoic ores came from

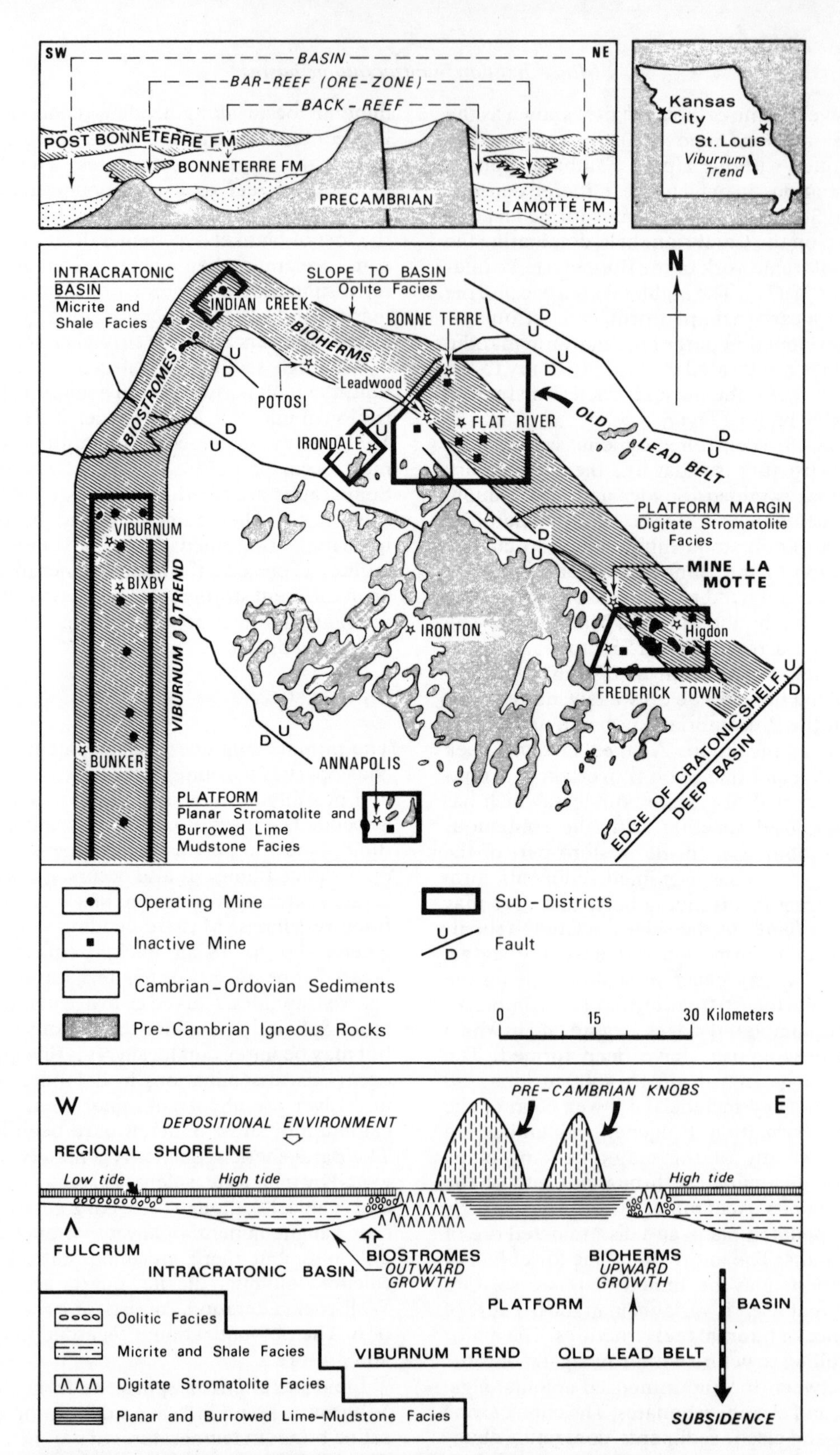

Figure 4.3. Major geological features of the Mississippi Valley-type lead district of southeast Missouri (after Snyder and Gerdemann, 1968). Depositional framework of the Upper Cambrian Bonneterre Formation (after Larsen, 1977).

the sandstone aquifer, the Lamotte Formation (Doe and Delevaux, 1972). The data indicate that the country rocks of the ore, the Bonneterre Dolomite, could not have been the source. The lead in the Old Lead Belt could have been gathered from about 30 000 km^2 of the underlying Lamotte Formation. The lead-isotope data from the Lamotte Sandstone indicate that the source area for the lead had an age of 1200–1300 Ma, similar to that of the lead in the Bonneterre Dolomite deposits (Köppel and Saager, 1976).

The lead of the ore deposits has an extremely high $^{206}Pb/^{204}Pb$ ratio, averaging 20.81 (Doe and Delevaux, 1972). Such highly radiogenic lead has been referred to as Joplin or J lead, and it cannot be used for radiometric dating because it gives a negative or future age for the mineralization.

The localization of the ores in southeast Missouri has been greatly influenced by the positive Precambrian basement of the St. Francois Mountains (Kisvarsanyi, 1977). The basement provided an aquiclude surface that guided the brines towards structurally, lithologically and chemically favourable traps in the overlying porous and permeable predominantly calcareous sediments. The brines also leached metals from the Precambrian rocks to generate a unique copper–nickel–cobalt radiogenic lead-rich mineral paragenesis. One may conclude that the connate water, carrying the metals, circulated widely through the underlying Lamotte Sandstone, a permeable aquifer, and flowed upwards into structural highs of the Bonneterre carbonate to be deposited.

The Pennines of England

The north–south hilly backbone of England has seen lead-mining since the Roman occupation. Mining is centred on three domed-up regions—the Alston and Askrigg blocks of the north, and the Derbyshire Dome of the Peak District to the south (*Figure 4.4*). It is estimated that to date the lead production from the north Pennines has been about 3 million t, and from the south, half a million t. Fluorite production has been about 1.5 million t from the north and 2.6 million t from the south. The extensive literature of the southern Pennines is reviewed by Ford, T.D. (1976), and that of the northern Pennines by Ineson (1976), while a comparison with the Irish base-metal deposits is given by Evans, A.M. (1976). A more condensed review is to be found in Dunham, K. *et al.* (1978).

Carboniferous sedimentation was thin over the Pennines, little more than 1200 m in the northern domed part, where the strata overlie basement highs (Dunham, K. *et al.*, 1978). The Pennines experienced Hercynian (Carbo-Permian) north–south tension that opened fractures trending between 65° and 115°, permitting long-continued epigenetic mineralizing fluids into the fracture systems to form the ore deposits. The main ore bodies are known locally as *rakes*, which are the filled, generally vertical, wrench fault fractures. Most of the rakes are horizontally slickensided, illustrating the wrench movement. *Scrins* are open joint fillings. Whereas most rakes trend east–west, the scrins trend northwest–southeast to northeast–southwest. Pipes represent the fillings of palaeo caves and voids between large boulders, or fillings of solution breccias, and *flats* are veins lying nearly parallel to the bedding.

In the northern Alston and Askrigg blocks, the mineralization is confined to the Viséan and Namurian host rocks and only a few deposits penetrated the Coal Measures. The Great, or Main Limestone, 20–25 m thick, lies at the base of the Namurian and carries the highest proportion of the ore. To the south in Derbyshire, the Namurian shales are barren, and most of the mineralization is in the underlying Viséan limestones.

Galena is the principal ore. In the north it contains 42–400 ppm, and in Derbyshire 25–30 ppm silver. Sphalerite has been mined in parts of the Alston block and also in Derbyshire. Cadmium has been recovered. Chalcopyrite, chalcocite and covellite are localized. The ores may also contain pyrite, marcasite and pyrrhotite along with rarer minerals. Fluorite is the main present-day product. In the north it occurs as mauve, green and yellow crystals containing yttrium, europium and other rare earths. In Derbyshire the commonest variety is colourless while the dark-purple Blue-John variety appears to have been coloured by hydrocarbons.

Barite is widespread, coarse grained in the Alston block, fine grained in Derbyshire. Witherite, siderite and ankerite occur in the north, and cerussite forms secondary deposits. Quartz is common as a wall-rock alteration, and sulphides have migrated into the limestones.

The mineralized fractures follow a very regular pattern. In the Alston block they trend 65° and 100°. They represent normal faults with displacements up to 15 m. These fractures are near-vertical in the limestones, but have dips as low as 45° in soft shales. Open spaces occur only where they are vertical, therefore the ore shoots have a much greater longitudinal than vertical extent. All the major fluorite bodies lie on east–west sections,

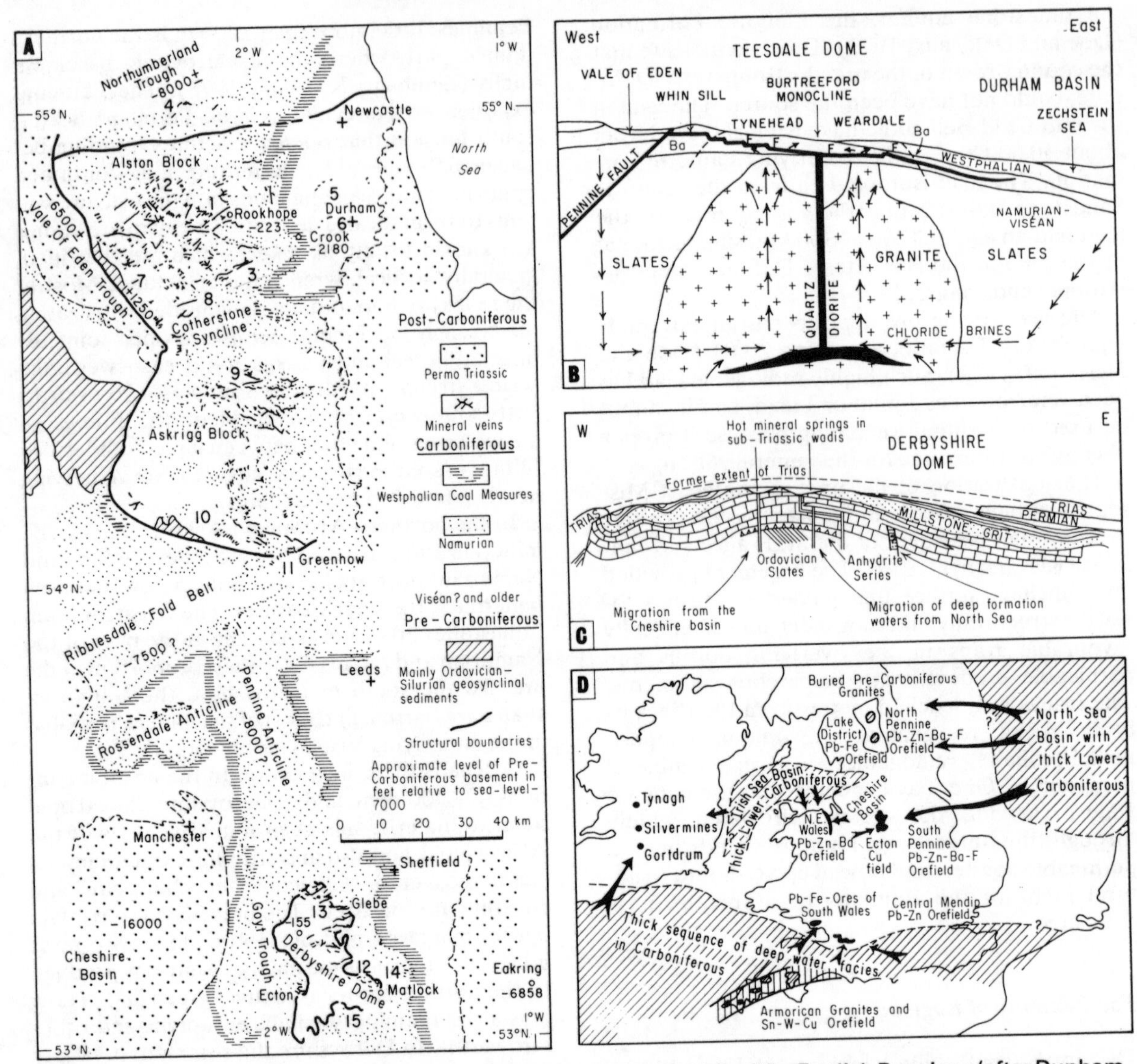

Figure 4.4. (A) Distribution of Post Carboniferous vein mineralization of the English Pennines (after Dunham, K. *et al.*, 1978). (B) Genetic model for the northern Pennines (after Solomon *et al.*, 1971) and (C) for the southern Pennines (after Ford, T.D., 1976). (D) The possible migration routes for formation waters from the basins to the marginal highs during the Carboniferous Mississippi Valley-type ore deposition cycle (after Ford, T.D., 1976). Names of mine districts: (1) Rookhope valley; (2) Nenthead; (3) Teesdale; (4) Settlingstones; (5) South Moor; (6) New Brancepeth; (7) Silverband; (8) Lunedale; (9) Swaledale; (10) Grassington; (11) Greenhow Hill; (12) Millclose; (13) Great Rake – Deep Rake; (14) Masson Hill; and (15) Golconda.

whereas the westnorthwest–eastsoutheast sectors carry no fluorite and obviously were closed during the mineralization (Greenwood, 1977).

In the Askrigg block, the fracture pattern trends eastsoutheast and east–west. In Derbyshire, the preferred mineralized horizon is the uppermost limestone bed beneath the Namurian shale and above the first lava or tuff. The ore shoots are long and shallow, but some have a width up to 20 m. Some deposits may represent filled karst features or metasomatic pipes (Mason, J.E., 1974). The age of faulting is not known, but it has a clear overlap with the mineralization (Ford, T.D., 1976). The temperatures of mineral deposition were 70–140°C and the inclusions are highly saline (up to 30% salts by weight).

Isotope dating indicates mineralization at peaks around 270 Ma (Early Permian), 235 Ma (Late Permian), continuing to 180 Ma (Late Triassic). The $^{206}Pb/^{204}Pb$ ratios are closely confined around 18.4, not nearly as radiogenic as the Mississippi Valley deposits of North America, but radiogenic enough to cast doubt on the validity of the dates obtained. Nevertheless they appear to be consistent with the geological evidence. In the north Pennines, the age of mineralization is constrained by the fact that it post-dates the Whin Sill (265 Ma). Some traces of mineralization are also found in the Upper Permian Magnesian Limestone. The sulphate originated from the Lower Carboniferous connate waters (Solomon *et al.*, 1971). Many of the mineral deposits contain hydrocarbon compounds, mostly paraffins bearing a close similarity to petroleum obtained from the East Midlands oil fields.

The mineral fluids obviously migrated from the host rocks into the opening fractures. It is therefore surprising to find a lateral rather than a vertical zoning. The Alston block has a concentric pattern; central zones have veins filled with fluorite, surrounded by a zone filled with barite, witherite and calcite (*Figure 4.4*). Fluid-inclusion studies indicate that the central fluorite zone was deposited at 220°C, with a temperature gradient outwards to 100°C. Galena occurs throughout the fluorite zone, but is most abundant near the fluorite–barite transition, and sphalerite also occurs sporadically here. In the barite fringe, sulphides diminish.

In the Askrigg block, fluorite is again restricted to the limited inner areas. Outer and higher zones carry barite and witherite, giving way to calcite, aragonite and dolomite. Copper occurrences in the fluorite zones are the epicentres of distribution. The model is that hot fluids rose at a few centres and spread outwards upon reaching laterally extensive fracture channels in the Carboniferous domes. The yttrium enrichment in the fluorite can be used to identify the feeding channels.

The Derbyshire field is zoned westwards from the contact between Namurian shale and massive limestone (*Figure 4.4*). Fluorite gives way laterally to mixed fluorite–barite and eventually barite and calcite are dominant. In the fluorite zone, the fluid inclusions indicate a temperature of deposition of 103–165°C (Rogers, P.J., 1977). The richest galena deposits are towards the eastern margin of the field. The distribution suggests that the source of the mineralizing fluids lay beneath the shale to the east, and oil borings in the east have revealed fluorite at depths of 800 m, supporting the westerly migration paths of *Figure 4.4* (D).

GENETIC MODEL

The historical development of hypotheses for the Pennine mineralization is an interesting lesson in the changing pattern of thought with time. A gravity survey by Bott and Masson-Smith (1957) led to the establishment of an underlying Weardale granite (*Figure 4.4*), and it was thought that the deposits were related to hydrothermal mineralization from this granite source. To test the theory, the granite was drilled but found to have a date of 380 Ma and thus to pre-date the overlying Carboniferous rocks (Dunham, K.C. *et al.*, 1965). Nevertheless this Weardale granite coincides strikingly with the distribution of fluorite mineralization in the Alston block. Fluid-inclusion studies (Sawkins, 1966) then showed that the fluorite was deposited from hypersaline brines, five to six times more concentrated than sea water and with a higher potassium/sodium ratio. The sulphur-isotope $\delta^{34}S$ values in the Alston block showed a range much wider than acceptable for any igneous–hydrothermal hypothesis. The range of sulphur isotopes in barite from the outer zones is so small, however, to suggest a sulphate origin from a homogeneous source, probably shallow ground water of an early Carboniferous age. Therefore the most plausible theory suggested a mingling of deep-seated, chloride-bearing formation waters, which had travelled up through the Weardale granite, with cooler sulphate waters already present in the Carboniferous host rocks. Either a higher-than-present temperature gradient or penetration of the deep waters to at least 8 km, to attain a temperature of about 200°C, before resurgence would have been required. For the source of the water, the adjacent deep basins of Northumberland, Bowland and the North Sea are obvious candidates. Ineson (1976) has summarized the various models by having the hot connate hypersaline brine coming from the deeper basins, moving through the granite and other basement rocks and leaching out the metals, then rising into the overlying carbonates where mineralization was precipitated by mixing with cooler meteoric water (*Figure 4.4*). As to the source of the elements, silicon, magnesium and iron can be shown to have been leached from the 290-Ma old Whin Sill, and leaching of the underlying granites would have provided the lead, barium, zinc and fluorine. The granite contains 20–80 ppm lead, 30–55 ppm zinc, 590–880 ppm barium and 820–2270 ppm fluorine. It is not known what the source could have been for the Derbyshire and Askrigg blocks. Sulphide sulphur could have come from the reduction of sulphate in the formation waters, but it is also common in the Carboniferous shales. Hypersaline brines can be invoked as the extracting medium, rather like the Red Sea model given earlier in this chapter.

The hypothesis of basin-margin migration of deep formation waters fits the facts better than any other (Ford, T.D., 1976). The principal migration paths of the formation waters migrating up-dip from the basins into the highs is shown in *Figure 4.4*. By early Permian times the Lower Carboniferous deposit of the North Sea basin was probably buried by at least 4 km of Upper Carboniferous sediments, and by Triassic times there could have been up to 6 km of sediments. As a result of compaction pressure and an appropriate geothermal gradient, migration of the deeply buried and modified connate water would have been facilitated westwards into the regional highs around the basin margin (*Figure 4.4*). Repeated fault formation would have facilitated the movement of the waters. The active wrench faulting would have assisted the movement of the waters through the basement rocks. Where did the waters go after depositing their minerals in the rakes and scrins? The demineralized waters could well have found their way to the surface as hot springs or into the Triassic ground water, and it is known that the Triassic sandstones do contain a scatter of small subeconomic mineral deposits all around the Peak District.

The Lower Carboniferous deposits of Ireland (Evans, A.M., 1976) are associated with eastnortheast or northeast-trending faults, and they are clearly epigenetic. The mineralizing solutions have ascended along the faults and spread out into the country rocks. General models can be put forward to relate them to the other brine-derived Carboniferous deposits of England (*Figure 4.4*). However, just as in the northern Pennines, there may be a genetic relation to buried granites (Evans, A.M., 1976).

Pine Point, Northwest Territories, Canada

The Pine Point zinc–lead mining area is located on the south shore of the Great Slave Lake (*Figure 4.5*). There are about 40 known ore bodies scattered over an area of about 50 × 10 km. The individual ore bodies range from a few thousand to over 15 million t. Ore reserves are estimated to be about 41 million t averaging 2.4% lead and 6% zinc. Most of the ore occurs in the Middle Devonian coarse dolomitic Presqu'ile Barrier Reef, but more than half the known reserves occur outside this coarse dolomite. The geology has been described by Beales and Jackson (1968), Jackson, S.A. and Folinsbee (1969) and Skall (1975). The Devonian Pine Point complex formed a barrier to the great Muskeg–Prairie evaporite basin of North America (described in Chapter 5), separating it from the open sea to the northwest. Fore-reef, reef and back-reef facies are recognizable in the elongate platform bounded on the north by the Mackenzie Shale basin and on the south by the Muskeg–Prairie evaporite basin (*Figure 4.5*). During the development of the barrier reef, the carbonate platform was frequently exposed, then followed by renewed growth. In the Middle and Late Devonian periods a marine transgression buried the palaeo-karst features.

The ore bodies are characterized by rather simple mineralogy: galena, crystalline and colloform sphalerite, pyrite, marcasite and occasional pyrrhotite are associated with dolomite and calcite gangue. An unusual feature is the virtual absence of fluorite and barite. Most ore bodies fill vugs and cavities in the limestone and their margins are sharp. Massive ore bodies have a greater vertical than horizontal extent, but tabular ore bodies a greater horizontal than vertical extent. The massive bodies have a higher grade and a lead/zinc ratio of approximately 1:1.7, whereas the tabular ones have a ratio of 1:2.6. Colloform sphalerite mainly occurs in the massive, and rarely in the tabular bodies. The mineralization was associated with collapse structures and rarely penetrates into the unbrecciated limestone.

The mean $\delta^{34}S$ value of +20.1 per thousand parts of ore is remarkably similar to that for the evaporite anhydrite of the Muskeg–Prairie basin to the southeast (Sasaki and Krouse, 1969). The sulphate in formation brines from nearby strata of the same age also shows the same isotope composition. Therefore the most plausible source for the Pine Point sulphur appears to be the Middle Devonian sea-water sulphate, probably transported laterally as connate brines out of the evaporite basin into the barrier reef.

The lead is not nearly as radiogenic as the Mississippi Valley deposits of the Tri-state and Missouri. Lead isotope $^{206}Pb/^{204}Pb$ ratios have been determined to be closely grouped around 18.3 (Cumming and Robertson, 1969). The Pine Point lead is therefore remarkably similar to that of the Pennines of England. The isotopes indicate an age of mineralization of 250–275 Ma, appreciably younger than the host rocks.

The cycle of mineralization began with dolomitization, followed by deposition of lead–zinc minerals, then followed by the precipitation of calcite (Fritz, 1969) (*Figure 4.5*). The temperature of

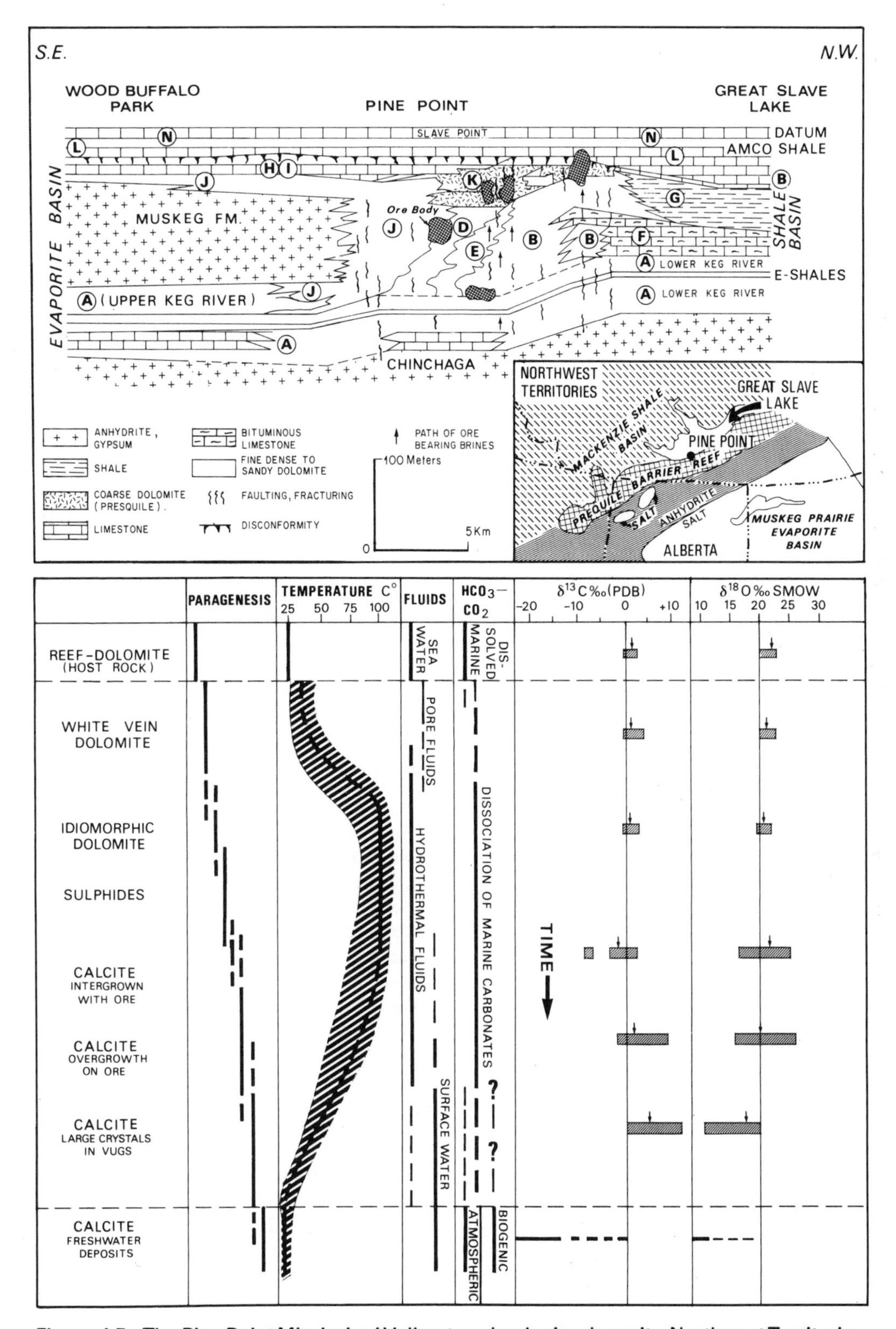

Figure 4.5. The Pine Point Mississippi Valley-type lead–zinc deposits, Northwest Territories, Canada. Top: stratigraphic cross-section to show the post-Middle Devonian emplacement locations of the sulphides (after Skall, 1975). Facies: (A) marine platform; (B) off-reef; (D) organic barrier; (E) clean arenite; (F) tentaculites; (G) Buffalo River Shale; (H) Gastropod; (I) Amphipora; (J) south flank back-reef; (K) Presqu'ile; (L) green shaly; (N) tidal flat member. Inset: location map after Jackson and Folinsbee (1969). Bottom: changes in carbon and oxygen isotopes with the cycle of mineralization (after Fritz, 1976).

formation was confined within the range 51–97°C (Roedder, 1976) and decreased slowly towards the end. Oxygen-isotope studies indicate that the source of solutions was the formation waters. The mineralization ended with the influx of isotopically lighter surface waters into the reef system, as shown by the gradual decrease in ^{18}O content of the late calcite (Fritz, 1969). The carbon originated from the dissolution of marine carbonates by the hydrothermal fluids.

The Pine Point deposit offers an excellent example of Mississippi Valley-type mineralization, in which escaping basinal brines from the Muskeg–Prairie evaporite basin carried metals outwards and upwards into the Presqu'ile Barrier Reef complex, made more porous by karstification and brecciation during periods of emergence. The mineralized brines mingled with hydrogen sulphide-rich limestone formation waters, resulting in the lead–zinc deposits. Nowhere is there any igneous association, and it must be inferred that the metals were carried far from the evaporite basin centre, which may have had similarities with the Red Sea brine formation.

Appalachian Zinc–Lead Deposits, Eastern USA

The major differences between these and the Mississippi Valley deposits is that the Appalachian deposits occur in rocks that have undergone post-mineralization orogenic deformation. They occur scattered along the Appalachians from Tennessee to Pennsylvania, thence into Newfoundland and Nova Scotia. A general review of their geology has been given by Hoagland (1976). There are three main categories. (1) The most important deposits are essentially lead-free, represented by the east-Tennessee districts of Copper Ridge and Mascot–Jefferson City and central Tennessee; Timberville, Virginia; Friedensville, Pennsylvania; and the Newfoundland occurrence in the Great Northern Peninsula. All these occur within Lower Ordovician dolomitic formations. (2) A second class of major economic importance is characterized by zinc with subordinate lead, with an example at Austinville–Ivanhoe, Virginia, localized at a facies change in breccia and reef material of Lower Cambrian age. (3) The third class is represented by Gays River, Nova Scotia, where zinc–lead mineralization occurs in a reef complex developed on a major unconformity at the edge of a large evaporite basin. The carbonate host is of Mississippian age, and the deposit has very strong resemblance to the Pine Point deposit of northwest Canada.

All the deposits occur in shallow-water marine carbonate sediments. The ore is commonly found in dolomitized limestone facies at or near the interface between limestone and dolomite. The reefs are interpreted as having occurred at or near the edge of the continental shelf. Karst-related breccia zones are prominent features of the mineralized areas of Tennessee.

The sphalerite is characteristically very pure, the iron content averaging about 0.5%. The gangue is predominantly dolomite with some calcite. Silica is commonly present as thin silicified selvages along limestone–ore contacts, as local intergrowths of black jasper and as bands or patches within the ore breccias. Barite, fluorite and chalcopyrite are rare in the Appalachian deposits.

In the Austinville–Ivanhoe area of Virginia, the ore zones extend over a distance of about 10 km nearly parallel to the 60° regional strike (Brown, W.H. and Weinberg, 1968). Their footwalls are conformable with the bedding, but the hangingwalls commonly cut across the bedding, with transgressions up to 30 m or more. Dip lengths of the bodies can be measured from 15 m to 120 m. Strike lengths range from 30 m to over 600 m. Major faults affect the extent of the bodies. Some are confined to one side of a fault only, and others enlarge greatly at the intersections of two faults. The sphalerite may be both dark-brown or honey-coloured. The very dark variety is nearly always in contact with pyrite. The host rock is always dolomite. An interesting feature in the neighbourhood of the ore bodies is the change of the dark-grey or black dolomite to light-grey, as if it had been bleached. The ore textures are described as mosaic breccia (accounting for about 43% of the total ore), rosette ore (31%), rubble breccia (15%) and disseminated ore (11%).

The Friedensville mine of Pennsylvania lies only 120 km west of New York City. Its geology has been described by Callaham (1968). The ore is lead-free with a sphalerite/pyrite ratio of about 3:1. The ore is localized in Lower Ordovician solution collapse breccias developed during the erosion interval marking the unconformity between Middle Ordovician formations. The sulphides commonly coat breccia fragments and occur as bands parallel to the bedding and within the matrix of the breccia, and also as dissemination in the matrix. The sphalerite is frequently deformed (Hoagland, 1976). The pyrite is usually deformed and fractured, and chalcedony intergrown with calcite occurs as fracture filling in the shattered pyrite. Repeated folding and deformation is indicated by the fact that the chert layers that fill the fractures have been deformed. Bleaching of the hangingwall dolomite is common. The exact significance of the bleached host rock is however unknown, because it has not been studied outside the mining regions. Roughly strata-bound limonite ores occur in the same mining region, in the Cambrian and Ordovician formations, both stratigraphically below and above the zinc deposits. Some had pyritic root zones and are regarded as oxidized derivatives of pyritic deposits localized in solution collapse breccias and formed during Ordovician time. They have a relation to the zinc mineralization for some of the iron ore has high zinc contents.

The most important zinc deposits of Tennessee are those of the Mascot–Jefferson City district (Crawford and Hoagland, 1968). The ore occurs as irregular bodies in Ordovician dolomite. Frequently they are perpendicular to the stratification. Distinctive types of ores may be described as coarse rock matrix breccias, fine rock matrix breccias and inflation breccias. The breccias are not invariably mineralized, indeed more are barren than mineralized. Ore breccia containing fragments of ore with later generations of superimposed sphalerite testify to the continuity of brecciation and permeability within the ore zone during a prolonged period of mineralization. Sphalerite is the only primary metallic mineral of economic significance.

The lead isotope $^{206}Pb/^{204}Pb$ ratios range from 18.48 to 19.33 (Sangster, 1976), distinctly different from those of the Mississippi Valley. Thus, even though geological features of the deposits of the folded Appalachian Mountains are markedly similar in many respects to those in the Mississippi Valley, the lead isotopic compositions suggest that at least the galenas have a different origin from the J-type leads of the Mississippi Valley (Heyl *et al.*, 1966). It is worthwhile noting that the Appalachian deposits represent the miogeoclinal equivalents of the coeval volcanogenic copper–lead–zinc massive sulphide deposits of Maine and Quebec, Bathurst in New Brunswick and Newfoundland to the north. These deposits are considered to be the results of volcanic exhalations into the Cambro-Ordovician seas. Lead in volcanogenic deposits is non-radiogenic, and similar to that in the Appalachian zinc–lead deposits. The source of the lead of the Appalachian deposits may therefore be a volcanic arc now buried beneath the younger sediments to the east (Sangster, 1976).

The Appalachian Mississippi Valley-type deposits can be regarded as resulting from large-scale metallized brine migration outwards and upwards towards the northwest from the Appalachian geosyncline, moving up-dip during compaction of the thick sedimentary sequence towards the Appalachian shelf, where the brines precipitated predominantly zinc sulphide in the shelf limestones, which had been made porous through dolomitization and brecciation during periods of uplift and karstification. The brines did not penetrate the Precambrian basement and hence the deposits do not contain J-type radiogenic lead.

Fluorite Deposits

Fluorite is predominantly deposited epigenetically in a large variety of styles and range of temperature. There is no single characteristic geological mode of occurrence, but there exists a close genetic relationship to the Mississippi Valley-type of deposition. In many cases there is direct evidence that alkaline and acid igneous rocks may have been the major suppliers of the fluorine to the brines that deposited the ore.

The commonest depositional environments are listed by Grogan *et al.* (1974) as: fissure veins in many kinds of country rocks, notably carbonates and granites; stratiform or *manto* deposits in carbonate rocks; replacements in carbonate rocks along contacts with intrusive and extrusive acid igneous rocks; stockworks or fillings in sheared or tectonically shattered zones; carbonatites and alkalic igneous complexes; residual concentrations formed by the weathering of less-rich primary deposits; fillings of karstic features such as caves and breccia pipes of explosive, collapse, or solution origin; and as a component of the gangue of other economic mineral deposits.

The most important types of deposits are in fissure veins and as manto or stratiform bodies, and some of the type localities are described in this chapter. The fluorite is frequently but not always associated with galena, sphalerite and barite. The South African deposits are referred to in Chapter 6, where the fluorine may have been supplied by the Bushveld granite. The Spanish fluorite deposits of the Astorias district of the north are predominantly fissure vein fillings trending southeast or northeast in Carboniferous limestone, in karstic solution cavities and also as bedded deposits in a Permo-Triassic limestone–sandstone conglomerate sequence, where there is no known igneous association. The Osor vein, near Gerona in the Pyrenees, is an east–west fissure in porphyritic granite.

Mexico

Mexico has been the world leader in fluorite production since about 1956, and about 90% of its production is exported to the USA. The main producing states are San Luis Potosi, Guanajuato, Coahuila, Chihuahua and Durango. In the San Luis and Guanajuato districts, the contact between the Lower Cretaceous Cuesta del Cura limestone and Tertiary rhyolites has been favourable to the deposition of large chimneys and cave fillings, averaging 65–85% fluorite (Pickard, 1974). The Las Cuevas fluorite deposit of the Zaragoza–Rio Verde district accounts for over 7% of world total fluorite production (Ruiz *et al.*, 1980). Mineralization occurred 30 Ma ago. Important production comes from narrow mantos and veins in the upper section of the Cretaceous Georgetown limestone. Mantos are flat-lying bedded or strata-bound ore bodies. They occur at the contact between the uppermost Lower Cretaceous Georgetown limestone and the disconformably overlying Del Rio shale or Buda limestone (Kesler, S.E., 1977). The deposits were formed by replacement followed by open-space filling and they consist of fluorite with minor calcite, quartz, and local celestite and gypsum; sulphides are extremely rare. Wall-rock alteration, in haloes about 30 cm wide around the mantos, include fluoritization and local silicification. The primary inclusions in the fluorite contain saline liquid (8–18% sodium chloride equivalent by weight, with sodium/potassium ratios ranging from 2.5 to 3.14). The inclusions also contain petroleum material. The temperatures of homogenization range from 110°C to 330°C with a strong maximum at 150°C. The Las Cuevas deposit has fluid inclusions that indicate a range of 60–130°C (Ruiz *et al.*, 1980). Evidence for boiling is seen in many deposits (Kesler, S.E., 1977).

The deposits are spatially associated with Mid-Tertiary rhyolite, which is anomalously high in fluorine, and some of the chimneys, veins and contact deposits are immediately adjacent to these rhyolites. One of the contact deposits contains fluorite with polyphase inclusions containing salinities of up to 40%, the homogenization temperatures being as high as 370–430°C. The sulphur-isotope data indicate that the host Del Rio Formation was not the source.

The favoured hypothesis for the origin of the manto deposits is given by Kesler, S.E. (1977) as follows: they formed when a zone of formation water, trapped beneath the Del Rio Shale, was intersected by the rhyolite magma, which contributed heat, fluorine and possibly magmatic water and sulphur to the formation water zone. On a regional scale, the northern Coahuila fluorite district appears to be related to a zone of Mid-Tertiary alkaline igneous rocks that extends from the Big Bend area of the USA to Tampico in Mexico. In Chihuahua and Durango, narrow high-grade veins occur in the Middle-to-Late Tertiary volcanic series. Fluorite is the major gangue material in the lead–zinc veins of the Parral district.

The close association of the Mexican fluorite deposits with rhyolite and alkaline igneous rocks, combined with the occasionally high homogenization temperatures of the fluid inclusions, lend credance to a magmatic affiliation postulated for other fluorite districts, especially in Kentucky.

Southern Illinois–Kentucky, USA

More than three-quarters of the fluorite produced in the USA has come from this district, with substantial amounts of lead and zinc by-products. The geology of the deposits has been reviewed by Trace (1974) and Grogan and Bradbury (1968). The deposits occur in a predominantly carbonate sequence of Mississippian age. Highly altered Permian mica peridotite and lamprophyre dykes and sills intrude the strata, and the mineralization post-dates them.

The fluorite is localized either as veins filling complex northeast-trending steeply dipping fault zones, or as bedding replacements principally along three horizons in Mississippian strata (*Figure 4.6*). The faults are predominantly normal, with dips ranging from 45° to 90°, resulting in a northeast-trending pattern of horsts and grabens (Hook, 1974). The fluorite veins are fissure fillings along faults and fault breccia zones. A typical vein is lenticular, pinches and swells erratically, and is commonly a mixture of fluorite and highly variable amounts of calcite and country-rock fragments. Most veins range from about 1 m to 3 m in width. The average height of the ore shoots is 30–60 m; they usually extend 60–120 m. There is an increase in the fluorine content of the illite of shales neighbouring a fissure fluorite body, from about 0.10% 100 m from the fault to about 0.50% close to the fault (Daniel and Hood, 1975). Magnesium and potassium rock contents also increase towards the fault.

Most of the bedding replacement deposits occur in the Cave-in-Rock area of Illinois. They are stratiform elongated bodies trending northeast. The textures are either well banded, imperfectly banded, or of breccia. Ore bodies are commonly 15–60 m wide, 1.5–6 m thick and extend along strike up to 3 km. Fluorite is the principal mineral; sphalerite, galena and barite are uncommon. Coarse quartz may be abundant in some deposits to the extent of rendering them commercially

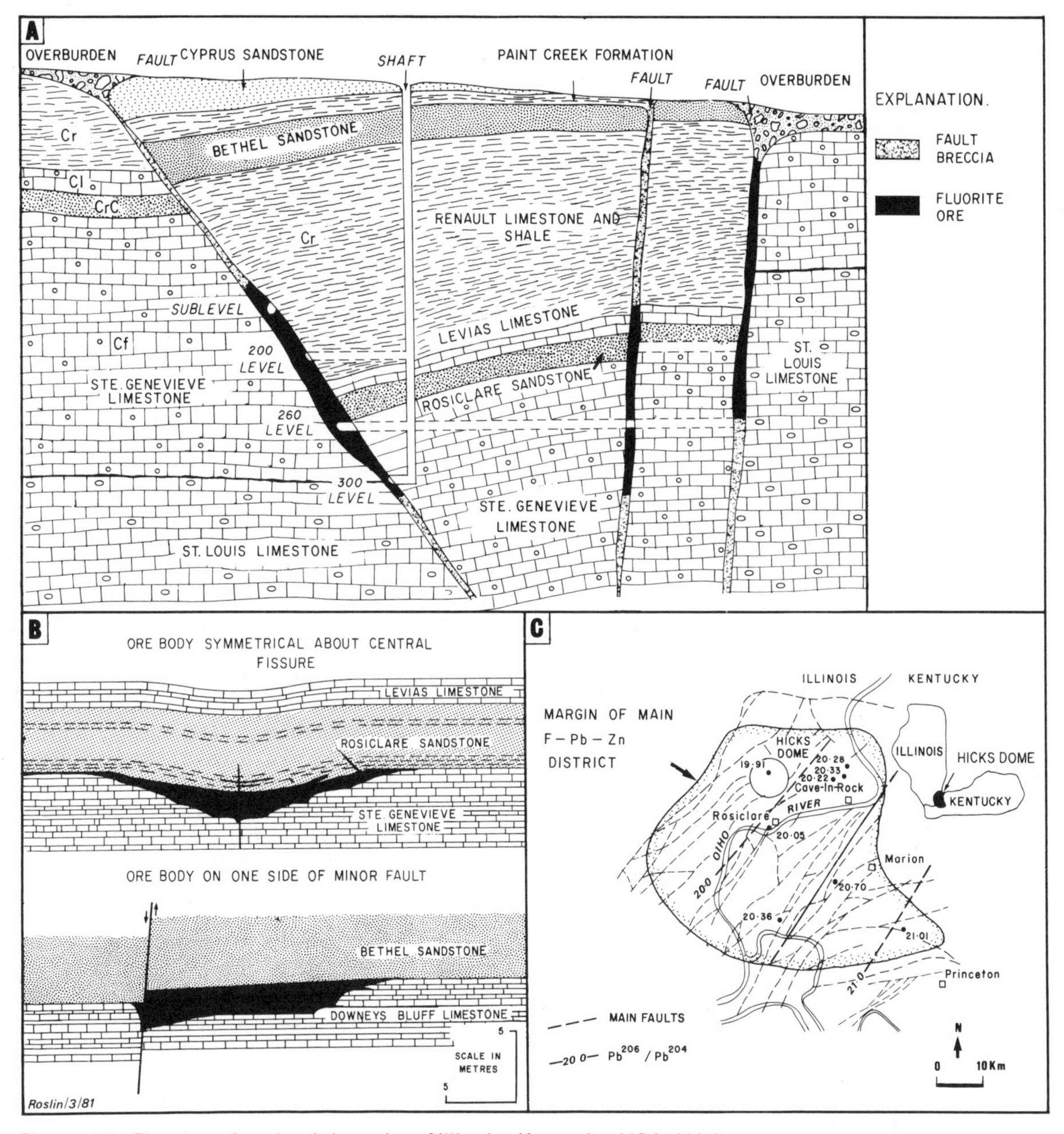

Figure 4.6. Fluorite–zinc–lead deposits of Illinois–Kentucky, USA. (A) Cross-section of the Davenport area, Kentucky, showing the vein fluorite deposits (after Trace, 1974). (B) Cross-sections of the Cave-in-Rock area, Illinois, showing bedding replacement or manto fluorite deposits (after Grogan and Bradbury, 1968). (C) Distribution of $^{206}Pb/^{204}Pb$ isotope ratios across Hicks Dome area of Illinois (after Heyl *et al.*, 1974).

worthless. A general genetic sequence for the bedded deposits is: calcite, followed by yellow, white and purple fluorite, chalcopyrite, quartz, sphalerite, galena, late purple fluorite, bitumen, barite and witherite. Calcite deposition overlaps the whole sequence.

Hicks Dome is a structural high that trends northeast across the area. It represents an alkalic cryptovolcanic centre consisting of explosion breccias that are mineralized with fluorite and barite and contain abnormally high concentrations of thorium, beryllium and rare earths as well as lead, zinc, titanium and niobium minerals. The deposits occur in breccia pipes that have sill-like extensions. The breccia clasts are mainly of Devonian and Ordovician sedimentary rocks. Associated with the breccias are alkalic dykes and breccia pipes dated at 265 Ma.

The least radiogenic lead occurs in galena in and near Hicks Dome, and the most radiogenic lead towards the southeast edge (Heyl *et al.*, 1974) (*Figure 4.6*). This suggests a local magmatic contribution from the mantle through the cryptovolcanic pipes, largely absent outside the dome. The lead-isotope zonation closely follows the zonation of silver, antimony and copper in the galena, which are much more enriched near the dome than elsewhere in the district.

Fluid inclusions in the early yellow fluorite of the Cave-in-Rock district closely resemble oil-field brines, with salinities around 15% sodium chloride equivalent by weight. The later quartz stage has salinities of only 9.3%, and probably represents a mixing of formation connate brines with meteoric water (Hall, W.E. and Friedman, 1963). The fluorite-deposition temperatures have been closely determined within the range 142–145°C (Roedder, 1976).

Although some geologists favour a syngenetic genesis for the deposits (Amstutz and Park, 1967), the majority favour an epigenetic origin in which the elements were carried up by hot brine formation water moving out of the large adjacent sedimentary basin. The circulation was probably driven by deep-seated igneous plutons. The fluids flowed along faults and aquifers in the underlying sedimentary strata to be localized in their present positions within the fault zones, fault breccias, or along bedding planes forming manto deposits. The ^{208}Pb enrichment in the cryptovolcano of Hicks Dome indicates a localized minor magmatic component to the lead in the galena. This component mingled with the main basinal brines, which were later diluted with meteoric fluids to form a three-component ore fluid system (Heyl *et al.*, 1974).

Barite

Commercial deposits of barite may be subdivided in a four-fold classification, and many mining regions show a spectrum of deposit-types: syndepositional bedded deposits; strata-bound replacement deposits; vein deposits; and residual deposits. Barite is an integral component of the Mississippi Valley-type deposits, and many of the mining regions of that ore association also produce barite. However there is sufficient evidence that some bedded deposits are truly syndepositional with the strata. It seems likely therefore that the hydrothermal brines that migrated up-dip towards the regional highs to form strata-bound replacement and fissure-filling deposits of the Mississippi Valley-type must also have exhalated elsewhere on to the sea floor to cause barite enrichment of the sediments being deposited. It is likely in North America that during Devonian and Mississippian times the hydrothermal fluids, which exhalated on to the basin floors to give syngenetic deposits, must also have deeply circulated into older rocks around the basin margins to give Mississippi Valley-type epigenetic deposits. In addition, barite is rather easily concentrated into residual deposits as a result of deep chemical weathering, especially if the country rocks are limestones.

North America is the greatest producer of barite, and shows the complete spectrum of depositional environments. Major production has been obtained from Arkansas, Missouri, Georgia and Nevada. A review is to be found in Brobst (1958, 1975). Barite is produced from Mississippi Valley-type deposits both in England and Germany.

Bedded Deposits

This category is of growing economic value. The dark-barite beds range from a few centimetres to more than 15 m thick and are extensive along strike. They are commonly intercalated with dark chert, siliceous shale and siltstone. Some are massive, but others laminated. The barite grains are usually less than 0.1 mm. Rossettes and nodules of barite are common in some beds and in the adjacent formations. Many beds are of 50–95% barite. The chief impurity is fine-grained quartz. The black-barite layers characteristically contain several percent organic matter and hydrogen sulphide is given off when the ore is broken. This is thought to be a product of decay of organic matter and reduction of sulphate by anaerobic bacteria. Fatty acids and hydrocarbons have been found in some.

In the USA the bedded deposits are found in strata of Devonian or Mississippian age in Nevada and Arkansas. The delicate sedimentary textures indicate a synsedimentary origin. The source of the barite is unknown. Dunham, A.C. and Hanor (1967) suggested that it has been leached from the feldspars of deeply buried igneous rocks of the calc-alkaline series, and feldspathic metamorphic basement rocks could also be a likely source. Connate brines moving out of a basin and circulating through the basement would end up emanating into the regions of contemporaneous sedimentation, as in the Red Sea (*Figure 4.1*).

ARKANSAS

The bedded barite deposits in the Mississippian Stanley Shale of Arkansas contain three small-scale sedimentary features that indicate their syn-sedimentary nature (Zimmerman and Amstutz, 1964). They are: continuous barite beds of variable thickness and grade; lenticular and worm-like bodies; and nodules of pure barite. All these patterns may be associated locally and there are gradations between them.

The larger barite nodules occur as well-defined beds within silty shale. They frequently overlie silt or shale beds barren of barite. There is a simple rhythmic sequence of beds of shale, silt, shale with lenses and nodules of barite, and beds of dense barite. The barite nodules vary from perfectly spherical to ellipsoidal or even worm-like or sausage-like. The trend of the sediments around the nodules is typical of sedimentological and diagenetic features. Local bedding around nodules often contains barite laminae, or worm-like to sausage-like bodies, often with a core of pyrite. The barite layers exhibit sedimentary or diagenetic flowage with small-scale intraformational drag folds. The nodules often show a recrystallized radiating pattern, with barite rods radiating from a common centre or from a central concentric oolitic core in which one of the zones consists of pyrite. These sedimentological and diagenetic features indicate that the barite was in the sediments before diagenesis.

NEVADA

Large high-grade bedded barite deposits occur in and near East Northumberland Canyon, Nye County, south of the Shoshone Range, central Nevada. The barite beds consist of massive and rosette rock, and minor amounts of conglomeratic barite rock, interlayered with dark chert and shale of Ordovician age (Shawe *et al.*, 1969). The barite beds are from 60 cm to 15 m thick. The dark-grey barite rock contains 71–94% barium sulphate; silica and dark carbonaceous materials are the principal impurities. Phosphatic nodules are common in the barite conglomerate. Delicate radiolaria tests form the nuclei of many barite rosettes, indicating space for barite growth before sediment compaction. Barite intraformational conglomerate is evidence of submarine erosion and redeposition during barite sedimentation. Graded bedding and the lateral continuity and conformity of barite beds in the enclosing chert and shaly mudstone beds, together with phosphatic nodules and organic material, all strongly affirm a marine sedimentary origin for the barite.

Some barite deposits in the Shoshone Range contain rhythmic layering of several types. One type includes interlayered barite and clay seams with iron oxide grains and stains where the barite beds gradually change in thickness upwards. There are top-and-bottom features: the bedding of shale layers around barite nodules, the gradation upwards of clay seams containing iron stains and grains to barite with fine-grained quartz, and the lateral gradation of clay seams into stylolites. Chert and very fine-grained quartz are interlayered with the barite. The thickest barite beds occur where the chert and shaly layers are thinnest. These features are taken to suggest a syngenetic process for the barite deposition (Zimmerman, 1969).

Replacement Deposits

Epigenetic replacements result in strata-bound barite deposits, which may closely resemble bedded syndepositional barite. In many cases it is extremely difficult to distinguish between the two origins because diagenetic recrystallization of a syngenetic deposit can be expected to cause recrystallization and replacement of sedimentary features such as fossils, as well as the migration of barite into the neighbouring rock strata.

Barite is distributed in the USA only in large-scale uplifted structural features, either domal or fold belts, suggesting that the fluids that carried the barite moved upwards and outwards from the adjacent deeper basins. Although any particular barite deposit never shows any demonstrable relationship to igneous bodies, it is likely that feldspar-rich granitoids and metamorphic rocks in the basement may have been the ultimate source of the barium, leached by the hydrothermal fluids from the feldspars (Dunham, A.C. and Hanor, 1967).

NEVADA

Bedded barite deposits are contained in the Devonian Slaven Chert of the Shoshone Range, central Nevada. The scattered localities are closely related to bedded chert. Individual deposits range from just over 1 m to about 600 m long. The outcrops are elongated parallel to the host bedding. Commonly they form large irregular bodies associated with smaller pods separated at distances of 100 m or more. Pre-mineral and post-mineral faulting and folding and irregular replacement of the beds makes the boundaries irregular (Ketner, 1965). The deposits are of interbedded barite, chert and minor limestone. There are sharp bedding contacts between the barite and chert and there are no altered wallrocks. The barite rocks contain microscopic chert and unreplaced limestone remnants, carbonaceous matter and small amounts of pyrite and iron oxide. Veinlets of barite, quartz and calcite transect many barite beds. Most of the barite is nearly as fine grained as the chert. Laminae are commonly less than 1 mm. Separations between the laminae resemble stylolites in their jagged cross-section and dark colour. Where the barite has recrystallized, it is white and the laminated texture is obliterated. It is instead massive and bladed. Like the host-rock chert, the barite beds contain shaly intervals of two to several centimetres, giving the rock a slabby fissility.

Fossils have been replaced by barite, giving the ore an augen-like texture. The barite is commonly dark-grey, just like the enclosing chert, but is locally white. It is apparent that the barite replaced both chert and limestone, for some contacts between barite and host rock go directly across the bedding.

Devonian chert, metaquartzite and greenstone in the Southern Independence mountains of northeast Nevada are mineralized by barite (Ketner, 1975). The barite is not bedded; it ranges from pure white and massive to dark-grey irregular mixtures of barite, white metaquartzite, black chert and greenstone. There is abundant evidence to show that the deposit was a result of replacement of the host rock, for example, the transection of the host-rock textures by blades and rosettes of barite in greenstone, chert and quartzite. All grades of replacement are found, from rocks that contain small isolated barite crystals to rocks composed almost wholly of interlocking barite with isolated interstitial remnants of the host. In some cases the barite has crystallized parallel to the chert bedding, but has broken some of the beds apart by the force of crystallization. Some barite fills fractures.

It appears that the barite was introduced over lengthy periods of time. Thus in some cases it looks as if the barite was introduced soon after deposition when the chert was still soft. However in other cases, the barite was certainly brittle and recrystallized, since quartz-filled fractures in the chert are offset by barite veins.

NOVA SCOTIA, CANADA

The Mississippian strata contain gypsum and anhydrite-rich horizons, as well as epigenetic barite and base-metal sulphides (Boyle *et al.*, 1976). The barite occurs in two styles: one is of reddish-to-greyish massive cryptocrystalline barite, as in the Magnet Cove deposit, which is a large pipe-like mass localized in a complex brecciated drag fold zone between two fault systems (*Figure 4.7*). The barite and associated sulphides are formed by replacement and fracture filling of limestone, conglomerate, fissile limestone and anhydrite. The principal minerals are barite, siderite, pyrite, marcasite, galena, sphalerite, chalcopyrite, tennantite, proustite, pearceite, acanthite, manganese oxides and haematite. Supergene minerals include limonite, manganese oxides and malachite. The barite bodies exhibit no obvious relationship to any igneous body. The other type of barite occurs as veins and irregular pods principally in limestones, and it is invariably coarsely crystalline.

The cryptocrystalline barite is thought to have been a replacement of gypsum and anhydrite, whereas the coarse-grained barite was introduced as barium and sulphate-bearing solutions filling vein voids. The replacement of anhydrite and gypsum is supported by the isotope data (Boyle *et al.*, 1976). The barite, anhydrite and gypsum have similar $\delta^{34}S$ values ranging from +12 to 31 per thousand. On the other hand the coarsely crystalline barite has a value of +12 to +18, similar to the sulphate of deep circulating brines ($\delta^{34}S$ + 11 to 15). The lead–zinc–fluorite–barite deposits of the northern Pennines of England have similarities to those of the Walton–Cheverie area, Nova Scotia. These bodies probably resulted from rising connate brines charged with sulphate, fluoride, barium, lead, zinc and copper.

Vein and Cavity-filling Deposits

The barite fills faults, joints, bedding planes, breccia zones, karstic solution channels and sink structures, and the deposits are clearly epigenetic. Most of the vein and cavity fillings are dense grey to white, and they may be associated with metallic sulphide ore minerals. The deposits have sharp contacts with the country rocks. Deposits in collapse and sink structures are common in Missouri and in the Appalachian states; they are rich, but usually small. Upon exposure the deposits may be enriched by deep chemical weathering to give residual deposits (*see* page 66). They are clearly epithermal, and may be likened to the Mississippi Valley-type of deposition that is found in the Pennines of England or the Appalachians of North America, being confined to fillings of open structures, predominantly in carbonate country rocks.

IRELAND

The Ballynoe deposit near Silvermines, County Tipperary, Ireland, is a tabular conformable gently dipping body in limestone 3–5 m thick with reserves of more than 2 million t of barite, located in Lower Carboniferous strata. A general account of

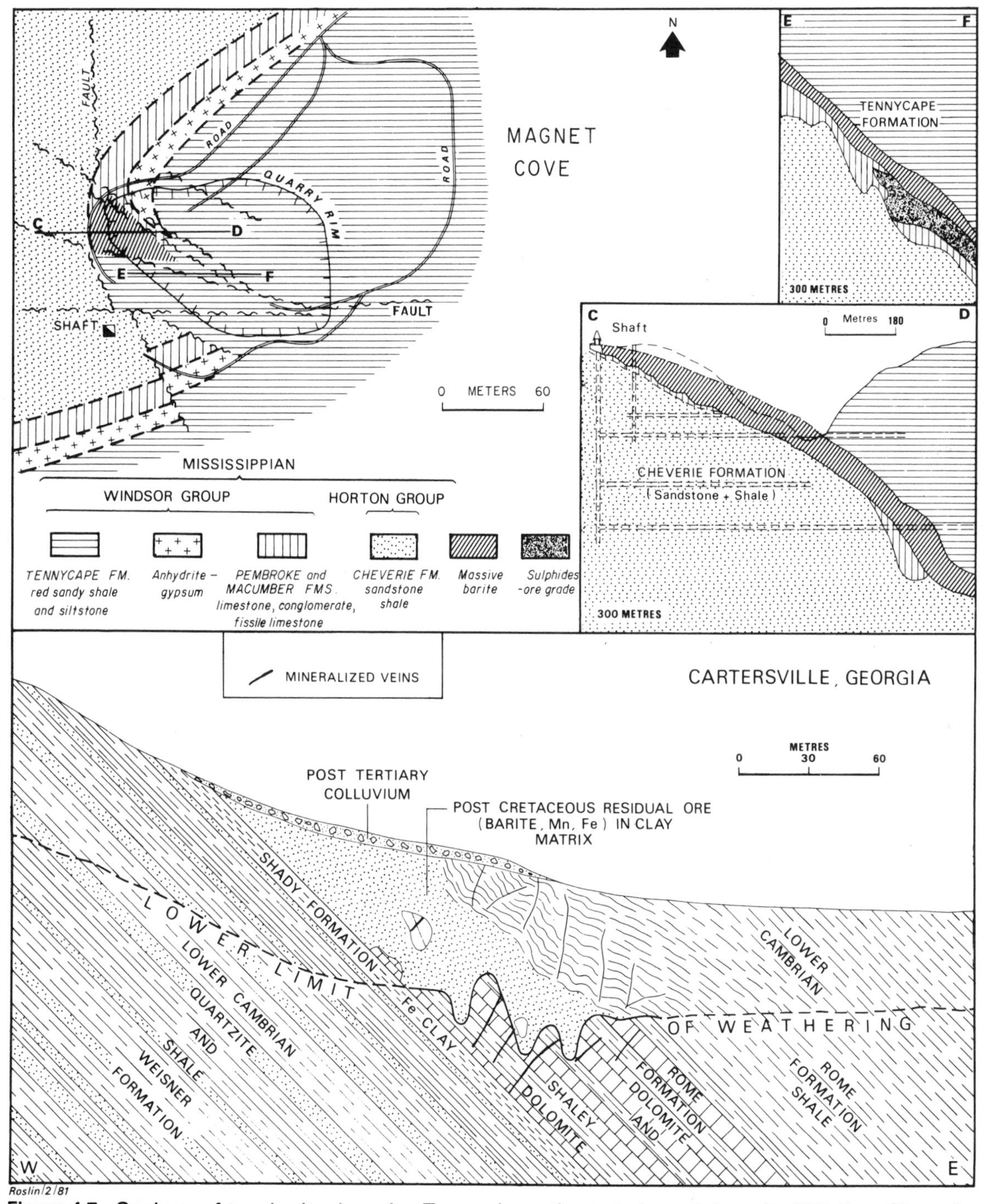

Figure 4.7. Geology of two barite deposits. Top: epigenetic strata-bound deposit of Walton–Cheverie, Nova Scotia, Canada (after Boyle *et al.*, 1976). The barite ore is genetically associated with a base-metal sulphide body. Bottom: the residual barite–clay deposit of Cartersville district, Georgia (after Kesler, T.L., 1950). The unweathered host rock contains fault fillings of sulphides in a barite matrix and stratiform specular haematite in the Shady Formation.

the Silvermines geology and mineral deposit is given by Taylor, S. and Andrew (1978), with summaries by Williams, C.E. and McArdle (1978) and Evans, A.M. (1976). Fluid inclusions suggest a temperature of deposition around 250°C, and the $\delta^{34}S$ values for the barite are around + 15 to 19 per thousand. They are thus coincident with values for evaporites of the Carboniferous basins. The deposits are generally held to be epigenetic, deposited by hot brines that entered the area via major faults, but it is quite likely that the barium and even other base metals may have been leached from granites in the basement.

Arkansas

White and clear barite occurs as the cement of some Cretaceous conglomerates in Arkansas. In this case the barite is epigenetic and may have its source in bedded barite deposits in the underlying Palaeozoic formations occurring in the neighbourhood.

Kentucky

The now exhausted barite deposits of central Kentucky occurred in fault fillings on the Jessamine Dome. The north–northeast-striking veins are of fissure and fault breccia filling. The mineralogy is barite, calcite, sphalerite, galena and fluorite in decreasing order. The host rocks are Middle Ordovician limestones. The deposits have been described by Plummer (1971). Fluid inclusions are extremely saline (5–35% sodium chloride equivalent by weight), and the temperature of deposition was 72–132°C. It is clear therefore that the deposits are epigenetic, and the mineralizing fluids were basinal brines migrating up-dip to the domal structure. It is suggested that the time of mineralization was early Tertiary, following uplift of the domal structure and development of the present-day topography. It is likely that the barite was deposited as a result of interaction between the mineralized brines and the regional meteoric ground water. This is indicated because the barite deposits are closely associated with the present-day ground-water table.

Southeast Asia

Strata-bound and fault-filling barite are common in the Middle Ordovician Wumbye Formation limestones in the eastern highlands of Burma (Goossens, 1978). Mining is carried out near Maymio, east of Mandalay. A new large deposit at Bawsaing is closely associated with argentiferous galena, found within strongly folded Lower-to-Middle Ordovician carbonates of the Wumbye Formation. When exposed, the galena is converted to cerussite. The barite body occurs beneath residual barite contained within clay. The barite also occurs as some fracture fillings in limestone, but generally it is stratiform. The barite contains only traces of galena. Although the deposits are lenticular, they all occur in the same stratigraphic horizon. A large deposit has also been concentrated in the nose of an overturned anticline. Barite also occurs as the principal gangue in the stratiform galena deposits of similar age that are mined nearby. Because of their restriction to the same stratigraphic horizon, these barite and base-metal deposits may originally have been syngenetic, but many have been transported into favourable host rocks, and therefore may be regarded as Mississippi Valley-type.

In Thailand there are at least 15 barite localities. Most occur as veins and cavity fillings along local structures, and some are of the residual-type (Asnachinda and Pitragool, 1978). Some of the veins occur in igneous rocks, but most are in limestones. The strata-bound deposits are in a variety of ages of strata from the Lower Palaeozoic to the Mesozoic.

The main barite deposit of Peninsular Malaysia is at Tasek Chini, where it forms a filling of a steeply dipping fault zone, 3.5 m wide. There are also many localities of residual barite (Hosking, 1973a), but the generally high iron content of Malaysian barite makes it less attractive than Thailand barite for local oil-industry needs. The central peninsular area where barite is mined is also known for its many residual deposits of manganese, and therefore bears a resemblance to the occurrences of Georgia (*see* page 67).

Residual Deposits

When limestones containing barite deposits are subjected to deep chemical weathering, the barite strongly resists weathering, whereas the limestone and other country rocks may be preferentially dissolved—an excellent mechanism for beneficiation of barite ore. The deposits are of irregular or rounded blocks of barite in residual clay, well known from Missouri, Tennessee and Georgia. Most of the residual barite is white, and translucent to opaque. It occurs as mammillary, fibrous, or dense masses but subhedral to euhedral crystals are also found. Most of the barite masses are

3–15 cm in diameter. Chert, jasperoid and druzy quartz are common associates. Small amounts of pyrite, galena and sphalerite occur in the barite. The matrix is of incompletely weathered rock fragments and yellow-to-red illitic clay. The grade is very variable, from 10% to 20% barite. The depth of the deposits depends on local weathering conditions. In Georgia some terminate 45 m beneath the surface, however in Missouri they terminate less than 5 m beneath the surface. The bedrock of residual deposits usually contains veins filled with barite, fluorite, quartz and calcite, and locally pyrite, sphalerite and galena, but usually the grade is too low to continue mining into the hard rock.

Georgia

The Cartersville district of Georgia, 65 km northeast of Atlanta, is one of the richest barite provinces of North America. Its geology has been described by Kesler, T.L. (1950). The region is known for its residual deposits of barite, manganese oxides, brown iron, ochre and umber and its primary bedded deposits of specular haematite. Primary ore and gangue minerals were deposited in faults formed in fractured calcareous rocks of the Lower Cambrian formations. The minerals include pyrite, galena, sphalerite, chalcopyrite, enargite, tennantite, vein quartz, carbonates and barite. The barite commonly encloses all the sulphides. The specular haematite does not occur in the vein and breccia deposits, and is confined to beds in the Shady Formation (*Figure 4.7*). The deposition of ores in the fault zones began before faulting had ended, for the primary minerals have been brecciated and enveloped in subsequently deposited jasperoid. The jasperoid is of fine-grained quartz, which replaced most of the earlier formed vein carbonate and part of the wall rocks of the fault zones and breccia fragments.

Intense weathering has formed secondary residual deposits – the basis for the barite industry. Barite occurs as lenticular streaks, massive fine-grained rounded blocks and whitish dull masses with iron-stained surfaces within residual clays (*Figure 4.7*). Barite in the residual deposits is always accompanied by angular boulders of jasperoid. The barite is of aggregates of curved crystals, enclosing small amounts of pyrite, altered to limonite. Fluid-inclusion studies of the residual barite (Rife, 1971) have shown that it was originally deposited in the veins and breccias at temperatures of 126–297°C, indicating a hydrothermal depositional history of fracture filling. Most of the larger residual barite and manganese deposits become exhausted at depths of 15–30 m.

Tennessee

Tennessee has barite deposits in a range of host rocks ranging in age from Precambrian to Mississippian. The deposits have been reviewed by Maher (1970). The most productive deposits are residual, in a clay matrix, overlying the Early Ordovician Kingsport Formation. However barite also occurs in veins, in breccia fillings and there are some bedded deposits. All these may give rise to residual deposits on weathering, especially if the host is a carbonate rock. Some deposits contain galena, fluorite, sphalerite and pyrite along with the barite, and therefore have a similarity to the Mississippi Valley-type. However barite is not a conspicuous gangue mineral in the major base-metal sulphide deposits of Tennessee.

Many deposits in the Valley and Ridge province of east Tennessee are composed of barite cementing angular fragments of host rocks. The Sweetwater deposits are known to be of this breccia-filling type, but the breccia itself is not sufficiently barite-rich to be mined. The weathered residual ores overlying the breccias form the main ore material. Weathering has removed the rock material and preferentially concentrated the barite. The most common impurity in the barite is quartz. Many deposits contain barite masses in the residual clay that are larger than those observed in the bedrock. It therefore appears that the original barite has been dissolved and reprecipitated in the weathering cycle. This probably resulted from resolution of barite by acid generated by oxidation of associated pyrite, local migration and reprecipitation (Maher, 1970).

The thickness of the ore-bearing residual clays ranges from a few metres to more than 30 m, with an average of 10 m. The ratio of barite/clay matrix ranges from 1:2 to 1:20 and for the majority of mines is around 1:13. Wherever mining has exposed bedrock, it contains veins and breccia masses cemented by barite, fluorite and pyrite with local concentrations of sphalerite and galena, but not of high enough grade to warrant mining.

It is possible that barite deposits result from epigenetic introduction of barium expelled from deeply buried feldspathic rocks, so that the hydrothermal solutions must have moved deep into the igneous basement via important regional fault systems. It is also likely that some of the hydrothermal solutions emanated through fractures into Mississippian seas to form submarine-exhalative syngenetic deposits. However the Ordovician deposits are epigenetic and predominantly in fissures and fault breccias. The deposition was therefore as open-space fillings.

Sedimentary Ironstone Deposits of Shallow Epicratonic Seas

There is an important category of sedimentary iron formations, which Kimberley (1978b) has called 'sandy, clayey, and oolitic, shallow-inland-sea iron formations'. They may be more generally referred to as ironstones. This is the main Phanerozoic-type of iron deposit. The ironstone is normally quite oolitic and unbanded. Some, like the Silurian Clinton ironstones of the USA, do not appear to be oolitic in hand specimen, but thin sections show them to be characteristically oolitic. The European ironstones are distinctly oolitic in hand specimen. The main ferriferous minerals that precipitate to form oolitic sedimentary ironstones are goethite, haematite, limonite, siderite, chamosite, and, to a lesser extent pyrite, iron-containing chlorite and magnetite. Common non-ferriferous minerals include detrital quartz, calcite, dolomite and apatite. Authigenic chert may occasionally be present but is not characteristic as in the Precambrian banded-iron formations. Common metal concentrations are 34% iron, 0.6% phosphorus and 0.08% vanadium. Manganese is more variable but the content is mostly between 0.1% and 0.5%.

Sedimentary features of the ironstones indicate deposition in very shallow agitated waters, with such features as ooids broken and regrown, ripped up clasts of penecontemporaneously lithified oolitic sediments, large-scale cross-bedding, abundant ripple marks and rain imprints. The fossils also indicate shallow water, and they are frequently replaced by ferriferous minerals. Ubiquitous trace fossils indicate contemporaneous burrowing in the iron-rich sediments. In some cases, enclosed wood and seeds may be replaced by ferriferous minerals.

It is apparent that the major iron ore deposits of the world (such as the Lake Superior banded-iron formations, and the Clinton and Minette ironstones of Europe) were deposited in epicratonic or pericratonic basins. They formed in very shallow basins, which were separated from the open sea by barriers of oolite and interclast sand or by tectonic islands (Dimroth, 1975). The terrigenous sediments associated with iron ores are all extremely mature, indicating that the eroding land mass had little relief and was deeply weathered. Although the great Silurian Clinton Group and the Jurassic ironstones of Europe were deposited in inland seas that probably were not entirely landlocked, it is possible to suggest the Aral Sea of central Asia as a good present-day analogue. Although landlocked, and all drainage is from the surrounding continent, the present-day Aral Sea water is quite salty (Dickey, 1968). The waters are not modified sea-water, but intermediate between river and ocean water, poorer in sodium and chloride and richer in calcium, magnesium and sulphate than sea water. The bottom sediments of the Aral Sea include sands, silts, marls, clays, delta clays and oolitic formations. Approximately 4.5 million t of iron are estimated to be brought into the Aral Sea each year by the great Amu Darya and Syr Darya rivers, 3% in solution, the rest in suspension. In addition to forming part of some of the component clay minerals, the iron occurs in the bottom sediments as gelatinous hydrated goethite. Hydrotroilite occurs as round lumps especially in the organic-rich layers. Grains of quartz, coated with iron oxide, have been blown in from the surrounding desert. The iron content of the sediments varies from 0.5% in calcite sands to 3.5% in the clays and marls (Dickey, 1968).

Iron oolitic formations are not distributed evenly throughout the Phanerozoic period, but are concentrated at specific times. Periods of important ironstone formation were the Ordovician and Silurian (for example, Clinton, Wabana, Normandy, Brittany, Thuringia, Bohemia, Sardinia, Wales and western Morocco), Late Carboniferous (Scotland and England) and the Jurassic and early Cretaceous (Germany, Lorraine, England, Chile). Between these periods, iron oolite formation is rare. This indicates that their formation was dependent on specific regional influences, one of which was the prevailing climate (Wopfner and Schwarzbach, 1976).

Iron oolites formed preferentially in epicratonic marine or landlocked sea basins. According to Borchert (1960), hydroxides and oxides form in shallow aerated waters, followed by siderite, chamosite and ultimately pyrite with increasing depth of water. This facies zonation corresponds to the vertical zonation of the sea water, comprising in descending order oxygen, carbon dioxide and hydrogen sulphide zones. The iron is carried into the basin of sedimentation, and may be incorporated in the minerals of the sediments, but the iron is mobilized from the sediments by waters with increasing carbon dioxide concentration. This model is the most convincing for sediments of the Black Sea or Aral Sea. Such conditions are thought to have been prevalent during ice-free periods when, owing to the absence of polar ice caps and cold climate and oxygenated bottom currents, a predominance of black-shale facies was developed on the sea floors. Some authors argue that the iron was derived from laterized land surfaces, eroded and transported by rivers. The most important factor for ironstone development is a low rate of clastic supply (Talbot, 1974), but it is possible that erosion and transportation of iron was from a variety of rocks exposed on the land surface around the basin margins. Hot humid climates are considered most favourable. Present-day occurrences of chamosite are limited to within 10° of the equator, and it is clear that it forms only where the sea temperature exceeds 20°C.

Haematite and other ferric compounds are the only iron minerals that can exist in true equilibrium with depositional waters (Curtis and Spears, 1968). However, ferrous compounds (pyrite, siderite and chamosite) are stable beneath the sediment/water interface in a basin. The ironstones are chemical deposits in an environment in which clastic imput was diminished, because the hinterland had long been reduced to a peneplain on which there was intense chemical weathering or laterization. The ironstones apparently accumulated in shallow marine basins in intertidal zones close to the shore, and the sedimentology and fauna indicate an oxidizing regime. One could therefore expect limonite to be stable in such an environment, and pyrite and siderite to be stable only beneath the water/sediment interface. This appears to be the case from studies of ironstones. Chamosite is more difficult to explain for it appears to have been a primary mineral, but low E_h values are necessary for its formation. Since diagenetic processes are important in the evolution of the ironstone textures, the chamosite now present may not represent the initial precipitate. A mixed gel could have been the initial precipitate, and the chamosite may have recrystallized from it during diagenesis.

Some ironstones have been subjected to reworking and redeposition, and in Europe they have been called detrital iron ores (Zitzmann and Neumann-Redlin, 1977). Some sedimentary ore fields have a range from purely oolitic ironstone to partly reworked detrital-type. In some cases the ironstone was deposited in elongated valleys resulting from the karstic erosion of limestone areas.

Middle Silurian Clinton Group Ironstones

The Clinton Group forms the lower half of the Niagaran Series in the central Appalachian basin. A major review is to be found in Hunter, R.E. (1970) and details of field occurrences in Muskatt (1972). It is mainly a shale formation, with current-bedded sandstones common in the east and carbonates in the west (*Figure 4.8*). The ironstones for which the Clinton Group was famous are mainly oolitic haematite beds, but oolitic chamosite is common. There is conclusive evidence that the oolites were formed by precipitation of iron compounds on grains while they were being transported across the sea floor. The Clinton cross-bedding is dominantly of the torrential-type or planar-type, typical of a fluviatile environment (Muskatt, 1972). There is a uniformity of current direction from the land towards the northwest throughout Clinton time. The ripple marks trend 358°, indicating the palaeo-strike during deposition. The heavy minerals in the Clinton Group of central New York indicate that the strata were principally derived from the erosion of low-grade metasediments with some contributions from acid igneous rocks.

The Clinton Group may be divided into several sedimentary facies (*Figure 4.8*). A silica-cemented sandstone facies occupies the easternmost parts of the basin. It includes sandstones and conglomerates, and cross-bedding is common, indicating a consistent regional current direction in a northwesterly direction towards the deeper basin (Muskatt, 1972). This facies received very little iron sedimentation, and it appears to represent fluvial channel deposits. The silica-cemented sandstone facies grades upwards and westwards into a red argillaceous sandstone facies. Its environment appears to have been deltaic or alluvial floodplain. Iron occurs in the red argillaceous sandstones mainly as finely crystalline haematite associated with the argillaceous matrix. The chlorite of the matrix is also iron-rich. Small amounts of ankeritic dolomite, siderite and pyrite occur in some green beds. It is uncertain whether the red colouration is due to post-depositional oxidation of iron-bearing chlorite or to the preservation of fine-grained iron oxide derived from soil of the source land to the east. The red argillaceous sandstone facies grades westwards into one dominated by greyish-red haematite-cemented sandstone. Trace fossils are common, and other fossils indicate a marine environment. Chamosite, collophane pellets, ripple marks and cross-bedding are common. The presence of *Lingula* indicates a neritic environment, which may be interpreted as tidal flats to shallow sea below low tide.

The sandstones are cemented principally by quartz overgrowths and by haematite and iron-rich chlorite. It seems likely that the haematite was precipitated as oolitic coatings on sand grains before their final deposition, followed by crystallization, or solution and reprecipitation after burial; but some of the haematite may have been precipitated after the sand deposition. Haematite remained stable after burial evidently because the E_h of the interstitial solutions was kept high by the

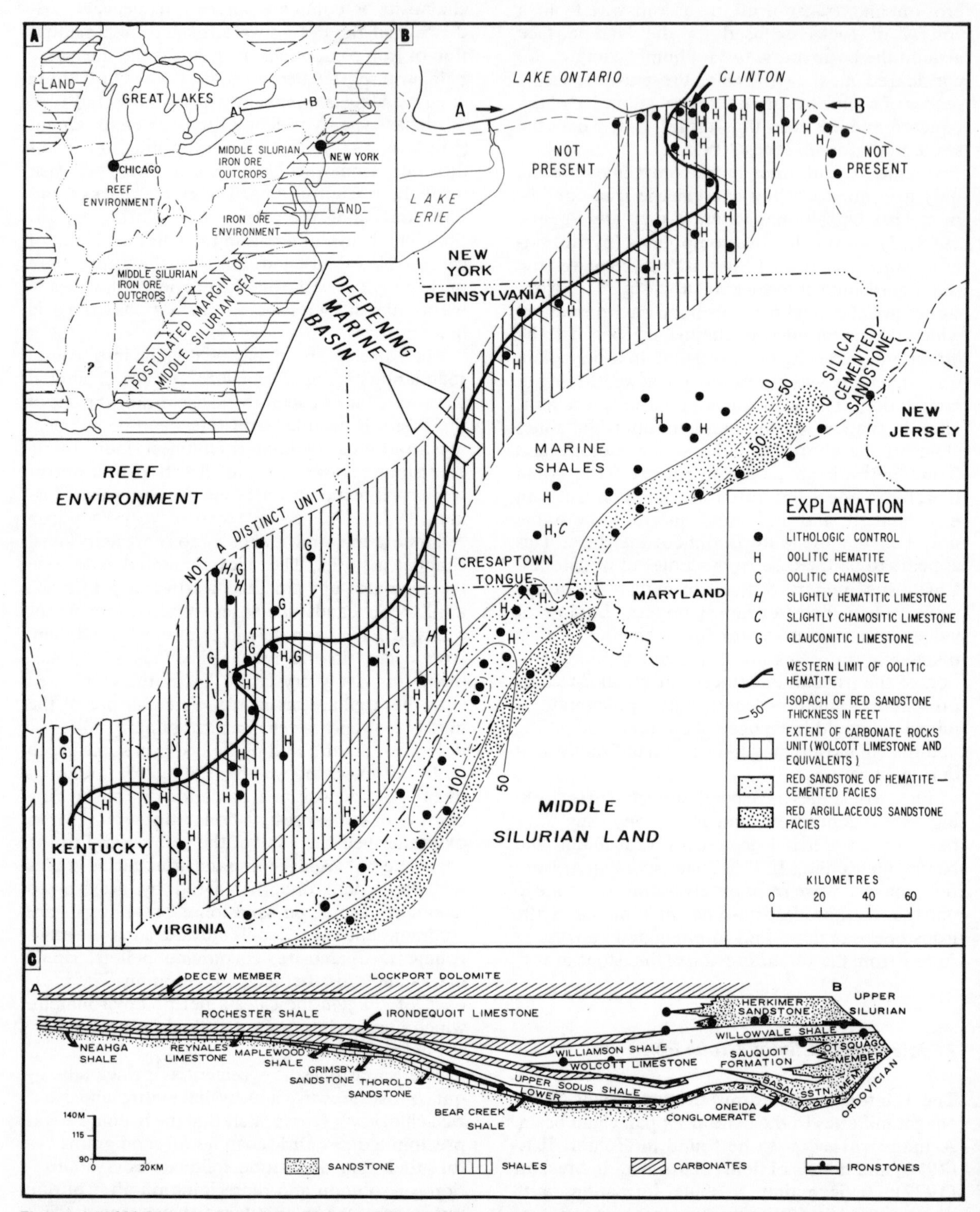

Figure 4.8. (A) Distribution of the Silurian Clinton Group sedimentary ironstone ores of the USA. Relationship of the shallow-sea environment to the land masses (after Wolf, 1976a). (B) Lithofacies map of the Upper, Lower and Middle Clinton Group (after Hunter, R.E., 1970). (C) East–west cross-section from southeast Lake Ontario through the town of Clinton (after Hunter, R.E., 1970).

large amount of ferric oxide within the sediment.

Chlorite-cemented sandstone may locally take the place of the red argillaceous sandstone and haematite-cemented sandstone facies. Small-scale ripple marks are less common in this facies, but large-scale cross-bedding is common. The cross-bedding has both a basinwards (northwesterly) direction and shorewards directed towards the southeast. These bimodal cross-bedding patterns have been attributed to ebb and flow tidal currents. It appears that the quartz grains were initially chlorite-coated before deposition, and the chlorite has eventually ended up as the cement during diagenesis.

The above sandstones intertongue westwards with greenish-grey and subordinate greyish-red shales containing thin beds of definitely marine sandstone, siltstone and carbonate. The iron content is contained in chlorite, but some is as illite and glauconite.

West of the green shale facies, and intertonguing with it, is a facies of limestone, dolomite and subordinate calcareous shales. The most common rock-type is biocalcarenite and coquinal limestone and their dolomitized equivalents. These rocks were deposited in the deeper water, far from the terrigenous input into the basin. Very little iron was deposited in this facies, but some glauconite and chamosite occur locally.

The Upper Clinton Group is characterized by a westwards transition from the silica-cemented to a carbonate-cemented facies. Many of the carbonate-cemented sandstones contain marine fossils. The cross-bedding is predominantly basinwards (northwestwards). The amount of iron is quite small, but thin beds of ironstone do occur. The Rochester Shale overlies the Clinton green shale facies (*Figure 4.8*). Illite and chlorite are the main minerals. It formed in deeper water farther offshore than the carbonate-cemented sandstones, and it has a very low iron content.

The Ironstones

Ironstones occur in the haematite-cemented sandstone facies, green shale facies, carbonate facies and carbonate-cemented sandstones of the Clinton Group. Most of the ironstones are of oolitic haematite. Most of the oolitic nuclei are of quartz sand grains, or carbonate fossil grains. Fossils are typically completely or partially replaced by iron minerals. All the oolitic ironstones are cemented mainly by calcite and dolomite. Thin beds of oolitic chamosite are commonly associated with oolitic haematite and overlying or underlying shale, and they represent deeper-water conditions than the oolitic haematite.

The horizons at which most of the ironstones are concentrated coincide with major faunal breaks in the Clinton Group, and at least some of the horizons are represented by unconformities in parts of the basin.

Source of the Iron

The preponderance of iron in the eastern shallow facies of the basin, and the low iron content in the deeper-water facies towards the west, clearly indicate that the iron was derived from the east presumably from the same land surface that was eroded to give the clastic materials that infilled the basin. The land surface being eroded probably was of deeply weathered low-grade metamorphic and some igneous rocks, which had not been widely uncovered until Clinton Times. The Clinton Group sediments are not particularly feldspathic, indicating that the eroded land mass was probably deeply weathered. A low-topography peneplain can explain the low sedimentation rates and the deeply weathered regolith could have contributed an abundant iron supply to both the drainage and ground-water systems. If, on the other hand, the hinterland had been orogenically uplifted, causing rapid erosion, the rock materials would not have been deeply weathered and the iron would therefore be dispersed in thick sedimentary sequences characterized by an absence of ironstones. The transportation of iron from the land to the basin may have been as ferrous iron in solution in river water of low E_h and pH.

Iron Deposition

The low-iron content of the silica-cemented sandstone facies indicates that this was the environment of iron transportation, not deposition. Whether or not the iron was carried by rivers or by ground water that was discharged subaqueously near the shoreline, precipitation took place soon after the iron-bearing water reached the marine environment. The intermingling of the fresh with the marine water caused precipitation before currents could distribute it uniformly throughout the Clinton Basin. Rapid precipitation is to be expected when water, having the low pH and moderately low E_h necessary for high solubility of iron, is introduced into a body of water, such as a shallow marine basin that has a higher pH and E_h (Curtis and Spears, 1968).

Europe

The environment of deposition of the Jurassic iron oolites was an epicontinental marine basin, which, during the Jurassic, covered large parts of central Europe, England, and most of the North Sea. The basin was subdivided by land masses such as the Rhenish Massif, and by peninsulas and shoals (*Figure 4.9*). Oolitic ironstones formed in proximity to these land areas, and embayments and re-entrants appear to have been the preferred loci of their formation (Wopfner and Schwarzbach, 1976).

ENGLAND

Westphalian ironstones

Ironstones occur in all the major coal fields of Britain, with the main deposits in the Midland Valley of Scotland, South Wales, north Staffordshire and Yorkshire, but none is now of economic interest (Slater and Highley, 1977). They usually lie above the coal seams, consist of siderite and are referred to as 'clayband' ores, or as 'blackband' ores where there is a significant proportion of carbonaceous material. These ironstones are presumed to have formed in fresh-to-brackish coastal swamp water. Their nodular character is due to diagenetic accretion.

Jurassic and Cretaceous ironstones

Sedimentary ironstones provide the major source of iron worked in Britain today. Summaries of the deposits are to be found in Hemingway (1974a, b), Slater and Highley (1977) and Dunham, K. *et al.* (1978).

The principal deposits of Jurassic ironstone occur in the Lower and Middle Jurassic.The ore field extends from the Cleveland Hills to beyond Banbury, a distance of over 300 km (*Figure 4.9*). The ironstones are low grade (20–35% iron) and occur as a series of disconnected flat-lying lenticular beds. The are of limited vertical thickness (less than 10 m), but of considerable lateral extent, particularly along the strike direction, where ore horizons may be traced over tens of kilometres. Along dip, each ironstone passes laterally into ferruginous sands and clays. The deposits are characterized by small rounded ferruginous oolites, cemented by an iron-rich matrix. They occur in limestones, sandstones and mudstones, and frequently show a cyclical development. The principal ore minerals are chamosite, siderite and limonite. The chamosite (30% iron) occurs as a primary constituent of the oolites and matrix. The oolites are of concentric layers of chamosite around cryptocrystalline chamosite or fragments of earlier oolites. They may show diagenetic changes to opaline silica, kaolinite, siderite, pyrite and sometimes magnetite. Siderite occurs as a matrix component in both oolites and mudstones. Limonite may be a primary constituent of the oolites, but invariably formed as a result of surface exposure of the ironstones. Limonitization frequently has enhanced the grade of the ore. Calcite occurs as shell debris, or as a matrix component. Quartz is present as detrital sand. Clay is associated with the chamosite and limonite.

The ironstones originated in shallow water along the margins of a marine sedimentary basin (*Figure 4.9*). The formation of oolites is ascribed to agitation of the sediment by constant current action, indicating a near-shore environment. The frequent association of oolites with estuarine sediments suggests their deposition close to river mouths.

Following the prevailing continental conditions of Britain during the Permo-Triassic, the Jurassic represents an incursion of marine conditions, when epicontinental seas covered much of southern England. Local intracratonic down-warping led to the formation of the basins in which the ironstones were deposited. In the Yorkshire ironstones, the oolites contain thick-shelled fossils such as oysters and pectens. The Main Seam of the Cleveland Ironstone contains abundant cross-bedding. Erosion on many scales is frequent, and ranges from the absence of a distinctive bed, to the presence of ironstone intraclasts within a seam (Hemingway, 1974a). The deposits were therefore formed in shallow water near-shore where wave action was constantly experienced.

The Oxfordian oolitic ironstones of Wiltshire have been described by Talbot (1974). The environment of deposition has been interpreted from the sedimentology and fossil assemblage to be lagoon or near-shore embayment. The iron was derived from the eroding deeply weathered land mass, and rivers transported the iron together with the clastic material. The single most important factor responsible for the appearance of oolitic ironstones was the reduction of clastic supply from the low relief land. The iron may have been transported in colloidal suspension or on the surface of clay minerals (*Figure 4.9*). The suspensions coagulated rapidly upon reaching the salt water just offshore, forming chamosite oolites from gel. The lagoons or embayments probably represented hypersaline areas in the seas as shown by their restricted faunal assemblage. Such hypersaline conditions obviously favoured the rapid precipitation of the iron (Curtis and Spears, 1968).

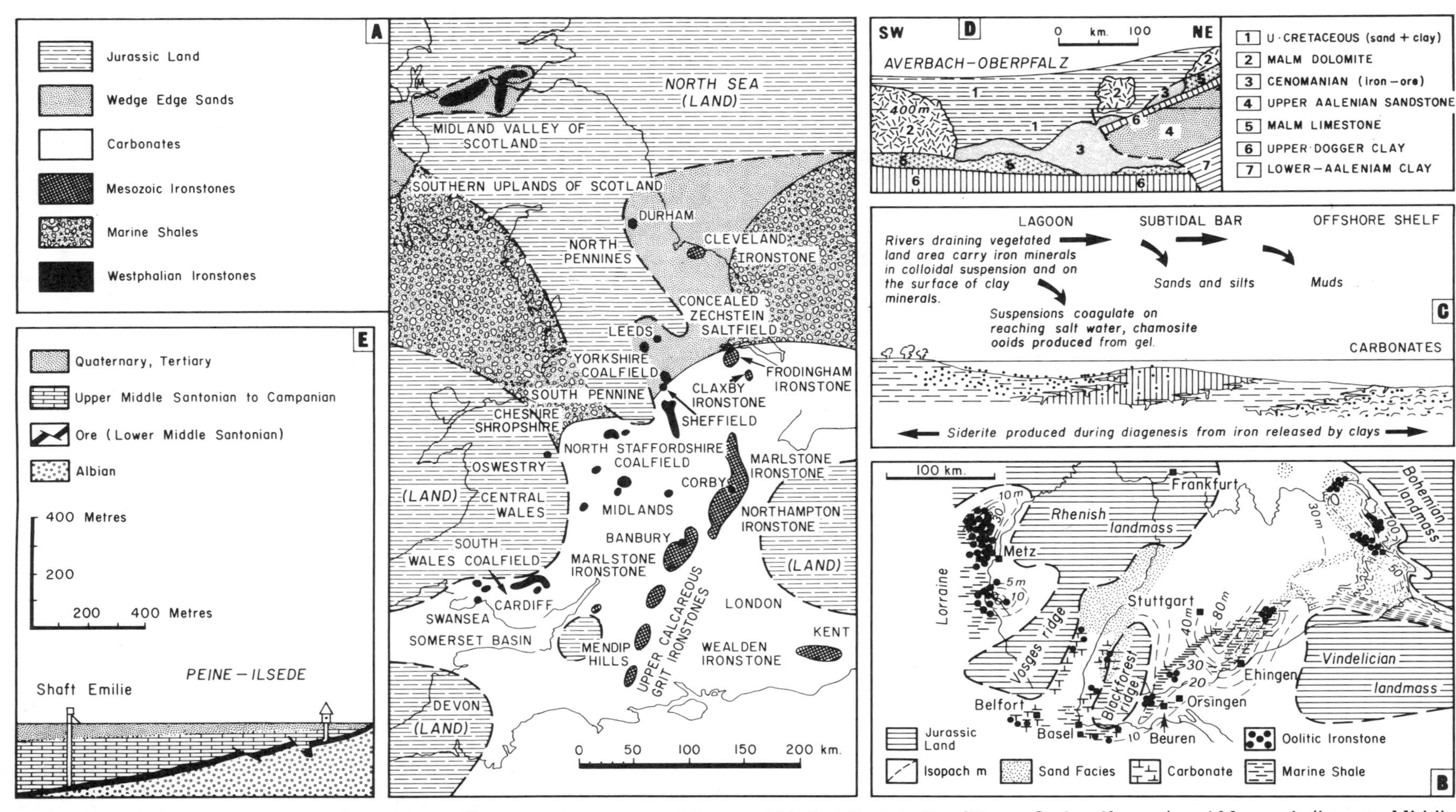

Figure 4.9. Palaeo-geographic setting of the sedimentary ironstones of Europe. (A) The Westphalian (Upper Carboniferous) and Mesozoic (Lower–Middle Jurassic) ironstones of Britain (based on Dunham, K. *et al.*, 1978). (B) The Liassic of Lorraine (France) and the Dogger (Middle Jurassic) of Germany (based on Wopfner and Schwarzbach (1976) and Ziegler, P.A. (1975). (C) Depositional model for the Upper Oxfordian Upper Calcareous Grit ironstones of Wiltshire, England, after Talbot (1974). (D) Cross-section of the Upper Cretaceous karst-related iron ore of Oberpfalz (near Bayreuth, Germany). (E) The Peine–Ilsede detrital reworked iron ore deposits of the northern Harz foreland (after Neumann-Redlin *et al.*, 1977).

FRANCE

The most important iron ores of France are the marine sedimentary oolitic ironstones of Lower Jurassic age of the Lorraine region, where they are referred to as minette ores (*Figure 4.9*). Similar ores also occur in the Ordovician of Normandy and in the lower Devonian of Brittany. The ore deposits have been comprehensively reviewed by Horon (1977) and Bubenicek (1964). The minette deposits of Lorraine are formed of oolitic ironstones composed of limonite, with chlorite, siderite, quartz and calcite in a clay matrix. The ore beds usually range from 31% to 37% iron, with silica usually less than 20%, and calcium oxide ranging from 5% to 20%. The ironstones are closely associated with shale, mudstone and sandstone or limestone (Zitzmann and Neumann-Redlin, 1977).

In the Lorraine district there are 12 mineable beds at the Lower/Middle Jurassic boundary, restricted to several northeast-trending synclinal axes. The enclosing rocks are compacted arenaceous material—sands composed of oolites, quartz, grains and shell fragments. They are characteristically cross-bedded or laminated. The minette ore was therefore deposited in a littoral environment under the influence of waves (Bubenicek, 1964). The mean diameter of the detrital grains increases from the base upwards through the ferriferous sequence. They are therefore a result of marine regression, leading to the deposition of coarser and more littoral material on older sediments of finer grain size. The intervening transgressions resulted in little sedimentation, but considerable reworking of the sediments occurred.

Large-scale cross-bedding is characteristic of the ore. An erosion surface bounds the ore at the top, indicating that deposition of the ore bed was followed by subaerial exposure and erosion, perhaps after an interval of deltaic deposition. The mudstones underlying the ore horizon are interpreted as shallow marine muds and the oolitic ironstone as an oolite bar.

GERMANY

Mesozoic iron ores of the minette-type are by far the most important in Germany. The geology of the deposits has been reviewed by Neumann-Redlin *et al.* (1977). They range in age from early Jurassic to Late Cretaceous, the Cretaceous being the most important. The Jurassic ores are mainly oolitic brown ironstone. The Lower Cretaceous are mixed oolitic and detrital and the Upper Cretaceous are detrital. The Jurassic ores are mainly beds of goethite, haematite, siderite and chamosite. In the Middle Jurassic (Dogger), marine sedimentary ores occur in the entire Swabian–Franconian Alb, and in the Upper Rhine Graben.

The Upper Cretaceous ores account for more than 60% of the country's production. Lenticular, tectonically disturbed bodies of brown iron ore and white siderite occur in the Oberpfalz, near Bayreuth in Bohemia. These are special ores. They lie in dolines, pockets and troughs in karstified Malm limestones or close to the large Saxonian fault (*Figure 4.9*). Apparently the iron ore was precipitated from the drainage system that was directed through lengthy erosional valleys in the karstic land surface.

Upper Cretaceous detrital ores occur at several localities in northern Germany and in the Harz foreland. These detrital ores are called Peine-type, after the locality in the Harz (*Figure 4.9*). They are reworked and enriched Lower Cretaceous clay ironstone formations that contained phosphorite concretions, and occur in moderately dipping synclinal strata whose structure is related in part to salt diapirs.

Detrital ores consist of conglomeratic, ungraded and unrolled ore lumps in a calcareous, marly or argillaceous matrix, depending on the kind of parent rocks. The whole is partly cemented by limonite, formed from the oxidation of siderite in the oxygenated water near the coast during deposition. The ores originate from iron-rich ironstones that were decomposed, removed and accumulated in near-shore regions, and were partly reworked. They were redeposited in troughs and depressions. In the Peine–Ilsede district of the Harz foreland, they occur in beds 2–20 m thick. They overlap the Lower Cretaceous formations as a basal conglomerate formed of reworked Lower Cretaceous clay ironstone and phosphorite concretions. The Peine ores are, on average, 24–33% iron, 6–18% silica, 17–27% calcium oxide, 0.5–4% manganese and 0.8–2% phosphorus. Some sedimentary ores are in part oolitic and in part reworked detrital.

Fresh-water Ferromanganese Deposits

Fresh-water ferromanganese deposits occur in lakes and streams that occupy local and regional depressions in cratonic glacial terrain. Formerly called bog iron ores, they are small and heterogeneous and are now of no significance. The subject has been reviewed by Callender and Bowser (1976). They occur at the more northerly latitudes of North America and Eurasia, and are associated with glacial deposits. Glaciated areas are usually poorly drained, and the ground-up material in glacial till is easily weathered under cool temperate climatic conditions. Many lakes and bogs occupy depressions in glaciated terrain, and contain ferromanganese oxides. Ferromanganese nodules and concretions are found in the lake sediments. The ferromanganese occurs as coatings and crusts and distinct concretions surrounding a variety of geological materials, on sand grains, cobbles, boulders and glacial clay. The iron/manganese ratio of the deposits varies widely, but mostly lies within the range 10–0.6. The generally low minor-element content of fresh-water ferromanganese oxides is a result of rapid nodule accretion rates, and metal–organic complexation/diagenetic fractionation of transition metals in the lacustrine environment. The mineralogy of the nodules is similar to marine nodules, with birnessite, todorokite and goethite being the most common. Psilomelane, the barium-rich manganese oxide, is abundant in the Lake Michigan ferromanganese deposits (Rossman *et al.*, 1972).

Two major sources of dissolved metals appear important in the development of fresh-water ferromanganese concretions: lake and river water overlying the nodule substrate, and interstitial water/ground water percolating through the substrate. Once aqueous solutions containing iron, manganese and other less abundant transition metals encounter suitable nuclei in an oxidizing environment, ferromanganese oxides of varying iron/manganese ratios and minor-element content precipitate. The iron and manganese originate from soil-weathering products. In temperate climates, iron and manganese minerals in the soils are attacked by aqueous solutions containing dissolved gases and organic matter.

The predominant pH range for streams and rivers draining glaciated terrain is between 5 and 7.5, which would allow a minimum concentration of aerated river water of 0.007 $mg\,l^{-1}$ iron and 100 $mg\,l^{-1}$ manganese. Where river water mixes with lake water, the river water that contains the metals and organic matter is diluted by an infinite volume of fresh water, which causes the inorganic and metal–organic complexes to break down. The river/lake water interface is therefore important in decreasing the total dissolved manganese and iron. If ground water rich in dissolved metals enters a lake, the more oxidizing (pH 8) conditions cause oxidation–precipitation of ferromanganese from this ground-water inflow.

The surfaces of marine nodules are usually covered with an active faunal and microfloral community, suggesting that biological precipitation is important. This may also be the case in fresh-water nodule precipitation. It is also possible that the ferromanganese is remobilized into the upper sediments during burial and diagenesis (Callender and Bowser, 1976).

The Quaternary meadow and white iron ores of central Europe were once of economic importance. These bog iron ores contain 55–75% free water. When dried they contain 40–50% iron. With increasing depth, they usually pass from limonite to ferrocarbonate (Zitzmann and Neumann-Redlin, 1977).

Uranium Mineralization in Continental Basins

The strata-controlled epigenetic uranium deposits are predominantly found in intermontaine basins that developed on cratonic platforms. Generally they occur well inland from the continental margins, but the fluviatile coastal environment of large river embayments may also be important. These deposits were formerly held to be syngenetic, but an epigenetic origin has been demonstrated. The epigenetic theories fall into two categories—hypogene and supergene (Fischer, 1968). The hypogene concept supposes deep-seated hydrothermal activity. The supergene theory is now more widely held. It attributes the deposits to moving ground water. The age of mineralization is difficult to determine, but takes place long after the sediments have lithified. Langford (1977) favours a surficial origin operating under continental conditions in a savannah-like climate. This type of uranium deposition forms the major reserves of the USA, and the deposits of Wyoming, Colorado and Texas offer the best examples of sandstone-type uranium mineralization (*Figure 4.10*).

Sandstone Uranium Fields of the Western USA

Most deposits are tabular layers that lie nearly parallel to the bedding, and for that reason are called peneconcordant deposits (Finch, 1967). The ore minerals mainly fill the pore spaces of the host rocks, but they partly replace the sand grains, plant fossils, and accessory and cementing materials of the rocks. In some deposits the uranium occurs in asphaltic material that impregnates and partly replaces the sandstone. Other deposits are

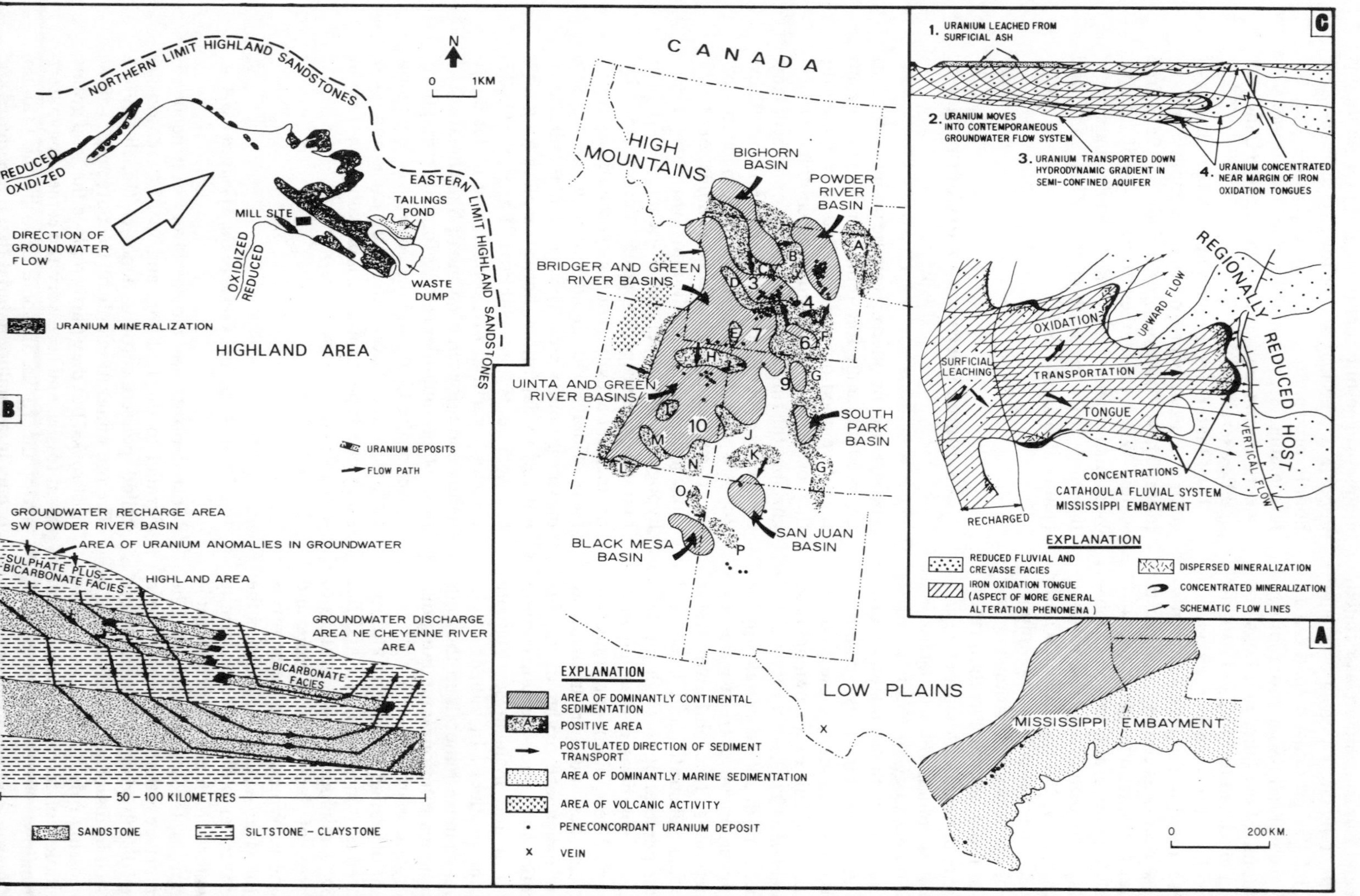

Figure 4.10. (A) Distribution of uranium deposits in sandstone of Eocene Age in Western USA (after Finch, 1967). *Positive areas:* A = Black Hills uplift; B = Bighorn Mountains; C = Owl Creek Mountains; D = Wind River Range; E = Rock Springs Uplift; F = Laramie Mountains; G = Front Range; H = Uinta Range; I = San Rafael Swell; J = Uncompaghre Up-warp; K = San Juan Mountains; L = Kaibab Up-warp; M = Circle Cliffs Up-warp; N = Monument Up-warp; O = Defiance Up-warp; P = Zuni Up-warp. *Basins:* 3 = Wind River Basin; 4 = Shirley Basin; 6 = Hanna Basin; 7 = Washakie Basin; 9 = North Park Basin; 10 = Green River Basin. (B) Plan (upper) and cross-section (lower) of the Highland area of the Powder River Basin (after Dahl and Hagmaier, 1974) to show how the deposits are formed by moving ground water. (C) Cross-section (upper) and plan (lower) of the Mississippi Embayment deposits (after Galloway, 1978) to show their relationships to ground-water movement patterns.

mainly localized along fractures that are discordant to the bedding and are referred to as vein deposits. In these, the uranium mineralization occupies fracture partings and impregnates the wall rocks. The deposits range from small masses only a few metres to hundreds of metres across containing a million tonnes of ore or more.

Nearly all peneconcordant deposits are in beds of Devonian age or younger and they correlate with the evolutionary development of woody land plants, which occur in abundance as fossils in most of the host rocks (Finch, 1967). Most of the deposits are in lenticular sandstone beds that accumulated from fresh-water streams in one of three environments: shallow depressions in foreland belts that lie between the stable interior of the continent and intracratonic basins (*Figure 4.10*); intermontaine basins; or coastal environments near shorelines. Drainage was poor in each of these environments, and as a result the formation pore waters remained in the beds, at least until structural deformation caused draining or flushing of these solutions. Through reaction with the rocks, these pore solutions probably became enriched in dissolved metals. Tuffaceous horizons are common in the rock sequences and these could have furnished uranium and vanadium, though erosion of the land surrounding the basins made the main contribution.

Finch (1967) has compiled data on 4600 uranium deposits in the USA, and 4300 of these are of the penconcordant sandstone-type. He found that 97% of these occur within continental sediments; the remainder are in near-shore or mixed marine–continental sequences. The deposits are restricted to sandstone lenses at one stratigraphic zone. In addition to uranium, the deposits contain vanadium and copper.

Mineralogy

Uraninite and coffinite are the principal uranium minerals. The uraninite occurs as hard black lustrous grains, or dense structureless and microbotryoidal masses. Coffinite and uraninite mainly replace fossil wood and the sandstone immediately adjacent to it, but some is disseminated throughout the pores of the sandstone. The primary vanadium minerals are also black; these are haggite and the more abundant montroseite, and they occupy pores in the sandstones and also replace sandstone grains and fossil wood. The vanadium silicate roscoelite and vanadium-bearing chlorite coat sand grains and fill the pores of sandstones.

Chalcopyrite, bornite, chalcocite and covellite occur as discrete particles or intergrowths and mostly replace the pyrite in plant fossils. Pyrite or marcasite always occur and form as much as 10% of some ores.

The chief yellow secondary minerals are hydrous vanadates, carnotite and tyuyamunite. The chief secondary green minerals are metatorbernite and metazeunerite; these are associated with goethite, lepidocrocite, jarosite, haematite, calcite, gypsum and manganese oxides. The ore minerals are always very fine grained.

There are many detailed summaries of the sandstone-type uranium deposits, notably by Finch (1967) and Rackley (1976). The deposits of the Colorado Plateau have been summarized by Fischer (1968) and those of Wyoming and South Dakota by Harshman (1968). A condensed summary is to be found in DeVoto (1978).

Nature of the Sandstone Hosts

The strata-controlled deposits with the greatest known reserves of uranium occur in formations that include both oxidized and reduced lithofacies. Such formations are called varicoloured because of their red, green and grey components. The most prominent of these formations include the Triassic Chinle Formation of northeast Arizona, the Jurassic Formation of northwest New Mexico, the Eocene Wind River, Wasatch and Battle Spring formations of Wyoming, and the Miocene and Pliocene Catahoula Tuff and adjacent formations of south Texas.

Organic carbonaceous material is rich in pyrite in the reduced facies. Thick units dominated by reduced facies alternate with thick units dominated by oxidized facies. They record periods of time when the climate varied from humid tropical to hot arid, and this is called a savannah climate, which Bailey and Childers (1977) regard as important for the formation of sandstone-type uranium deposits. In a uniformly humid climate, chemical weathering is enhanced, but uranium and humate are not systematically leached out of the eroding weathered highlands.

Langford (1977) has amplified the argument of climatic control. Tabular deposits are confined to Devonian or younger continental sedimentary sequences. This is clearly because the development of reducing conditions in terrestrial clastic deposits is impossible without a well-developed land flora, which evolved only from the Devonian onwards. The introduction of humate from the land plants into the basins of deposition create the reducing conditions necessary for the deposition of the uranium and vanadium minerals.

Several different configurations of ore deposits

have been described, and they can be related to the flow of ground waters through the formations.

Trend deposits

These are large tabular deposits within sandstones and conglomerates, with the following characteristics. (a) Ore bodies are distributed along mineralized belts or trends that may have several kilometres length. (b) The ore bodies are generally tabular in habit and oriented subparallel to the gross stratification. (c) Commonly the individual ore bodies are elongated parallel to the long dimension of the enclosing sandstone or conglomerate bodies, and the palaeo-drainage pattern controls the trend of mineralization (*Figure 4.10*). However, in a complex of interconnecting channel sandstone bodies, the trend of mineralization might cross the palaeo-drainage, suggesting some other dominant control. When the trend of mineralization crosses the palaeo-drainage, the individual ore bodies are typically amoebic in plan with little preferred orientation (Childers and Bailey, 1979). (d) Uranium mineralization occurs in grey-to-black reduced rock, which is usually carbonaceous. Most of the carbonaceous material is amorphous and finely disseminated, and it coats grains and partially fills interstices between the sand grains. (e) Uranium ore bodies commonly form the nuclei of larger volumes of grey reduced rock; this contrasts sharply with the more extensively developed red haematite-stained or white-bleached sandstone or conglomerate. These reduced rocks, which enclose ore bodies, are interpreted by many geologists to be altered, because they are closely associated with the ore, and also because they are different in appearance from the typical oxidized unmineralized rocks farther removed from the sites of mineralization. (f) Trend deposits occur in formations with both oxidized and reduced facies of fluvial origin. Generally the oxidized facies is dominant and in many cases the uranium appears to have concentrated within oxidized rock that became reduced after the ore bodies were formed.

Childers and Bailey (1979) propose that most trend deposits are early diagenetic in origin. The most favourable environment for their formation is a large fluvial system under the influence of an extended savannah climate dominated by arid phases. In most of the tributary drainages with intermittent flow, fluctuating ground-water tables and high oxidation potentials result in ideal conditions for complete dissolution and removal of decaying organic matter and uranium. The complexes of humic substances and uranium, leached from the tributary drainages, were concentrated in the sands beneath the main trunk streams. The process of concentration might have been a gradual thickening of colloidal or gelatinous masses.

To illustrate this type of deposit, the *Big Indian Ore Belt* of the Lisbon Valley anticline of Utah has been selected. Its geology is described by Wood (1968). Uranium ore deposits in the Lisbon Valley area, near the centre of the Colorado Plateau, occur in an arcuate belt 24 km long by 2.4 km wide, on the east flank of the Lisbon Valley anticline (*Figure 4.11*). The size of the ore deposits is in the range 500–1.5 million t. Ore bodies average 1.8 m thick, are tabular, amoeba-shaped masses, concordant to the bedding, and are in the thickest and lowest sandstone unit of the Moss Back Member of the Triassic Chinle Formation. The host rock is predominantly of fluviatile calcareous, fine-grained to conglomeratic sandstone. Uraninite is the principal ore mineral, and fills the pores in the sandstone and partly replaces sand grains and fossil wood fragments. Intimately associated with it are small amounts of coffinite and the vanadium mineral montroseite and doloresite. Secondary uranium and vanadium minerals are found in areas of oxidation, but they are of no economic importance.

Most of the ore in the belt is within a 150-m elevation interval. The large ore bodies are hundreds of metres above the present-day water table, but more than 95% of the ore is of unoxidized uraninite. The spread of the ore bodies across the belt is appreciably wider at the northwestern and southeastern ends of the anticline, where the beds have a more gentle dip. The bedded ore deposits are displaced by Tertiary normal faulting. Uranium deposits appear to be more persistent or to occur more often where porous friable fine-grained Permian Cutler Sandstone units, bleached to light-grey, suboutcrop at the non-conformity under Moss Back sandstone, which was lithologically favourable for uranium. Small deposits of uraninite ore occur in the Culter Sandstone adjacent to the Moss Back ore deposits. Carnotite deposits also occur about 150 m laterally up-dip and at approximately the same elevation, but lower stratigraphically in the Cutler section than the ore occurrence subjacent to the Moss Back ore. Persistent bleaching, resulting from leaching of ferric oxide, is obvious both in the Chinle and Cutler formations in and around the ore and particularly along the non-conformity. Abundant coalified plant material occurs as lenses or aggregates in and above the basal ore sandstone or as carbon trash disseminated in mudstone beds above the ore sandstone. Wood (1968) suggests the fol-

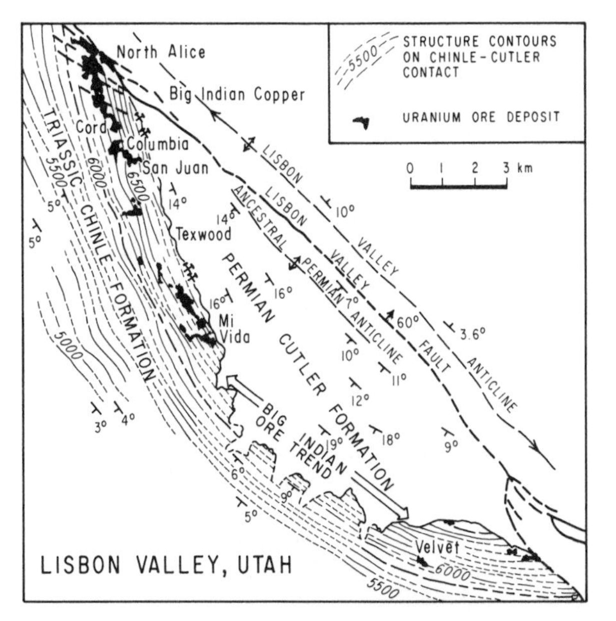

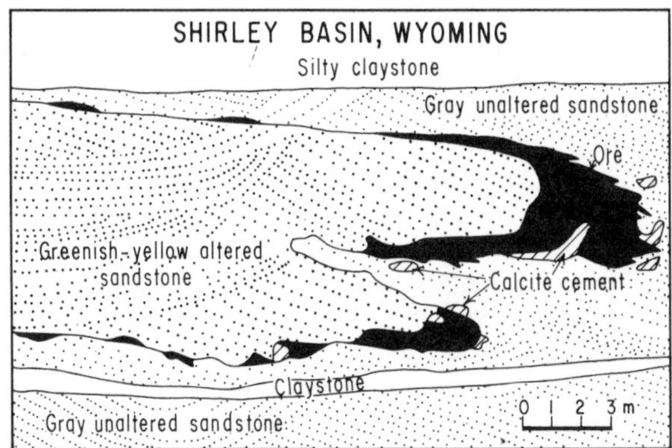

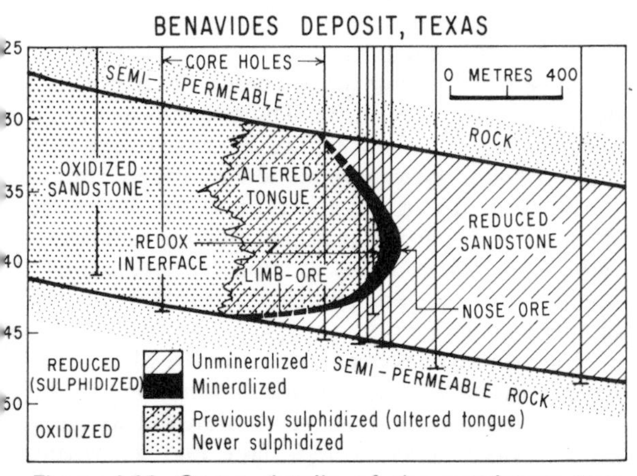

Figure 4.11. Some details of the sandstone-type uranium deposits of the USA. The Trend Deposits of the Big Indian Ore trend of the Lisbon Valley, Utah (after Wood, 1968). The roll-front deposit of the Shirley Basin, Wyoming (after Harshman, 1968). The roll-front deposit of Benavides, Texas (after Reynolds, R.L. and Goldhaber, 1978).

lowing sequence to explain the mineralization: uranium, probably derived from the Chinle Formation by diagenetic processes, was transported in connate ground water, moved by compaction and deposited under reducing conditions. Uranium was emplaced around the crest of the ancestral anticline, prior to the Laramide orogeny. During the orogeny, the Tertiary Lisbon Valley anticline was induced upon the Permian anticline, and penecontemporaneously was uplifted and faulted. The Big Indian Ore Belt was elevated to approximately its present position. The ore deposits were positioned by either a water/gas interface or by a connate/vadose water interface near the crest of the ancestral Permian anticline (*Figure 4.11*).

Other notable trend deposits are those of the Ambrosia Lake Trend (Kelley *et al.*, 1968) and the Uravan Mineral belt (Motica, 1968).

Roll-front deposits

Roll fronts and roll-front uranium deposits occur marginally to alteration within sandstones and conglomerates, and the alteration is believed to have been formed by the mineralizing solutions (Harshman, 1968, 1972). The most common characteristic of alteration associated with roll fronts is oxidation of iron minerals. In some areas, altered sandstones in outcrop are more intensely oxidized than their unaltered weathered equivalents. The mineralizing solutions frequently appear to have had higher oxidation potentials than those that have been produced by recent weathering. Solution banding similar to Liesegang rings (concentric colour banding cross-cutting the bedding) is another common feature associated with the alteration. These bands consist of various tints of red, yellow, rusty-brown and other colours. Solution bands probably reflect slight changes in the chemical composition of the ground water during mineralization. Solution bands near a roll front or other geochemical interface frequently parallel the interface itself.

Colour zonation is common in alteration complexes. Red haematite-staining may characterize parts of an alteration complex, whereas yellow-brown or rusty limonite-staining may dominate other parts, and white-bleached sandstone may characterize still other parts of the complex.

In the subsurface, below the zone of weathering, unaltered sandstone, equivalent to the altered complex, is typically grey, pyritic and carbonaceous (Childers and Bailey, 1979). In some areas, the altered sandstone is found to have been secondarily reduced after mineralization was completed. This is interpreted as occurring when the ground water became stagnant after mineralization and abundant humates yielded hydrogen sulphide reductants to the ground water.

Alteration complexes or tongues of alteration (*Figure 4.11*) vary considerably in size. They may be tens of kilometres in length and several kilometres in width. Childers and Bailey (1979) feel that there are two types of roll fronts. (a) Those associated with both oxidized and reduced formations, which may have been produced during the early diagenesis of the formations—this category produces roll fronts by solutions that moved in the same directions as the streams that deposited the host sands. (b) Those that occur in uniformly reduced formations, frequently occurring beneath unconformities—these are interpreted as having been formed beneath erosion surfaces during a time at which the climatic conditions were favourable for solution, transport and uranium concentration. They may have formed long after the host rocks were lithified, are truly epigenetic, and the roll fronts may have no relationship to the palaeostream directions.

One of the best examples is provided by the *Shirley Basin* of Wyoming, described by Harshman (1968, 1972). The Wind River Formation of Eocene age is the host rock for large high-grade uranium deposits. The major deposits are in a northwest-trending belt of sandstones that were deposited in stream channels and were lithologically favourable for uranium accumulation. The reserves of the Shirley Basin are estimated at 22 000 t of U_3O_8. The palaeo-topography influenced the character and position of the mineralized belt. The major ore deposits are in a well-defined belt that lies west of a prominent ridge on the pre-Wind River erosion surface. Sediments favourable for subsequent metallization were deposited locally in some of the tributary-stream channels 5–16 km west of the buried ridge.

The major deposits within the belt are at or near the margins of large tongues of altered sandstone formed in the most permeable parts of thick sandstone beds (*Figure 4.11*). At least two tongues of altered sandstone are present in the belt, one of which is 8 km long, 5 km wide and 21 m thick; the other is somewhat smaller.

Ore bodies consist of a few hundred tonnes to several hundred thousand tonnes of ore containing from 0.1% to 2.0% U_3O_8. The ore mineral is uraninite associated with pyrite, marcasite, haematite and calcite. Deposition is believed to have taken place when neutral to slightly alkaline oxidizing uranium-bearing ground water, on entering the basin, encountered a reducing environment. The reductant may have been hydrogen sulphide produced from the decay of plant material in the strata.

The roll-type deposits are crescentic in cross-section and elongate along the margins of large tongues of altered sandstone (*Figure 4.11*). The alteration was produced by the ore-bearing ground water. Elements were added to and removed from the host sandstone (Harshman, 1974), in bands or zones that show a remarkably consistent pattern. Uranium, selenium, vanadium, molybdenum and ferrous-bearing minerals have been deposited outwards from the edge of an altered tongue in exactly the same sequence as required by their redox potentials. These elements transport in neutral to slightly alkaline solutions. Elements such as copper, which are mobile in acid solutions only, are not present in the ore in significant amounts. The deposition is related to a drop in E_h that reached a minimum with deposition of jordisite at a considerable distance from the edge of the oxidized tongue.

Uranium deposits in the southern *Powder River Basin* of Wyoming are excellent examples of large roll-type ore bodies (Dahl and Hagmaier, 1974). The host rocks are Palaeocene sandstones deposited at point bars by a meandering stream. The source of the uranium is thought to be tuffaceous and arkosic debris indigenous to the sedimentary sequence containing the host rocks. The largest deposits of highest grade occur near the distal margins of permeable, slightly dipping sandstones where they grade laterally into organic-rich siltstones, claystones and lignites deposited in back-swamp or flood-basin environments (*Figure 4.10*). The deposits are epigenetic, formed by precipitation of uranium from ground-water solutions that moved through the host rocks from a recharge area southwest of the deposits towards a discharge area northeast of the deposits. The deposits are large because the host rocks are extensive and the ground-water system remained relatively stable. Pyrite formed early in and around the host units through a biogenic process utilizing sulphate-reducing bacteria. This was important in establishing a permissive geochemical environment for the later ore genesis.

Roll-front deposits also occur in the Mississippi Embayment. *The Catahoula Formation* of the Texas Gulf Coast Plain consists of fluvial channel-fill, crevasse-splay, flood-plain and lacustrine facies (*Figure 4.10*). Large volumes of volcanic ash were deposited during the Late Oligocene and Miocene epochs.

Galloway (1978) gives the following sequence of events in uranium distribution. (1) Uranium was leached from the volcanic ash soon after deposition by pedogenic processes and moved into shallow ground-water circulation cells. (2) Oxidizing uranium-enriched waters entered semi-confined aquifer sands in areas of regional recharge. (3) The geometry of the flow of ground water was determined by the three-dimensional facies and structural framework of the aquifer system. (4) Uranium was concentrated by aqueous geochemical gradients as discrete mineralization fronts that were closely associated with iron oxide fronts. Uranium was preferentially deposited where facies changes or faulting induced cross-stratal flow from permeable fluvial channel facies into interbedded or less permeable overbank facies (*Figure 4.10*). (5) Post-mineralization decrease of regional and local ground-water flux has resulted in re-equilibration of large parts of the aquifer with the regionally reducing subsurface environment.

In the reduced rock in front of and enveloping the altered tongue (*Figure 4.11*), pyrite and marcasite have almost completely replaced detrital titanomagnetite and titanohaematite. Subsequent oxidation produced limonite pseudomorphs (Reynolds, R.L. and Goldhaber, 1978). Iron oxide and titanium oxide 1-km up-dip from the roll front were never sulphidized: titanomagnetite and martite constitute nearly half the heavy mineral fraction and limonite is absent. Oxidized rocks 200 m closer to the nose of the roll front, however, are nearly devoid of titanomagnetite and contain abundant limonite. The redox interface that bounds the altered tongue (*Figure 4.11*), therefore, moved more than 200 m but less than 1 km. Detrital iron oxide and titanium oxide are increasingly sulphidized in the down-dip direction. Organic carbon content is low throughout the host sandstone, which implies that bacteria were not involved in the generation of the sulphides. The hydrogen sulphide for the initial sulphidization may have been derived from oil and gas in formations that underlie the Catahoula Tuff (Reynolds, R.L. and Goldhaber, 1978), perhaps infiltrating via a fault that occurs 1.5 km down-dip from the ore body. The introduction of sulphur can be divided into two stages by isotope studies (Goldhaber *et al.*, 1978). The early sulphur entered the aquifer via the fault, and the later sulphur was remobilized by ground water that formed the uranium roll-front ore body.

Not all uranium remains in sedimentary strata, and the hydrothermal solutions may infiltrate fractures in underlying or overlying formations. An example of this is now given.

Schwartzwalder Deposit, Front Range, Colorado

The deposit is 13 km northeast of Golden, and has been described by Young, E.J. (1979). The total production is about 5 000 t of U_3O_8. The ore tenor was 0.6% but is now 0.35%. The deposit is in the metamorphic Precambrian Idaho Springs Formation, less than 0.8 km from the contact with Palaeozoic sedimentary rocks. The mines are either on or near northeast-trending Precambrian faults. The Illinois Fault strikes north and contains the main vein. The ore is entirely fracture-controlled. The pitchblende is accompanied by pyrite and minor galena, jordisite, molybdenite and sphalerite. The gangue is quartz, calcite, ankerite, siderite, red haematite and alkali feldspar. The following factors are critical to the ore localization (Young, E.J., 1979). (1) There is structural complexity and multimovements along the fault system. (2) Many of the surrounding sedimentary rocks are uranium-rich. Some of these strata were thrust deep under the Precambrian Formations along the west-dipping Golden reverse fault. Coal and asphalt in these formations contain uranium. The ultimate source of the uranium is thought to be volcanic and pyroclastic rocks in the Cretaceous–Tertiary Denue Formation. (3) The Ralston dyke, of age 62 Ma, possessed the heat necessary to drive hydrothermal circulation.

The deposits are estimated to be 60 Ma old, and to be meteoric hydrothermal. The hydrothermal solutions infiltrated the fracture systems in the Idaho Springs Formation. The meteoric–hydrothermal origin is likely because: pitchblende in the deposit is rich in molybdenum but poor in thorium and rare earths, and is typical of sedimentary pitchblende; the pitchblende unit-cell dimensions are more characteristic of sedimentary rather than magmatic pegmatite pitchblende; and muscovite in the Idaho Springs Formation close to the uranium mineralization still retains its Precambrian potassium/argon age, indicating that it was not affected by the mineralization, which was accordingly of low temperature.

It appears that meteoric water moved down-dip along bedding planes in the sedimentary strata that had been tilted during the Laramide tectonism, leaching uranium from the rocks, and percolated into fractures and into the Golden Fault zone. Subsequently the heat source of the 62-Ma old Ralston dyke drove a hydrothermal circulation cell through the rocks and the fluids especially concentrated along the Rogers and Illinois faults.

5

Other epicontinental sea deposits

Some commodities are economically concentrated within both intracratonic basins, and equally miogeoclinal or shelf environments along the margins of major cratonic continents. The sandstone uranium deposits could be included here, but they have such a strong affinity for intracontinental basins that they have accordingly been included in the previous chapter. 'Epicontinental' is used here in the sense of including both basins that lie within, as well as those that lie along, the shelf of continental masses.

Phosphorites and anhydrites may occur equally in intracratonic basins or along the margins of major continental masses, and so they are included in this chapter. However, some of the well-known phosphorite and anhydrite deposits of the world may appropriately have been described in the previous chapter.

Phosphorites

The world supply of phosphate is made up of approximately 82% from sedimentary deposits, 16% from igneous intrusions and 2% from guano accumulations. The igneous associations are described in Chapter 6. *Phosphorite* is a sedimentary rock containing 10% (by volume) or more of phosphate grains. *Phosphatic* denotes all sediments that contain between 1% and 10% phosphate grains, for example, phosphatic mudstone, phosphatic dolomite, etc. Major reviews of the extensive literature have been made by Cook (1976), McKelvey (1967), and Notholt (1980). Sedimentary phosphorites are of marine origin, but weathering and leaching may have led to post-depositional enrichment. High-grade phosphorite beds may contain as much as 30% P_2O_5 but mining grades may extend below 10% in favourable conditions.

Stratigraphic columns through some of the world's typical phosphate deposits are given in *Figure 5.1*, from which the wide range of thickness and rock associations can be seen. Some phosphatic beds are lenticular over distances of a few metres, others are remarkably persistent.

The non-phosphatic matrix, the thin interbeds and the associated sediments are rather variable. They include dark carbonaceous mudstone and shale, chert, limestone and dolomite, sandstone, diatomite and tuff. Phosphorites grade laterally into limestones and dolomite and chert or diatomite.

Distribution of Phosphorites

Phosphorogenesis was not restricted to any stage in the earth's development, and there are major absences of phosphate deposition in the Oligocene, the Triassic and from the Silurian to the Lower Carboniferous periods. The major episodes of phosphate deposition are: Miocene–Pliocene (14 Ma ago) occurring, for example, along continental margins such as those in Florida and Peru; Upper Cretaceous–Eocene (65 Ma ago) occurring in West Africa, North Africa and the Middle East; Jurassic (148 Ma ago) occurring in the USSR; Permian (250 Ma ago) occurring in the huge Western field of North America; Ordovician (464 Ma ago) occurring in Tennessee and Alabama; and Cambrian (542 Ma ago) occurring in Karatau in the USSR, and the Georgina basin of Australia. Precambrian deposits are more sparse, but significant. The known episodes are: Upper Proterozoic I (620 Ma ago) occurring in the Volta area; Upper Proterozoic II (700–800 Ma ago) occurring in China; and Middle Proterozoic (1200–1600 Ma ago) occurring in Australia at Rum Jungle and Broken Hill (Cook and McElhinny, 1979).

Phosphorogenesis undoubtedly depends on the distributions of land masses in relation to the oceans and the periodicity of occurrence may simply reflect times when there were suitable distributions of land and oceans at suitable latitudes. The

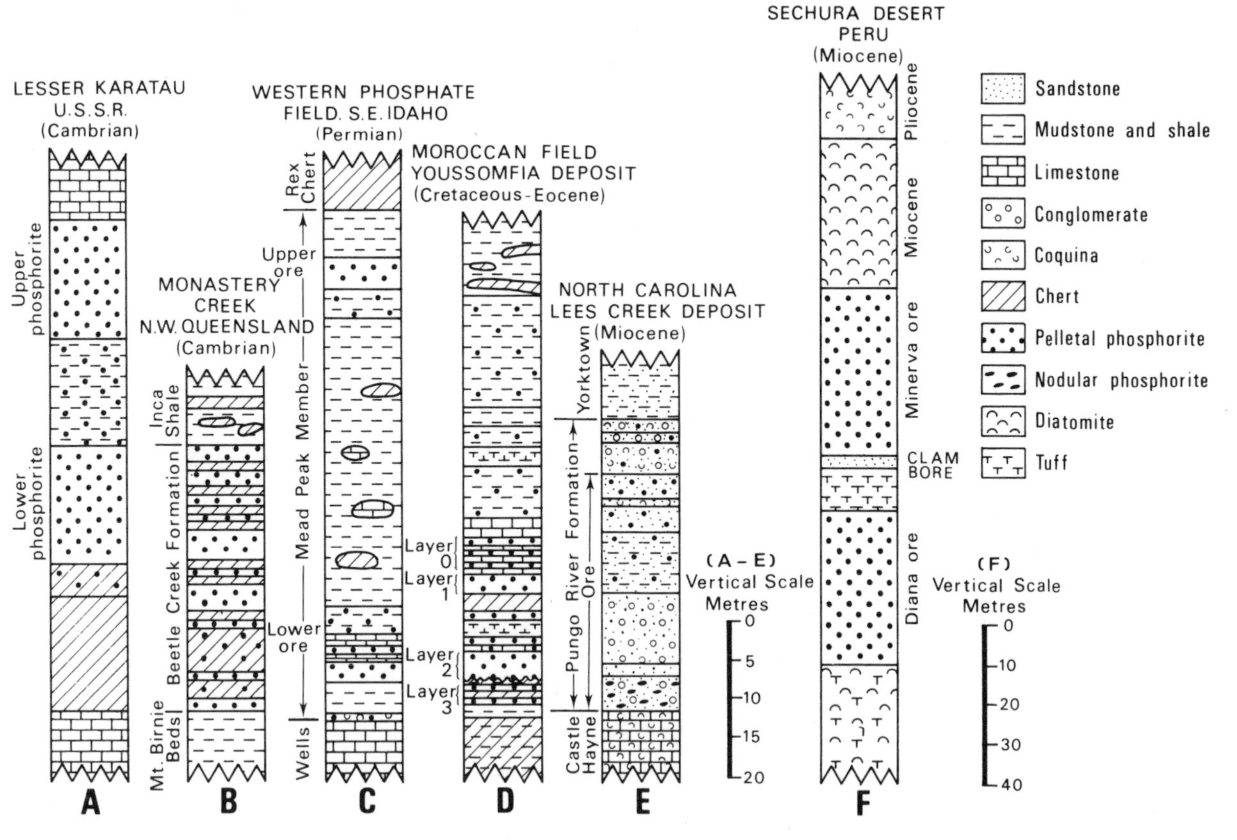

Figure 5.1. Stratigraphic columns for some typical phosphorite deposits. Based on Cook (1976) and Cheney, T.M. *et al.* (1979).

past history of the earth shows the existence of major geographic provinces. They include: the Late Precambrian province of central and east Asia; the Cambrian province of central and east Asia and of northern Australia; the very rich Permian province of western North America; the Upper Cretaceous–Eocene Tethys province of the Middle East and North Africa and South America; and the Miocene province of southeast North America.

Phosphate-rich sediments are present on modern continental shelves, for example, off southwest Africa, southeast North America and southern California. The present-day formation is beyond doubt, and has been discussed by Veeh *et al.* (1973), and the mechanism described in detail by Riggs (1979b).

Petrology and Classification

Marine deposits have three basic forms: (1) phosphatic nodules, (2) phosphatic grains and pellets, and (3) fine-grained clay-sized structureless phosphorite or phosphatic shale.

The details of phosphorite petrology and classification have been elaborated by Riggs (1979a). Orthochemical phosphorite is non-clastic clay-sized sedimentary material formed physiochemically or biochemically within the basin of deposition. Allochemical grains are larger than clay size and have also been formed in the basin of deposition by similar processes. The authigenic microcrystalline phosphatic mud (microphosphorite) that precipitates *in situ* either biochemically or physiochemically is classified as orthochemical phosphorite (*Figure 5.2*). If the microphosphorite mud is subsequently modified into discrete clastic particles such as pellets or nodules or oolites, then it is classified as allochemical phosphorite. The muds may be torn up by biological or physical processes to produce the intraclastic allochems or ingested and excreted by organisms to form pelletal phosphorites. If there is sufficient water energy, the muds may also aggregate around a nucleus grain or shell fragment to form oolites or pseudo-

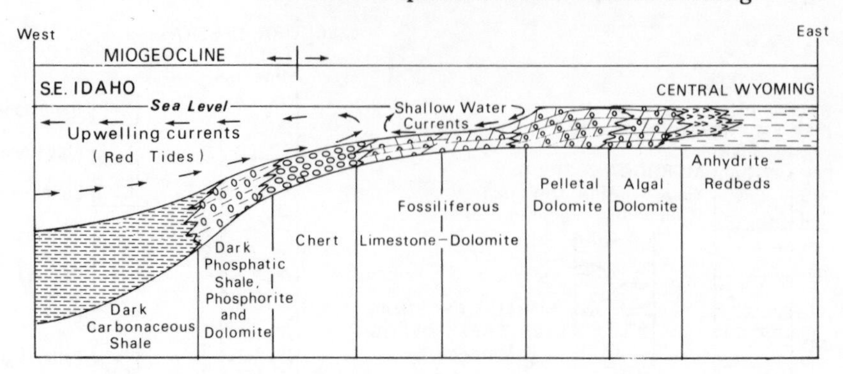

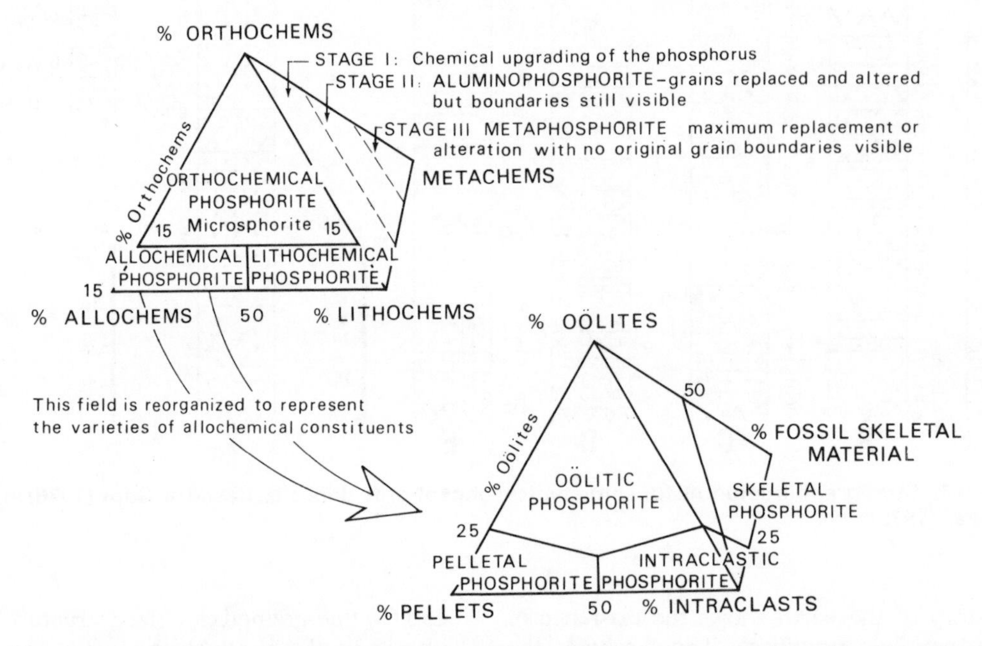

Figure 5.2. Top: schematic representation of the facies changes and origin of the Permian Phosphoria Formation of the western USA (after Sheldon, 1963). Bottom: classification scheme of phosphorites (after Riggs, 1979a).

oolites. A fourth type of allochemical grain is a fossil skeleton that drops down into the mud and may become phosphatized.

Lithochemical phosphorite constitutes older clastic fragments of similar composition, previously formed by similar physiochemical or biochemical processes, that have been eroded and transported as clastic grains into the basin of sedimentation. These form phosphorite grains, which have been reworked into younger sedimentary strata. Subaerial weathering processes chemically and mechanically change an exposed older phosphorite deposit, and eroded grains from such a source may be redeposited as metachemical grains. The total range of grains and rock-type classification is given diagrammatically in *Figure 5.2*.

Most phosphorites are composed of collophane, which is cryptocrystalline carbonate–fluorapatite, with a general formula $Ca_5(PO_4)_3(F,CO_3)$. The chemical variability has been reviewed by Tooms *et al.* (1969). Substitution of trace elements and the contents of fluorine and carbonate have a great influence on the agricultural use of the phosphate.

Uranium is an important element that may substitute in the apatite structure and many phosphorites can be a secondary source of uranium. Tooms *et al.* (1969) recorded an average content of 190 ppm uranium in phosphorites. The second most important source of by-product uranium is American phosphatic acid plants that treat phosphorites, principally in Florida (Finch *et al.*, 1973).

Some of the world's major oil fields have phosphatic source rocks. Powell, T.G. *et al.* (1975) have shown that although phosphorites commonly do not have excessively high concentrations of organic matter, a very considerable proportion of this organic matter is in a form of readily extractable hydrocarbons, and consequently phosphorites do constitute potentially important petroleum source rocks.

Types of Phosphate Deposits

The most widely used classification is fourfold: geoclinal or 'west coast' (for example, the western Phosphoria Formation field of the USA); platform or 'east coast' (for example, the Florida and North Carolina-type); weathered or residual (for example, the brown-rock deposits of Tennessee); and guano deposits on tropical islands (for example, Christmas Island).

Geoclinal (West-coast) Deposits

Some of these deposits have apparently been formed in basins associated with arc–trench systems, probably in fore-arc basins. They tend to be rather small and are commonly associated with thick volcanogenic sequences. They are generally uneconomic although the Sechura deposit of Peru (Cheney, T.M. *et al.*, 1979) is quite rich. It is different from other major marine deposits in that the sequence consists chiefly of phosphorite and diatomite and many of the beds are tuffaceous. The deposits of Sayan and central Tien-Shan in central Asia contain up to 200 million t of medium-grade ore.

The miogeoclinal deposits may have formed in basins overlying relatively inactive cratonic margins where there was little or no volcanic activity. They are of much greater economic importance. Deposits such as Karatau in the USSR and the western Phosphoria Formation field of the USA are important examples. They have a common association with chert, carbonaceous shales and dolomite. The phosphates are pelletal and extend over hundreds of thousands of square kilometres, with high grades of P_2O_5 throughout. They may be somewhat folded and faulted.

Platform (East-coast) Deposits

These occur in cratonic areas or along the passive borders of cratons (Atlantic margins). They are usually rich, cover hundreds of square kilometres, and are several metres thick. The best example is the Tertiary Atlantic coastal-plain deposits of Florida and North Carolina. This coastal-plain deposit represents the single most important world reserve, accounting for about 32% of world supply (Riggs, 1979a; 1980). Another is the Mishash deposit of Israel. Alternatively platform deposits occur as thin low-grade nodular deposits covering many thousands of square kilometres as in the Mesozoic Moscow Basin and the Ordovician of Central Australia.

Platform deposits are commonly associated with terrigenous sediments such as quartz siltstone and sandstone, or with carbonates, and have no chert. Gluaconite is sometimes abundant, and phosphatized coquinas occur in some. Shallow-water features such as cross-bedding are frequent.

A clear-cut distinction between the west-coast and east-coast types is not always possible and the Cambrian deposits of northern Australia have features of both categories, such as an abundance of chert, but they also have coquinas and cross-bedding features.

Weathered and Residual Deposits

Post-depositional modifications and weathering after uplift caused concentration of phosphate as a result of chemical and mechanical weathering. Chemical weathering produces high-grade ore by leaching out the calcareous and non-phosphate matrix to give, for example, the brown-rock deposits of Tennessee. Sometimes the phosphate may go into solution and be redeposited within the weathered profile, to give the white-rock deposits of Tennessee. Mechanically reworked deposits are exemplified by the Pliocene Bone Valley Formation of Florida in which the phosphate material is derived from the Miocene Hawthurn Formation. Such reworked phosphorites are commonly pebbly and nodular and mixed with coarse cross-bedded quartzose sandstones. They are typically lenticular and infill bedrock depressions. Where the phosphorite has been derived from a calcareous unit, the erosion surface at the top of the carbonate may be karst-like.

Guano Deposits

These deposits result directly or indirectly from accumulations of excreta of sea birds and to a lesser extent bats. Major occurrences are on the remote islands of the Pacific and Indian Oceans (White, W.C. and Warin, 1964). Nitrogenous guano contains a large part of the original nitrogenous organic matter of the excreta, which contains 22% nitrogen and only 4% P_2O_5. Phosphatic guano has matured and lost its nitrogen through decomposition, which proceeds rapidly. Modern guano contains 10–12% P_2O_5. Guano that has been leached by percolating waters may contain up to 32–40% P_2O_5.

Nitrogenous guano forms extensive thick white cappings to islands. The best known is Chincha island off Peru, which attains a thickness of 50 m.

Phosphatic guano consists essentially of calcium phosphates, formed by reaction of the guano with the underlying calcareous sands of the islands. It forms a flat layer of pale-brown to dark-brown or reddish-brown crust characteristic of sand cays and atoll islands. These crusts are seldom large and the P_2O_5 content is variable; they consist of collophane and calcite.

Phosphate rock deposits are few but important and rich and occur on elevated atolls and limestone and volcanic islands. Rainfall removes the phosphate and redeposits it within the underlying rock as collophane. Mostly the phosphatized rock is limestone, but examples of igneous and metamorphic rocks are known (McKelvey, 1967).

The most impressive example is Christmas Island (105° E 11° S), which is extensively mined. The island is of a sequence of basalts and limestones ranging from Eocene to Miocene (White, W.C. and Warin, 1964). The 143-km^2 island is covered with a thick blanket of mainly oolitic and pisolitic phosphate overlying the deeply etched limestone surface. Coherent phosphate and phosphatized limestone frequently contain up to 40% P_2O_5.

The Permian Phosphoria Formation of the Western USA

The Permian rocks of southwest Montana and adjacent parts of Idaho and Wyoming form the largest phosphorite field, well known from the publications of McKelvey *et al.* (1959), Sheldon (1963), Cressman and Swanson (1964) and Swanson, R.W. (1970, 1973). The Permian strata form part of a series of chiefly shallow-water Palaeozoic and Mesozoic marine sediments, as much as 2800 m thick in places, that are overlain by a similar thickness of continental sediments and volcanic rocks. The marine basin was formed directly on Precambrian basement. The Permian basin was open to the sea towards the west and southwest and shallowed to the east and north. The southwest part has been referred to as miogeoclinal and the northeast part as the platform part of the succession (*Figure 5.2*).

The phosphate ore is the Meade Peak and Retort members of the Phosphoria Formation. It is mainly phosphatic shale and oolitic and pelletal phosphorite that accumulated chiefly near the border zone between the cordilleran miogeocline and the eastern platform. These members are thickest in Montana and contain the richest phosphorus. The Meade Peak phosphorite lenses out to the east and north, but the Retort member is much more extensive, but also lenses out towards the northeast.

The phosphate members are of interlaminated phosphatic shale and phosphorite; they have an F/P_2O_5 ratio of approximately 0.1. The members contain, on average, 10–12% P_2O_5. The two phosphatic members contain more than 10^{10} t P_2O_5, 80% of which is in the Retort member. More than 6×10^9 t of rock contain more than 24% P_2O_5. Meade Peak and Retort phosphatic shale members contain roughly 1 part of uranium per 3000 parts of P_2O_5. Although it is low, uranium is a potential by-product of the treatment of the high-grade phosphate (Swanson, R.W., 1970), and a total amount of 4×10^5 t of uranium is present in the Phosphoria Formation phosphorite containing 24% or more P_2O_5.

Oil shale is also abundant interbedded with phosphatic shales, especially in the Dillon area (Swanson, R.W., 1970). The phosphate shale members are always associated with bedded cherts, which occur as the Rex and Tosi members of the Phosphoria Formation (Cressman and Swanson, 1964). The chert is composed of microcrystalline quartz and chalcedony mixed with clay and detrital quartz. Siliceous sponge spicules form more than half of the non-detrital silica.

The Shedhorn Sandstone is composed of well-sorted sandstone, derived from the sedimentary terrain outcropping around the basin. The shallow-water nature of the basin of deposition is clear from the rapid facies changes and lensing of the members. The different facies are related to water depth, as suggested by Sheldon (1963), and shown diagrammatically in *Figure 5.3*. Chert was deposited as a spicule ooze, mostly at depths of less than 50 m, and the massive chert was formed by solution of the siliceous tests. The dark mudstone of the phosphatic shale members was deposited below wave level in unoxygenated water. The source of the carbonaceous matter was marine plankton (Cressman and Swanson, 1964). The phosphate was introduced into the basin by cold up-welling waters from the deep ocean. Either chert or phosphorite was deposited, depending partly on the depth of water in the basin. With changing palaeo-geography, some waters became more saline and warmer and carbonates were deposited.

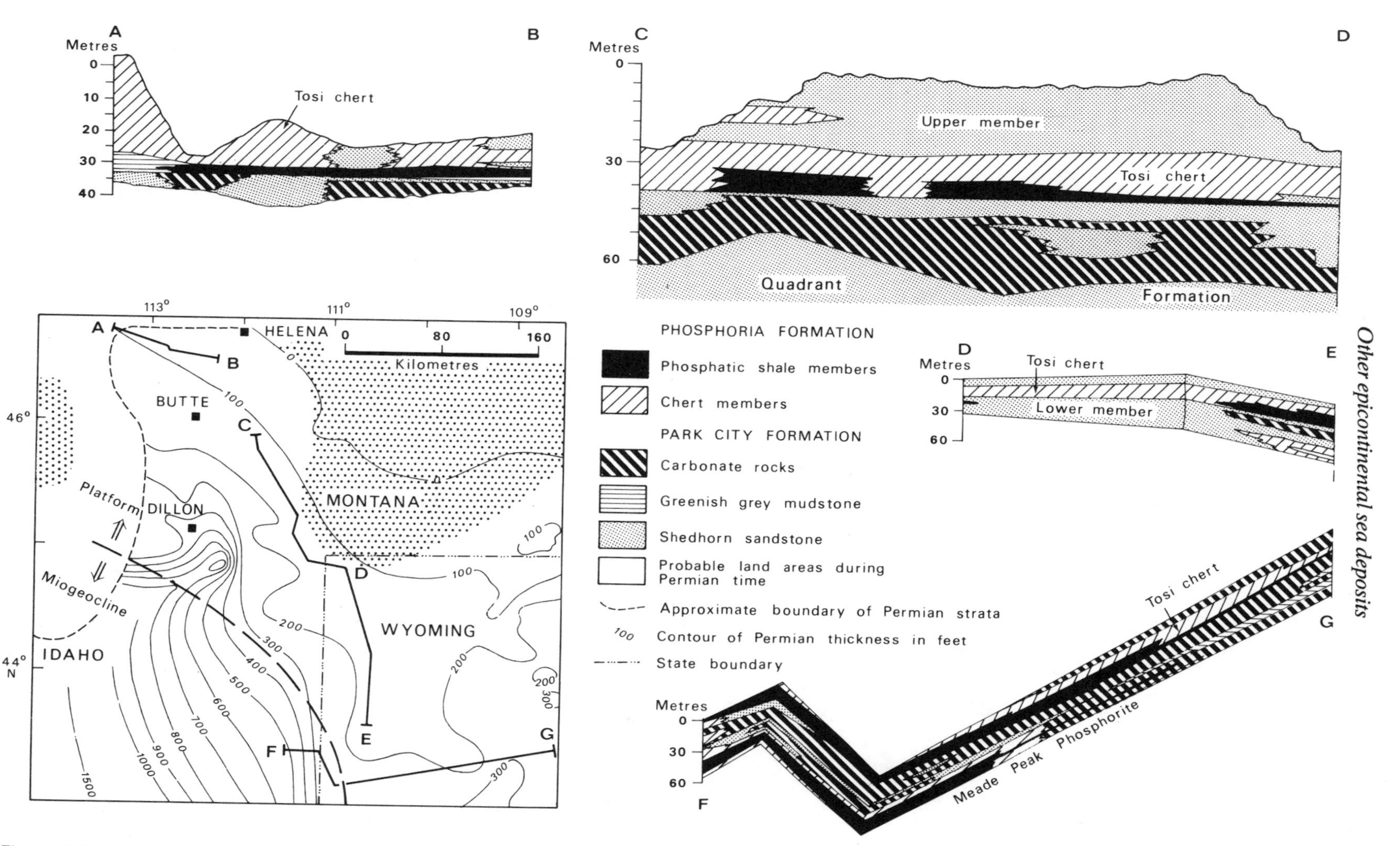

Figure 5.3. Permian rocks of southwestern Montana and adjacent areas. Thickness contours and distribution from Swanson, R.W. (1970). Cross-sections are simplified from Swanson, R.W. (1973), Cressman and Swanson (1964) and Swanson, R.W. (1970).

THE CAMBRIAN GEORGINA BASIN NEAR MOUNT ISA, NORTHERN AUSTRALIA

Early Middle Cambrian phosphorites were deposited in shallow near-shore marine environments varying from lagoonal, estuarine and littoral to intertidal. Some deposits show penecontemporaneous erosion by stream channels in addition to subaerial erosion during periods of regression. There are three types of phosphorite: mudstone phosphorite, replacement phosphorite and pelletal phosphorite. The latter two were formed by diagenetic phosphatization of carbonate skeletal sands, bioclastic and micritic limestones, and dolomites (Howard and Hough, 1979). The phosphatic mudstone, which is the main ore, is considered to be an orthochemical sediment. The phosphorite basin occupies a partly fault-controlled depression on the Precambrian land mass. Some of the mudstone phosphorites include thin beds that may contain up to 45% iron oxide (Fe_2O_3) and 6% manganese oxide (MnO).

The basin has three transgressive–regressive cycles, which resulted in interdigitating facies changes. The associated sediments are dolomite, limestone, chert and sandstone and the basin grades landward into a redbed sequence. There is a positive correlation between the P_2O_5 content of the phosphorites and uranium; however the uranium is more readily leached from outcrops than the phosphorus. The overall content is 1–86 ppm uranium.

Intense Cenozoic weathering and laterization has modified the outcrops. The exposed weathered phosphorite has been referred to as phoscrete, which caps high-grade phosphate mudstone and pelletal phosphorite. Some of the phoscretes may also have formed by replacement of the carbonates during the Cambrian period.

The Georgina Basin phosphorites are presumed to have been formed by the diagenetic phosphatic replacement of coquinas, carbonate sands and muds at the sediment/water interface by incoming phosphate-rich oceanic currents.

Genesis of Phosphorites

The majority of phosphorites were deposited at low latitudes within the zone from the palaeo-equator to about 40° N or S (Cook and McElhinny, 1979). The present-day ocean is nearly saturated with phosphate, but the contents are far from uniform. Warm surface waters of the ocean contain only about 0.01 ppm or less, whereas deep cold waters contain 0.3 ppm (McKelvey, 1967). The lowest phosphorus concentrations are at the surface and rapidly increase to a maximum at a depth of about 1000 m, with only a slight and variable decrease below that (Gulbrandsen and Robertson, 1973). Oceanic up-wellings are thought to be one of the major mechanisms for recycling the deep supply of phosphorus back to the surface. Such up-wellings result from major oceanic current divergences and represent areas of high primary organic productivity. Regions of up-welling are characterized by blooms of dinoflagellates (red tides) and abundant diatoms. Important coastal areas of present-day up-welling are along the west coast of California and South America, the west coast of south and north Africa, the Red Sea and the Arabian Peninsula. The major oceanic up-wellings are generally found in the trade-wind belts where surface waters are blown offshore and the offshore current is augmented by the Coriolis force (Riggs, 1979b).

The solubility of phosphate in up-welling water decreases as the temperature and pH increase. Hence apatite may be precipitated by organic and inorganic processes. Gastropod and pelecypod shells that were calcareous in life may become phosphatized upon burial in zones of up-welling (McKelvey, 1967).

The localization and formation of fossil phosphate deposits can therefore be expected to be related to the palaeo-positions of extensive coastlines in low latitudes in relation to the major oceanic currents. It is the complex interrelations between these that decide the positioning of the major up-wellings (Cook and McElhinny, 1979). However, the tectonic activity of the coastline is important. Maximum phosphorite formation occurs intermediately between total quiescence of a coastline, which will produce carbonate banks, and extreme mountain building, which will supply a deluge of terrigenous and volcanogenic sediments (Riggs, 1979b). But more important, the formation of phosphorite appears to occur during periods of changing tectonic activity, which apparently modify the regional chemical system of the bottom waters. The phosphorogenic system occurs simultaneously with and adjacent to other anomalous chemical sediments (*Figures 5.2, 5.3*) with which the phosphorites commonly mix and become interbedded as the chemical systems fluctuate with the tectonic periodicity. The different kinds of deposits thus oscillate laterally. The iron-rich and silica-rich sediments (glauconite, diatomite, chert) occur dominantly on the polar side of the phosphorite belt, while the magnesium-rich sediments (dolomites and clays) occur dominantly on the equator side (Riggs, 1979b).

Phosphate precipitates as the cold, chemically saturated and somewhat toxic up-welling bottom currents move onto the shallow shelf environments and across the structural highs. The phosphorite accumulates as a loose colloidal microcrystalline mud suspension. Under high-energy water conditions some of the muds are physically aggregated to produce oolites or pellets. The phosphate pellets are frequently subjected to the movement of currents, which thereby remove the finer-grained less-phosphatic materials and hence enrich the deposit in P_2O_5. Diatomites and phosphatic diatomites are found in areas subjected to much less current movement (Cheney, T.M. *et al.*, 1979).

Potash–Salt–Anhydrite–Sulphur Evaporites

Evaporites are important commercially for the chemical industry. It is clear that the greatest salt deposits of the world have a strong affinity for the interiors of subsiding basins (Sloss, 1969). There are three major types of geological setting for salt accumulation.

(1) Asymmetric fault-bounded intracratonic troughs (yoked basins), which receive their sedimentation from the craton highlands, are the sites of some large salt accumulations. The Plio–Pleistocene of the Dead Sea Graben is a good example. However, yoked basins are relatively ephemeral features, and throughout much of their duration are generally choked with clastic debris derived from uplift of the surrounding highlands as the basin sinks. Thus ancient yoked basin salt deposits are relatively rare.

(2) Basins on the miogeoclinal (passive) margins of continental cratons, for example, on either side of the Atlantic Ocean, where salt deposits are relatively infrequent in the stratigraphic record, perhaps because the miogeoclines generally had an open circulation system and unimpeded communication with the open ocean. Nevertheless miogeoclinal successions do contain the largest known salt accumulations. The Triassic or Jurassic salt deposits of the Gulf of Mexico Coast basin of North America are estimated to underlie an area of as great as 5×10^5 km^2, and the salt total is estimated to be nearly 2×10^5 km^3 in volume (Sloss, 1969). The Lower Cretaceous salt deposits of the Gabon and Congo basins were deposited on the miogeoclinal rifted coast of the African craton formed as South America drifted away from it. The Congo Basin contains a layer of almost undisturbed salt varying in thickness up to 1000 m. The Congo Basin is very rich in carnallite and the reserves of potash and magnesium are estimated to be several billion tonnes. Only sylvite is mined at present (De Ruiter, 1979).

(3) The most typical and most commonly found sites of salt accumulation are the large interior basins of continental cratons, where salt deposition was present in the vertical stratigraphic successions over long spans of time. These cratonic interior basins are likely to receive basin-centre evaporites, commonly including salt, whenever two conditions are simultaneously met: significant subsidence of the basin interiors in relation to the adjacent shelves; and absence of major volumes of terrigenous clastic material to choke the basin.

Commercially valuable potash salts such as sylvite and langbeinite generally occupy only small parts of evaporite basins, whereas non-commercial potassium-bearing minerals such as carnallite and polyhalite are of more widespread distribution in the basins (Goldsmith, 1969). Sylvinite is the name given to mixtures of halite and sylvite mined as a potassium ore.

The Middle Devonian deposits of the Great Plains region of North America are possibly the largest in the world. The Permian Zeichstein evaporites of Europe are equally famous and both will be briefly described.

The Middle Devonian Prairie Evaporite

The evaporites of the Muskeg or Prairie Formation, in western Canada, which extends into North Dakota and Montana, contain some of the richest potash deposits of the world. They also have direct relationships to the occurrence of oil.

The rocks were deposited in the elongated down-warped Williston Basin formed on the continental cratonic crust. The rock succession underwent little structural complication and correlation from well to well is easy. Isopachs of the Muskeg Formation show the deepest parts of the basin to lie generally along its central axis, with thicknesses in excess of 150 m near Regina and north of Edmonton (*Figure 5.4*).

The longitudinal cross-section (from Klingspor, 1969) indicates the progressive change in the basin from northwest to southwest (*Figure 5.4*). The basin was open to the sea in the northwest. An arcuate barrier carbonate reef complex extends from northeast British Columbia into the Northwestern Territories and skirts a platform of variable width where shelf carbonates dominate. Patch and pinnacle reefs are present on the platform in great numbers. Eastward and inward from the carbonate platform is a broad region where the

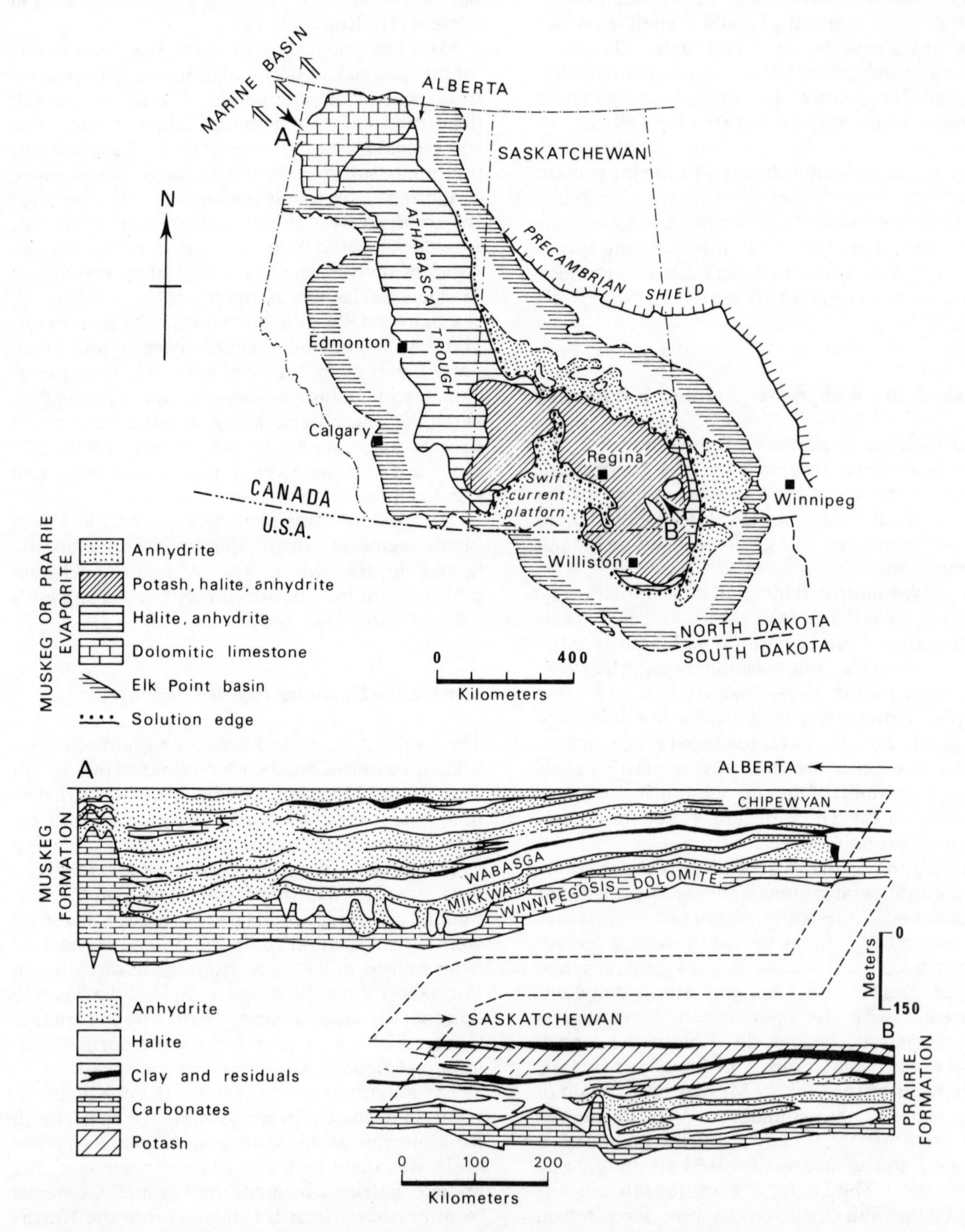

Figure 5.4. The Devonian Muskeg – Prairie Province of western North America, based on Worsley and Fuzesy (1979). Cross-sections are from Klingspor (1969).

Muskeg Formation is composed mainly of anhydrite, and carbonates decrease to less than 25% of the succession. The carbonate platform is known as the 'Carcajou Shelf', which constituted a broad barrier impeding the supply of sea water from the open sea southeastwards into the basin. The shelf was rich in reefs, banks and intertidal flats and, as the oceanic waters flowed over them, they underwent progressive evaporation and saturation.

The gentle slope from the shelf platform into the interior basin is characterized by a transitional facies, with finely varved halite as the dominant rock. This is best seen in the Athabascan Trough, which coincides with the axis of the basin. Anhydrite is present only in thin bands at the base of each salt cycle and in the rhythmic varves that divide all salt layers. The Athabascan Trough is flanked on the east and west by broad fringes of semi-saline and evaporite facies, which grade in western Alberta into shoreline, deltaic and fluvial sands.

Near the Alberta–Saskatchewan border, the halite begins to contain sylvite crystals, then thin sylvite beds indicate that southeastwards there was a higher degree of salt concentration (*Figure 5.4*). This hypersaline facies of the Prairie Evaporite extends through Saskatchewan. A sylvinite facies is widespread in the broad interior of the basin, where up to 10% of the salts consists of sylvinite potash beds. Carnallitic potash ores are common, although of erratic occurrence. Some local potash beds are extraordinarily thick, and represent sub-basins in which the concentrated brines experienced periods of maximum evaporation (Klingspor, 1969). Basement tectonics influenced both potash deposition and its preservation (Worsley and Fuzesy, 1979). The rich potash beds of Saskatchewan extend as far south as North Dakota and northeast Montana, where they occur at depths exceeding 1065 m (Anderson, S.B. and Swinehart, 1979).

Permian Evaporites of the North Sea Basin

The Zechstein salt and potash deposits form the basis of the major chemical industries of England, Germany and Poland. The North Sea Basin was well developed in the Carboniferous period and continued to subside through the Permian before the opening of the Atlantic Ocean. Progressive subsidence led to the establishment of the Permian Rotliegendes basins and the Mid-North Sea–Ringkøbing–Fyn High appeared as a major ridge separating the less well known northern basin from the Southern Permian Basin (*Figure 5.5*). The Rotliegendes evaporites were deposited in the Southern Permian Basin inland sea. Continued subsidence of the arid basin possibly below sea level resulted finally in the spectacular ingression of the Zechstein Sea, possibly through the northern North Sea (*Figure 5.5*). Thin shelf carbonates and sulphate sequence deposition was restricted to the basin margins; these graded into thick basin-filling halite series that reached a thickness in excess of 1000 m, both in the southern and northern Permian basins (Ziegler, W.H., 1975).

Oil-company drilling has provided excellent correlations of the Zechstein strata from the English coast of Yorkshire and Norfolk to Germany and Denmark (Brunstrom and Walmsley, 1969; Taylor, J.C.M. and Colter, 1975). There are four major evaporite cycles recognized in Germany, which can be correlated to England and Denmark with remarkable lateral persistence. The first three cycles began with a marine transgression and typically had well-developed carbonates at the base. An ideal marine evaporite cycle is as follows: (top) anhydrite; potash salts; halite; anhydrite; dolomite; limestone; siltstone; clay (base). The early part of the first cycle contains the distinctive Kupferschiefer (copper shale), which extends far under the North Sea from Germany and is equivalent to the English Marl Slate of Durham, easily recognized in well logs by its high radioactivity.

Halite and potash salts were deposited in the Zechstein 1 (Z1), or Werra cycle, in Germany near the Harz Mountains and in southeast Holland, but only anhydrite is found in the North Sea. Important potash beds occur near the top of the Z2, or Strassfurt cycle. The Z3, or Leine cycle, contains important carnallite and sylvite evaporites in the form of the Reidel and Ronnenberg Potash Beds. The Z4, or Aller cycle, is continental and lacks a carbonate phase, although a 30-cm carbonate phase is known from Denmark (Brunstrom and Walmsley, 1969). Potash beds are thickest and most consistent along the western margin of the Zechstein 3 and 4 basin, which extends on to the Yorkshire coast of England. The potash ores in both cycles are mainly of sylvinite in the onshore areas of Yorkshire. The layered Upper Potash of the Z4 cycle is less variable but generally of lower grade than the Boulby Potash of the Z3 cycle (Smith, D.B. and Crosby, 1979). Towards the North Sea Basin centre, the potash salts are highly variable in thickness and grade and have been generally affected by diapirism of the halite of the Z2 cycle. The Z3 Boulby Potash overlies the thick Boulby Halite. It is of sylvinite and contains rounded fragments of the rocks of the adjacent

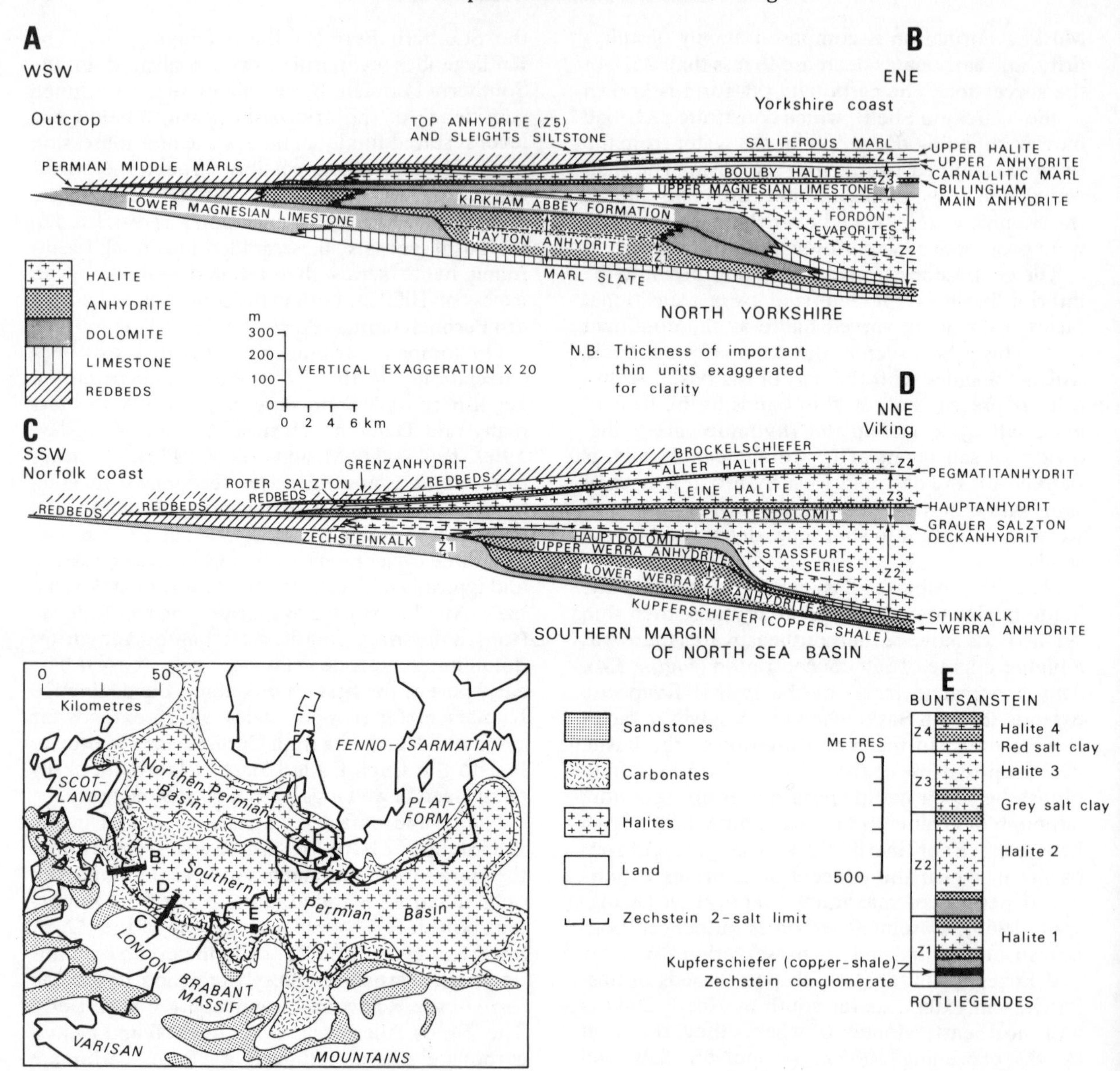

Figure 5.5. The Upper Permian Zechstein palaeo-geography (after Ziegler, P.A., 1975). Cross-sections after Taylor, J.C.M. and Colter (1975). Dutch/German border column after Brunstrom and Walmsley (1969).

beds (Woods, 1979). Its thickness varies from less than 1 m to as much as 20 m, and it has an overall high grade and simple mineralogy. The potassium chloride content may extend up to 45%. However, its high temperature (40°C), gas content and the tectonic deformation tend to cause mining problems.

The Kupferschiefer

The copper concentration in this horizontally persistent member of the Zechstein 1 cycle has supported a long history of mining in the region near Leipzig between the Harz and the Erzgebirge mountains. Its thickness ranges from slightly less than 40 cm to slightly greater than 80 cm (Jung and Knitzschke, 1976). The lowest unit is formed by basal conglomerate, sandstone and siltstone, which overlie unconformably a basement of crystalline rocks, low-grade metasediments and volcanic rocks. This unconformity is related to the spectacular incursion of the Zechstein Sea flooding over the Mid-North Sea–Ringkøbing–Fyn High, and inundating the inland sea and its numerous fringing volcanic centres that were active on the High and along the southern margin of the basin during the Rotliegendes. Some volcanic activity persisted on the Mid-North Sea High during Zechstein times (Ziegler, W.H., 1975).

The Kupferschiefer is a fine-grained, finely laminated claymarl or marlstone containing carbonaceous material. In general there is a difference between the normal basin facies and the red-spotted facies, coloured by haematite, which overlies topographic highs. The thickness can change rapidly from 0 to 58 cm within an area of sand bars.

All significant copper concentrations occur within deposits that are the equivalents of the redbeds of the Zechstein foreland beyond the rises in the basement topography. High E_h potentials correspond with the top of the basement highs, which are free from non-ferrous metals, whereas deposits that originated along the slopes from the highs accumulated under slightly negative to neutral E_h values leading to predominantly copper-rich sulphide precipitates. Strongly negative E_h values in the surface regions led to lead–zinc mineralization (Jung and Knitzschke, 1976). The horizontal and vertical sequence of copper, lead, followed by zinc deposition was formed by a normal marine transgression. The chemical agent that was responsible for the precipitation of the metals was hydrogen sulphide released during decomposition of the organic matter, but in particular from the reduction of sulphates during the metabolism of bacteria under stagnant euxinic water conditions.

The major metal contents of the Kupferschiefer are: copper (which ranges up to 29 kg t^{-1}), lead (up to 8.6 kg t^{-1}) and zinc (up to 18.5 kg t^{-1}). Important trace elements are silver (up to 190 g t^{-1}), cobalt (up to 160 g t^{-1}), nickel (up to 147 g t^{-1}), vanadium (up to 914 g t^{-1}) and molybdenum (up to 308 g t^{-1}).

Lateral changes of the paragenetic sequences within the facies changes from the basement highs towards the deepest parts of the basin are: (shallow) haematite-type; chalcocite-type; bornite/chalcocite-type; bornite-type; bornite/-chalcopyrite-type; galena/chalcopyrite-type; sphalerite-type; pyrite-type (deep). This sequence corresponds to the solubility products of the sulphides in decreasing order of the E_h value away from the basement highs. The haematite-rich parts are not commercial, but other mineral facies are.

The formation of the Kupferschiefer has been likened to the present-day Black Sea, whose muds are of sulphide sapropels (Degens and Ross, 1974). The supply of the abnormally high metal components to the Zechstein Sea is totally unknown, but there is an overabundance of theories, which have changed through time; these are reviewed by Jung and Knitzschke (1976). Volcanic activity was relatively widespread around the flanks of the lower Rotliegendes inland sea, preferentially along the North Sea High, and also along the southern margins of the sea, at the northern part of the Variscan foreland (Ziegler, P.A., 1975). It is hard to avoid the conclusion that the source of metals in the early part of the marine Zechstein Sea was this important phase of volcanic activity. The marine incursion into the basin may have reworked the metal concentrations related to these volcanoes. Volcanic activity continued in the North Sea High throughout Zechstein times, so that submarine exhalations may also have contributed to the metal enrichments. It has also been suggested that connate water passing through the basement of Late Variscan molasse formations may have leached out the non-ferrous metals and deposited them in the Zechstein Sea.

The Question of Water Depth

Modern marine evaporite deposits form in shallow, slowly subsiding basins, or embayments near the sea coast, in regions of arid or semi-arid climate. It is, however, very difficult to draw an exact analogy between these modern 'salina' deposits and the ancient evaporite deposits such as the Muskeg and Zechstein, although they are mineralogically similar (Schmalz, 1969). The outstanding anomalies are: ancient evaporites include thick basal sapropelic units that may contain economic base-metal sulphides; some anhydrite and clastic layers in the overlying salts are bituminous, and the strongly reducing conditions of the ancient basins, necessary to satisfy these features, cannot be reconciled with the modern 'salina' model; most ancient salt deposits show little or no evidence of emergent conditions during their active salt deposition, except in the very late stages, which makes them dramatically different from the 'salinas'; and ancient evaporites are characteristically very much thicker than their modern counterparts, for example, the Zechstein sequence includes more than 1100 m of evaporites.

To overcome these incompatibilities, Schmalz (1969) and other workers have proposed a deep-water model for ancient deposits. The salts were deposited in a basin several hundreds to thousands of metres deep. The deep basin had a restricted connection with the open ocean (like the present-day Mediterranean Sea). The deeper-water hypothesis offers a direct explanation of the petroleum, natural gas and the base-metal sulphide deposits, as well as providing a suitable hydrographic and chemical model for the salt deposition, not necessarily restricted to arid or semi-arid climatic conditions.

The most obvious contrast between a deep-sea model and the 'salina' model is the water depth. Whereas the 'salina' model implies that the brine depth above the salt is of a few centimetres to a few tens of metres, the depth of water in the deep-basin model is assumed to be approximately the same as the ultimate thickness of the deposit. The final thickness of the 'salina' is determined by the rate of subsidence of the basin floor, but subsidence of the deep basin will have negligible effect on the final deposit thickness (Schmalz, 1969). The deep-basin model seems to be preferred by most writers for the Zechstein Sea (Taylor, J.C.M. and Colter, 1975).

Sulphur

Roughly one-third of the world supply of sulphur is from elemental sulphur deposits within evaporite sequences. Sulphur is also obtained from volcanic fumaroles and separated in refineries from sulphur-bearing fossil fuels, or by roasting pyrite and pyrrhotite ores. Anhydrite and gypsum can be processed to yield sulphate-bearing fertilizers, or reduced to elemental sulphur.

Elemental sulphur deposits are widely distributed and occur in two geological environments: basins containing both hydrocarbons and evaporites; and zones of Cenozoic volcanism. The classification used in this chapter is after Ruckmick *et al.* (1979).

BIOEPIGENETIC SULPHUR DEPOSITS

These deposits are formed by anaerobic bacterial reduction of gypsum or anhydrite and oxidation of hydrocarbons in basins where evaporite beds are underlain by petroleum (Ruckmick *et al.*, 1979).

The only significant difference between the cap-rock and the strata-bound deposits is that the position of the anhydrite above the petroleum in the cap rock is caused by diapirism, rather than by sedimentary processes. In the salt domes, the residual anhydrite of the cap rock is accumulated by solution and removal of halite from rock salt by artesian ground water (*Figure 5.6*). Bioepigenetic sulphur deposits occur where joints and faults permit water, hydrocarbons and bacteria to rise into the evaporite or cap rock. In this environment, bacterial oxidation of hydrocarbons occurs to give carbon dioxide, followed by reduction of gypsum to hydrogen sulphide and alteration of gypsum to secondary calcite (Davis, J.B. and Kirkland, 1973). By dissolution and dispersion of both hydrocarbons and sulphate ions, circulating artesian meteoric water in contact with both hydrocarbons and gypsum provides an ideal environment for high levels of bacterial activity. Another require-

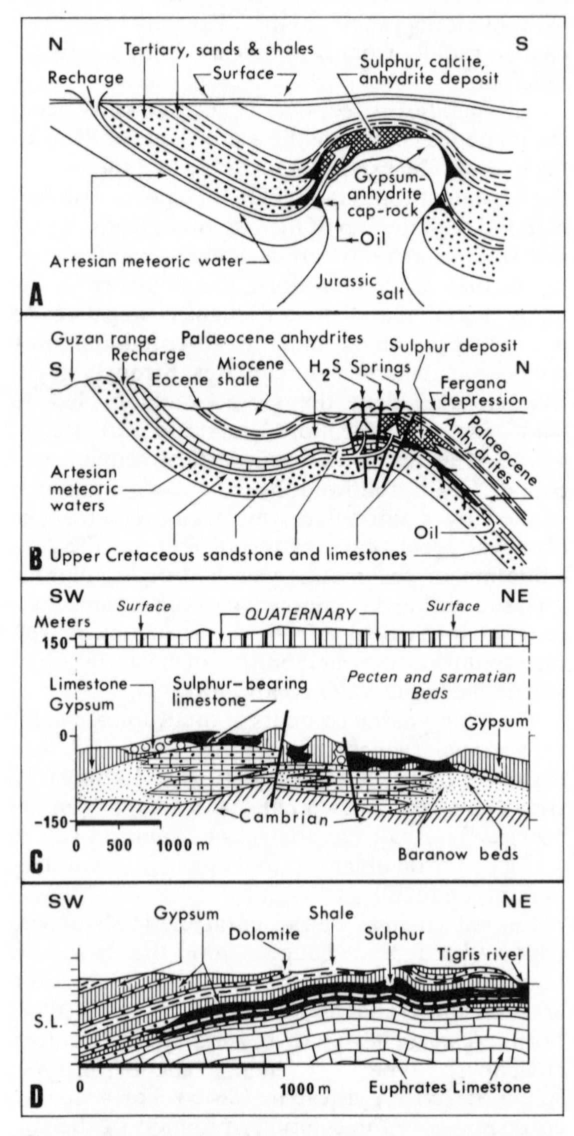

Figure 5.6. Details of some bioepigenetic sulphur deposits. (A) Gulf of Mexico coast salt–dome sulphur deposit (after Ruckmick *et al.*, 1979). (B) Shor-Su strata-bound deposit, Uzhbekistan, USSR (after Ruckmick *et al.*, 1979). (C) The Tertiary Tarnpbrzeg deposit, Poland (from Pawlowski *et al.*, 1979). (D) The Tertiary Mishraq deposit, northern Iraq (after Barker, J.M. *et al.*, 1979).

ment appears to be the presence of relatively impermeable beds, such as shales capping the system and allowing hydrogen sulphide sufficient residence time to be oxidized by sulphate ions and carbon dioxide.

Structural controls localize hydrocarbons and artesian circulation to result in the sulphur deposits. For cap-rock deposits, the structural control is the salt diapir itself. Structural control of the strata-bound deposits involves either faults and joints or flexures, or a combination of both. The Polish deposits (*Figure 5.6*) are related to sulphur mineralization formed in structural highs created by horst blocks in the Cis-Carpathian trough (Pawlowski *et al.*, 1979). The Shor-Su deposits of the Fergana depression of the USSR (*Figure 5.6*) are principally controlled by anticlinal structures. Mishraq, the very large deposit of Iraq (Barker, J.M. *et al.*, 1979) also occurs in a prominent anticline.

BIOSYNGENETIC SULPHUR DEPOSITS

These are sedimentary stratiform deposits that result from bacterial reduction of sulphate to hydrogen sulphide and oxidation of hydrogen sulphide to elemental sulphur by oxygenated surface waters, in lagoonal or otherwise restricted evaporite and euxinic marine or lacustrine basins. Sulphur-isotope and carbon-isotope analyses distinguish between biogenic sulphur deposits and non-biogenic deposits, but do not distinguish the epigenetic from the syngenetic (Davis, J.B. and Kirkland, 1979). Strongly stratiform deposits such as the Polish and Russian Cis-Carpathian deposits and the Iraq deposits (*Figure 5.6*) may indeed be primarily syngenetic with secondary localization by epigenetic processes. Pawlowski *et al.* (1979) believe, however, that the native sulphur deposits of Poland resulted from the epigenetic alteration of marine gypsum at structurally favourable sites, where the gypsum was up-faulted and overlain by impermeable clays. The hydrocarbons moved upwards from the lower strata.

VOLCANIC DEPOSITS

Sulphur deposition in the fumaroles of volcanoes is usually common but not commercial. However, they form the basis of the commercial supply of sulphur in Japan (Mukaiyama, 1970).

OXIDATIVE DEPOSITS

This category includes the deposits of hot springs. Such deposits are widespread but small. They were deposited in fractures and open spaces in a variety of rock-types and there is no evidence of a biogenic replacement associated with them.

THERMOGENIC DEPOSITS

At high temperature and pressure, liquid sulphur may be formed by oxidation of hydrogen sulphide derived from direct reaction between hydrocarbons and anhydrite, for example, in the deep Anadareo Basin of Oklahoma (Ruckmick *et al.*, 1979). However, this category is unlikely to be commercial since it is encountered in deep oil-well drilling only.

Uraniferous marine black shales

Many marine black shales, which are rich in organic matter, contain significant uranium contents, and they may provide a future uranium resource, though mining would present severe environmental problems. A review is given by Bell (1978), who tabulates uranium contents normally within the range 10–100 ppm. Normally black shales contain more than 2% organic carbon. Thin phosphatic layers and flat nodular phosphates are present in some. Almost all are characterized by very thin laminations persistent over several kilometres. Uraniferous black shale units are usually less than 5 m thick.

The uranium is always clearly associated with organic content, or with apatite in the case of phosphorite lenses or layers. Discrete uranium minerals have not been detected, except when the shales have been metamorphosed, and then disseminated fine-grained uraninite and thucholite are found.

The ideal home of black shales is a marine basin of restricted water circulation in which the bottom conditions are euxinic. The Black Sea is taken as the ideal present-day example (Richards, F.A., 1965). At depth it is highly anoxic. The deeper-water levels maintain a reducing and toxic environment as a consequence of removal of oxygen by settling of organic matter and a lack of aeration. It appears that living organisms may be effective in fixing the uranium in the black muds (Degens *et al.*, 1977).

One important point that emerges from the studies of the Black Sea (Degens and Ross, 1974) is that the rivers that drain into the Black Sea are not enriched particularly in uranium and it can be concluded that, in the case of a sea of this size, a strongly uraniferous source area is not necessary. The bulk of the uranium is dispersed in the shelf and peripheral areas of the Black Sea and Azov Sea.

The ancient environments, particularly for the Alum Shale of Sweden and the Chattanooga Shale of the USA, are clearly epicontinental and of shallow-water and probably cold-water deposition, rather like the present-day Baltic Sea (Bell, 1978).

The most important conditions for development of uraniferous black shales include the following: extremely slow sedimentation rates controlled largely by scarce input of terrigenous material; high organic carbon input; poor aeration of the bottom regions of the sea; the presence of hydrogen sulphide in the reducing environment; living biological fixation processes in the muds; and maintenance of a uranium concentration above 2 ppb in the waters.

At present, only the deposits in Sweden can be considered a uranium resource. Others may be found and the environmental problems connected with mining will probably be overcome.

Swedish Black Shales

Sedimentary rocks of early Palaeozoic age in southern Sweden, and farther north along the Caledonian Front, have horizons rich in uranium (Grip, 1978). In south Sweden, the Upper Cambrian alum shale contains hydrocarbons and finely disseminated uraninite, especially near Billingen mountain and Närke. At Billingen, a bed of bituminous shale, about 3.6 km thick, contains, on average, 300 ppm uranium, 680 ppm vanadium, 300 ppm molybdenum, 0.3% phosphorus pentoxide, 7% sulphur, 8% aluminium and 15–20% kerogen (Armands, 1972a). The shale is being mined and processed at Ranstad. The largest production was in 1966, when 425 000 t was extracted. The shale contains 10–15% pyrite and 12–17% organic carbon.

Farther north along the Caledonian Front, folded shales contain considerable uranium contents, but here they are of Ordovician age. The highest grades occur in the Lake Tåsjö area. An early Ordovician glauconite – phosphorite shale overlies the Alum Shale. Its average contents are 200–400 ppm uranium, 200 ppm vanadium, 0.1% rare earths and 4.0–9.3% phosphorus pentoxide (Armands, 1972b). Its thickness averages 5 m, but it has not yet been mined. The shale contains 9–24% apatite, 0.8–11% pyrite and 0.5–3.3% organic carbon.

Stacking the ore for a year to weather enhances the final recovery, because acids generated during the weathering of the pyrite enhance the removal of the uranium.

USA

Most of the identified resources are in the upper member of the Late Devonian to early Mississippian Chattanooga Shale of Tennessee, Kentucky and Alabama (Swanson, V.E., 1961). The member is 3.6–5.1 m thick over an area of 10 000 km^2, and averages 0.007% U_3O_8. Unfortunately concentrations in excess of 70 ppm are uncommon.

Mercury Deposits

All major deposits are mined for their mercury contents alone, and all contain cinnabar as the chief ore mineral. However, there is frequently an association with gold and antimony. Silica and carbonate minerals are common as introduced gangue, but pyrite and marcasite may also be common associates. The mineralization is of low-temperature deposition, from 50°C to 200°C. The low-temperature mobility of mercury and antimony is well illustrated by hot springs of western North America, in which mercury is undergoing present-day deposition. Some of these springs also simultaneously deposit antimony (Dickson and Tunell, 1968). All deposits are shallow, and were formed within 1000 m of the surface.

Undoubtedly mercury deposits have formed throughout geological time, but owing to their shallowness and low-temperature mobility in hot ground water, most mercury has been remobilized and deposited in young formations. Some of the major deposits are strata-bound and appear to have resulted from submarine hot-spring emanations directly into the basin of sedimentation, as at Almaden in Spain and Idrija in Yugoslavia. However, several major deposits are clearly epigenetic and localized along major fundamental faults, especially along the continental margins. Continental margin plutonism has mobilized mercury on a grand scale, as in the case of the Chinese deposits, in which there is a striking affinity for carbonate host rocks, accounting for over 80% of the deposits (Ikonnikov, 1975). In China, as in several other countries, mercury and antimony mineralization frequently overlap (Maucher, 1976), and the Chinese deposits are predominantly of cinnabar and stibnite, with accessory pyrite, orpiment and realgar. The principal mining regions are in the southern provinces of western Hupeh and western Hunan. Mercury may survive an erosion cycle, and cinnabar frequently occurs as placer deposits.

The rich deposits of the Bau mining region of Sarawak, Borneo, occurred as cinnabar–quartz–pyrite–marcasite and realgar fillings of breccia in sandstones and shales, but the greatest production came from eluvial cinnabar (Wilford, 1955). The mobilization of the mercury is probably related to a line of high-level acid plutons that characterized the active eastern continental margin of Sundaland, now represented by the Lupar ophiolite line. Epithermal mineralization within this Late Cretaceous active margin is represented by an association of mercury, antimony and gold (Hutchison and Taylor, 1978). Complex arsenical ores contain native arsenic, stibnite, pyrite, free gold and cinnabar. The total gold production was 36 000 kg, antimony ore 85 000 t and mercury 760 t, but the Bau Region is now largely exhausted.

The worldwide distribution of the major mercury deposits is strongly concentrated in two orogenic regions. (1) The circum-Pacific volcano–plutonic system, with important distributions in the Andes of South America (Petersen, 1972), and the western cordillera of North America, continuing through Japan, China, the Philippines, Borneo and eastern Australia (Bailey, E.H. *et al.*, 1973). (2) The northern borderlands of the Mediterranean and the southern margins of the Alpine fold belt. The world's three largest deposits are Almaden in Spain, Idria in Yugoslavia and Monte Amiata in Italy.

In summary, most deposits occur on the continent side of post-Jurassic subduction zones. It appears that the mercury was originally mobilized from the continental crust and concentrated in the eugeoclinal or intra-arc basins along the convergent active margins.

A major deposit-type is associated with serpentinite or ophiolite melange where it is in tectonic contact with either continental rocks or accretionary prism melange.

Although mercury deposits may be broadly classified as epicontinental, they appear to be confined to active convergent plate margins where there was sufficient geothermal gradients to remobilize the metal into younger formations. These formations may be the tectonic melange areas, or overlying basins in which strata-bound mineralization was introduced via major deep-seated faults.

The Almaden Mercury Deposit, Spain

Almaden is by far the largest deposit in the world. It produces about 2068 t (60 000 flasks) of mercury per year, and accounts for Spain's pre-eminent position in mercury production. The mining area is located in the Cuidad Real province, southwest of Madrid. The geology has been comprehensively described by Saupé (1973) and summarized by Mamen (1971) and Dixon (1979). The mine is worked at a grade of 2–3% mercury and has yielded 250 000 t of mercury. It seems probable that the reserves equal this.

The ore formation is a sequence of earliest Silurian orthoquartzites, interbedded with ripple-marked siltstones, generally dipping vertically to steeply towards the north (*Figure 5.7*). The mine area forms the southern limb of a syncline. The immediate footwall of the stratiform ore body is sometimes marked by a rather altered basalt flow. The ore formation consists of two or three quartzite members, separated by slaty siltstones. The quartzites are clean and medium-to-fine grained. In many places, the hangingwall of the ore formation is a Llandoverian graptolitic black shale. Above this lies a thick sequence of pyroclastic rocks.

Within the sandstones, the mineralization is in the form of lenses, corresponding to the more porous zones. They range from 3 m to 5 m thick, are worked to a depth of 385 m below surface, and extend along strike for a distance of up to 400 m. The ore is quartzite containing minor amounts of pyrite, kaolinite and organic matter. The principal ore mineral is cinnabar and small amounts of native amalgam. There are three varieties of ore: compact masses of cinnabar, cinnabar in veins and quartz sandstones impregnated with cinnabar. The cinnabar fills pore spaces and microfractures in the rock. The mercury content varies between 0.6% (cut-off grade) and 20%, and its distribution reflects variation in grain size and sedimentary structures. Some minor flexures and faults, mostly oriented north–south, disrupt the ore beds to a slight extent.

The association with volcanic rocks has generally been interpreted as genetic (Saupé, 1973). It is suggested that mercury was emplaced in the more porous parts of the sandstones, which acted as palaeo-channelways for subsurface-water flow. Dixon (1979) and Saupé (1973) imagined the mercury source to have been volcanic sublimates. Saupé (1973) assigned a sedimentary origin to the deposit based on his observations of apparent sedimentary structures in the cinnabar and stratification of quartz and cinnabar. The strata-bound nature within the quartzite horizons, and especially the cinnabar grains included between the detrital core of the quartz grains and their secondary silica overgrowths, clearly demonstrate a syn-

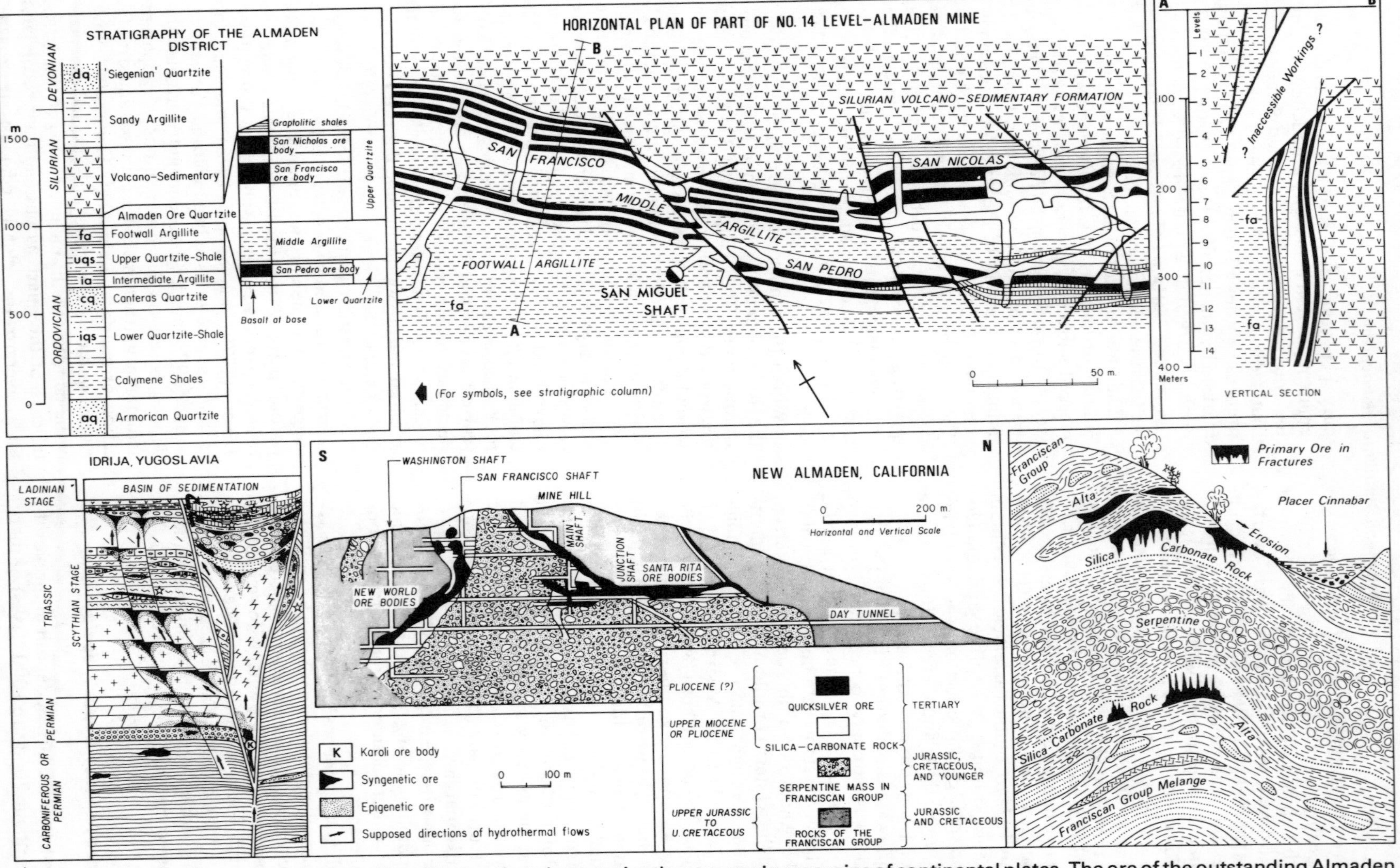

Figure 5.7. Various styles of mercury mineralization that characterize the converging margins of continental plates. The ore of the outstanding Almaden Mine of Spain is earliest Silurian and syngenetic (after Saupé, 1973). The ore of Idria, Yugoslavia, is partly syngenetic within Ladinian formations, but is also partly epigenetic (after Mlakar and Drovenik, 1971). The ore of the New Almaden mine of California is entirely epigenetic along the boundary zone between diapiric serpentinite and Franciscan melange (after Bailey, E.H. and Everhart, 1964).

depositional origin for the mineralization, which is thought to have been genetically linked to the Gothlandian volcanic activity within the sedimentary basin. Calvo and Guilemany (1975) have shown that the quartz, pyrite, calcite and dolomite of the ore bodies have a common sedimentary origin. They suggested that a molten cinnabar–mercury mixture moved through the rocks and impregnated the porous channelways. On cooling, the excess mercury segregated to form droplets of mercury within the cinnabar crystals. However, the carbonates intercalated within the ore sequence have isotopic constitutions, which suggest that they were derived from magmatic sources (Eichmann *et al.*, 1977), and that they probably owe their origin to the mafic igneous suite of the region.

The absence of mercury mineralization along the faults, and the presence of identical mineralization on both sides of the cross-cutting post-mineralization dykes, indicates that the deposit is not epigenetic (Saupé, 1967). These lamprophyre dykes, dated at 301 Ma, cut and roast the ore, so are distinctly post-mineralization. The nearest major intrusive is a granite, 25 km away, and this is not likely to have been genetically involved in the mercury deposit.

The problem of ore genesis has been discussed by Maucher (1976). One would expect that an initial synsedimentary concentration in the basin would have been in the Ordovician pelitic sediments that are rich in carbonaceous material. It is difficult to explain the restriction of the mineralization to the sandstones, unless an initial pelitic enrichment was mobilized by the regional flow of ground waters through the more permeable formations. The mineralization localization within the sandstones may therefore be a diagenetic redistribution within the basin.

Elsewhere in the world, strata-bound cinnabar often coincides with scheelite and stibnite. However, the Almaden deposit is unusual in that it is of cinnabar only. Stibnite is reported from the district, but only from vein deposits, and the scheelite deposits of the region are granite-controlled. Maucher (1976) suggests that it was possible that stibnite and scheelite were preconcentrated in the sediments but have been reconcentrated hydrothermally in the thermal aureoles of the granites.

The Idria Mine, Yugoslavia

This mining region lies in northwest Yugoslavia, and offers one of the best illustrations of repeated mercury remobilization. Cinnabar occurs in different clastic, pyroclastic and carbonate rocks of different ages (*Figure 5.7*). It has been found in all horizons of Carboniferous to Middle Triassic. The overlying Upper Triassic-to-Tertiary succession is barren. The geology of the mining region has been described by Mlakar (1964) and Mlakar and Drovenik (1971), and summarized by Maucher (1976).

Mlakar and Drovenik (1971) distinguish two phases of mineralization. The second phase coincides with the deposition of the Ladinian Skonca Beds and Langobardian pyroclastics. In this phase, the Carboniferous, Scythian and Anisian beds may also have received a second mineralization, but as the Langobardian conglomerate already contains mineralized pebbles of the Upper Scythian dolomite, a first mineralization phase must pre-date the Middle Triassic tectonic erosional unconformity. The ores of the older mineralization phase are controlled mainly by contacts between lithostratigraphic units; in addition, they also show epigenetic patterns. The ore in the Langobardian Sandstone, Skonca beds and tuffites is characterized by many sedimentary fabrics, including graded bedding, cross-bedding, intraformational unconformities and glide folds within the ore beds. The situation during the Ladinian Stage, as envisioned by Mlakar and Drovenik (1971), is given in *Figure 5.7*. At this stage, the strata-bound ore bodies were formed by submarine thermal springs within a subsiding basin. The deposition of the ore coincided with tectonic and volcanic activity, causing turbidity currents and slumping of the unconsolidated sediments, resulting in mechanical transportation and redeposition of the ores before and during diagenesis of the sediments.

The region is characterized by four superimposed thrust sheets (Mlakar, 1964). Most of the ore is formed by cinnabar that has replaced Permian and Triassic dolomite, but some occurs in shale and sandstones. Individual ore bodies are large, many having a plan area of over 2000 m^2 and extending through a vertical interval of 60 m. Ore has been mined to a depth of 400 m. Understanding of the fault system has led to the discovery of new major ore deposits offset by 3 km from the mine that has been operating for the past 450 years.

Because only sparse native mercury occurred in the uppermost thrust plate and rich cinnabar occurred in the plate below, it has been suggested that the ore was formed in Triassic time, before the thrust faulting, and was thereafter protected from erosion by the overthrust plate. The near-surface native mercury is believed to be the result of post-ore and post-thrusting upward migration of vapour from the main ore body. The grade of ore mined is within the range 0.1–0.2% mercury. Production from the area has exceeded 110 000 t.

The ultimate source of the mercury remains hypothetical (Maucher, 1976). Even if it is assumed that the greatest part of the mercury has been derived from known Permian and Carboniferous mercury-bearing sediments, older source beds cannot be excluded.

New Almaden Mercury District, California

This district, situated a few kilometres south of San Jose in Santa Clara County, California, has yielded about 38 000 t of mercury, or nearly 40% of the total production of the USA. Its geology has been described by Bailey, E.H. and Everhart (1964). The whole country-rock pattern of the region is one of a tectonically active melange belt. The Franciscan melange is cut by tabular serpentinite masses that resemble serpentinite melanges. That the cold serpentinite masses flowed upwards along fracture zones to form tabular diapiric intrusions, which even extruded on to the land surface, has been well documented by Dickinson (1966) for the California Coastal Ranges.

Tabular masses of serpentinite that have been intruded into the Franciscan melange as sheets of foliate serpentinite breccia are of particular interest because their altered parts contain the mercury ore bodies. Some of the serpentinite melanges are conformable with the structure of the Franciscan, but others cross-cut and occupy fault zones. The larger masses consist of blocks of unsheared serpentinite embedded in a matrix of sheared serpentinite. The smaller masses are entirely of sheared serpentinite.

Silica–carbonate rock is the host of the mercury mineralization (*Figure 5.7*). It has formed locally by hydrothermal alteration of the serpentinite and may occur as a thin peripheral shell although some thick serpentinite masses are totally converted. There are abundant relict textures inherited from the serpentinite. Its dominant minerals are quartz and magnesite. It is presumed that the alteration to quartz–carbonate rock took place during the earliest phases of the hydrothermal mineralization, probably in Pliocene times (Bailey, E.H. and Everhart, 1964). Steep north-trending to northeast-trending fractures traverse the silica–carbonate rocks only, and they appear to represent tension fractures.

The mineralization is of simple mineralogy. Cinnabar is the only economic mineral, although native mercury impregnates and enriches the ore. Small contents of sulphides include pyrite, stibnite, chalcopyrite, sphalerite, galena and bornite. The gangue is of quartz and dolomite with some hyocarbons. Some of the mercury deposits have been eroded to give alluvial placer deposits, which have been mined (*Figure 5.7*).

The deposits resulted from replacement of the silica–carbonate rocks along steep northeast-trending fractures and also fracture fillings. The replacements extended only a few centimetres outwards from the fractures, but was so complete that the rock contains more than 50% cinnabar. The ore bodies that have been mined were large and exceptionally rich. The largest was 60 m wide and 4.8 m thick and extended down-dip for 470 m. Ore bodies have been found from near-surface to as deep as 600 m. During the first 15 years of mining, the ore contained more than 20% mercury, but the grade has now declined to 0.5%.

The ore was thought to have been deposited during the Pliocene epoch by low-temperature (50–130°C) ground-water solutions rising along steep faults and shear zones along the contacts between the serpentinite and the Franciscan melange. The richest deposits formed along gently dipping contacts, where the ore solutions were trapped. Although most of the deposits are on top of a serpentinite mass, some also occupy the under surfaces. The mining district is now largely abandoned, but is considered still to have a future potential.

Bailey, E.H. *et al.* (1973) have pointed out that the main mercury belt of the USA trends through the California Coast Ranges and has yielded 85% of the country's production. Virtually all the productive deposits lies close to or generally just below the Coast Range Thrust, which represents a Late Mesozoic subduction zone. This major structure provides an exceptionally fine guide to where to look for mercury ore deposits. Guild (1978) has echoed the same sentiments, pointing out that mercury is principally concentrated in or near young structures more or less parallel to the continental border. Most of the deposits are near transcurrent faults and there is no obvious relationship to volcanism.

It is noteworthy that the large mercury mines of Palawan Island, Philippines, lie along the faulted contact between an ophiolite melange and a pre-Mesozoic continental microcontinent (Hutchison and Taylor, 1978). This mining region accounts for 1.4% of world production. Cinnabar is the chief ore mineral. Common gangue minerals are quartz, chalcedony, opalite, calcite, pyrite and marcasite. Yellow ochre results from cinnabar weathering.

Fissure veins, breccia fillings, replacements, cavity fillings and disseminations represent the ore textures. Free metallic mercury occurs as globules in the open fractures. It is clear that this deposit is of very low-temperature deposition from hot springs in a subaerial environment, and the age of deposition is Pleistocene to Recent (Fernandez, H.E., 1968).

6

Intrusive bodies emplaced in stable cratonic terrain

Under this heading are included some spectacular and economically extremely important, though rare, deposits. There is no limitation on the timing of the intrusions; they may be Precambrian or Phanerozoic. The only common feature is that the terrain into which they have been intruded is characterized by long geological stability. Such intrusions are absent from oceanic areas and restricted to continental shield areas.

They may be subdivided into: deposits in large-layered complexes, which may be further subdivided into sheet-like intrusions either without repetitive layering, for example, the Sudbury Basin, or with repetitive layering, for example, the Bushveld complex, and dyke-like intrusions, for example, the Great Dyke of Zimbabwe; and deposits in alkaline complexes including carbonatites and kimberlites, for example, Khibina.

The Sudbury Basin

This unique funnel-shaped nickel-bearing basic lopolith, composed of norite, micropegmatite and a so-called sublayer, has been dated at between 1680 and 2000 Ma, and it is the greatest repository of commercial nickel in the world. The lopolith outcrops as an elliptical ring of 1.5–6.5 km width, with long axis 59.5 km and short axis 27.5 km. It consists of concentric micropegmatite and norite members and discontinuous marginal facies, or sublayers that include quartz diorite dykes. The outer contact has a general inward dip (*Figure 6.1*). The copper–nickel–iron deposits occur around the periphery of the norite and in the dykes (Souch *et al.*, 1969).

The upper part of the funnel-shaped intrusion is of granophyre micropegmatite and plagioclase-rich rock (Peredery and Naldrett, 1975). The granophyre intrudes and engulfs the plagioclase-rich rock (Naldrett *et al.*, 1972), which is a phase of the upper part of the oxide-rich gabbro. The micropegmatite is thought to be a differentiate of the norite that collected at its top towards the centre of the lopolith structure.

The base of the norite consists of a thin layer of mafic norite that grades abruptly upwards into felsic norite, which is the principal rock-type, and referred to as the Nickel Irruptive. Felsic norite is in abrupt contact with the overlying gabbro phase, which is relatively rich in opaque minerals and apatite and forms the base of the zone of transition into the micropegmatite (Souch *et al.*, 1969). The average nickel content in the mafic norite is 170 ppm, in the felsic norite 60 ppm, but in the micropegmatite only 20 ppm.

The copper–nickel sulphide ores are associated with the discontinuous sublayer, which is a sulphide and inclusion-bearing silicate intrusive rock. Two facies can be distinguished in the sublayer (Naldrett *et al.*, 1972). They are: igneous-textured gabbro–norite and diorite; and metamorphic-textured felsic to mafic leucocratic breccia. Both facies contain iron–nickel–copper sulphides and xenoliths of the underlying footwall rocks and of exotic anorthosite–mafic–ultramafic cumulates (Pattison, 1979). The gabbro–norites show enrichment in quartz and alkali feldspar towards the footwall, while their pyroxenes exhibit iron enrichment. The sublayer is therefore upside-down in respect to the normal crystal settling differentiation model, suggesting that assimilation played a large part in its formation. Pattison (1979) holds that the gabbro–norites and the breccias are contemporaneous and both pre-date the main Sudbury Nickel Irruptive funnel-shaped intrusion. There is a strong spatial relationship between concentration of exotic xenoliths and concentration of iron–nickel–copper sulphides. A genetic link, though not proven, is very likely.

Disseminated sulphide in leucocratic breccia has

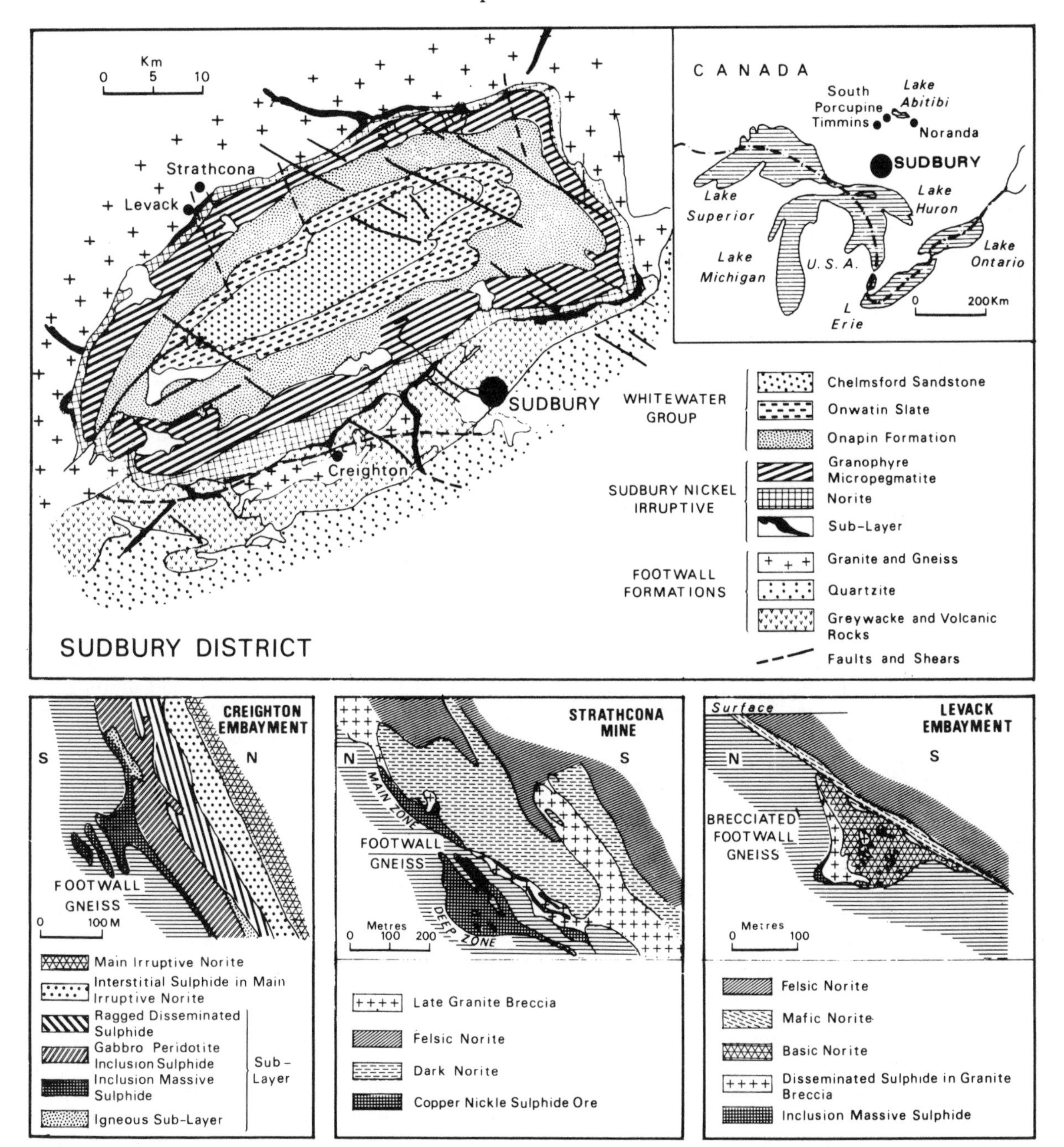

Figure 6.1. Simplified geological map of the Sudbury Basin, showing cross-sections of the Creighton mine of the South Range to illustrate the northwards-dipping footwall gneiss. The cross-section of the Strathcona and Levack mines of the North Range show the southwards dip of the footwall gneiss and the iron–nickel–copper ore in relation to the embayment structures (after Abel *et al.*, 1979; Pattison, 1979; Souch *et al.*, 1969).

a distinctive fragmental texture in contrast to the interstitial sulphide of the adjacent igneous sub-layer. The rounded-to-subangular sulphide fragments vary from 1 cm to 10 cm in diameter and there is no size sorting. This ore is clearly fragmental in character and smaller ore blebs of less than 1 cm may represent melted or abraded sulphide fragments. They occur in a matrix that is not brecciated and hence have been brought in from a pre-existing source (Pattison, 1979).

The vast bulk of the Sudbury ore consists of varying proportions of hexagonal and monoclinic pyrrhotite, pentlandite and chalcopyrite. Locally there are occurrences of pyrite, cubanite and millerite. There is a clear increase in Cu/(Cu + Ni) ratio towards the footwall and some iron concentrations show a marked increase of (Cu + Ni)/Fe ratio towards the footwall, especially where the ore projects into it. These zonations are interpreted as indicating subsolidus migration of mobile copper and nickel sulphides down a thermal gradient induced by the overlying norite magma lopolith (Naldrett and Kullerud, 1967). The overall Cu/(Cu + Ni) ratio, based on mine-production figures, averages around 0.5 for the whole of the Sudbury Basin. Individual mines, however, deviate widely from this mean, suggesting pre-emplacement differentiation or an inhomogeneous source for the ore. The host rocks are highly contaminated with xenoliths so that the original silicate magma, with which the sulphide was genetically related, is unknown. The ore often occurs in spatial association with embayment structures in the footwall (*Figure 6.1*). A complete section through an idealized sublayer occurrence would consist, from top to bottom, of: contaminated hybrid basal noritic Nickel Irruptive; igneous sublayer; leucocratic breccia; megabreccia; and brecciated Sudbury footwall formation.

The Sudbury Basin is unique and hence the hypothesis that it represents a hypervelocity asteroidal impact, with impact-induced triggering of the various intrusive events, is clearly acceptable. A spectacular cause is necessary for this unique deposit. The impact or astrobleme hypothesis is based on considerable evidence, as reviewed by Dence (1972), Dietz (1972) and French (1972). Although the structure may be impact-induced, there is no evidence that the ore is of cosmogenic origin. The greenstone belts of Ontario are well known for their copper, iron and nickel occurrences, and the most reasonable hypothesis is that the lithosphere beneath the Sudbury Basin is rich in these metals and the impact supplied the mechanism for their mobilization.

Two processes may explain the origin of the sublayer and its associated nickel ores (Pattison, 1979): segregation of a sulphide-triggered magma and its early intrusion along the contact between brecciated footwall rocks and the overlying Onaping Formation; or direct emplacement of sulphide-enriched impact-produced melt along the walls of the crater, the sulphides being produced from previous concentrations in basic magmatic country rocks, such as komatiites in greenstone belts. In both theories, the leucocratic breccias were formed by mechanical attrition of brecciated footwall rocks as the igneous sublayer was intruded.

Creighton Embayment

The Creighton Embayment (*Figure 6.1*) is a trough-like depression in the South Range footwall that extends downwards for at least 3 km. Mostly it is filled with sulphide and inclusion-bearing quartz-rich Irruptive norite. Like many embayment structures, it shows evidence of *in situ* gravity differentiation. Massive sulphides, with few xenoliths, occupy the embayment base and fill fractures in the footwall formation. They grade upwards by increase of inclusions and become known as gabbro–peridotite–inclusion sulphide and ragged disseminated sulphide ore (Souch *et al.*, 1969).

Strathcona Mine

Massive chalcopyrite and cubanite ore, with lesser pentlandite and pyrrhotite of grade 9.09% copper, 0.52% nickel and 34.3 g t^{-1} silver, occupies a system of fractures striking 68° and dipping 45° southwest in the footwall feldspathic gneiss (Abel *et al.*, 1979). The ore zone is 150–350 m beneath the Nickel Irruptive contact (*Figure 6.1*). The intrusion of a high-temperature copper-rich liquid, containing a crystallizing nickel-rich pyrrhotite phase, into a system of open fractures in a dilatant zone in the footwall is considered to be the most likely mode of formation.

Levack Embayment

The Levack Mine occupies the end of a large nose-like embayment of the sublayer rocks into the footwall gneiss (*Figure 6.1*). The sublayer consists of sulphide and inclusion-bearing norite, sulphide-bearing granite breccia and inclusion–massive sulphide ore. The sulphides in the basic norite occur as patches interstitial to the silicate minerals. The granite breccia ore occurs as blebs, pods and stringers and accounts for 65% of the ore mined (Souch *et al.*, 1969).

Bushveld Igneous Complex

Although there are several other layered igneous bodies—the Skaergaard of eastern Greenland,

Stillwater in Montana, the Cuillins of Skye—the Bushveld is unique in that it is the largest and the greatest repository of metals in the world. Its importance is illustrated by its 1975 production figures (Von Gruenewaldt, 1977): platinum group metals and gold 85 t, nickel 18 250 t, copper 10 800 t, chromite 2 075 378 t, vanadium pentoxide 19 002 t, magnetite 1 561 670 t, magnesite 27 000 t, fluorite 202 583 t, tin 5232 t and andalusite 77 149 t. The Bushveld complex is estimated to contain about 86% of world platinum-group metal reserves, 83% of the chromium, 64% of the vanadium and about 44% of the nickel (Von Gruenewaldt, 1979). The nickel ore is of very low grade compared with Sudbury, but the disseminated sulphide deposits can be economically exploited because of the associated platinum-group metals, gold and copper. The variety and amount of products in the Bushveld is accordingly staggering. The major mineral deposits occur over an area of about 30 000 km^2 in the Central Transvaal. The total estimated extent of the complex is 66 000 km^2, about 55% of which is covered by younger formations.

The voluminous literature on the complex has been summarized by Willemse (1969a), Hunter, D.R. (1976) and Von Gruenewaldt (1979). The vast resources are concentrated in several mineral-rich layers of which the Merensky Reef, the UG2 chromitite layer and the Platreef are the most important. Details of the famous Merensky Reef are given by Cousins (1969), Vermaak (1976) and Vermaak and Hendriks (1976).

All major basic plutons, like the Bushveld, show subhorizontal compositional layering over a wide range of scale, owing to a kind of magmatic 'sedimentation' or cumulate settling under gravity modified by convection and local currents within the magma chamber. The greater part of the pluton has differentiated and solidified from the base upwards, by sinking of early crystallized phases to the floor. By crystal cumulation, the solid floor progressively rises. Liquid remaining in the pore spaces between the cumulus crystals eventually solidifies as crystalline intercumulate material.

Visual conspicuous lithological layering is termed rhythmic, since it is characteristically repeated in vertical sections. There is also a continuous gradual change in the chemistry of the mineral phases from the base to the top of the pluton. This is referred to as cryptic layering. Cryptic layering is to be expected from the experimentally established order of liquidus temperature, a proof that the pluton progressively crystallized from the floor upwards.

Most writers today visualize emplacement of layered mafic plutons as a short-lived act of intrusion. The original magma is sealed within a chilled frozen skin, now represented by the chilled border-zone rocks, and slow cooling is ensured through slow loss of heat resulting from the blanketing of the overlying country rocks. The process of sedimentation of higher-temperature phases accumulating on the floor under gravity is interrupted by convection currents within the magma chamber. There are many enigmas, such as the great lateral persistence of, for example, the Merensky Reef, which implies a general absence of convection, but disturbances obviously did occur as shown by graded bedding, potholes in layers and boulder beds—in fact many features that are found in sedimentary rocks. The great similarity to sedimentary rocks forces one to bear in mind that these processes of 'sedimentation' took place at basaltic liquidus temperatures.

Tectonic Setting

The four mafic-to-ultramafic compartments that constitute the layered sequence of the Bushveld complex were emplaced 2095 Ma ago into the relatively undeformed sedimentary cover of the Kaapvaal Craton beneath the subaerially extruded Pretoria Series composed of leptite, andesite tuff, agglomerate and felsite (*Figure 6.2*). The loci of emplacement were controlled largely by the major structural features of the Kaapvaal Craton. The complex occurs at the intersection of the following prominent structural trends: Great Dyke of Zimbabwe, Pretoria–Zebediela anticline, Murchison and Amsterdam lineaments. The present-day configuration of the complex is controlled by a series of dome-like and basin-like features, which have characterized the Kaapvaal Craton since the Archaean era. These features developed in response to the interference of northnorthwest-oriented to eastnortheast-oriented anticlinal and synclinal warps (Von Gruenewaldt, 1979). For this reason, the theory that the complex is a product of meteorite impact (Hamilton, 1970) cannot be supported. The igneous layering and upwards differentiation trend, as shown by the cryptic layering, show no anomalies and, unlike the Sudbury Complex, there are no features that could have resulted from impact.

The Bushveld rocks will be briefly described under the following headings: the Transvaal System and its diabase sills; the layered sequence of mafic and ultramafic rocks; the epicrustal rocks; and the Bushveld Granite.

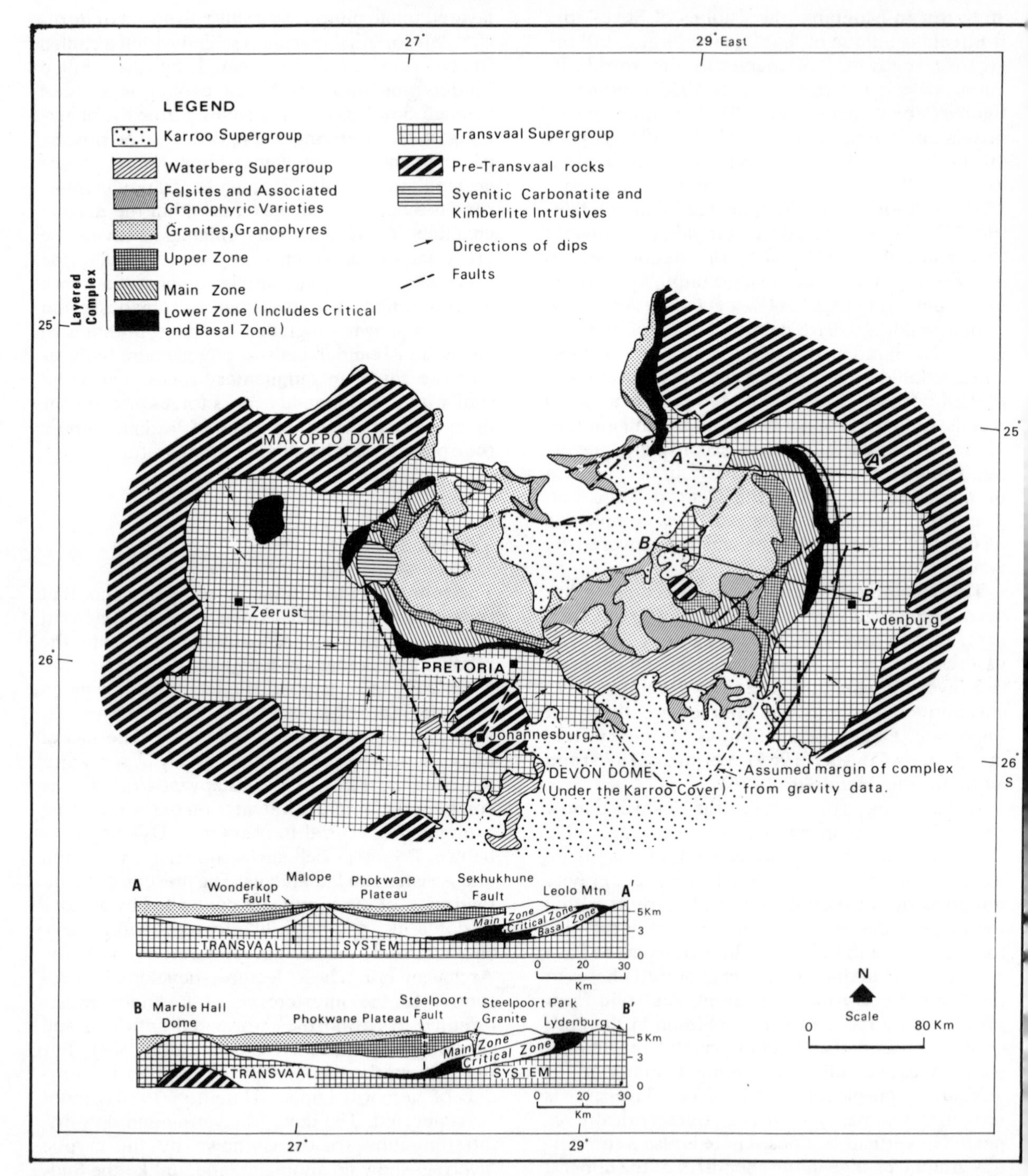

Figure 6.2. Simplified geological map of the Bushveld complex based on Hunter, D.R. (1976) and Vermaak (1976) with cross-sections drawn with the help of geophysical data (after Von Gruenewaldt, 1979).

The Transvaal System

The Transvaal System forms 85% of the country rocks of the complex. It is of shale, dolomite, andesite, quartzite, banded-iron formation, conglomerate and agglomerate, The andesite lava flows may be regarded as the precursors of the Bushveld-complex magmatism. Diabase sills extend into the Transvaal System. They comprise a range of plagioclase-rich and clinopyroxene-rich rocks, frequently altered to hornblende, chlorite and epidote assemblages. However, these sills may actually represent lava flows, but their details are poorly known.

FLUORITE

Most of South Africa's fluorite resources (116×10^6 t of ore with a calcium fluoride content greater than 25%) are found in the Bushveld complex; they are associated with the Bushveld Granite. Workable concentrations occur within the granite as pipe-like bodies, and vein-like and brecciated bodies in the leptite and felsite.

The fluorite at the Zeerust area is not related to the granite, but to the basement dolomite (Martini, 1976). The Upper part of the Precambrian Dolomite Series is mineralized with fluorite and minor lead–zinc. The deposits show similarities to the Mississippi Valley-type but have been subsequently metamorphosed in the contact aureole of the Bushveld complex. The locations are mostly controlled by the palaeo-porosity of the dolomite such as vuggy horizons and palaeo-karst (*Figure 6.3*). The fluorite district is localized on a palaeo-relief, which acted as an ore trap. The most attractive hypothesis for the fluorite formation, which is one of the largest of the world, is precipitation from solutions that derived fluorine from the Dolomite Series and the Pretoria Series during diagenesis. Nevertheless the geographic coincidence of this major stratiform fluorite district with

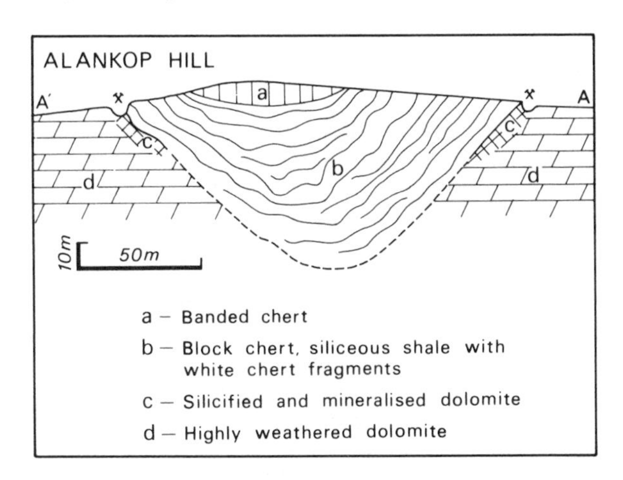

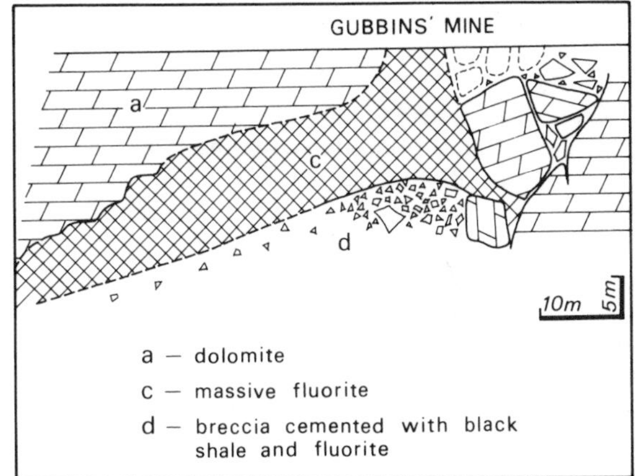

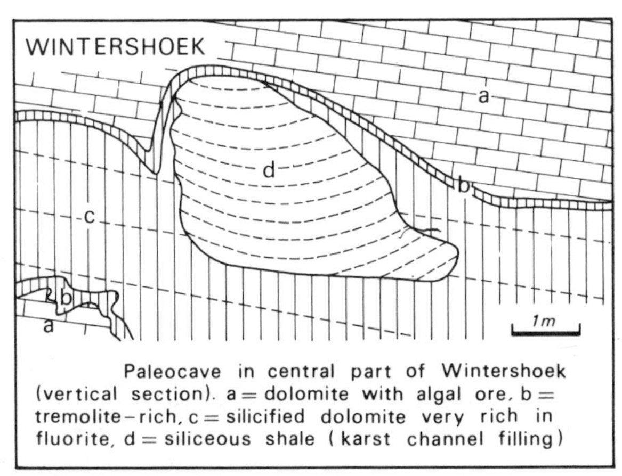

Paleocave in central part of Wintershoek (vertical section). a = dolomite with algal ore, b = tremolite-rich, c = silicified dolomite very rich in fluorite, d = siliceous shale (karst channel filling)

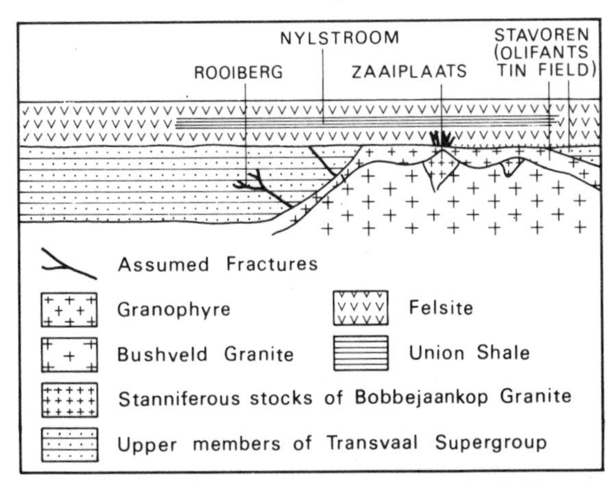

Figure 6.3. Examples of fluorite mineralization of the Marico district, South Africa (after Martini, 1976) and schematic representation of main locales for tin mineralization in the Bushveld epicrustal rocks (after Hunter, D.R., 1976).

the subsequent Bushveld intrusion and its rich mineralization must lead to speculation of a genetic connection. Perhaps the fluorine had a mantle source, coming up into the Transvaal rocks along major structural dislocations that subsequently would allow the mantle-derived Bushveld magma to gain similar access to the crust.

Total reserves of mineable fluorite ore, which lie under thin cover, with an average grade of 15% calcium fluoride, are 100–150 million t. The ore is of several types: *algal ore*, which occurs as irregular nodular masses and less commonly as laminae of fluorite in stromatolites; *blackspar*, which is finely crystallized black fluorite replacing dolomite (the black colour is due to carbon inclusions inherited from the dolomite); *banded spar*, which is irregularly banded, possibly deposited by the filling of voids in cavernous dolomite, although the ore bodies of this type are small; and *breccia spar*, which forms a cement of breccia bodies in the dolomite, forming pipe-like bodies, and is locally associated with lead–zinc mineralization. Fluorite is also associated with black chert around palaeosink holes now filled with impure black chert (*Figure 6.3*).

Other substantial deposits within the Bushveld complex, but not related to it in time, are related to the younger alkaline intrusions. There are three major deposits, one of which is unfortunately rich in apatite, which cannot yet be successfully separated commercially. These deposits give a further 10×10^6 t of fluorite reserves. The initial concentration was in the Precambrian Dolomite Series, and subsequent igneous events have remobilized it to form economic deposits of varying ages in the same general region.

All fluorite deposits in the Transvaal are confined within long discrete linear zones up to 30 km wide and 650 km long. These zones are aligned along the eastnortheast (Murchison) and north-northwest (Franspoort) trends, which are fundamental to the Kaapvaal Craton and have controlled the emplacement of the Bushveld complex. They have also clearly controlled the introduction of fluorine into the Dolomite Series and into the younger igneous centres. Many of the fluorite deposits are related to some of the numerous alkaline intrusions and alkalic complexes that are confined to these fundamental tectonic zones (Wilson, 1979). Within the individual fluorite-enriched zones, the commercial deposits have greatly differing ages. Taken as a whole, the ages range from about 2100 Ma to less than 125 Ma and the present-day ground water is fluorine-enriched. Fluorite is now being precipitated in some localities from warm springs.

The Layered Sequence

The layered sequence comprises a thickness of about 7620 m of mafic and ultramafic rocks. The major subdivisions are based on persistent marker horizons that are easily mapped (*Figure 6.4*). They are: the main Chromitite Seam, the Merensky Reef and the main Magnetite Seam. The Layered Sequence is subdivided into five zones, described from the base upwards.

THE CHILL ZONE (MARULENG NORITE)

These norites intrude transgressively into the Pretoria Series of the Transvaal Group. Its relationship to the overlying Basal-Zone ultramafic rocks is not clear, and in some places it may be missing. The Chill-Zone norite cannot be considered to represent the original undifferentiated magma of the Bushveld complex, because it is highly contaminated with sedimentary rocks, notably quartzite.

THE BASAL ZONE

This is the lowest unit of the complex. However, it does not always form the base, for the Main Zone abuts for long distances against the country rocks. The zone displays marked layering composed of thick units. The Hendriksplaats Norite has a strike length of more than 80 km; it contains numerous sedimentary-rock inclusions that yield interesting metamorphic assemblages. It is an anomalous unit, and is regarded as uplifted Maruleng Norite because it occurs only where the Maruleng Norite is missing from the sequence and it is underlain by quartzite.

In western Transvaal the Basal Zone forms a basin structure in which the other overlying zones are absent. The rocks are of pyroxenite and harzburgite containing chromitite seams, richer in aluminium and poorer in iron than in the higher stratiform series.

THE CRITICAL ZONE

This zone is layered on a grand scale and ranges from banded norite to variable successions of pyroxenite, norite, anorthosite and chromitite. Most of the chromitite seams occur in this zone, and the upper marker is the Merensky Reef. Chromite is a cumulus phase within the anorthosite suite of the zone. The zone is terminated by the two most complete macrocyclic units, namely the Merensky Reef and the overlying Bastard Reef, after whose deposition both chromite and olivine

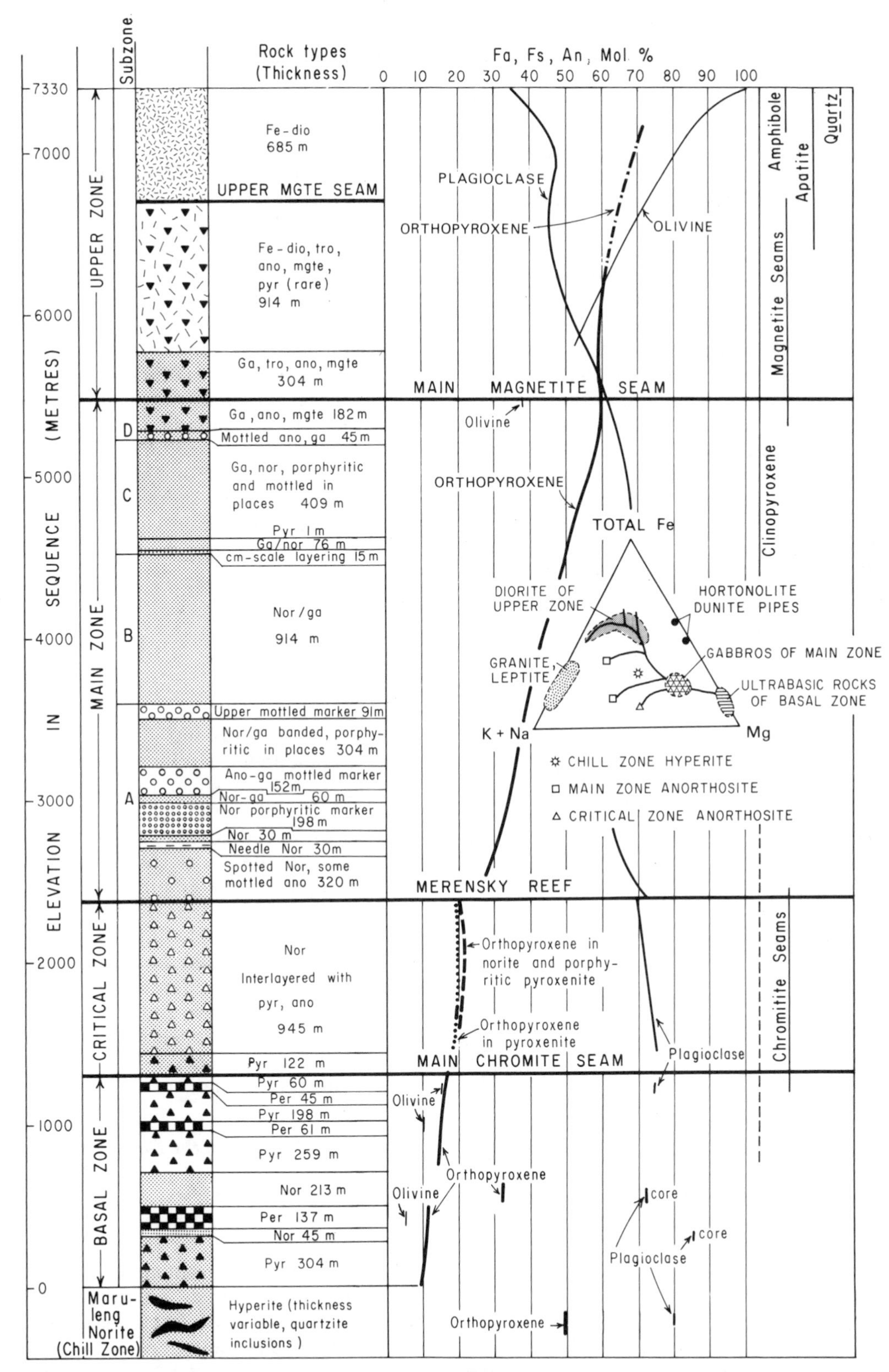

Figure 6.4. Igneous stratigraphy for the eastern-Transvaal section of the Bushveld complex, showing also the mineral trends in the cryptic layering and the differentiation trends as shown by an AFM diagram (after Willemse, 1969a). An, anorthite; ano, anorthosite; dio, diorite; fe-dio, ferrodiorite; fa, fayalite; fs, ferrosilite; ga, gabbro; mgte, magnetite; nor, norite; per, peridotite; pyr, pyroxenite; and tro, troctolite.

disappear as cumulus phases. *The Merensky Reef* is probably the most persistent magmatic–sedimentary layer of all layered complexes in the world. It is unique and represents a complete mineral grading of the maximum number of cumulus phases—chromite, locally olivine, orthopyroxene with some clinopyroxene and plagioclase. The apparently isolated magma segment, from which the reef differentiated, must have been reasonably close to a basaltic composition. Although its layering is uniform on a macroscale, individual layers in the noritic range are sometimes cyclic from melanorite to anorthosite.

The persistent lower-chromite seam mostly rests on a dimpled footwall surface of pure white anorthosite a few centimetres thick. It, however, disappears where the reef is disturbed by potholes and other irregularities. A noteworthy feature is the presence of melanocratic boulders within the underlying anorthosite, forming a marker horizon 28.5 m below the reef. It has been deduced that these boulders sank through the magma and disrupted the earlier-forming chromite cumulus layers to form the basal pegmatoid reef.

THE MAIN ZONE

The Main Zone has a great thickness of gabbroic rocks, noritic towards the base and gabbroic higher up. In addition to the obvious layering, cryptic layering is shown by the upward progressive change in orthopyroxene from En_{75} near the base to En_{40} at the top (*Figure 6.4*).

THE UPPER ZONE

The Upper Zone is well known for rock-type alterations. It has 20 seams of magnetite ore as well as anorthosite and troctolite layers. Cryptic layering is spectacular, and the olivine grades upwards from Fa_{54} to Fa_{100}, while plagioclase becomes sodic upwards, ranging from An_{60} at the base to An_{34} at the top. The strong iron enrichment of the Upper Zone is shown by the diorites in the AFM ($A = Na_2O + K_2O$; F = total iron as FeO; M = MgO) diagram (*Figure 6.4*).

PEGMATITE PHASES

Ultramafic pegmatoids are prominent in the Layered Sequence. The following are the different types. (1) Bronzite pegmatoids form pipe-like bodies in the Basal Zone; they contain nickeliferous sulphides. (2) Hortonolite dunite and dunite are closely associated with platiniferous pipes. Dunite forms the outer zone and hortonolite dunite the core of the pipes, but some are purely of dunite. The three platiniferous pipes are called Onverwacht, Mooihoek and Driekop. They occur in the Critical Zone. (3) Diallagite pegmatoid forms irregular masses, anastomosing veins and pipe-like bodies. They transgress the igneous layering and are quite extensive, occurring in the Critical Zone, Main Zone and Upper Zone. (4) Magnetite pegmatoid—by a greater concentration of magnetite, the diallagite pegmatoid passes into massive magnetite iron ore. (5) Vermiculite pegmatoid forms pipe-like bodies of pure vermiculite.

The initial Bushveld magma is deduced to have been close to a magnesium-rich tholeiitic basalt. This conclusion is based on the dominance of the complex by the Upper and Main Zones, relative to the ultramafic rocks of the Critical and Lower Zones. The rocks of the Upper and Main Zones contain average magnesium oxide values typical for normal basalt.

Epicrustal Rocks

Extensive areas in the central portion of the complex are occupied by the Bushveld Granite, which was subconcordantly intruded 1950 Ma ago into felsite, granophyre, leptite and microgranite as a crudely stratiform sheet roughly 2800 m thick. The granite does not normally transgress the layered rocks and is separated from them by leptite, microgranite, partly feldspathized quartzite and granophyre. The granite is therefore also essentially basin-shaped in form and its roof is mostly of felsite and granophyre.

The Bushveld Granite occupies too large a volume to represent a differentiated product of the mafic Bushveld magma, and one is forced to think of it as a product of anatexis of the sedimentary formations of the Transvaal System and older underlying formations.

Economic Deposits of the Bushveld Complex

PLATINUM-GROUP METALS AND ASSOCIATED GOLD, NICKEL AND COPPER

By far the most important mineral deposits are the sulphide ores. Three extensive deposits are known—the Merensky Reef, the UG2 Chromitite Layer and the Platreef.

Merensky Reef

This layer has a thickness variable from a few metres to 26 m. On average it consists of a cumulate orthopyroxene–chromite pegmatoid containing post-cumulus feldspar and clinopyroxene, with local olivine-rich and quartz-rich facies. The average thickness mined is 0.8 m and its average dip is 12°. The geothermal gradient of 20°C km^{-1} restricts mining to a maximum depth of 1200 m below surface. The base metals are present as pyrite, pyrrhotite, pentlandite and chalcopyrite.

The precious metals are associated with the base-metal sulphides. Idiomorphic braggite, cooperite and laurite, with minor sperrylite, are the main platinoid minerals, Platinum–iron alloy occurs commonly intergrown with other base-metal sulphides. A significant amount of platinoids occurs isomorphously within the pyrrhotite, pentlandite and pyrite. The precious metals have a clear preference for occurrence at the contact of the base-metal sulphides with the silicate–oxide gangue (Vermaak and Hendriks, 1976).

UG2 Chromitite Layer

This layer occurs 150–300 m below the Merensky Reef. Its thickness varies from 90 cm to 150 cm. Many of the chromitite layers, especially those higher in the sequence, contain small quantities of sulphides associated with platinum-group minerals. The highest concentration occurs in the UG2 layer. Eighty percent of the platinoid minerals occur along the grain boundaries or in association with interstitial sulphides. The UG2 layer is comparatively sulphur-poor and hence its nickel and copper contents are less than those of the Merensky Reef and Platreef. The platinoid elements content of the UG2 layer is almost twice that of the Merensky Reef, but its chromite content is of low quality.

The Platreef

The Platreef is of harzburgite at the base followed by a pegmatitic feldspathic pyroxenite, overlain by a porphyritic pyroxenite. It is up to 200 m thickness and is richly mineralized over thicknesses between 6 m and 45 m. It is proposed that the sulphide mineralization may be related to the transgression of the Layered Sequence of the complex across a floor of dolomite. Sedimentary sulphur from the dolomite was incorporated in the magma to assist in the separation of an immiscible sulphide liquid. The mineralization is of variable intensity. Some details of the reefs are given in Table 6.1. The grains of platinum-group minerals in the Merensky Reef and the UG2 Chromitite Layer are so small that they could not have settled sufficiently rapidly to become concentrated together with the chromite crystals, which are far larger. In the Merensky Reef they could have settled in the form of composites of sulphide liquid and platinum-group minerals (Hiemstra, 1979). However in the UG2 layer, even the sulphide particles are far too small to have settled. It is suggested that the sulphide grains adhered to settling chromite crystals and thus were carried down with them to form cumulates.

CHROMITE

The commercial chromite is confined to the Critical Zone. It occurs as cumulate layers varying in thickness from a few centimetres to more than 2 m. Details are given by Cameron, E.N. and Desborough (1969). The chrome-ore resources are calculated to be 2300 × 10^{6} t to a vertical mining depth of only 300 m. The resources can be multi-

Table 6.1 Platinoids, nickel and copper contents (% by weight of precious metals in the ore, unless otherwise stated) of the platinoid ores of the Bushveld and other complexes (from Cousins and Vermaak, 1976; Naldrett, 1973)

Element	*Merensky Reef*	*UG2 Chromitite*	*Platreef*	*Great Dyke*	*Sudbury (average)*
Pt	59	41	42	57	40
Pd	25	34	46	29	44
Ru	8	12	4	5.6	5
Rh	3	9	3	4.3	4
Ir	1.0	1.9	0.8	—	1.2
Os	0.8	1.7	0.6	0.6	1.0
Au	3.2	1.3	3.4	7.8	4.8
Platinoids + Au resources (kg × 10^{6})	18.15	32.50	12.24		
Ni (weight% in ore)	0.18	Variable	Variable	—	1.5
Ni resources (t × 10^{6})	4.3	3.8	14.7	—	9.7
Cu (weight% in ore)	0.11	Low	Variable	—	1.3
Cu resources (t × 10^{6})	2.65	—	7.30	—	8.53

plied by a factor of 10 if deeper mining were contemplated. The Critical Zone is divisible into chromitic and non-chromitic intervals. Layers of chromitites occur within the former. These layers have chromite as the sole cumulus phase. Most chromitic intervals consist of chromitic pyroxenite or anorthosite, or both, but two consist mainly of chromitic dunite and harzburgite. The layering in the Critical Zone, both visible and cryptic, is grossly a result of progressive magmatic differentiation, as shown by the upwards change from predominant pyroxenite to predominant norite and anorthosite, and the upwards increase in the chromite of total iron, iron/magnesium, and iron/chromium, increase of iron/magnesium in the orthopyroxenes and Ab/An in the plagioclase. Departures from this trend as well as major chemical and petrological disconformities indicate a complex igneous–sedimentary history.

It is deduced that an increase of the oxygen fugacity is responsible for the formation of massive chromitite layers in the Bushveld complex (Snethlage and Klemm, 1978). The chromitites are a result of cumulate magmatic sedimentation. Currents within the magma are considered to have had an important role in the process of repetitions, but the lateral persistence of chromitite layers for distances up to 65 km is hard to reconcile with active magma current action.

Vanadiferous Magnetite

Vanadiferous magnetite occurs as plug-like bodies and as cumulate layers. The former are pegmatitic, composed of coarse-grained diallage, olivine and amphibole, which dilute the ore below commercial grade. Only the Kennedy's Vale plug-like mass has been mined; it contains about 2% vanadium pentoxide and has yielded 1×10^6 t of ore.

The real magnetite layers are encountered in the uppermost 300 m of the Main Zone and in the lower 1200 m of the Upper Zone. Twenty-six seams have been identified. The uppermost is about 9 m thick and is composed of disseminated magnetite in anorthosite. The main seam, 1220 m lower down, is 1.4–2.6 m thick massive magnetite, exposed for a strike length of 120 km. It contains, on average, 1.6% vanadium pentoxide. The ore minerals are magnetite, ilmenite, ulvite, maghemite and martite (Willemse, 1969b). The vanadium obviously takes the place of titanium in the minerals. The lowest layer contains 2% vanadium pentoxide and 14% titanium dioxide, and the uppermost layer about 0.3% vanadium pentoxide and 18–20% titanium dioxide. The vanadium pentoxide contents of the magnetite plugs seem to correspond approximately to that of the cumulate layers in the vicinity.

At present only the main magnetite layer, 1.8 m thick, is mined. The vanadium pentoxide resources of this layer are estimated to be 17×10^9 kg in 2×10^9 t of ore.

Tin

Six tin fields occur within the granitic hub of the Bushveld complex (*Figure 6.3*). Thirty-six percent of the production has been from ore bodies in the granite, 56% from exogranite deposits in sedimentary rocks of the Transvaal Supergroup and 8% from deposits in felsite (Hunter, D.R., 1973). The cassiterite deposits occur as endogranitic pipes and as primary disseminations, or as exogranite fissure veins, fault breccias or replacement bodies (Wilson, 1979). The endogranitic deposits are located in the roof of stocks of miarolitic granites that intrude the stratiform Bushveld Granite. The stocks are depleted in barium, strontium and europium and enriched in rubidium and caesium. Pipe-like ore bodies are variable in shape, roughly cylindrical and up to 12 m in diameter. A typical pipe has a core of sericitized and chloritized granite carrying disseminated cassiterite and variable amounts of pyrite, fluorite and scheelite, surrounded by shells of red feldspar, tourmaline, white quartz and red silicified granite.

The granites are presumed to have resulted from anatexis of the basement rocks of the Bushveld complex that occurred 1950 Ma ago. The fluorine, which is greatly enriched in these rocks (*see* 'Fluorite' section, page 107), behaved as a flux as the mass of the Bushveld complex caused the basin to depress, resulting in basement anatexis, fluxed by the fluorine (Wilson, 1979). The fluxed magmas were highly volatile and became enriched in tin and other trace metals by volatile stripping, or volatile scavenging. This hypothesis explains the undeniable coincidence of the late-stage tin-containing granites with the fluorite-enriched linear tectonic zones and also indicates that the relationship is genetic.

The cassiterite deposits are therefore related to late-stage granite intrusions derived by anatexis brought about by the selective fluorine-fluxing of sialic crustal rocks in fluorine-rich zones within ancient long-lived linear tectonic zones in the Kaapvaal Craton. These ancestral zones are seen on the surface by the linear distribution of fluorite deposits, by the high fluorine content of the ground water and by the young alkaline intrusions.

The six tin fields have produced 90 000 t of tin, all from underground mining.

ANDALUSITE

Andalusite deposits in the metamorphic aureole of the Bushveld complex are mined along its northeast and southwest margins. Large crystals are developed in hornfels. Resources of high-grade andalusite (more than 55% aluminium oxide (Al_2O_3) and less than 1.5% total iron as Fe_2O_3) are estimated to be around 10×10^6 t. Extensive andalusite deposits occur in soils overlying the hornfelsed shale, as well as in alluvial sands in the rivers that drain the metamorphic aureole terrain.

MAGNESITE

Weathering of the ultrabasic igneous rocks of the complex has given several localities of magnesite. Mining operations have allowed the resources to be estimated at about 10×10^6 t (Von Gruenewaldt, 1977).

Great Dyke of Zimbabwe

The Great Dyke is a north-trending linear body 480 km long with an average width of 5.8 km (*Figure 6.5*). The intrusion has been dated to be at least 2530 Ma old. Reviews of the literature are to be found in Bichan (1970) and descriptions of the chromite deposits in Worst (1964) and Bichan (1969). For most of its length it was intruded into a basement complex of Archaean granites. The Dyke consists of four layered lopolithic complexes each of which has a cyclic sequence of ultramafic rocks overlain by a gabbro capping. The major one is the Hartley complex, near Salisbury. The structure of each complex is broadly gentle synclinal in which the layering dips inwards towards the dyke centre at angles as steep as 25° near the outer contacts to as shallow as 5° in the central areas. The dyke itself has steep contacts with the country rocks. Rocks in the immediate vicinity of the contact zone have been sheared and desilicified. Xenoliths of country rocks are very common in the gabbro. There is a complete absence of a marginal chill zone.

A typical vertical sequence from the top downwards is (*Figure 6.5*): gabbroic rocks; 143 m of pyroxenite with olivine in the basal part; 46 m of picrite and harzburgite; chromite seam no. 1 of 25 cm thickness; 24 m of harzburgite; chromite seam no. 2 of 35 cm thickness; underlain by pyroxenite (Bichan, 1969).

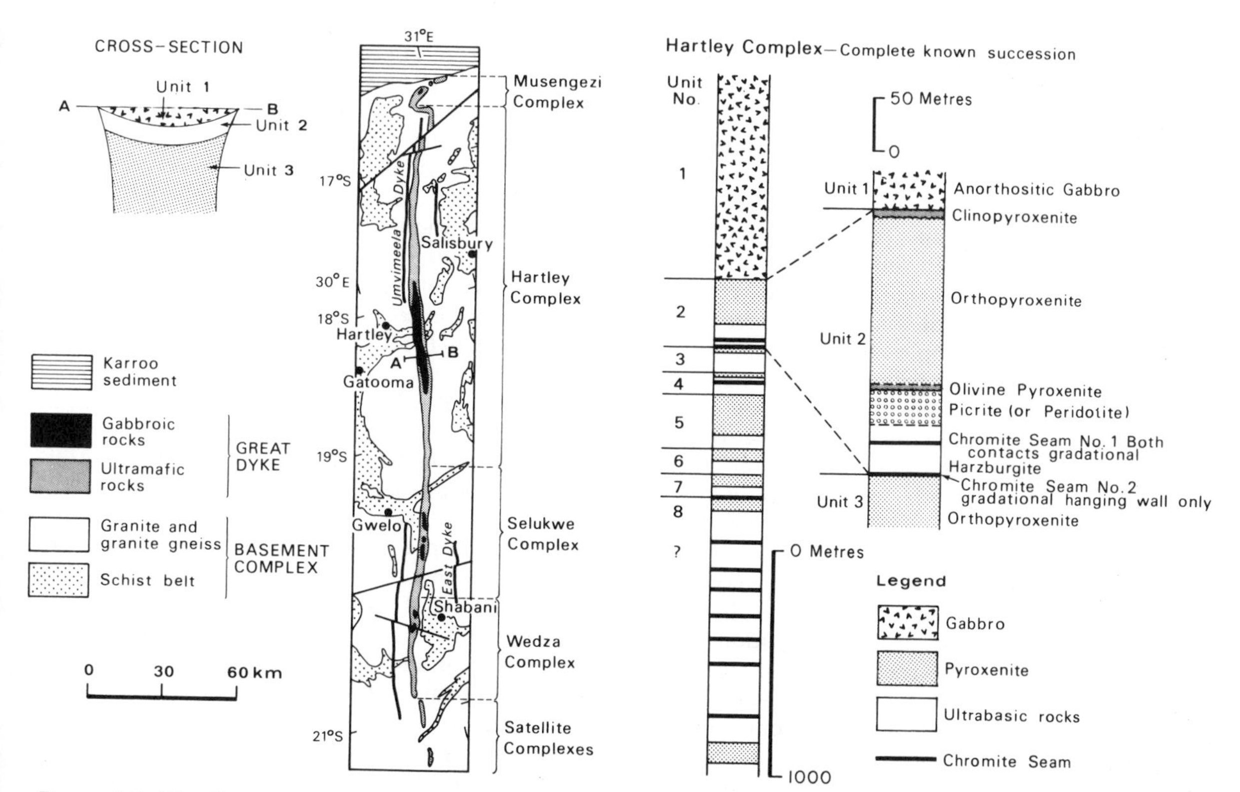

Figure 6.5. The Great Dyke of Zimbabwe with details of igneous layering and chromite in its Hartley complex (after Bichan, 1970).

The resources of the Dyke computed to a depth of 150 m are 190 million t of metallurgical chromium ore (with 48% Cr_2O_3, chromium/iron 2.8) and 350 million t of chemical/refractory ore (49% Cr_2O_3, chromium/iron 2.3). This makes it one of the largest repositories of chromium in the world. In the north, fine-grained eluvial chromite is mined. The reserve of eluvial ore is estimated to be about 60 million t (De Kun, 1965).

It is suggested that the Dyke was formed by pulsatory injections of mantle-derived magma of a tholeiitic affinity as the Archaean crust fractured. The pulses followed each other rapidly and the chromite layers (*Figure 6.5*) are a result of cumulate settling.

Alkaline igneous rocks

The alkaline igneous rocks are insignificant with regard to their volume. However, they continue to intrigue geologists because of their rarity and unusual geochemistry and mineralogy, and therefore they may be important economically. The literature of alkaline rocks far outweighs their importance as crustal-formers. Rock nomenclature has been uncontrolled and many rock names have been coined from individual localities. The complicated nomenclature will deter anyone not familiar with these rocks, therefore a glossary is presented in *Table 6.2* of the names used in this section.

There are strong affinities between alkaline ultrabasic rocks, carbonatites and kimberlites. Some of the links are: the alkalic ultrabasic rocks of ring complexes occur in the same tectonic environment as kimberlites, and they intrude stable continental cratons; the ring complexes often intrude carbonatite intrusions; carbonatites are sometimes associated with central-complex kimberlites; many kimberlites contain much primary carbonate; with only a few exceptions, the $^{87}Sr/^{86}Sr$ ratios of the alkaline rocks vary within the range 0.704–0.710, values lower than the average crustal rocks through which they intrude, so it is therefore concluded that most alkaline rocks could not have formed by anatexis of crustal rocks (Powell, J.L. and Bell, 1974)—the alkaline magmas may have differentiated from mantle-derived basaltic magma, but in many cases contamination from crustal and other mantle rocks is certain; and the intrusions are commonly high-level volcanogenic phenomena, including diatremes, cone sheets, ring-dykes and pipes.

The principal rock-types developed are carbonatites, nephelinites, ijolites, pyroxenites and fenites. Fenites are the hallmark of the alkaline rock complexes. Kimberlites also occur in the carbonatite nephelinite association but they are more rare. Igneous complexes displaying carbonatite and nephelinite rocks are present in all the continental cratonic areas of the world. The typical association is absent from oceanic regions. Alkaline rocks have been well documented by LeBas (1977), Sørensen (1974), Heinrich (1966) and Tuttle and Gittins (1966).

Apatite Deposits of Igneous Origin

Phosphate deposits of igneous origin are of great importance. In 1977 they contributed 19 million t of apatite concentrate (18% of the world phosphate-rock supply). Mining centres exist: in South America, at Araxa and Jacupiranga; in Africa at Sukulu, Dorowa, Glenover and Palabora; and in Europe, in the Kola Peninsula at Khibina and Kovdor in the USSR and Sokli in Finland (Notholt, 1979).

The apatite in the alkaline igneous complexes is clearly of igneous origin, but an apocryphal theory of their formation is given by Mannheim (1979), called the 'coprogenic impact theory', whereby an artificial satellite composed of organic wastes, from an advanced civilization belonging to an unknown planet, escaped from its orbit and eventually impacted with the earth on the Kola Peninsula.

In the Kola Peninsula of the USSR (*Figure 6.6*) alkaline central-ring complexes of different ages and composition intrude Archaean gneisses and migmatites close to local fracture zones in the Baltic Shield (Gerasimovsky *et al.*, 1974). The alkaline intrusions vary in age from middle Proterozoic, through Caledonian to Hercynian. The latter Palaeozoic episode includes the world's two largest alkaline complexes of Khibina (1327 km^2) and Lovozera (650 km^2).

KHIBINA, KOLA PENINSULA

The Khibina ring complex, 16 km south of Murmansk, was intruded 290 Ma ago into Archaean gneiss. The complex has a diameter of 40 km and is polyphase of the central type (*Figure 6.6*). There are steep contacts with the country rocks. The contact rim is of metasomatic fenite and albitite. The first intrusive phase was of nepheline and alkali syenite and nepheline syenite porphyry in the form of small bodies in the peripheral zone of the complex. Phase 2 was of massive khibinite (*Table 6.2*), phase 3 trachytoid khibinite and phase 4 rischorrite, which forms a complex ring structure. Ijolite, melteigite and urtite are the members of phase 5 and they form the well-known strati-

Table 6.2 Glossary of alkaline igneous rock names

Albitite: albite syenite, often aplitic
Alkali pyroxenite: pyroxenite containing sodic pyroxenes or minor feldspathoids
Alkali syenite: syenite containing sodic pyroxenes and/or amphibole and/or feldspathoid ($<$ 10% of the total felsic minerals)
Alnöite: alkali lamprophyre containing melilite, biotite, olivine, titanaugite, nepheline, perovskite, garnet and calcite
Carbonatite: carbonate-rich ($>$ 50% carbonate) rock of apparent magmatic derivation or descent
Fenite: alkali syenite formed by contact metasomatism around alkaline intrusions (process is called *fenitization*)
Foyaite: nepheline syenite, characterized by a foyaitic texture, that is, intergranular texture, because of arrangement of plates and tables of microperthite
Ijolite: nepheline–pyroxene rock, mesocratic, feldspar-free
Jacupirangite: alkali pyroxenite containing titanaugite, titanomagnetite, nepheline, apatite, perovskite and melanite
Khibinite: nepheline syenite, leucocratic with granular or trachytoid textures, eudialyte-bearing
Kimberlite: serpentinized and carbonated mica–peridotite characterized by high-pressure minerals often occurring as diatremes
Lamprophyre: ultrabasic and basic dyke rocks rich in alkalis, iron and magnesium; they contain biotite, and/or barkevikite, with augite, olivine, and rarely melilite; fledspars are plagioclase or orthoclase; some are chemically similar to nephelinite
Melteigite: melanocratic rock composed mainly of alkali pyroxene and nepheline
Nepheline syenite: group name for feldspathoidal syenite with nepheline as the main feldspathoid
Nephelinite: volcanic rock composed mainly of nepheline and augite; devoid of feldspar and olivine
Okaite: hauyne-bearing intrusive rock composed mainly of melilite and augite; feldspar-free and olivine-free
Peridotite: ultrabasic rock consisting mainly of olivine and pyroxene; feldspar typically is absent
Pyroxenite: ultrabasic rock composed essentially of enstatite or augite or both
Rischorrite: nepheline syenite having nepheline crystals poikilitically enclosed in microcline perthite; also contains biotite, aegirine, augite, etc.
Shonkinite: melanocratic rock composed of alkali feldspar, feldspathoids and more that 60% mafic minerals (pyroxene, olivine, biotite)
Sövite: biotite–calcite carbonatite
Syenite: alkali feldspar rock, often with minor oligoclase, sometimes quartz or nepheline-bearing; leucocratic with biotite, hornblende and augite; it is called alkali syenite when it contains sodic minerals
Trachytoid texture: texture characterized by large phenocrysts of alkali feldspar that exhibit flow alignment
Urtite: leucocratic volcanic rock containing mainly nepheline ($<$ 70%) and augite; devoid of feldspar and olivine; differs from ijolite in its smaller amount of mafic minerals

form complex that contains the world's largest apatite–nepheline ore deposit (Gerasimovsky *et al.*, 1974). Phase 6 is of medium-grained aegirine–nepheline syenite and phase 7 of foyaites. The final stages are of numerous alkalic pegmatites and lamprophyre dykes. The phase-5 rocks contain an annular zone of apatite–nepheline rock sandwiched between a lower zone of urtite and an upper zone of poikilitic nepheline syenite (richorrite). The ore body is up to 200 m thick and several kilometres long, dipping 30–45° towards the centre of the complex. It has an arcuate outcrop extending 11 km along the surface (Notholt, 1979). Its depth has been proved to hundreds of metres. The apatite–nepheline ore contains 40–90% apatite (20–45% phosphorus pentoxide P_2O_5). The apatite reserve equals 2000 million t and the annual production is 10–18 million t. It is also economic to mine ore containing less than 20% P_2O_5. Fluorine (about 1.4% in apatite), rare earths (about 0.8%) and strontium (about 1%) are also extracted from the apatite ore. Nepheline is also concentrated (Semenov, 1974).

The ore body has fold structures, fissure, shear and breccia zones, along which weathering has progressed. The weathered rock is also mined as ore.

Individual apatite deposits grade from high-grade apatite ore to apatitic ijolite and urtite, generally containing 15–75% apatite (6–31% P_2O_5), 10–80% nepheline, 1–25% aegirine and 5–12% sphene. The fluorapatite contains up to 11.42% strontium oxide and 4.90% rare-earth oxides, RE_2O_3 (Notholt, 1979).

JACUPIRANGA, BRAZIL (48° W 23° S)

The Jacupiranga alkalic complex is oval-shaped, covering an area of 65 km^2, intruded at the contact between Precambrian mica schists and granodiorite (Melcher, 1966). The major rock-types are Jacupirangite, pyroxenite, peridotite, ijolite, fenite and nepheline syenite. Carbonatite occupies a small central portion of the complex of less than 1 km diameter (*Figure 6.7*).

Chemical weathering of the carbonatite has produced a 1–20-m thick soil regolith, which averages 22% P_2O_5 and 26% iron oxide, Fe_2O_3. The phosphate, as apatite, is mined and concentrated. The annual production is 50 000 t of concentrate averaging 39% P_2O_5. The apatite is little affected by the weathering and appears as the same ovoid crystals as occur in the fresh carbonatite.

Baddeleyite and pyrochlore also occur in the

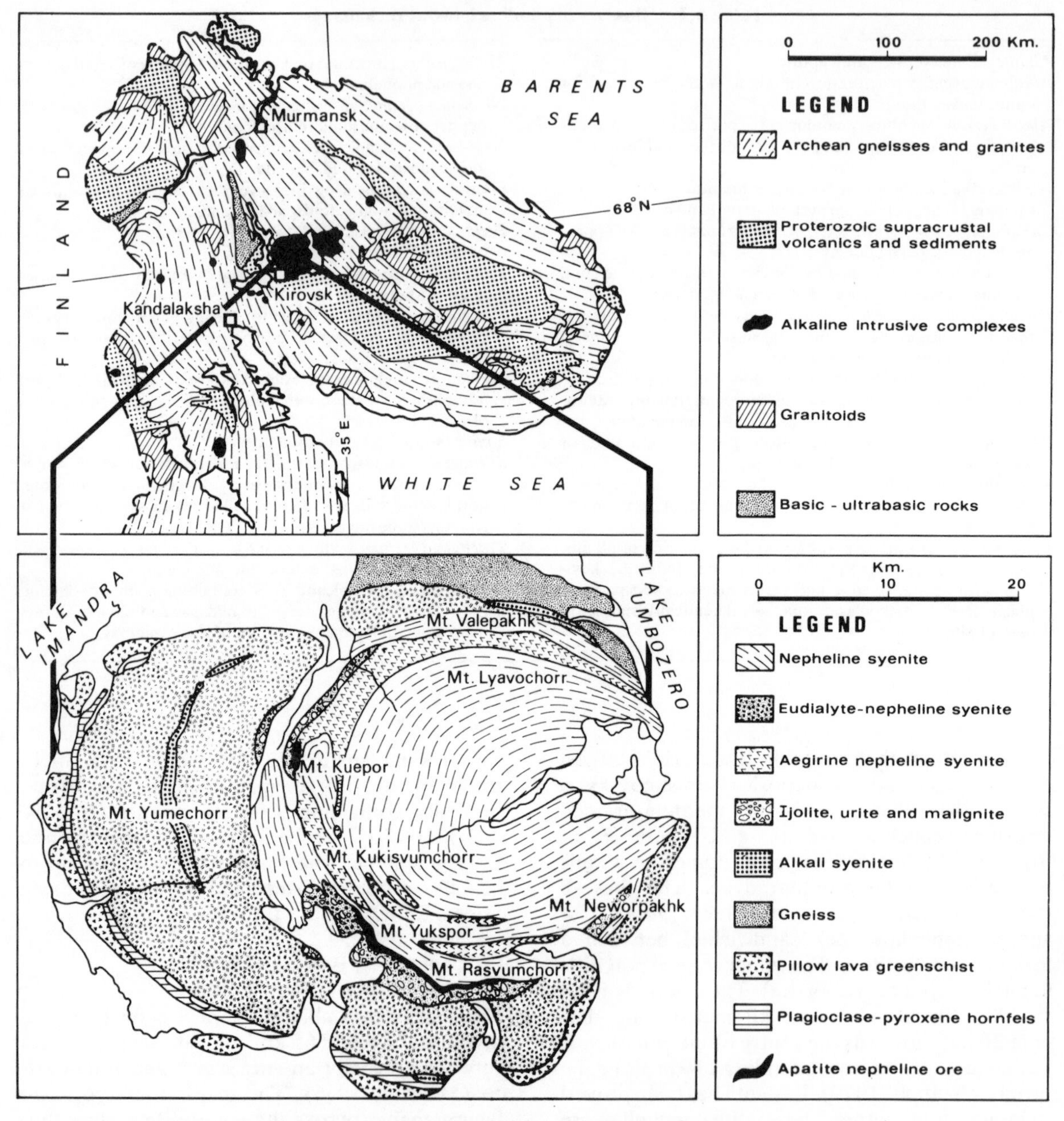

Figure 6.6. Simplified geological map of the Kola Peninsula, USSR with details of the Khibina alkaline massif to show the setting of the apatite–nepheline ore (after Gerasimovsky *et al.*, 1974).

soil, but the concentration is not economic. The unweathered carbonatite is now also mined for its apatite.

Mineralization in Carbonatites

Carbonatites constitute an immense reserve of niobium, chiefly as pyrochlore. Some also contain major accumulations of rare earths and apatite (Semenov, 1974).

Fifty carbonatite complexes occur in North America in northern Ontario and western Quebec (Erdosh, 1979). All occur along recognized major tectonic features; their ages range from 120 Ma to 1700 Ma. They all contain apatite in the carbonatite phase, amounting to 5–25% and karstic leach-

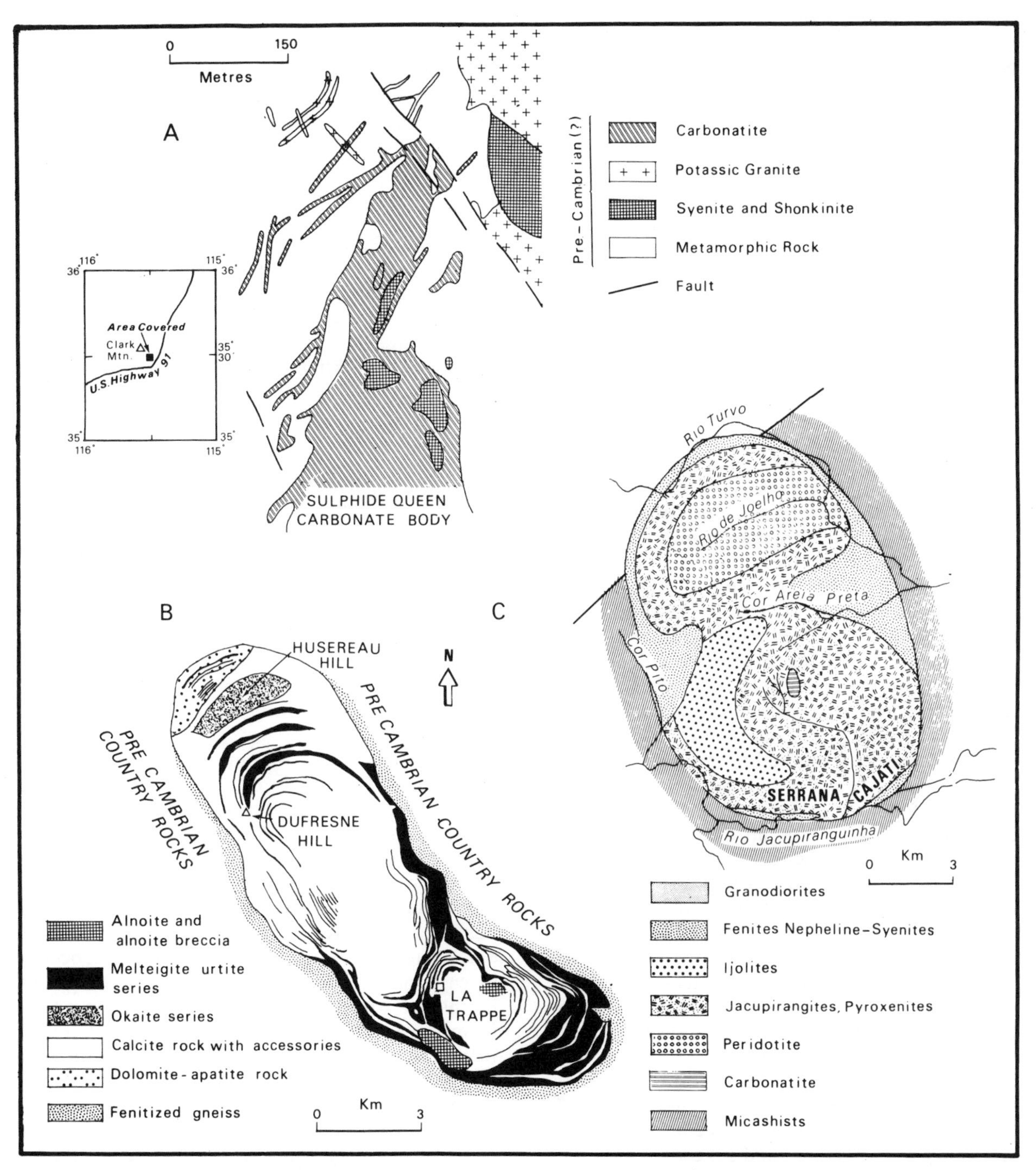

Figure 6.7. Some details of three carbonatite complexes. (A) The Mountain Pass carbonatite of California, which is mined for rare earths (after Olson, J.C. and Pray, 1954), (B) the Oka alkaline complex of the Montreal area of Canada, mined for niobium (after Gold *et al.*, 1967) and (C) the jacupiranga alkaline complex of Brazil, mined for apatite (after Melcher, 1966).

ing of the carbonate may have formed residual economic concentrations.

NIOBIUM

The important minerals are pyrochlore, niobium-containing perovskite, columbite and epistolite. Pyrochlore deposits are especially connected with carbonatitic derivatives of ultramafic alkaline massifs (Oka, Quebec) and less commonly with nepheline syenites and alkali syenites (Lake Nipissing, Ontario) (Rowe, 1958).

Oka, Quebec (74° W, near Montreal)

Extending eastwards from Montreal, seven hills of alkaline rock form the Monteregian alkaline petrographic province (Gold, 1967; Gold *et al.*, 1967). The age of the small stocks, necks, plugs, dykes and sills is 110 Ma, or early Cretaceous. Carbonatites of seven different types are associated in a double-ring structure intrusive into Precambrian gneisses and anorthosites. The rock-types include okaite–jacupirangite, ijolite, urtite and alnöite. The $^{87}Sr/^{86}Sr$ ratio is 0.7040. The complex has two centres, that in the southeast is about 1 km diameter and there is a larger elliptical centre (1 × 3 km) in the northwest (*Figure 6.7*). Around these cores multiple ring-dykes, arcuate dykes, and cone sheets of silicate rock and carbonatites are complexly interleaved. Metasomatism or fenitization is marked. The sequence of rock development has been deduced by Gold *et al.* (1967) as follows. (1) Fenitization of the gneissic cover rocks, followed by an emplacement of an early carbonatite phase as dykes and ring-dykes, and ijolitization of the enclosed country rocks. (2) Intrusion of okaite–jacupirangite rocks as arcuate dykes mainly in the northern part of the complex. (3) Intrusion of a pyrochlore carbonatite, followed by monticellite carbonatite. (4) Intrusion of ijolite dykes followed by solid-state flow in the carbonatite. (5) Hydrothermal activity along fractures, producing biotite and enrichment of the carbonatite in thorian pyrochlore. (6) Dolomitic carbonatite dykes, and explosion breccia formation. (7) Lamprophyre dykes. (8) Emplacement of alnöite and alnöitic breccia pipes and dykes.

There are several large niobium ore bodies of tabular-to-lensoid shape. The chief host rock is an aegirine augite–biotite–pyrochlore sövite. Pyrochlore occurs in concentrations up to 20%. The approximate range of composition is 1–20% pyrochlore, 75–80% calcite, 6–10% pyroxene, 3–4% apatite, 3–4% magnetite, 3–4% mica and 1–2% pyrite. The ore shoots range in length from 3 m to 520 m with a width of 3–60 m. The pyrochlore is euhedral to subhedral, variable in composition (40–56% niobium pentoxide, Nb_2O_5). The niobium content apparently varies inversely with the thorium and uranium contents. The variety betafite (with 32–35% Nb_2O_5 and 8–8.2% thorium dioxide, ThO_2) occurs in biotitized ijolite or in sövite at altered ijolite contacts. Niobian perovskite has high niobium contents (25–47% Nb_2O_5); it occurs in altered okaite.

The Sokli carbonatite complex of Finnish Lapland is related to the same structures that controlled the emplacement of the Khibina and Lovozera complexes of the Kola Peninsula. The pyrochlore and apatite–magnetite mineralization has been described by Vartiainen and Paarma (1979).

RARE EARTHS

Rare earths are particularly concentrated in alkaline rocks. Pegmatites rich in lovchorrite–rinkolite have been exploited in the Khibina massif of the Kola Peninsula (Semenov, 1974) and the rare-earth potential of African carbonatites was discussed by Deans (1966). The greatest concentration of rare earths is at Mountain Pass mine (35.47°N, 115.53°W) in California.

Mountain Pass, California

The geology of the Mountain Pass area of San Bernardino County is of Precambrian potassic alkaline igneous rocks, carbonatites and potassic granites intrusive into Precambrian metamorphic rocks (Olson, J.C. and Pray, 1954; Olson, J.C. *et al.*, 1954). The rare earths are chiefly cerium, lanthanum and neodymium, occurring mainly in bastnaesite, a rare-earth fluorcarbonate. The rare earths are rich in europium. The bastnaesite occurs in carbonate veins. About 200 veins, composed mainly of carbonate and barite, range in width from 0.3 m to 2 m. They cut across the Precambrian foliation; some are emplaced in fractures in shonkinite–syenite intrusions. The biggest carbonatite body is the Sulphide Queen (*Figure 6.7*), which is a carbonatite body of 200 m maximum width, by 730 m long. The carbonates calcite, dolomite and ankerite form more than 50% of the vein. Barite averages 20–25%. Bastnaesite forms 5–15% of the rock and locally reaches 60%. The shonkinite, syenite, granite and the metamorphic rocks are all Precambrian. The carbonate dyke rocks are presumed to have differentiated from the same magma as supplied the potash-rich dyke rocks. Shonkinite is the oldest potash-rich intrusive, and syenite and granite are younger (Olson, J.C. and Pray, 1954). Even younger are dykes of fine-grained shonkinite. Potassic igneous dykes occur throughout the region. The carbonate rocks cut, and are therefore younger than, the potassic igneous rocks.

The Sulphide Queen carbonate rock vein consists of 40–75% calcite, 15–50% barite and 5–15% bastnaesite. Samples from the ore body average 6.89% RE_2O_3 (with a range up to 38.92%). The carbonatite dykes carry up to 18.64% RE_2O_3 with an average of 5.17%. The estimated reserves for the district are 25 million t of carbonatite with 5–10% RE_2O_3, 20–25% barium sulphate and less than 8% thorium dioxide. Europium is also recovered commercially.

Uraniferous Peralkaline Plutons

The uranium content of some peralkaline intrusives is higher than that of carbonatites. However, at Palabora, South Africa, a carbonatite is host to a major disseminated copper deposit, which also produces by-product phosphates and uranium. The ore contains 0.7% copper, 0.01% thorium and uranium amounting to 0.004% in uranothorianite and badelleyite (Von Backström, 1974).

THE ILIMAUSSAQ INTRUSION, GREENLAND

The 1180–1150 Ma old intrusion of south Greenland covers an outcrop area of 150 km^2 and is composed of an early augite syenite followed by agpaitic nepheline syenites (Sørensen *et al.*, 1978).

Laminated lujavrite forms part of the complex; it is composed of microcline, albite, nepheline, sodalite, alkali amphibole and alkali pyroxene, with accessory sphalerite, britholite, eudialyte, monazite, pyrochlore, lithium-containing mica, steenstrupine and villiaumite. It generally contains more than 50 ppm uranium, 50 ppm thorium and 3000 ppm rare earths. In a number of places, large masses contain more than 200 ppm uranium, 600 ppm thorium and 10 000 ppm rare earths.

In the northernmost part of the intrusion, uranium-enriched lujavrites are in contact with strongly contact-altered volcanic rocks, which have been mineralized and contain 100–3000 ppm uranium and 300–15 000 ppm thorium. The reserves in the main area are 5800 t uranium, contained in 18 600 000 t of ore with an average 310 ppm uranium. There is 2.6 times this amount of thorium.

Kimberlite and Diamonds

Kimberlite is a very rare, potassic, ultrabasic, hybrid igneous rock that occurs as small diatremes, or as dykes or sills of limited extent (*Figure 6.8*). Its porphyritic appearance is due to megacrysts of olivine, enstatite, chrome–diopside, pyrope, picro-ilmenite and phlogopite set in a fine-grained matrix of which serpentine, carbonates, phlogopite, magnetite and perovskite form the major part (Dawson, 1967, 1971).

Kimberlites may be classified on the basis of their content of five major minerals: diopside, monticellite, phlogopite, calcite and serpentine. Olivine is always abundant and because a distinction cannot easily be made between olivine derived from the kimberlite magma and xenocrystic olivine, its presence is not used in the classification (Skinner, E.M.W. and Clement, 1979).

Many of the megacrysts are derived from fragmentation of mantle-derived blocks that are embedded in the kimberlite and are in various stages of reaction with the matrix, which may or may not contain diamond. Most kimberlites lack diamond. Even the most diamondiferous usually contain about 1 ppm of widely dispersed diamonds. Inclusion studies from the Finsch kimberlite of South Africa (Gurney *et al.*, 1979) suggest a diamond crystallization temperature of 1100°C at a pressure close to 50 kbar. It is assumed that kimberlite resulted from the partial melting of phlogopite–dolomite–peridotite (Wyllie, 1979) within the diamond stability field. The abundant carbon dioxide generates an initial melt of carbonatite–kimberlite affinities that preferentially dissolves calcium. When the silicate phases in equilibrium with the melt become depleted in calcium, conditions become favourable for diamond formation (Gurney *et al.*, 1979).

Kimberlites contain two generations of olivine, one xenocrystal and of high temperature (Fo_{91-94}), the other occurring in the ground mass (Fo_{88-89}). Phlogopite may also be distinguished as either xenocrystal or kimberlitic magmatic (Elthon and Ridley, 1979). The picro-ilmenite grains are always mantled by perovskite and titanium, magnesium, chromium and aluminium magnetites.

Kimberlite dykes infill major deep-seated fractures that cross the cratons on a geometric pattern. The kimberlite that infilled these fractures was a fluid that penetrated to within 2.3 km of the surface before there was an explosive breakthrough with subsequent formation and infilling of the high-level diatremes by a gas–solid fluidization process (Dawson, 1971). The kimberlite magma was probably emplaced at low temperatures, con-

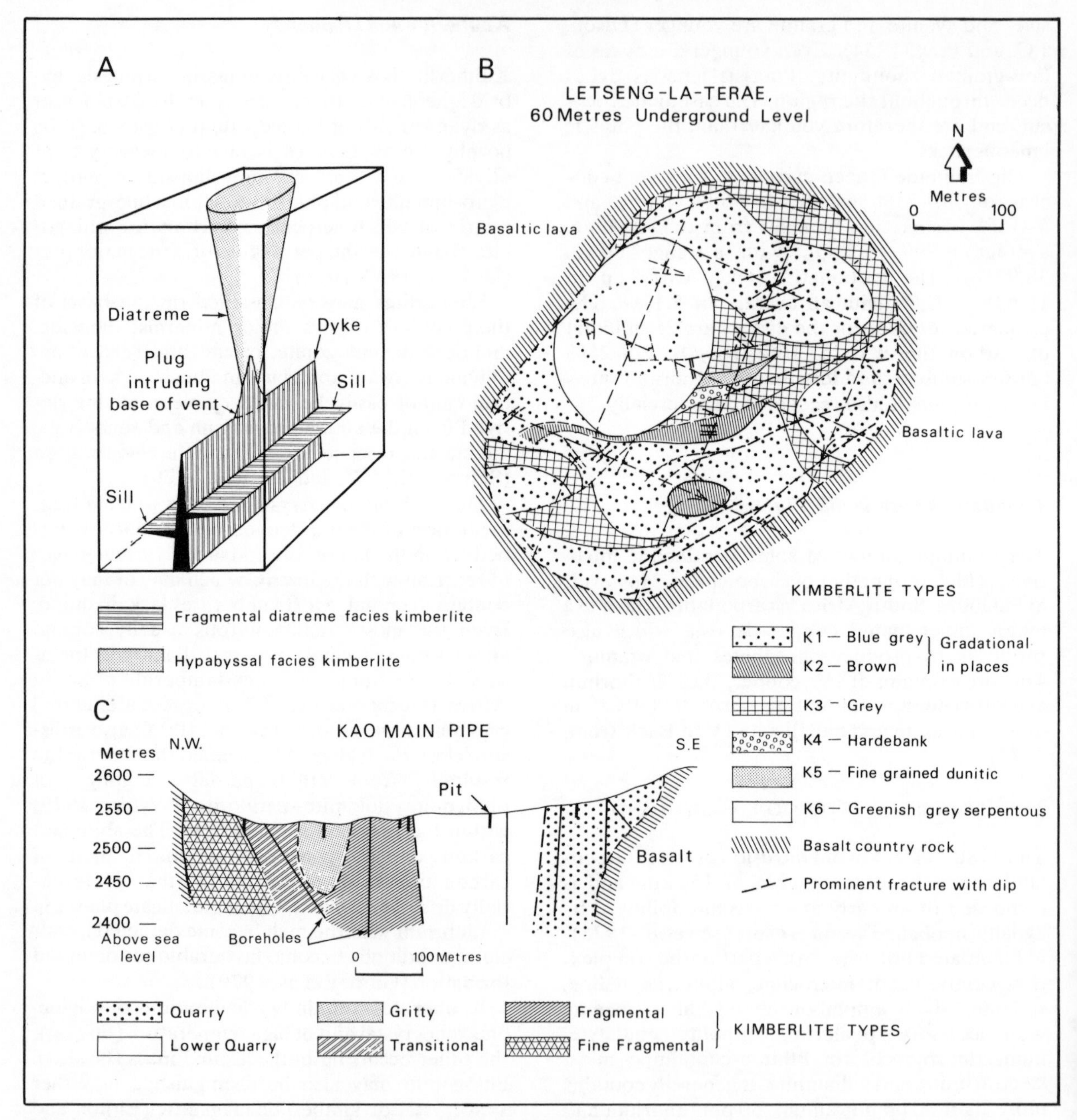

Figure 6.8. Schematic relationship between kimberlite diatremes and deeper-seated dykes and sills (after Dawson, 1971), and the geology of two Lesotho kimberlites mined for diamonds, the Letseng-la-terae (after Bloomer and Nixon, 1973) and the Kao deposit (after Rolfe, 1973).

sidered to be about 600°C (Mitchell, R.H., 1973). From their formation at 200 km in the mantle to their emplacement in the upper crust, kimberlites incorporate xenoliths and xenocrysts resulting from the dismemberment of xenoliths. The crystallization is closely tied to the dissolution of carbon dioxide and acquisition of water during the ascent. The magma is continuously saturated with carbon dioxide, which must degas from the melt during ascent in the depth interval 100–80 km, and this degassing would assist the eventual explosive eruption of the kimberlite. The gas evolution would assist crack propagation (Wyllie, 1979). Diatremes probably occur in regions of the continental crust where the rocks at depth are rich in carbon dioxide (Anderson, O.L., 1979). A crack passing through such a region would collect abundant carbon dioxide gas, which would accumulate at its tip. This allows the crack to accelerate upwards. The crack becomes unstable and the speed approaches the shear velocity of sound as the crack breaks through to the surface (Anderson, O.L., 1979). The argument of Wyllie (1979) that the most likely source of kimberlite in the mantle is a periodotite, enriched in carbon dioxide and water, would also provide the proper environment for the rapid upwards propagation of cracks to permit their intrusion.

The silicate kimberlite melt, plus cumulate crystals (including diamonds) and the remaining volatile phases are therefore rapidly emplaced into the upper crust as kimberlite bodies. The coarse peridotite and eclogite xenoliths are from shallow mantle overlying the proto-kimberlite magma chamber, sampled on the way upwards. The development of carbon dioxide overpressure not only provides an upwards migrating mechanism but also leads to the commonly altered nature of the kimberlite matrix. The trend towards heavier $\delta^{18}O$ values in fragmental kimberlites is thought to be due to the inter-reaction between the rising kimberlite magma with meteoric water in the upper crust (Kobelski *et al.*, 1979).

Kimberlites can vary widely in texture, from the fragmental varieties found in diatremes, to massive varieties found within the hypabyssal dykes and the rare sills. Both petrographically and chemically, kimberlite is a hybrid rock, formed by incorporation of crystals mainly derived from fragmentation of upper-mantle garnet lherzolite into a matrix that has strong affinities with carbonatite. Garnet lherzolite and to a lesser extent eclogite are the main upper-mantle xenoliths. Kimberlite is the only known primary ore of diamond. Oxidation and graphitization may account for the absence of diamonds from most kimberlites, but the graphite found in most is considered to be a primary mineral, so that the majority of kimberlites are thought never to have contained any diamonds (Robinson, D.N., 1978).

Kimberlite bodies usually occur as clusters. Diamonds found in adjacent pipes may be similar in property or equally probably may be remarkably different. Some surface enrichment in diamond content over kimberlite pipes undoubtedly results from weathering and removal of the matrix. Some kimberlites have relatively richer upper layers, composed of tuffaceous or brecciated kimberlite in which the diamonds could have been concentrated by loss of volcanic dust. In hypabyssal kimberlites the diamond content is maintained with depth. An exception is the De Beers Mine, Kimberley, where diminishing diamond contents are attributed to the thinning out in depth of the relatively richer central core of the diatreme.

Diamonds are generally many thousands of times more abundant in the eclogite xenoliths than in the kimberlite matrix (Robinson, D.N., 1978). This strongly suggests that they are xenocrysts derived from disaggregated eclogite, and not an intrinsic part of the kimberlite magma. Peridotite xenoliths are more common than eclogite xenoliths. Diamonds have also been found in them, especially in garnet lherzolite, which is the main kind of xenolith. Nodules of monomineralic ilmenite, garnet, diopside, enstatite and olivine also occur. Some garnet nodules contain diamond, but they are thought to be part of the eclogite.

Lead-isotope dating of the mineral phases included within diamond crystals indicate that the diamonds are much older than the kimberlites that contain them. Hence the diamonds are xenocrysts picked up like the xenoliths by the kimberlite.

The weight of evidence suggests that the diamonds were formed deep within the mantle under thermodynamically stable conditions. The favoured hypothesis proposes that kimberlite is formed by incipient melting of parental garnet lherzolite in which phlogopite is an additional phase. The initial melt liquid, plus fragments of the parental five-phase garnet lherzolite, is what is known as kimberlite (Dawson, 1971).

The most obvious reason for the primary diamonds being practically restricted to kimberlites is that these diatreme rocks are the only ones that originated at sufficient depth to have frequently sampled the diamond-bearing regions underlying the continental crust. Kimberlites not bearing diamonds may have too shallow roots. Since diamond is the high-pressure polymorph of graphite, the apparently greater potential for kimberlites in continental-shield areas to contain substantial amounts of diamond can be related to the relatively cool geothermal gradient beneath the shield areas.

Small diamond crystals have been recovered directly from dunite and even basalt. The cryptocrystalline diamond aggregate called carbonado has never been found in kimberlite. It is very common in Brazilian placers. Its fine crystallite size and its isotopic composition suggest derivation from organic carbon; its inclusions resemble crustal metamorphic minerals. It has been suggested that carbonado formed as a result of subduction of organic carbon-rich sediments at trench areas (Robinson, D.N., 1978). Diamonds have also been found in iron meteorites and ureilites.

From their distribution pattern, it can be shown that kimberlites are virtually confined to the old Precambrian cratons and that diamond-bearing kimberlites exist on the older cratonic nuclei that have not been deformed for the past 2500 Ma (Dawson, 1971). Prior to 1870, all known diamond deposits were considered to be detrital. Soon thereafter, it became apparent that a deposit being mined near Kimberley, South Africa, was in volcanic material, which was named kimberlite. Kimberlite had also been worked for diamonds for many centuries at Majdgawan, India, and in 1930 it was determined to be kimberlite-like. The known occurrences today are in South and Central Africa, Siberia, India, North America and Australia.

Kimberlites are widely distributed throughout the USA where they have an age spread from Middle Palaeozoic to Middle Tertiary (Meyer, 1976). Individual occurrences may be related to local structural trends. Diamonds have been found in the occurrences in Arkansas and Colorado–Wyoming, and elsewhere they are found in placer deposits. Many placer occurrences are related to moraines of the last Ice Age and hence they cannot be traced to kimberlite sources.

Although Moore, A.C. (1973) stated that there are no authentic descriptions of kimberlites in Australia, there are ultrabasic lamprophyric dykes in the northern Flinders Ranges that have carbonatite affinities. More recent descriptions by Stracke *et al.* (1979) confirm the existence of post-Proterozoic kimberlites in southeastern Australia. Their emplacement is probably related to fracture patterns connected with the early stages (80–60 Ma ago) of transform faulting during the beginning of the separation of the Tasman Sea. Evaluation of kimberlite pipes in the remote Kimberley area of the far-north of Western Australia has indicated that 31 pipes have been found at Ellendale and promising quantities of diamonds have been sampled. The largest stone recovered was 5.9 carats (1.18 g). In pipe A, 563 carats were recovered from 10 629 t of kimberlite, and in pipe B, the recovery was 653 carats from 8164 t of kimberlite. At the near-by Argyle prospect, placer diamonds are being recovered from gravel deposits. The Kimberley area is now in production and will prove to be the richest in the world. However, it contains few gemstone-quality stones.

The world's major gemstone fields are in South Africa, Namibia, Tanzania Plateau, West Africa and the USSR. The main economic diamond deposits of the USSR (alluvial and kimberlite pipes) are the Yakutian region lying between 62° and 66° N and 110° and 120° E (Davidson, 1967). Kimberlite dykes are relatively more common in the outlying fields than in the central ones, where pipes predominate. The kimberlites show a wide age spread, and the field evidence suggests that there were periods of kimberlite volcanism both in the Palaeozoic and Mesozoic. The diamond morphology varies from field to field. For example, in the Malaya Batuobiya group, 72% are octahedral and 8% dodecahedral, and in the Daldyn–Alakit group 23% are octahedral and 62% dodecahedral. The remainder are of transitional forms.

Age

The kimberlite blows at Igwisi Hills, central Tanzania, are recent. The diatremes of the Colorado Plateau are Tertiary. Southwest Africa, South Africa, the Siberian Shield and central USA have kimberlites of several ages, with major intrusion episodes in the Cretaceous. Diamonds are found in Precambrian conglomerates or greywackes in Brazil, South Africa, West Africa, India and Australia, and subsequent kimberlite activity of Precambrian age is found in some of these areas. The evidence is that cratons characterized by kimberlites have repeated kimberlite intrusions throughout geological time. Potassium–argon ages of kimberlites frequently are older than the country rocks in which the diatremes occur. This supports a neo (Bemmelen, 1970a), from the Kamchatka Peninsula and also from the island of Malaita, Solomon Islands (Hutchison, 1975). The alnöitic breccia of Malaita island is different from the kimberlites of continental settings as shown by the chemistry of the silicate minerals (Dawson *et al.*, 1978).

Although kimberlite is virtually confined to continental areas, diamond-bearing breccia pipes are known from the Meratus Mountains of South Borneo (Bemmelen, 1970a), from Kamchatka Peninsula and also from the island of Malaita, Solomon Islands (Hutchison, 1975). The alnöitic breccia of Malaita island is different from the kimberlites of continental settings as shown by the chemistry of the silicate minerals (Dawson *et al.*, 1978).

Lesotho, Southern Africa

Lesotho has 17 pipes, 21 dyke enlargements or blows and over 200 dykes of kimberlite; a high proportion are barren. Large diamonds of superb quality occur at Letseng-la-terae (Bloomer and Nixon, 1973) and at Kao (Rolfe, 1973); these rank among the worlds ten largest pipes.

Letseng-la-terae (29° E 29° S)

The main pipe (area 15.9 ha) and the satellite pipe

(4.7 ha) occur 3100 m above sea level. The soft blue kimberlite is overlain by 1 m of yellow ground and then basaltic gravel. The pipes penetrate approximately 1.5 km of Karroo Stormberg basaltic lavas (*Figure 6.8*). The main pipe tapers downwards, but the satellite pipe expands to at least 150 m depth.

The K1 and K3 kimberlites are soft, friable, tuffaceous, varying from bluish-grey (K1) to greyish-green (K3) with a relative density of 2.58, K2 is brown and harder and less tuffaceous. K4 is greenish-black, fine sugary-textured, tough and compact with a relative density of 3.00. K5 is poor in xenoliths and rich in fresh olivine. K6 is the main diamond-bearing kimberlite; it is soft, friable, greenish-grey and brown with phenocrysts of altered olivine in a fine-grained matrix.

Inclusions in the kimberlites consist of Triassic–Jurassic Karroo sediments and lavas, Archaean gneisses, granulites, lherzolite nodules and monomineralic nodules. The diamonds are unusual for their virtual absence of octahedra. Most are irregularly shaped, broken dodecahedra and cleavage chips. Brown and faint-yellow colours predominate (Bloomer and Nixon, 1973).

Kao (28°40′ E 29° S)

Kao main pipe (area 19.8 ha) and satellitic pipe (3.2 ha) decrease in size with depth. Six types of kimberlite are mapped (*Figure 6.8*). The Quarry kimberlite probably represents the first blow during intrusion of the main pipe; it is grey and relatively abundant in ilmenite and diamonds. The lower Quarry kimberlite came next; it is khaki-green and composed of fresh olivine and pyroxene. The Gritty kimberlite came next; it contains mica in a brown friable matrix, rich in xenoliths. The blue-green Fragmental kimberlite contains more zircons than the others. The final intrusion was of Fine Fragmental kimberlite, which is similar to the Fragmental but contains more rounded xenoliths. Diamond contents vary from 3 carats to 18 carats (0.6–3.6 g) per 100 t of kimberlite. Deep-seated nodules include basement gneisses, granulite and eclogite, lherzolite and harzburgite (Rolfe, 1973). The diamonds range from octahedra to dodecahedra, the proportion of octahedra decreasing with the diamond size.

Cratonic Rift-related Tin–(Niobium–Tantalum–Tungsten) Mineralization

Intracratonic rifts, for example, the East Africa Rift and the Baikal Rift, are considered to result from crustal thinning over a mantle hot-spot. They develop on cratonic lithospheric plates that are either stationary or moving slowly with respect to the underlying mantle plume (Burke, K.C. and Wilson, 1976). Rifts radiating outwards from hot-spots are characterized by a fault trough with updomed flanks. Associated volcanoes are characteristically alkaline, and they erupt rhyolitic and undersaturated alkaline lavas and tuffs, underlain by plutonic rocks including granites. The granites of such within-plate tectonic settings are characterized by very high niobium contents (170 ppm niobium on average) as compared with the granites of volcano–plutonic arcs of destructive plate margins (15–20 ppm niobium, on average) (Pearce and Gale, 1977).

With continued development, tholeiitic basalt is injected along the axial zone and the rift splits longitudinally into two halves separated by a spreading ocean rise system, as in the case of the Red Sea. Many rifts fail to develop into ocean rises, perhaps as a result of the continent starting to move with respect to the mantle hot-spot (Burke, K.C. and Wilson, 1976) and such rifts are preserved as failed intracontinental rifts or aulacogens, characterized by a great thickness of continentally derived sediments (for example, the Belt Basin of northwest America). They commonly pass at one end into a major delta (for example, the Benue Trough–Niger Delta). Granite rocks on the rift flanks are usually preserved because the tectonic setting again becomes cratonic with the extinction of the hot-spot activity.

The Precambrian mineralization within the Beltian aulacogen is described in Chapter 10. Tin mineralization is known from several areas that are interpreted as ancient failed rifts resulting from hot-spot activity (Sillitoe, R.J., 1974; Mitchell, A.H.G., 1979). Such tin deposits fit the category described by Taylor, R.G. (1979b) as anorogenic. The granites are not associated with fold or orogenic belts and have been emplaced epizonally in major zones of fracturing of cratonic shields. The igneous bodies are small oval or circular-ring complexes. The minor stocks and ring-dykes are often capped by sheets. Minor volcanic rocks are associated with them. Groups of ring complexes show strong linear alignments, related to the rifting and hot-spot track. The rocks associated with the mineralization are predominantly granite, microgranite, or rhyolite and alkali granite. Both plutonic and porphyry textures are characteristic. Mining is commonly from placer deposits eroded from the complexes and primary lodes have minor significance. Examples are to be found in Nigeria, Rondonia (Brazil) (*Figure 8.12*) and Missouri.

Tin and Associated Mineralization

Tin in rift-related settings is associated with granitic rocks mostly emplaced during the earliest stages of development of intracratonic rifts that failed to develop into ocean basins. Pre-intrusion host rocks are usually either metamorphic rocks of an earlier orogeny or rift-related continental sediments and silicic volcanic rocks slightly older than the mineralized epizonal plutons (Mitchell, A.H.G., 1979).

The mineralized igneous rocks are high-level granitic bodies in the form of subvolcanic ring complexes, although some mineralization occurs in the overlying intruded volcanic rocks. The tin-bearing plutons are mostly biotite granites, but also include syenites and peralkaline granite. The plutons are anorogenic and there is an absence of deformation or regional metamorphism associated with their intrusion. Pegmatites, tourmaline and beryl are uncommon. The ore minerals are predominantly cassiterite, with lesser wolframite and columbite–tantalite. Topaz, fluorite and lepidolite are common. Mineralization is usually disseminated in the roof zones of the volcanogenic plutons and is related to post-emplacement hydrothermal albitization, greisenization and quartz veining.

Rondonia, Brazil

Twenty-five tin-mineralized bodies of the Late Proterozoic granite complexes of sizes variable up to 25 km diameter have been mapped (Priem *et al.*, 1971); they are predominantly of biotite granite, with lesser microgranite, rhyolite and silicic breccias. The granites have a high $^{87}Sr/^{86}Sr$ initial ratio of 0.718, which supports the argument of Hutchison and Chakraborty (1979) that tin in such anorogenic granites has been remobilized from the former pegmatites and metamorphic rocks of the cratonic continental crust. There is no basis for supporting the hypothesis of Sillitoe, R.H. (1974) that tin in economic amounts can have a direct mantle source. The plutons are epizonal bodies intruding high-grade metamorphic and volcanic rocks of the older basement and they form subvolcanic ring complexes. To the north, the granites intrude sediments within a graben structure. Deformation is anorogenic and limited to faulting and fracturing. The tin is mined from alluvial deposits derived from the primary granite mineralization. Cassiterite occurs in greisenized zones and in quartz–topaz veins within the granites and in the contact zones of the country rocks. Wolframite and columbite–tantalite are locally associated with the cassiterite.

St. Francois Mountains, Missouri

Mineralization at the Silvermine area of Missouri has been interpreted by Lowell, G.R. (1976) in terms of hot-spot and incipient northeast-trending intracontinental rifting of Proterozoic age. The mineralization occurs in unmetamorphosed epizonal alkalic to peralkaline granites associated with extensive rhyolites. The igneous activity has been dated at 1500 Ma. Quartz veins in the contact zones between the Silvermine granite and the rhyolites contain wolframite, chalcopyrite, sphalerite and argentiferous galena, as well as sub-economic amounts of cassiterite. Other minerals in the veins include fluorite, pyrite, topaz and lepidolite.

Nigeria

The Precambrian basement complex of southwest Nigeria consists of migmatite, metasediments including schists and quartzite, meta-igneous rocks and the older granites that include pegmatites and charnockites. Radiometric dating indicates two major metamorphic–igneous events, 1950 Ma ago and a younger Pan African event 600 Ma ago (Matheis, 1979). Cassiterite, with minor columbite–tantalite, is mined from the placer deposits that overlie and surround the Precambrian pegmatites, which are related to the Pan African Older Granite suite (Jacobson and Webb, 1946). There are two distinct events in the mineralization process. The first consisted of the emplacement of simple pegmatites composed of quartz, microcline, oligoclase, muscovite, biotite, garnet and schorl tourmaline. The second phase is represented by intense albitization accompanied by economic mineralization of cassiterite, columbite–tantalite and such accessory minerals as beryl, lepidolite and more tourmaline. The alluvial and eluvial deposits have been eroded from the pegmatite bodies.

However, the Precambrian tin accounts for only 5% or less of the total Nigerian production, the other 96% is won mainly from aluvial and eluvial mines associated with Jurassic ring-complex volcanogenic granites. The presence of tin in the Precambrian rocks indicates that the basement, through which the Younger Granites intrude, is enriched in tin, supporting the conclusion of Hutchison and Chakraborty (1979) that the source of the Jurassic epizonal granites was the tin-enriched pegmatite zones in the Precambrian basement. This is reinforced by the $^{87}Sr/^{86}Sr$ ratios of the Jurassic granites, which may be as high as 0.7134. Of particular interest is Bowden's (1970) lead-isotope studies, from which he has concluded that the Jurassic granites with 'old lead ages' cannot have differentiated from the upper mantle, but must have an anatectic source in mineralized continental crust. Alexiev (1970) and Bowden and Turner (1974) have postulated, on the basis of rare-earth characteristics and petrographic evidence, that the tin concentrated in the Nigerian granites is related to post-magmatic albitization.

The tin-bearing granites of the younger tin fields are concentrated in a 200 km wide north–south-trending zone, centred on the Jos–Bukuru Complex of the Central Plateau (Buchanan *et al.*,1971, MacLeod *et al.*, 1971). The mineralization is associated with ring complexes of which 40 are known, ranging in outcrop area from 1500 km^2 to 2 km^2. The age of the granites systematically decreases from Late Triassic in the north to Middle Jurassic in the south. To the north in Niger, the ring complexes are of Carboniferous and Silurian age (Bowden *et al.*, 1976). The ring complexes of Nigeria lie west of the Benue Trough. The most intense mineralization in both the Precambrian pegmatitic deposits and the Jurassic granites lie close to the intersection of a prominent east-northeast lineament of the Precambrian pegmatites with the north–south lineament of the Jurassic ring-complex plutons (*Figure 6.9*). These lineaments are of Precambrian parentage and have controlled the break-up of Gondwanaland (Wright, 1970).

Nigeria produces annually about 10 000 t of cassiterite mainly from alluvial fields of the Jos Plateau. Cassiterite occurs only in association with the Jurassic biotite granites and is richest in those that have the highest degree of hydrothermal albitization. The richest concentrations are in the roof zones of the plutons impounded beneath an impermeable roof of country rocks. The richest deposits are therefore in those plutons that have recently been unroofed and shallowly eroded. Ten of the sixty ring complexes provide over 90% of the Nigerian output and that of Jos–Bukuru produces 75% of the output (*Figure 6.9*). The cassiterite is dispersed throughout the higher levels of the biotite granite and is not concentrated in either quartz veins or in greisen zones. Most of the cassiterite is strongly zoned and dark-brown to black. Some of the aluvial and eluvial deposits have been covered over by Cenozoic basalt flows that are invariably laterized (*Figure 6.9*). The Jurassic granites are also the source of the columbite–tantalite. Nigerian annual production has been as high as 3000 t of concentrate from the tin-bearing alluvium, but annual production has declined. The mineral occurs as an accessory in the biotite granite and like cassiterite is not related to greisen zones or quartz veins. It is fine grained, always less than 5 mm, and is highest in concentration where the granite has been albitized. The average content is about 45 $g\,t^{-1}$ of granite. The mineral is invariably close in composition to the niobium end-member with the Nb_2O_5/Ta_2O_5 ratio always in excess of 5:1. In contrast, the columbite–tantalite in the Precambrian pegmatites shows a wide range over the spectrum columbite–tantalite.

Wolframite occurs in association with the cassiterite but production is decreasing. The mineralization is sporadic and it is not profitable to mine except as a by-product of cassiterite. It is never found in the alluvial deposits.

The Jos–Bukuru complex has the widest exposure of biotite granite (*Figure 6.9*). It is a volcanogenic structure with gently dipping sheeted components, the majority of which are mineralized with cassiterite, columbite–tantalite and wolframite. The good river drainage system has led to numerous economic placer deposits widely distributed around and over the biotite granite outcrops. The primary mineralization is highly disseminated. Much of the biotite granite has been greisenized but the mineralization is not associated with greisenization, but with post-magmatic albitization, although they have no lode concentrations. In some areas, for example, at Rayfield and Passakai, the decomposed granite is mined by hydraulic methods for its disseminated columbite. These deposits contain more than 15 000 t of columbite. The high columbite values are confined to an intensely altered and metasomatized zone in the granite, decomposed to depths of 60 m. The altered granite is enriched in albite and pale-green lithium mica, and riddled with pegmatitic veins and knots, coarse-quartz segregations and late quartz–epidote veins.

Pegmatite Mineralization

Many pegmatites occur within amphibolite-facies metamorphic terrane and are apparently not related to later intrusive granites. They may be considered as the crystallized products of anatectic melts produced *in situ* or in the neighbourhood during the metamorphism. Good examples are to be found in the economically important pegmatites of the Precambrian greenstone belts. On the other hand, pegmatites frequently occur spatially and genetically related to the roof cupola zones of granite plutons, occurring either within the roof zone of the granite or injected into the country

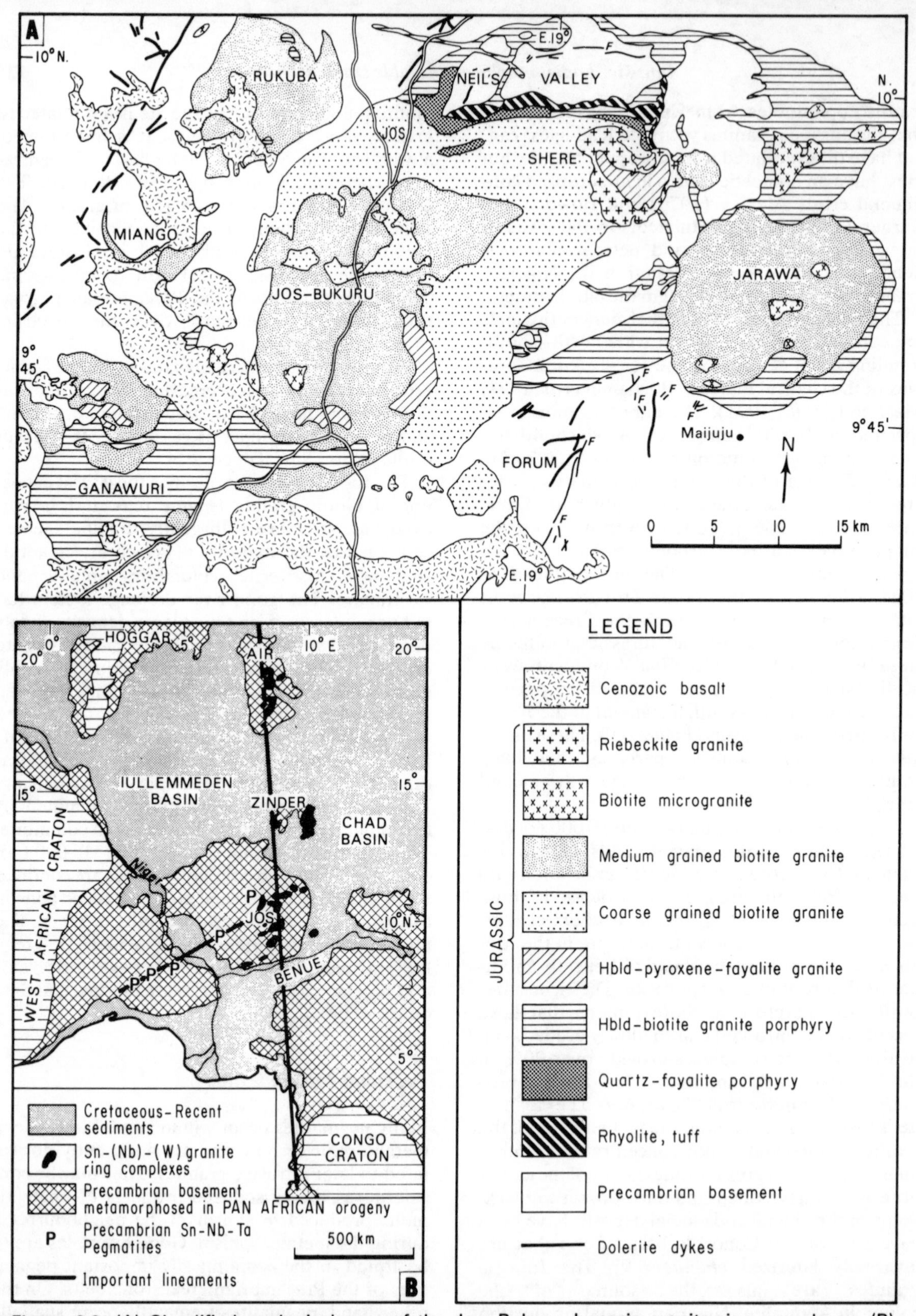

Figure 6.9. (A) Simplified geological map of the Jos–Bukuru Jurassic granite ring complexes. (B) Structural trends of the Pan African (600 Ma) pegmatites and of the Jurassic biotite granite and alkaline granite ring complexes of Nigeria (after Matheis, 1979). The Sn–(Nb–Ta–W) mineralization is associated with the biotite granite (after McLeod *et al.*, 1971).

rocks. Examples of this occurrence are to be found throughout the earth's history, from Precambrian to Cenozoic.

Pegmatite magma, distinguished by a high content of dissolved water, can be formed either through partial anatectic melting of metamorphic rocks or as a rest-liquid in a cooling igneous pluton The crystallization of a pegmatite body can occur directly from a hydrous silicate melt to give a coarse-grained or pegmatitic granite. It can also form from an exsolved aqueous fluid concomitant with a silicate melt to yield giant-textured pegmatite or aplite, or from an aqueous fluid with an absence of silicate melt to give a wide variety of late-stage products (Jahns and Burnham, 1969). A large number of publications bear witness to the fact that a transition exists between pegmatites, aplites and quartz veins.

Following this observed spectral sequence, Varlamoff (1972) has classified granite-related pegmatites on the basis of their occurrence in Central and West Africa (*Figure 6.10*). Type 0: composed of microcline, plagioclase, biotite and magnetite. Type 1: microcline, plagioclase, biotite and small amounts of quartz. Type 2: biotite, black tourmaline, abundant graphic textures between quartz–feldspar and quartz–tourmaline. Type 3: muscovite, biotite; black tourmaline is scattered throughout the pegmatite. Type 4: muscovite and black tourmaline are scattered throughout the pegmatite. Type 5: abundant muscovite (of non-commercial quality), quartz, microcline, small rare prisms of beryl; characterized by a beginning of albitization. Type 6: large prisms of beryl. Large crystals of amblygonite or spodumene or both, columbite–tantalite and microlite. Frequently well zoned with a quartz zone or large quartz concentration. Moderate albitization and greisenization. In some cases, green tourmaline and lepidolite. Type 7: partially or completely albitized; contains quartz, spodumene and muscovite; greisenization is subordinate. In the Congo, Rwanda Burundi and Uganda, this type contains commercial cassiterite, columbo–tantalite and white beryl. Type 7–8: quartz veins and large microcline crystals with some muscovite and cassiterite. Type 8: quartz veins with muscovite and cassiterite. Type 9: quartz veins with cassiterite and/or wolframite, or scheelite.

In high-level granite plutons, quartz veins and pegmatites are located within the roof zone of the granite pluton. In deeper-seated plutons, pegmatite types 7 and 6 are in the cupolas of the roof rocks (*Figure 6.10*). In deep-seated plutons, quartz types 0 and 1 are within the granite (Varlamoff, 1972).

Precambrian tin provinces are usually characterized by a preponderance of pegmatites. Schuiling (1967a,b) ascribed this fact to the level of erosion, based on his observations that Cenozoic tin deposits are in volcanogenic settings (for example, Mexico), Mesozoic are associated with the roof zones of granite batholiths (for example, Peninsular Malaysia) and Precambrian deposits are predominantly in pegmatites associated with migmatites and greenstone terranes (for example, Zimbabwe). However, these major differences may reflect the changing thermal regime of the earth's crust and it is not simply a matter of level of erosion, for many Mesozoic fold belts have been more deeply eroded than some stable Precambrian cratons.

The best-known Precambrian pegmatite deposits are in Southern and Central Africa, Nigeria, Brazil and Australia. Eluvial–alluvial concentrations have provided most of the tin–niobium–tantalum production, which is small compared with the Mesozoic granite batholith-associated mineralization. The lithium, beryllium and caesium minerals are obtained from the primary deposits.

SWAZILAND (SOUTHERN AFRICA)

Alluvial tin-mining ceased in 1966, by which time 10 400 t of tin had been recovered (Hunter, D.R., 1973). The stanniferous pegmatites occur in a granite–greenstone complex dated at 2000 Ma. Mineralized pegmatites are confined to a 5-km wide linear northwest-aligned belt and appear to be related to albitization (Davies, D.N., 1964). The cassiterite is notable for its coarse euhedral bipyramidal crystals. The cassiterite of pegmatite deposits is characteristically of a bipyramidal habit and the prism is usually suppressed (Hosking, 1979). Magnetite is a common accessory, with lesser amounts of yttrotantalite, monazite, garnet, fluorite and beryl. The cassiterite content is always highest where the pegmatites are magnetite-rich.

BIKITA (ZIMBABWE)

The most important pegmatite area in Southern Africa is at Bikita (*Figure 9.1*) located at the eastern extremity of the Fort Victoria greenstone belt in Zimbabwe (Martin, 1964). Shallow dipping pegmatite sheets, dated at 2360–2650 Ma, are now mined for lithium and beryllium. Like other stanniferous pegmatites in Zimbabwe, the deposits are not large. However, the Bikita Pegmatite is one of the largest deposits of lithium,

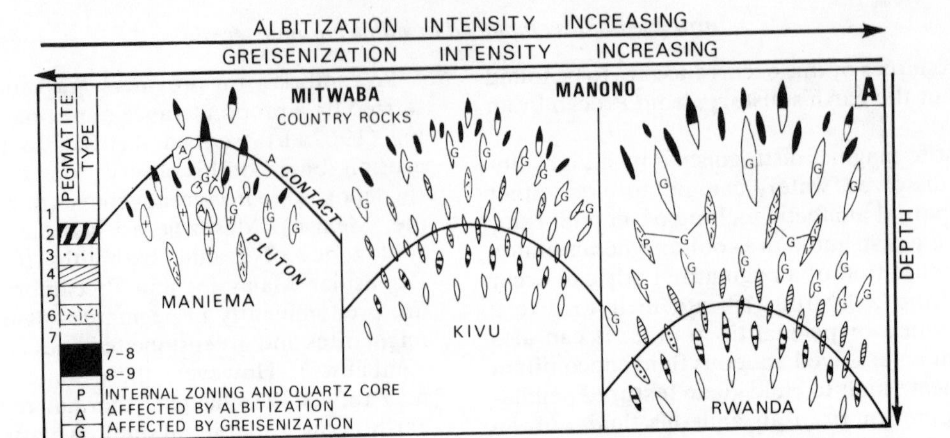

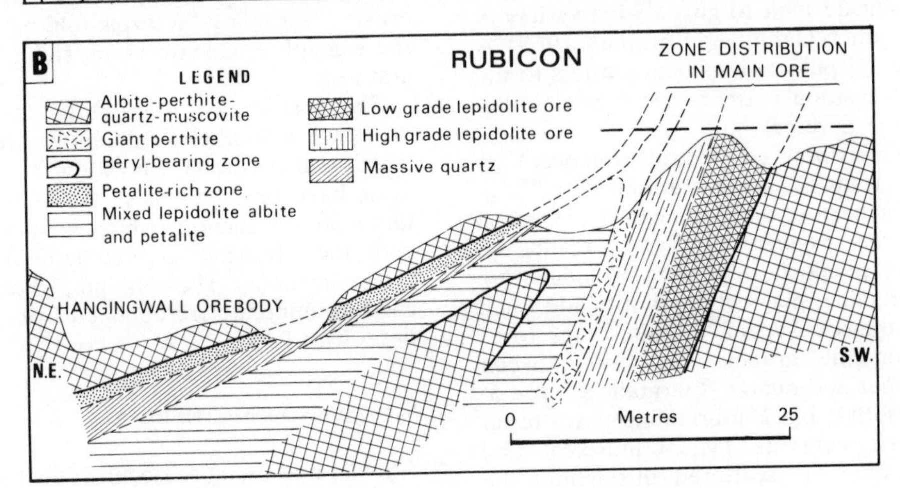

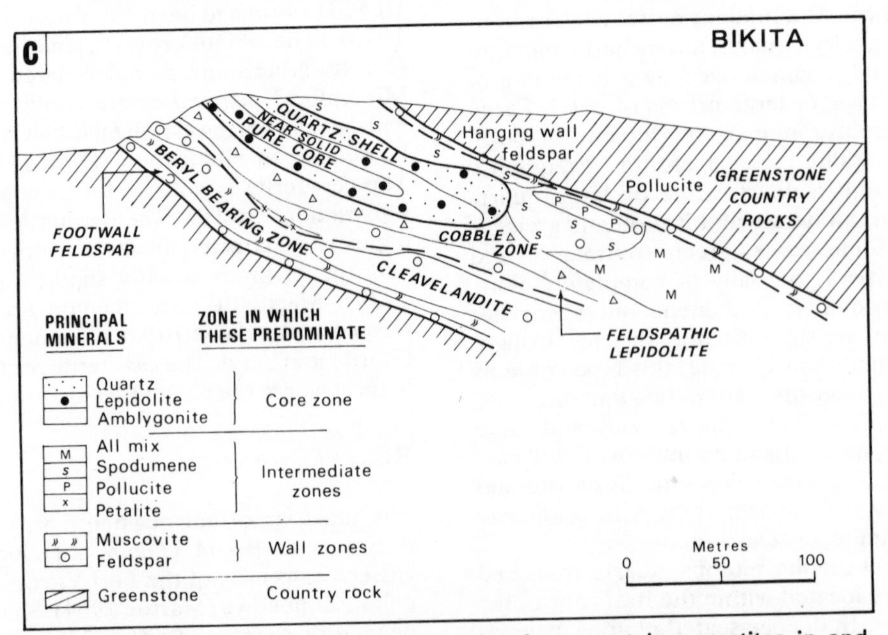

Figure 6.10. (A) Diagrammatic distribution of rare-metal pegmatites in and around granitic cupolas, Congo and Rwanda (after Varlamoff, 1972). (B) Idealized cross-section of the Rubicon pegmatite, Southwest Africa (after Roering and Gevers, 1964). (C) Cross-section through the main Bikita pegmatite, Zimbabwe (after Cooper, D.G., 1964).

caesium and beryllium in the world. The main pegmatite is about 2 km long, 30–365 m in outcrop width, with a true width of 45–60 m. It is well zoned, but very irregular (Cooper, D.G., 1964). No two cross-cuts expose the same sequence inside the wall zones, so that every cross-section is uniquely different (*Figure 6.10*).

The zonation bears some similarity to the ideal sequence of Cameron, E.N. *et al.* (1949) for North American pegmatites. The important differences are that the three most economically important zones – the lepidolite–quartz lenses, the petalite–feldspar zone, and the pollucite zone – are all unknown to the classification of Cameron, E.N. *et al.* (1949).

The most valuable parts of the pegmatite are in the core zone (*Figure 6.10*). The minerals of major economic importance are given by Cooper, D.G. (1964) as petalite, lepidolite, spodumene, pollucite, beryl, eucryptite and amblygonite. Cassiterite, tantalite and microlite were disseminated in marginal pockets in quartz-rich zones in large masses of lepidolite greisen, but are now mined out.

KAMATIVI (ZIMBABWE)

Tin-bearing pegmatites, dated at 2100 Ma, occur within basement muscovite–biotite–quartz schists (Fick, 1960). Some pegmatites are vertical or steeply dipping and may cut or follow the schistosity, while shallow dipping pegmatites are conformable with the schists. They can vary from 0.3 m to 9 m in width. Some contain up to 12% lithium oxide, Li_2O, with a 5–10% total of spodumene, amblygonite, cassiterite and tourmaline. The pegmatites not bearing lithium contain more cassiterite and tourmaline. The cassiterite is usually black subhedral with a grain size up to 40 mm.

GRAVELOTTE–MICA

The second largest pegmatite region of Southern Africa is the Gravelotte–Mica pegmatite field, located adjacent to the Murchison greenstone belt of the Kaapvaal Craton (*Figure 9.1*). It contains important deposits of mica as well as beryllium, lithium, tantalum, columbium, corundum, feldspar and silica. Emeralds together with beryl are mined from the pegmatites that intrude the biotite schists.

SOUTHWEST AFRICA (NAMIBIA)

Lithium and beryllium pegmatites, associated with 510 Ma old granites, occur within the Precambrian Damara System in the Karibib district of Southwest Africa (Roering and Gevers, 1964). The economic pegmatites are situated within larger bodies of pegmatitic granite that occur as bosses, dykes and irregular bodies. The Rubicon ore body has an uninterrupted zone of lithium and beryllium ore extending for 976 m along a northwest trend. The average thickness of the economic zone is 30–45 m (*Figure 6.10*). There is a symmetrical pattern of zonation, idealized as follows: outer pegmatite zone consisting of albite, perthite, quartz and muscovite; beryl zone containing cleavelandite (albite), muscovite, quartz, beryl and frondelite; lithium-ore zone—in the hanging-wall it is petalite-rich, and in the footwall it is lepidolite-rich; and quartz-rich core.

In some of the pegmatite deposits, there is a zone between the lithium zone and the quartz core that contains cleavelandite–beryl–columbite–(tantalite)–cassiterite.

AUSTRALIA

There are several minor pegmatite tin deposits in the cratonic metamorphic terrain of Australia. They have been reviewed by Taylor, R.G. (1979a). The Pilbara tin field, west Australia, contains two main types of pegmatite mineralization: lithium–(tin, niobium–tantalum, beryllium) pegmatitic dykes that intrude the metamorphosed basement; and tin–(tantalum–niobium, beryllium) low-angle vein swarms of pegmatite-type around the margins of 3050 Ma old granites and pegmatites. The deposits are minor but have led to local alluvial placer deposits.

The 1700 Ma old cassiterite-bearing pegmatites of the Broken Hill region are emplaced parallel to the bedding of the phyllites and schists (Katz and Tuckwell, 1979). The cassiterite occurs as irregular masses at the intersection of cross-fractures with the pegmatites. Greisenization is apparent in the cassiterite zones.

7

Batholith-associated mineralization

General Features of Batholiths

Batholiths are characteristic features of cordilleran volcano–plutonic arcs at destructive plate junctures or collision belts paralleling orogenic suture zones. Important mineralization is associated with the granitoid phases of the batholiths, which vary considerably in character and composition with tectonic setting.

Two contrasting granitic series can be readily identified, which appear to characterize different types of orogenic belts. One is a compositionally expanded calc-alkaline series, the other is compositionally restricted and predominantly granitic. The difference is most clearly expressed in terms of the gabbro–diorite/tonalite–granodiorite/granite proportions of the plutonic arc suite; the former is characterized by a proportion near 15:50:35 and the latter by one near 2:18:80 (Pitcher, 1979). The compositionally expanded series has been called the calc-alkaline plutonic association by Lameyre (1980), and it spans the quartz/alkali feldspar/plagioclase fields (*Figure 8.2*) of granite, granodiorite and monzonite, to diorite and gabbro (Streckeisen, 1976). The compositionally restricted series has been called leucogranite by Lameyre (1980). The series is confined to the fields of alkali granite and granite. Its restricted composition, nature of xenoliths and tectonic setting imply that it results from the anatexis of a continental basement of metasediments, metavolcanic rocks and older granitoids (Lameyre, 1980), whereas the calc-alkaline series may have a mantle source.

In discussing these two orogenic series, the twofold subdivision of Chappell and White (1974) and White, A.J.R. and Chappell (1977) is useful and has been applied widely (Chappell, 1978).

The S-type plutonic suite is compositionally restricted to granite and adamellite. Normally the sodium oxide is less than 3.2% in rocks that contain approximately 5% potassium oxide and decreases to less than 2.2% in rocks that contain about 2% potassium oxide. The molecular ratio $Al_2O_3/(Na_2O + K_2O + CaO)$ exceeds 1.1 and usually S-type granites contain more than 1% C.I.P.W. (Cross – Iddings – Pirsson – Washington) normative corundum. They range to types rich in silica, but their major-element plots on variation diagrams tend to be irregular. S-type granites are usually rich in biotite and devoid of hornblende. A continental crustal source is implied by their initial $^{87}Sr/^{86}Sr$ ratios of 0.706 or greater; xenoliths contained in them are of metasediments. Tin and tungsten mineralization appears to be confined to the more silicic S-type granites, indicating that tin is recycled from buried continental crust by anatectic events (Hutchison and Chakraborty, 1979).

In contrast, the I-type plutonic suite is high in sodium. Normally sodium oxide exceeds 3.2% in felsic rocks decreasing to more than 2.2% in more mafic varieties. The molecular ratio $Al_2O_3/(Na_2O + K_2O + CaO)$ is less than 1.1 and I-type granites contain C.I.P.W. normative diopside or less than 1% normative corundum. They are not compositionally restricted and there is a broad range from felsic to mafic with a peak in the tonalite field (*Figure 8.2*). The spectrum in any one batholith system shows a regular interelement variation as displayed on near-linear variation diagrams. I-type granitoids have an igneous source as shown by their content of hornblende-rich xenoliths and they may have a mantle source as indicated by the low $^{87}Sr/^{86}Sr$ ratios of 0.704–0.076. The rocks themselves usually contain hornblende and sphene. There is no association with tin, but rather with porphyry-type copper, iron and gold deposits. Ongoing research is showing that tungsten also has an important association with I-type granitoids in addition to its association with the S-type.

Some plutonic arcs are composite and contain both types of granitoid; for example, the Japanese island arc and the eastern volcanic–plutonic arc of the Malay Peninsula. Usually the S-types are earlier in the intrusive sequence and often have a strong secondary foliation truncated by later I-type intrusions that are either massive or have a dominant primary foliation (Chappell and White, 1974).

The S-type granites characteristically contain ilmenite and monazite and have low magnesium/iron ratios in the biotites. They are referred to as the ilmenite series (Ishihara, 1977). They were generated in the middle-to-lower continental crust by anatexis of the carbon-bearing metamorphic and sedimentary rocks that caused the magma to have a low oxygen fugacity.

The I-type granitoids contain magnetite and sphene and have high magnesium/iron ratios in the biotite. They are referred to as the magnetite series (Ishihara, 1977). They were generated in the upper mantle and have not been contaminated by carbon-bearing crustal rocks, and hence were formed under a higher oxygen fugacity.

S-type and I type can occur separately or together. Some orogenic belts, such as the Mesozoic of the Andes, are characterized almost exclusively by the compositionally expanded I-type with initial $^{87}Sr/^{86}Sr$ ratios near 0.7042, accompanied by significant volumes of gabbro and associated in time and space with andesite and ignimbrite (Pitcher, 1979).

Other orogenic belts like the Main Range of the Malay Peninsula (Hutchison, 1977, 1978c; Ishihara *et al.*,1979) are exclusively of S-type granite, devoid of a direct volcanic association and strongly compositionally restricted. In contrast, the eastern-belt granites of the Malay Peninsula have mixed S and I characters, and include a volcanic and gabbro–tonalite association.

In the Mesozoic cordillera of the western USA, the plutons are distinctly I-type on the west with initial $^{87}Sr/^{86}Sr$ ratios as low as 0.702 in western Oregon. These ratios progressively increase eastwards to 0.707 in Montana (Armstrong, R.L. *et al.*, 1977). The edge of the craton can be taken as the western limit of Precambrian outcrops, which coincludes rather well with the granite strontium-isotope contour line of 0.704. All plutons to the west of this contour are definitely I-type and east of the line they show an intermediate character, but there are no truly S-type granites in the cordillera and this may explain the absence of tin.

The Caledonian granitoids of Scotland and Ireland are distinctly intermediate between I-type and S-type (Pitcher, 1979), which again might explain the absence of tin.

It is clearly demonstrated by the data that geographical position related to the cratonic margin, lithospheric depth and crustal thickness control the nature of the granite magmas. It has been observed (Lipman *et al.*, 1971; Dickinson, 1975b) that the potassium contents of the plutonic rocks steadily increase inwards from the consuming plate margin towards the cratonic interior, and appear to have a positive correlation with depth to the underlying Benioff Zone. However, the exact significance of this is open to speculation and the nature of the intruded crust plays some role.

Batholith-Type and Orogenic Association

There are three contrasting types of orogens (Zwart, 1967; Pitcher, 1979). Their general features may be summarized as follows (*Figure 7.1*).

Alpinotype orogenies

Features are island-arc volcaniclastics and lavas deposited in oceanic trenches, crustal-shortening involving thrusting with nappes predominant and high-pressure regional metamorphism with wide progressive zonation; ultrabasic rocks are abundant. Granitoid batholiths are characteristically absent.

Andinotype Orogenies

Features are island-arc or continental-margin volcaniclastic rocks and lavas deposited in troughs of eugeoclinal-type located within the continental lip and paired with belts of shelf facies clastics; there is little crustal shortening but vertical movements are dominant, with open drape-folding that lacks cleavage; there is regional burial metamorphism. Compositionally expanded disharmonious batholiths exist with important basic plutonic and andesitic volcanic association. I-type granitoids with crustal involvement occur only in the later stage of evolution; $^{87}Sr/^{86}Sr$ initial ratios are less than 0.706, including restite material from subcrustal sources.

Hercynotype Orogenies

Features are non-volcanic continentally derived sediments in intracratonic basins of miogeoclinal-type, crustal shortening with upright folds and cleavage and low-pressure metamorphism with prograde zonation; ultrabasic rocks are rare. Compositionally contracted, harmonious, granitic

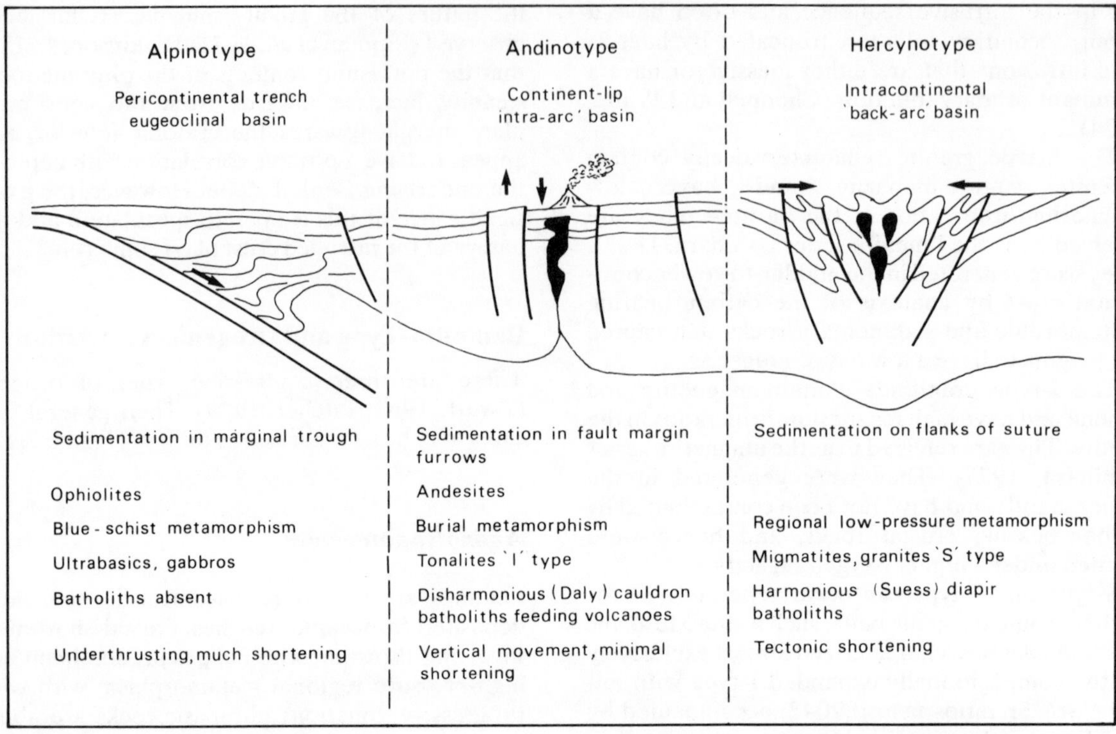

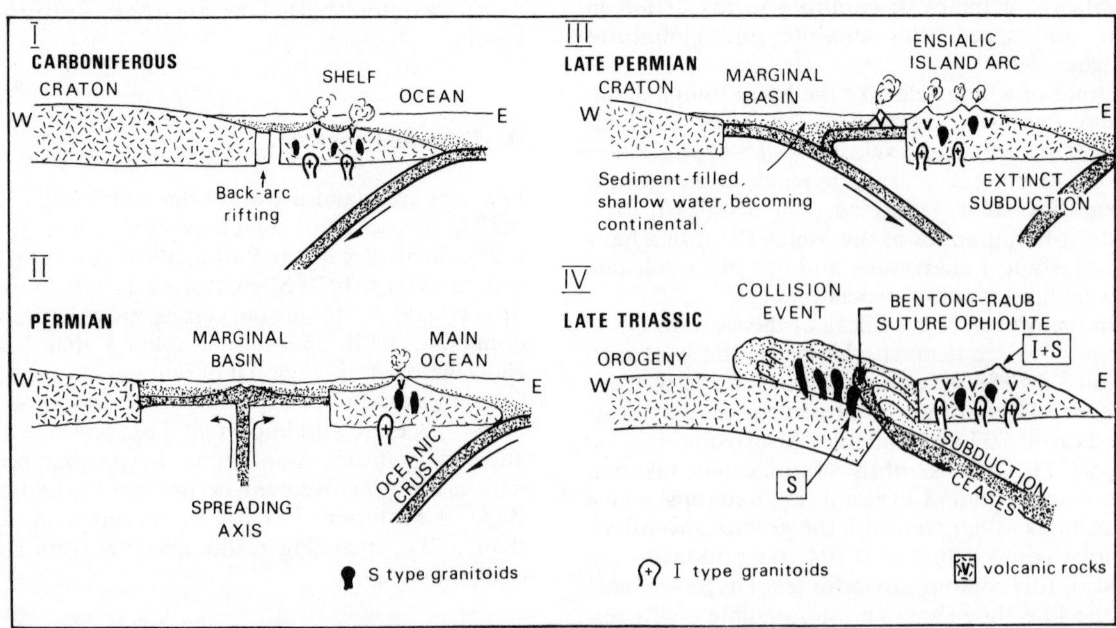

Figure 7.1. Top: schematic classification of orogen-types (after Pitcher, 1979). Bottom: the evolution of the paired granitoid belts of the Malay Peninsula (after Hutchison, 1977, 1978c), shown in diagrammatic cross-sections.

batholiths exist with only minor basic associations and generally lack contemporaneous volcanics. The $^{87}Sr/^{86}Sr$ ratios are greater than 0.706 with a considerable range; batholith variation diagrams usually fail to show linear trends; inherited xenocrysts derived from recycled crustal rocks are characteristic.

This threefold subdivision undoubtedly over-simplifies the tectonic spectrum of orogenesis since each mobile belt is unique. However, the division does show the range of the spectrum. It is only in the Andinotype orogeny that there is a space–time relationship between plutonism and volcanism. The relatively dry magmas of the Andinotype rise into the higher crust by way of important vertical faults. Even here, however, it is not yet clear what the relationship is between the batholiths and the volcanic rocks. It may be that silicic ignimbrites are vented from plutonic-based calderas, whereas the andesites are extruded from the deep fissures.

In the Hercynotype orogen, the batholithic magmas are wet and do not rise and are impounded within unfractured folded and metamorphosed country rocks. The batholiths are deep seated within the folded basin and may be classified as mesozonal as opposed to the epizonal Andinotype. Some of the Hercynotype granites have been recognized by Mitchell, A.H.G. (1979) as collision-related. The mechanism envisioned is attempted subduction of a continental plate as a basin closes (*Figure 7.1*), resulting in crustal thickening and anatexis; there are examples in the Himalayas, southwest England and the Main Range of the Malay Peninsula.

In some regions there may be closely paired orogenic belts. For example, Hutchison (1978c) has presented a model for the two batholith belts of the Malay Peninsula (*Figure 7.1*).

Relationships Between Granitoids and Hydrothermal Deposits

Residual silica-rich liquids collect preferentially under cupolas or bosses in the roof zones of granitoid batholiths during advanced stages of their crystallization (Emmons, 1933). The detailed behaviour of these liquids, and whether or not the country rocks fracture to allow their escape, and when they fracture, is clearly related to depth of the granitoid emplacement (Hesp and Varlamoff, 1977; Varlamoff, 1978) (*Figure 7.2*). The pegmatite style of mineralization is related to deep-seated granitoid emplacements at depths of the order of 10 km (lower mesozone or catazone). However, following Emmons (1933), pegmatites can be emplaced higher in the crust if fracturing or faulting intersected the roof at an early stage before the granitoid evolved to a hydrothermal system. Towards the end of crystallization, the residual liquid is depleted of granitic constituents and, if fracturing occurs at a late stage, hydrothermal deposits are formed. If the roof does not break during crystallization of the granitoid, the cassiterite and wolframite will form pipes within the granitoid or it will be finely disseminated throughout the roof zone. Granite cupolas associated with greisenization are characterized by a limited vertical extent of veining into the country rocks, and are mesozonal with a depth of emplacement around 8 km (*Figure 7.2*). Epizonal emplacements at about 4–6 km are associated with vein systems characterized by long down-dip persistence, and some of the fractures that are now filled with hydrothermal vein deposits may have even reached the surface. The subvolcanic intrusions, at depths down to 2 km, are associated with rhyolites, ignimbrites and volcanic vents.

Within this spectrum of depth of emplacement, there is a characteristic spectrum of mineralization and associated alteration (*Figure 7.2*). In particular, sulphides of zinc, lead, copper and pyrite are associated with tin and tungsten only in the epizonal and upper mesozonal granitoids, and albitization and greisenization with mesozonal emplacements only. The figure is a generalization, however, and the complexity of geological phenomena may result in many exceptions to the scheme.

The paired orogenic tin belts of the Malay Peninsula (Hutchison, 1977, 1978c) offer a good illustration of the change in style of mineralization in response to tectonic environment. The Main-Range Triassic granite batholith was mesozonally emplaced into folded greenschist facies metasediments. The granites are of the S-type ilmenite series. The hydrothermal solutions were impounded by the unfractured country rocks so that there are no tin lodes or veins with a considerable dip extent (Hosking, 1977). The commonest type of deposit is a greisen-bordered vein swarm confined sharply to the contact zone in the roof of the cupolas and bosses, with cassiterite restricted to a maximum of about 100 m down-dip.

In sharp contrast the epizonal eastern Permo-Triassic belt granitoids are of mixed ilmenite and magnetite series with well-developed andalusite and cordierite-bearing thermal aureoles. The overlying country rocks fractured easily, so that the hydrothermal solutions repeatedly filled the fractures to form Cornish-type lodes of considerable strike and dip length. Usually the cassiterite is concentrated approximately 500 m from the granite and extends outwards for another 500 m (Hosking, 1977). The Phuket region of Thailand is characterized by pegmatites that have been facilitated on their upward movement into the epizone by major strike slip faults, providing an exception to the scheme of *Figure 7.2* in which pegmatites are displayed as catazonal.

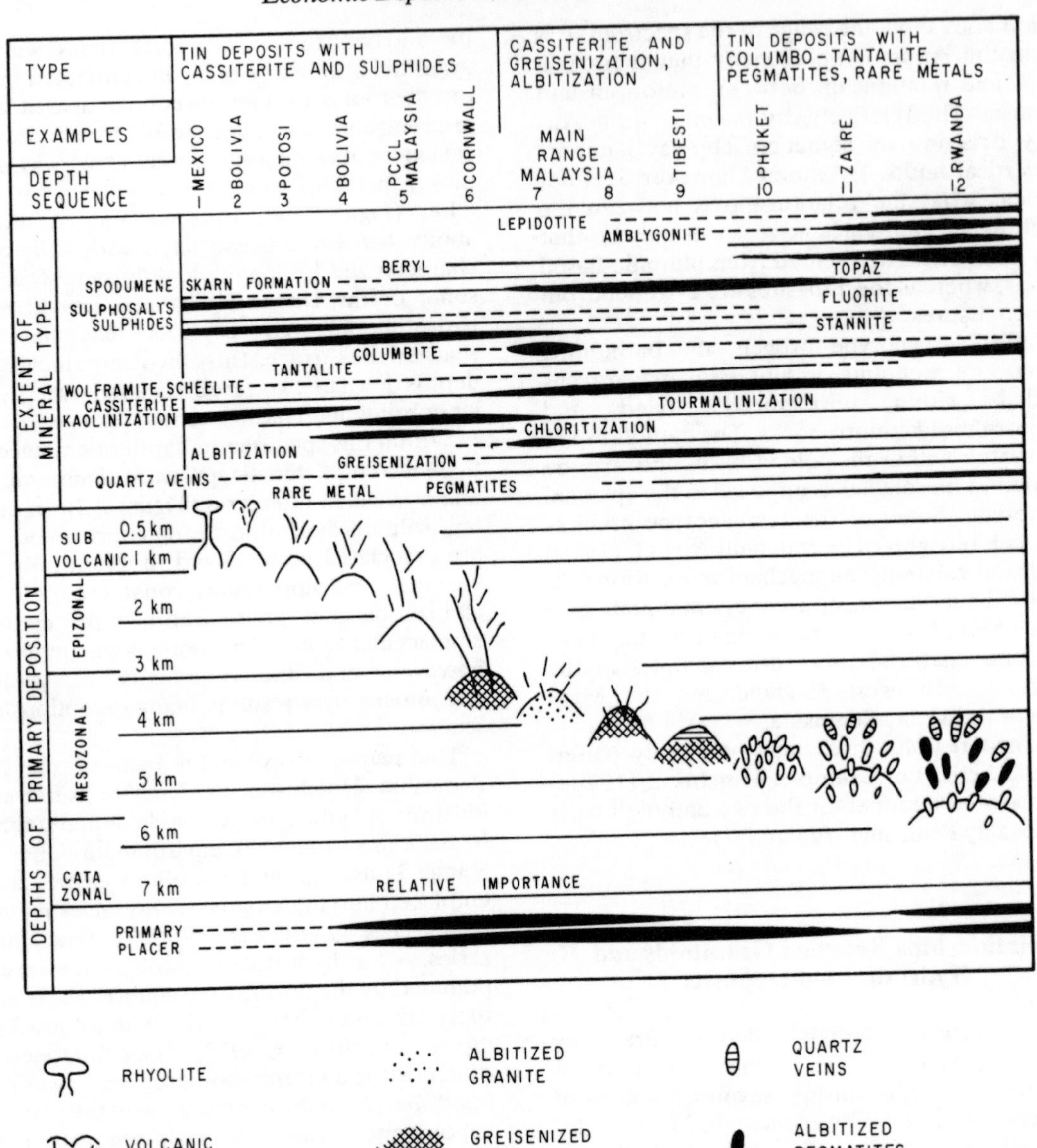

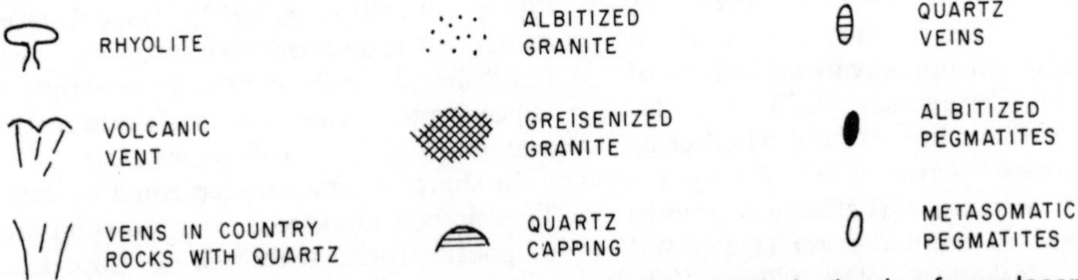

Figure 7.2. Classification of tin deposits according to their depth of emplacement of the associated granitoids (after Varlamoff, 1978), showing also the associated mineral associations and types of alteration and relative importance of the deposits. The depth scale of Varlamoff is too shallow. A more generally acceptable scale puts the base of the epizone at 6.2 km and the base of the mesozone at 10.8 km.

Temperature of Hydrothermal Formation and Alteration

Fluid-inclusion studies of hydrothermal vein deposits have shown that cassiterite crystallized at the higher-temperature range in the evolution of a quartz-rich hydrothermal vein deposit simultaneously with greisenization of the neighbouring granitoid or skarnization of country rocks, since both processes have been equated (Shcherba, 1970). In southern Brittany, France, Charoy and Weisbrod (1975) found the cassiterite temperature of crystallization to be around 350°C. In Czechoslovakia, the range of temperature was determined as 370–426°C (Durisova, 1978). In Cornwall (Moore, F. and Moore, 1979) it was determined as 360–380°C. In Finland, the fluids that deposited the cassiterite had a temperature of 260–390°C and a salinity of 3–17% sodium chloride equivalent by weight (Haapala and Kinnunen, 1979). At these temperatures, meteoric waters can be shown to be the dominant source of the hydrothermal fluids during greisenization and cassiterite deposition (Sheppard, S.M.F., 1977a). Later-stage kaolinization associated with the mineralized veins took place when the temperatures waned to about 150°C. In all vein systems, quartz may be of several generations, spanning a temperature range 400–120°C.

In most greisen-bordered veins, tungsten is deposited within the range 270–320°C (Ivanova, 1978), a range in exact agreement with the work of Casadevall and Rye (1980), who show also that the veins were formed from meteoric water. However the sulphur is of deep-seated origin. The lower-temperature deposition of wolframite, when compared with cassiterite, appears to have been triggered by a major influx of meteoric water into the circulating hydrothermal system. This may form the basis of an explanation why individual lodes in a tin–tungsten province rarely carry both cassiterite and wolframite. Although both metals are ultimately of magmatic origin, perhaps both from the same magma, the tungsten may be leached away by lower temperature meteoric–hydrothermal solutions to be redeposited elsewhere at lower temperature (Taylor, H.P., 1979).

The Malay Peninsula

The paired orogenic batholith belts of Peninsular Malaysia and Thailand have been analysed by Mitchell, A.H.G. (1977) and Hutchison (1978c) The two main tin granitoid belts are separated by a central basin filled with a Palaeozoic and Mesozoic folded sedimentary sequence that shallowed from marine in the Permo-Triassic to continental by the Jurassic (Hutchison, 1982). The interpretation of this central basin as representing a marginal sea is supported by its positive gravity anomaly, its sedimentary succession and its association with submarine volcanism. Mineralization in the central basin is of gold and base metals as would be expected from the absence of an underlying cratonic sialic crust (Taylor, D. and Hutchison, 1979). In contrast, the two fringing granitoid belts are characterized by negative gravity anomalies.

The Triassic western or Main-Range tin belt has mesozonal characteristics. The granites were emplaced in phyllitic greenschist facies metasediments with no thermal aureoles. They are of the S or ilmenite series, their alkali feldspar is of maximum microcline, and they are of restricted granitic composition. The associated tin deposits are closely confined to the granite–country rock contact, as summarized by Hosking (1977). There are no tin lodes or veins with persistent dip length, although some hydrothermal pipes in marble may be extensive. However, Emmons (1933) equated such pipes with a deep-seated environment in which the batholith roof did not fracture. The commonest type of deposit is a greisen-bordered vein swarm within the granite along the roof zone of the cupolas and bosses. In these, the cassiterite is restricted to a maximum of about 100 m downdip. Rich massive lodes of irregular shape and size are characteristically impounded at the granite–country rock contact. They have a similarity with the *carbonas* of Cornwall.

The Permo-Triassic epizonal eastern-belt tin granitoids contain orthoclase and intermediate microcline, and are surrounded by andalusite and cordierite-bearing thermal aureoles. The plutonic suite includes abundant tonalite and a significant proportion of gabbro, but the major volume is of granodiorite, and there is a strong intermediate-to-rhyolitic volcanic association. The granitoids are both of the magnetite (I) and ilmenite (S) series, rather sporadically mixed (Ishihara *et al.*, 1979). In this, there is a similarity with the Tasman Geosyncline (Chappell, 1978) and the ensialic arc of Japan (Ishihara, 1977), which is now separated from its former cratonic attachment by the Sea of Japan marginal basin. The granitoid cupolas are covered with country rocks that fractured readily and the hydrothermal solutions frequently filled the reopened fissures to form Cornish-type lodes of considerable strike and dip dimension (Hosking, 1977).

The eastern Malay Peninsula separated during the early Palaeozoic from its craton by the opening of the central marginal basin (Hutchison, 1978c), which subsequently closed because of extinction of spreading and reversal of subduction polarity, with final Late Triassic collision of the island arc against its ancestral craton to give the S or Hercynotype Main-Range granite batholith (*Figure 7.1*).

The two orogenic batholith belts of the Malay Peninsula contain tin and tungsten deposits but of different character, reflecting the differences in their tectonic history. Whereas the eastern belt represents an Andinotype volcano–plutonic arc, isostatically stable since the end of the Triassic period, the Main-Range belt was formerly deep seated and tectonically depressed and has been isostatically uplifting since the Triassic to cause unroofing of the granite cupolas and bosses of the batholith during the Late Tertiary period, resulting in the impressive Quaternary placer deposits of the giant Kinta Valley, Kuala Lumpur and Bangka tin fields.

The Eastern Volcano–Plutonic Arc

PAHANG CONSOLIDATED CO. LTD. MINE (SUNGEI LEMBING, MALAYSIA)

This mine is situated about 20 km west of Kuantan, and is generally referred to as PCCL. The geology has been described by Fitch (1952). A thick monotonous sequence of Viséan-basal Namurian shale with subordinate siltstone and sandstone generally dips 30° towards the east-northeast. At the mine location the rock sequence is intruded by a ridge of Permian granite (Hutchison and Snelling, 1971). The strata are fossiliferous and ripple-marked. The thermal aureole around the granite contains chiastolite (Fitch, 1952). The mine has workings to a depth of 340 m. The tin lodes are steeply dipping mineralized faults that have been repeatedly reopened. The principal lodes strike due east and also eastsoutheast and the common dip is 70–90°. Most lodes lie about 600 m from the top of the granite. As in Cornwall, England, the mineralization does not extend down into the granite, although the faults and quartz veins continue downwards. The zone of mineralization occupies a band about 600 m wide that lies parallel to the granite contact (*Figure 7.3*). There is a sharp distinction between this style and that of the Main Range, in which the lodes occur within the contact zone of the granite cupolas.

Much the same type of mineralization occurs down-dip for at least 600 m, indicating that the temperature gradient was very gentle and that meteoric water circulating through the country rocks played a dominant role. The lack of obvious zoning in individual lodes is a feature of all Southeast Asian tin deposits (Hosking, 1970b, 1973a) and in this they differ markedly from those of southwest England. As in Cornwall, cassiterite of several generations can be recognized. The broad paragenesis appears to be: cassiterite, followed by arsenopyrite, pyrite, pyrrhotite, sphalerite (with exsolved chalcopyrite), chalcopyrite, sphalerite (without exsolution) and galena. Quartz was introduced at various times. A large amount of iron-rich chlorite was formed immediately after the cassiterite. The frequent reopening of the lodes led to a commonly brecciated texture. Further complications result from pre-mineralization and post-mineralization faulting, but the structural history of the PCCL mine is not fully resolved. In several cases, the lodes follow propylitized dacitic dykes along part of their extent, and these dykes have played a role in localizing the lodes, yet the dykes themselves do not appear to be mineralized, presumably because they remained unfractured.

Fluid-inclusion studies (Foo, 1977) indicate that the repeated opening of the lodes resulted in fluid boiling. The quartz deposition occurred in various stages, at 272°C, 180°C, 148°C and 70°C.

Wolframite has been produced nearby in the past at Kuala Sungei Perong (*Figure 7.3*) and also from the granitic area at Ulu Sungei Reman, where it occurred in vein quartz along with tourmaline and chlorite (Fitch, 1952). The amount was always small compared with the cassiterite, and although occurring in the same general district, wolframite and cassiterite do not occur together in the same lode.

BUKIT BESI (TRENGGANU–PAHANG, MALAYSIA)

The Permo-Carboniferous shales are locally calcareous and the sequence contains minor sandstone and limestone. The sequence has been folded and subjected to low-grade metamorphism, then further deformed and contact-metamorphosed by Permian granitoid intrusions (Bean, 1969). The marble and calcareous shale horizons were converted to skarn containing garnet, pyroxene and amphibole assemblages. Near the granite contact, calcareous bodies were converted to amphibolite containing magnetite and pyrrhotite.

Iron-ore bodies all lie close to the granite. The ore has formed by replacement of shale. The bodies are irregular in shape and are composed of magnetite, secondary martite and haematite enclosing relicts of unreplaced country rock. Some of the ore bodies are tabular and distinctly strata-bound, and these formed the major ore bodies, now mined out. They contain significant pyrite. They also occur as replacements of roof pendants within the main granitoid outcrops. The skarn ore is of massive magnetite in a gangue of garnet and diopside and it forms tabular bodies (Bean, 1969).

Ore bodies were preferentially formed where the sediments dip towards the granite contact, facilitating the ingress of hydrothermal fluids. Locally there occurred marked concentrations of cassiterite owing to replacement of the host rocks around fracture zones. At Sri Bangun, the cassiterite is associated with iron oxides in the shales. At Batu Tiga, the ore consists of early magnetite, followed by cassiterite and scheelite, chlorite, pyrrhotite, chalcopyrite and late pyrite (Hosking, 1973a). These ores developed in a calc-silicate host and the deposition was controlled by fault zones, and impounded by overlying metaquartzite.

Apart from these tin concentrations, a certain amount of tin was present throughout the iron ore, of average content 0.07% (Bean, 1969). This tin is structurally held in the magnetite.

Hosking (1973a) feels that the iron ore should not be called skarn because most of it is not in a calc-silicate hornfels. Furthermore the cassiterite introduction was controlled by mineralizing fluids ingressing along fault zones. The deposit is therefore hydrothermal and metasomatic. Hosking (1973a) also holds that all the Malaysian pyrometasomatic deposits are hydrothermal and happen to have developed within a favourable calc-silicate hornfels host, but not exclusively. However, these fine distinctions in ore-body classification are unnecessary, for Shcherba (1970) has included skarn and amphibolitization under the hydrothermal heading of greisen. The different types of rock alteration and mineralization simply reflect different responses of the variety of country rocks to the ingress of hydrothermal fluids.

It is unrealistic to imagine that the same granitoid that provided the tin also gave the iron. The volcano–plutonic belt contains abundant diorites as well as granite. Either one, acting as a heat source, could have caused circulation of meteoric water throughout the granite as well as the diorites and sediments to mobilize both iron and tin and deposit them around the granitic heat source. It is reasonable to assume that an ilmenite-series granite is the source of the tin and a magnetite-series diorite that of the iron, but the later tin granite may have been responsible for driving the hydrothermal activity. The possibility also cannot be ruled out that the sedimentary rocks were iron-enriched during Permo-Carboniferous volcanism contemporaneous with their deposition.

PELEPAH KANAN (KOTA TINGGI, JOHORE, MALAYSIA)

This opencast mine hill is composed of massive magnetite, largely altered to martite, resting on bedded metaquartzite and calc-silicate hornfels, which is intruded from beneath by porphyritic biotite granite. The calc-silicate rocks have resulted from metamorphism of a sequence of calcareous shales, siltstones and mudstones containing the following mineral assemblages: quartz–epidote – sericite, quartz – diopside – epidote – sphene and quartz–diopside. Near the granite contact there is a garnetiferous hornfels and some rocks contain cordierite porphyroblasts (Hosking, 1973a).

Mineralized veins, of thickness from 30 cm to less than 2 cm, intrude the metasediments usually concordantly, but some are vertical and form a stockwork. The veins are of variable mineralogy, containing the following paragenesis: alkali feldspar, followed by biotite–chlorite, coarse cassiterite, scheelite(?), lollingite, acicular cassiterite, ankerite, siderite, haematite and fluorite. Quartz is abundant and overlaps the above sequence. Although feldspathic, the veins are not pegmatite and there are two generations of cassiterite (Bean, 1969). The earlier cassiterite is characterized by intense colour-zoning, which is a most striking feature; later needles are never zoned. The cassiterite is ferromagnetic and its magnetism is destroyed by heating to 830°C for 15 minutes (Flinter, 1960). The darker the cassiterite zone, the greater its magnetic susceptibility. Generally the magnetization appears to be due to hydrated ferrous stannate. However, the magnetic character is also due in part to magnetite inclusions within the cassiterite (Grubb and Hannaford, 1966).

It is generally held that this contact deposit is a hydrothermal iron–tin replacement body developed above the granite contact beneath an impermeable cap rock that impounded the fluids (Hosking, 1973a). However, one must conclude that the sediments were enriched in iron before the contact metamorphism, for their mineralogy includes the iron-rich amphibole grunerite. Hydrothermal introduction of tin mineralization from the biotite granite is therefore imprinted upon and mixed with an earlier iron-enrichment, perhaps hydrothermally leached from a diorite source by the hydrothermal fluids circulating and being driven by the granite heat source.

The iron-ore deposit, which is large, has proved of little value because the tin content is too high for smelting. The weathered skin of the iron ore has been mined for its cassiterite but the recovery is poor.

BILLITON (INDONESIA)

The peneplained island of Billiton is composed of east–west-striking Carbo-Permian highly inclined alternating beds of sandstone and clay–slate, intruded by a Triassic sequence of gabbro, granodiorite and granite, dated by a rubidium/strontium isochron at 213 Ma (Jones, M.T. *et al.*, 1977). The granitoids show some contact-metamorphic effect on the country rocks (Adam, 1960).

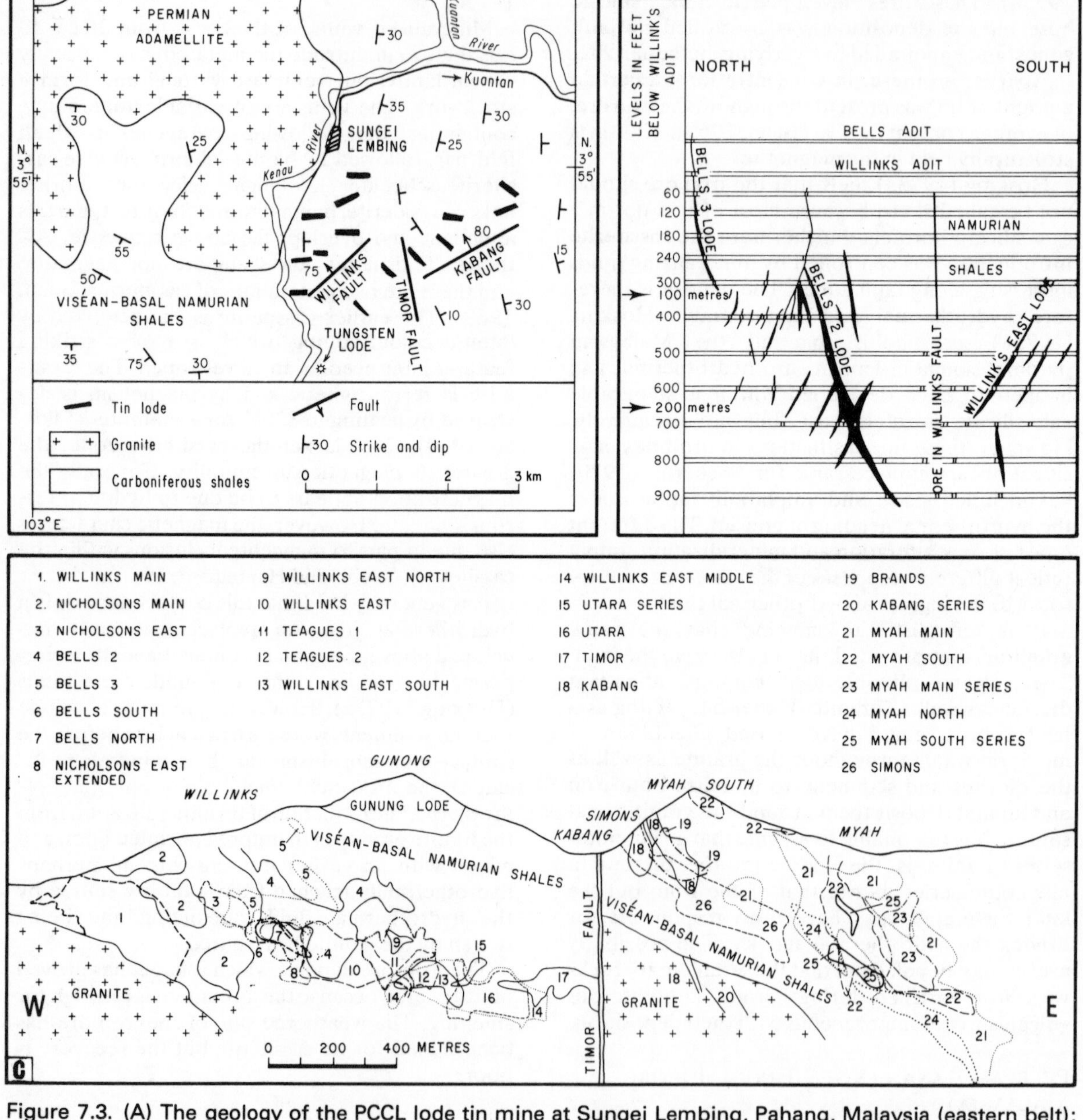

Figure 7.3. (A) The geology of the PCCL lode tin mine at Sungei Lembing, Pahang, Malaysia (eastern belt); based on Fitch (1952) and Hosking (1937a), simplified geological maps; (B) vertical north–south section showing the lodes in relation to the Willinks fault; (C) east–west cross-section with the lodes projected on to the plane of the section showing that lodes lie parallel to and away from the granite contact.

The Klappa Kampit mine reached a depth of 300 m. The lodes are entirely within the sedimentary rocks and no granite has ever been encountered in the mine. The most important lodes are conformable with the sandstone bedding and have been called 'bedding plane lodes'; they are of considerable lateral extent. Just as at PCCL in Malaysia, no vertical zoning has been observed. The chief minerals are pyrite, pyrrhotite, arsenopyrite, chalcopyrite, cassiterite, magnetite, chlorite, galena and quartz. All lodes are rich in iron and sulphur. Large parts are almost entirely of magnetite. The lodes are devoid of wolframite, though neighbouring mines have fissure veins containing wolframite but no cassiterite (Adam, 1960). Some localities have a similarity with skarn mineralization and the assemblage includes amphibole, pyroxene, garnet and fayalite; in addition, there is tourmaline, zircon, biotite and fluorite. Hosking (1977) has suggested that the bedding plane stratiform lodes may be of volcanic parentage. This problem is discussed further below.

Lower-grade tin mineralization occurs along major strike slip faults and within fracture fillings in the sandstone. The initial strontium ratio of the granite (0.7139) clearly indicates a continental crustal source and the galena has yielded lead-isotope data suggesting that the lead is also from such a source (Jones, M.T. *et al.*, 1977).

A general review of the geology and mineralogy of the tungsten vein deposits of the eastern belt is to be found in Hosking (1973b).

Main-Range Belt (Malaysia)

The Kinta Valley, around Ipoh town, is the greatest alluvial tin field in the world. The placer cassiterite has been deposited close to the contact of the Palaeozoic strata, which are predominantly calcareous, with the Triassic Main-Range granite (Ingham and Bradford, 1960). The valley has given several interesting hard-rock deposits, now largely mined out, and it also displays a regional geochemical zoning (*Figure 7.4*). Tungsten occurrences all lie very close to the granite, whereas those of lead–zinc are in the metasediments (Hosking, 1973a). Some of the primary mineralization styles are described below.

Tin Skarns

The intrusive contacts between the Triassic granites and the Permian and older marble formations have provided a favourable environment for hydrothermal skarn deposition of tin and tungsten.

If tin is introduced at an early or high-temperature stage in the contact phenomena, it may be incorporated in andradite and axinite (Hosking, 1973a) even up to 2% tin. If the introduction is somewhat later, malayaite, a late-stage hydrothermal silicate isostructural with sphene, is deposited. At Batang Padang (Kinta Valley) malayaite is associated with varlamoffite (Alexander and Flinter, 1965), where it occurs in collapsed pipe-like deposits in the limestone floor of an alluvial tin mine. The malayaite alters to varlamoffite, which is a hydrous oxide of tin and results from the supergene oxidation of stannite. It is intimately intergrown with cassiterite. In all the skarn deposits, cassiterite and sulphides replace dolomite and tremolite, and stannite replaces cassiterite.

At the Hiap Huat mine, Kuala Lumpur area, malayaite rims sphene and is associated with vesuvianite, diopside, epidote, alkali feldspar, chlorite, calcite, lollengite and quartz. At Templer Park, Kuala Lumpur, malayaite, scheelite and pyrite occur in a vesuvianite–garnet–diopside–wollastonite–hedenbergite–quartz skarn. Malayaite and scheelite also occur in skarn elsewhere in the Kuala Lumpur area in the Sungei Way and Pudu Ulu mines. However, not all tin skarns contain malayaite. Those at Telok Kruen consist of cassiterite, arsenopyrite, bismuth and scheelite with a gangue of calcite and tremolite.

Several irregular chimney-shaped hydrothermal skarn deposits have been found in the Kinta Valley, referred to as pipes. All are fissure fillings or metasomatic replacements of limestone. The hydrothermal fluids have presumably originated from the granite and have carried sulphur, arsenic, tin, copper and iron to form the pipe-like bodies consisting of calcite, arsenopyrite, cassiterite, chalcopyrite, bornite and pyrite. In some cases the lodes contain tremolite, fluorite, quartz, fluoborite, phlogopite and talc. No pipe has been traced into the granite because of mining difficulties.

The most important stanniferous skarn deposit ever discovered was the Beatrice Pipe, Selibin, Kinta Valley (Ingham and Bradford, 1960). The body was an irregular pipe-like shape of length 22 m, width 6 m, elongated in the direction 15°E. Thirty metres to the south, a continuation dipped towards 196°, became steeper with depth and almost ended up vertical (*Figure 7.4*). The pipe abruptly ended against granite at a depth of 120 m at a faulted limestone–granite contact (Willbourn, 1932). The total length of the pipe exceeded 260 m. The amount of cassiterite varied considerably; it frequently occurred in bands parallel to the border of the pipe with the limestone and as irregular crystal clusters filling gaps between arsenopyrite. The gangue was mainly tremolite and arsenopyrite, with minor fluoborite and phlogopite. Talc, fluorite and dolomite occurred in small amounts mostly in the margins. Chalcopyrite, pyrrhotite and bornite accounted for the 5% or less copper content. Most of the tin occurred as cassiterite, but some stannite was present. The pipe yielded 9000 t of cassiterite concentrate. It was found under the alluvium as oxidized ore composed of rusty iron oxides, scorodite, arsenopyrite

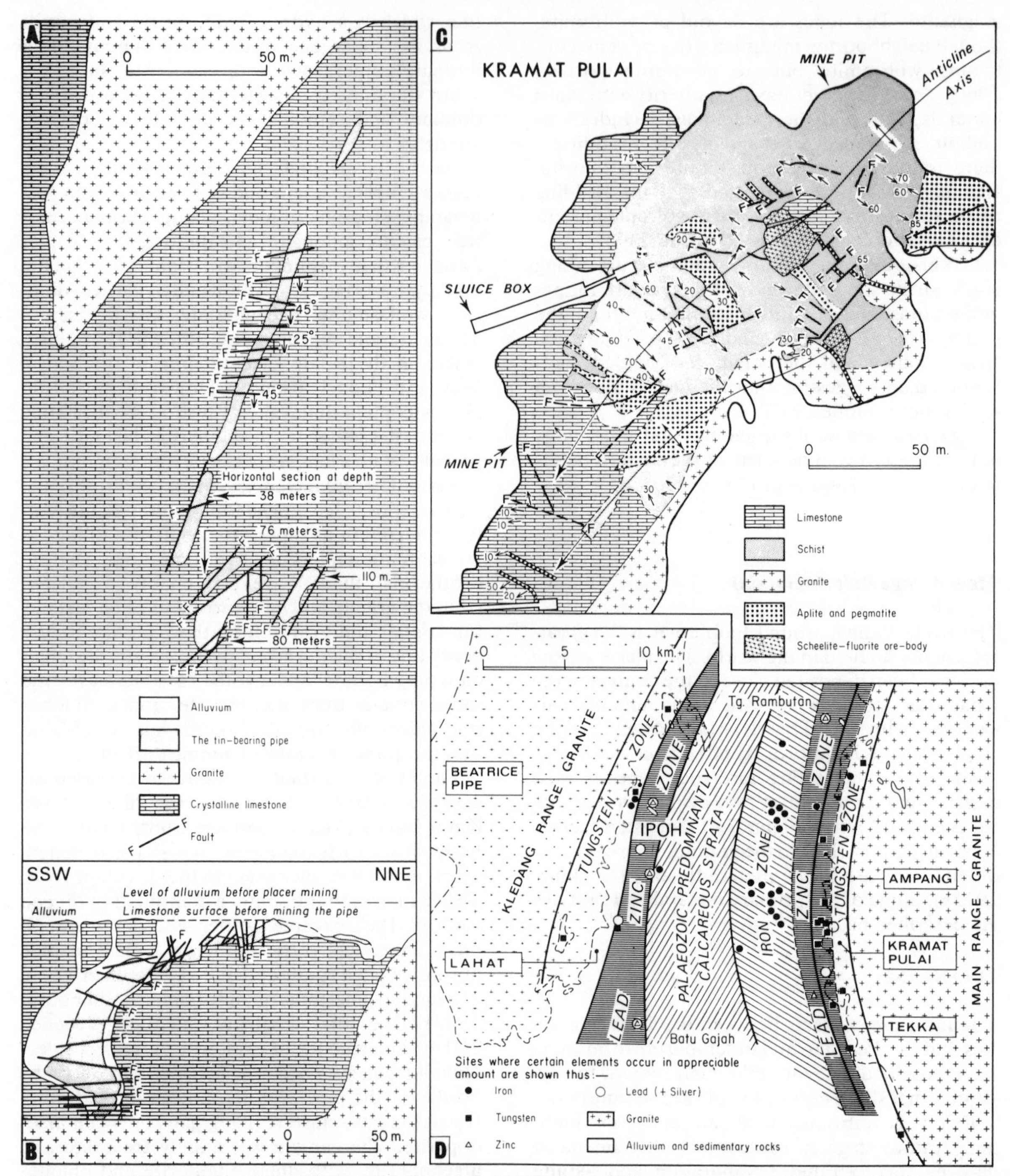

Figure 7.4. Ore bodies of the Kinta Valley, Malaysia. (A) Plan, and (B) cross-section of the Beatrice Pipe, after Willbourn (1932). (C) The Kramat Pulai scheelite–fluorite mine (after Willbourn and Ingham, 1933). (D) Metal zonation in the valley (after Hosking, 1973a).

and cassiterite, changing into unoxidized sulphide ore at depth.

The Lahat pipe of the Kinta Valley was mined to a depth of 100 m. It consisted mainly of angular pieces of cassiterite and iron oxides cemented by calcite. Only below 90 m was the ore found to be unbrecciated and to consist mainly of arsenopyrite. The brecciation and oxidation has resulted from solution of the limestone and solution-sinking beneath the overlying alluvium. The output of cassiterite concentrate was 1500 t. The Sin Nam Lee pipe at Ampang, Kinta Valley, is in dolomite. The gangue is of dolomite and tremolite. The ore is of cassiterite, arsenopyrite, pyrite, stannite, chalcopyrite, jamesonite and tennantite (Hosking, 1973a).

TUNGSTEN SKARNS: THE KRAMAT PULAI SCHEELITE MINE (KINTA VALLEY)

At the peak of its production in 1936, this mine had an annual output of more than 1400 t, and was the largest in the world, making Malaya the third largest tungsten producer, after Burma and China.

The main ore body lay under 7 m of alluvium at a distance of 30 m from the Main-Range granite (*Figure 7.4*). It is an irregular body in Palaeozoic schist interbedded with and faulted against limestone. The rocks are traversed by dykes and veins of aplite and pegmatite (Willbourn and Ingham, 1933). The veins vary from 60 cm to 3.8 m in width; they dip steeply with a strike perpendicular to the granite contact, bearing 90–115°, and are often deeply weathered. The ore is almost entirely of fluorite and scheelite of a coarse pegmatitic texture. Gangue amounts to less than 1% and is chiefly quartz. The country rocks carry the metamorphic minerals garnet, vesuvianite, axinite, tremolite and actinolite. Tourmaline and cassiterite occur in association with the pegmatites (Ingham and Bradford, 1960).

Folding and faulting occurred before and after deposition of the ore. Anticlinal folds plunge 8° towards 240°. Most of the ore was taken from the No. 1 ore body. Its scheelite content ranged from 3% to 15%, and occupied the centre of a limestone anticline under schist, which is cut by an aplite dyke. It is a hydrothermal replacement of limestone as shown by wollastonite and bustamite in the ore. Ore body No. 2 also occupied an anticline at a fault contact with granite (*Figure 7.4*).

A number of faults, which displace the main ore body and pass into the schist, are also filled with scheelite–fluorite ore of coarsely pegmatitic texture. The ore is therefore a low-temperature hydrothermal deposit (Willbourn and Ingham, 1933). The schist had impounded the hydrothermal fluids but some leak-away fractures were healed by scheelite–fluorite ore.

In addition to its production of scheelite, Kramat Pulai mine has produced more than 8000 t of cassiterite. Other by-products were corundum, diaspore, fluorite, galena, yttrotungstite, ferberite, wolframite, stolzite, tungstite, stibnite, sphalerite, arsenopyrite, pyrite, pyrrhotite, tremolite, beryl and grossularite (Ingham and Bradford, 1960). The corundum was usually found as pebbles and boulders in clay at the limestone–granite contact. The ferberite occurred along with yttrotungstite in a fault fissure in weathered granite.

Elsewhere in Malaysia, scheelite may occur in deposits in which there is no wolframite. But wolframite may occur in association with scheelite as at Chenderong, where it is also associated with cassiterite, and at Ulu Langat scheelite locally cements fractured cassiterite in greisen-bordered veins (Hosking, 1973a).

BRECCIA ORES

Although of minor importance compared with the greisen-bordered vein swarm and the skarn-type of deposits, breccia ore is an interesting category. The exact mechanism by which rocks are broken during hydrothermal activity is not understood. It is held by Sawkins (1969) that a lot of brecciation may be chemical. Reactions between the hydrothermal fluids and fine-grained silica and dolomite may produce something akin to the cement–aggregate reaction or alkali reactivity that results in expansion and cracking of concrete. Many authors attribute brecciation to the accumulation of magmatic water trapped beneath the cooled rind of the apical region of a pluton (Norton, D.L. and Cathles, 1973). The impounding of the fluids in the roof zone is done by an impermeable capping-rock formation; their eventual escape causes brecciation. The vapour release along tensional cracks can lead to the formation of a breccia pipe. As shown by Burnham and Ohmoto (1980), second or retrograde boiling causes a separate hydrothermal phase to separate. Whereas the increase in pluton volume caused by this at an emplacement depth of 2 km is as much as 50%, causing explosive fracturing of the roof to form porphyry-type mineralization, the volume increase at 5 km depth is less (about 15%) but may still be sufficient to moderately fracture the roof zone of a deep-seated batholith to form localized breccias. The majority of breccia-type ores are xenothermal, consisting of superimposed high-temperature and low-temperature mineralization.

In many countries xenothermal deposits are related to volcanic centres, as in Japan (Nakamura, 1976). However, the Southeast Asian xenothermal or telescoped vein deposits are related to the roof zones of granites.

Tin breccia

Two tin-bearing much-weathered breccia ore bodies within granite, lensoid-shaped of dimension approximately 40 m × 12 m, were mined in the Yap Peng mine, Sungei Besi, Kuala Lumpur (Yeap, 1978a). The breccia consists of angular-to-rounded rock fragments, 1–50 cm long, composed of sericitized and kaolinized granite, and vein quartz, embedded in a cassiterite-bearing brown clayey matrix. The surrounding granite country rock is kaolinized, silicified and tourmalinized. The matrix of the breccia was mineralized by quartz, beryl, cassiterite, arsenopyrite, pyrite, sphalerite and chalcopyrite, filling the spaces between and sometimes replacing the rock fragments. Manganese was also introduced as rhodochrosite and rhodonite. A late hydrothermal stage deposited chalcedony, minor fluorite and opal as replacements of parts of the breccia (Yeap, 1978a).

Tungsten breccia

The unique Khao Soon deposit is the largest in Peninsular Thailand with a monthly production of 180 t. The funnel-shaped breccia pipe cuts altered metasedimentary rocks of the Upper Palaeozoic Kanchanaburi Series, which underlie the Ratburi Limestone. The rock sequence includes quartzite, greywacke and phyllite. The 300-m deep mine occurs near both the axis of a north–south-trending anticline and a north–south fault (Arrykul *et al.*, 1978). The single 400 m × 800 m ore body exhibits lenses, pockets and pipe-like shapes. The ore is of fragmented angular altered country rocks cemented by a matrix of ferberite, some quartz and rarely pyrite. Stibnite and marcasite occur as accessories in the upper parts. Silicification and pyritization are common in the rock fragments and in the wall rocks. Ferberite–quartz veins occur in the hornfelsed quartzite and phyllites. Ferberite and stibnite may also be found in vugs in the mudstone country rocks. The overlapping of the higher-temperature wolframite by later stibnite indicates a xenothermal classification. The ferberite contains on average, 5% manganese tungstate and hubnerite occurs, but only very locally, in the deposit (Ishihara *et al.*, 1980). It is therefore likely that the deposit may have formed in a near-surface environment and meteoric water may have played an important role. Nevertheless the temperature significance of the wolframite composition is still under discussion (Hsu, 1976).

VEIN SWARMS IN AND NEAR THE GRANITOIDS

Vein swarms in the outer zones of granite plutons are widespread and undeniably formed the main source of detrital cassiterite in the Quaternary alluvium as a result of deep kaolinization and erosion. In the Kinta Valley, the best *in situ* mined areas are in the Kledang Range near Menglembu and around Tekka (*Figure 7.4*) where the deeply weathered granite and its vein system are mined directly by hydraulic methods.

Most of the deposits are of thin stringers of cassiterite and cassiterite–quartz transgressing the granite or as small lodes of quartz, cassiterite, pyrite and arsenopyrite near the contact with the country rocks. The granite is invariably greisenized, tourmalinized and kaolinized in proximity to these veins.

One of the most common types of ore is referred to as 'streaky bacon'; it consists of a great number of very thin parallel veinlets cut by thicker quartz veins. The 'streaky-bacon' zone penetrates the granite in the form of a lode often more than 4 m in width. It contains cassiterite, brown biotite, tourmaline, pyrite, magnetite, chlorite and muscovite, as at Menglembu, where the veinlets strike east-northeast. The veinlets cut straight through crystals of feldspar and biotite in the granite, without displacing or altering them, and there is no mineralization in the adjacent granite. The thicker veins, or leaders, occur singly or as two or three parallel veins separated by a few centimetres or several metres of intervening granite. The leaders and adjacent granite are often very rich in cassiterite (Scrivenor, 1928)

The Tekka mine

This mine, 13 km southeast of Ipoh, has been described by Riley (1967), Loew *et al.* (1969) and Teh (1978). A plexus of quartz veins cuts highly weathered granite. In the mining area the granite intrudes a sequence of limestone and schist. An early set of fractures strikes 320° and they contain mineralized quartz veins ranging from 15 cm to 30 cm in width. These consist mainly of cassiterite (some tantaliferous), wolframite, tourmaline, topaz and arsenopyrite in a quartz matrix.

A later set strikes 30° with a 5-cm average width, broadening to 15 cm where they intersect the earlier set. They contain no wolframite, are sulphide rich, but contain no cassiterite. They are composed of quartz, stannite, arsenopyrite and other sulphides.

The veins represent a good example of a telescoped xenothermal deposit, containing both high-temperature cassiterite, wolframite and columbite–tantalite, associated with low-temperature galena and stibnite. A general paragenetic sequence is as follows: hypogene minerals are loellingite, followed by arsenopyrite, cassiterite, sphalerite, stannite, chalcopyrite, kobellite and enargite; the supergene minerals are covellite and varlamoffite (Leow *et al.*, 1969). Only the cassiterite is of economic importance. Where the veins are numerous and closely spaced, the granite is silicified, tourmalinized and greisenized, and arsenopyrite and cassiterite occur along the joint surfaces. The oligoclase in the granite is frequently replaced by topaz (Riley, 1967). Topaz and muscovite occur in all the veins except those that are of pure quartz.

It has been suggested that this xenothermal deposit could not have formed in the deep-seated mesozonal environment proposed by Hutchison (1977) for the Main-Range granite, and that it may be related to a younger granite emplaced when the region was uplifted (Teh, 1978). Indeed the Main Range continues at the present time to be hot, as witnessed by the numerous hot springs associated with faults in the granite. The Main Range has been dramatically uplifted since the Triassic period (Hutchison, 1978c) and lower-temperature hydrothermal circulation through the rocks can be expected to follow the uplift and unroofing of the granites. This theory would require that the mineralization continued over a long period of geological time, and that a high temperature existed in the Triassic, waning to low temperature by the Tertiary.

In addition to quartz veining in granite, a considerable part of the Kinta Valley has mineralized veins extending into the country rocks, mainly the schists. They are usually quartz veins containing cassiterite associated with pyrite or arsenopyrite and/or tourmaline. They are generally conformable with the foliation of the schist. These veins are frequently associated with stanniferous aplite or pure quartz veins (Ingham and Bradford, 1960).

IRREGULAR REPLACEMENT ORE BODIES OR 'CARBONAS'

Several extremely rich replacement ore bodies were worked in the Triassic granite not far from the contact with marble in the bedrock of the deep alluvial mine of Sungei Besi, Kuala Lumpur area. They have been likened by Yeap (1978b) to the irregular shaped massive lodes of Cornwall, called *carbonas* (Hosking, 1964, 1970a). They are of short vertical extent, less than 30 m and have no depth extension; their long axis is aligned at 20°. These massive lens-shaped lodes have a paragenetic sequence of quartz, followed by tourmaline, cassiterite, arsenopyrite, pyrite, fluorite, sphalerite and chalcopyrite. They have been mostly oxidized by supergene processes.

Granite wall-rock alteration includes intense greisenization, tourmalinization, silicification and kaolinization. The hydrothermal solutions are thought to have been impounded beneath an anticline of marble, and localized at the intersection of two sets of fissures, trending 290° and 20°. Fluid-inclusion studies on quartz and fluorite suggest initial boiling of the solutions at about 330–450°C and deposition of the ore within the range 197–230°C (Yeap, 1978b).

Phuket, Thailand

Coarse-grained early Cretaceous porphyritic granite, dated by rubidium/strontium ratios at 115 Ma, intrudes the Palaeozoic Phuket Group and Ratburi Limestone (Garson *et al.*, 1975). Fine-grained two-mica granite has a rubidium/strontium age of 56 Ma. The Phuket Group has been contact-metamorphosed in aureoles up to 2 km wide. Cordierite hornfels is common and andalusite and sillimanite have been recorded. All the granites have been affected by early Tertiary potassium/argon rejuvenation resulting from uplift along the major northeast-trending Phangnga and Khlong Marui faults. Pegmatites, dated at 52 Ma and 57 Ma, were intruded into fracture zones parallel to these major faults.

The Phuket region produces more than half the total cassiterite production of Thailand, mainly from sea dredging and on-land alluvium. Cassiterite occurs as disseminated small grains in the biotite granite and in thin quartz stringers along its outer margins, but is not found in the other granitoids. The highly weathered granite is mined directly by hydraulic methods. The grade is variable within the range 0.05–0.5% cassiterite. Associated minerals are monazite, ilmenite, yttrotantalite, ilmenorutile and garnet; wolframite occurs locally.

It is probable that most of the alluvial cassiterite was eroded from zones of stockworks at the margins of the biotite granite plutons. These zones extend up to 1–2 km from the igneous contact and consist of tourmalinized hornfels with a network of narrow veins, dykes and sheets of quartz, aplite, pegmatite and fine-grained granite. This style of mineralization is akin to the epizonal Eastern Belt of Peninsular Malaysia. The veins contain grains and lumps of cassiterite varying from less than 1 mm to several centimetres long. Associated minerals include wolframite, scheelite, pyrite, arsenopyrite, yttrotantalite, columbite, ilmenorutile, manganese oxides and tourmaline. Late veinlets rich in muscovite and chlorite are locally common.

The cassiterite from Phuket is the most consistently strongly pleochroic, from intense red to pale-brown or colourless, of all the Southeast Asian districts, and this has been ascribed to a tantalum content (Hosking, 1977).

Two distinctive types of pegmatite are known: mica–tourmaline and lepidolite. Some are up to 15 m thick, some contain tin mineralization, but others are barren. Those at Reung Kiet and Bang I Tum are amongst the largest known unzoned pegmatites (Garson *et al.*, 1970). Both are 1 km long and 20 m wide and trend 220°, parallel to the major Phangnga fault zone with dips steeply to the southeast. They are composed of aggregates of quartz, albite and pink lithium muscovite. The main accessories are tourmaline, ilmenite and cassiterite, which form large grains usually anhedral but sometimes bipyramidal. At Tantikovit, a zone of deeply weathered near-vertical lenses of kaolinized pegmatite up to 10 m wide can be traced for 300 m. It is pure white and consists of quartz, albite, muscovite and tourmaline, together with cassiterite, ilmenite and wolframite.

The lepidolite pegmatites are assemblages of quartz, lepidolite, lithium muscovite and albite (Garson *et al.*, 1970), but they contain accessory lithium tourmaline, beryl, cassiterite, chryosoberyl, ilmenorutile, tantalite, microlite, yttrotantalite, yttrobastnaesite, xenotime, monazite, orthite, spessartite and topaz. Their normal tin-content range is 0.1–0.2%.

A significant curiosity of the Phuket region is the detrital diamonds in the tin-bearing alluvium. They are thought to be Precambrian and to have eroded from the Carboniferous Phuket Group pebbly mudstone (Garson *et al.*, 1975).

The Cretaceous Phuket tin granites are all of the ilmenite series (Ishihara *et al.*, 1978, 1980) and therefore are thought to have been derived from anatexis of an underlying continental crust. The belt of plutons can be related to eastwards subduction from the Bay of Bengal at a time when the Burmese and Sumatran volcanic arcs were continuous before the opening of the Andaman Sea marginal basin. The volcanic arc must then have lain parallel to and immediately to the west of the tin belt (Mitchell, A.H.G., 1977; Hutchison, 1978c).

The Tungsten–(Tin) Deposits of China

Southeast China is the richest tungsten metallogenic province of the world. The theory of Hutchison and Chakraborty (1979) that tin is mobilized from the deep continental crust by polycyclic events, leading to its concentration in acid plutonic rocks, can be equally applied to tungsten. One could argue that China is rich in tungsten (and tin) deposits because of the tectonic and magmatic events spanning the range 430–95 Ma ago, which have been favourable to the final concentration of these metals in the more evolved acid igneous rocks (Xu *et al.*, 1980; Yan *et al.*, 1980). The continental crustal origin of tungsten is further supported by the fact that the granites of southeast China are dominantly S-type and predominantly of the ilmenite series (Takahashi *et al.*, 1980). Whereas the Yanshanian compositionally restricted S-type granites of Nan Ling are associated with tungsten–tin mineralization, the contemporaneous I-type compositionally expanded series of the Middle and Lower Yangtze Valley, of implied mantle origin, is associated with iron, copper, molybdenum, zinc and lead mineralization (Wang *et al.*, 1980). This supports the finding of Ishihara (1980) for Japan in which tin and tungsten mineralization is found preponderantly associated with ilmenite-series granites and molybdenum, copper, lead, zinc, silver and gold with magnetite-series granitoids.

It was the youngest igneous cycle, the Yanshanian, that produced multistage high-level batholiths surrounded by thermal aureoles, with which the famous Nan Ling Range tungsten deposits are associated, throughout Jurassic time. However, they were preceded by more deep-seated Caledonian and Hercynian episodes of plutonic activity that played important early roles in the metal concentration history (Xu *et al.*, 1980). The cordilleran margin of China can be related to a northwest-dipping Benioff Zone that has persisted throughout most of the Phanerozoic up to the Cretaceous period. The granites exhibit a zonal distribution trending northeast and roughly parallel to the Benioff Zone. The locus of granitic magmatism migrated with time from the northwest to the southeast (Wang *et al.*, 1980). The Nan Ling Range accounts for more than 50% of the Chinese annual tungsten production of more than 22 000 t, but important deposits also occur in south Kiangsi, north Kwang Tung and southeast Hunan. The geology of these important deposits has been summarized by Ikonnikov (1975), and the classification is widely applicable throughout the whole region including the deposits in Burma, Thailand and Peninsular Malaysia (Hutchison and Taylor, 1978).

Quartz Vein Deposits (Southern Kiangsi-type)

These are the most important in the region; they fill fissures in the granite and country rocks. Groups of thin veins are most common. The veins dip steeply, some tens to hundreds of metres long (the longest is more than 1000 m). Their width is from a few millimetres to a few metres. The veins are greisen-bordered. This category may be subdivided as follows.

TUNGSTEN–TIN IN GREISEN

The adjacent granite is highly greisenized. The gangue of the veins is of quartz, muscovite, secondary feldspar, tourmaline, topaz and beryl. The main ore minerals are cassiterite and wolframite, accompanied by scheelite, chalcopyrite, molybdenite, pyrite, galena, sphalerite, arsenopyrite, pyrrhotite and monazite. Ore grade is low.

WOLFRAMITE–QUARTZ

These are the most important type throughout the region of east and Southeast Asia. They occur extending out from the top parts of acid plutons as steeply dipping veins or vein groups. Quartz is the chief gangue forming 80–95% of the veins, accompanied by feldspar and muscovite. Tourmaline, topaz, beryl or fluorite and arsenopyrite are common. In addition to wolframite, there can be cassiterite, molybdenite, bismuthinite, pyrite, chalcopyrite, sphalerite, galena and scheelite. Their grade is high. Wolframite is the major ore mineral, and the scheelite has no value.

SCHEELITE–WOLFRAMITE–QUARTZ DEPOSITS

These are related to granite, monzonite, or granodiorite. The ores occur as veins with a gangue of quartz, some plagioclase, calcite, barite, muscovite and tourmaline. Scheelite and wolframite are the main ore minerals, with more scheelite than wolframite. Associated minerals are cassiterite, pyrite, chalcopyrite, galena, sphalerite, molybdenite and arsenopyrite. They are small deposits.

TUNGSTEN–ANTIMONY–GOLD DEPOSITS

These occur as gently dipping veins. Quartz is the main gangue, accompanied sometimes by barite and calcite. In addition to scheelite, the ore contains stibnite, pyrite and native gold. They are medium-sized deposits.

Skarns of Scheelite (Southeast Hunan-type)

These are second in importance in China and occur near granitoids within 700–1000 m of the contact. The host rock is limestone, dolomite, calcareous shale and schist. They are of irregular lenses or pockets and are of variable size. Garnet and diopside are the main gangue, with scheelite, cassiterite, molybdenite, bismuthinite, chalcopyrite, sphalerite and galena. The main ore mineral is scheelite.

Wolframite–Scheelite Reticulated Veins (East Kwantung-type)

The ore bodies occupy reticulated fissures in Mesozoic sandstones near the granitoid contacts. The veins are evenly mineralized and form huge deposits. Quartz is of less importance to the sulphides. The most common are pyrrhotite, arsenopyrite, pyrite, bismuthinite, galena and chalcopyrite. The veins also contain cassiterite. Silicates are muscovite, sericite, biotite, topaz, andalusite, tourmaline, chlorite and quartz. The major ore minerals are wolframite and scheelite, which forms 25–45% of the tungsten oxide content. Both the wolframite and the scheelite are fine grained (0.02–10 mm).

The tungsten deposits usually show superimposed multistage mineralization, because the granites are multistage. In any one deposit, a close association of different types and generations of ore can be observed (Yan *et al.*, 1980). For example, wolframite veins may cut through scheelite skarn; many veins are composite. Wall-rock alteration may also be multistaged or telescoped. It is deduced that mineralization may have spanned tens of millions of years. Nevertheless, the following stages can be discerned (*Figure 7.5*).

EARLY SILICATE STAGE

This stage is characterized by skarn formation, alkali feldspar and greisenization. Mineralization may be disseminated or may form commercial

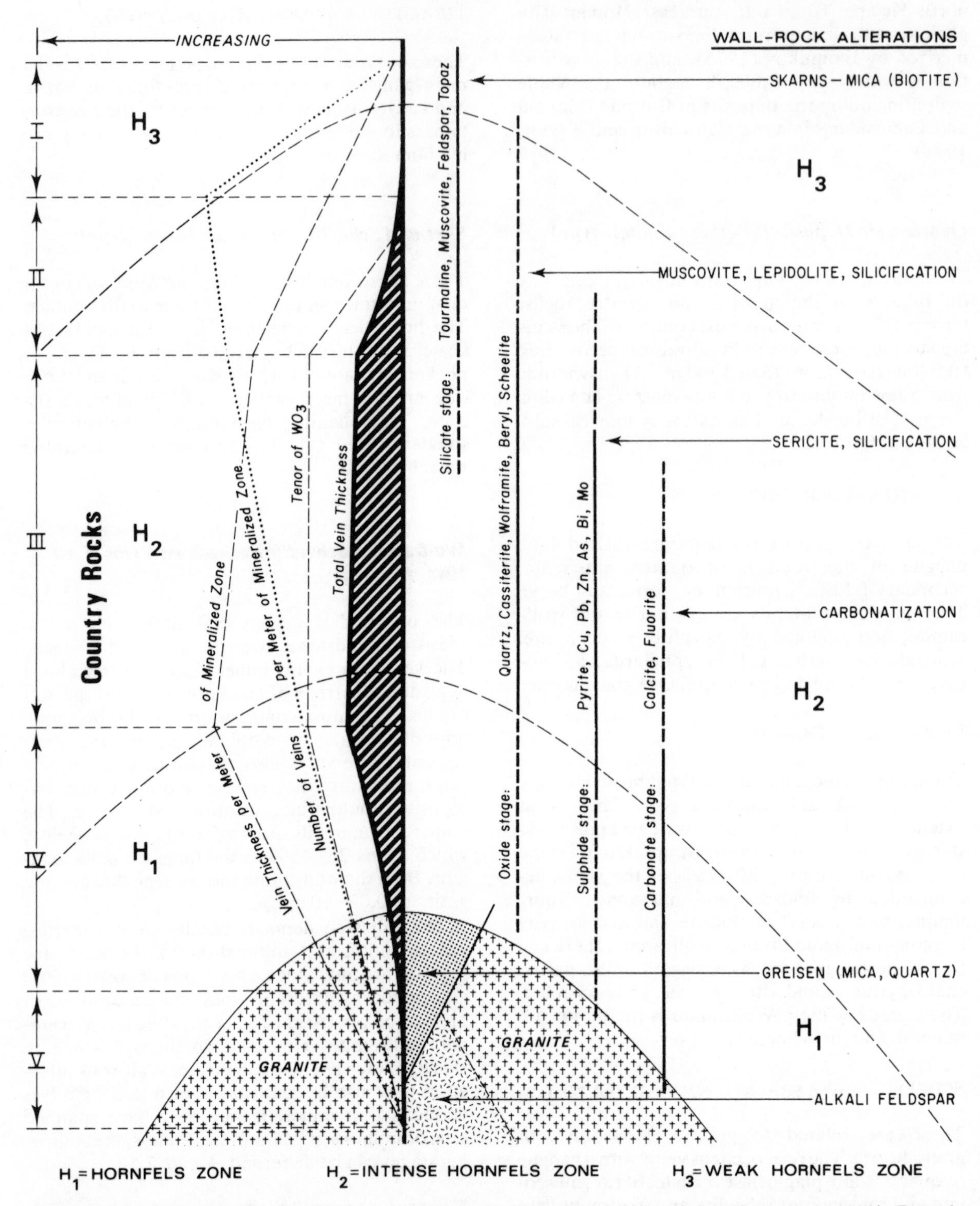

Figure 7.5. Vertical zoning pattern for tungsten vein deposits of China (after Yan *et al.*, 1980). Zone I: mica–quartz veinlets zone. Zone II: densely distributed veinlets zone (poor in mineralization). Zone III: veinlets and vein zone (principal commercial ore). Zone IV: veins zone (zone of commercial ore). Zone V: sparsely distributed vein zone (barren).

tungsten–tin deposits. The temperature range was 500–400°C.

OXIDE STAGE

This is the main ore-forming stage for tungsten. The mineral paragenesis is: molybdenite, followed by (molybdenum-bearing scheelite), wolframite–quartz, lithium-containing muscovite–beryl–cassiterite–wolframite–(feldspar)–quartz and zinnwaldite – cassiterite – wolframite – sulphides – quartz. The temperature range was 400–330°C.

SULPHIDE STAGE

This stage represents declining intensity of tungsten–tin mineralization. The main minerals are (cassiterite – wolframite) – scheelite – sulphides – quartz. The temperature range was 330–185°C.

CARBONATE STAGE

This is characterized by weak tungsten–tin mineralization. The paragenesis is fluorite, scheelite, pyrite, quartz and carbonates. The temperature range was 185–129°C, as determined by fluid-inclusion studies (Yan *et al.*, 1980).

Wall-rock alteration changes with style of mineralization (*Figure 7.5*). At the silicate stage, skarn formation and feldspar development predominate. At the early wolframite–molybdenite stage, biotite is characteristic. At the main wolframite stage, greisenization and silicification are important with extensive development of tourmaline, topaz, muscovite and fluorite. At the sulphide stage, sericitization, chloritization and silicification are important, while the carbonate stage is characterized by carbonatization.

Vein-type Zoning

Reversed vertical zoning is commonly observed in the tungsten vein deposits (Yan *et al.*, 1980), summarized in *Figure 7.5*. In the top part (I), the main minerals are muscovite and quartz. Sometimes topaz, zinnwaldite and tourmaline may be important, with or without molybdenite, cassiterite and wolframite. The minerals are usually fine grained. Wolframite forms tabular or acicular crystals. Comb-structured quartz veinlets and planar mica veinlets are developed, and they usually indicate underlying mineralized zones at greater depth.

In the upper part (II), oxides and silicates are important, such as zinnwaldite, muscovite, feldspar, tourmaline and topaz, as well as beryl, cassiterite, wolframite and quartz. Sulphides such as arsenopyrite, molybdenite, bismuthinite and pyrrhotite may also occur. The ores are commonly banded. Wolframite occurs inside an outer zinnwaldite band, and it is associated with cassiterite, beryl and topaz. Wolframite crystals are tabular. The ore bodies are composed of small veins and veinlets or zones rich in veinlets ('streaky bacon').

In the middle part (III), wolframite, cassiterite and other oxides predominate. Sulphides increase in quantity and variety. The gangue is often clearly banded. Miarolitic cavities and brecciated structures frequently occur. The wolframite is of large tabular crystals unevenly distributed. This is the richest part of any vein system (*Figure 7.5*).

In the lower part (IV) the economic mineral content diminishes, but sometimes scheelite, stannite and helvite appear. In the bottom part (V), the tin and tungsten minerals disappear almost completely. However, fluorite, ankerite, siderite, rhodochrosite, calcite and other carbonates increase in amount. The veins disappear with depth.

A newly recognized porphyry-type of tungsten mineralization (Yan *et al.*, 1980) is discussed in Chapter 8.

The southern limit of the Nan Ling Range is in Hong Kong, where the late Yanshanian uppermost Jurassic 140-Ma old Needle Hill granite contains wolframite-bearing quartz veins; they are principally of quartz, but contain feldspar and biotite. The ore minerals are wolframite, galena, sphalerite, bismuthinite, cassiterite, pyrite and others, including fluorite, but there is no scheelite (Davis, S.G., 1964).

As seen from the descriptions of the Chinese deposits, the tungsten mineralization commonly overlaps with that of tin. Hutchison and Taylor (1978) pointed out that in the region of China and Southeast Asia there is a strong tendency for tungsten to occur farther within the continent where the continental crust is presumably thicker, whereas tin occurs in a zone lying outside the main tungsten belt. Thus in the Malay Peninsula, tungsten is of greater economic importance northwards towards Thailand and Burma, whereas tin is more important southwards towards Malaysia and Indonesia. Likewise the tin belt lies closer to the east coast of China,and the tungsten belt more inland, but they largely overlap, and both metals must have been brought up by granitoid emplacements originating in the underlying crust, then concentrated hydrothermally. This view is amplified by Xu *et al.* (1980) who concluded that the tungsten, tin, niobium and tantalum were originally in the continental basement strata and remobilized by S-type granites, whereas the Chinese iron, copper, lead, zinc and gold deposits were brought from the mantle by I-type granites.

Yunnan is the largest producer of tin in China. The most important district is at Kochiu, 190 km south of Kunming. This is a unique pipe-like deposit with no analogies in the whole of Southeast Asia, and it is fantastically rich. The ore occurs in irregular bodies in limestone and contains cassiterite, limonite and haematite with a little lead, zinc, copper and arsenic. The average grade is 2.4–5.0% tin. Reserves are of the order of 1 million t of tin.

Cornwall, Southwest England

This ancient mining district is estimated to have produced a total of 2.5×10^6 t of tin and 2×10^6 t of copper. At present china clay (kaolinite) is the major product. Excellent descriptions of the mineral deposits are to be found in Hosking (1964, 1970a) and a general review in Dunham, K. *et al.* (1978) and Jackson, N.J. (1979).

Most of the mineralization lies close to the roof of an elongate Late Palaeozoic (Armorican) granite batholith of which five major cupolas are exposed (*Figure 7.6*), intruding through a marine and deltaic succession of folded Devonian-to-Carboniferous slates, shales, sandstone, limestone and volcanic rocks, referred to generally as *killas*. During cooling of the magma, the complex pattern of joints, resulting from an earlier deformation, was reactivated to provide channels for acid porphyry dykes called *elvans*, and metalliferous mineral veins. Radiometric dating (Halliday, 1980) has outlined the following sequence: about 295 Ma ago emplacement of the major plutons, 285+ Ma ago formation of mineralized pegmatites, 280–285 Ma ago greisenization and 275–280 Ma ago intrusion of porphyry dykes; about 270 Ma ago was the major polymetallic mineralization episode. In some cases the fractures were reopened and filled several times, so that the above sequence is but a generalization.

Types of Deposits

The classification most widely accepted is by Hosking (1964, 1970a).

Pyrometasomatic Deposits

These are few and relatively unimportant. The best examples are around the Dartmoor granite where a zone of pyrrhotite–arsenopyrite–spalerite–pyrite–chalcopyrite mineralization is developed within garnet–diopside–actinolite–wollastonite–axinite skarn in the Carboniferous slates and chert. Copper was worked near Okehampton. Farther west a similar sulphide association contains tin in the form of malayaite ($CaSnSiO_5$).

Pegmatites

These occur as scattered small lenses close to granite contacts but are insignificant. At South Crofty mine they carry wolframite and arsenopyrite.

Greisen-bordered Vein Swarms

These form a host of subparallel quartz veinlets in the granite with walls of greisen or narrow selvages of chlorite rosettes, tourmaline needles, feldspar and fluorite. The central quartz vein rarely exceeds 0.15 m but it carries most of the cassiterite, wolframite, arsenopyrite, chalcopyrite, chalcocite and stannite. The greisen walls contain finer-grained cassiterite and suphides but no wolframite. When the veins occur in the granite, the granite between the greisen borders is kaolinized. The best examples are at Cligga Head and St. Michaels Mount, where the order of mineral deposition is cassiterite, wolframite, lollingite, arsenopyrite, stannite, followed by chalcopyrite. Quartz deposition overlaps all the above ore minerals, and was deposited over the range 380–280°C, while most of the quartz was deposited at 360–380°C (Moore, F. and Moore, 1979). Meteoric waters are considered to be the dominant source of the hydrothermal fluids during greisenization (Sheppard, S.M.F., 1977b).

The veins may continue a short distance into the country slates as normal lode veins, but they may also pinch-out within the granite. Most vein swarms dip steeply. Where they are mined from opencast pits, the grades range is 0.15–0.40% tin and at Mulberry more than 1×10^6 t of ore was mined.

Stockwork Deposits

These networks of narrow ramifying mineralized structures may occur within any country rock-type. Some are closely associated with major hydrothermal veins, others have no such connection and exhibit preferential ore deposition along one set of joints. The veins contain cassiterite, wolframite and arsenopyrite.

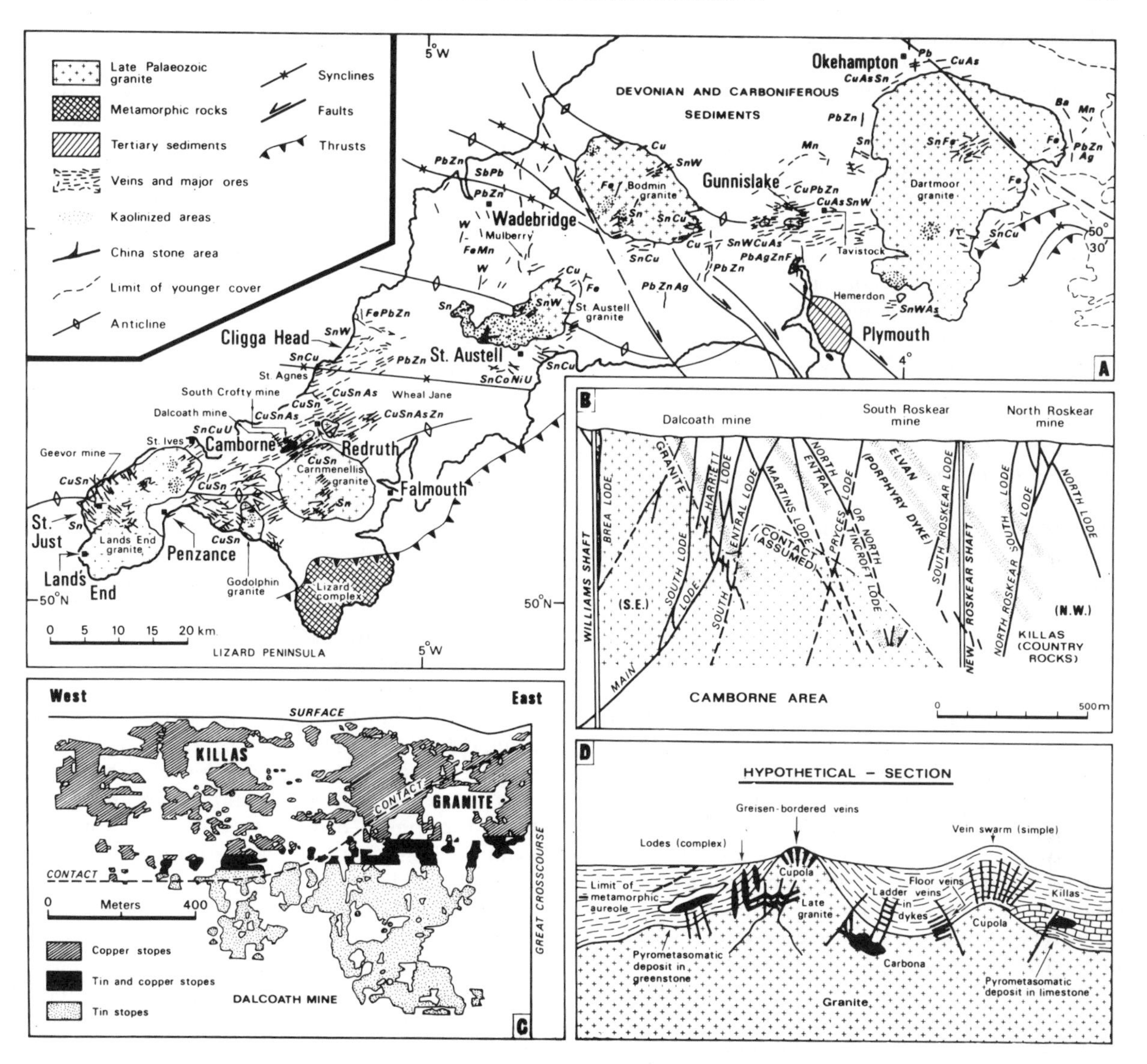

Figure 7.6. (A) Distribution of ores and veins in the Armorican granites and Devonian–Carboniferous country rocks of Cornwall (after Dunham, K. *et al.*, 1978). (B) Cross-section of the Dalcoath mines of the Camborne area to show the relationship of the lodes to the porphyry dykes (elvans) and the granite (after Hosking, 1964) and (C) the zoning of tin and copper ores (after Hosking, 1964). (D) Schematic cross-section to show the range of major types of tin deposits in southwest England (after Hosking, 1970a).

Fissure Veins of Cornish-type Lodes

These have yielded most of the Cornish production. They are tabular and generally less than 2 m wide. Some have a dip extent up to 900 m and have been followed along strike for up to 6 km. Most dip from vertical to 60°, but there are flat-lying examples such as in Wheal Jane mine. The dip and strike is variable, and veins may merge and cross each other. They are regarded as filling a conjugate fracture pattern in the country rocks. There is no consistent relationship between the granite-shape and the vein trend, and there is no radial pattern around a pluton. Cross-veins, perpendicular to the lodes, are usually filled with barren quartz.

Fissure veins have complex banded structures indicating a history of polyphase deposition and repeated fracture opening. Internal brecciation and recementation may occur.

Banded and braided veins usually include layers

of barren gangue material, while the ore minerals are confined to a few bands. Breccia veins consist of fragmented wall rock cemented by ore and gangue minerals that occupy faults and shear fractures. Replacement veins appear to be formed by the outward migration of hydrothermal fluids from a narrow fissure to replace wall rocks, shear zones, or selected horizons within the metasediments.

Fissure vein systems can claim to be amongst the world's major producers of copper, tin and arsenic, and numerous examples have been fully documented by Dines (1956). The main fissure vein lodes have the following mineral paragenesis (Hosking, 1970a): cassiterite, wolframite, arsenopyrite, pyrrhotite, sphalerite, stannite, followed by calcopyrite. Lower-temperature veins farther from the granite heat sources may have the following general sequence: pitchblende, argentite, galena, sphalerite, pyrite, siderite, tetrahedrite, stibnite, followed by haematite. These veins are complex and contain a larger variety of minerals.

Pipe and Carbona Deposits

These are uncommon, usually small, occur within the granite and are associated with fissure veins. Pipes are vertical cylindrical, about 130 m high and 4 m across, consisting of a 1-cm wide centre of vein quartz and clay surrounded by altered granite containing cassiterite and some sulphides and fluorite. Carbonas are similar but have an irregular shape and attitude and contain tourmaline.

Vertical zoning is apparent in some deeper-worked mines, notably in the Camborne district at the Dalcoath mine (*Figure 7.6*). Generally the sequence (from the granite outwards) is: oxides of tin with tungsten in the upper parts; zone of sulphides of copper intermixed with tungsten and tin; zone of sulphides of lead and silver; zone of sulphides of antimony; and zone of carbonates of iron and manganese. However, there are so many exceptions to this zonation that it is only a generalization, and the zones overlap and are of variable thickness. It is generally held that the zoning results from temperatures declining outwards and upwards from the granite heat source towards the land surface. Since the hydrothermal solutions are based on meteoric water (Sheppard, S.M.F., 1977b) and are not magmatic, the circulation of these solutions through the country rocks can be expected to result in a mineral zonation away from the granite heat source. However, repeated mineralization episodes and reopening of the veins has caused complexities in this simple picture.

The Economic Minerals

The only tin ore is cassiterite, of which wood tin (a colloidal layered intergrowth of cassiterite and quartz) (Moh and Hutchison, 1977) is an uncommon variety (Hosking, 1970a). The cassiterite is usually accompanied by arsenopyrite, with or without chalcopyrite, chalcocite, specularite, sphalerite and wolframite. Stannite (Cu_2FeSnS_4) is widely distributed, especially in the greisen-bordered veins. The ores of copper are chalcopyrite and chalcocite. The only arsenic is in arsenopyrite. Wolframite occurs in the early quartz–tourmaline and quartz–feldspar veins. Scheelite is known from several high-temperature veins and from the pyrometasomatic deposits. The only zinc is in sphalerite.

With the basic assumption that the Lizard Complex represents a suture zone as the Rheic Ocean closed by southwards collision of the pre-Devonian basement of Wales and Devon with Europe, Mitchell, A.H.G. (1974) proposed that the Cornish granites result from crustal thickening caused by underthrusting along a southwards-dipping Benioff Zone. His model is identical to that for the Main-Range granite of the Malay Peninsula (Mitchell, A.H.G., 1979). However, Bromley (1975) expressed scepticism because the model does not explain why the granites on both sides of the suture zone (Cornwall on the north; Portugal, Spain, Brittany and the Massif Central on the south) should both be tin-bearing. However, Hutchison and Chakraborty (1979) argued that tin becomes enriched in granites if cyclic anatectic events mobilize it from a suitable continental basement, so that no meaningful analysis can be made about the distribution of tin since these basement rocks nowhere outcrop. However, the analogy with the Malay Peninsula is a good one, for two distinct tin-belt granites are separated by a suture zone and a sedimentary basin (Hutchison, 1978c).

Pasto Buena, North Peru

Tungsten and base-metal mineralization, with a significant silver content, occur in narrow near-vertical quartz vein systems that span several hundred metres on either side of the upper intrusive contact of a 9.5 Ma old quartz monzonite stock emplaced in a Jurassic–Cretaceous shale and sandstone sequence (*Figure 7.7*). The stock has four pervasive and roughly parallel alteration zones from depth upwards: alkalic; phyllic–sericitic; argillic; and propylitic (Landis and Rye, 1974). Greisen assemblages of zinnwaldite, fluor-

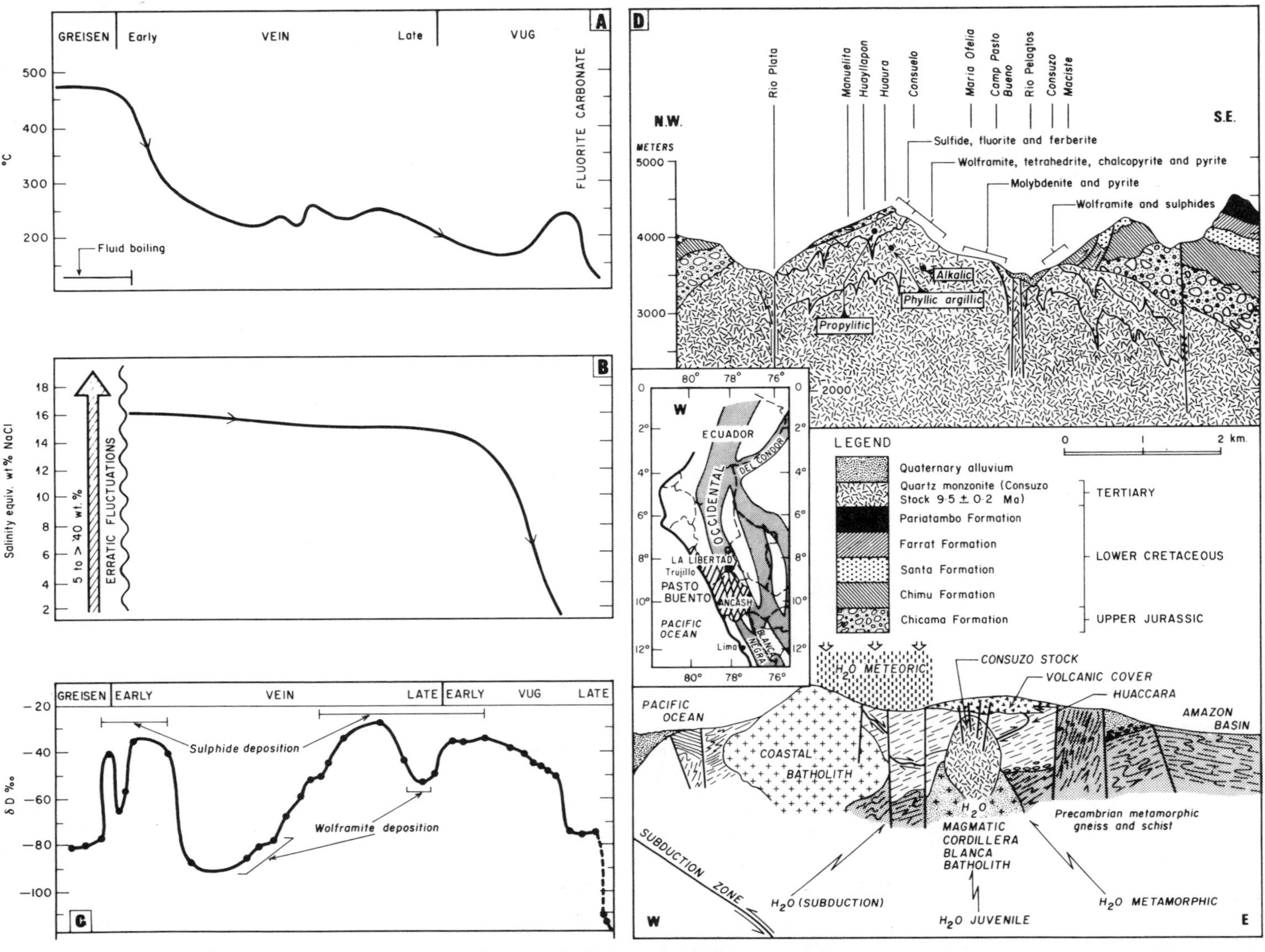

Figure 7.7. Geology of the Pasto Buena tungsten–base-metal vein deposit of northern Peru. Graphs to show: (A) the range of temperature of mineralization based on fluid-inclusion studies; (B) salinity of the fluids; and (C) their δD values. (D) Cross-sections show the major locations around the roof zone of the Consuzo stock and the bottom diagram suggests the origin of the waters that were responsible for the hydrothermal system (after Landis and Rye, 1974).

ite, pyrite and minor topaz and tourmaline occur around the quartz veins in the phyllic zone.

The principal vein minerals are wolframite, tetrahedrite/tennantite, sphalerite, galena and pyrite in a gangue of quartz, fluorite, sericite and carbonates. The major wolframite early phase is of thick stubby brown blades of huebnerite (75% manganese tungstate) randomly oriented in quartz veins. The greisen stage accounts for about 60–70% of the mineral deposition; the later vein stage for 25–35%, and the later vug stage for less than 5%. Calcite, rhodochrosite and dolomite are confined to the late-stage vugs.

Fluid-inclusion studies indicate that greisen and early-vein-stage fluids were very saline (more than 40% sodium chloride equivalent by weight), of high temperature (400–500°C) and of magmatic derivation. The subsequent vein-stage fluids were in the temperature range 175–290°C with a salinity of 2–17% sodium chloride equivalent by weight. Boiling occurred in the greisen and early vein stage only. The patterns for δD_{H_2O} and $\delta^{18}O_{H_2O}$ values for the hydrothermal fluids indicate that mixing of meteoric water and possibly metamorphic or connate water with magmatic water accounted for the vein-stage mineralization. The meteoric water circulated deep into the hydrothermal plumbing system, which was driven by the heat source of the Consuzo stock.

Wolframite deposition was associated with episodes of meteoric water influx that are reflected in temperature, salinity and δD values of the water in the fluid inclusions. The dips in the δD and $\delta^{18}O$ graphs (*Figure 7.7*) indicate that the tungsten mineralization is somehow triggered by the influx of the relatively dilute meteoric water (Taylor, H.P., 1979). It is not certain, however, whether the tungsten is carried by the meteoric water or that its precipitation is simply brought about by the abrupt fluctuations in temperature, salinity and the other chemical variables that accompanied the mixing of meteoric and magmatic water. Sulphide mineralization is however associated with water of magmatic derivation. Carbon-isotope studies indicate that the carbon was derived both from a sedimentary and magmatic source. The sulphur is of mantle origin. Although most of the components of the ore deposits are ultimately of magmatic origin, clearly significant volumes of meteoric and other water were involved in the vein stages of the hydrothermal deposition. The most plausible sequence of events is; early magmatic–hydrothermal biotite alteration and greisenization; influx of saline connate water into the cooling stock during the vein-formation stage because of the establishment of a convection system around the heat source; mixing of the saline connate water with fresh ground water that entered through the deep fracture system; and influx of meteoric–hydrothermal water, even lighter than the present-day local meteoric water (Taylor, H.P., 1979).

Panasqueira, Portugal

These vein deposits are related to Hercynian 290 Ma old granite, and form the leading source of tungsten in Europe, outputting annually about 1250 t of wolframite and 62 t of cassiterite concentrates. The geology and mineralization has been detailed by Kelly and Rye (1979).

A large number of nearly horizontal ferberite–quartz veins cut sharply across the steep bedding and foliation of the greenschist facies pre-Ordovician Beria Schist (*Figure 7.8*). The schists are contact-metamorphosed and spotted in an aureole around the granite. The vein openings were created by vertical dilation of pre-ore joints, post-dating the metamorphism, thought to have been caused by erosional unloading of the country. The mineralization is related to one or more greisenized granite cupolas of a larger deeper batholith. A highly silicified cap occurs at the roof of the cupola, but its mineral content is subeconomic. Economic vein zones are laterally extensive but restricted vertically to a zone of 100–300 m thickness.

The vein paragenesis is complex but four distinct phases can be recognized: oxide silicate stage —principally of quartz, muscovite, tourmaline, topaz, arsenopyrite, cassiterite and wolframite; the main sulphide stage—quartz, muscovite, arsenopyrite, pyrite, pyrrhotite, sphalerite, chalcopyrite, stannite, apatite and galena; pyrrhotite alteration stage—pyrite, chalcopyrite, marcasite and siderite; and the late carbonate stage—chlorite, dolomite and calcite. The spread of the solutions and deposition of these stages through the vein system is complicated and not yet fully resolved.

The veins fluids were sodium-chloride-dominated, with deposition within the range 360–230°C. The closing carbonate stage was at 120°C or lower. Sulphur-isotope and oxygen-isotope studies (Kelly and Rye, 1979) indicate that the fluids were predominantly of meteoric water in the carbonate veins, but the higher-temperature tin–tungsten deposition was by deep-seated magmatic water. The deposit represents a mixing of magmatic with meteoric water as the system cooled beyond the sulphide stage. The flat vein system is cut by steep faults that contain

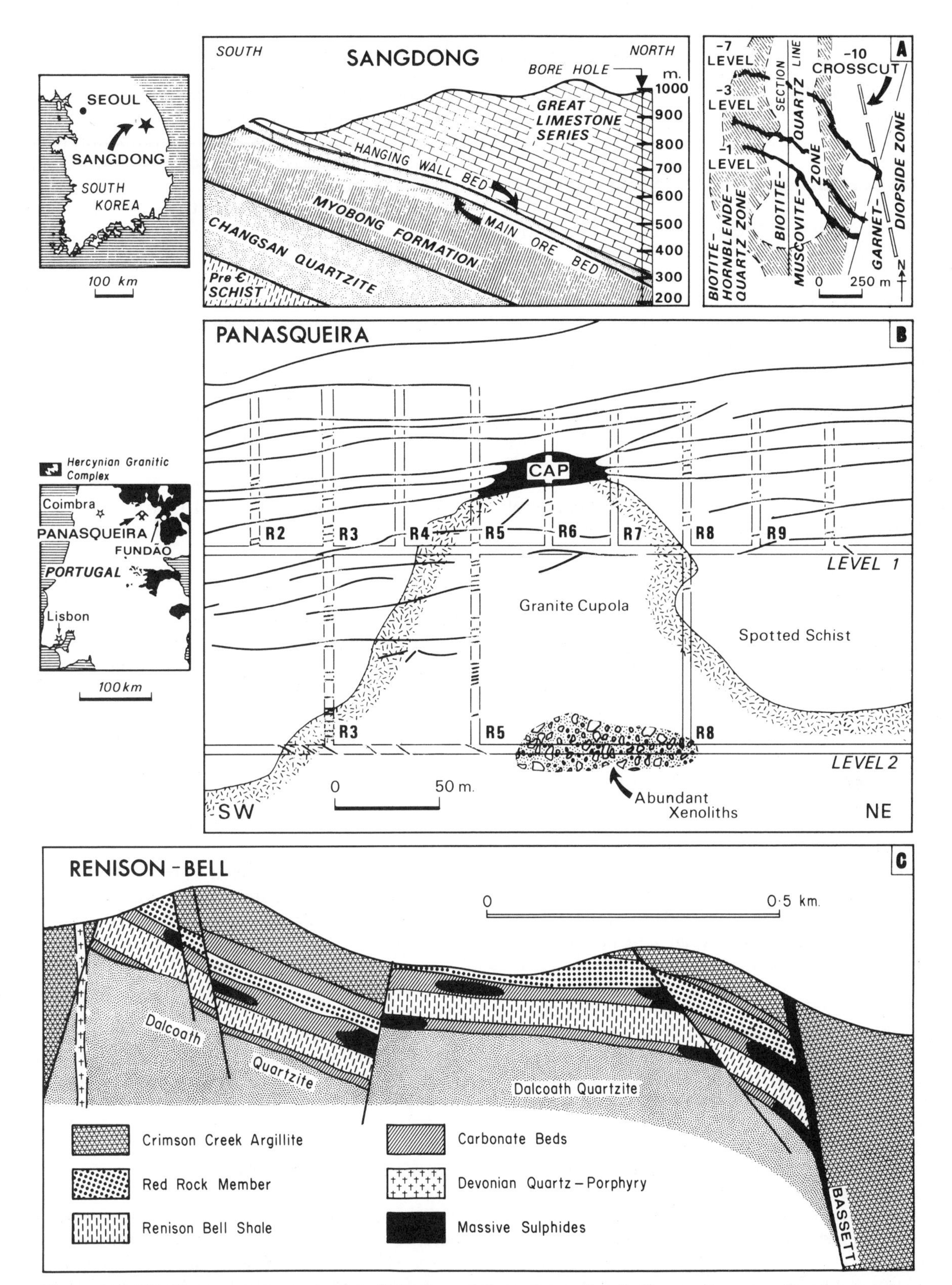

Figure 7.8. (A) The strata-bound scheelite mine of Sangdong, South Korea. Cross-section from John, Y.W. (1963) and lateral zoning pattern along the mineralized horizon after Farrar *et al.* (1978). (B) The tungsten–tin flat-lying Hercynian vein system of Panasqueira, Portugal (after Kelly and Rye, 1979). (C) Cross-section of the Renison tin mine, Tasmania, Australia (after Newnham, 1975).

economic sulphide mineralization, deposited at 100°C by meteoric water.

Hamme Tungsten District, North Carolina

This district of North Carolina and Virgina is the largest quartz–wolframite veins regions in the USA. More than 50 veins occur, but none is now mined. The veins occur at the border between the Lower Palaeozoic Vance County pluton and greenschist facies meta pelities (Foose *et al.*, 1980). The principal ore mineral is huebnerite, accompanied by scattered pyrite, sphalerite, galena, chalcopyrite and tetrahedrite in a gangue predominantly of quartz, with minor fluorite, sericite and carbonate.The prismatic huebnerite crystals are oriented perpendicular to the vein-layering. Some have terminations with cappings of sulphides that have concentric growth zones.

The deposits have resulted from open-space filling of planar faults or fractures and were subsequently deformed and metamorphosed (Foose *et al.*, 1980). The veins were deposited from meteoric water before the 550°C metamorphism (Casadevall and Rye, 1980). The primary mineralization took place at 260–320°C and the sulphur had a deep-seated origin.

Strata-bound Tungsten and Tin Deposits

It is so customary to regard tungsten (and tin) mineralization as having been introduced into country rocks by hydrothermal solutions emanating from granitic plutons that there has been a tendency to regard strata-bound tungsten (and tin) deposits also as hydrothermal and having been introduced selectively into favourable beds. However, the extensive nature of some of the bedded examples leads to an alternative hypothesis—that the mineralization was synsedimentary and caused by submarine exhalations. The debate goes on, and the argument for tin and tungsten synsedimentary deposition is by no means universally accepted.

Recent evidence from worldwide occurrences shows that tungsten may have been deposited by volcano–sedimentary proceesses that result in stratabound concentrations (Stumpfl, 1977). These deposits frequently also contain minor amounts of base-metal sulphides. Sedimentary textures are well preserved in many strata-bound tungsten deposits and they apparently survive metamorphism to amphibolite facies. Granulite facies metamorphism results in changes in texture that obliterate the original sedimentary textures. However, manganese haloes around strata-bound deposits frequently survive amphibolite facies metamorphism, so that they can be of use in exploration (Stumpfl, 1979a).

The following strata-bound deposits have been cited by Stumpfl (1977) as possibly submarine-exhaltive.

Namaqualand (Southwest Africa)

Tungsten ore occurs as concordant quartz–ferbertie veins in the 130 m thick wolfram schist, regionally metamorphosed 1213 Ma ago. The ferberite is accompanied by minor amounts of chalcopyrite, pyrite, scheelite and molybdenite. The stratabound horizons can be traced over 40 km along strike (Stumpfl, 1979b).

Eastern Alps

Strata-bound scheelite ore occurs in a volcano–sedimentary succession of Lower Palaeozoic age in the province of Salzburg. The Felbertal ore body extends 2500 km along strike. Many of the Alpine scheelite deposits are associated with basic submarine volcanic rocks. At Kleinarltal, the scheelite ore is bound to carbonate rock (Höll *et al.*, 1972).

Norway

Strata-bound deposits occur in the Bindal area of north Norway. Scheelite-bearing skarn is strongly strata-bound in a hornblende–biotite gneiss.

Zimbabwe

Tungsten mineralization occurs widely in the 1950 Ma old Piriwiri and the 2900 Ma old Bulawayan systems. Scheelite skarns are strata-bound and associated with metavolcanic schists.

Sangdong Mine, South Korea

This important ore deposit is the largest and most perfect example of a strata-bound tungsten deposit. The geology has been described by John, Y. W. (1963) and Farrar *et al.* (1978). Most of the ore is restricted to a single strata-bound ore body, the 'Main Vein', some 3.5–5 m thick located in folded quartz-rich early Cambrian slate (*Figure 7.8*). The ore body has been traced for over 1500 m along both strike and dip. It strikes 75° and dips 15–30° towards the northeast. The ore body is characterized by a network of randomly orientated quartz veins (John, Y. W., 1963). This mine produces over 70% of the Korean tungsten output. Its interpretation as a volcanogenic–synsedimentary deposit is strengthened by the absence of granites and other plutonic bodies from the mine neighbourhood.

The fine-grained disseminated scheelite is associated with silicate mineral assemblages (*Figure 7.8*). A central zone, 3000 m wide, is dominated by quartz, biotite and muscovite. Flanking this is a zone with abundant hornblende, quartz and minor biotite. The outer zone is of diopside–garnet skarn. Closely spaced scheelite and sulphide-bearing quartz veins are abundant, particularly in the central micaceous zone. Ore grade decreases from 1.5% to 2.5% tungsten oxide (WO_3) in the mica zone, from 0.3% to 1.5% in the hornblende zone, to subeconomic grades containing less than 0.5% in the skarn zone.

The major minerals (hornblende, biotite and muscovite) and whole rock slate have been dated by potassium/argon methods at 83 Ma (Farrar *et al.*, 1978). Thus the rock sequence was metamorphosed in the Cretaceous period. The nearest granite (4 km southeast of the mine) has beed dated at 1762 Ma, therefore it clearly is not genetically related to the deposit. It is possible that the deposit was strata-bound and of Cambrian age (Maucher, 1972), associated with basic volcanism. A similar origin has been argued for the strata-bound tungsten deposits of the eastern Alps (Höll *et al.*, 1972).

The Main-Vein diopside–garnet zone formed at 460±25°C. Farrar *et al.* (1978) favour the deposit to be hydrothermal and related to an Upper Cretaceous buried granite that has not yet been discovered. The evidence is ambiguous, and the Sangdong deposit may have been of Cambrian age, modified and recrystallized during the Cretaceous, or it may have been hydrothermally introduced in this very strata-bound manner into the Cambrian sediments during the Cretaceous igneous and hydrothermal events.

King Island, Bass Strait, Australia

The south-dipping stratiform scheelite deposits of King Island (40° S, 144° E) are 5–40 m thick (Danielson, 1975), containing reserves of 7 million t of ore averaging 0.75% WO_3. The mine series is of a succession of hornfels and skarn of Late Precambrian to early Cambrian age, metamorphosed by late Devonian to early Carboniferous granite. The fine-grained (0.05–0.2 mm) scheelite occurs in and along the margins of andradite garnet in the skarn horizons. The ore concentrate contains 0.01% tin, but cassiterite has never been detected. Sulphides such as pyrite, arsenopyrite, pyrrhotite, chalcopyrite and molybdenite are also present.

The deposits have generally been recognized to result from selective metasomatic replacement of appropriate dolomitic horizons, the hydrothermal solutions emanating from granite plutons. However, their bedded nature and the absence of granite from the mining area may indicate a synsedimentary origin, related to the tuffaceous beds that occur in the mine sequence. The metamorphic aureole superimposed by the Devonian–Carboniferous granites makes it difficult to prove whether the ore bodies were Cambrian syngenetic or Devonian–Carboniferous metasomatic. This deposit well displays the quandary presented by strata-bound mineralization, such as tungsten, which is traditionally regarded as hydrothermal and not exhalative.

The possibility that the bedded lodes of the Klappa Kampit tin mine of Billiton, Indonesia, may be of volcanogenic synsedimentary origin has been suggested by Hosking (1977), as discussed above. However, this is not the only locality that exhibits strata-bound tin deposition. The tin fields of Tasmania, Australia, offer some of the best examples of strongly strata-bound and possibly synsedimentary tin deposition and Hutchinson (1979) has developed the argument that they had a submarine-exhalative origin.

Renison–Bell Tin Field

The ore bodies are either fault-controlled or strata-bound; the latter invariably lie adjacent to mineralized fault zones, which may have provided feeder channelways for the hydrothermal solutions, so that selective replacement hypothesis may be acceptable here (*Figure 7.8*). The mineralized carbonate horizons occur in a sequence of quartzites, shales, volcanic rocks and tuffs or younger Precambrian to Upper Cambrian age (Newnham, 1975).

A 2-m wide microdiorite dyke and a 5–10 m wide quartz porphyry dyke intrude the mine area, and Devonian-to-Carboniferous granites lie outside of the mining area.

Typically sulphides form 60–75% of the strata-bound ores, mainly of pyrrhotite, with minor pyrite, chalcopyrite and arsenopyrite. The gangue is of quartz, calcite, fluorite, tourmaline, talc and chlorite. Cassiterite occurs both in the sulphides and gangue, and has an average grain size of 50–100 μm. The ore averages 1.3–1.8% tin.

The fault-controlled ore bodies extend up to 2000 m along strike, and down to 800 m along dip, with average thickness of 5–15 m. Post-ore movement on the faults is evidenced by fractured grains and second-generation pyrite.

This deposit may be equally well interpreted as selective Devonian–Carboniferous metasomatic mineralization along faults extending into favourable carbonate beds, or one can hold that the mineralization was originally in the Cambrian carbonate beds, subsequently remobilized by metamorphism and faulting, and driven by the granite heat source.

Cleveland Mine

The Upper Precambrian to Cambrian succession consists of a basic volcanic unit, micaceous sandstone and grey argillite, which contains the tin mineralization (Ransom and Hunt, 1975). The succession is in greenschist facies grade of metamorphism. The ore occurs as strata-bound lode lenses, and it may be argued that these are the original sedimentary metal-rich lenses, unrelated to later structural controls. The inferred ore reserves are 2:8 million t containing 1.02% tin and 0.43% copper (Cox, R. and Glasson, 1970).

The sulphide-rich mineralized lenses are composed of complex combinations of pyrrhotite, pyrite, marcasite, cassiterite, quartz, carbonate, fluorite, actinolite, chlorite and tourmaline. Accessory minerals are arsenopyrite, sphalerite, haematite, stannite and tetrahedrite (Ransom and Hunt, 1975). The controls of mineralization are apparently entirely stratigraphic, restricted to discrete layers within the well-defined shale and chert sequence.

Mount Bischoff Tin Ore Body

This ore body yielded 56 000 t of tin metal from 5.5 million t of ore. The bulk came from a subhorizontal ore body whose base is strictly conformable with the underlying shale (Knight, 1975). The Late Proterozoic succession includes shale, quartzite and dolomite. It is the dolomite beds that are mineralized. The ore is of more than 50% sulphide, mainly pyrrhotite, pyrite and marcasite, together with talc, quartz and coarse-grained carbonates.

A 1.7 km × 2.7 km zone of interconnecting quartz porphyry dykes occurs around the mineralized zone, dated at 349 Ma. Sills from these dykes intrude the ore. Locally the porphyries are intensely enriched in topaz, and they contain cassiterite in the vicinity of the stratiform ore bodies. Several quartz–pyrite–cassiterite lodes cross-cut the footwall shales and sandstones, some worked to a depth of 330 m, with an average width of 65 cm, and a strike extension of 600 m. This ore deposit was formerly considered to be a hydrothermal replacement of the selective dolomitic horizons, emanating from the quartz porphyries. However the strictly conformable nature argues that it is synsedimentary, and the age of mineralization therefore late Precambrian. The intrusion of the early Carboniferous porphyry dyke plexus could have remobilized the mineralization in a hydrothermal system centred on the dykes.

Iron Contact–Hydrothermal Deposits

Compared with the Precambrian banded-iron formations, contact deposits are of lesser worldwide significance. They predominantly occur in skarn at the contact with a variety of iron-rich plutonic rocks, such as diorite. Magnetite–haematite ore bodies are widely distributed throughout the volcano–plutonic circum-Pacific arc systems. The contact metamorphic, pyrometasomatic, and/or replacement deposits are economically more important than other varieties. They are cross-cutting irregular bodies of magnetite, superficially oxidized to haematite. There are hundreds distributed from east Australia, Malaysia, China, Japan, Kamchatka, on the west Pacific, through Canada and the USA to South America on the east (Park, 1972).

Many of the bodies occur against limestone, associated with skarn, but other country rocks such as andesite–breccia are also favourable. There is a strong structural or stratigraphic control on the movement of the hydrothermal fluids that deposited the bodies. The associated plutons range from granodiorite to diorite and gabbro.

Many of the deposits, particularly in South America have a close association with magnetite–apatite veins. In Chile, apatite-bearing dykes are very extensive, and some have been mined for phosphorus (Park, 1961). Therefore, in some cases it appears as if volatile-rich ore magma must have segregated and intruded or even extruded, as at Laco, Chile (Park, 1961). This bears a similarity to the Proterozoic Kiruna-type ore (Frietsch, 1978). However, most of the deposits were formed by replacement, rather than by direct injection. Park (1972) concludes that the iron of the circum-Pacific may have been remobilized from deeper crustal sources, such as Precambrian banded-iron formations in many cases. The polycyclic igneous events of the destructive plate margins have caused a remobilization of the iron, and major faults may have played important roles in the upward migration of the ore magmas and hydrothermal solutions. The mobilized iron could have been emplaced in many different forms, but all would have been greatly modified by continuing hydrothermal activity in the volcano–plutonic arc. Hence deposits are not confined to any one age or style, and the circum-Pacific mineralization extends throughout the Phanerozoic.

The following examples are chosen to illustrate briefly the vast circum-Pacific province.

Philippines

The pyrometasomatic iron deposits occur in volcano–plutonic–sedimentary sequences of age variable from the Cretaceous to the Miocene (Bryner, 1969). The host rocks are predominantly limestone, and they especially lie at their contact with volcanic rocks. Most of the ores are magnetite, though some contain an oxidized haematite zone.

At *Larap*, the ore lies immediately below a thick volcanic sequence. This is the largest and oldest mining region of the country, and has produced over 16 million t of iron ore. The deposits are of several lenticular bodies that are conformable with the sediments. They contain magnetite, and magnetite–pyrite zones. Each body is surrounded by skarn and is adjacent to syenite porphyry sills and dykes of thickness up to 30 m, which were probably responsible for the sedimentary–volcanic contact, and most are replacement deposits.

At the *Thanksgiving* mine, the ore is unusual, but serves to illustrate that not just iron can be concentrated in the contact–hydrothermal deposits. The mine comprises a number of small sulphide ore bodies, extracted for their gold, silver, zinc, copper, lead, cadmium and sulphur contents (Callow, 1967). The ore is massive sulphide, mainly of sphalerite and pyrite, with minor chalcopyrite, galena, arsenopyrite, magnetite, haematite, chlorite, calcite, quartz and garnet. Tellurides and native gold are scattered through the ore.

The ore occurs as veins and pods, mainly along the contact between diorite porphyry dykes and limestone, and a few are entirely within the limestone. They vary from 3 cm to 150 cm in width, and are 30 m or more in length. The diorite porphyry is considered to be the source of the metals and of the hydrothermal activity. The contacts are characterized by zones of garnet–clinozoisite skarn, and the deposition is considered to have been in the temperature range of 350–400°C (Callow, 1967).

Bukit Ibam Iron Mine, Peninsular Malaysia

This ore body formed a single lens with maximum dimensions of 930 m × 240 m × 60 m. It is in part concordant with and in part discordant to the surrounding rocks and is cut by faulting subparallel to the ore lens (*Figure 7.9*). The ore is mainly strata-bound between an overlying sequence of Triassic rhyolite tuff and igrimbirte of thickness about 300 m, and underlain by a sequence of siliceous hornfels, impure limestone and siliceous schist, which has been intruded by a hornblende diorite. An interesting feature of the diorite is its absence of magnetite, in contrast to other plutonic bodies exposed in the neighbourhood (Taylor, D., 1971). The ore occurs within the metamorphic aureole of the diorite. At the peak of its production in 1965, the annual output was more than 2 million t of grade 60–64% iron.

The primary ore can be divided into an enriched upper oxidized zone of haematite–magnetite, and an underlying protore containing well-crystallized single crystals or aggregates of magnetite (up to 10 cm diameter) in a green matrix composed of actinolite, anthophyllite, talc and serpentine. Copper, zinc and bismuth were present in minor amounts, copper (0.08%) being mainly in the limonitic ore along the hangingwall in a zone of secondary sulphide mineralization. Zinc (0.07%) was also mainly in the limonitic ore, and also in chloritic rocks along the hangingwall. Bismuth (average 0.06%) was in the friable haematite–magnetite ore. The ore is confined to a zone within a sheath of sheared and chloritized rocks. The diorite is not in contact with the ore, but occurs at a depth of less than 60 m below it. The major fault that cuts the ore body is itself mineralized and the overlying altered rhyolitic rocks contain irregular patches of specularite–pyrite mineralization.

Although Taylor, D. (1971) considered that the ore body was contemporaneous with the rhyolitic tuffs, the Cornwall-type mineralization (*see* p. 159) holds many similarities, including the contact metamorphic mineral assemblage and the sub-

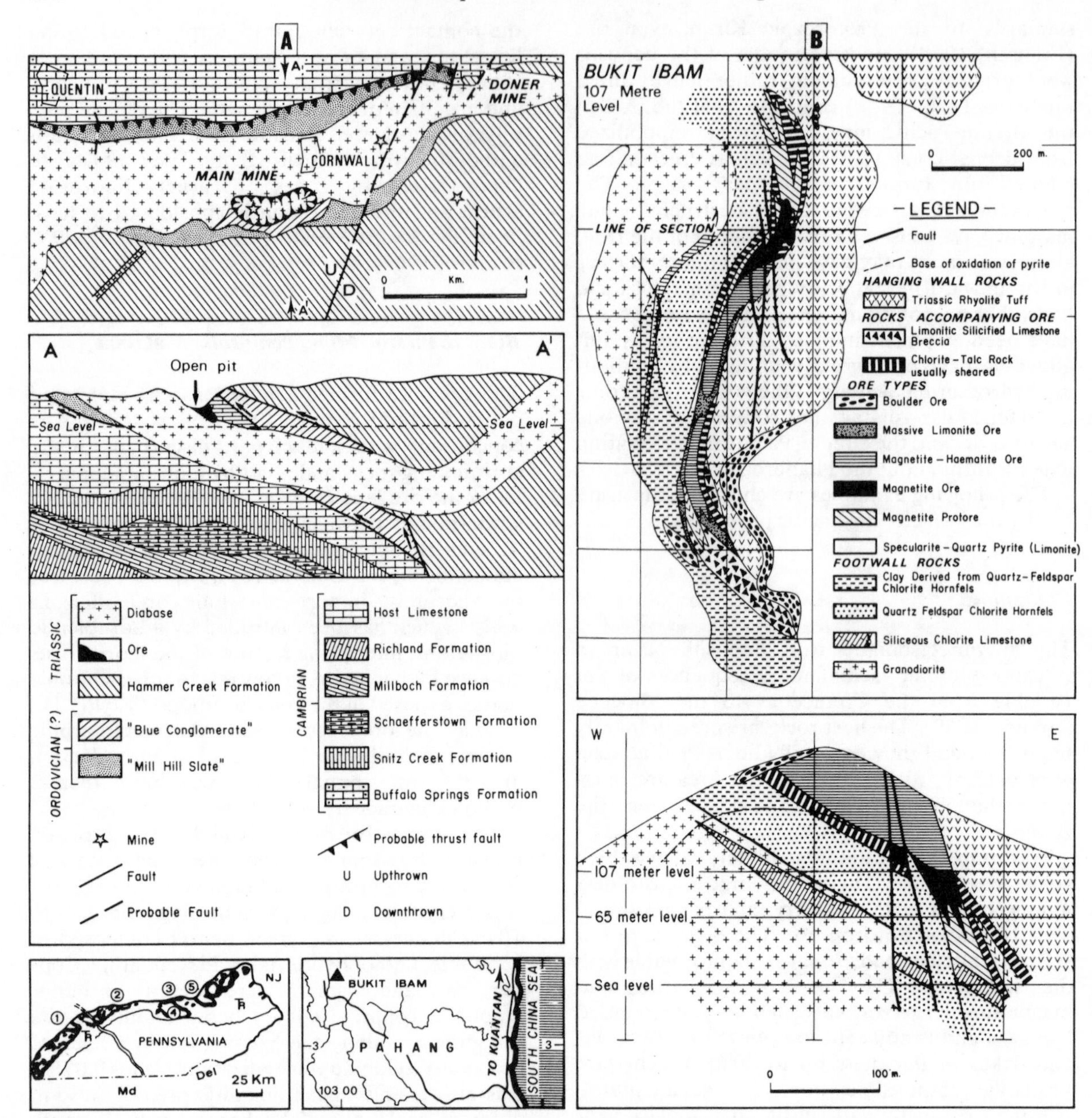

Figure 7.9. (A) The Cornwall (Pennsylvania) contact magnetite deposit and its relation to the other diabase–magnetite ore bodies of the Triassic belt. (1) Dillsburg, (2) Cornwall, (3) Wheatfield, (4) Morgantown–French Creek, (5) Boyertown (after Lapham, 1968). (B) The Bukit Ibam (Peninsular Malaysia) strata-bound structurally controlled replacement magnetite ore body, possibly also of Triassic age (after Taylor, D., 1971).

sequent hydrothermal metasomatic changes. It seems therefore appropriate to apply the model of Eugster and Chou (1979) to this deposit, and to suggest that the iron has been removed from the underlying diorite and deposited in a favourable stratigraphic setting beneath the thick rhyolite tuff formation by a hydrothermal circulation cell. The succession underlying the rhyolite tuff contains limestone horizons, and the ore could have replaced a favourable limestone bed. Indeed, many of the smaller ore deposits in the Bukit Ibam vicinity are related to limestone beds and skarns. The serpentinization of the protore matrix favours the interpretation of hydrothermal alteration of a previous metamorphic assemblage within the aureole of the diorite, which could have included pyroxenes. The mine is now abandoned, but the serpentine continues to be used for local carving of souvenirs under the trade name 'Dara Jade'.

The numerous other iron deposits of Peninsular Malaysia have been reviewed by Bean (1969). Most of them were mined as oxidized supergene altered material and the underlying primary protore was never discovered. From the few that were mined into the protore, Hosking (1973a) concluded that all were mesothermal hydrothermal ores. The significant association with tin mineralization, as at Bukit Besi and Pelepah Kanan (described p. 137), indicates a hydrothermal origin in which the fluids were leaching both iron and tin, perhaps both from diorite and granite, or at least from a range of plutonic rocks in the volcano–plutonic arc.

China

There are two major types of contact ore deposition (Ikonnikov, 1975). *The Tamiao*-type of deposit is genetically related to gabbro and anorthosite. Irregular veins and lenses of ore occur in the contact zones, frequently controlled by faults. The ore contains 30–40% iron, more than 10% titanium dioxide and less than 1% vanadium pentoxide. The type locality is at Tamiao in the Luanping district, 150 km northeast of Peking. The ore is of magnetite, titanium–iron ore and haematite, associated with sulphides, sphene, rutile, apatite, quartz and calcite, and occurring as bodies 100–500 m long, by 10–30 m wide, within gabbro and anorthosite at the contact zone. This type of deposit has similarities with the well-studied Cornwall-type magnetite deposit of Pennsylvania (*see* right). *The Tayeh and Shihlu*-types occur in the contact zones of acid-to-intermediate plutons against limestones. Their shape is irregular and they have all the typical features of skarns. The main minerals are magnetite and haematite, associated with sulphides, with a significant copper and cobalt content. The ore grade is usually high. In the Tayeh district of Hupeh, 85 km southeast of Wuhan, ore veins and lenses occur at the contact between granodiorite and Triassic limestone. The ore is both of haematite and magnetite, containing 60–65% iron and enough cobalt and copper for commercial extraction.

The Shihlu-type of deposit is exemplified on the island of Hainan. The irregularly shaped ore bodies occur at the contact between granite or quartz diorite and hydrothermally altered dolomite or sandstone. The country rocks are strongly sericitized and silicified. Haematite is the chief mineral, and magnetite increases with depth. The grade is within the range 58–63% iron.

Japan

The iron deposits have been reviewed by Mitsuchi (1952). The Kamaishi mine is the most important and the Shinyana lens the largest, of dimensions 400 m long, by 80 m wide, with a depth extension of 550 m. Granodiorite intrudes Mesozoic limestone, shales and sandstone. Skarn is developed along the contact and lenticular bodies of magnetite occur adjacent to the skarn. Minor haematite, chalcopyrite, bornite, pyrite, pyrrhotite, sphalerite, galena, gold and silver are disseminated in the magnetite.

Canada

Magnetite is a common constituent of skarns in southwest British Columbia (Sangster, 1969). Most of the intrusive rocks on Vancouver Island are older than the magnetite deposits. On Texada Island, just south of Vancouver Island, a sequence of limestone, schist, quartzite and volcanic rocks is intruded by quartz diorite. Both the country rocks and the diorite are replaced by extensive skarn zones containing massive magnetite. Diorite dykes also cut the magnetite. The ore contains considerable amounts of pyrite, calcite, quartz and up to 1% copper. The deposition of the ore was by replacement processes.

Cornwall, Pennsylvania

Although not in the circum-Pacific Belt, this well-known deposit has revealed a possible mechanism of hydrothermal deposition that may be widely applied to contact deposits of volcano–plutonic arcs.

The Cornwall magnetite mine was the oldest continuously operating mine in North America; it began in 1742. The geology and mineralization have been described by Spencer, A.C. (1908), Lapham (1968) and Lapham and Gray (1973). The metallization occurs as a limestone replacement above the essentially conformable Triassic south-dipping portion of a saucer-shaped diabase sheet (*Figure 7.9*). Thermal metamorphism of the carbonate-bearing country rocks largely to calc-silicates is ubiquitous. Potassium metasomatism is confined to the ore zones. Associated magnetite–actinolite mineralization replaces all earlier metamorphic and metasomatic minerals and sometimes even the diabase chilled margin. The generalized paragenetic sequence of the ore is: calcite, plagioclase, diopside, garnet, epidote, tremolite, haematite, talc, serpentine, mica, diopside, orthoclase, actinolite, pyrite, magnetite, chlorite, pyrrhotite, chalcopyrite, sphalerite, followed by zeolites, hydroxides and sulphates.

Most important to the understanding of the Cornwall-type deposits is the introduction of new materials into the hornfels zone around the diabase. Diabase granophyre is enriched in pink iron-rich orthoclase, which is also present in sections of the ore zone.

Content temperatures were estimated at 600–670°C. Ore emplacement occurred after the diabase was fractured and at a lower temperature of about 500°C. Eugster and Chou (1979) propose a model for the genesis of the ore bodies that can be applied generally to magnetite contact–hydrothermal deposits. The model is based on convecting supercritical aqueous chloride solutions, driven by the hot diabase. Acquisition of ferrous chloride by the hydrothermal solution is accomplished at depth (at the hot end of the circulating hydrothermal cell) as a two-stage process. Firstly, hydrochloric acid is formed by conversion of muscovite to alkali feldspar, and then iron-minerals such as magnetite, pyrite, or ilmenite in the rocks are dissolved to yield ferrous chloride. The source rock for this iron may be a basic igneous rock such as diabase or an intermediate rock such as diorite, both of which are usually iron-rich. If one looks at the differentiation trends of igneous rock suites in AFM diagrams (*Figure 6.4*), then the peak of the iron enrichment is usually reached in the dioritic or andesitic members of a plutonic arc suite. The ferrous chloride solutions circulate to the cold outer margins of the circulating cell, where they come in contact with beds of limestone. Magnetite and pyrite are precipitated during reaction with the calcium carbonate and the solutions become enriched in calcium chloride and carbon dioxide, some of which may return downwards with the return flow of the convection cell. By this process, there is no difficulty in transporting large amounts of magnetite or pyrite from a source rock to a skarn formation. The neutralization of the hydrothermal solutions by carbonate formations is supported by the worldwide affinity of magnetite deposits for carbonate formations.

This new interpretation of the Cornwall-type deposits is interestingly similar to an early hypothesis of Spencer, A.C. (1908), who wrote: 'They have formed by the more or less complete metasomatic replacement of sedimentary rocks by iron minerals precipitated from heated solutions set into circulation by the invading diabase'. The model proposed for the Cornwall-type deposits can be widely applied throughout volcano–plutonic arcs, especially in the circum-Pacific Belt, which is rich in suitable iron-rich diorite or andesite.

Uranium Deposits Associated With Granitoids

The large uranium deposits of the world occur in continental sedimentary basins or within Precambrian regionally metamorphosed terranes. However, the S-type leucogranites, which represent the anatectic partial melts derived from continental sedimentary formations, are also important hosts of significant uranium mineralization. The uranium resources of France, for example, are predominantly within Hercynian granites, and S-type granites elsewhere in the world are now being carefully considered as suitable hosts for uranium hydrothermal mineralization (Moreau, 1977; Rogers, J.J.W. *et al.*, 1978). The plutons are located in regions of thickened sialic crust that have been metamorphosed, and the granites have been called collision-related (Mitchell, A.H.G., 1979). Moreau (1977) considers that only Hercynotype granites may be fertile for uranium mineralization because they result from anatexis of continental sediments within marginal basins that have been filled with sediments entirely derived from continental erosion. I-type granites may occur where the continental crust is thin; they do not have a continental source, and are low in uranium.

In S-granite plutons the uranium content is usually high and occurs mainly as microcrystals of non-thorian uraninite. The uranium has been further concentrated by hydrothermal processes into veins and fracture zones within and close to the granite batholiths. Fehn *et al.* (1978) have shown that the heat generation in large plutons, which contain abnormal concentrations of uranium, can drive rather intense hydrothermal convection cells. Pitchblende is apt to be deposited in response to reduction of U^{6+} to U^{4+} and to the cooling of the solutions. These conditions are most likely to be met during vertical movement of the solutions along steeply dipping fracture zones in the plutons, which facilitate the easy upwards movement of the fluids.

A correlation is expected between anomalous

enrichment in uranium, tin and tungsten, all of which elements have an affinity for S-type plutons. This is clearly shown in the European Hercynian granite chain. For example, the leucogranites of the western Limoges region of the Massif Central of France have higher uranium and tin values than the cratonic granites of the massif of Guéret farther east. The major uranium deposits occur within the granites of St. Sylvestre, which have uranium contents of 8–21 ppm and tin contents of 8–240 ppm, whereas the granites of Guéret contain only 3–8 ppm uranium and 3–7 ppm tin.

The tin–tungsten-rich granites of Peninsular Thailand have high uranium contents of 5–57 ppm (average 16 ppm) and this contrasts with the I-type granites of Japan, which have only, on average, 3 ppm uranium (Ishihara and Mochizuki, 1980). Rogers, J.J.W. *et al.* (1978) developed the argument that a high initial $^{87}Sr/^{86}Sr$ ratio is important for a granite, indicating that it results from anatexis of continental sialic crust, if it is to be considered a potential source of mineralization. High initial strontium ratios are characteristic of Hercynotype granites.

Many authors have noted that granites play an important role in supplying uranium to younger overlying sedimentary strata, and some regions of the continental crust have been uranium provinces since early Precambrian time. The old granites contain anomalously high thorium values and the granites may have lost their uranium into circulating hydrothermal systems. Isotopic evidence shows that many of the Wyoming granites of North America lost most of their uranium during the Cenozoic era (Stuckless, 1979) so that they have served as a source of uranium to the sedimentary deposits of that region. In the Bokan area of Alaska, uranium mineralization is related to Mesozoic peralkaline albite–riebeckite granite (MacKevett, 1963). There are four types of mineralization: primary dissemination of uranium minerals in the later granitic phases caused by magmatic–hydrothermal activity; primary mineralization in aplites and pegmatites associated with the granite; epigenetic vein and fracture fillings with some replacements; and secondary hydrothermal deposition in the pores of overlying clastic sedimentary rocks.

The Hercynian Granitic Deposits

The middle European Hercynian chain extends from Spain to the Ertzgebirge, through Cornwall, Brittany, Vendée, the Massif Central, the Vosges and the Black Forest (*Figure 7.10*). There is a clear association between the granites and uranium, tin and tungsten, and the region constitutes a metallogenic province for these elements.

Massif Central

There are four major regions in France that contain hydrothermal uranium deposits: Morvan, Limousin, Forez and Vendée. They are all located within Hercynian granites, and three of them are in the Massif Central. The granites have anomalously high uranium contents (15–20 ppm) and frequently contain tin and tungsten deposits within their aureoles. In the two mica–granites, 50–70% of the contained uranium occurs as microscopically disseminated uraninite (Moreau, 1977). In contrast, the uranium in the biotite granite is predominantly contained in accessory minerals such as zircon, monazite and apatite. The age of the primary mineralization averages 250 Ma (Rich *et al.*, 1977). Hypogene mineralization occurs as pitchblende and minor coffinite. It is accompanied by pyrite, marcasite, galena, sphalerite and chalcopyrite. Generally the deposits are shallow, occurring within a few hundred metres of the surface, and they die out with depth. The ore occurs as uraniferous silicified material in fault zones, fracture zones, stockworks, small veins and mineralized zones within *episyenite* (alkali metasomatized carbonated and desilicified granite). Uranium mineralization is particularly intense where these veins intersect lamprophyre dykes, which obviously have impounded the hydrothermal fluids (*Figure 7.10*).

The French Massif Central is composed of metamorphic and granitic rocks. The oldest metamorphic formations are believed to be Precambrian, and metamorphism to sillimanite grade, accompanied by anatectic granites, has been dated at 526 Ma. The granites that constitute half of the outcrops are, however, of Hercynian age, but they contain xenoliths of Precambrian gneisses and anatectic granites of the older infrastructure. Hercynian magmatism began in the Upper Devonian period with andesitic tuffs. Then several granites were emplaced in the Carboniferous. Whole rock isochrons have been obtained at 335, 325, 310, 272 and 270 Ma, and the initial strontium isotope ratios vary between 0.704 and 0.718 (Vialette, 1973).

The Forez deposits

The Bois Noirs–Limouzat deposit is one of the best examples of granite hydrothermal mineralization in the European Hercynian chain, and has

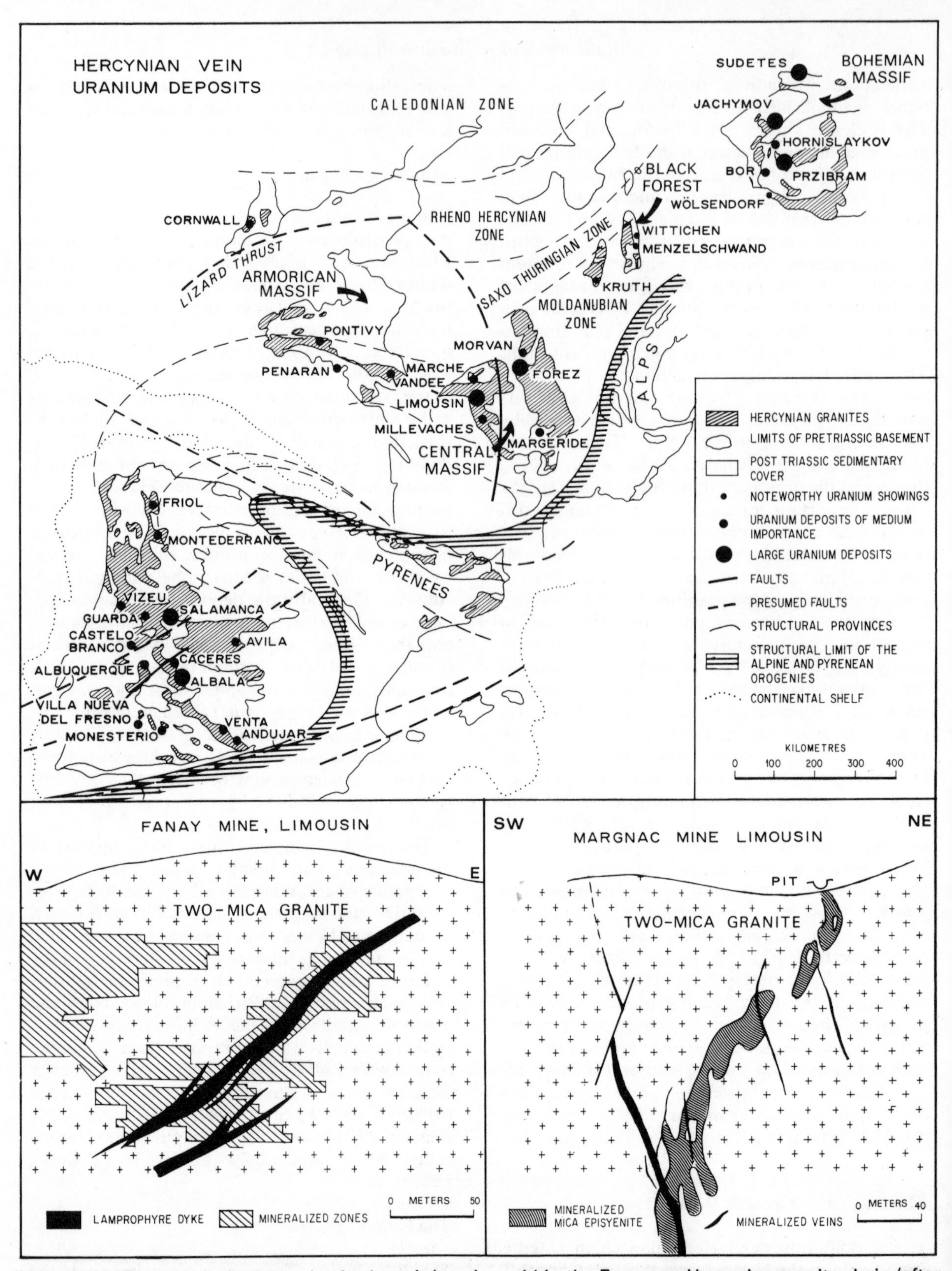

Figure 7.10. Top: the hydrothermal vein deposit location within the European Hercynian granite chain (after Cuney, 1978). Bottom: cross-sections of the Fanay and Margnac mines of the Limousin district of the western Massif Central of France (after LeRoy, 1978).

been described by Cuney (1978). It contained about 6400 t of uranium metal, but unfortunately closed in 1980 because no further important deposits could be found in the district; it was mined to a depth of 400 m. Two types of mineralized structures are found within the granite: type-I is of east–west-trending veins that dip 70–80° to the south. They vary from tens of centimetres to a few metres in thickness. Larger veins are filled with breccias cemented by quartz–pitchblende primary ore and iron sulphides. The mineralization is discontinuous along strike. Ore grades are generally between 0.5 and 5% metal by weight (Cuney, 1978). The type-II structures trend 135–165° and dip 20° to the south. The shear directions contain ore zones of thicknesses ranging from a few metres to tens of metres and they contain mainly secondary hexavalent uranium minerals (torbernite, etc.) and sooty pitchblende paint coating the joints. Age determination on the sooty pitchblende gives a Tertiary age. The grade is only 0.1–0.3% uranium metal, but type II accounts for more than half the uranium of the deposit. The veins occur within the Bois Noirs granite.

The following sequence of events, which led to the uranium mineralization, is summarized from Cuney (1978): the Bois Noirs granite resulted from anatexis of uranium-rich sediments near granulite-facies conditions. The temperature of anatexis is estimated at 800°C. The magma was syntectonically emplaced along east–west structures into a non-metamorphic higher crustal environment. The pluton cooled inwards, and a late-phase magmatic fluid collected and migrated upwards towards the core and altered the primary minerals orthoclase, oligoclase and biotite to microcline, albite and chlorite, respectively. The alteration of the primary minerals sphene, zircon and monazite resulted in the partial liberation of their uranium contents. This igneous–hydrothermal fluid migrated outwards to precipitate uraninite and its deposition was associated with pervasive greisenization of the granite.

The pluton of the Bois Noirs continued to experience hot conditions throughout the span of 335–270 Ma years and this stimulated hydrothermal circulation cells. The enrichment of the pluton in radioactive elements would also have provided an additional heat source.

Water of meteoric origin, mixed with carbon dioxide from a deep source, produced the remobilization of the uranium contained in the primary uraninite. The east–west foliation of the granite was reopened by tectonic activity, which produced open zones filled with porous crushed granite breccias. The fluids were rich in carbon dioxide and the uranium was transported in the form of uranyl monocarbonate and significant amounts of hydrogen sulphide were present in the solutions. The fluid pressure was nearly lithostatic in the deep zones (265–775 bar) and became progressively hydrostatic in the breccia zones (100–300 bar), where the pitchblende was deposited. The carbon dioxide concentration reduced from 3 mole% to less than 1 mole%. The uranyl carbonate complex was destabilized and hexavalent uranium was reduced by oxidation of the reduced sulphur species in the solution. Marcasite and pyrite deposition alternated with pitchblende.

The stages in the complex mineralization have been summarized from Cuney (1978) as follows. Stage I: quartz + pitchblende + marcasite + pyrite. Filling temperatures were estimated by fluid-inclusion studies at around 100°C and salinities around 1.7% sodium chloride equivalent by weight. Stage II: quartz + pyrite + haematite + coffinite + galena. Filling temperatures were determined to have been around 280°C with a salinity of 8.5% sodium chloride equivalent by weight. Stage III: quartz + chalcopyrite occurred over a declining temperature range from 150°C to 50°C with a constant 13–15% sodium chloride equivalent by weight. Stage IV: chalcopyrite + fluorite + calcite + whewellite + magnetite. Filling temperatures of the calcite are 90–160°C. The fluids contained carbon dioxide and hydrocarbons and the salinity was 7.3–15.6% sodium chloride equivalent by weight. Stage V: paragenesis is mainly of quartz + bismuthinite with chalcopyrite, with a filling temperature of about 150°C. The high salinity was of 23% sodium chloride equivalent by weight. The above stages I–V were deposited in type-I structures.

At a much later event, pitchblende was altered to coffinite without new uranium introduction. These recrystallization remobilizations were related to Oligocene uplift of the horst structures and the uranium was remobilized into the type-II structures in the form of hexavalent uranium minerals and sooty pitchblende (Cuney, 1978). This gives the stage VI mineralization, which is of black uranium oxides + hexavalent uranium minerals + manganese hydroxides of low deposition temperature, probably less than 50°C, and the fluids were of very low variable salinity, 19–1.5% sodium chloride equivalent by weight.

Poty *et al.* (1974) consider that the presence of carbon dioxide and hydrocarbons in the fluid inclusions is important in the uranium genesis. If the carbon dioxide-rich fluids were too highly oxidized, there would be a deposition of haematite and destruction of pitchblende during the hydrothermal phase. If the equilibrium is displaced towards reducing E_h values, there will be no remobilization. Fluids were always close to the pyrite–haematite boundary, on the pyrite side during deposition and on the haematite side after deposition. There is therefore in the Massif Central deposits a common association of pitchblende and pyrite.

Limousin

The La Crouzille uraniferous district of the Haut Limousin, 20 km north of Limoges, is the main French intragranite uranium district, with more than 22 500 t of proven uranium metal. The host rock is the St. Sylvestre two-mica granite, which is surrounded by metamorphic formations. The major metamorphic–anatectic episode has been dated at 360–380 Ma (Duthou, 1977). The geology of the main mining districts of Margnac and Fanay has been described by LeRoy (1978). There are two distinct granite facies in the region—the Brame, which is foliated and contains biotite and sillimanite, and the St. Sylvestre granite, which contains both muscovite and biotite. The determined age of 315 Ma is now regarded as a re-equilibration age when muscovite formed, and the actual intrusion took place 350–360 Ma ago (LeRoy, 1978). The existence of the two distinct granites is thought to be due to deuteric processes of muscovitization and albitization operating within the same plutonic complex. These granites are called 'fertile' (Moreau, 1977), to indicate that they contain more uranium than the normal two-mica granites. The St. Sylvestre massif contains pegmatites and quartz veins with tin, tungsten and gold.

An early magmatic–hydrothermal alteration of the two-mica granite is referred to as episyenitization. In feldspar episyenite the quartz is gradually leached, the muscovites are altered to feldspar and the biotites altered to chlorite and alkali feldspar. Mica episyenitization is related to the intersections of north–south and east–west fractures, and appears as irregular pipes. All the granite components except muscovite disappear, but some orthoclase may remain (LeRoy, 1978).

In contrast to the earlier ductile deformations, three fracture directions, north–south, east–west and northwest–southeast, developed. Intrusions of lamprophyric dykes were intruded at 285 Ma into the fractures.

The mineralization is both of the vein-type and the disseminated-type (*Figure 7.10*). The veins occupy breccia zones, and the mineralization consists of pitchblende, pyrite and microcrystalline haematitic quartz, cemented by muscovitic and argillic minerals and black oxides. The following paragenetic sequence has been described by LeRoy (1978): (1) quartz + pyrite; (2) spherulites of pitchblende (up to 4 cm diameter), followed by pyrite; (3) microcrystalline haematitic quartz that contains pyrite and haematite; (4) brecciation followed by quartz and marcasite, and pitchblende altered to coffinite—rare galena was deposited; (5) violet fluorite followed further brecciation—pink barite was deposited, containing numerous inclusions of pyrite, marcasite and haematite, and calcite ends the succession.

The disseminated mineralization represents nearly 40% of the ore of Margnac and 30% of the Fanay mines. It is mainly in mica episyenite pipes and veins. The mineralogy is the same as the veins. Pitchblende is the main ore and coats the rock forming minerals.

It is also clear that the lamprophyre dykes acted as impounding bodies to the further migration of the hydrothermal fluids, and their control of the mineralization is especially seen in the Fanay mine (*Figure 7.10*).

The following mineralization sequence has been deduced by LeRoy (1978), supported by his fluid-inclusion and ore textural studies, supplemented by data from Poty *et al.* (1974): the lamprophyric activity produced localized heat sources that heated up the meteoric waters, which had infiltrated the plutonic complex, to form a convecting hydrothermal system. The fluids were channelled into the faults and fractures of the east–west and north–south system causing the micaceous episyenitization of the granite. The earlier feldspar episyenitization is considered to have occurred during earlier plastic deformation of the pluton.

Pitchblende and pyrite mineralization immediately followed the lamprophyre emplacements and the episyenitization, and a minimum age of 275 Ma is generally accepted. The mineralization was precipitated from carbon dioxide-rich fluids by the unmixing of complex carbon dioxide–water mixtures following a drop in pressure within the fault and episyenite zones. The temperature of the solutions in the episyenite of the Margnac mine was approximately 345°C (Poty *et al.*, 1974).

After pitchblende was deposited, the fluid became progressively more water-rich during the precipitation of microcrystalline quartz and haematitization of the pyrite. The deposition of quartz plus marcasite and the transformation of pitchblende to coffinite took place when temperatures fell from 330°C to 140°C, related to the progressive decay of the hydrothermal heat source. The sequence ended with the deposition of fluorite, barite and calcite, beginning at 135°C and continuing to a much lower temperature.

Supergene modifications into sooty pitchblende and other secondary minerals date from about 30 Ma ago and continue to the present.

OTHER MAJOR HERCYNIAN DEPOSITS

As shown in *Figure 7.10* the other major deposits are in Spain and the Bohemian Massif. A summary of some of these deposits is to be found in Rich *et al.* (1977) and a fuller account in Ruzicka (1971, 1978). In all these regions the granites are anomalously enriched in uranium, so that the region constitutes a uranium metallogenic province. For example, in southwest England (Cornwall) the granites contain a uranium average of 12 ppm, while the Lands End Granite of Cornwall contains an average 17 ppm (Simpson *et al.*, 1977). Only the Massif Central contains classic intragranite uranium deposits; elsewhere the veins are exogranite. For example at Salamanca, the most important mineralization occurs within Lower Palaeozoic pelitic schists. At Příbram, Czechoslovakia, the ore occurs in veins within shales, siltstones and greywackes around a Lower Carboniferous granite. At Jáchymov, southwest Erzegebirge, the pitchblende ore occurs in veins within scapolitized and metamorphic rocks that contain abundant sulphides, graphite and biotite, but veins within the granite itself are usually uneconomic. As with many uranium localities throughout this region, tin mineralization is associated with the greisenized granites of the Erzegebirge (Ruzicka, 1971).

Spokane Mountain Region, Washington State

The two deposits of Midnite Mine and Spokane Mountain occur in the contact zones between the Cretaceous Loon Lake porphyritic quartz monzonite batholith and the Togo Phyllite of the 1300 Ma old Precambrian Beltian Prichard Formation. Both mines occur along the northnortheast-trending faulted Deer Trail anticlinal structure. The deposits occur in a large roof pendant of deeply weathered metapelitic rocks of the Togo Formation that overlies the batholith.

SPOKANE MOUNTAIN DEPOSIT

This deposit occurs 61 km northwest of Spokane. It averages 2 m thickness with a grade of 0.25% U_3O_8 and lies 46–152 m below surface. It has a linear trend 70–80° with a 305-m minimum length and a 46-m average width (Robbins, 1978). Pitchblende, in veinlets or as coatings on fractures, is associated with iron sulphides and is localized mainly in the unoxidized zone of the graphitic phyllite and chlorite schist of the Togo Formation within 30 m of its contact with the quartz monzonite. The deposit could have been syngenetic within the Togo Formation, but became localized around the granitoid contact by Cretaceous hydrothermal activity. However, most of the final deposition is related to deep weathering and ground-water movement related to the Cenozoic topography. It differs from the Midnite Mine in that the mineralization extends into the granitoid along the continuous fracture zones (*Figure 7.11*). It is also possible that the Cretaceous granitoid introduced the uranium into the contact zone, and this deposit has been likened to many of the exogranite Hercynian deposits of Europe (Moreau, 1977).

MIDNITE MINE

This mine is situated 56 km northwest of Spokane. Essentially all primary mineralization is found in the metasedimentary rocks, close to the intrusive contact (*Figure 7.11*). The contact zone is highly fractured. Pitchblende occurs as disseminations along foliation planes, as replacements and fracture fillings. Uranium bodies cross-cut sedimentary lithologies without noticeable effect (Nash and Lehrman, 1975). The ore bodies are of pitchblende stockworks or irregular mineralized masses, which are generally elongated parallel to the intrusion contact and often occupy topographic lows in the contact topography (*Figure 7.11*).

Below the oxidized zone, pitchblende and some coffinite occur with pyrite and marcasite, and small amounts of pyrrhotite and molybdenite. The inconspicuous gangue includes quartz, carbonate and iron oxides. Ore-associated alteration of the country rocks is not pronounced. The isotopic age of the pitchblende mineralization is 102–108 Ma.

The origin of these deposits is thought by some (for example, Robbins, 1978) to be originally concentrated in the Precambrian sediments and subsequently hydrothermally concentrated during the Cretaceous granitoid emplacement. When viewed from the European point of view (Moreau, 1977) the deposits are regarded as having their source in the Cretaceous Loon Lake batholith, and the impermeable nature of the Togo Phyllite prevented the upwards escape of the hydrothermal fluids to form the highly impounded nature of their concentration.

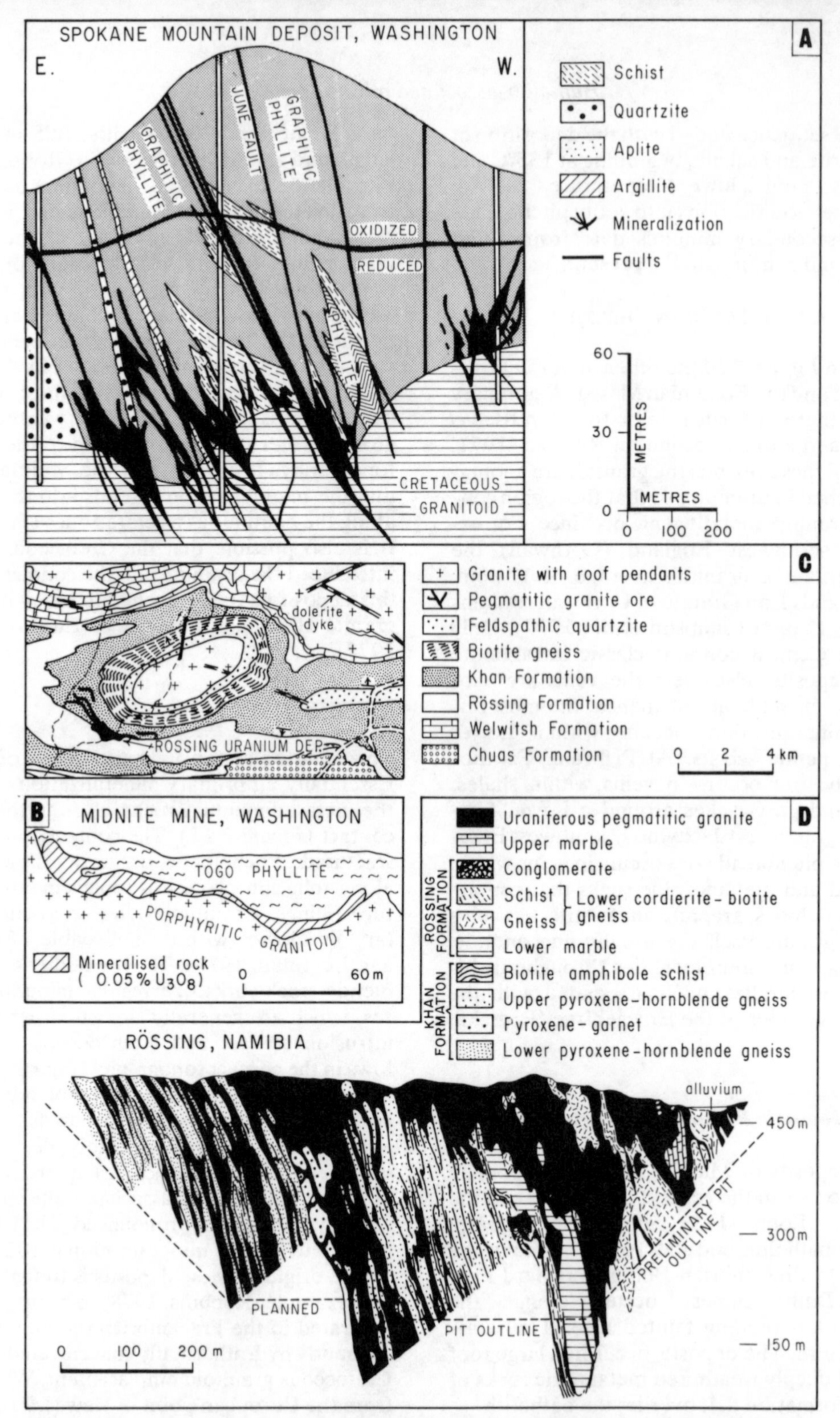

Figure 7.11. Granite-related uranium ore deposits. (A) The Spokane Mountain deposit (after Robbins, 1978) and (B) the Midnite Mine (after Nash and Lehrman, 1975) are both in Washington State, USA. The mineralization in both cases is granite-related, at Midnite Mine it is exogranite, but at Spokane Mountain it penetrates the granite. (C,D) The Rössing deposit of Namibia is the world's largest uranium mine. The mineralization is within pegmatitic granite that has formed anatectically from a sequence of high-grade schists and gneisses (after Berning *et al.*, 1976).

Rössing, Namibia

Rössing is the type example of what Armstrong, F.C. (1974) and McMillan (1977b) unfortunately call 'porphyry'-type uranium deposits. There is no justification whatsoever for this description; they are in no way like porphyry-copper deposits, and porphyries play no role in their genesis. Rather they are metamorphic–anatectic and the uranium was concentrated from the original sedimentary sequence into anatectic melts formed during high-grade regional metamorphism. The granitic melts have remained largely within their metamorphic envelope in a catazonal environment. This kind of deposit lends credence to the granite-related deposits of the Hercynotype terrains, in which the granite melts have migrated upwards from their anatectic source into mesozonal or even epizonal unmetamorphic regions of the crust. In the Rössing area, the granitoids remain within their metamorphic envelope as a migmatite formation. The Mary Kathleen deposit of Queensland, Australia, has also been cited (for example, by Derry, 1980) as igneous-related, but there is no evidence for this, and it is accordingly described along with other Proterozoic deposits.

The Rössing deposit lies 65 km northeast of the coastal town of Swakopmund. Cheney (1981) quotes a reserve of greater than 130 000 t U_3O_8 with an average tenor of 0.035% U_3O_8. The deposit lies within the central zone of the Late Precambrian Damaran orogenic belt. The country rocks belong to the Kahn and Rössing formations, which have been metamorphosed to pyroxene–hornblende gneisses, pyroxene–garnet gneiss, marble and cordierite–biotite gneiss. Anatexis produced various types of granites about 510 Ma ago. Triassic dolerite dykes are common throughout the area. The mine geology has been described by Berning *et al.* (1976).

The deposit occurs in a migmatitic zone in which uraniferous alaskitic granite shows both concordant and subcordant relationships with the gneisses. The country-rock sequence comprises metasediments and metavolcanic rocks. Contact metamorphic effects are evident in metasediments adjoining the alaskite intrusives, and skarns are developed adjacent to marble.

Tight vertical or slightly overturned folds trend northeast–southwest. The Rössing ore body is situated along the northern limb of a complex synclinorium developed between the domal structure of *Figure 7.11* and the Kahn metasediments.

The syntectonic igneous rocks are variously named pegmatite, potash granite or alaskite by different workers; they constitute the main ore. Completely barren or subeconomic alaskite continues along strike beyond the limits of the mining area. The alaskite is preferentially emplaced in the pyroxene–garnet gneiss–amphibolite and amphibolite–biotite schist–marble–cordierite–biotite gneiss units.

Uraninite is the dominant primary ore mineral. Its grain size ranges from a few micrometres to 0.3 mm. It is included within quartz, feldspar and biotite, but also occurs interstitially and along cracks in these minerals. Monazite is closely associated with the uraninite. Betafite is of minor amount and occurs as inclusions in quartz and feldspar. Zircon, apatite and sphene are always associated. Pyrite, chalcopyrite, bornite, molybdenite, arsenopyrite, magnetite, haematite, ilmenite and fluorite occur regularly in small amounts. Of the secondary uranium minerals, beta-uranophane is the most abundant. Uraninite contains about 55% of the uranium of the deposit, betafite about 5% and the secondary minerals about 40%.

Berning *et al.* (1976) propose the following origin for the ore, and clearly it is the only acceptable interpretation: the uranium was initially syngenetic within the sedimentary–volcanic rock sequence before regional metamorphism. High-grade regional metamorphism was accompanied by extensive anatexis. The uranium was preferentially concentrated in the anatectic melt of alaskite composition. The localization of the uranium-enriched alaskite coincides approximately with the original extent of enriched Precambrian sediments. Anatexis outside this area resulted in normal alaskites that contain only subeconomic uranium concentrations. A considerable amount of melt remained and crystallized within the metasediments from which it was formed, but there was obviously some upward magma migration to cause thermal metamorphic overprints on some of the metasediments.

8

Epigenetic deposits of volcanic and epizonal plutonic association

This chapter describes a spectrum from epithermal vein deposits, genetically related to near-surface volcanogenic fracture systems, to higher-temperature vein and dissemination deposits associated with epizonal plutons, which have intruded older volcanic and sedimentary formations (*Table 8.1*). The petrological and ore details vary with tectonic setting, whether in ensimatic island arcs or in volcano–plutonic arcs within ensialic cratons. The host magmas are predominantly of mantle origin, presumed to have risen from a Benioff Zone, but the metals in the ore deposits need not all be cogenetic with the magma, for large-scale circulation of meteoric hydrothermal waters, driven by the magmatic heat source, would have scavenged metals from the country rocks. The mineralization is usually connected with highly differentiated silicic and alkalic fine-grained porphyritic epizonal intrusives belonging to polyphase igneous complexes.

Porphyry-copper Deposits

Porphyry-copper deposits are large low-grade (average 0.4% copper) deposits of disseminated copper and iron sulphides, often associated with gold, molybdenum and silver, which occur in and around the roof zones of cupolas within the deep dioritic to granodioritic roots of calc-alkaline volcanoes, both in island arcs and on cratonic continental margins. The mineralization is grossly disseminated over large volumes of rock. Mining is from opencast pits of large size, and the whole mineralized zone is removed as ore by large-scale bulk mining. The proportion of sulphide localized along continuous megascopic fractures varies greatly between deposits and within a deposit. Microscopically it can be shown that the apparently disseminated copper sulphides are fracture-controlled. Breccias are commonly intrusive into the host rock and their matrix may be clastic, igneous or mineralization, but sometimes they may post-date the mineralization (Hollister, 1975, 1978; Gustafson, 1978).

Economic Significance

Porphyry-copper deposits provide 52.4% of the world's 351×10^6 t of copper resources. Canada accounts for 8% (British Columbia 5.5%), USA 21.4% (western USA 18.6%), Central America 5.4% and South America 25.8% (Sutherland Brown and Cathro, 1976).

Among the giant porphyry deposits are: Bingham, Morenci, San Manuel–Kalamazoo and Safford (USA), La Caridad (Mexico), El Teniente and Chuquicamata (Chile), Cerro Colorado (Panama), Sar Cheshmeh (Iran), Panguna (Bougainville–Papua New Guinea), Atlas on Cebu, Sipalay on Negros Occidental (Philippines), Valley Copper (Canada) and the Kal'makyr deposit of the Almalyk region (USSR) (*Figure 8.1*).

Classification

The classification into volcanic, hypabyssal and plutonic is based on the depth of formation, which may vary from as shallow as 500 m to as deep as 5 km below surface (Nielson, 1976).

Plutonic porphyry deposits occur within or at the margins of large plutons centred in areas of complex multiphase intrusive activity (Nielson, 1976). They lack obvious concentric zoning of alteration–mineralization, which is more structurally controlled than in the higher-level varieties, and occur throughout the Mesozoic–Cenozoic in the circum-Pacific, for example, at Butte (Montana), Chuquicamata (Chile) and Tapadaa (Sulawesi).

Table 8.1 Epigenetic disseminated and vein deposits of volcanic and epizonal plutonic association

Type	*Tectonic setting*		*Depth of formation (m)*	*Alteration–mineralization and temperature*	*Remarks on occurrence*
	Island arc[a]	*Cratonic arc*[a]			
Porphyry copper					*Island arcs:* associated with quartz monzodiorite and granodiorite. Important Au, negligible Mo *Cratonic:* associated with quartz monzonite and granite. Important Mo, minor to negligible Au
(a) plutonic	x	X	5000–3000	Early magmatic hydrothermal, 450–600°C	
(b) hypabyssal	X	X	3000–1000		
(c) volcanic	x		500–1000	Late meteoric hydrothermal, 300–450°C	
Porphyry molybdenum		x	1000–2000	Identical to porphyry copper	Associated with rhyolite porphyry. May contain W and Sn
Skarn	x	x	500–5000	Similar to porphyry copper	Occurs where porphyry copper intrusions are in contact with limestone country rocks
Epithermal vein					Commonly contain Cu, Fe, Pb and Zn sulphides in addition to the precious metals. Veins are in fractured rocks, caldera, or radial fissures. The volcanic association may be obscure in the lower temperature veins. Categories (a), (b) and (c) occur in the same settings as porphyry coppers. Category (d) never does
(a) Ag (Au)	x	X	100–1000	Substantially meteoric water alteration–mineralization, 200–300°C	
(b) Au (Ag)	X	x			
(c) Au (Ag) tellurides	X	x			
(d) Au (Sb) polymetallic	x	X			
Porphyry tin		x	50–200	Variable, 150–550°C	Commonly associated with silver. Tin may be as cassiterite or as wood tin

[a] X major occurrence, x minor occurrence.

Hypabyssal porphyry systems are the most common throughout the world. The mineralization is genetically connected with small hypabyssal, roughly cylindrical, composite and dominantly porphyritic intrusions in the form of cupolas, plugs, breccias and dykes. Mineralization is restricted to specific intrusive phases within the hypabyssal complex. Alteration–mineralization is broadly concentric and centred on the intrusions. This most common type of deposit has been well described by Lowell, J.D. and Guilbert (1970). Excellent examples are provided by Bingham (Utah), Cerro Colorado (Panama) and Panguna (Bougainville). Depth of formation ranges from 500 m to 2 km beneath the surface (Nielson, 1976).

Volcanic porphyry systems represent the near-surface expression of the hypabyssal systems (Sutherland Brown, 1976). The plutons have risen high enough to intrude coeval volcanic formations. The shape of the intrusive bodies is variable from cylindrical, irregular, to anastomosing dykes and sills. The bodies are commonly relatively small and the mineralized zone may be large in comparison (Sutherland Brown, 1976); breccias are common. Red Mountain (Arizona) and El Salvador (Chile) are good examples (Nielson, 1976).

Host-rock Characteristics

Intrusions within island arcs, which are genetically related to porphyry-copper deposits, are poorer in alkali feldspar than their counterparts that occur within ensialic cratonic plates (*Figure 8.2*).

Although most of the mineralized island-arc intrusions are quartz diorite, there are more potassic intrusions including quartz monzonite and even quartz syenite. All these are lower in potassium than those occurring in cratonic continental mar-

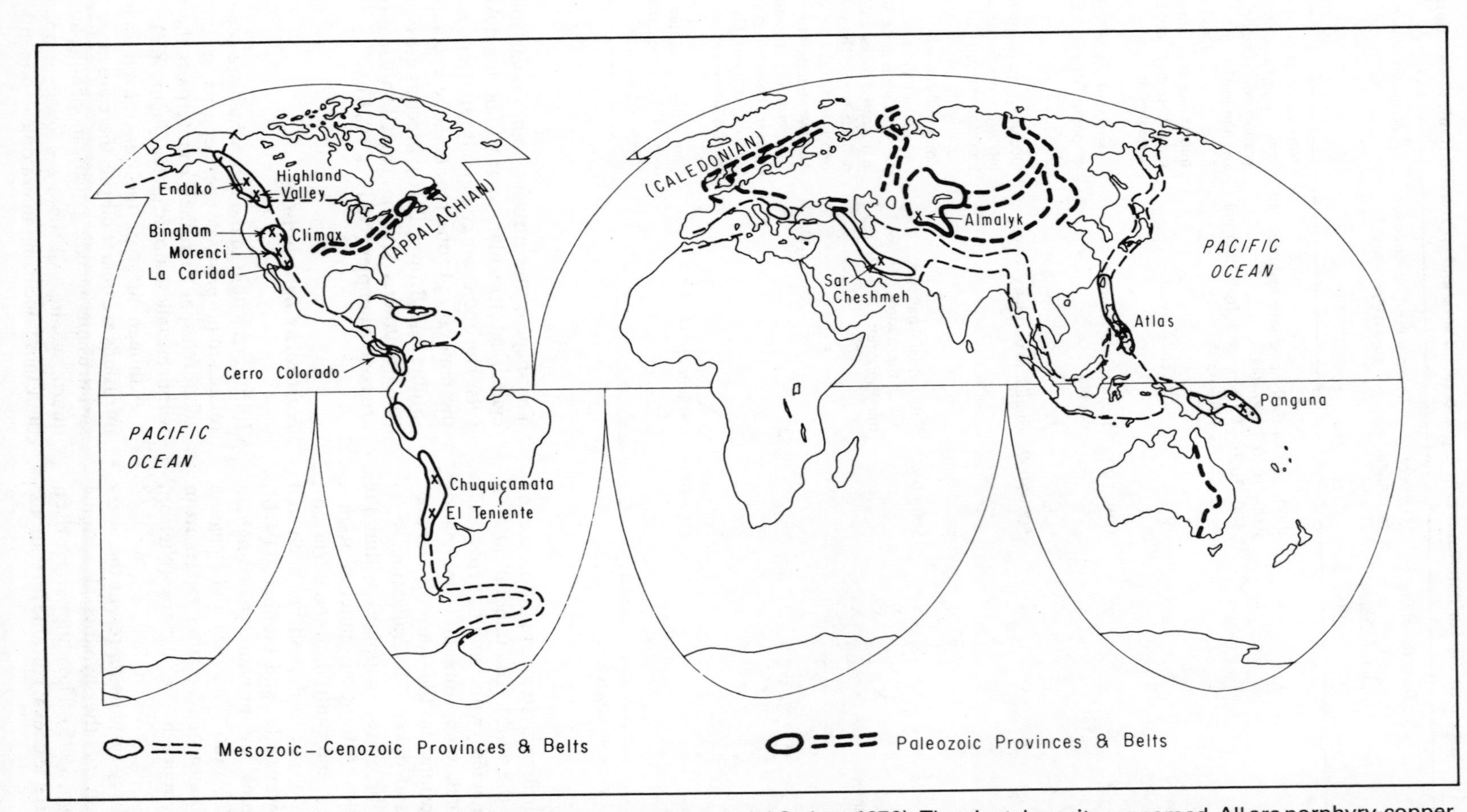

Figure 8.1. Porphyry provinces of the world (after Sutherland Brown and Cathro, 1976). The giant deposits are named. All are porphyry-copper deposits except Climax and Endako, which are porphyry molybdenum.

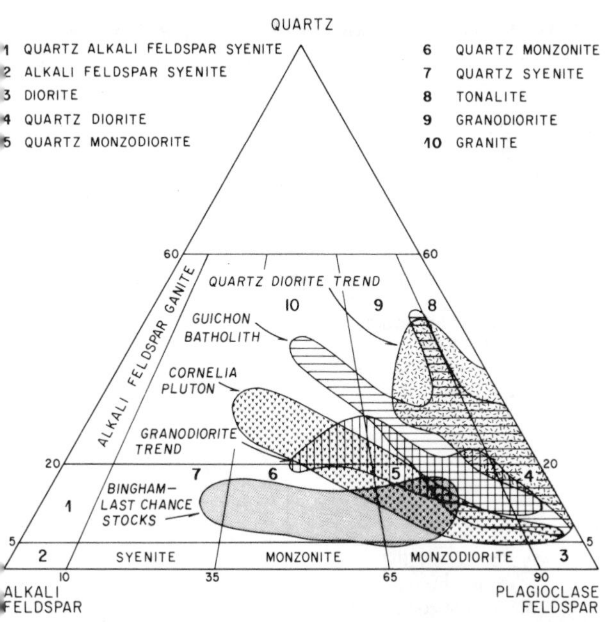

Figure 8.2. The composition of the porphyry stocks that host porphyry-copper deposits (after Kesler, S.E. *et al.*, 1975). The quartz-feldspar model classification is after Streckeisen (1976). The quarts diorite trend characterizes ensimatic island arcs. The trends directed towards alkali feldspar enrichment characterize stocks that have intruded cratoric crust.

gins, and island-arc quartz monzonites lack alkali feldspar phenocrysts (Kesler, S.E. *et al.*, 1975). There is a close and essential link between porphyritic intrusions and disseminated copper mineralization (Gustafson, 1978). The important textures, which are related to porphyry-copper mineralization, range from coarse seriate (subporphyritic) to distinctly porphyritic, and fine equigranular phases are commonly present. The rock sequence is characterized by a multiplicity of intrusive events and variety of textures.

A common feature of porphyry-copper deposits, from both island-arc and continental cratonic setting, is their spatial and temporal relation to an intrusive body that is among the most differentiated (both quartz-rich and alkali feldspar-rich) and among the youngest. This implies that differentiation plays a part in the origin of these deposits and in the characteristics of the igneous rocks with which they are associated (Creasey, 1977). It has been argued, from the viewpoint of copper partitioning and coordination within the minerals, that wherever there are composite intrusions, as is the rule for porphyry-copper deposits, the more aluminous intrusives are the better targets for copper mineralization (Feiss, 1978).

Stable Isotope Studies as an Indicator of Derivation

The $^{34}S/^{32}S$ ratios of porphyry-copper sulphides suggest an upper-mantle origin (Rye and Ohmoto, 1974). The lead-isotopic composition of the sulphides is similar to that of the associated igneous rocks. The conclusion is justified that the pluton and the copper may have come from the same source, and that the upper crustal material through which the magma passed did not contribute significantly to either (Doe and Stacey, 1974).

The plutonic rocks associated with the mineralization have initial $^{87}Sr/^{86}Sr$ isotopic ratios that are compatible with their origin in the mantle, with limited contamination from the crust (Hedge, 1974). Initial strontium isotope ratios in intrusive systems from island arcs are all lower than 0.705. On the other hand, initial ratios for plutons that have intruded cratonic continental plates exceed 0.705 and may be as high as 0.710. Rubidium/strontium ratios of the intrusive rocks also appear to differ, though less clearly. Island-arc values average about 0.2 and cratonic values 0.3–0.4 (Kesler, S.E. *et al.*, 1975).

Comparison between Island-arc and Continental Deposits

The major differences between deposits occurring in island arcs and those situated in continental cratonic crust have been well summarized by Saegart and Lewis (1976), who have described the Philippine deposits, and for comparison Lowell and Guilbert (1970) have described the American western cordillera. The comparison is given in *Table 8.2.* The differences, especially the strontium isotope ratios and the gold and molybdenum contents, may reflect contamination from the crust through which the mantle-derived magmas rose. Alternatively the mantle wedge above the Benioff Zone may be different between island arcs and cratonic continental margins, though why this may be so is unknown. Contamination from the crust is the simplest and most readily supported hypothesis.

Hydrothermal Alteration–Mineralization

Host rocks are invariably altered to secondary mineral assemblages. Major differences in these assemblages reflect host-rock composition. For example, purely igneous country rocks lead to

Table 8.2 Comparison of island-arc (Philippine) and cordillera (American) porphyry-copper deposits

Characteristic	*Philippine (island arc)*	*American (cratonic continental margin)*
Age (pluton and mineralization)	Palaeocene to Late Miocene	Predominantly Laramide (Late Cretaceous to Early Eocene)
Source pluton	Quartz diorite to diorite	Quartz monzonite
Initial $^{87}Sr/^{86}Sr$ ratio	<0.705	>0.705
Host rocks	Cretaceous to Mid Tertiary metavolcanic, metasedimentary and keratophyres	Precambrian to Cretaceous sediments and metasediments
Ore-body geometry	Tabular	Vertical cylinders
Ore distribution (% in source pluton versus % in host rock)	>50%/<50%	70%/30%
Distribution of deposits	Clusters	Clusters
Average ore tonnage	133×10^6	136×10^6
Average hypogene ore grade of copper	0.46%	0.45%
Alteration in ore zones	Potassic (biotite, chlorite, epidote, sericite, alkali feldspar)	Potassic to phyllic
Alteration in fringe areas	Propylitic (chlorite, epidote, calcite, sericite, clay)	Phyllic, argillic, propylitic
Zonation of alteration assemblages	Overlapping, indistinct	Distinct zones
Hypogene sulphides	Chalcopyrite and minor bornite	Chalcopyrite and minor bornite
Sulphide mode of occurrence	Quartz–sulphide veinlets with lesser dissemination	Disseminated and fracture filling
Weight% sulphides in ore zones	2–6% (average 3%)	3–10%
Pyrite/chalcopyrite in ore zones	0.8/1 to 4/1 (average 2/1)	3/1 to13/1
Supergene enrichment	Negligible	Common
Oxidation and leaching	Shallow (partial oxidation to 100 m depth)	Shallow to deep
Associated gold	Average 0.30 g t^{-1}	Minor
Associated molybdenum	0.007% Mo	0.015% Mo
Associated magnetite	Abundant (average 3%)	Sparse
Anhydrite or gypsum	Common as moderate amounts (often post-mineralization)	Common minor amounts
Intramineralization to post-mineralization dykes	Abundant, narrow and thick	Few, narrow dykes
Post-mineralization structures	Abundant shatter fractures	Major fault displacements

silicified assemblages, whereas if carbonates form a part of the country rocks, skarns may develop (Theodore, 1977). Alteration may be local, and partial or complete. There is always a zonal arrangement of the alteration–mineralization (Gustafson, 1978), but it may be imperfect in the plutonic porphyry variety. Superposition and replacement of successive alteration assemblages is the rule. A ubiquitous sequence is the replacement of the original magmatic minerals with biotite and alkali feldspar, accompanied by sulphide deposition characterized by a high copper/iron ratio. This earlier alteration is followed by later pyritic assemblages with lower copper/iron ratios in which biotite and alkali feldspar are destroyed (Gustafson, 1978).

The hydrothermal alteration usually includes a potassic core in which orthoclase and biotite are formed. This core is surrounded by a phyllic zone in which quartz–sericite–pyrite assemblages are formed. This is succeeded by an argillic zone in which clay minerals and pyrite are formed. The final outer zone is propylitic, in which chlorite is the characteristic alteration product (*Figure 8.3*). The alteration zones are concentric but often incomplete. The ore minerals, deposited during the alteration, are copper and molybdenum sulphides (Lowell, J.D. and Guilbert, 1970).

The potassic zone may consist of either an orthoclase–biotite or an orthoclase–chlorite zone, and in a few deposits both may occur (Hollister, 1975); in rare cases, the potassic zone may be absent. The phyllic zone is always present and contains the greatest (up to 20% of the rock) development of pyrite. Although this is called the sericite–quartz–pyrite zone, other sericite-like minerals such as pyrophyllite may be important. The argillic zone is characterized by kaolinite, montmorillonite and pyrite; it may be missing in some deposits.

The propylitic zone, characterized by chlorite and pyrite, is invariably present. Hypogene copper and molybdenum sulphides may occur together or separately in the potassic and in the phyllic zones. Copper appears to be richer at the interface between the potassic and phyllic zones. Copper rarely occurs in the argillic zone, but when it does it is accompanied by other sulphides.

Cause of the Alteration-Mineralization

Oxygen-isotope and hydrogen-isotope studies of hydrothermal biotites, from the potassic zones, show that their D/H ratio and their ^{18}O values are identical to normal magmatic biotites (Sheppard, S.M.F. *et al.*, 1971). Hence the potassic alteration is a result of hydrothermal activity caused by magmatic water (*Figure 8.3*). The temperature of this hydrothermal alteration has been estimated in the range 450–600°C.

In contrast, the isotope data from the sericite and clay minerals of the other zones require a large component of meteoric or saline formation water, and the temperature of formation in the phyllic zone is estimated at 300–450°C (Sheppard, S.M.F. *et al.*, 1971).

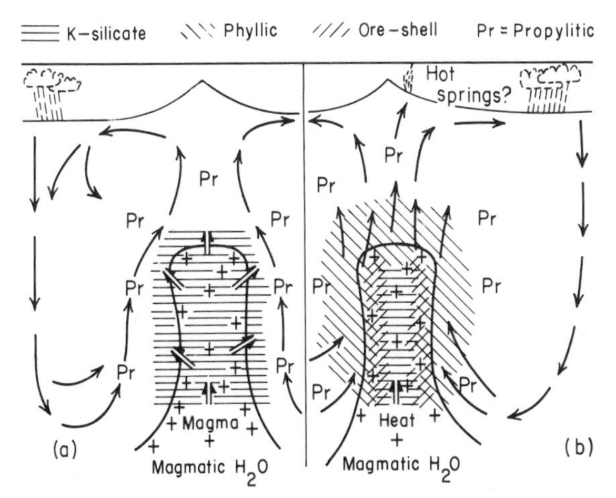

Figure 8.3. Schematic cross-section through a porphyry-copper deposit to show the interaction between magmatic–hydrothermal and meteoric–hydrothermal alteration–mineralization: (a) early stage; (b) late stage (after Sheppard, S.M.F., 1977a).

The copper mineralization is not necessarily a result of the igneous differentiation from diorite. A wide range of orogenic calc-alkaline magmas contain sufficient iron and copper to be capable of giving rise to porphyry-copper deposits (Gustafson, 1978). The critical factor appears to be the evolution of the volatile fraction as the magma rises. Only magmas that have avoided the separation and loss of the carbon dioxide-rich fluid phase may reach shallow crustal levels of 2–3 km with enough sulphur and accompanying metals to produce concentrations in the cupolas (Gustafson, 1978). It is clear from isotope studies that the potassic zone alteration was caused entirely by magmatic hydrothermal fluids with no contribution from ground water or meteoric water. A closed system, which prevented escape of this hydrothermal fluid to the surface, seems to have been necessary (*Figure 8.3*).

Outside the potassic core zone, a meteoric–hydrothermal circulation was induced externally, which led to phyllic, argillic and propylitic zones of alteration and transfer of metals (*Figure 8.3*). Both magmatic and meteoric systems were simultaneously present in a developing deposit, but the external system persisted and overlapped on to the core magmatic system as the temperatures waned and the volcano became solfataric (Hollister, 1975).

The Panguna Deposit of Bougainville

This example has been selected to illustrate the characteristic features of a typical porphyry-copper deposit (*Figure 8.4*). The geology has been described by Baldwin *et al.* (1978). The Oligocene–Lower Miocene volcanic country rocks (Panguna Andesite) were intruded by a quartz diorite 4.5 Ma ago. This was followed by porphyries, breccia formation and biotite–granodiorite intrusion. The magmatic hydrothermal–mineralization system was then inundated by hydrothermal ground water. Post-mineralization intrusion, 3.4 Ma ago, of the Biuro granodiorite and a leucocratic quartz diorite, was followed by extrusion, 1.6 Ma ago, of the Nautango andesite.

Commercial production began in April 1972, and, up to September 1977, 167.8×10^6 t of ore were produced, with an average grade of 0.68% copper, 0.90 ppm gold and 1.96 ppm silver. Intrusive and angular collapse breccias contain higher grades, 0.73–1.21% of copper. The dominant copper mineral is chalcopyrite; bornite is significant only locally. Within the ore body, 60% of the

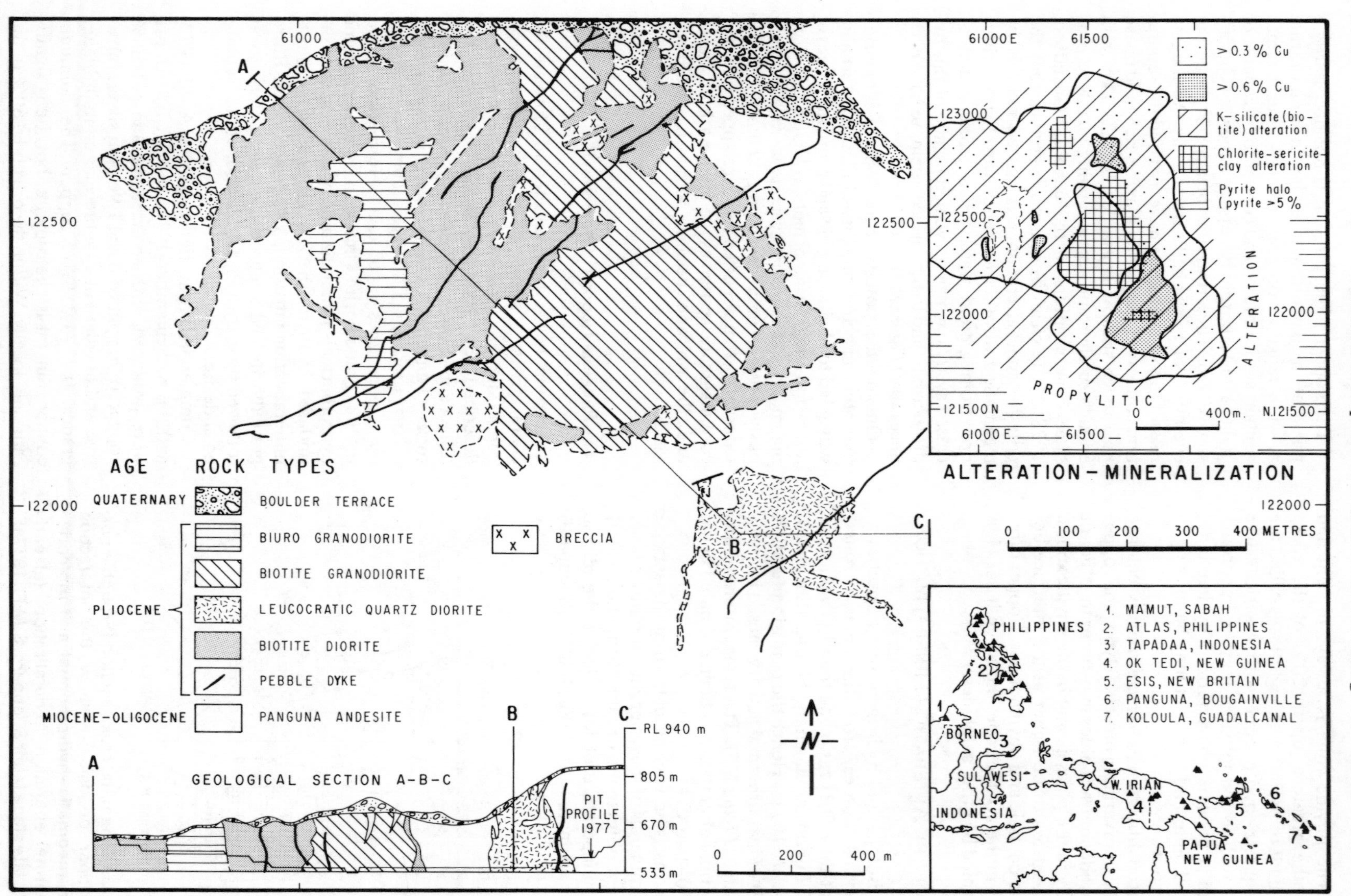

Figure 8.4. The Panguna porphyry-copper deposits of Bougainville, Papua New Guinea. Geology is based on Baldwin *et al.* (1978). Alteration details are based on Ford, J.H. (1978).

copper sulphides are in the andesite, 35% in the intrusives and 5% are disseminated. The high-grade ore zones always contain quartz and sulphide veining, except for the breccias, in which the matrix is mineralized. Mineralization is mainly on fractures, and the ore grade falls rapidly with decreasing amount of quartz veining.

The average magnetite content of the ore body is 2.7% by weight. Molybdenum is low, averaging 72–80 ppm in the intrusives, and 33 ppm in the Panguna andesite (Baldwin *et al.*, 1978).

ALTERATION–MINERALIZATION

There were three broad alteration events, as the inner magmatic–hydrothermal system declined, and was eventually disrupted by the influx of the external heated ground-water system (Ford, J.H., 1978): early—causing the formation of amphibole and magnetite; main—causing the formation of potassium silicates; and late—a feldspar-destructive stage. The early and late stages were not associated with mineralization.

The details of the alteration are: *amphibole–magnetite*—this replaces primary pyroxene and forms veinlets and fracture coatings in the Panguna andesite; *potassium silicate*—secondary biotite, chlorite, alkali feldspar, magnetite and anhydrite are formed and sulphides are deposited; *chlorite–sericite (phyllic)*—this alteration grades outwards from the biotite zone; *sericite–clay (argillic)*—this represents a pervasive alteration; and *propylitic*—this causes the formation of chlorite, epidote and pyrite.

Mg/(Mg + Fe) ratios in both the hydrothermal amphibole and biotite increase towards the ore zone. Biotitization occurred at 370–540°C. The magmatic–hydrothermal ore fluids were highly saline. The alteration–mineralization is similar to all other porphyry-copper deposits. The mineralized potassium silicate core zone coincides with high potassium, copper, copper/sulphur, rubidium and rubidium/strontium values and decreased levels of calcium, manganese, strontium, potassium/rubidium, zinc and lead.

An outward temperature decline was the main control on the alteration–mineralization zonation, as evidenced by: increasing $\delta^{18}O$ values of the vein quartz away from the intrusives, with an outward thermal gradient of about 30°C/100 m; and increasing Mg/(Mg + Fe) ratios in the hydrothermal biotite towards the middle of the ore body.

Fluid-inclusion studies by Eastoe (1978) lead to the following conclusions: copper mineralization was under a pressure of 200–300 bar and at 350–700°C. Assuming a hydrostatic pressure system, this is equivalent to a depth of porphyry-copper formation of 2–3 km.

The copper and iron sulphides, quartz, anhydrite and haematite were deposited by dense boiling magmatic–hydrothermal solutions of 70% salinity. The dissolved salts were rich in sodium chloride and potassium chloride. The copper content may have been up to 1900 ppm in the liquid. The zone of cooling and dilution of the salt-rich liquid corresponded roughly with the 0.3% copper contour of the ore body (*Figure 8.4*).

The system cooled below 400°C before undergoing renewed mineralization, which accompanied the intrusion of the porphyritic stocks.

Ground water of less than 10% salinity inundated the ore body at 400°C and deposited quartz–pyrite, pyrite–clay and sphalerite–pyrite veins at about 300°C.

In summary, a magmatic–hydrothermal system was surrounded by a cooler meteoric–hydrothermal system. The latter collapsed inward as the former waned below 400°C (*Figure 8.3*).

Copper-Bearing Skarns

Many porphyry-copper cupolas have intruded wholly or partly into limestone country rock. The hydrothermal–mineralization solutions have interacted with the limestone to produce copper skarn ore.

The following are examples of porphyry-copper deposits that contain significant amounts of metal-bearing skarn: Bingham (Utah), Morenci–Metcalf (Arizona) and Cananea (Mexico). At Cananea, the skarns are primarily lead–zinc-bearing (Livingston, 1973), the others are copper-bearing.

The best-studied skarn deposit of this type is at Ely, Nevada (James, L.P., 1976). A generalized cross-section of the deposit (*Figure 8.5*) compares the differences of the alteration–mineralization in the porphyry and the limestone country rock. Wall-rock limestone, adjacent to secondary biotite–alkali feldspar assemblages (potassic zone), is replaced by skarn that contains massive andradite–diopside rock. The post-magmatic fluids, which produced the potassic alteration assemblages in the porphyries, commonly reacted with the carbonates to yield andradite–diopside-bearing skarn. There is disseminated chalcopyrite in the adjacent altered porphyry. However, the bulk of the copper in the skarn was deposited as pyrite–chalcopyrite veins, which cut the andradite–diopside rock. The mineralized veins are surrounded by envelopes of actinolite–

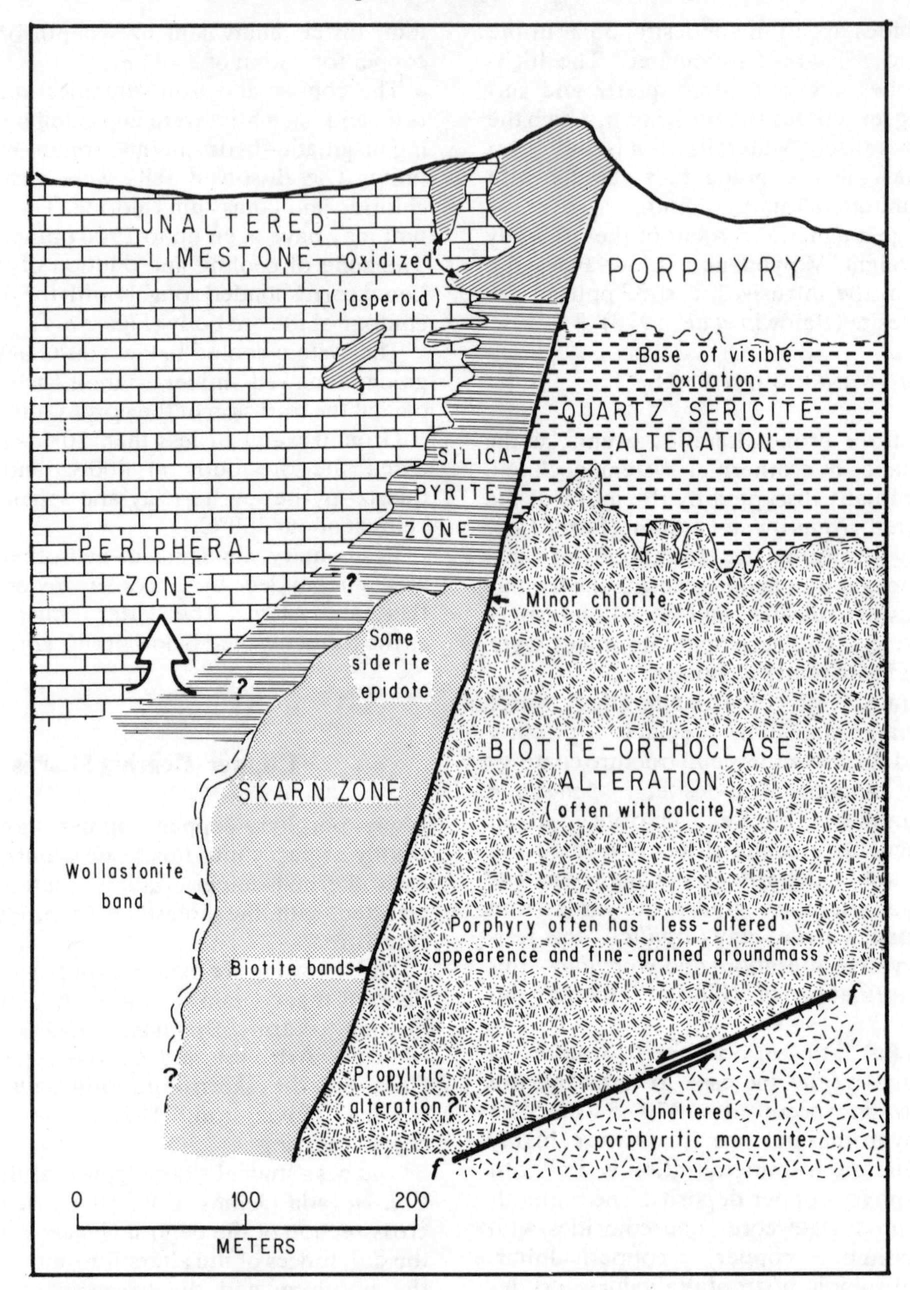

Figure 8.5. Generalized cross-section showing the geometry of the alteration–mineralization in limestone and porphyry country rocks at the Ely porphyry-copper deposits, Nevada (after Theodore, 1977).

calcite–quartz–nontronite alteration, which indicate a clay–sulphide stage superimposed on the earlier calc-silicate skarn (Theodore, 1977). The mineral zones in the skarn generally lie parallel to the porphyry contact.

Adjacent to the quartz–sericite alteration zone in the porphyry, the limestone contains secondary calcite–pyrite–nontronite–quartz assemblages (*Figure 8.5*), formed simultaneously with the adjacent zone in the porphyry.

The alteration of carbonate to skarn may laterally range widely from deposit to deposit, and skarn formation usually extends far beyond the economic limits of the hypogene ore. Abundant pre-mineralization structures in the epizonal environment provide channelways readily accessible to the circulation of the skarn-forming hydrothermal fluids. In several deposits, additive iron metasomatism has been preceded by widespread magnesium metasomatism, and the mineral zonation is formed primarily by infiltration. In others, the skarn seems to have formed by a combination of infiltration and diffusion as the carbonates reacted with surrounding silicates to yield skarn.

Detailed paragenetic and fluid-inclusion studies at Copper Canyon, Nevada (Theodore, 1977) suggest that saline non-boiling fluids, within the temperature range 450–550°C, were initially involved in the crystallization of the anhydrous silicate minerals in the skarn. The fluids became highly saline and were boiling, within the temperature range 310–390°C, as the skarn evolved through its sulphide and hydrous silicate stages.

Porphyry-Molybdenum Deposits

Porphyry-molybdenum deposits supply 95% of the world's annual production of 90 700 t of metal. Total identified world molybdenum resources amount to 2.68×10^7 t.

Molybdenum is also produced as a by-product to copper from the porphyry-copper deposits of cratonic setting. When molybdenite is produced as a by-product from a copper porphyry, the MoS_2 grade may range from 0.016% to 0.029% (Soregaroli and Sutherland Brown, 1976). The bulk of the world's production comes from the giant post-Laramide (20–40 Ma) deposits of Climax and Henderson mines in Colorado and the Jurassic Endako mine in British Columbia. In the Canadian cordillera, 97% of the molybdenum produced is from porphyry deposits. Significant molybdenum mineralization is almost exclusively associated with porphyritic granitic magmas, but a small production comes from pegmatite, vein and skarn deposits.

Intrusive stocks, genetically related to porphyry-molybdenum deposits, are leucocratic silicic porphyritic granite or microgranite. The intrusions are more silicic than those related to molybdenum-bearing copper porphyries. Euhedral-to-subhedral phenocrysts of quartz, alkali feldspar, and more rarely sodic plagioclase and biotite, are characteristic. Intrusions are composite. Later phases become progressively leucocratic. The mineralization–alteration is characteristic and similar to that described for copper porphyries. Mineralized zones are controlled by fractures developed in and around genetically related intrusive stocks. Most are as stockwork zones, elliptical to annular in plan and cylindrical to irregular in a vertical dimension.

Quartz is ubiquitous in all molybdenum deposits. Alkali feldspar may accompany it and may form up to 30% of the veins. Together quartz and feldspar account for 90% of the gangue. Molybdenite is the only ore mineral, but wolframite may accompany it.

Urad and Henderson Mines

The stockwork molybdenite ore bodies are related to an Oligocene rhyolitic subvolcanic centre called the Red Mountain Complex, which consists of 15 separate magmatic phases (Wallace *et al.*, 1978). The major phases are all highly differentiated and are composed of quartz, alkali feldspar, albite and minor biotite.

The Urad ore body is cylindrical. The fissure system in which it developed consists of a set of cone sheet fractures caused by a quartz porphyry intrusion. Half the ore body is within the Red Mountain Complex, half within the Precambrian Silver Plume granitic country rocks. The mine is now exhausted; it produced about 13×10^6 t of ore of average grade 0.348% MoS_2. The Urad ore body was deposited before the Henderson and at a shallower depth.

The Henderson ore body is much larger and is an umbrella-shaped stockwork system entirely within the Red Mountain Complex. It contains more than 300×10^6 t of ore (Wallace *et al.*, 1978).

Hydrothermal alteration–mineralization consists of the following zones, which progress outwards: alkali feldspar-enriched zone; silicified zone; sericite–quartz–pyrite zone; and argillic zone.

Because it was so strongly differentiated, the Red Mountain Complex is believed to have a continental crustal origin and to have therefore derived its molybdenum from the crust.

The Urad–Henderson deposits are remarkably similar to those of Climax (*see* p. 178), except for the absence of tungsten and tin. Wallace *et al.* (1978) suggest that these elements originated in the Climax deposit from the abundant Precambrian metamorphic basement rocks through which the porphyries have been emplaced. Such rocks are lacking at Urad–Henderson.

The Endako Deposit, British Columbia

The deposits of the Canadian cordillera range in age from 20 Ma to 141 Ma. The largest deposit is at Endako, and its geology has been described by Kimura *et al.* (1976). Present production is at 24 000 t of ore per day containing 0.15% molybdenum. The ore body occurs in the Jurassic Topley composite alaskite, granite and quartz–monzonite intrusions into early Mesozoic and Late Palaeozoic sedimentary and volcanic rocks. Hydrothermal mineralization–alteration includes alkali feldspar envelopes on veins and fractures, quartz–sericite pyrite envelopes on veins and fractures, and pervasive kaolinization of the Endako quartz–monzonite.

The Climax Mine, Colorado

The detailed geology of the Climax mine, Colorado (*Figure 8.6*) has been described by Wallace *et al.* (1968) and the alteration–mineralization by Hall, W.E. *et al.* (1974). This mine is unique; its annual production consists of 2552 t of molybdenum from molybdenite concentrate oxidized to MoO_3, 736 t of WO_3 from huebnerite concentrates, 45 t of tin from cassiterite concentrates, 16 t of pyrite and 27 t of monazite. Pyrite, however, is the main opaque mineral, averaging 2–3%, but it is mainly discarded as uneconomic.

The Oligocene rhyolite porphyry Climax stock was intruded into Precambrian schists and gneisses of the Idaho Springs Formation in four separate phases, each intruding the former at successively lower levels, and each accompanied by alteration–mineralization. The fourth, and last phase, showed almost no commercially exploitable mineralization. The important Mosquito fault, of 2400 m throw, separates Palaeozoic sedimentary rocks from the Precambrian country rocks of the Climax mine (*Figure 8.6*). The alteration zones are spacially related to each molybdenite ore body. They include a silica zone below, a potassium silicate zone that approximately coincides with the ore, and overlying quartz–sericite–pyrite–topaz, argillic and propylitic zones (*Figure 8.7*).

The isotopic and fluid-inclusion studies (Hall, W.E. *et al.*, 1974) support a model in which the ore bodies were formed by mixing of magmatic water with later meteoric water, just as in porphyry-copper deposits (*Figure 8.3*).

The mineralization is of very fine-grained molybdenite, which occurs along with quartz in fractures in the Climax porphyry and contiguous host rocks. Pyrite mineralization forms a broad zone, predominantly above and peripheral to the molybdenite ore, as quartz–pyrite–sericite veinlets that contain finely disseminated fluorite and topaz.

Tungsten mineralization, in the form of huebnerite, coincides with the pyrite zone. However, the tungsten mineralization may in places overlap downwards on to the molybdenite ore zone. Small crystals of cassiterite are set in vugs, or are embedded in pockets of sericite in the quartz–sericite–pyrite veinlets. The tin dissemination appears to be confined to the tungsten zone.

Most of the Ceresco ore body has been removed by erosion and glaciation, which exposed the Upper ore body (*Figure 8.6*).

Discussion

Porphyry-molybdenum deposits present an intriguing problem. A complete spectrum does not exist between porphyry-copper and porphyry-molybdenum deposits. Porphyry-copper deposits contain significant molybdenum only when they occur within cratonic continental plates, never in wholly ensimatic island arcs. In which case, one might argue that the molybdenum content is scavenged from the sialic cratonic basement by the circulating ground-water–hydrothermal system, during alteration–mineralization centred around the hot pluton. The problem is exacerbated by the fact that porphyry-molybdenum deposits characteristically show an absence of copper. Their plutons are therefore unlikely to have come from the same Benioff-Zone source as do the porphyry-copper plutons. This is further supported by their more highly differentiated silicic nature. Yet the porphyry nature of both copper and molybdenum deposits is virtually identical.

The close association of tungsten and tin, which are considered to be elements that have been concentrated in continental crust by polycyclic events involving anatexis (Hutchison and Chakraborty, 1979), together with the absence of copper, which is mantle-derived, and the more strongly differentiated granitic nature of the host plutons, strongly favours the hypothesis that porphyry-molybdenum deposits are derived from continental crust. On the other hand, there is universal agreement that the host plutons of porphyry-copper deposits have a mantle source (Taylor, D., 1973), and their differences in respect of their molybdenum and gold contents, between cratonic and island-arc settings, may result from the differences in the country rocks from which the circulating meteoric–ground-water hydrothermal systems have scavenged these elements.

Perhaps the clue to porphyry-molybdenum deposits lies in the dramatic ^{18}O depletions of many of the hydrothermal fluids at Climax (Hall, W.E. *et*

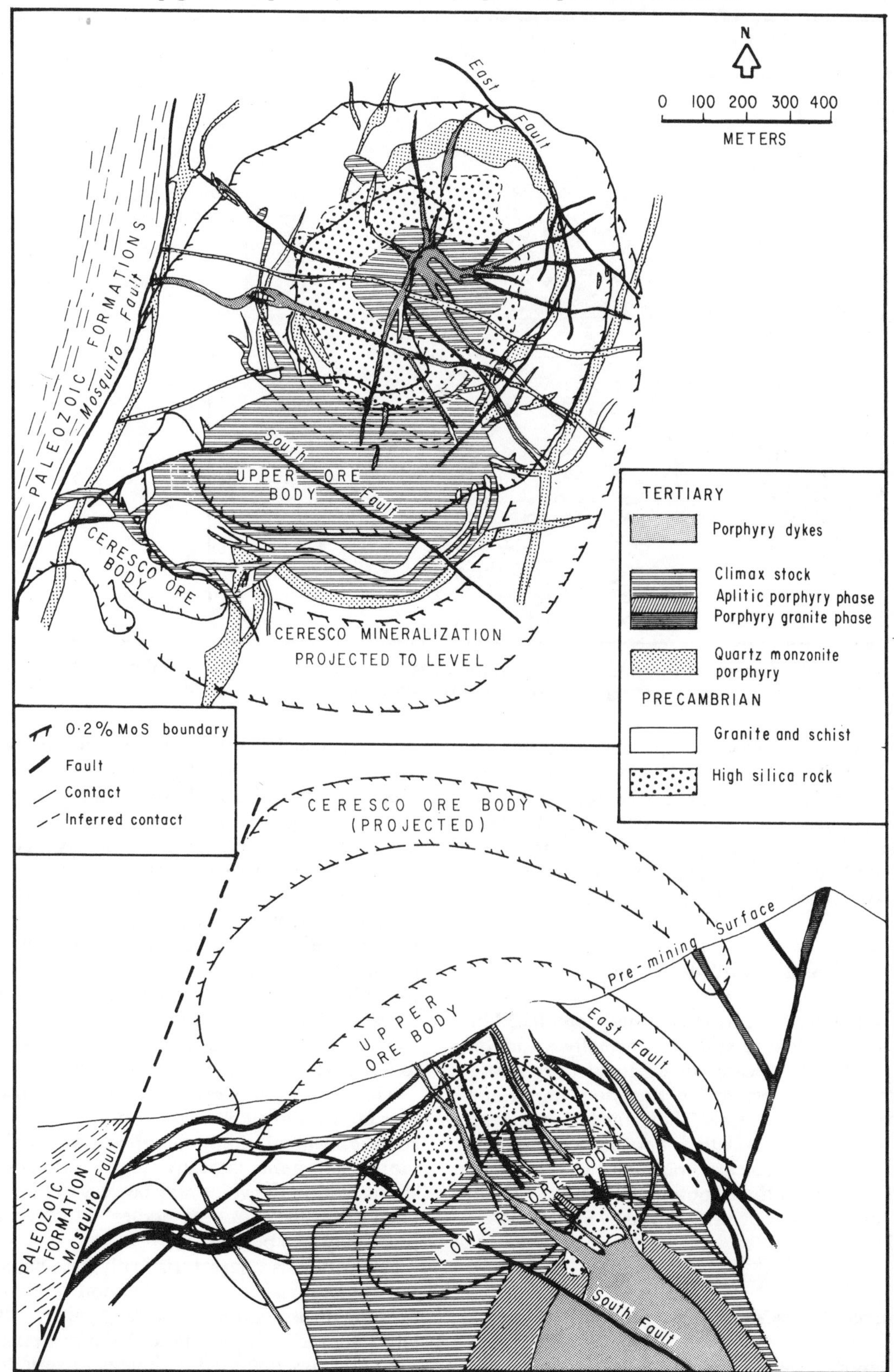

Figure 8.6. Upper: aereal geology of the Climax porphyry-molybdenum mine, Colorado. Lower: east–west cross-section through the mine area (after Wallace *et al.*, 1968).

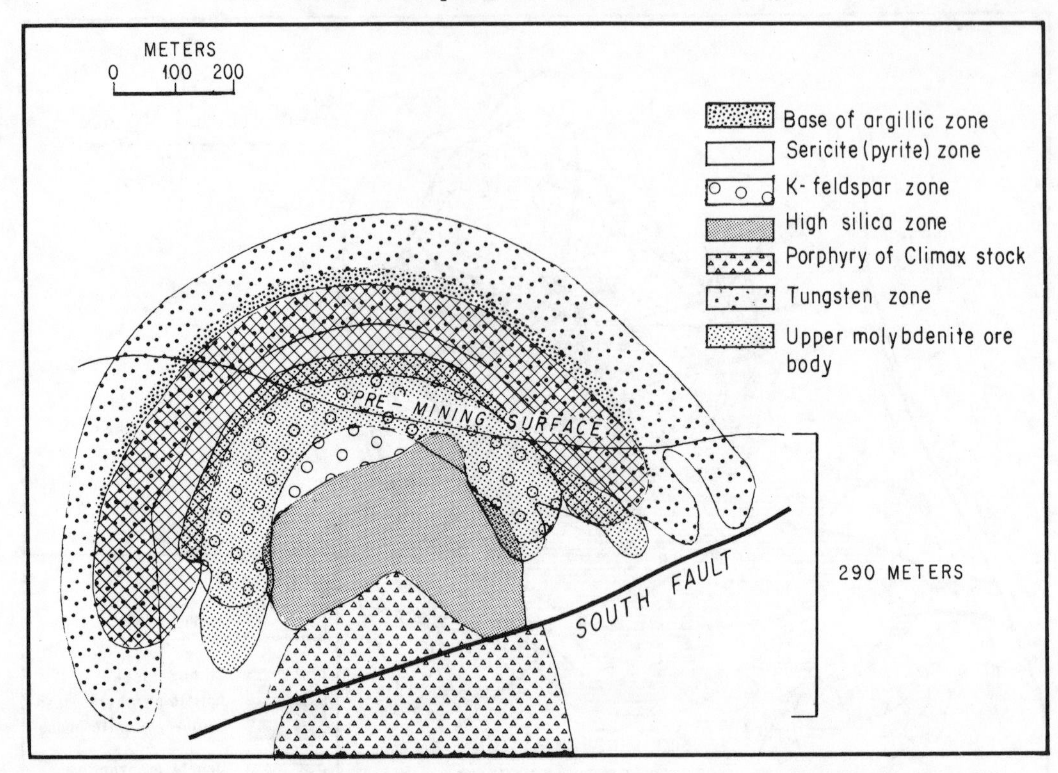

Figure 8.7. Details of the hydrothermal mineralization of the Upper Ore Body of the Climax mine as seen in a cross-section through the mine (after Hall, W.E. *et al.*, 1974).

al., 1974), greater than in any porphyry-copper deposit (Sheppard, S.M.F., 1977a), which dictate that enormous quantities of heated meteoric waters circulated around the high-level rhyolitic pluton. Hence the molybdenum, tungsten and tin could have been scavenged from a wide zone of the surrounding continental cratonic rocks.

Epithermal Vein Deposits

Vein mineralization, hydrothermally emplaced in near-surface fracture systems, normally within or associated with volcanic or epizonal porphyritic formations, has resulted in the most important source of precious metals. They may be associated with base metals, and the deposits occur in the upper levels of volcano–plutonic arcs both in island-arc and cratonic continental settings. Their main features have been reviewed by Sillitoe, R.H. (1977).

Characteristic Features

The deposits were emplaced as veins in or associated with volcanic or epizonal plutonic sequences, which vary from andesitic to rhyolitic. Most are of Mid-to-Late Cenozoic age, but Mesozoic examples are fairly common. The generally young age of formation means that many of the mines are in regions of present-day high geothermal gradients, and hot springs may frequently be closely associated with the mining regions.

Most of the vein systems are related to subaerial volcanogenic fracturing, and cauldron subsidence is important though not necessary. In many cases the mineralization has followed these subaerial volcanogenic fracture systems quite far from the source pluton, so that no observable relationship to intrusive rocks can be demonstrated.

The veins occupy pre-existing tension fractures and faults, and they occur in branching patterns, commonly increasingly complex upwards; they are more persistent laterally than vertically. Ore shoots within the veins may be very rich, and have been called 'bonanza ore bodies'. Stockworks and pipes may also be present.

Alteration is characteristically associated with the mineralization. Propylitization is commonly present on a regional scale, and not directly related to the vein formation. Characteristically, the veins are bounded by haloes of sericitization and silicification, which commonly contain adularia. Alternatively a more advanced argillic alteration may be formed, in which the ore zones and veins have haloes of clays, alunite, pyrophyllite and diaspore.

Mineralogy

The deposits principally contain gold and silver in various proportions but also commonly contain base metals. The main gangue mineral is quartz, commonly in the form of chalcedony, accompanied by variable amounts of pyrite, calcite, dolomite, barite, adularia, rhodochrosite, rhodonite and fluorite. The silver occurs as argentite or acanthite and as arsenical or antimonial sulphosalts. The gold occurs native, but is usually argentiferous, although it may occur as tellurides.

Textures

Druzy cavities, filling, comb, colloform, crustiform and intermineral breccias are common textures of the ores. They suggest deposition under low confining pressure. Both the ores and the gangue minerals are fine grained.

Depth of Formation

The shallow, 100–1000 m, depth of the ore deposition is emphasized by the abrupt downwards termination of the mineralization or by the downwards increase of base-metal content at the expense of the precious metals. Ore shoots in veins have therefore rarely been worked for more than 600 m vertically.

Age Relations

The mineralization is generally emplaced a few million years after formation of the host volcanic rocks. This is illustrated by numerous examples of radiometric dating studies. At Guanajuato, Mexico, volcanism ended after 32 Ma and the principal vein-type silver–gold mineralization, dated by its adularia, occurred between 27.4 and 30.7 Ma (Gross, W.H., 1975). In the Hauraki region of New Zealand, the host-rock andesites give dates of 17 Ma, while the hydrothermal alteration associated with the mineralized veins give an age ranging from 2.5 Ma to 7 Ma (Robinson, B.W., 1974).

Nature of the Mineralizing Solutions

Fluid-inclusion studies on a number of deposits confirm that the hydrothermal mineralizing fluids were of low salinity and low temperature (200–300°C). The deposits are truly epithermal, and these studies show that the early estimate of 200°C by Lindgren (1933) was somewhat low.

There is a suggestion that salinities for veins that carry base metals ($>2\%$ sodium chloride equivalent by weight) are somewhat higher than those that carry precious metals ($<2\%$ sodium chloride equivalent by weight). There is also a tendency for epithermal veins that carry base metals to have formed at higher temperature, within the range 320–370°C (Rye and Sawkins, 1974).

Origin of the Metals

The majority of the examined veins possess low $\delta^{18}O$ values, indicating a substantial meteoric water component in the ore-depositing fluids (Sillitoe, R.H., 1977). However, in some deposits, there was a definite contribution from magmatic water to the hydrothermal mineralizing solutions. This was particularly the case at the Tui mine of New Zealand, where the δD value obtained from fluid in galena strongly suggests a magmatic derivation (Robinson, B.W., 1974).

In the green tuff region of Japan, the ore fluid that gave rise to both the Kuroko deposits (*see* Chapter 3) and the associated epithermal vein deposits, from evidence of similar mineralogy and sulphur-isotope and lead-isotope ratios (Ohmoto and Rye, 1974), is thought to be the same. The fluid has been shown to be of at least 75% sea water in the Kuroko ores. Therefore, in epithermal vein systems that connected with the sea in island-arc settings, it can be assumed that hydrothermal fluids dominated by sea water may have been responsible for the mineral deposition.

It appears that the ultimate origin of the ore in epithermal vein deposits may differ with tectonic setting. In the case of ores from the Tui mine of New Zealand (Robinson, B.W., 1974) and the Guanajuato mine of Mexico (Gross, W.H., 1975), the source of the metals has been proposed to be the underlying sialic basement, on the basis of sulphur-isotope and lead-isotope studies. In contrast, the ores from the green-tuff region of the island arc of Japan must be mantle-derived (Ohmoto and Rye, 1974).

The nature of the basement exerts strong control on the metal contents of the veins. In the Central American epithermal vein belt, those deposits that occur within the cratonic part of Guatemala and Honduras are uniformly rich in silver relative to gold. In contrast those in Costa Rica, Panama and Nicaragua, which are not underlain by the craton, are uniformly rich in gold relative to silver (Kesler, S.E., 1978). Similarly in North America, the deposits of Nevada and Colorado, which formed in cratonic sialic crust, are pre-eminently silver-rich, while those lying outside the craton in California are gold-rich (*Figure 8.8*).

Distribution of the Deposits

The worldwide distribution is rather similar to that of porphyry copper (*Figure 8.1*). In many cases, they occur close to porphyry-copper deposits, as in Taiwan and Mexico, but in others only epithermal precious-metal vein deposits and an absence of porphyry-copper deposits characterize the volcanic arc, as in Sumatra (Hutchison and Taylor, 1978). This is obviously related to level of erosion, for deeper erosion is necessary to expose porphyry-copper deposits. The gold–stibnite vein deposits, however, do not occur in the same orogenic belts as porphyry-copper deposits. As has been noted by Stanton (1972) and Taylor, D. and Hutchison (1979), they are related to Mesozoic epizonal granite porphyries, which occur within continental areas. Their tectonic setting is not as clear as the other epithermal vein deposits.

North-American Cordillera

The western cordillera of North America (*Figure 8.8*) represents the greatest concentration of silver–gold vein deposits in the world. It includes such well-known districts as the San Juan field of Colorado (*see* p. 183), the Great Basin of Nevada, the Mother Lode of California, and the great silver deposits of Mexico.

In Nevada, the mining areas of Bullfrog, Goldfield and Silver Peak districts are spacially related to caldera ring fractures (Albers and Kleinhampl, 1970). The Bodie district is genetically related to dacitic plugs. The pre-eminent deposits of Bodie, Aurora, Tonopah, Comstock Lode and Goldfield are all related to the final stages of andesite–dacite volcanism (Silberman and McKee, 1974). Goldfield has produced 3110 t of silver and 130 t of gold from veins that were emplaced 20 Ma ago in earlier (30 Ma) caldera ring fractures (Ashley, 1974). The mineralization is in a large area of advanced argillic alteration in which the veins

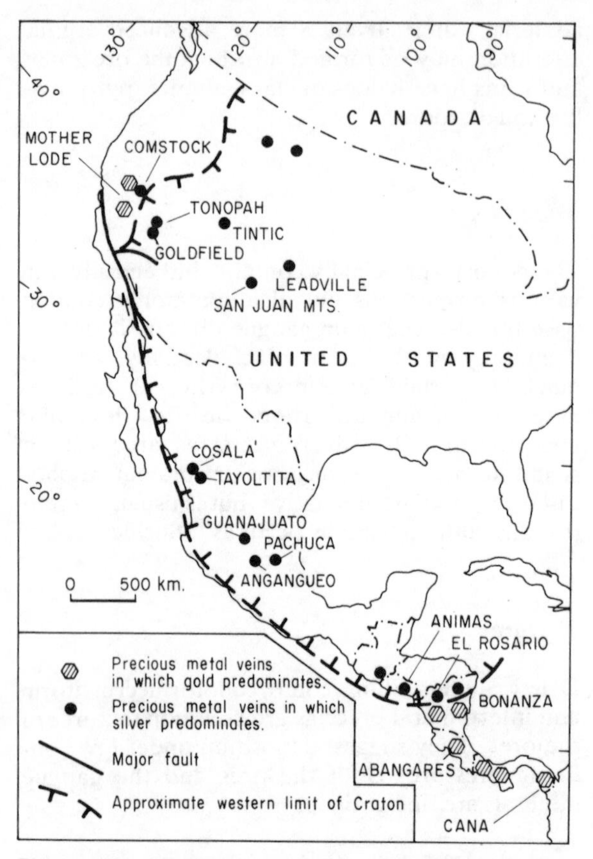

Figure 8.8. Distribution of precious-metal epithermal vein deposits in the western cordillera of North America. Silver is important where deposits lie within the craton. Outside the craton, gold is important. There is an intermediate zone in west Nevada where some deposits are gold-rich, others silver-rich (after Sillitoe, R.H., 1977; Guild, 1978; Kesler, S.E., 1978).

carry irregular patches and pipes of sulphides, sulpho-salts and native gold. The Tonopah district has produced 3110 t of silver and 59 t of gold from replacement veins, characterized by quartz–sericite–adularia alteration within andesite lavas and breccias. The principal minerals include argentite, polybasite, sphalerite, galena, chalcopyrite, pyargyrite and pyrite (Nolan, 1935). The low ^{18}O values for the Tonopah and Bodie deposits imply that meteoric waters were overwhelmingly the dominant source for the hydrothermal mineralizing fluids (Sheppard, S.M.F., 1977a). Oxygen-isotopic studies of the Comstock Lode (Taylor, H.P., 1973; O' Neil and Silberman, 1974) indicate that low-salinity meteoric solutions played a dominant role in the transportation and deposition of the mineralization. In the majority of deposits, the large scale of the meteoric system, and resulting voluminous zone of propylitic alteration, have overwhelmed and largely obliterated the earlier magmatic–hydrothermal stage, which may have been critical for the initiation of such deposits (Sheppard, S.M.F., 1977a).

California has been the source of more than 3290 t of gold. The most productive areas have been Grass Valley, Nevada City, Alleghany, Sierra City district, the Mother Lode belt, and the East and West Gold Belts (Clark, W.B., 1970). Most deposits are peripheral to the Uppermost Jurassic to Cretaceous Sierra Nevada batholith. Dodge and Bateman (1977) suggested that the role of the batholith was limited to supplying heat to a circulating meteoric water system, which mobilized the metals from the volcanogenic and metasedimentary country rocks. The major gold-mining districts are concentrated along or near the Melones Fault zone and in the East Gold Belt. They lie within metamorphic country rocks. The deposits along the Melones Fault constitute the famous Mother Lode (*Figure 8.8*). Deposits close to the batholith generally contain abundant pyrrhotite, whereas pyrite increases away from the batholith. Ores of the East Belt are more complex than the Mother Lode, and contain appreciable arsenopyrite, chalcopyrite, galena, sphalerite and iron sulphides (Clark, W.B., 1970).

The Mexican belt of Cenozoic volcanic rocks, extending from the border with the USA 2400 km to Mexico City, contains the greatest known concentration of silver in the world. The famous mines of Pachuca and Guanajuato have both yielded more than 30 000 t of silver. Pachuca was responsible for a production of 37 324 t, or 6% of the world's silver, commencing in the sixteenth century. It also produced 190 t of gold, and important amounts of lead, zinc and copper. Most of the veins occur in Tertiary andesites, which are overlain by rhyodacitic-to-rhyolitic pyroclastic formations, 32–23 Ma old (Sillitoe, R.H., 1977). Fractures, formed during the uplift of the Sierra Madre Occidental, are the location of the Oligocene-to-Pliocene epithermal veins. Argentite (and acanthite) and the ruby silvers (pyrargyrite and proustite) are always deposited later than the base-metal sulphides, and gold is concentrated in the final phases (Wisser, 1966). Deposits close to the interface with the subvolcanic basement are richer in silver, carry abundant base-metal sulphides but only minor quartz, and are narrow. In contrast, the veins higher in the volcanic pile are richer in gold and quartz, poorer in base metals, and are broader. Mineralization extends vertically over 700 m, and zones of albitization, silicification, sericitization and carbonatization fringe the veins (Sillitoe, R.H., 1977).

There is an obvious relationship between the gold/silver ratios of the epithermal deposits and their tectonic setting. Silver is important only where the mineralization occurs on old cratonic plates (*Figure 8.8*). The gold fields of California lie west of the margin of the Precambrian craton (Guild, 1978), and the gold deposits of southern Central America lie beyond the southern margin of the craton (Kesler, S.E., 1978). It may therefore be concluded that silver is of continental character, whereas gold is of oceanic or mantle derivation. The base metals do not discriminate, and occur in both settings.

The San Juan Mountains of Southwest Colorado

The San Juan volcanic field is an erosional remnant of a huge volcanic province that covered the southern Rocky Mountains in the Oligocene (*Figure 8.9*). The volcanic formations rest on a basement of metamorphic, sedimentary and igneous rocks that range in age from precambrian to Tertiary. The importance of this famous mining district is shown by the total production, from the four mines of Hinsdale, Ouray, San Juan and San Miguel, from 1873 to 1964: 237 t of gold, 499 t of silver, 99 137 t of copper, 635 654 t of lead, and 309 560 t of zinc (Burbank and Luedke, 1968). Tertiary volcanic activity started about 35 Ma ago, and extended for 5 Ma with the building of andesitic-to-rhyodacitic stratovolcanoes. About 30 Ma ago, activity changed to silicic ash flows and tuffs, derived from 17 collapsed calderas during a 4-Ma interval. Intermediate lavas were extruded after and during the caldera formation. The calderas were resurgently domed, during which time felsic intrusions were emplaced around their margins. About 25 Ma ago, intermediate volcanism gave way to bimodal mafic–silicic volcanism, and between 25 Ma and 5 Ma ago, the Lake City caldera developed. Minor alkalic and silicic intrusives were partly controlled by earlier caldera structures (Steven, 1975; Steven and Lipman, 1976).

Minor vein deposits were emplaced in the cores of the 35–30-Ma old stratovolcanoes, for example, at El Queva, and mineralized veins were emplaced, associated directly with the resurgence of the calderas (Lipman *et al.*, 1976). Apart from the minor mineralization, about 95% of the mineral production was derived from caldera ring faults, outward-extending radial fractures, and graben faults along the resurgent domes (*Figure 8.9*), all produced during subsidence and resurgence of the

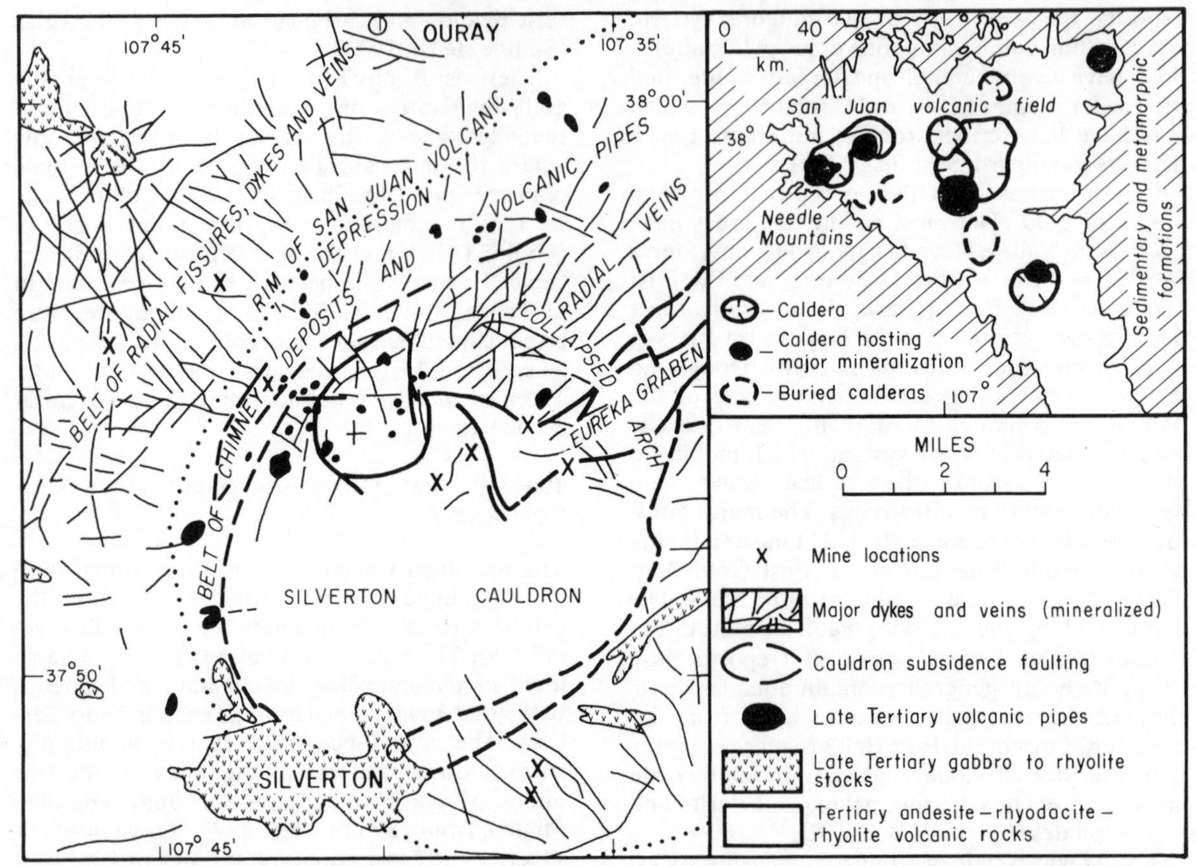

Figure 8.9. Structural map of the Silverton cauldron and its relation to the San Juan volcanic field, Colorado, showing the location of the epithermal silver–gold mineralization (after Burbank and Luedke, 1968).

30–26-Ma old calderas (Lipman *et al.*, 1976). Mineralization was, however, considerably later than the caldera development, and was related to the silicic magmas belonging to the bimodal suite.

The ore deposits possess epithermal characteristics, and occur in breccia pipes, associated with small quartz latite porphyry stocks, in the Red Mountain district and in the Silverton caldera fault zone. The common sulphides are pyrite, sphalerite, galena, chalcopyrite and tennantite. In the higher-grade ores, argentiferous tennantite, tetrahedrite, and proustite form the greatest proportion of the sulphides. Arsenopyrite may occur. Tellurides of gold and silver are sparse. Free gold, in association with quartz and adularia, was introduced late. The common gangue is of quartz, barite, calcite, ankerite and fluorite. There are complex manganiferous ores, which contain rhodonite, rhodochrosite and tephroite. They occur mostly in relatively barren bodies (Burbank and Luedke, 1968). Enargite is rare in the vein deposits, but of common occurrence in the breccia pipe or chimney deposits.

Radiometric dating of the mineralization

Epithermal ore deposition took place over the same time span as the associated volcanism, but the richest ore deposits were emplaced later than the formation of the calderas, particularly the Silverton caldera, about 5–15 Ma later than the termination of the caldera formation (Lipman *et al.*, 1976). Adularia from the Camp Bird vein was dated at 10.5 Ma, some 17 Ma later than the age of the host fractures. Alunite, accompanying the mineralization in the Summitville district, was introduced 6.4 Ma after the subsidence that formed the caldera.

Fluid-inclusion studies

Studies by Nash (1975) provide evidence for ore deposition within the range 315–249°C, from fluids carrying 7.9–0.1% sodium chloride equivalent by weight. The base-metal and gold-deposition stages were near 290–280°C, and salinities were less than 2%. Liquids of salinities 1.6–0.2% sodium chloride equivalent by weight were deposited in the range 308–216°C. No systematic variation in temperature, or salinity, was documented over the 1100-m vertical interval in the veins.

Alteration

Large volumes of volcanic rocks have been propylitized before the ore veining. The alteration increases with depth. The altered rocks range from low-grade quartz–chlorite–calcite assemblages, to albite–epidote–chlorite assemblages analogous to hornfels. This propylitic alteration preceded vein formation (Burbank and Luedke, 1968). Alteration, related to the vein deposits, is commonly restricted to a few metres from the margins of the veins. Pyrite, quartz, sericite, clay minerals, calcite and chlorite are typical alterations. Silicification is particularly characteristic of the walls of some chimney deposits. The chimneys commonly lie in or near strongly leached and altered volcanic rocks of the cauldron rim zones. Over large areas, the rocks are altered to aggregates of quartz, dickite and other clay minerals, pyrophyllite, zunyite, diaspore and alunite. All the original iron has been converted to pyrite.

Stable isotope studies

Quartz from an ore vein in the Gold King mine, near the centre of the Silverton caldera, and dickite from a vein in the Ouray area both have low ^{18}O values. These data support the general conclusion, made by Taylor, H.P. (1974), that the epithermal vein deposits were formed 'from hydrothermal fluids that contained a significant component of meteoric ground water'. The hydrothermal system in the radial fissure zone, outside the caldera, was much richer in meteoric ground water, or had a higher water-to-rock ratio, than the system prevalent within the caldera (Matsuhisa, 1978).

Western Pacific

Numerous epithermal vein deposits also characterize the western margin of the Pacific. Gold is more important than silver, since many deposits occur within ensimatic island arcs.

The Khakandzha gold deposit of the Chukotka peninsula of eastern Siberia consists of gold–argentite–polybasite mineralization, in rhyolite and andesite, associated with intense silicification and formation of adularia. The nearby Karamken gold–silver deposit consists of veins of quartz, adularia and pyrite, chalcopyrite, sphalerite, galena, marcasite, freibergite, tennantite, sulphostannates, selenides, sulpho-salts of silver, argentiferous pyrite, electrum and silver within the cauldron faulting system of a dacitic volcano (Borodaevskaya and Rozhkov, 1977).

In Japan, epithermal fissure-filling gold–silver, lead–zinc and manganese veins are associated with and are considered to be transitional to the Kuroko massive sulphide deposits (Lambert and Sato, 1974), but they also occur in subaerial volcanic sequences (Nishiwaki *et al.*, 1971). Radiometric dating of adularia indicates that not all are contemporaneous with the Kuroko deposits, and some are as young as 3 – 8 Ma (Sato, 1974).

The Chinkuashih gold–luzonite–enargite epithermal vein deposit of northeast Taiwan produced during its peak in 1935, 2–2.5 t of gold, 6000–7000 t of copper, and silver increased to 2 t in 1961. The ore bodies are related to north–south faulting in Pleistocene dacite and sandstone (Huang, 1955). The ore minerals are pyrite, enargite, luzonite and native gold, and the gangue is of quartz, barite, alunite, wurtzite and kaolinite. Crustification and druzy textures are common. There is a marked outwards zonation of alteration from silicification, through argillation to chloritization. Where the grade of ore is high the pyrite occurs as octahedra, and where low as cubes; elsewhere it occurs as pyritohedra (Tan, 1976). Fluid-inclusion studies indicate a temperature of formation within the range 228–305°C (Folinsbee *et al.*, 1972).

The Lepanto ore deposit of the Philippines has produced 27 000 t of ore annually with an average grade of 3.0% copper and 5 ppm gold (Gonzales, 1956). It is very similar to the deposit of Taiwan; both were formed at a depth of less than 500 m. A major difference is that the gold at Lepanto is in the form of tellurides, while at Chinkuashih it is mainly native or alloyed with silver (Tan, 1976). At Acupan mine, hydrothermal breccias carry native gold (Bryner, 1969).

Epithermal gold mineralization at Tavua, Fiji, has produced 124 t of gold from fissure-filling veins of gold–silver tellurides and auriferous sulphides, where north–south-trending faults have intersected a caldera ring fault system (Denholm, 1967).

The Hauraki region of North Island, New Zealand, has produced 1360 t of gold, from silver–gold and base-metal veins, within fractures in Late Cenozoic andesites and rhyolites. The Waihi mine has produced 124 t of gold from quartz veins in dacite (Williams, G.J., 1965). Late Miocene-to-Pliocene precious-metal mineralization on Great Barrier Island consists of quartz veining in andesite. The veins are enclosed by haloes of silicification and argillic and propylitic alteration. They contain early pyrite and marcasite, followed by base-metal sulphides, and then sulpho-salts, of which the ruby silvers are dominant, and finally by selenides and stibnite (Ramsay and Kobe, 1974). This mineralization can be considered to be continuing to the present day, for the hot spring waters have a high silver content.

Gold–Antimony Association

Whereas the gold–silver and gold–quartz associations may occupy the same tectonic setting as porphyry–copper deposits, but at a higher structural level in the volcanic pile, the gold–stibnite, stibnite and stibnite–gold–scheelite veins never occur in association with porphyry-copper deposits (Stanton, 1972, Ch. 17; Hutchison and Taylor, 1978).

It appears that epithermal mineralization involving antimony is characteristic of regions that are underlain by continental crust, though rather attenuated in the case of Peninsular Malaysia.

NORTHWEST BORNEO (SARAWAK AND KALIMANTAN)

The region is marked by important Late Cretaceous (80–85 Ma) epizonal intrusive activity. The calc-alkaline belt is marked by epithermal mineralization with gold, antimony and mercury in Sarawak and gold in west Kalimantan. Much of the gold has been from placers, but the best primary production has been from Bau, near Kuching, Sarawak. The Bau mineralization is related to a line of minor Tertiary dacitic and granodioritic stocks, trending northnortheast perpendicular to the tectonic zones, emplaced in the Cretaceous Bau Limestone. These could have provided the heat source to drive the hydrothermal circulation responsible for the mineralization. The mineralization consists of complex arsenical ores containing native arsenic together with other arsenical species, stibnite, pyrite and a little free gold. The quartz-rich veins contain more free gold. Silver is subordinate. Fluid-inclusion studies indicate the temperature of formation of the mineralization to be within the range 140–250°C (Wong, 1979).

Total gold production was 36 t, mainly extracted between 1900 and 1921. The gold occurs in fissures within the limestone, but much of the production was from residual and eluvial deposits associated with solution cavities in the limestone surface. Most of the gold production from Kalimantan has been from placer deposits, but coarse gold has been mined from quartz veins in hornfels at Gunung Mas.

The region has produced about 85 000 t of antimony ore but the yield has rapidly declined in importance. The main deposits were at Paku, where stibnite–quartz ore occupied faults at a Jurassic–Cretaceous limestone–shale contact. Mineralizing solutions appear to have migrated upwards through the limestone until they met the overlying impermeable shale (Wilford, 1955).

Most of the ore mined from the Bau area was from eluvial and residual deposits, concentrated as masses within solution hollows in the limestone following weathering and erosion of the shale–limestone contact. Mining of the primary veins, which continue downwards into the limestone, was largely unprofitable because of rapid and irregular thinning of the veins and the expense of draining the mines in this region of high rainfall.

Mercury was mined as cinnabar within breccia zones in the shales and also from eluvial deposits, giving a total production of 750 t (Lau, 1970).

MAINLAND SOUTHEAST ASIA

Gold, antimony and minor mercury mineralization occurs in a narrow zone within Permo-Triassic sedimentary and volcanic rocks on the east side of the Main Range of Peninsular Malaysia (Hutchison and Taylor, 1978). Only gold has been of importance, with one mine, the Raub Australian Gold Mine, yielding over 80% of the recorded production.

Raub Australian Gold Mine

From 1889 to 1938 this mine produced 21 t of gold. When it went into disuse around 1955, the gold-mining industry of the Malay Peninsula became insignificant. The ore bodies are hydrothermal replacements and fissure fillings located in zones of compressional faulting, known as the Eastern and Western lode channels (*Figure 8.10*). The minerals in the lode channels are quartz, gold, pyrite, arsenopyrite, stibnite, scheelite and traces of chalcopyrite and cerussite. Silver is generally absent. The stibnite occurs as large masses, and the gold frequently occurs within quartz, stibnite, or arsenopyrite. The veins are largely conformable within the folded Triassic calcareous shale. Quartz lodes also occur as discontinuous replacement bodies within the shale. There is no demonstrable relationship with igneous intrusions. However, Richardson (1939) implied that the deposits were

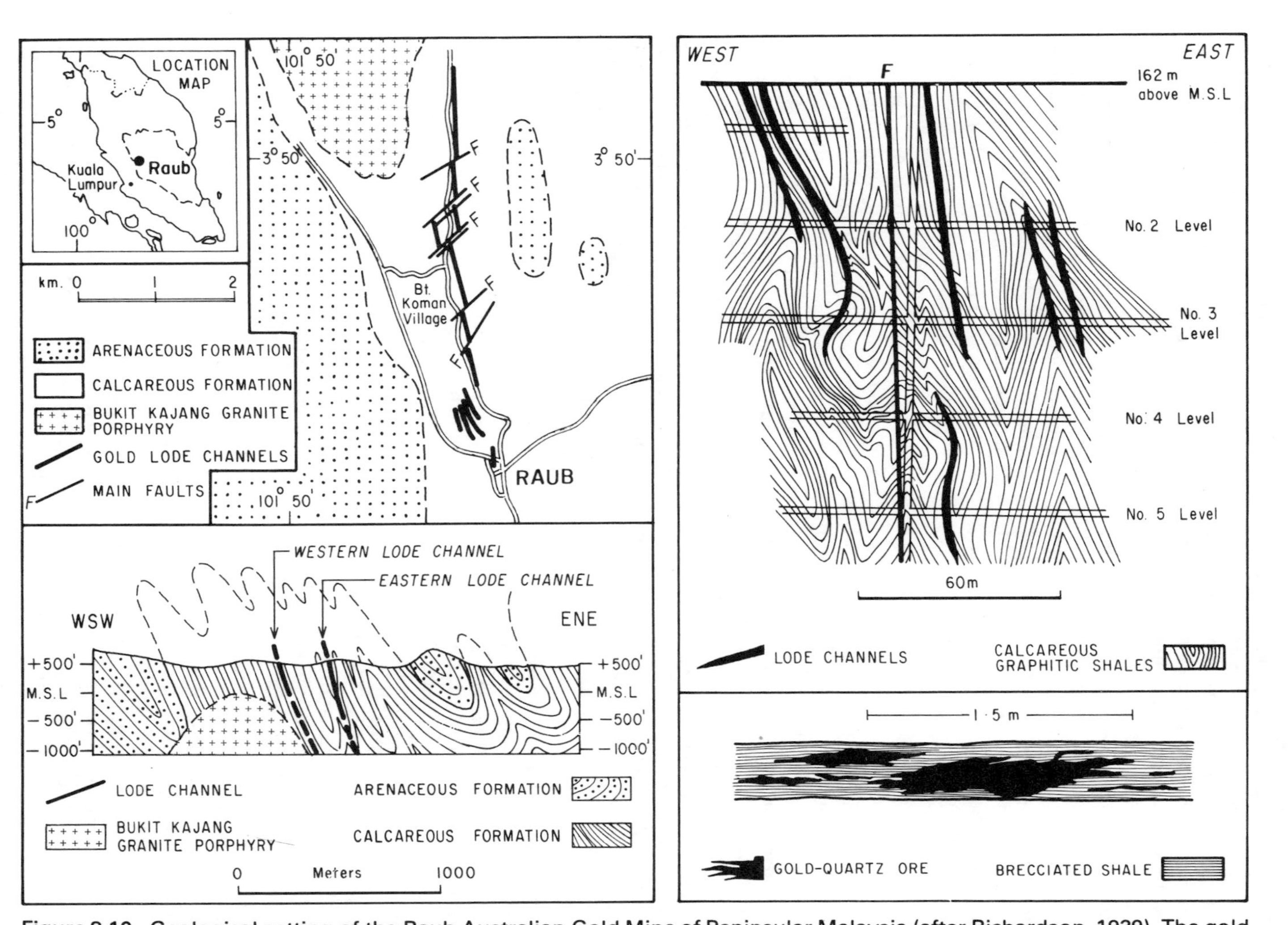

Figure 8.10. Geological setting of the Raub Australian Gold Mine of Peninsular Malaysia (after Richardson, 1939). The gold and stibnite occurs in quartz veins (upper right) or in discontinuous replacements or 'sweat-outs' (lower right) in the Triassic sediments. A genetic relation to the quartz porphyry has been suggested.

genetically related to the Bukit Kajang rhyolite porphyry, which is probably of Late Triassic age. However most of the lodes resemble typical 'sweat-out' veins and pods of quartz that commonly occur in low-grade metamorphic assemblages. It is likely that the lodes and quartz veins resulted from mobilization of the economic elements from the country rocks by circulating meteoric hydrothermal solutions. If the Bukit Kajang rhyolite porphyry did play a role, it was only to raise the regional temperature and facilitate the meteoric water circulation. Farther north along the same tectonic line, a lode deposit was formerly mined at Toh Moh in south Thailand. About 1.8 t of gold was extracted from about 200 000 t of ore (Hutchison and Taylor, 1978).

Antimony

Antimony occurs at many places in Thailand and Burma east of the Shan Scarp. Annual production has been below 100 t in Burma with the annual Thai production increasing recently to over 7000 t. All the deposits appear to be small quartz–stibnite veins unrelated to igneous intrusions, and all are related to fault zones. Some occur within brecciated zones in Palaeozoic limestones. The proposed genetic relation with Triassic granites is unlikely because many of the deposits occur in fault zones that are of Tertiary age (Tantisukrit, 1975).

China is considered to have greater reserves of antimony than any other country. The main deposits are in the south, predominantly in Hunan, which holds more than 60% of the nation's total. The Hsikuangshan district is the premier mining region. The ore contains stibnite in a gangue of quartz and occurs in sandstone. It is of low grade with 6–9% antimony on average. The district produces over 25 000 t of metal antimony annually (Ikonnikov, 1975). Other regions in Hunan are at I-yang, where quartz–stibnite veins cut schist and shale, and in the Meicheng district. In addition to the pure antimony ores of China, many of the tungsten and tin deposits also contain antimony in the same hydrothermal systems.

As well as the main quartz–stibnite veins, there is also an important stibnite–galena–arsenopyrite replacement in carbonate country rocks (Juan, 1946). The antimony mineralization usually occurs in regions of important shearing and brecciation.

Bolivia is the world's major producer of antimony (*see* Appendix). The northern province of the Cordillera Real (*Figure 8.11*) is dominated by Mesozoic epizonal granite batholiths surrounded by Palaeozoic country rocks that have been metamorphosed in thermal aureoles. The southern province is a higher-level Cenozoic volcanic province. In both regions, the stibnite veins are widely spread far to the west and east of the igneous chain (Ahfeld, 1967); they occupy old faults and form lenticular bodies concordant within the non-metamorphic Palaeozoic shales. The gangue is usually of quartz. The stibnite is generally fine grained because of recrystallization and is often accompanied by ferberite and gold. The southern region is richer in stibnite than the Cordillera Real. The stibnite is considered to be older than the young volcanism, because near La Puerta, west of Potosi (*Figure 8.11*), stibnite pebbles are abundant within Lower Cretaceous conglomerates. Most of the antimony mineralization is held to be Tertiary or Late Mesozoic. Generally the antimony veins are areally separated from the zinc–lead veins, but transitions are known.

Antimony does occur within the tin veins closer to the igneous chain, but not as stibnite. It is contained in the complex lead–tin sulpho-salts jamesonite and boulangerite, but stibnite is known from the latite stocks in south Lipez. Stibnite also forms veinlets in young deposits, which are more common in the north; it consists of druzy deposits of acicular crystals accompanied by pyrite, but the deposits are uneconomic. They are associated with wood tin and were deposited from thermal ground water. A hot spring terrace at Uncia is of antimony-bearing psilomelane, opal and barite.

The epigenetic antimony mineralization continues northwards through Peru (Petersen, 1965), although those of tin and tungsten do not. The main economic elements in the Peruvian sector of the Cordillera are iron, copper, arsenic, antimony, lead and zinc, while silver and bismuth are also economically important. The mineralization is related to centres of igneous activity or with persistent longitudinal faults. Limestone contacts have acted as particularly favourable places for mineralization deposition.

Volcanogenic Tin Deposits

This minor kind of tin mineralization is interesting because it occurs in the same tectonic setting as porphyry-copper and epithermal metal vein deposits along convergent continental margins. However the distribution of tin is much more erratic than copper, lead and zinc, and the presence of tin in cordilleran margins may have resulted from its remobilization from pre-existing deposits already present in the continental crust

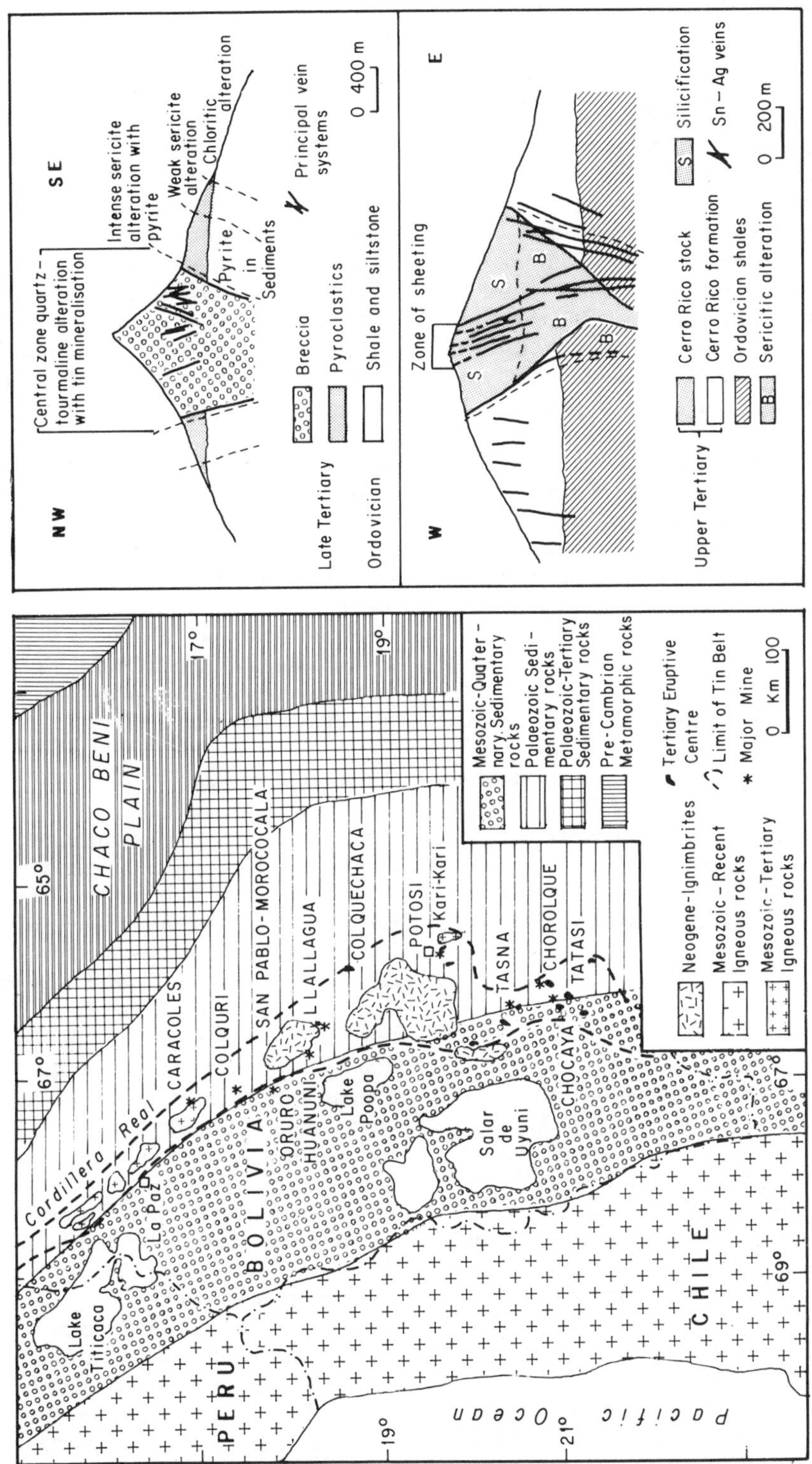

Figure 8.11. Regional geology of the Central Andes showing the Bolivian tin belt. Right upper: schematic cross-section through the Cerro Chorolque. Right lower: schematic cross-section through the Cerro Rico, Potosi (after Grant *et al.*, 1977).

before the active igneous arc was initiated. This argument is further developed under the descriptions of the Bolivian tin deposits, given below.

MEXICO

The major occurrence is in a 900-km belt of western Mexico, but the annual output is restricted to a few hundred tonnes of tin concentrate. The ore occurs as short, narrow fissure veins, low-grade disseminations and breccia fillings in Mid Tertiary rhyolite (Foshag and Fries, 1942). Tin-bearing breccias are found in the upper parts of some steeply flow-banded plugs of vesicular rhyolite. The metal is present as incrustations of cassiterite and wood tin, and is always accompanied by specularite. Mimetite, tridymite, quartz, chalcedony, opal, sanidine, white mica, zeolites, topaz, fluorite, bismuth, magnetite and limonite also occur. The chief alteration mineral is montmorillonite. The tin was deposited close to eruptive vents from hydrothermal fluids of temperatures claimed to be below 150°C, during the waning stages of volcanism (Pan and Ypma, 1974).

USA

Similar volcanogenic tin deposits occur in Mid-to-Late Tertiary rhyolites at Taylor Creek, New Mexico, and Lander County, Nevada. In contrast to the Mexican deposits, the presence of bixbyite and pseudo-brookite in Taylor Creek suggests that the tin mineralization was generated from a gas phase at temperatures higher than 500°C, immediately after deposition of the host rhyolite (Lufkin, 1976).

BOLIVIA

Bolivia accounts for about 14% of the world's tin production, but unlike many major producing countries, most of the mining is of primary vein deposits. General accounts of the deposits are given by Ahfeld (1967), Turneaure (1971) and Rivas (1979). The geological setting of the active volcano–plutonic convergent margin is illustrated in *Figure 8.11*. Geological and geophysical evidence indicates that the Precambrian rocks, which form the Brazilian Craton, extend westwards beneath the full width of the Andean Orogen (James, D.E., 1971; Clark, A.H. *et al.*, 1976). This basement is overlain by a thick Lower Palaeozoic clastic marine sequence that was deposited in intracratonic basins before the active cordilleran margin developed. Depending on the particular reconstruction of Pangea, it might be hypothesized that the west coast of South America is a rifted margin that separated Antarctica and Australia from South America. Although most reconstructions show the west coast of South America as always being a margin, this is probably not true for there is a remarkably good fit with Antarctica, and the Bolivian re-entrant looks like a triple junction. By analogy with Nigeria, a failed arm or aulacogen, similar to the Nigerian Benue Trough, might be suspected here, though there are no suggestions of this in the literature.

The tin belt is a well-defined metallogenic province that extends 800 km throughout the length of the Eastern Cordillera of Bolivia (*Figure 8.11*). In addition to tin, there is a significant production of tungsten, silver, zinc, lead, antimony, bismuth and gold, mostly from polymetallic vein deposits (Petersen, 1970; Turneaure, 1971). There is a clear genetic relationship to cordilleran Cenozoic acid igneous activity, in which there were two broad episodes—Triassic–Jurassic, when granitic and granodioritic plutons were emplaced in the Cordillera Real to the north (Evernden *et al.*, 1977), and in the Miocene, when there was high-level plutonic and volcanic activity in the central and southern regions (Grant *et al.*, 1979). The tin belt can be conveniently divided into two contrasting regions (Ahfeld, 1967).

The Cordillera Real

This northern region is more elevated and more deeply eroded than the south. The plutonic arc is of batholiths of Triassic-to-Lower Jurassic and Oligocene-to-Miocene granodiorite and granite. The plutons are surrounded by well-developed contact aureoles (*Figure 8.12*) in the Lower Palaeozoic country rocks (Rivas, 1979). The plutons are associated with vein-type tin–tungsten deposits, which are numerous, but account for only a small part of Bolivia's tin production. This province contains no porphyry-type deposits because the tectonic level of igneous emplacement was deep epizonal, estimated to have been at depths of 2000–4000 m (Kelly and Turneaure, 1970).

The Late Triassic batholiths have well-zoned deposits: (1) tin-bismuth, with an outer zone of copper–lead–zinc–silver; (2) tungsten–antimony–mercury; (3) gold (Turneaure, 1971). Most tin deposits are within the batholiths or the contact aureoles. The veins are mainly of cassiterite, wolframite, or scheelite. Pyrrhotite and sphalerite are abundant, and quartz is the principal gangue mineral. Tourmaline is common in the altered country rocks. Of minor importance are chalcopyrite–wolframite veins, lead–zinc–

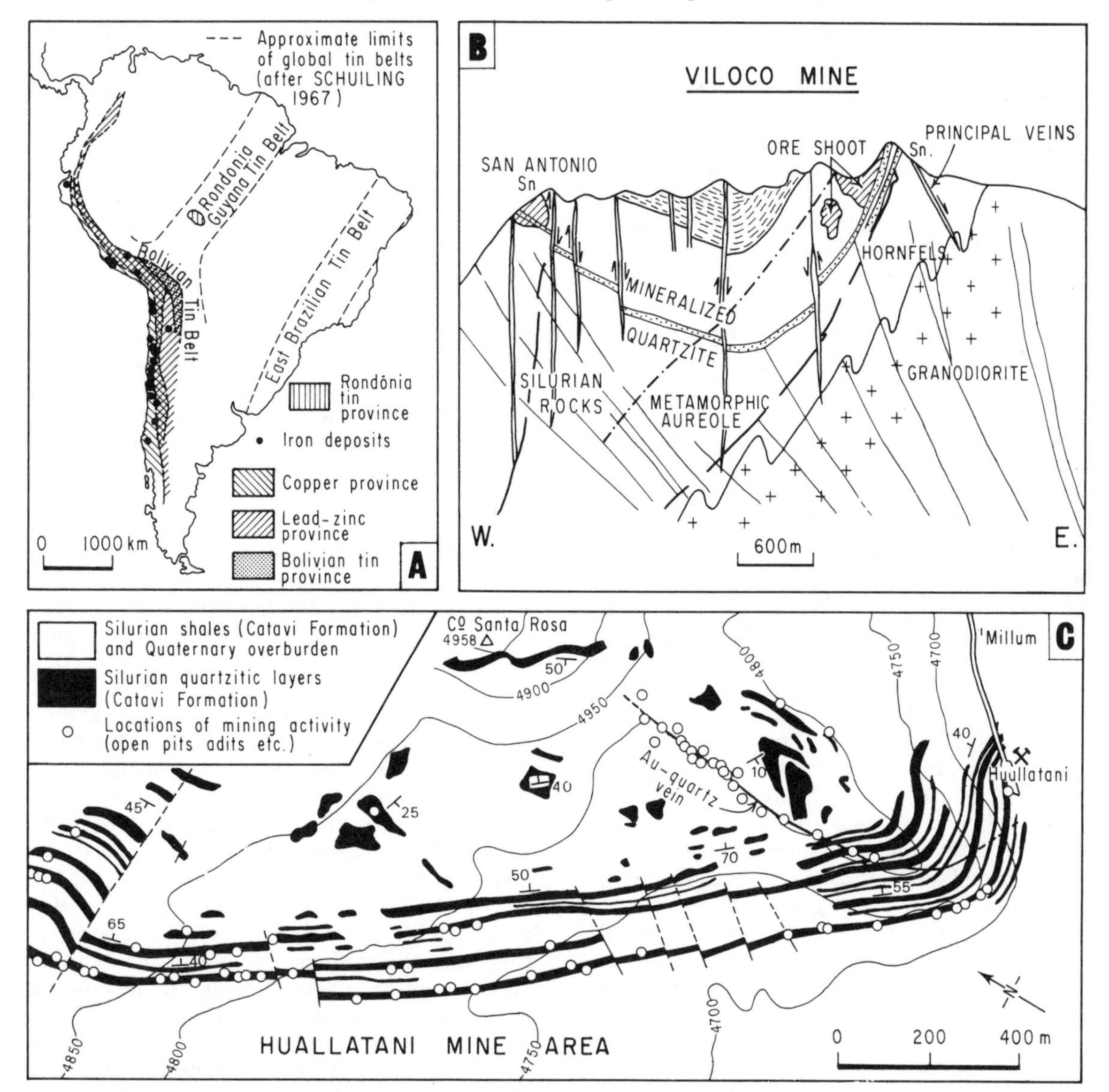

Figure 8.12. (A) The metallogenic belts of western South America (after Schuiling, 1967a; Petersen, 1972; Sillitoe, R.H., 1976). (B,C) The fissure and strata-bound tin deposits of Bolivia, as illustrated by the Viloco and Huallatani areas (after Schneider and Lehmann, 1977; Rivas, 1979).

antimony deposits, gold–quartz veins and tin-bearing pegmatites (Turneaure, 1971).

The cassiterite–wolframite mineralization consists of the following. (1) An early vein stage, which fluid-inclusion studies have shown to have been deposited within the range 530–300°C from solutions that had salinities as high as 46% sodium chloride equivalent by weight (Kelly and Turneaure, 1970). The deposition of cassiterite and wolframite was accompanied by active boiling of the saline hydrothermal solutions. Salinities progressively declined throughout the following stages. (1) An early apatite formation at 400–300°C. (2) A base-metal sulphide stage between 400°C and 260°C. (3) Hypogene alteration of pyrrhotite to pyrite and marcasite to siderite, at 260–200°C. (4) Late minerals were deposited at 200–133°C forming crusts and veinlets of siderite, fluorite and hydrous phosphates, some of which represent supergene alteration at temperatures below 70°C.

A remarkable feature of the Cordillera Real is the frequent occurrence of strata-bound lodes, which resemble the bedded lodes of Klappa Kampit, Billiton, Indonesia, yet most of the veins are clearly epigenetic and occupy fissures transverse to the batholith contacts with the metamorphosed country rocks.

Viloco is a good example of this (*Figure 8.12*). The deposit is located in a 200-m wide north–south-striking syncline of Silurian lutites near the contact with biotite granodiorite. Quartzites on both sides of the syncline are stratiformly mineralized. The rocks are cut by faults perpendicular to the igneous contact, and the mineralized veins are fault fissure fillings. This deposit is famous for its collection and quality of large black euhedral cassiterite crystals. The mineralization begins in the upper part of the hill with a little molybdenite, sometimes as flakes disseminated in the granodiorite, then continues near the contact, with veins of wolframite, scheelite, tourmaline, quartz, pyrrhotite, jamesonite, arsenopyrite, siderite and principally cassiterite. Finally the veins degenerate at the outer edges of the mineralized zone to marmatite. The veins pinch out 300–400 m below surface. In addition to veins within the country rocks, cassiterite veins within tension fractures in the granodiorite are now being mined (Rivas, 1979).

About 20 km north of La Paz, two of the most striking examples of strata-bound tin deposits occur at Kellhuani and Huallatani (*Figure 8.12*). Both are bound to Silurian quartzite, and the mining activities around the hillsides clearly define the sandstone beds (Schneider and Lehmann, 1977). The quartzites range in thickness from a few centimetres to 30 m and are irregularly mineralized, from rich to subeconomic. At present, mining is restricted to small-scale mines because of mineral-dressing problems, but large-scale mining will result in this region becoming a major tin-producing centre. These so-called manto-type deposits can be traced along the strike of the Silurian formations over a 250-km distance up to the Poopa region. The mineralized sandstones alternate with black shales. Just as in the Viloco mine, mineralization also occurs along fault zones, but the deposits are largely strata-bound. Schneider and Lehmann (1977) interpret these manto-type tin deposits as Silurian palaeo-placers, implying that the tin of this region was already in the Lower Palaeozoic continental formations, and has been remobilized in the volcano–plutonic Mesozoic arc to give rise to the later igneous-related deposits. The deposits consist of fine-grained cassiterite, tourmaline and abundant pyrite and haematite. This simple mineralogy contrasts strongly with the complex mineralogy of the clearly epigenetic vein deposits of the region, which consist of sphalerite, galena, cassiterite, pyrite, stannite, fluorite, wolframite, chalcopyrite, siderite and bismuthinite. Thin sections of the mineralized quartzites show metamorphic recrystallization textures. Green tourmaline appears as an apparently clastic component enriched along some bedding planes. Cassiterite occurs as fine dissemination or fills small fissures on a millimetre–centimetre scale. Concordant cassiterite concentrations are rare, so that a doubt remains as to whether these are truly syngenetic palaeoplacers, or that the cassiterite has been epigenetically emplaced in the sandstones. Ore from the Monte Blanco Mine, however, strongly suggests that the cassiterite was confined to certain bedding planes of a sandstone that has been metamorphosed to quartzite (Schneider and Lehmann, 1977).

The southern volcanogenic region

In the less elevated and less deeply eroded region south of the Santa Vera Cruz pluton, the igneous rocks form small isolated porphyry stocks, dykes and larger eruptive complexes. There are also extensive areas of ignimbrite and tuff, and only one plutonic body—the Kari-Kari batholith near Potosi (*Figure 8.11*). The volcanogenic tin province has been described by Grant *et al.* (1977, 1980). There are three distinct periods of igneous activity (Grant *et al.*, 1979). In the north from Oruro to Potosi, the eruptive complexes, the Kari-Kari batholith and associated volcanics are dated at 23–19 Ma (Lower Miocene). These overlap in time with the southern plutons of the Cordillera Real. From Potosi southwards, the igneous ages are Middle Miocene (17–12 Ma). The ignimbrite sheets gives ages between 8 Ma and 6 Ma and post-date the important tin mineralization. Most of the tin production comes from vein-type deposits within or at the margins of the acid igneous eruptive complexes of Tertiary age (*Figure 8.11*). At some of these there is also a potentially important low-grade porphyry-type mineralization, consisting of disseminated, veinlet-controlled and breccia-controlled cassiterite and sulphides. The mineralization is developed within or peripheral to stock-like bodies of rhyodacite porphyry and breccia that fill volcanic vents. The igneous rocks are pervasively altered, giving assemblages of quartz–tourmaline, tourmaline–sericite, and sericite–clay, often zoned outwards and upwards from the eruptive centre. Low-grade cassiterite mineralization is dispersed throughout the inner zones. In most of the porphyry-type deposits, the early dispersed mineralization is cross-cut by a system of veins that carry high-grade ore, as at Cerro Chorolque and Potosi (*Figure 8.11*).

The porphyry-tin deposits bear many resemblances to porphyry-copper deposits (Sillitoe, R.H. *et al.*, 1975). Important features include: the mineralization is centred on small (typically 1–2 km^2) porphyry stocks of acid-to-intermediate composition; the stocks are subvolcanic, and were emplaced beneath or within the vents of subaerial volcanoes; hydrothermal breccias are a prominent feature, although in some cases the stocks are large breccia pipes—typically there were several pulses of porphyry intrusion and brecciation; the stocks and their wall rocks have undergone intense hydrothermal alteration, which was feldspar-destructive, and sericite-rich and tourmaline-rich assemblages predominate. Zonation—both lateral and vertical—is evident; cassiterite is accompanied by sulphides, particularly pyrite, but also including stannite, chalcopyrite, sphalerite and arsenopyrite, and in some cases the deposits are zoned, with a pyrite-rich low-grade halo surrounding a low-pyrite higher-grade core; and the pervasive alteration and porphyry mineralization is independent of the higher-grade vein ores.

Geologic and stable isotope studies suggest a genetic model involving an initial supply of heat, fluid and ore components associated with the emplacement of the porphyry magma at shallow depths of 1000–2000 m in the vents. Earliest hydrothermal activity produced widespread explosive brecciation, and this was followed by intense rock alteration. Fluid-inclusion studies show that the hydrothermal systems were initiated by highly saline magmatic fluid at temperatures in excess of 500°C. When temperatures fell to 450°C, the pervasive quartz–tourmaline alteration, fracturing, and hornfelsing of the country rocks took place. Intermittent boiling of the fluids occurred; widespread disseminated cassiterite was deposited during this phase. The earliest stages of growth of quartz–cassiterite veins took place around 400°C from a highly saline fluid, of about 40% sodium chloride equivalent by weight. Vein growth continued when temperatures fell below 200°C and salinity decreased. The bulk of the cassiterite deposition appears to have taken place within the temperature range 250–300°C, accompanied by a major decrease of salinity of the mineralizing fluids (Grant *et al.*, 1977). The final stages of waning temperature and salinity probably represent a mixing of saline magmatic fluid with cool dilute meteoric water.

The data from Llallagua mine may be summarized as follows. (a) Pervasive sericite alteration and disseminated mineralization took place within the range 500–400°C with salinities between 18% and 25% sodium chloride equivalent by weight. This activity was accompanied by intermittent boiling of the fluids. (b) An early cassiterite–sulphide vein stage was deposited between 400°C and 300°C at salinities declining to about 8% sodium chloride equivalent by weight. The latest sulphide–carbonate vein stage was deposited at temperatures declining from 300°C to 100°C with low salinity fluids, less than 8% sodium chloride equivalent by weight. During the above declining hydrothermal temperature range, cassiterite deposition was restricted in the temperature range 480–300°C (Grant *et al.*, 1980). Cassiterite is the only mineral mined; it is accompanied by quartz, bismuthinite, wolframite, monazite, apatite, pyrrhotite, pyrite, marcasite, arsenopyrite, stannite, sphalerite, siderite, vivianite and other phosphates, representing the range from magmatic hydrothermal to low-temperature meteoric hydrothermal (Rivas, 1979).

Most porphyry-style cassiterite and sulphide deposition was associated with sericitic alteration at moderate temperatures and fluid salinity. Later, owing to structural adjustments probably related to the cooling of the underlying plutons, major fracture systems formed, and tapped the deeper magmatic sources. Hydrothermal circulation and ore deposition were then confined to the major fractures, resulting in high-grade vein ores that cross-cut the earlier pervasive alteration and mineralization (Grant *et al.*, 1980).

Discussion

The ultimate source of the tin in the Bolivian and western cordilleran belt of North America may be presumed to be the same as that of the associated rhyolite or rhyodacitic stocks. Sillitoe, R.H. (1976) has suggested that because the high silica rhyolites of the Sierra Madre Occidental, Mexico, have low initial strontium ratios of 0.704–0.705, their source, together with their tin, could be subcrustal and subduction-related. He also argued that the Bolivian tin was released from a depth of about 300 km on the shallow-dipping Benioff Zone (Sillitoe, R.H., 1976). However, although the Benioff Zone may play some role in the evolution of mantle-derived magmas, the magmas themselves are produced within the mantle wedge well above the Benioff Zone, at depths calculated to be around 100–120 km beneath the surface (Whitford *et al.*, 1979). If the role of the Benioff Zone in magma production is unclear, it is even more so in regard to the genesis of tin in the overlying crust. The heat flux from the magmas, which have risen into the crust, may be the only link between tin and the mantle. Clark, A.H. *et al.* (1976) rejected Sillitoes's hypothesis, citing the fact that there is a complete absence of tin minerals in the Neogene porphyry-copper centres of the Sierias Pampeanas, a region rich in Palaeozoic tin deposits. There is also a sudden transition in northern Argentina from these deposits to the tin polymetallic centres of the main tin belt, and this sudden transition does not coincide with any known changes in the subduction pattern. Clark, A.H. *et al.* (1976) tentatively conclude that the great concentration of tin deposits in the Andes reflects a contribution from a persistent anomaly, which has been drawn on periodically since the early Palaeozoic. They suggest that this enriched zone is in the upper mantle rather than in the lower crust, though the arguments are not compelling.

If Sillitoe, R.H. (1976) is correct and tin is contributed in economic amounts directly in one magmatic cycle to the crust from the mantle, then all volcano–plutonic arcs of convergent plate margins should be endowed by tin deposits. It is clear that tin does come from the mantle in single magmatic cycles, as in some of the porphyry-copper centres of the Philippines, for example, but the contributions from the mantle are insufficient to give tin deposits and the levels of contribution to the base-metal deposits are restricted to less than about 5 ppm. The absence of economic deposits of tin, tungsten, silver and molybdenum from wholly ensimatic island arcs strongly argues that these elements are remobilized from sialic crust. Gold and base metals show no preference for ensimatic island arcs or continental cordillera. Hence they are considered to be subduction-related and to have been brought up in significant amounts in one or a few magmatic cycles directly from the mantle by the ascending magmas. Thus, porphyry coppers are equally to be found along continental margins or in wholly ensimatic island arcs. The subduction-related magmas provide high heat flow regimes that are capable of driving large-scale hydrothermal fluid circulation cells within the crust, which leach and concentrate the economic elements. It appears necessary to conclude that the rising of hot mantle-derived magmas results in localized anatexis of continental crustal materials and that any element already present in the crust, if it has an affinity for acid magmas, will be preferentially concentrated in the granitic magmas. Such anatectically derived crustal magmas will be characterized by higher initial strontium-isotope ratios, but the actual ratios will depend on the materials that were anatectically melted. One may also suggest that both mantle-derived and crustal-derived magmas will drive the large-scale hydrothermal fluid circulation cells that scavenge such metals as tin, tungsten, silver and molybdenum from the continental crust. The absence of economic deposits of these metals from wholly ensimatic island arcs is because the crust was directly derived from mantle materials and was not sufficiently differentiated by polycyclic orogenic events to have a continental affinity. Hutchison and Chakraborty (1979) conclude that continental crust becomes progressively more differentiated with respect to tin by polycyclic orogenic events, involving metamorphism, anatexis and related processes. As a result of these processes, it becomes characterized by a highly irregular tin distribution.

When the mantle-derived magmas rise into ensimatic island-arc crust, or even into continental crust that is less evolved, as indicated by its low initial strontium-isotope ratios, the plutons set up hydrothermal circulation cells that scavenge gold and base metals from the country rocks, but such regions will be low in economic deposits of tin, tungsten and silver. The nature of the continental crust before it became an active volcano–plutonic arc system is a fundamental question in the metallogeny of active plate margins. Based on these arguments, the problem of the restriction of tin largely to Bolivia in the South American cordillera will be considered.

Metallogenic distribution in the Andes

The metallogenic distribution map of the South American cordillera (*Figure 8.12*) is reproduced here from Petersen (1972). It is notable that the copper, lead–zinc and iron provinces extend fairly well along most of the active cordilleran margin. One might therefore infer that these metals were introduced into the crust by igneous events that originated in the mantle. The tin belt is strikingly impersistent, and largely confined to Bolivia, with a slight extension into Peru. Why is there no continuity? One might argue like Clark, A.H. *et al.* (1976) that the upper mantle in this region is anomalously enriched in tin. But why not argue that it was the crust that was enriched, and that igneous events that originated in the mantle have provided the necessary heat flux to mobilize the older tin into the Mesozoic and Tertiary igneous centres? Schuiling (1967a,b) clearly showed that tin is restricted to several well-defined belts within the continental crust. In South America, one might define a broad Precambrian belt extending northeast through Rondonia, in Brazil, to Guyana. It is remarkable that the southwestwards extension of this belt intersects the Andean cordillera in the Bolivian tin belt (*Figure 8.12*).

Schneider and Lehmann (1977) develop the argument that the stratabound manto-type tin deposits within Silurian quartzites provide the missing link in the metallogenic development of the Bolivian region. The tin was already in the continental crust of a Precambrian belt extending from Guyana, through Rondonia, to Bolivia. A significant remobilization took place during the Lower Palaeozoic, before the cordilleran margin was active, and whatever this event was, it gave rise to palaeo-placers in the Silurian Catavi and Llallagua quartzites. Mesozoic and Tertiary events related to the cordilleran plate margin provided the heat necessary to remobilize these into epigenetic vein and porphyry-type tin deposits. Thus it might be concluded that the Bolivian tin is inherited from the Precambrian cratonic basement.

9

Archaean-style mineralization

Archaean Tectonics

A highly characteristic feature of the Archaean crust is the existence of elongated greenschist facies mafic–ultramafic igneous rocks and associated metasediments (greenstone belts) within more extensive areas of granites and gneisses in the shield areas of all continents. These are especially well displayed in the shields of North America, India, Australia and the Zimbabwe Craton (*Figure 9.1*).

Archaean-type greenstone belts persisted into the Lower Proterozoic, for example, in the Guyana Shield prior to the Trans-Amazonian orogeny (Gibbs, 1980). Therefore there was not an abrupt global change in the crust–mantle system at the end of the Archaean, and Archaean-style mineralization may persist into the Lower Proterozoic in some parts of the world.

Greenstone belts are exceptionally richly mineralized and hence their tectonic evolution should be considered. The following summary is largely based on Glikson (1976a,b).

Firstly it is important to note the principal features of the greenstone belts.

Geotectonic Setting

They occur as outliers within intrusive granites and the belts are up to several hundred kilometres in length (*Figure 9.1*).

Intrusive Assemblages

They consist of: subvolcanic mafic-to-ultramafic sills; tonalitic, trondhjemitic and granodioritic early plutons; and granite and syenite late plutons.

Volcanic Assemblages

They consist of bimodal mafic–felsic suites including abundant ultramafic volcanic rocks (komatiites). Basalt–andesite–rhyolite cycles occur in the Canadian Shield. Pillowed basalts are common. Alkaline volcanics are rare, but do occur at high stratigraphic levels. Acid volcanic lenses and pyroclastic rocks are common.

Komatiites constitute an igneous suite of volcanic and hypabyssal rocks ranging from dunite and peridotite to basalt. Ultramafic komatiites are rich in olivine and are spinifex-textured (like intergrown blades of coarse grass) massive or pillowed lava flows, occurring as bedded volcaniclastic rocks or as small dykes and sills (Arndt *et al.*, 1979). Peridotitic komatiites can be recognized and defined by the presence of this characteristic spinifex texture (Nesbitt *et al.*, 1979). Their chemistry suggests that large amounts of mantle melting are involved. There is no generally acceptable definition of basaltic komatiites, except that they are high-magnesium basalts. Post-Archaean komatiites are extremely rare. Komatiites were erupted subaqueously, and probably are of submarine origin. Nowhere is there any evidence of subaerial eruption. The rate of eruption was great. The scarcity of interflow sediments indicates that eruptions frequently occurred one after the other in fairly rapid succession.

Chemistry of the Volcanic Rocks

Low-potassium oceanic-type tholeiites are very common. Sodium-rich rhyolites and dacites are also very common. The tholeiites have flat rare-earth element patterns but the acid volcanics have highly fractionated patterns.

Sedimentary Assemblages

Greywacke–shale assemblages predominate. Other formations are carbonaceous shale, chert, banded-iron formations and conglomerates. There are minor carbonates and quartzites. Pure shale units are very rare.

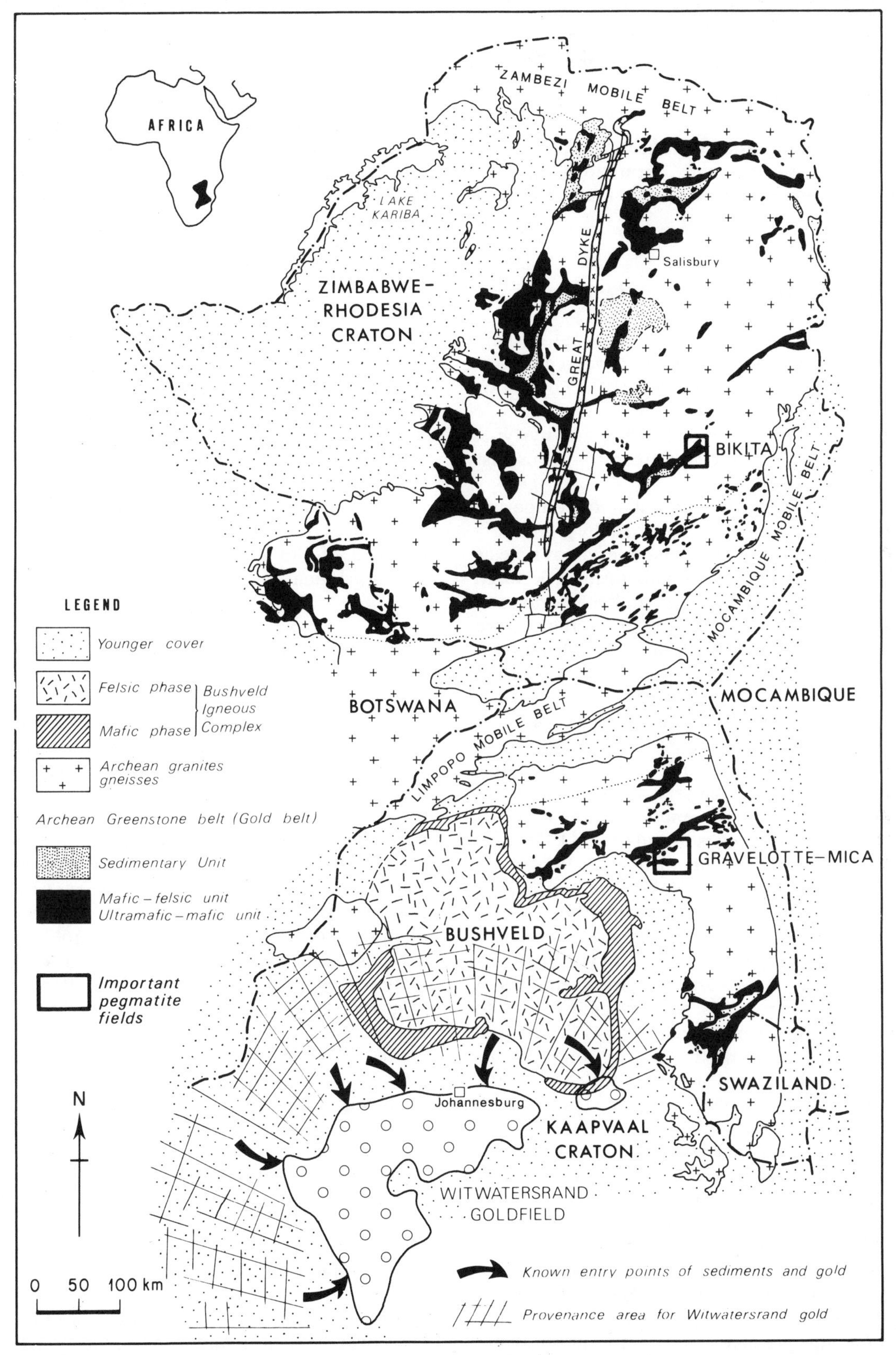

Figure 9.1. The Archaean granite-greenstone terrane of the Zimbabwe and Kaapvaal Cratons showing the Great Dyke, Bushveld complex and the Witwatersrand fold field (after Anhaeusser, 1976b).

Successions

In ascending stratigraphic order the succession is: ultramafic–mafic volcanic formations, mafic–felsic volcanics, greywacke–shale units, polymict conglomerate.

Typical Structures and Metamorphism

The greenstone belts occur as synclinal outliers down-folded and down-faulted between granites. Anticlines are commonly accompanied by major faults. Strike faults are common. The tectonic deformation is mainly of the gravity-type. Metamorphism is of the low-pressure greenschist facies.

Greenstone belts can accumulate over a period of 50–100 Ma and are accompanied by granodiorites and granites over the same interval of time as the volcanic activity. Most belts formed between 3700 Ma and 2500 Ma ago, and the formation of the belts occurred in not more than three generations, occupying less than half of the total Archaean time. There is no general agreement on their origin.

Greenstone belts may be regarded as ophiolitic assemblages initially formed on oceanic crust. They may have initiated as rift or graben structures or something akin to Phanerozoic marginal basins. The following proposed genesis is given by Glikson (1976a,b) and Collerson and Fryer (1978): (1) high geothermal gradients of at least 90°C km^{-1} resulted in widespread granulite facies crust; (2) initial rifting of an oceanic ophiolite mafic–ultramafic crust caused by subsidence of the crust above a rising mantle diapir (*Figure 9.2*); (3) partial melting of eclogite and/or mafic granulites of the subsiding crustal segments giving rise to plagiogranite and tonalite diapirs; (4) differential vertical movements of the plagiogranite diapirs and intervening segments of ultramafic–mafic crust accounting for the linear troughs in which the sediments accumulate; (5) a worldwide amphibolite-facies anatectic metamorphic episode about 2600 Ma ago gave rise to younger potassium granites and syenites, which rose and spread out at high crustal levels (*Figure 9.2*), their source being most likely the older plagiogranites.

Arth (1976) showed that a greenstone belt of northeast Minnesota formed 2750–2700 Ma ago. The $^{87}Sr/^{86}Sr$ initial isotope ratios for all rocks including the greenstones and the granites are lower than 0.7009. These low ratios do not allow the formation of any of the rocks from a pre-existing granodiorite or more potassic crustal rocks. The proposed model is that all the rocks are of mantle partial-melting origin, with possible anatexis of greywackes of the greenstone belt. Greenstone belts therefore represent processes of upper-mantle differentiation rather than remobilization of older crust.

The partial melting of the upper mantle to form greenstone–granodiorite belts is envisioned to have taken place at spreading ridges or at mantle hot-spots in oceans and marginal basins (Burke, K. *et al.*, 1976). They also suggest plate tectonic activity rather similar to Phanerozoic arc-trench consuming plate margins as important in producing the tectonism of the belts. The marginal basin hypothesis is widely favoured for Archaean greenstone belts and Tarney *et al.* (1976) draw a close analogy with the Cretaceous marginal basin development of south Chile (*Figure 9.2*). The major difference, of course, was the higher geothermal gradient of the Archaean, which led to the ultramafic komatiitic volcanic formations.

Mineralization Associated with Greenstone Belts

Greenstone belts are perhaps the most highly enriched in economic metal deposits and of a wider range of type than any other known tectonic setting. One might argue that the peak of mineralization was attained in the Archaean. The types of mineral deposits may be summarized as follows. (1) *In the volcanic successions of greenstone belts*—consisting of iron formations of the Algoman-type and manganiferous sediments interstratified with the volcanic rocks; gold–quartz and gold telluride deposits segregated after deformation of the greenstone belts, often after periods of granite emplacement; and sulphide deposits of copper–zinc–iron, mainly related to intermediate-to-acid volcanic centres. (2) *In the ultrabasic–basic intrusive and extrusive rocks*—consisting of nickel sulphide deposits associated with the ultrabasic volcanic rocks of the greenstone belt.

The most important types of deposit will now be detailed.

Iron Formations

Ironstone is a chemical sedimentary rock that contains over 15% iron, and iron formation is a mappable rock unit composed mostly of ironstone. Iron formations may be classified as either cherty or chert-poor. The extensive literature is full of

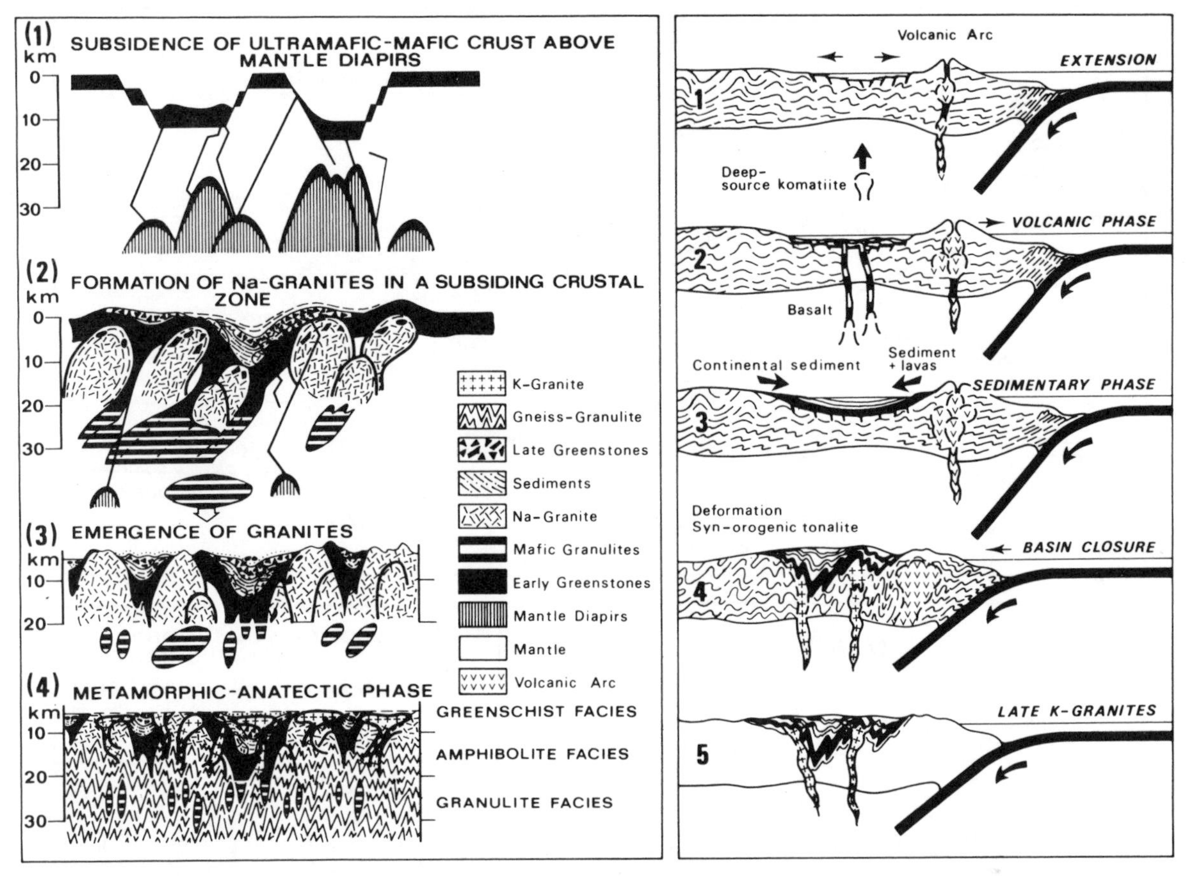

Figure 9.2. Left: model of Archaean greenstone belt evolution after Glikson (1976a,b). Stages: (1) rifting over mantle diapirs; (2) partial melting of eclogite or amphibolite to give sodium granites; (3) vertical movement of granite diapirs to give the greenstone troughs; (4) anatexis at 2600 Ma to give younger granites and metamorphism. Right: marginal basin model of Tarney *et al.* (1976) showing a sequence of back-arc rifting, extension, folding and igneous activity by analogy with the Cretaceous of south Chile.

names that have a strong geographical flavour, for example, rocks within the cherty-iron formations may be named jaspilite, taconite, itabirite, and so on. Kimberley (1978b) has proposed a palaeo-environmental classification based on the physical characteristics of the ironstone and the associated rocks. His *shallow volcanic platform iron formation* is also widely known as the Algoman-type (Gross, G.A., 1973). This type is predominantly of Archaean age and connected with greenstone belts and its close association with the acid volcanic and pyroclastic rocks leaves no doubt that the iron and silica were derived from the volcanic sources (Eichler, 1976). Most if not all Archaean iron formations are attributed to a volcanic-exhalative source. Greenstone belts are characterized by iron formations in the Canadian Shield (Goodwin, 1973b), in parts of the Australian Shield (Trendall, 1973), the USA (Bayley and James, 1973), Southern Africa (Beukes, 1973), South America (Dorr, 1973a), the USSR (Alexandrov, 1973) and India (Krishnan, 1973).

Thickness of the iron formations is usually a few metres to tens of metres but may exceed 300 m. Subjacent volcanic rocks commonly display chemical alteration. Banding is characteristic of iron formations to the extent that they are commonly referred to as banded-iron formations but this feature may be absent or poorly developed, for example, in the chert-poor iron formation of the Helen section (*Figure 9.3*). The most widespread evidence for shallow-water deposition is the bedded occurrence of ripped-up clasts of banded ironstone between undisturbed beds. Pyritic mudstone with over 10% free carbon is commonly interbedded with ironstone, and sulphur-isotope and carbon-isotope ratios indicate organic fractionation for the Helen deposit. Siderite, pyrite and

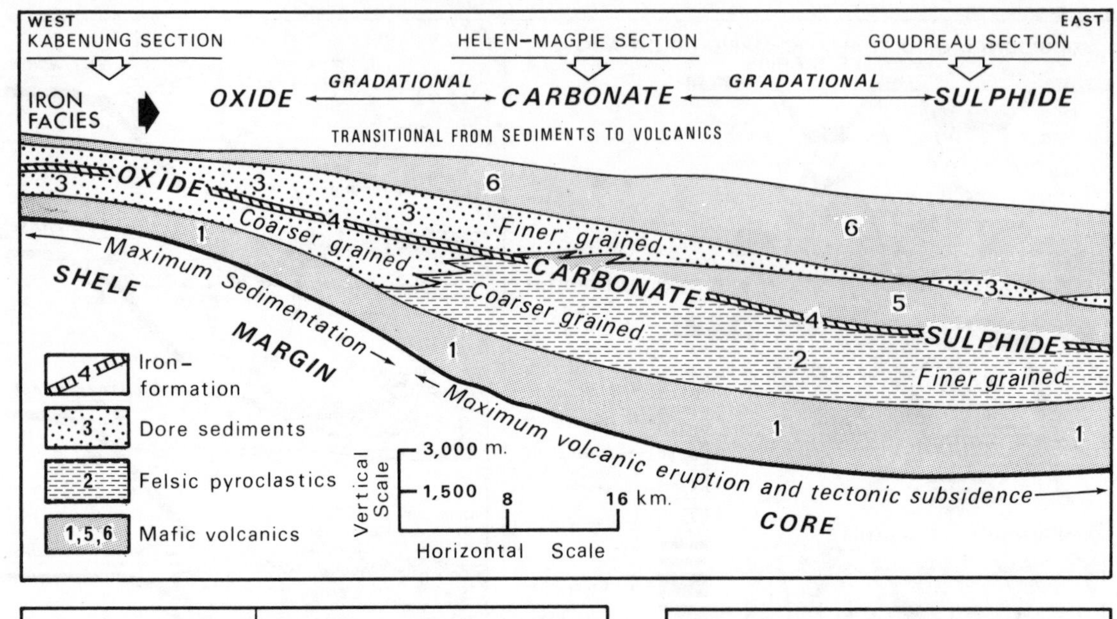

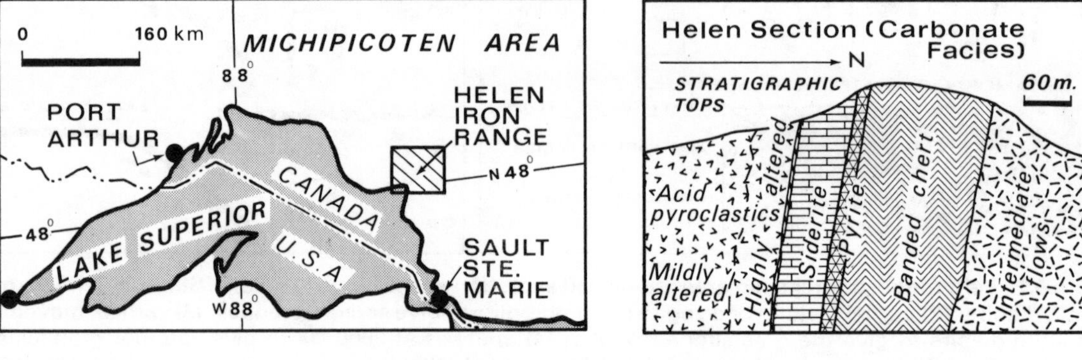

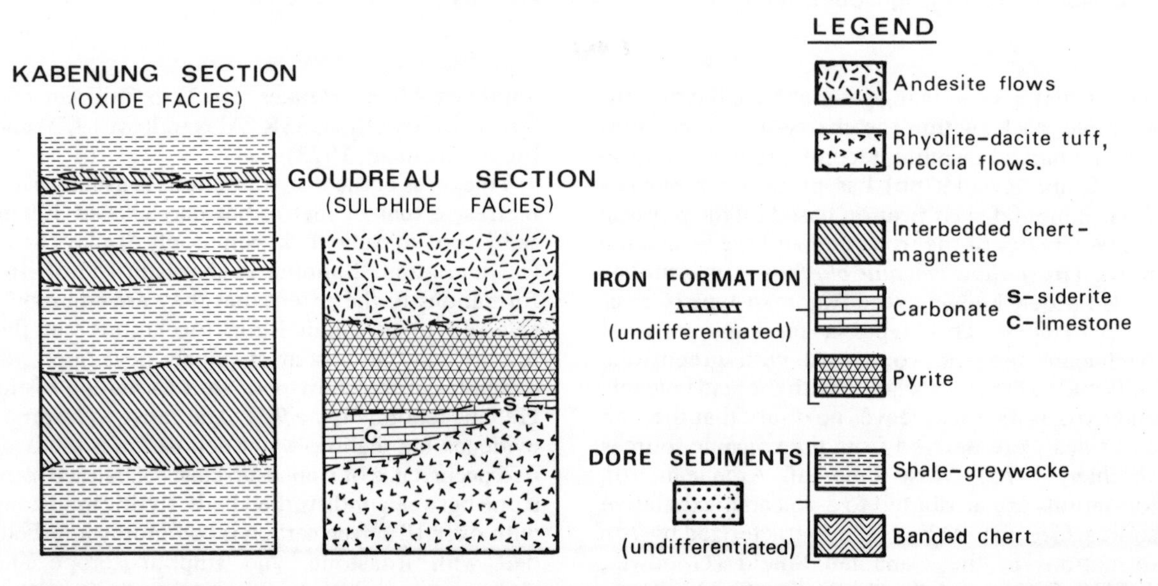

Figure 9.3. Details of the Algoman-type iron formation of the Michipicoten basin, northeast Lake Superior. East–west cross-section and typical stratigraphic sections of the different facies (after Goodwin, 1973a).

pyrrhotite are abundant relative to other iron formations. Typically, siderite or magnetite bands alternate with chert but the ironstone may locally be massive siderite or pyrite. The Algoman-type of iron formation is chemically heterogeneous and phosphorus may be locally concentrated while stratiform base-metal sulphides may occur as facies variations (Gross, G.A., 1973). Europium is often anomalously concentrated. This may be an indication of the genetic relationship to plagioclase-rich volcanic rocks. The iron formations are characteristically non-oolitic, though oolitic texture is not totally absent. The palaeoenvironment is interpreted to be a volcano largely eroded or collapsed down to wave base or built up to shallow-water depth, with a platform around a small island or a shallow rim protecting a lagoon (Quade, 1976; Kimberley, 1978b).

Some Archaean iron formations may be classified as deep-water type (Kimberley, 1978b); they are characteristically iron-poor and associated with much more abundant ferriferous chert. The deposit near Savant Lake, Ontario, consists of magnetite–chert ironstone occurring within and/or above the uppermost pelitic unit of an ideal turbidite sequence. The generalized palaeo-environment of the deep-water type is on a deeper-water slope closer to a submarine volcano than to the continental margin.

MICHIPICOTEN BASIN, LAKE SUPERIOR, CANADA

The 112 km × 48 km basin of north central Ontario consists of mafic and felsic volcanic rocks, clastic sedimentary rocks and iron formation, collectively known as the Michipicoten Group (older than 2500 Ma), now complexly folded, faulted and intruded by mafic and felsic dykes, stocks and batholiths (Goodwin, 1973a,b).

The stratigraphic succession, variable from 300 m in the west to 12 000 m in the east, contains six distinctive volcanic and sedimentary facies (*Figure 9.3*). The lowest unit is mainly of andesite flows showing pillowed structure, and dyke intrusions, and it is continuous across the basin. In the eastern and central basin this lowest unit is overlain successively by thick piles of felsic pyroclastic rocks, iron formation, mafic volcanic rocks, thin discontinuous clastic sediments and still younger mafic volcanics. In the west part of the basin, the lowest volcanic unit is overlain successively by a great thickness of clastic sediments with intercalated iron formation and younger mafic volcanic unit. In common with many Archaean volcanic assemblages, the rocks here are intermediate between the tholeiitic and calc-alkaline series. They correspond with modern island-arc and adjacent oceanic crust assemblages. The felsic pyroclastic formations represent highly explosive eruptions, and together with their coeval plutons they represent the sites of clusters of vents that rapidly produced thick volcanic piles.

Iron formations occur in volcanic rocks in the central and eastern part of the basin. Typically they occur at the prominent contacts between felsic eruptive masses and mafic flows. In the west, the lateral equivalents are enclosed within clastic sediments.

The iron formation of volcanic association contains a threefold arrangement of, in descending order, banded chert, sulphide and carbonate members (Helen and Goudreau sections). The banded chert is commonly 30–60 m thick, reaching a maximum of 304 m at Helen range. The sulphide member is commonly 3–9 m thick but attains a maximum of 36 m at Goudreau (*Figure 9.3*). The carbonate member is lenticular, composed mainly of siderite with minor pyrite, magnetite, pyrrhotite and silicates, commonly less than 60 m and reaching a maximum of 106 m at the Helen iron range. It is sideritic in the basin centre but becomes calcareous eastwards.

The iron formation of sedimentary association is of thick cherty units composed of thinly interbanded chert, siliceous magnetite and jasper (haematitic chert). Individual zones may reach a thickness of 180 m. It is typically intercalated with greywacke and shale.

The iron facies are gradational across the basin from oxide, through carbonate, to sulphide facies (*Figure 9.3*). It is apparent that the oxide facies is largely coincident with a shallower-water sedimentary environment, which is partly of conglomerate in the west. The carbonate facies is of deeper-water association, and the sulphide facies of the deepest water towards the east.

The iron formations are viewed as direct volcanic products because: the major iron concentration (siderite and pyrite) overlie the thickest felsic volcanic piles, hence the iron is thought to be exhalative into the water from felsic vents; iron was deposited rapidly at the end of a mafic-to-felsic volcanic cycle implying an end stage of igneous differentiation as the source of the iron; and deposition of the iron formation was associated with pervasive leaching of silica from the freshly deposited felsic pyroclastic rocks to form chert.

It is important to note that the predominantly cherty iron oxide facies, which is the most commonly recognized type of iron formation, is best developed in the sedimentary environment to the west, on the slopes of the basin some distance from the volcanic-exhalative sources. The transition from formations directly associated with felsic volcanism to those laterally associated with a sedimentary sequence is an important clue to the Proterozoic banded-iron formations where no direct association with felsic volcanic rocks can be demonstrated.

Synvolcanic Nickel Sulphide Deposits

Nickel sulphide deposits occur in Archaean greenstone belts, in differentiated mafic intrusions into cratonic areas, and in ophiolite complexes (Naldrett and Cabri, 1976). Here, synvolcanic deposits that are largely restricted to Archaean greenstone belts are of concern. Within these belts, nickel sulphide deposits may be associated with komatiitic volcanic flows, tholeiitic volcanic flows, or some deposits may not have any clear komatiitic or tholeiitic parentage, probably because tectonism and metamorphism have caused variable amounts of redistribution of the ores. A detailed review of the deposits was given by Naldrett (1973).

The volcanic-associated iron–nickel–copper sulphide deposits are associated with ultramafic flows or shallow sills within sequences of komatiitic and associated tholeiitic volcanic and volcanogenic metasedimentary rocks (Groves *et al.*, 1979). They are equated with the volcanic-type ores of Barrett *et al.* (1977) and Naldrett (1973).

The sulphide ores occur at or towards the base of peridotitic komatiitic lenses, of thickness up to about 150 m but generally 30–50 m thick. They most probably represent the basal portions of the ultramafic lava flows. This is well documented by Muir and Comba (1979) for the deposits of the Abitibi greenstone belt of Ontario, northeast of the Great Lakes (*Figure 9.4*). In the Dundonald deposit of Abitibi, the extrusion of peridotitic lava was interspersed with sedimentation of carbonaceous pyritic chert and lapilli tuff. An immiscible nickeliferous sulphide melt existed as droplets carried in suspension within some of the flows. Subsequently the sulphide droplets were concentrated by a combination of gravitational settling and by entrapment within depressions or structural embayments on the palaeo-surface on which the lava was extruded (Muir and Comba, 1979). The sulphide melt concentrated and formed a liquid layer that ran along and formed the base of the flowing magma. A portion of the sulphide melt descended into the porous underlying volcano–sedimentary formation to form what is called 'inter-flow ore'. Although most komatiitic nickel sulphide ore occurs at the base of a flow, some occurs at the base of what have been described as sill-like feeders to such flows. A typical mineralized flow consists of a lower zone of massive sulphides overlain by a zone of net-textured ore comprising olivine crystals and interstitial sulphides, overlain by weakly mineralized peridotite grading upwards into unmineralized peridotite. Heat loss through the bottom of the flow caused the sulphide liquid to freeze upwards thus producing the massive ore zone (Usselman *et al.*, 1979). At the same time olivine accumulated on the sulphide liquid owing to heat loss through the top of the flow. The ever-growing olivine cumulus pile forced some of the olivine crystals into the sulphide liquid resulting in the net-textured ore (the so-called 'billiard ball model').

The Western Australia Deposits

The discovery at Kambalda led to a realization that this was a new kind of nickel–copper deposit, and subsequent exploration led to the discovery of similar deposits at Windarra, Wannaway, Nepean and other places in the general Kargoorlie region of Western Australia (*Figure 9.4*). All the deposits are related to highly magnesian phenocryst-rich komatiitic ultramafic flows, or subvolcanic sills, occurring within greenstone belts. Despite an amphibolite facies metamorphic overprint, many primary features can be seen (Groves *et al.*, 1979). Lack of wall-rock alteration beneath the basal massive ores and comparative geochemistry of massive ores and along-strike volcanic-exhalative metasediments indicate that most ores did not form by an exhalative process. Ore association with the ultramafic units, consistent nickel–copper ratios of 10–15, abundant platinum-group elements and contained ferrochromites indicate initial crystallization of sulphides and spinels from immiscible oxysulphide liquids. The sulphur contents of the mineralized and unmineralized ultramafic units, and the isotope ratios, favour a model of eruption of phenocryst-rich melt carrying sulphide droplets, which concentrated by gravity settling. Metamorphic modification and remobilization of the ores are common features.

The Kambalda Deposit

The ores and host rocks have undergone lower amphibolite facies metamorphism (Barrett *et al.*, 1977) and are cut by late-metamorphic felsic dykes. Most ore shoots occur at the base of the ultramafic sequence, in contact with the footwall

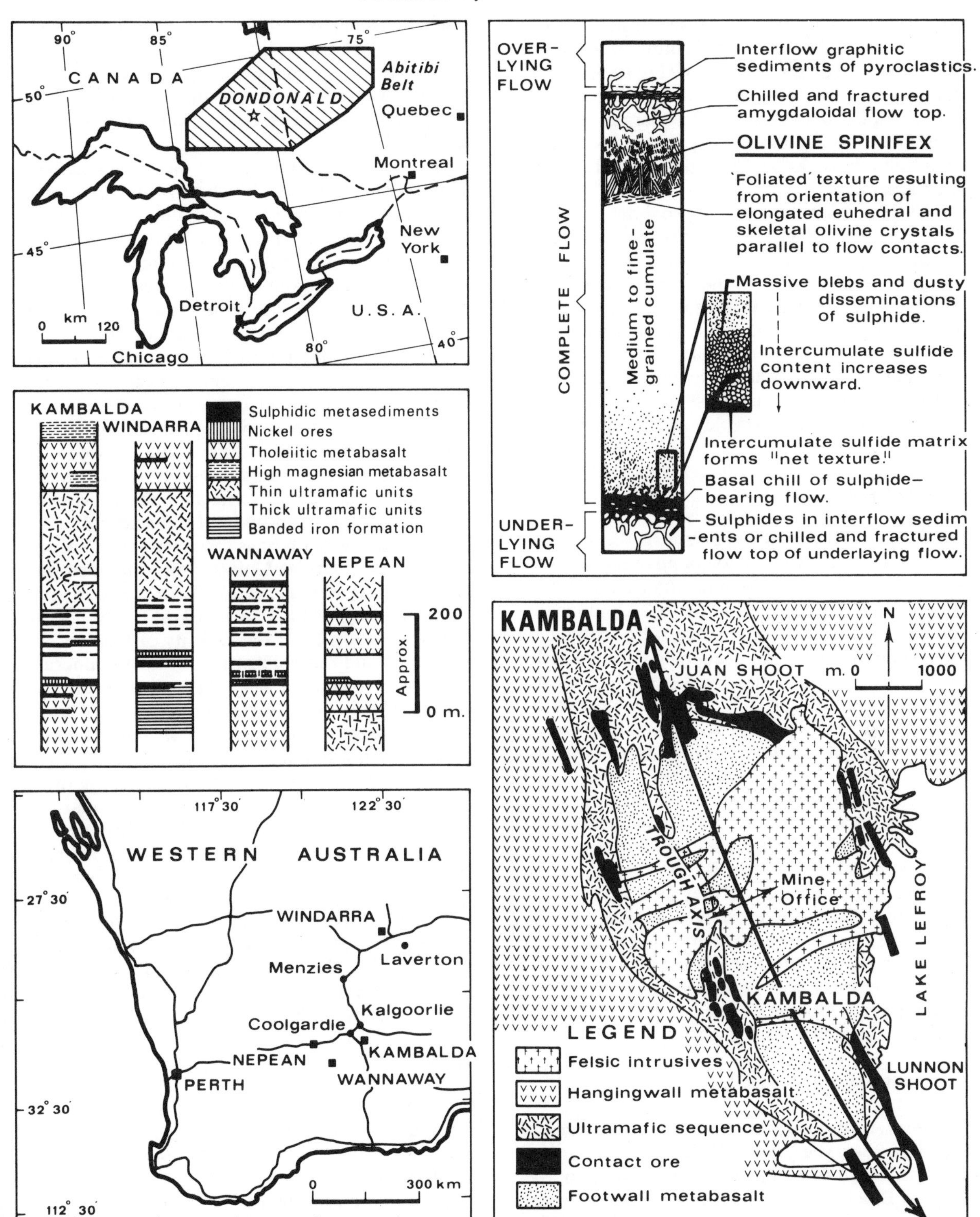

Figure 9.4. Typical stratigraphic sequences of Archaean greenstone belts that contain synvolcanic nickel sulphide ore bodies in western Australia (after Groves *et al.*, 1979) and in the Abitibi greenstone belt deposit of Dundonald (after Muir and Comba, 1979). Simplified geological map of the Kambalda deposit of Western Australia from Barrett *et al.* (1977).

metabasalts, and normally consist of massive ore overlain by a thicker and more continuous layer of matrix ore and disseminated ore (*Figure 9.4*). Less than 10% of the total contact ore has been displaced along shears or at the margin of felsic intrusions to positions entirely within metabasalt or ultramafic rocks and/or faulted contacts between them.

The Juan Main Shoot is approximately 800 m long and up to 200 m wide and is confined by faults or pinch-out structures into the wall rocks. The ore sequence is 0.5–210 m of massive ore, composed of pyrrhotite, pentlandite, pyrite and chalcopyrite, commonly overlain by matrix or disseminated sulphides in a talc–carbonate matrix. The ores have undergone a complex tectonic and metamorphic history in which the ductile behaviour of massive ores and sulphur diffusion played a prominent role (Barrett *et al.*, 1977).

Gold in Greenstone Belts

Gold mineralization is important in the Archaean and shows restrictions of occurrence that appear to be related to the maximum temperature attained in the crust (Watson, 1976). Although the source of the gold is now universally thought to have been the mafic and ultramafic volcanic rocks of the greenstone belts, the workable deposits were concentrated after the changes induced by the rise of granites into the surrounding greenstone belts. The optimal thermal conditions for remobilization are thought to have been 450–550°C (upper greenschist facies), while a lower temperature of 300–400°C favoured subsequent redeposition in quartz lodes and veins. Insofar as crustal temperature is related to depth, the presence or absence of gold may be determined by depth of subsequent erosion (Laznicka, 1973). However, there is little correlation between age of a deposit and depth of erosion. Many Archaean greenstone belts have been subjected to only a few kilometres of erosion of their overburden since they were formed.

WESTERN AUSTRALIA

Western Australia consists of two major shields, the Pilbara and the Yilgarn Blocks. Granites in the Pilbara Block and along the western margin of the Yilgarn Block are 3100–3000 Ma old. However, the important mineralized eastern margin of the Yilgarn Block is related to a widespread metamorphic episode at 2600–2700 Ma (Binns *et al.*, 1976). Both blocks are mainly of acid gneisses and granite with narrow elongate troughs, or greenstone belts, filled with acid, basic and ultrabasic komatiitic volcanic rocks. The metamorphic grade varies from prehnite–pumpellyite to mid-amphibolite facies. Most of the ore bodies occur within narrow zones of the volcanic rocks (Warren, 1973).

Gold previously dominated the mining activity in both the Archaean blocks. In the Kalgoorlie region, the most important mining centre (*Figure 9.4*), the gold is associated with acid dykes and sills, and occurs in gold–sulphide, gold–quartz and gold–telluride lodes. The Kalgoorlie Golden Mile is proposed as an Archaean volcanogenic-type deposit, modified and redistributed along later structures (Tomich, 1974). Rich pyrite–telluride and weakly pyritic quartz–carbonate gold ores are associated with the end phase of basic volcanism and generally confined to the broad interface zone between the submarine volcanic formations and the later acid volcanic rocks. There is virtually no ore associations within the main body of the acid volcanics. There are all gradations between massive and banded base-metal sulphides, vein sulphides, and disseminated sulphides, just as in the Canadian Archaean deposits (Tomich, 1976). The range of ore-type results from the severity of tectonic deformation and remobilization of the ore minerals. The basic volcanic rocks, which contain the disseminated sulphides, are therefore regarded as the original host rock for both the base metals and the gold-bearing sulphides and tellurides.

The geological setting of the lodes in other mining areas of the Yilgarn Block is similar, although a gold–arsenic association is also common to the north. In the Pilbara Block most of the gold occurs in quartzose lodes, commonly with pyrite, but in the east antimony is also common (Warren, 1973). It is generally held that the gold and the base metals were originally in the basic and ultrabasic submarine volcanic formations, mobilized and redistributed during metamorphic and tectonic evolution of the greenstone belts.

CANADIAN SHIELD

Canada has a significant gold production, about three-quarters of which is from gold–quartz veins and the rest from associated volcanogenic massive sulphide deposits (*see* p. 207).

The most important ore-mining region is around Abitibi Lake, northern Ontario, in the district of Porcupine (48° 30′ N; 81° W), 200 km north of Sudbury. At Porcupine, the gold, silver and scheelite occur in quartz veins within altered Keewatin lavas and lesser amounts within quartz porphyry.

The Keewatin lavas form the oldest rocks of the region; they are mainly of andesite and rhyolite and are conspicuously associated with banded-iron formation and tuff. Most of the ore is associated with andesite, dacite and basalt, some of which is spherulitic and pillowed (Dunbar, 1948).

Algoman intrusive rocks include granite, quartz porphyry and albitite. The quartz veins that carry gold, silver and scheelite also contain pyrite and pyrrhotite.

Out of a total of 36 mines in the Porcupine region, Hollinger Mine accounts for nearly half the total production; its geology has been described by Jones, W.A. (1948). From 1909 to 1945 the mine produced 424 t of gold and 81 t of silver. The ore bodies are of quartz veins, stringers, lodes and stockworks. The veins range from 30 cm to 21 m width and average 3 m. The host rock is of tightly folded Archaean basic-to-intermediate lava flows of Keewatin age; they are schistose and hydrothermally altered (*Figure 9.5*). Lens-shaped porphyry stocks intrude the lavas. The porphyry is also schistose and is therefore pre-metamorphism. A series of prominent fault planes strike northeast.

The rocks in the main-ore zone are highly schistose and hydrothermally altered. The strong regional tectonic deformation makes it difficult to differentiate the alteration events. The lavas in the ore zone are rich in chlorite, carbonate and muscovite. Three periods of mineralization have been identified (Jones, W.A., 1948): the quartz–ankerite vein system followed by quartz–scheelite–tourmaline associated with albite; the main quartz–ankerite system with which the sulphides arsenopyrite, pyrrhotite, chalcopyrite, sphalerite, galena and the tellurides of gold are associated; and a feeble quartz–calcite phase later than the gold mineralization.

A disseminated copper deposit occurs in a lens of felsic schist of volcanic origin in the Timmins area about 1.5 km northeast of the Hollinger Mine (Davies, J.F. and Luhta, 1978). The copper–molybdenum–gold mineralization and hydrothermal alteration have many characteristics of Tertiary porphyry-copper deposits. However, the host rock appears to be entirely within a volcanic sequence of the greenstone belt and no intrusive source can be identified, hence any similarity to a porphyry-copper deposit is accidental. The hydrothermal alteration associated with the mineralization is characteristic of most greenstone belts and may be related to the regional metamorphic events that mobilized the metals in and from their original volcanic host rocks.

About 170 km east of Porcupine, the Noranda region contains both massive sulphide deposits and gold-bearing quartz-veins. The former are by far the most important; they contain pyrite, pyrrhotite, chalcopyrite, sphalerite and gold in volcanic breccia and tuffs directly associated with rhyolite and andesite (Goodwin, 1965). Situated 120 km southeast of Porcupine, the Kirkland Lake area contains gold in fissure quartz veins and stockworks in granite, in shear zones and in rhyolite and felsite tuff. Argentiferous galena is associated with sphalerite, quartz and calcite at volcanic–sedimentary contacts. In the central part of the area a commercial banded-iron formation is also present (Goodwin, 1965).

The majority of the mineral deposits in the Porcupine, Kirkland Lake and Noranda region are concentrated in volcanic rocks. The metals are considered to have been originally hosted in the volcanic rocks, and later remobilized by tectonic and metamorphic events.

SOUTHERN AFRICA

Gold mineralization is preferentially developed in the volcanic-rich units and in the banded-iron formation of the Archaean greenstone belts (*Figure 9.5*). The basaltic and peridotitic volcanic rocks of the Lower Ultramafic Unit were the primary host of most of the gold mineralization of the Southern African greenstone belts (Anhaeusser, 1976a). The subsequent concentration of gold deposits in areas rimming granite intrusions must have come about from remobilization that resulted from the heat effects of the granites during regional metamorphism and tectonism.

The Zimbabwe and the Kaapvaal Cratons (*Figure 9.1*) have yielded radiometric ages within the range 2600–3600 Ma and the Great Dyke is 2530 Ma old.

The stratigraphy of the greenstone belts may be simply stated as a Lower Ultramafic unit (Sebakwian), superimposed by a mafic-to-felsic unit (Bulawayan), then a Sedimentary unit (Schamvaian), which changes upwards from argillaceous to arenaceous. The distribution of gold in these groups has been reviewed by Anhaeusser (1976a,b) and Fripp (1976a,b). Of the 100 larger mines in Zimbabwe (*Figure 9.5*), thirty are of stratiform gold mineralization associated predominantly with the Sebakwian group; seven are in more massive but stratiform sulphide deposits, with no stratigraphic restriction; fifty-six are in quartz lodes, predominantly in Bulawayan and Shamvaian groups, but also occurring in the Sebakwian group; and seven are strata-bound disseminated deposits, predominantly in the Bulawayan and Shamvaian groups (Fripp, 1976a).

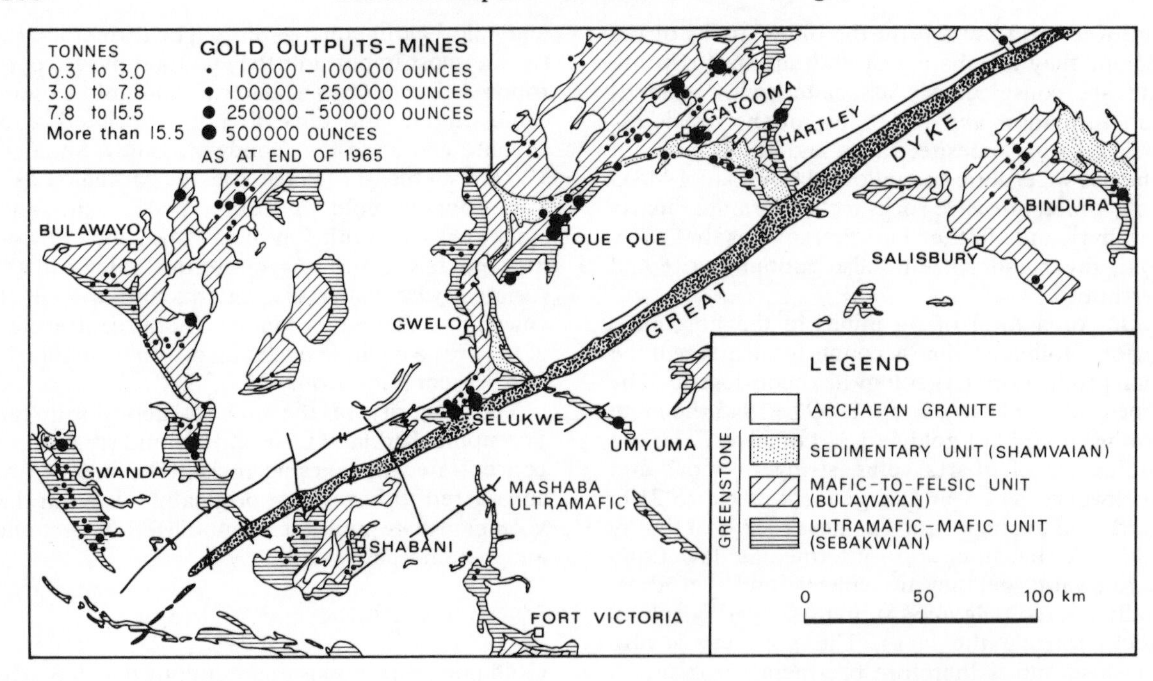

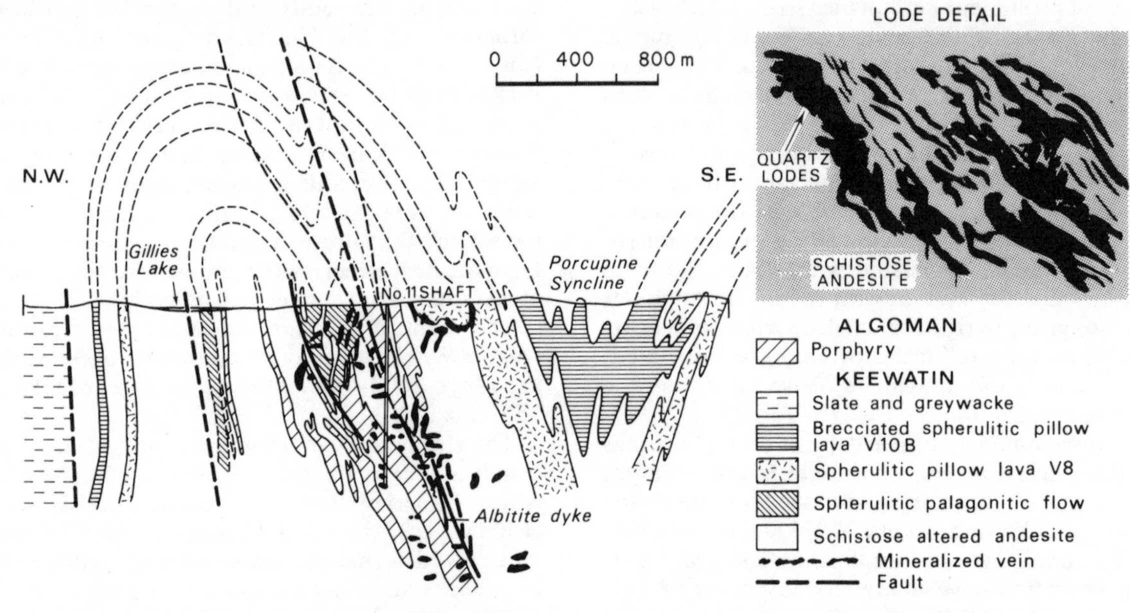

Figure 9.5. Bottom: geology of the Hollinger Mine of the Porcupine region of the Abitibi greenstone belt of Ontario (after Jones, W.A., 1948). Top: distribution of gold mines in the greenstone belts of part of the Zimbabwe Craton (from Anhaeusser, 1976b).

Stratiform deposits

The gold occurs in several thin beds of banded-iron formation that are interlayered with mafic and felsic water-deposited tuffs. The ore bodies are stratiform and confined to sulphide beds and mixed sulphide–carbonate facies of the iron formation, which consists of layers rich in quartz (chert), arsenopyrite, pyrrhotite and ankerite. The gold occurs as minute 50-μm grains of native gold within arsenopyrite crystals and the ore grade averages 11 ppm ($g\,t^{-1}$). Most of the deposits occur within the Sebakwian group. Individual gold-bearing beds are generally less than 5 m thick. In addition to gold, the stratiform iron formations contain copper, lead and silver. The typical ore is of alternating layers of well-layered carbonaceous chert and sulphides. In some mines the layers are folded. The gold and the sulphides are definitely pre-metamorphic and pre-folding and of synsedimentary origin. A volcanogenic model is favoured for the banded-iron formation (Fripp, 1976b), in which the deposits are considered to be submarine chemical precipitates deposited from active fumaroles.

Strata-bound massive sulphides

Although rare, there are a number of strata-bound gold deposits comprised of intermixed cherty sulphidic and chloritic material up to 5 m thickness containing evidence of soft sediment deformation.

Lode deposits

This is by far the most important type. The veins vary from less than 1 m to about 5 m and extend along strike as far as 2 km. They are of quartz, accompanied by lesser carbonate and sulphides, mainly pyrite with lesser pyrrhotite, sphalerite, galena, arsenopyrite, stibnite, chalcopyrite and scheelite. The gold is coarse grained. Gold tellurides and aurostibnite may be present. The gold is usually associated with the pyrite, chalcopyrite, stibnite and galena.

The vein wall rocks have been propylitized, extending from less than 1 m to several metres from the veins. The common alteration minerals in the mafic rocks are chlorite, epidote, tremolite, albite and some disseminated sulphides have been deposited. Most of the veins occur within thick mafic–felsic volcanic rocks of the Bulawayan unit. The distribution is structurally controlled, usually parallel to the axial planes of the F_1 deformation. Many veins are also folded and boudinaged. It is clear that hydrothermal solutions have been responsible for carrying the gold and the sulphides from their original host rocks into the vein systems. Metamorphism of the greenstone belts in the upper greenschist facies is considered conducive to remobilization of gold and final deposition is considered to have taken place at 300–400°C. Granites, intrusive into the greenstone belts, have provided suitable local heat sources to drive hydrothermal systems, and several mines in the Zimbabwe Craton are concentrated around granite stocks.

Folding and faulting of the greenstone terrain has had a powerful control on the final deposition of the quartz–gold lodes and veins. Since the rocks have generally undergone several periods of deformation, the fissure systems are complex and vary from mine to mine.

Strata-bound disseminated mineralization

Some mines extract clastic sedimentary rock that contains a few percent of disseminated pyrite and arsenopyrite. Such deposits are relatively thick and are mostly in the Bulawayan and Shamvaian units.

Massive Base-metal Sulphides

The latest studies of the massive pyritic base-metal sulphide deposits in volcanic rocks suggest that they are of volcanogenic origin, formed in recurrent episodes of sea-floor fumarolic activity during prolonged periods of subaqueous volcanism (Hutchinson, 1973). Such deposits are common throughout the geological history of the earth. The type that was common in the Archaean is the zinc–copper pyrite; this contains only minor amounts of lead but has important gold and silver contents, usually accompanying the zinc-rich sulphides. Gold is relatively more important than silver. However, the relative distribution of gold and silver is erratic and cannot be explained. The host rocks of the massive sulphides may range from basalt to rhyolite, having affinities intermediate between the tholeiitic and calc-alkaline series (Hutchinson, 1973). The lavas were deposited in successions ranging up to 12 000 m in thickness. At the base, the successions are of broad basaltic platforms, probably formed from widespread fissure eruptions from deep fissures into the mantle. Subsequent volcanism became increasingly felsic and domal in structure, and the sulphide deposits show a genetic relationship to volcanic domes and fumarolic breccia pipes. The intercalated sediments include cherts and the various facies of iron formations, siliceous tuffs and immature volcanogenic greywackes. The rock successions are tightly folded and usually metamorphosed to greenschist facies.

The best examples are the great base-metal deposits of the Keewatin greenstone belts in the Lake Superior province of the Canadian Shield including the Noranda deposit of the Abitibi greenstone belt (*Figure 9.6*).

Massive sulphide deposits of pyrite and pyrrhotite are relatively few in Southern Africa (Anhaeusser, 1976a). The Iron Duke mine north of Salisbury and other deposits in the Shamva–Salisbury greenstone belt represent the

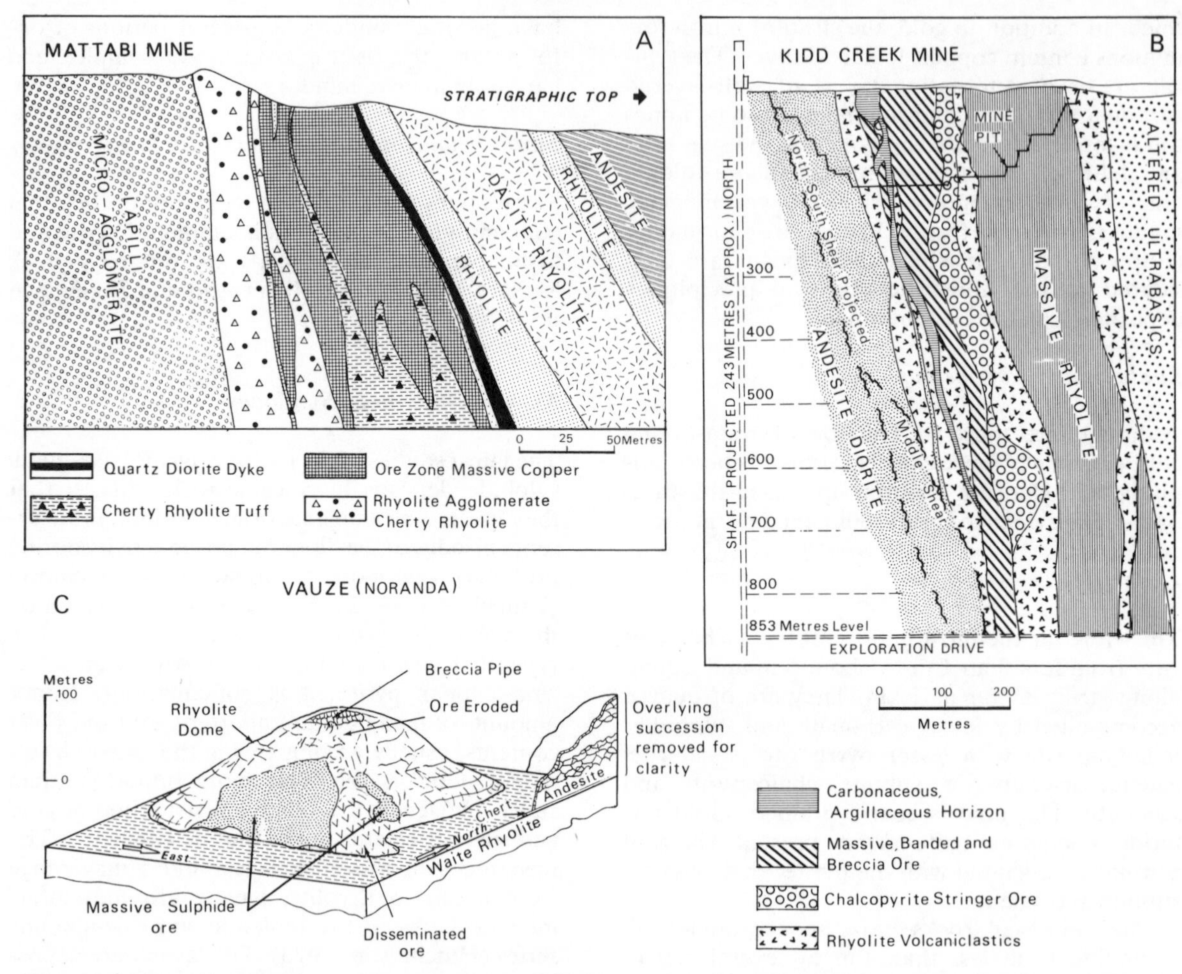

Figure 9.6. Cross-sections through two important Archaean Canadian Cu–Zn–Fe massive sulphide deposits at: (A) Mattabi, 180 km north of Lake Superior; (B) Kidd Creek mine, Noranda district of the Abitibi greenstone belt on the Ontario–Quebec border. (C) Detailed location of the massive sulphide deposit on the rhyolite dome after the strata have been restored to horizontal to show the original setting, and after removing the overlying unmineralized andesite (Vauze mine, Noranda District). The ore is considered exhalative from a fumarole breccia pipe and to have flowed down the steep flank of the submarine volcanic dome (after Franklin *et al.*, 1975; Spence, 1975; Walker *et al.*, 1975).

product of volcanic-exhalative activity during the deposition of the Shamvaian unit.

Abitibi Belt

The Abitibi greenstone belt, which straddles the Ontario–Quebec boundary of Canada, is the largest single greenstone terrain of the Canadian Shield. The felsic volcanic units contain all the productive massive base-metal sulphide deposits.

The best-known occur at Noranda, summarized by Spence and De Rosen-Spence (1975). Massive copper–zinc sulphide deposits overlie chloritic pipes of alteration and occur at or near the top of rhyolitic formations and many are associated with primary volcanic features such as lava domes and explosive breccias (*Figure 9.6*). They show a zoning of copper/zinc ratios and there is evidence of fragmentation of some massive sulphides prior to their having been covered by later flows of andesite or rhyolite. They are attributed to submarine volcanogenic processes that formed sulphide sinters over hot springs. There are three main types of massive sulphide: pyrrhotite–pyrite–chalcopyrite–sphalerite, gold and silver; pyrite–sphalerite, with gold and silver; and pyrite bodies with low zinc, copper, gold and silver contents.

The Vauze massive sulphide deposit of the Noranda area (*Figure 9.6*) has an underlying zone of disseminated sulphide, and it lies at the same andesite–rhyolite contact as the neighbouring deposits in the Noranda district (Spence, 1975). The breccia pipes in the rhyolite suggest that the sulphide bodies are a result of explosion from the pipe area. The sulphides must have formed as hard sulphide sinters in restricted lenses on the steep sides of a volcanic dome. The deposits were later covered by subsequent andesite flows.

The Kid Creek massive sulphide deposit (*Figure 9.6*) lies concordantly within a steeply dipping rhyolitic volcaniclastic pile, which is overlain by unmineralized andesite. The ore contains 9.75% zinc, 1.52% copper, 0.40% lead and 134.4 g t^{-1} silver. The silver occurs native, and within accessory acanthite, tetrahedrite and tennantite. Tin may reach 3% locally, occurring as cassiterite. The ore is syngenetic sedimentary and related to explosive subaqueous rhyolite volcanism.

The Mattagami Lake ore body (195 km northeast of Noranda) is a strata-bound massive iron–zinc–copper sulphide deposit of Archaean age within rhyodocite and tuff, and overlain by andesite (Roberts, 1975). The ore is mainly of pyrite and sphalerite, with some pyrrhotite and magnetite. The enclosing volcanic rocks are in greenschist facies and have been folded. The primary sedimentary features of the layered pyrite–sphalerite ore suggest that it has been emplaced by volcanic–exhalative processes in a subsiding volcano–sedimentary basin during the final stages of rhyolitic activity.

Other Canadian Deposits

The Mattabi massive sulphide ore body (180 km north of Fort William and the north coast of Lake Superior) occurs conformably in steeply dipping, moderately metamorphosed, Archaean rhyolite (Franklin *et al.*, 1975). Rhyolite agglomerate and dolomitic lapilli tuff underlie the ore body (*Figure 9.6*). Hydrothermal alteration in the underlying rhyolite agglomerate includes vein-type chalcopyrite–pyrite ore. The alteration zone includes siderite and chloritoid. Both copper and zinc occur in the siderite alteration zone. The ore body contains 11 700 000 t, with 7.6% zinc, 0.91% copper, 0.84% lead, 97.8 g t^{-1} silver and 0.2 g t^{-1} gold. The deposit differs from those at Noranda in the abundance of siderite and the paucity of chlorite. The siderite is probably due to a replacement of dolomite by rising metal-rich aqueous solutions. The zonal arrangement of chlorite and silicification is similar to that in the Japanese stockwork Kuroko deposits.

The Flin Flon massive zinc–copper–iron deposit (on the Saskatchewan–Manitoba border at 54° 46′ N) occurs in rhyolite and dacite, overlain by andesite (Koo and Mossman, 1975). Disseminated ore occurs in andesite breccia beneath the massive ore. The ore bodies are pyritic, with sphalerite and chalcopyrite. The ore deposition is pre-greenschist facies metamorphism. Relict structures indicate that they were syngenetic with the submarine acid volcanic flows and deposited along with the volcanic formations and the minor associated sedimentary rocks, and later folded and metamorphosed.

The Cobalt–Gowganda Copper–Nickel–Arsenic–Silver Ore Region, Ontario

The Archaean Keewatin intermediate-to-mafic flows and interstratified sediments contain strata-bound base-metal sulphides. The Cobalt–Gowganda area lies 120 km northnorthwest of North Bay, which in turn is 283 km north of Toronto. Pyrite, chalcopyrite, pyrrhotite, sphalerite and galena are the main minerals, with minor arsenopyrite, marcasite and tetrahedrite. Rich concentrations are largely confined to the sediments, usually cherty and commonly carbonaceous, which occur as bands in the Keewatin Archaean volcanic sequence (Jambor, 1971). Some bands contain several percent of sulphides. The deposits are associated with the Archaean volcanism. The deposits were intruded by minor mafic dykes and salic plutons associated with the Kenoran orogeny (2490 Ma ago), but the stratiform character of the Archaean mineralization was not disturbed by them.

Erosion of the Archaean volcano–sedimentary formations dispersed base-metal sulphides into the lower Huronian sedimentary sequence, especially in the basal strata adjacent to the Archaean–Proterozoic unconformity. The sulphides in the laminated sediments show a distinct monomineralic trend in some groups of laminae. Some laminae contain only sphalerite, some galena, and others only yellow sulphides. The amounts of sulphide decrease upwards in the Huronian sequence.

The Archaean and Huronian rocks were intruded 2155 Ma ago by the Nipissing diabase sheets and this resulted in silver ore being mobilized into veins, small fractures, faults and joints that cut the Archaean, Huronian and Nipissing formations. Ore production is greatest when the host rocks are Huronian sediments, but the Nipissing host also gives good ore from its fissures.

Some veins have cobalt contents up to 8.52%,

nickel up to 24.60%, antimony up to 5.50%, arsenic up to 64%, silver up to 9.48%, copper up to 0.21% and iron up to 13.20%. Bismuth is low (less than 0.19%), and uranium is extremely low (less than 0.3 ppm) (Petruk, 1971b). Native silver is the chief ore mineral. It is accompanied by a complex cobalt–nickel–arsenic–silver assemblage in a gangue predominantly of dolomite and calcite. The ore veins are steeply dipping to vertical and vary in width from 1 cm to over 30 cm, with an average of 5 cm. Some veins are over 305 m long, but all the ore in the veins is localized in ore shoots. A length of 60 m and depth of 30 m would be considered a good shoot. Small shoots are often rich in silver. The sequence of mineral deposition in the veins has been described by Petruk (1971a). The gangue of quartz and chlorite was first deposited within rising temperatures from 200°C to 550°C. The early ore minerals are nickel arsenides (Rammelsbergite) at around 550°C. The main sulphide deposition was at 520–470°C. Late sulphides are mainly of silver sulphide deposited below 400°C and down to 220°C. Cross-cutting veins of stephanite–quartz were deposited below 195°C. The late quartz was deposited between 200°C and 145°C. Finally, late sulphides continued to be deposited down to temperatures as low as 60°C.

It is deduced that the vein ore at Cobalt was derived principally from the Keewatin sulphide-rich interflow sediments, with a small contribution from the Archaean flows themselves. Mobilization of the ore elements probably resulted from the high heat influx when the Nipissing diabase sheets were emplaced. Deformation during the emplacement caused fissures that were infilled by the hydrothermal ore solutions (Boyle and Dass, 1971). Here, therefore, is an interesting case of an Archaean volcanogenic ore deposit being later remobilized by epigenetic hydrothermal processes.

Badham (1976) expressed the view that the complex silver–nickel–cobalt arsenide ore-type resulted from the mixing of felsic magma with basic magma formed above a shallow-dipping subduction zone in the interior parts of a plate margin.

Antimony in Greenstone Belts

The antimony-bearing horizon of the Murchison Range greenstone belt of the Kaapvaal Craton is one of the largest antimony deposits of the world. This deposit is apparently unique, for greenstone belts are not known elsewhere for their antimony contents. The Murchison Range association with a volcano–sedimentary sequence in a deformed basin may, however, have analogies elsewhere in the world. The Cretaceous strata-bound stibnite deposit of Schlaining, eastern Alps, occurs mainly within a 6-m thick ore schist horizon, and has been interpreted by Höll and Maucher (1976) as submarine hydrothermal. Höll (1977) and Maucher (1976) have also drawn attention to the association of stibnite with basic submarine volcanics in strata-bound mineral formations elsewhere in Europe, and the greenstone mineralization of South Africa may therefore not be unique in its tectonic setting.

The Murchison Range Antimony Line

The Murchison Range of the eastern Transvaal is a volcano–sedimentary rock assemblage belonging to the Archaean Swaziland Sequence of the eastern Kaapvaal Craton, lying about 400 km northeast of Johannesburg. The geology has been described by Muff (1978). The greenstone belt has a uniform northeast–southwest-trending strike and is surrounded by a basement complex of granites (*Figure 9.7*). Small exposures of aplitic muscovite granite are closely associated with the coarse-grained sedimentary rocks of the Antimony Line. They are highly mylonitized, and hence pre-date the regional deformation and metamorphism.

The Murchison Range greenstone belt underwent strong deformation accompanied by greenschist facies metamorphism, but the grade is slightly higher along the margins bordering the basement granites, and biotite becomes more conspicuous, and is accompanied by porphyroblasts of garnet and kyanite.

The Antimony Line is a mineralized zone consisting of carbonate–schist, talc–schist, cherty carbonate, chert, banded-iron formation and small lenses of quartzite (*Figure 9.7*). It stretches over a distance of approximately 50 km, and has a width ranging from 50 m to 100 m. It is paralleled by quartzites of the Antimony Bar, which runs approximately 50 m north of the mineralized zone.

The antimony-bearing horizon can be traced over a distance of about 20 km. The deposits are restricted to carbonatic rocks, which form a well-defined horizon in the central part of the Range. Interlayered with the carbonate horizons are grey, massive chert bands, siliceous carbonate lenses and cherty carbonate layers. The richest mineralization is associated with siliceous carbonate rocks and the ore is invariably accompanied by quartz and carbonate gangue. The contacts of the carbonate horizons are commonly talcose.

The important deposits of the Alpha and

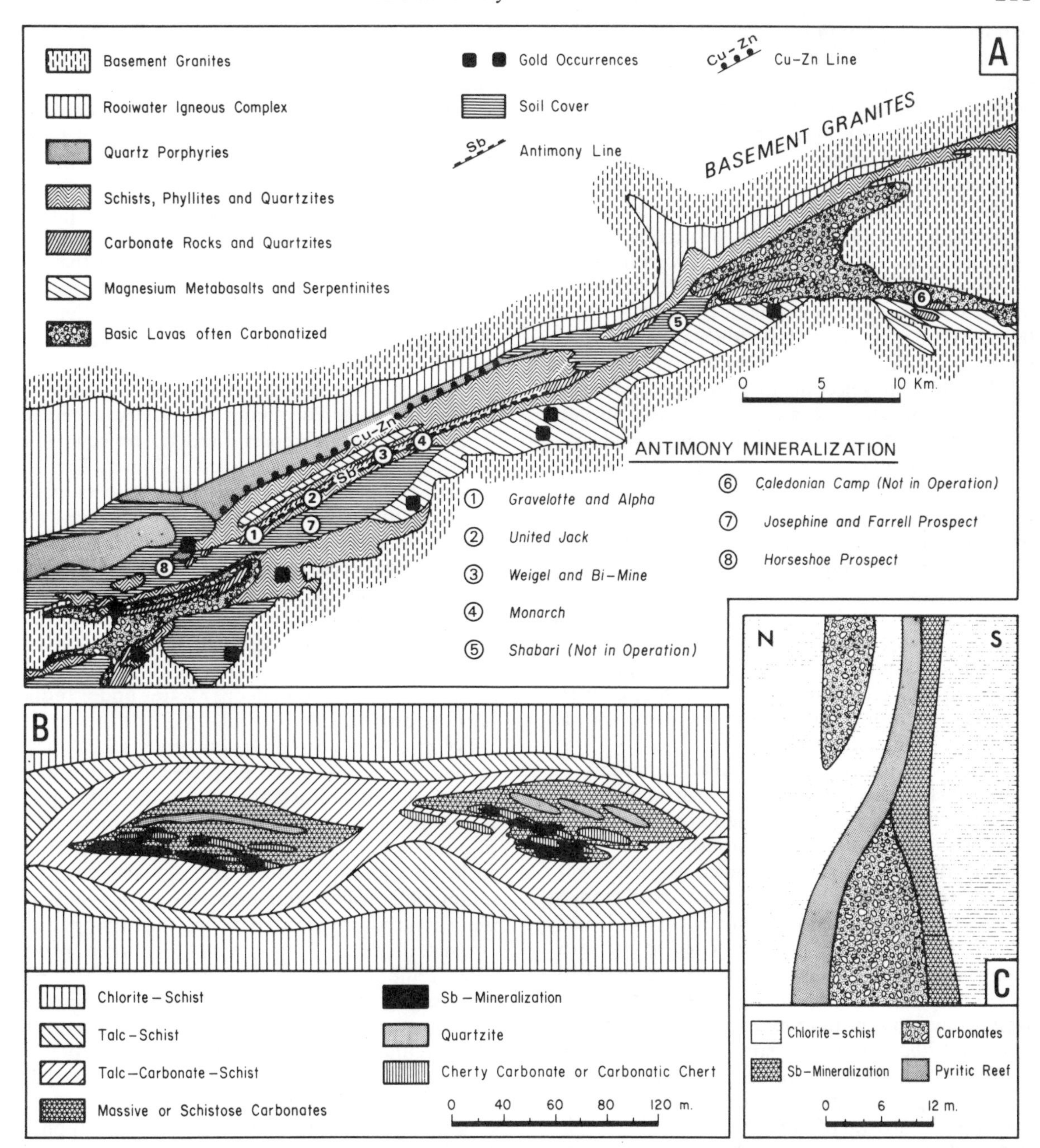

Figure 9.7. The stibnite deposit of the Murchison Range greenstone belt, Kaapvaal Craton, South Africa. (A) Simplified geological map of the Murchison Range; (B) idealized map of a portion of the Antimony Line showing the relationship of the mineralization to cherty carbonate or carbonatic chert rocks; (C) sketch of the Antimony Line at the Weigel Mine, showing the Pyritic Quartz Reef (Tetrahedrite Reef) and the Stibnite Reef (Antimony Reef) (after Muff, 1978).

Gravelotte mines occur near an important parasitic fold in the Antimony Bar. The ore deposits occur in steeply dipping lodes parallel to the formation strike. These mines account for most of South Africa's production, which in 1979 amounted to 11 960 t of metal.

In the Weigel Mine, the Antimony Line consists of two parallel reefs, which in places are in contact (*Figure 9.7*), although they may be separated by up to 15 m. The northern reef is called the Pyritic Quartz Reef and the southern the Antimony Reef. Commonly the Pyritic Quartz Reef consists of banded-iron formation, approximately 1–3 m wide and 20 m long, but in other areas it is represented by white vein quartz. The Stibnite Reef is a narrow zone of chloritic quartz carbonate schist interlayered in carbonatic chlorite schist. Usually several parallel stibnite veins of 1–5 cm width conform to the lithology.

There are three main ore-types in the greenstone belt. (1) Stibnite ore, which is almost monomineralic. It may contain rare native antimony and more common gold, and fills and cements fractures and breccias in the wall rocks of the Antimony Line. (2) Berthierite ore, which contains a wide variety of sulphides. It is mainly of berthierite, and is accompanied by chalcostibnite, ullmanite, gersdorffite and occasionally stibnite and tetrahedrite. (3) An iron sulphide-rich ore, composed of pyrite, arsenopyrite, chalcopyrite, sphalerite, tetrahedrite and gudmundite. There is evidence that much of the berthierite ore was originally of this type, but was subsequently overprinted. The iron sulphide-rich ore is best seen in the iron formation.

All the ore textures are clearly of metamorphic origin. The ore minerals, quartz and carbonate were reconstituted during the regional metamorphism of the greenstone belt. The metals were therefore present in the volcano–sedimentary pile before regional deformation.

The Murchison Range also contains a Copper/Zinc Line, which can be traced along the whole of the quartz porphyry horizon (*Figure 9.7*) and the sulphide mineralization is obviously genetically related to the acid volcanism.

The region has also mined mercury from a mercury–antimony association. Muff (1978) concludes that the metals were enriched in the sediments as submarine hydrothermal ore deposits. The association with mercury is important, and it is well known that both mercury and antimony can be deposited from hot springs (Dickson and Tunell, 1968). Since there are no unconformities within the greenstone volcano–sedimentary pile, the deposits are not of subaerial hot-spring origin, but rather are submarine–hydrothermal, and probably genetically related to the igneous activity represented by the quartz porphyry or the basic lavas.

Model of Archaean Mineralization

There is general acceptance of the volcanogenic–exhalative origin of Archaean mineral deposits within the marine basins of greenstone belts. The model shown in Figure 9.8 is based on Goodwin and Ridler (1970), Hutchinson *et al.* (1971), Anhaeusser (1976b) and Fripp (1976a,b).

The early ultramafic–mafic unit of a greenstone belt is known to be the ultimate source of nickel sulphides, for they occur at the base of komatiitic flows as settled accumulations of immiscible sulphide liquids. They may have subsequently mobilized into fissures produced during the folding of the greenstone belts. Most of the gold is undoubtedly ultimately from the mafic–ultramafic unit, for gold occurs within olivine in the Kaapvaal Craton. However, it has largely been mobilized during later events.

Most of the volcanic-exhalative deposits are related to explosive submarine activity, especially to fumaroles now represented by breccia pipes forming the tops of felsic volcano domes. The iron formations and the associated stratiform gold–silver sulphide deposits are linked genetically to these exhalative processes. Deposits close to the exhalative centres (proximal) gave rise to an oxide facies, whereas the distal deposits have no clear genetic relationship to the volcanic rocks and occur associated with sedimentary formations such as carbonate and clastic rocks. They are of sulphide facies, formed in deeper water under reducing conditions.

Folding and accompanying greenschist facies metamorphism and the emplacement of granite stocks played important roles in the subsequent remobilization, especially of the gold deposits. The plutons acted as heat centres that drove large-scale circulation of brine through the volcano–sedimentary pile, later depositing gold and sulphides in fissures created during the folding of the greenstone belts. Optimum mobilization occurred when temperatures were in excess of 500°C and optimum deposition in lodes and quartz veins occurred when temperatures waned below 400°C. The circulation of hot brine created a situation analogous to Phanerozoic porphyry-copper deposition, with extensive propylitization of the country rocks. Thus many of the gold and sulphide deposits are truly epigenetic, though their actual metal source, in particular the gold, is ultimately thought to be the mafic–ultramafic unit of the greenstone belts.

The massive zinc–copper–iron sulphide deposits, associated with silver and/or gold, are cogenetic with fumarole activity associated with the felsic explosive phases of the greenstone belts. Sulphide exhalations accumulated as sinters at the vents or plastered the steep slopes of the lava or pyroclastic submarine domes, later to be covered over predominantly by andesite flows.

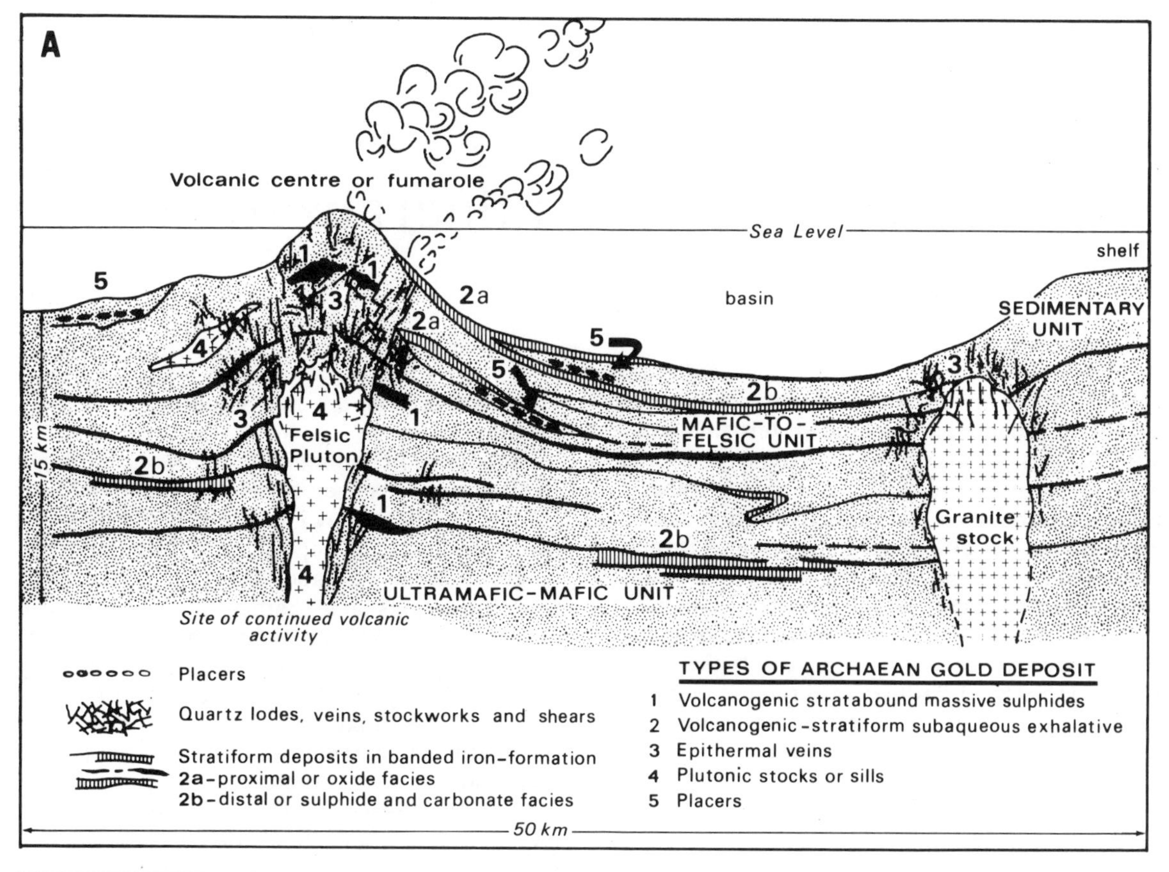

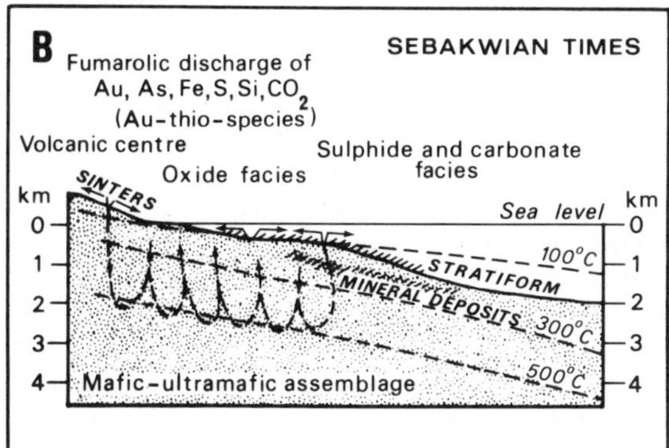

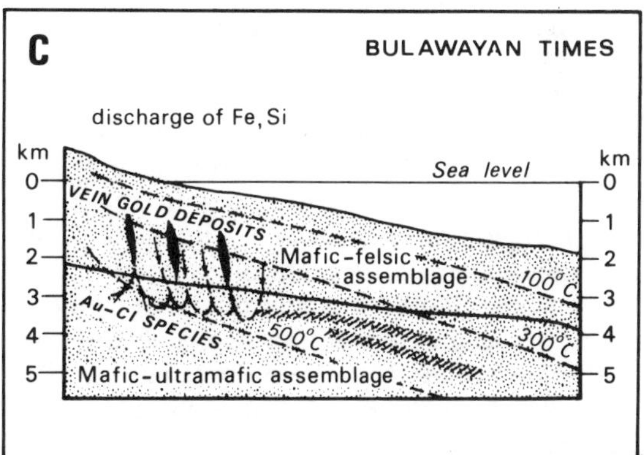

Figure 9.8. (A) Schematic diagram of an Archaean volcano–sedimentary greenstone belt basin showing the possible relations of mineralization to the volcanic-exhalative activity (from Anhaeusser, 1976b). (B) Model of stratiform mineral deposition of gold and sulphides associated with iron formation in the early phases of a greenstone belt. (C) Later lode or vein deposition in fractures formed by folding of the greenstone belt. The convecting brines reached their optimum depositional stage for gold at less than 400°C (after Fripp, 1976a,b).

10

Proterozoic-style mineralization

Iron formations and massive or stratiform sulphide deposits are characteristic of the Proterozoic. They differ from such deposits in the Archaean in that often there is no direct association with volcanic formations, though an indirect association can be invoked with volcanic rocks in contiguous formations. They occur in large epicontinental basins close to Archaean cratons and are assumed to have a volcanic-exhalative origin, though more distal from the eruptive centres than the typical Archaean deposits. Hydrothermal uranium deposits typically occur within high-grade metasediments associated with major unconformities.

Iron Formations

In his palaeo-environmental classification of iron formations, Kimberley (1978b) noted that the vast majority of early Proterozoic deposits may be categorized as metazoan-poor, extensive, chemical sediment-rich, shallow-sea iron formations. They are typified by occurrences in the Lake Superior region of Ontario and Minnesota. They are closely allied petrographically and compositionally to the shallow volcanic platform iron formations that characterized the Archaean and that are directly related to greenstone belts. The Proterozoic deposits probably have a similar genetic relationship, but are distal, or far removed from the volcanic centres, hence their volcanic connection is implied rather than demonstrable (Eichler, 1976).

The great iron formations of the world appear to have been deposited sometime between 1900 Ma and 2500 Ma ago (Goldich, 1973). This was the greatest epoch of iron deposition that the world has known. These Proterozoic formations occur worldwide; their thicknesses commonly exceed 100 m and in Western Australia and southwest USSR they exceed 2000 m. They are generally laterally extensive. In eastern Canada they can be correlated for a distance of 1100 km. However, a few are really quite restricted in extent.

Pronounced banding in the ironstone is a highly characteristic feature; the average band thickness is 1 cm. Oolitic texture is common in North American deposits, but is not universal. Mineral assemblages in diagenetic and low-grade metamorphic iron formations are characterized generally by fine-grained quartz or chert, iron phyllosilicates (greenalite, chamosite, chlorite, minnesotaite and stilpnomelane), iron-bearing carbonates (siderite, ferrodolomite, ankerite and calcite), iron oxides (haematite and magnetite), iron sulphides (pyrite) and sodium–iron amphiboles (riebeckite and crocidolite) (French, 1973). The grunerite is of metamorphic origin. Bands rich in these minerals typically alternate with recrystallized chert (Floran and Papike, 1975).

The generalized palaeo-environment is thought to be a shallow continental shelf with minimal relief, adjacent to a low-lying continent with intervening terrigenous sediment traps. The typical great areal extent is inconsistent with a model of lacustrine environment. Moreover most sedimentary rock associations are seemingly marine and the successions commonly include dolomite and chert.

The best-studied Lower Proterozoic deposit is that of the Mesabi and Gunflint Iron Formations, west of Lake Superior, which, in addition to its great economic value, is relatively unmetamorphosed and their primary sedimentary features are well preserved.

The Lake Superior-type of iron formation, characteristic of the Proterozoic, forms blanket deposits in epicontinental marine environments along the margins of old cratonic areas and they are generally devoid of volcanic rocks (Eichler, 1976).

Biwabik–Gunflint Iron Formation

The Biwabik Iron Formation of the Mesabi district of Minnesota and its continuation into Ontario as the Gunflint Iron Formation, has a strike length of 320 km. Near the Canadian–USA border it is interrupted and contact-metamorphosed by the Duluth Gabbro (*Figure 10.1*). The Iron Formation ranges in thickness from about 90 m to 240 m and the average is 120 m (Goodwin, 1956). It is structurally simple, nearly flat-lying with an average southeast dip of 5°. Local folds and brecciation do, however, occur. The geology, petrology and ore mineralogy have been summarized by Bayley and James (1973), Morey (1973) and Floran and Papike (1975).

The Iron Formation occurs within the Animikie Formation, dated radiometrically at between 1980 Ma and 1900 Ma, and is underlain by quartzite and impure argillite of the Pokegama Formation and grades up into the 1000-m thick Rove of Virginia Formation of interbedded argillite and greywacke.

The formation has traditionally been divided into four informal units: Upper Slaty, Upper Cherty, Lower Slaty and Lower Cherty. However, none of the rocks is strictly slaty; rather they are laminated and well bedded. In this Lake Superior region, the name *taconite* has been applied to a rock dominantly of chert with iron silicates, magnetite or haematite, in other words sedimentary iron ore. Carbonate may also be present. In general, taconite that is not composed of carbonate is coarse grained and granular, whereas taconite composed of carbonates and/or iron silicates is fine grained and laminated.

The cherty units are characterized by chert alternating with layers containing magnetite, siderite, ankerite and the iron silicates greenalite, minnesotaite and stilpnomelane. A full account of the mineralogy is given by Floran and Papike (1975). The chert contains mineralogically complex iron-rich granules containing widely different proportions of iron silicates, chert and magnetite. Some are rimmed by haematite; many display tiny internal spherules, possibly of organic origin. Magnetite occurs as discrete layers or as diagenetic replacements of silicate granules. The layering of the granule-bearing rock is commonly wavy and irregular and the centimetre thick layers rapidly lens out. In places the granule-bearing layers grade into oolites that are coated with haematite. Mengel (1976) showed that the size concentration of the iron-rich granules indicates that they were originally detrital and deposited into the basin of sedimentation. The major facies changes in the iron formation were recognized by Goodwin (1956), who described the deposition and repetition of various facies in the Lower and Upper members.

The tuffaceous nature of the Upper Shale member shows the importance of volcanic activity, and Goodwin (1956) suggested that the iron and silica were of volcanogenic origin. There are abundant penecontemporaneous deformations. The variety of these features indicates fluctuations in water level and also that water turbulence was common. Reworking of sediments was common and at least some of the iron formation was deposited subaerially (Walter, 1972).

Two fundamentally different kinds of iron formation can be recognized: cherty iron formation, consisting of granules, ooliths and interstitial cements; and banded or slaty iron formation, composed of matrices of fine-grained material with internally structureless laminae.

Greenalite associated with cherty quartz and minor minnesotaite are the dominant minerals of the granules. Stilpnomelane and haematite are less common. Recrystallized calcite, ankerite and siderite occur locally as cement and as replacement minerals. The most common cement is quartz. Iron silicate and siderite matrices are major constituents of the slaty iron formation, which also contains considerable amounts of secondary calcite and silicate. Stilpnomelane and chamosite are locally abundant in slaty rocks as pseudomorphs after shards. The chemistry of the greenalite, minnesotaite and stilpnomelane are consistent with textural and compositional data, supporting a primary origin (Floran and Papike, 1975).

Sedimentary features include oolitic texture, stromatolite structure, cross-bedding and beds of ripped-up clasts of banded ironstone (Gross, G.A., 1972). Much of the chemical sedimentary rock contains less than 15% iron and may be named ferriferous chert. The variety of sedimentary features is largely independent of iron content. Algal structures, composed of chert, have their internal details defined by iron oxide. They commonly occur in conglomeratic zones. In Ontario the algal structures form zones of 3 m × 1 m by about 60 cm height. The chert contains extremely well-preserved microfossils.

The Lower and Upper slaty members consist of dark thinly laminated iron formation, interlayered with cherty iron formation. The laminated rock is made up of chert, magnetite, iron silicate and siderite.

The average major element chemical composition of the Biwabik iron formation is (% by weight): silica 46.4, iron oxides (Fe_2O_3) 18.7, (FeO) 19.71, magnesium oxide 2.98, calcium oxide 1.60 and carbon dioxide 6.90.

The $\delta^{18}O$ and δD values suggest that the chert appears to have been formed from fresh waters at temperatures similar to those of Phanerozoic iron

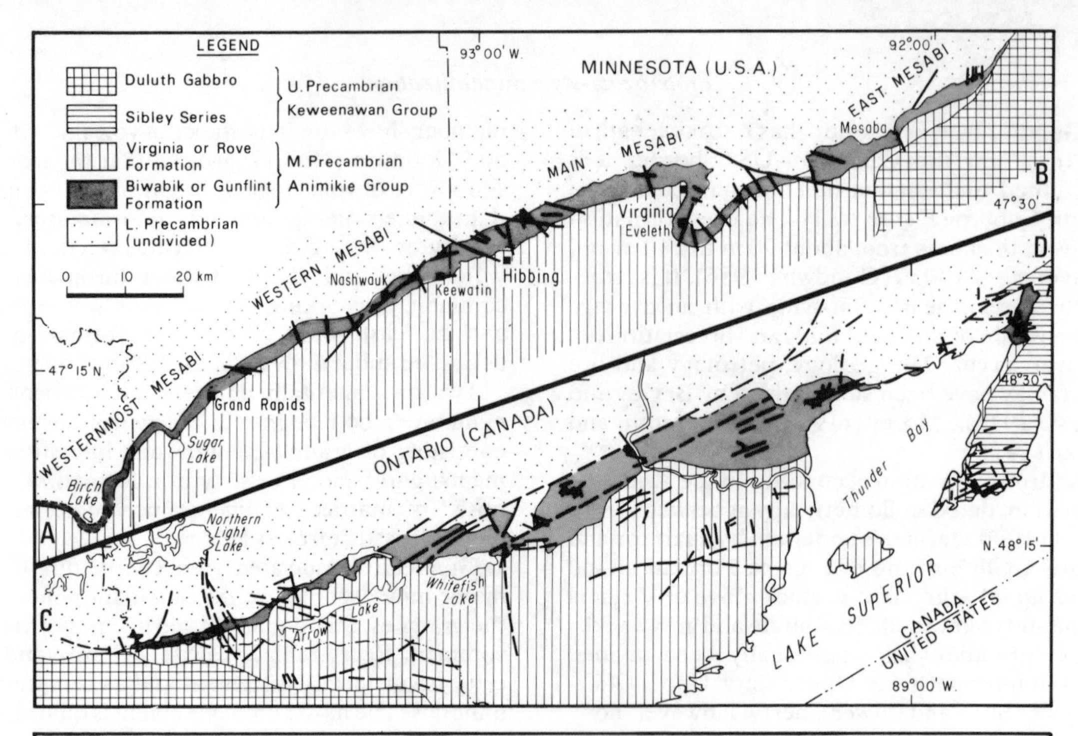

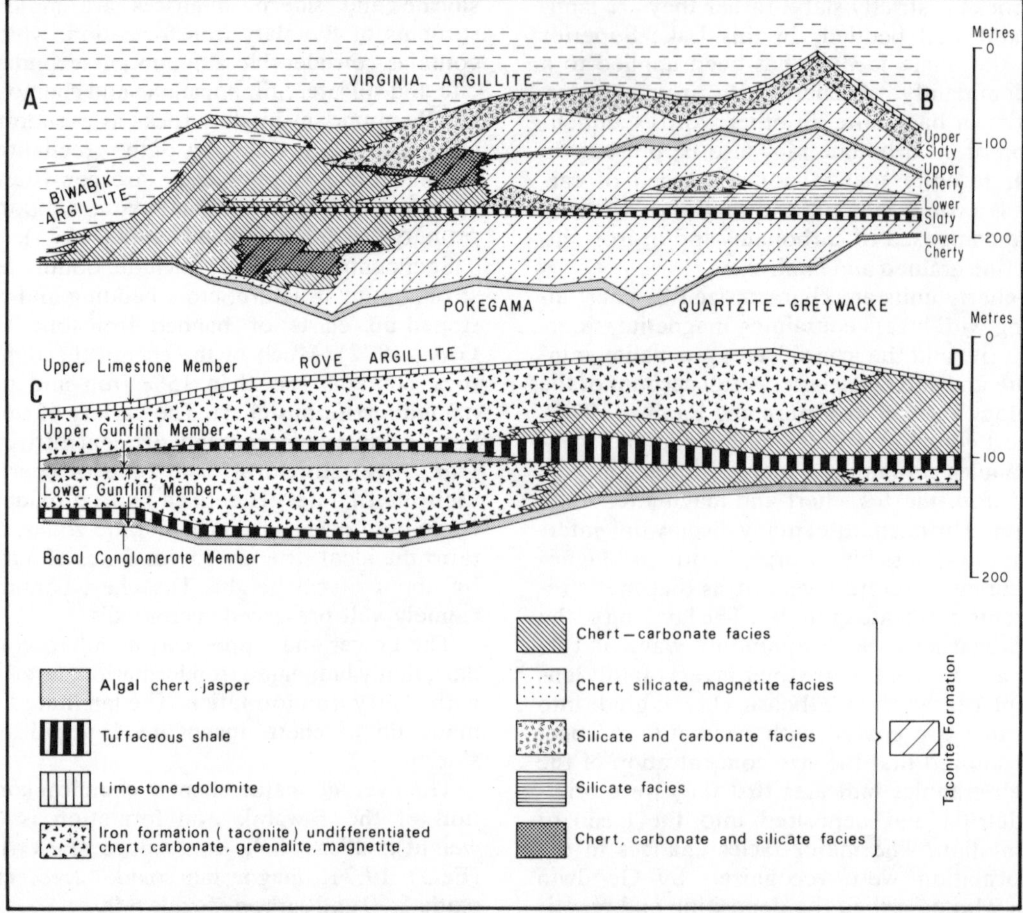

Figure 10.1. Top: generalized geological maps of the Mesabi and Gunflint ranges of the Lake Superior Precambrian Iron Formations. Bottom: stratigraphic cross-sections showing the facies differentiation (after Morey, 1973).

formations. The iron ores are of two types: magnetite-bearing taconite, and secondarily enriched ores.

Magnetite occurs as either disseminations of individual octahedra, aggregates of octahedra, or layered clusters of grain aggregates. Very fine-grained magnetite, occurring both as granules and in the matrix as disseminated and diffuse crystals 5 μm or less in size, is probably primary. Definite secondary magnetite euhedra, 0.05–0.1 mm in size, commonly replace other iron silicates and surround fine-grained siderite in the thin slaty bands associated with cherty taconite. Often an inner core of greenalite or minnesotaite is surrounded by a rim of magnetite.

Thick layers from which the magnetite can be mined occur in the Mesabi Range. The ore bodies commonly occur within the cherty members. The slaty units are usually too thin to mine. Any bed or part of a bed may be designated an ore body, depending on its grade of contained magnetite. The magnetite ores occur in two principal areas. The main Mesabi district is between Nashwauk and Mesaba and also east of Mesaba. The eastern district ores are modified by metamorphism caused by the Duluth Gabbro.

The oxidized ore bodies range in shape from fillings along narrow fissures, through channel-type, to blanket-type deposits. There is little doubt that they result from secondary oxidation and leaching of original iron formation. The original minerals were oxidized to haematite and goethite, while the non-ferrous elements were leached out. These secondary ores may either have formed by weathering or by hydrothermal action but the weathering hypothesis is most favoured. All these ores are related to an erosion surface, which in general is the present bedrock surface. The ores underlie glacial drift or a thin veneer of Cretaceous strata.

Supergene-enriched Iron Formations

Both in the Guyana Shield of South America and the Liberia Shield of West Africa, several metamorphosed iron formations have been strongly enriched by secondary supergene processes related to the formation of the Cretaceous Peneplain of Gondwanaland (Gruss, 1973). One deposit in each of these Precambrian shields will be briefly discussed to illustrate the similarity of the deposits and the processes that enriched them. They are now far separated, but were once contiguous (*Figure 10.2*).

CERRO BOLIVAR (VENEZUELA)

The deposits represent the relicts of a synclinorium of itabirite-bearing metasediments (dated at 2700–2900 Ma). The deposit has a strike of 20 km and outcrops up to 750 m wide. In this area an itabirite formation (metamorphosed quartz–haematite banded-iron formation) has a 200-m sedimentary thickness and is isoclinally folded. The ore has been described by Ruckmick (1963). The formation of high-grade ore from itabirite is due primarily to a removal of silica by rain and ground water and a relative upgrading of iron and aluminium, just as in the formation of laterite. The typical weathering profile is shown in *Figure 10.2*. The weathering residues of zones A and B form the supergene high-grade ore. These high grades rarely occur in coastal areas. The formation of the high-grade ore began about 24 Ma ago and is related to the old Gondwana peneplain, about 700–750 m above sea level. This peneplain dips at 1° towards the north and is covered by younger sedimentary formations.

The iron ores belong to the brown-type and black-type of weathered high-grade ores. The unweathered (primary) zone D consists of fine-banded to coarse-banded (0.05–2.0 cm) itabirite containing 39% iron and 42% silica on average, with a grain size of approximately 0.05–0.15 mm. The main iron minerals are magnetite and specularite. The fresh rock is overlain by a 10-cm thick zone C of soft itabirite. It contains siliceous fine ore (50–62% iron) and soft itabirite ore (45–55% iron). Zone B, of approximately 100 m depth is of black and brown supergene high-grade ore, brown fine ore (62–64% iron) and black fine ore (66–68% iron). The brown fine ore occurs on the rims of a syncline, and the black fine ore generally in the centre. About two-thirds of the reserves in zone B are of brown fine ore.

NIMBA (LIBERIA)

The Nimba deposit is considered to be the largest iron mine in Africa (Gruss, 1973). The high-grade ore is related to itabirite (2500 Ma old), which is fine banded (0.5–5 mm) and fine grained (0.03–0.1 mm). The sedimentary thickness of the itabirite ranges from 250 m to 400 m and the sequence is isoclinally folded (*Figure 10.2*). The main ore minerals are magnetite and haematite. The deposit has been deeply weathered causing supergene upgrading of the ore. There were two upgrading cycles: Cretaceous Gondwana peneplain at 1300 m above sea level; and an older peneplain at 1600 m above sea level.

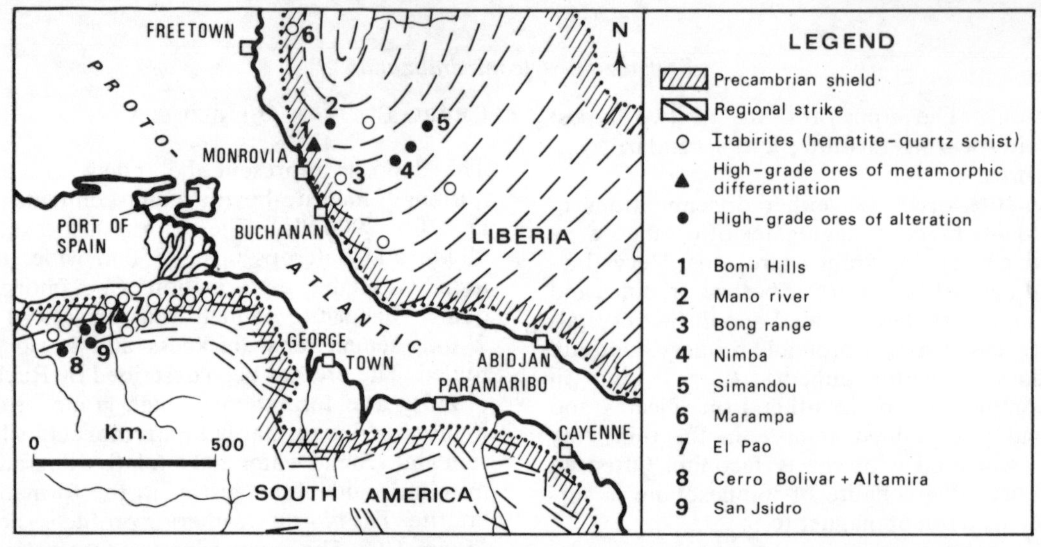

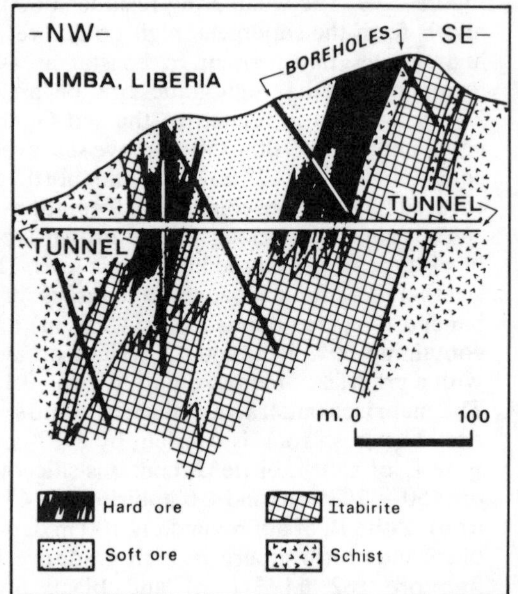

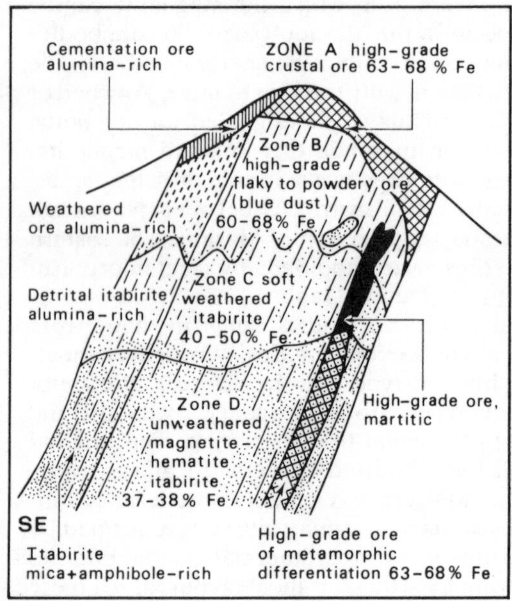

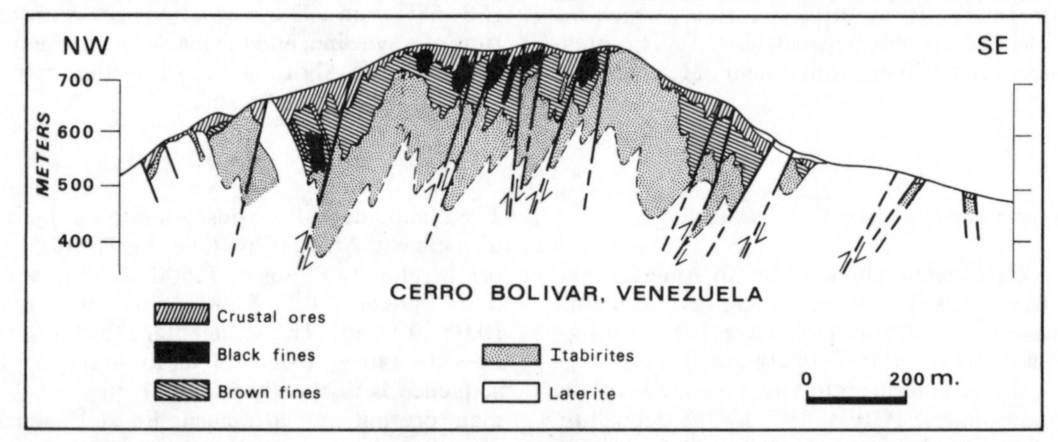

Figure 10.2. Distribution of the Precambrian itabirite (metamorphosed haematite-rich iron formation) ores in the Liberia and Guyana Shields (top); cross-sections through the Cerro Bolivar (bottom), and Nimba deposits (left centre); and typical weathering profiles of the deposits related to the Cretaceous peneplain of Gondwanaland (right centre) (after Gruss, 1973).

The high-grade ore of the Gondwana peneplain usually forms flat caps, reaching a depth of 75–100 m. The cementation ores (*Figure 10.2*) of zone A are 2–5 m thick on average. The brown ores of zone B may reach a thickness of 100 m. The fresh itabirite of zone D contains 38% iron and 42% silica on average. Compared with the brown high-grade ores bound to the Gondwana peneplain weathering episode, the blue high-grade ores of the older period of weathering do not show the typical weathering profile of *Figure 10.2*. They are of limited areal extent, but may extend to a depth of 600 m.

Kiruna-type Iron Ore Bodies

Iron ores of the Kiruna-type are unusual formations apparently restricted both in time and space. They are principally found in the Precambrian of Sweden and Missouri (USA), but why they should be so restricted is impossible to explain. They are characterized by a high concentration of iron and appreciable phosphorus content. The Swedish ores occur in acid volcanic rocks dated within the range 1605–1635 Ma, and those in Missouri with volcanics of age 1300–1500 Ma (Frietsch, 1978).

There are scattered deposits throughout the Baltic Shield similar to those at Kiruna. Magnetite–apatite–actinolite ores are also found in the circum-Pacific volcano–plutonic arc (*see* Chapter 7) and they have been likened to Kiruna-type, with examples in Mexico, Chile and Peru, ranging in age from Mesozoic to Recent.

The Kiruna-type ores consist of magnetite, less commonly haematite, and fluorine-bearing apatite is a characteristic accessory. The minor gangue is of actinolite and occasionally diopside or calcite. Titanium, manganese and sulphur are extremely low. The ores occur as massive bodies that are elongate, tabular or irregular in shape. The ore bodies may be either conformable or disconformable with the country rocks, which are usually of porphyritic keratophyre, rhyolite or ignimbrite, and are alkali-rich, consisting of 6–11% by weight of sodium oxide plus potassium oxide. In general, soda-rich keratophyres predominate, but some varieties are potassic. The high potash content may be related to hydrothermal alteration.

A genetic relationship between the magnetite ore and the volcanic rocks is widely accepted, and liquid immiscibility in an underlying magma chamber is the most reasonable theory to explain the association of ore and felsic igneous rocks. As a basic or intermediate magma cools and fractionates by gravity settling of the high-temperature crystals (olivine and pyroxene), the supernatant magma is expected to follow a trend towards strong iron enrichment (*see* example of the Bushveld complex in *Figure 6.4*). The strongly iron-enriched magma may further fractionate dramatically by liquid immiscibility. The iron-rich liquid will initially separate as droplets within the silicate liquid. The droplets will gradually coalesce and sink towards the floor of the magma chamber because of buoyancy forces, causing the silicate melt to migrate upwards. Subsequent tapping of the roof of the magma chamber will result in abundant keratophyre or rhyolite volcanic formations. The same fracture system can realistically be expected to tap periodically the lower horizons of the magma chamber, causing a release of the ore magma, which will rise up to intrude the cogenetic acid volcanics. In the case of the Baltic Shield, it appears that the ore magma was periodically tapped and was extruded as lava flows between periods of silicate magma eruptions. Therefore, depending on whether the ore magma remains intrusive within the volcanic country rocks or vents onto the surface, the results can be plutonic intrusions from depth, predominantly disconformable, or conformable extrusions on the acid volcanic magma flows.

The deposits of southeast Missouri, as exemplified by Pea Ridge (*see* p. 222) are clearly intrusive into the acid volcanic country rocks, but the Swedish examples (*see* below) appear as interflows within the keratophyre. The controversy regarding the origin appears to be unnecessary, for the stratiform nature of the Kiruna ore reflects the extrusive nature of the ore magma, whereas the Missouri examples are plutonic intrusive and therefore disconformable. The present controversy is summarized by Parák (1975) and Frietsch (1978). The magmatic nature of the ores seems clear, for frequently they contain xenoliths of country rock (silicate volcanics), and magnetite lava flows are known from the circum-Pacific at El Laco (Park, 1972). It is doubtful if the ores can be regarded as exhalative-sedimentary, as Parák (1975) suggests. They presumably represent the result of crystallization of ore magma, rather than the result of hydrothermal exhalations.

The Swedish Ores

Apatite-bearing iron ores are found in a large number of deposits, all within the volcanic formations of the Kiruna–Arvidsjaur Complex. In northern Sweden the volcanic rocks are dated at 1605–1635 Ma, whereas in southern Sweden they are 1900 Ma old. The main ore bodies occur in a restricted area around Kiruna and Gällivare. The total reserves are estimated to be 3.4×10^9 t. Individual deposits are large and have an average content of 52% iron. The deposits of Kiruna form the largest concentration of ore in Sweden. The total area underlain by ore is nearly 700 000 m^2 and reserves are estimated to be 2×10^9 t. The geology of the deposits has been described by Geijer and Ödman (1974) and summarized by Grip (1978). The ores are variously interpreted as magmatic intrusive, a view championed by Frietsch (1978) or volcano–sedimentary exhalative (Parák, 1975).

The ores are of primary magnetite containing variable amounts of apatite. In several deposits, the magnetite has been oxidized to haematite. At the Kiirunavaara deposit, the oxidation is related to superficial weathering, but in other deposits it is a result of hydrothermal activity, associated with wall-rock alteration.

The phosphorus content of the iron ores varies between 0.1% and 5.5%. The apatite generally contains 2.5–3.5% fluoride or chloride and is usually evenly disseminated throughout the ore, but may occasionally occur concentrated in bands. The apatite-bearing ores occur within porphyritic keratophyre, partly as extensive sheets and partly as irregular networks of veins in the country rocks (called breccia ore). In the Kiirunavaara deposit, there are dykes of porphyry that are comagmatic with the country rocks. Some of the dykes pre-date and others post-date the magnetite ore, clearly indicating that the magnetite ore is coeval with the acid-to-intermediate volcanism that dominates the region. The oldest rocks of the region are mainly greenstones derived from basalts and andesites, associated with acid volcanics, tuffs, graphite schists, carbonates and jaspilites, associated with copper mineralization. The greenstones (>1950 Ma old) are overlain by conglomerates up to 600 m thick, followed by keratophyre 1200 m thick. This in turn is unconformably overlain by quartz keratophyre, which often exhibits volcanic layering; it is associated with interlayers of agglomerate and conglomerate. The overlying Lower Hauki rocks are of sediments and volcanics, overlain in turn by phyllites and sandstones with conglomerate horizons. The regional strike is northeast and the succession dips towards the southeast at angles between 30° and 90° (*Figure 10.3*).

The large ore bodies occur in two main horizons. The giant Kiirunavaara and the smaller Luossavaara ore bodies both occupy unconformities between keratophyre and quartz keratophyre (*Figure 10.3*); they consist mainly of magnetite with small contents of apatite and actinolite. The keratophyre that overlies the Kiirunavaara and Luossavaara ore deposits contains abundant ore fragments, making up to 50% of the hangingwall rock; most fragments are angular. At Luossavaara, the hangingwall of the main ore also contains conglomerate, in which cobbles of pure magnetite ore, of several decimetres diameter, are found in a polymict conglomerate.

The stratigraphically higher contact between the quartz keratophyre and the Lower Hauki Complex contains a group of bodies called the Per Geijer ores. Faults frequently separate and cut the ore bodies.

The giant Kiirunavaara ore body has sharp contacts with overlying and underlying keratophyres, but the footwalls frequently contain fragments of keratophyre, commonly brecciated and transected by dykes and veinlets of ore. Ore breccias are common at the contact between the keratophyre and the porphyry dykes. The ore body is about 4000 m long and its width varies from 20 m to 200 m. The ore is known to continue to a depth of at least 1500 m; it is fine grained. Magnetite is dominant, but haematite occurs locally, and some bodies of haematite may be totally enclosed by magnetite. Fluorapatite is the main gangue, but actinolite, calcite, diopside and biotite occur in small amounts; sulphides are rare. There is a sharp but irregular contact between apatite-rich and apatite-poor ore zones. The phosphorus content generally increases with depth.

The Per Geijer ores have been affected by metamorphism. Silicification, sericitization and chloritization have occurred in the wallrocks and the ore is martitized.

The controversy between those authors that regard the ores as volcano–sedimentary exhalative (for example, Parák, 1975) and those that traditionally regard them as magmatic intrusive (for example, Frietsch, 1978) hinges on the nature of the ore breccias. The occurrences of ore-forming veins and dykes intersecting the host rocks clearly indicates an intrusive nature. The contact between the host rocks and the ore is always sharp, irrespective of whether it occurs as a large ore body or as small veins. In many cases there are irregular fragments of the country rocks enclosed in the ore. Ore breccias, in which the ore occurs in veins and dykes anastomosing in various directions and often forming a network, are typical of the Kiruna ores. In some localities, the wall rock is broken up by the ore into angular pieces along joint planes, suggesting that the ore was injected forcibly.

On the other hand, the close association of the

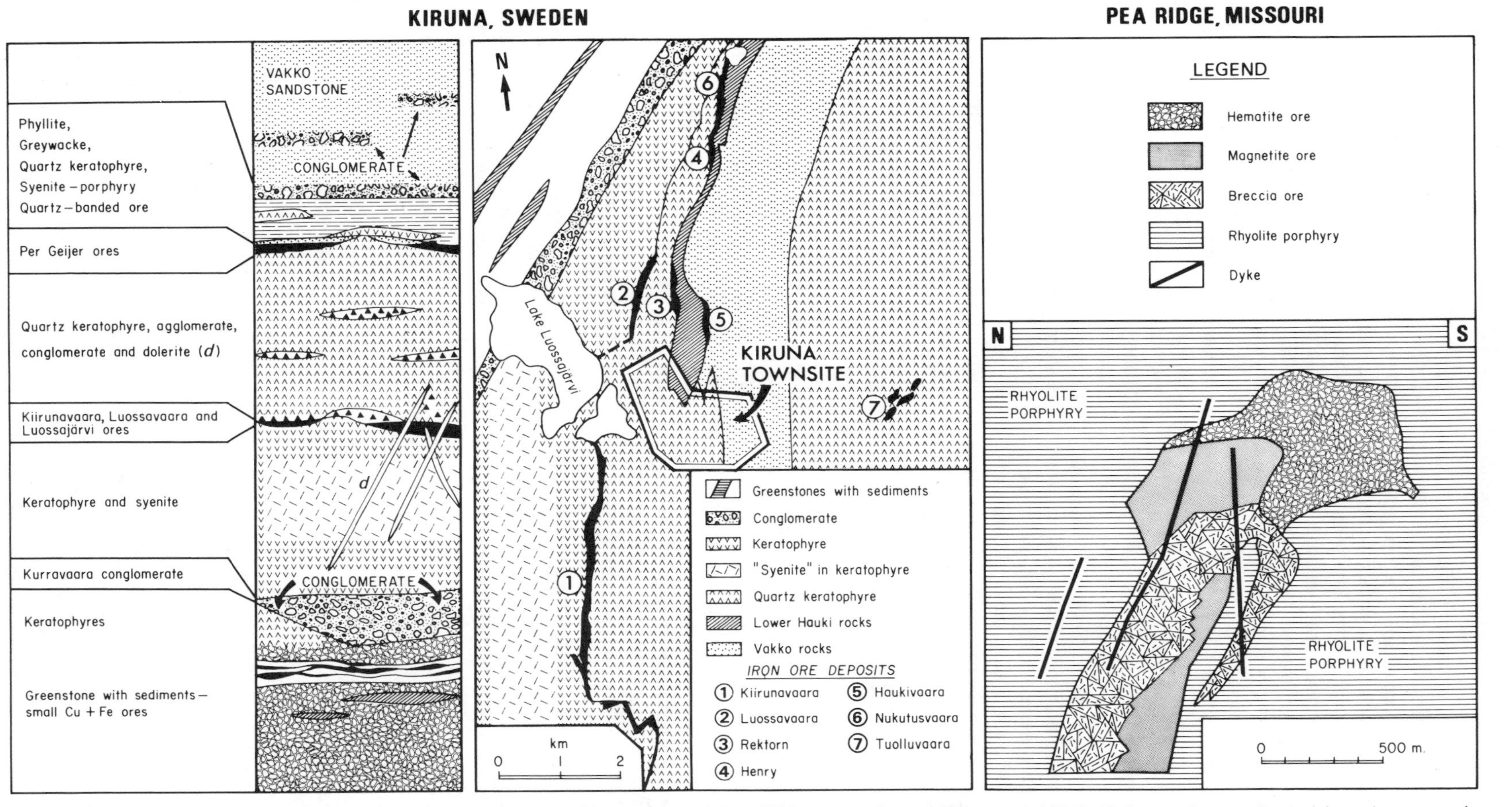

Figure 10.3. Magnetite–apatite ore deposits of Kiruna, Sweden, and Pea Ridge, southeast Missouri, USA. Schematic stratigraphic column and simplified geological map of the Kiruna area (after Parák, 1975). North–south cross-section of the Pea Ridge iron ore body (after Emery, 1968).

major ore bodies with unconformities, and the occurrence of ore clasts in conglomerates overlying the ore horizons, suggest that the ore magma may have been extruded and subjected to erosion before the successive keratophyre flows. In summary it appears that the magnetite–apatite ores of Kiruna exhibit both intrusive and extrusive phenomena characteristic of a volcanogenic environment, but there are no convincing data to indicate that they are exhalative.

Pea Ridge, Missouri

The Precambrian rhyolitic St. Francois Mountain area of southeast Missouri contains ten known iron occurrences and three major ore deposits (Snyder, 1969). The iron mineralization may be conformable with the foliation of the rhyolitic lavas or more commonly may be sharply disconformable and the plutons extend and widen with depth. The Bourbon body is a good example to illustrate their intrusive nature. It is predominantly disconformable and widens with depth. It also has an apophysis called the 'Handle', which is a tabular body extending from the main intrusion and injected conformably parallel to the foliation of the country-rock volcanics. The whole deposit therefore resembles an igneous disconformable stock with a sill extending outwards from it.

Haematite is the predominant mineral only at the Precambrian outcrops, in shallow deposits, or in the upper parts of deep deposits; it is clearly related spatially to the Precambrian outcrop pattern (Snyder, 1969). Magnetite is the dominant, and often the only, ore mineral in the deeper deposits. Titanium is insignificant. Copper, fluorine, phosphorus and sulphides are present in all deposits but their distribution is erratic.

Some deposits, notably Pea Ridge, Iron Mountain and Pilot Knob, contain breccia ore, and appear to represent forcible injections of an ore fluid viscous enough to support blocks of rhyolite in the form of xenoliths (Snyder, 1969).

The Pea Ridge iron-ore deposit, near Sullivan, Missouri, lies about 40 km due north of the Viburnum Trend Mississippi Valley lead deposits (*Figure 4.3*). The geology of the mine has been described by Emery (1968) and the whole iron mining region by Snyder (1969). Pea Ridge is one of the most outstanding examples of a magnetite ore that has been intruded from depth into Precambrian 1300–1500-Ma old rhyolite and ignimbrite. The ore body is a dyke-like mass predominantly of magnetite that extends from an unknown depth to within 400 m of the present land surface (*Figure 10.3*); it dips nearly vertically, intersects the layering of the volcanic country rocks at high angles, and the contacts are sharp. Faulting, both pre-ore and post-ore, appears to have had only a minor consequence and there appear to be no structural controls on the location of the ore body, which forced its way upwards in the manner of a magmatic pluton.

The ore body is composed of magnetite with specular haematite, quartz–apatite and pyrite occurring as accessory primary minerals. Specular haematite occurs in appreciable quantities on parts of the cap and footwalls of the ore body.

A rhyolite breccia ore zone is situated on the hangingwall side of the massive magnetite in the eastern part and on both walls in the west (*Figure 10.3*). This variety of ore consists of 30–45% magnetite containing embedded fresh and angular rhyolite fragments of a variety of sizes. In the west, the xenoliths include mafic metamorphic rocks, which are unknown as country rocks in any of the mine workings. The xenolith rhyolite, though resembling the rhyolite of the mine-level country rocks, frequently contains metamorphic minerals such as actinolite and epidote. This fact, together with the basic xenoliths, indicate that they are brought up from depth, and that the magnetite ore was viscous enough to have transported them upwards. The ore was not therefore hydrothermally deposited, and the breccia ore cannot be regarded as *in situ* hydrothermal brecciation.

Hangingwall quartz–actinolite and footwall quartz–haematite zones occur in certain rhyolite horizons and sericitization has affected parts of these zones. Emery (1968) concludes that the Pea Ridge ore body formed essentially by a single injection of a magmatically differentiated magnetite magma, from which late-stage fluids resulted in hydrothermal activity causing the amphibole, haematite and sericite alteration of parts of the deposit.

Bedded Manganese Ore

The geology of ancient manganese deposits has been summarized by Roy (1976). The stratiform deposits of India, Brazil and Ghana are traditionally well known and the recently explored deposits of Kalahari in the Cape Province of South Africa should assume increasing importance.

Several major non-volcanogenic deposits are intimately associated with banded-iron formations in Proterozoic basins. The association in Brazil has been described by Dorr (1973a, b) where stratigraphically controlled beds and lenses of manganese oxide ores up to 3 m thick and more than 100 m in length are included in the metamorphosed iron formation of the Minas Series at Minas Gerais.

Kalahari Manganese Field (28° S, 23° E)

Some of the largest reserves of manganese in the world (over 800 million t) are found in the Kalahari Manganese Field. The geology has been summarized by Beukes (1973) and Button (1976). In this region, the Ongeluk Lava, dated at 2224 Ma old, is overlain by a succession of chemical sediments, predominantly iron formation (*Figure 10.4*). The banded-iron formation conformably overlies the lavas and it contains thin jasper beds identical to those interbedded with the underlying lava flows.

Within the basal 100 m of the iron formation, up to three layers of primary manganese ore are developed over a strike length exceeding 50 km. They are conformable with the iron formation and limestone sequence (*Figure 10.4*). The banded-iron formation consists of alternating bands of jasper (0.25–3 cm thick) and haematite (0.25–1.00 cm thick). The manganese bands are distinctly layered and wholly conformable with the enclosing banded-iron formation, even in areas of deformation—a conclusive indication of their sedimentary origin. They are of manganiferous and calcareous sediments, bedded, sometimes pseudo-oolitic, and frequently streaked by thin wisps of white carbonate. The lowest manganese layer is up to 25 m thick and it developed about 30 m above the Ongeluk Lava.

All but one of the deposits is covered by younger sediments, including calcrete and wind-blown sand of the Kalahari desert. Beneath this cover, supergene-enriched derivatives of the primary ore are mined. The supergene enrichment fingers out with depth to primary manganese sediment. However, the bulk of the ore mined is not appreciably modified by surficial processes. The ore is mineralogically complex, especially because it may be locally intruded and metamorphosed by bostonite dykes and sills. The principal minerals are braunite, bixbyite, rhodochrosite, hausmannite and jacobsite.

The original sediment probably consisted of oolitic carbonate in a matrix of gelatinous manganese and iron hydroxides, manganese and iron carbonate, and in places hydrous silicates of magnesium and iron. Diagenesis resulted in the crystallization of wad, cryptomelane, braunite, jacobsite, rhodochrosite and calcite. Subsequent metamorphism and later supergene and hydrothermal activity produced secondary minerals. The differences in composition between the manganese bands and the iron bands of the enclosing iron formation, which contain generally less than 1% manganese oxide (MnO), suggest that very complete separation of iron and manganese took place during deposition.

The major control on the mineralization was the original sedimentological environment. Beukes (1973) concluded that the deposits are somehow genetically related to volcanic activity associated with the Ongeluk Lava.

The iron and manganese oxide facies represent a more proximal facies and the limestone, iron formation and manganese carbonate represent a more distal facies. The manganese was probably held in solution in the sea water. Chemical deposition was brought about by the onset of more oxidizing conditions in the basin of sedimentation, far removed from detrital sediments (Button, 1976).

Madhya Pradesh and Maharashtra, India

In India, syngenetic manganese formations are best developed in the Sausar Group (846–986 Ma metamorphic age; 1700–2000 Ma sedimentary age) that covers the districts of Nagpur and Bhandara (Maharashtra) and Chhindwara and Balaghat (Madhya Pradesh) in Central India. The manganese formations extend as an arcuate belt for more than 210 km with an average width of 32 km (at about 79° E, 22° N). The Sausar Group is of a miogeoclinal quartzite–carbonate sequence. Igneous rocks are extremely rare. The geology has been summarized by Roy (1973, 1976). The manganese formations are entirely stratigraphically controlled. They are particularly associated with the pelitic Mansar Formation. The manganese formations are beds of manganese oxide ore and manganese silicate rocks known as gondite (regionally metamorphosed spessartite–quartz rock, with or without other manganese silicates). They are intimately interbedded with the enclosing pelitic metasediments. The ores have been metamorphosed to various grades in the greenschist and amphibolite facies.

The deposits were originally laid down as manganese oxides, progressively reduced during metamorphism to form braunite–bixbyite–hollandite – jacobsite – hausmannite – vredenburgite assemblages. The interbedded gondites were originally of manganese oxides intermixed with quartz and clay. On metamorphism

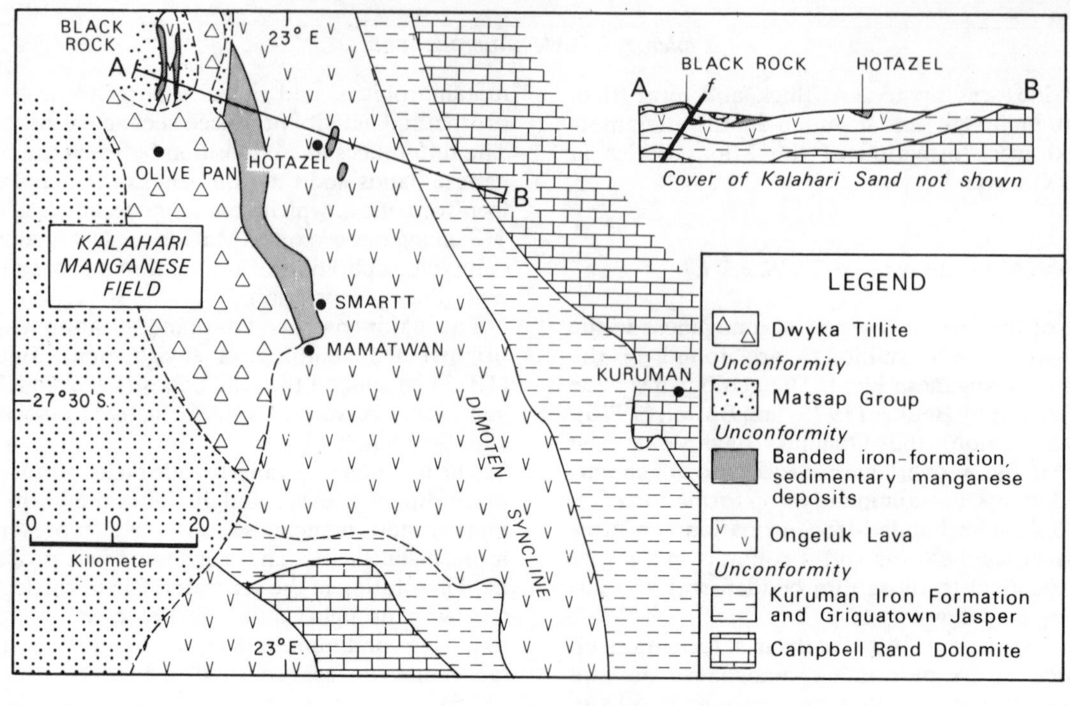

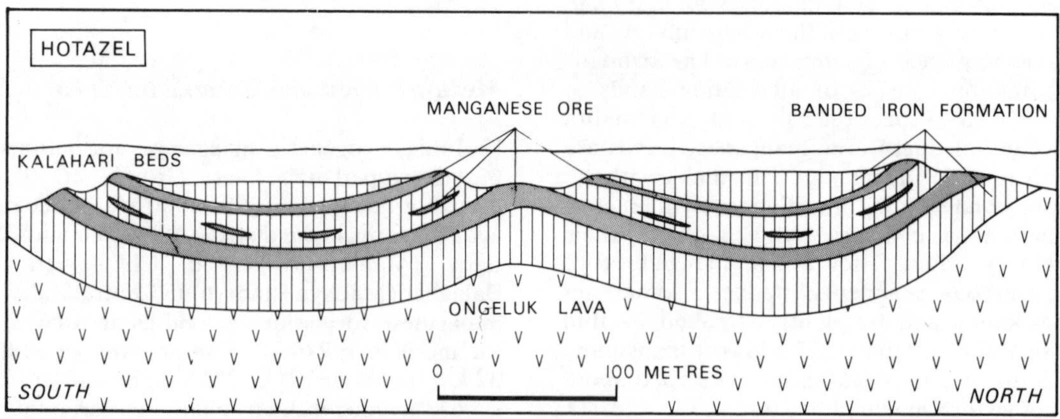

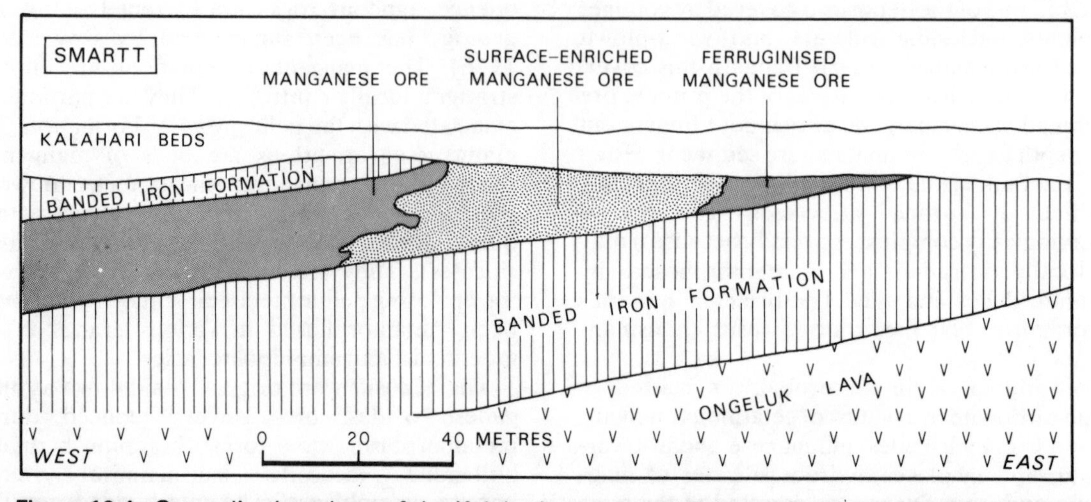

Figure 10.4. Generalized geology of the Kalahari manganese fields of Southern Africa and cross-sections showing the relation between the Precambrian bedded manganese ore and the banded-iron formation (after Beukes, 1973; Button, 1976).

they resulted in spessartite–rhodonite–quartz assemblages, often containing manganiferous pyroxenes, amphiboles and micas.

West Africa

Nsuta, Ghana, is the world's principal producer of high-quality manganese, situated near the gold-mining town of Tarkwa. The ore bodies occur in a sequence of phyllites 114–122 m thick in the Precambrian Birrimian System (2200–2400 Ma old). The phyllite zone includes graphitic schist, tuff, greywacke and gondite. It is deformed into near-isoclinal folds trending 15° E.

The gondite beds (quartz–spessartite rock) are formed by metamorphism of manganiferous sediments (Sorem and Cameron, 1960). The upper Birrimian System includes basic-to-intermediate lava flows and tuff metamorphosed to greenstones, interbedded with phyllites. The horizon of manganese phyllites and gondites is 450–600 m below the top of the Birrimian System and in the upper part of the phyllite group. Lenticular ore bodies and pipe-like veins are up to 300 m long and 30 m thick. The ore bodies are strongly modified by surface weathering and supergene enrichment. Replacement and cavity-filling ore-types are the most common. 'Nsuta MnO_2' and cryptomelane are the most common manganese oxides and pyrolusite and lithiophorite are common. The mineralogy of the ore is consistent with a supergene origin, although there is no doubt that the manganese deposits were originally of sedimentary origin.

The Gabon manganese ores at Moanda (1° S, 13° E) are mined from a supergene-enriched capping of high-grade ore. The annual production is 1.6 million t and reserves will last for 150 years at this rate (Weber, F., 1973). The residual supergene ore blanket is developed on a plateau directly overlying the non-metamorphic Francevillian Series (1740 Ma old) sedimentary sequence, which includes some spilite horizons. Drilling through the manganese laterite crust has revealed that the underlying black shales contain, on average, 15% manganese, and this has been enriched by supergene processes to 45%.

Massive Sulphide Deposits

Proterozoic massive sulphide deposits are typified by the large silver–lead–zinc deposit at Sullivan, British Columbia (Hutchinson, 1973; Sawkins, 1976). The deposits occur in settings characterized by very thick sedimentary sequences composed largely of continentally derived clastic sediments. In contrast to other types of sulphide deposit, they exhibit little or no relationship to volcanism or plutonism, so that they have long been something of a genetic puzzle. They exhibit extremely well-defined mineral banding.

The two most important Sullivan-type deposits are Sullivan itself (Freeze, 1966) and the Mount Isa deposit of Queensland (Bennett, E.M., 1965). The Sullivan ore body has been cited by Mitchell, A.H.G. and Garson (1976) as an example of mineralization related to an aborted rift zone, or aulacogen, situated on the edge of a continental plate. For this reason, and to develop the argument, the mineralization of the Belt–Purcell basin of western North America will be discussed as a whole.

Mineralization in the Belt–Purcell Aulacogen

The Belt–Purcell basin represents a slowly subsiding re-entrant at the western edge of the North American craton, beginning about 1500 Ma ago and persisting till about 900 Ma ago. The basin was not strictly graben-shaped, and at times had a triangular shape (*Figure 10.5*). Harrison *et al.* (1974) drew attention to its resemblance to an aulacogen. Sears and Price (1978) accepted this hypothesis and matched the Belt–Purcell basin with a counterpart on the Siberian craton. These aulacogens formed as the Archaean craton split up about 1500 Ma ago. The splitting can be imagined along arms combining in a triple junction. Whereas the major opening progressed along two of the arms, a third arm begins to split, but fails to achieve complete separation. This failed arm is what is referred to as an aulacogen. The western limit of the North American craton can be taken as the limit of Precambrian outcrops. It may also be more precisely defined as the line of 0.704 initial $^{87}Sr/^{86}Sr$ isotope ratios for cordilleran Mesozoic and Cenozoic igneous rocks (Armstrong, R.L. *et al.*, 1977). Isotopic ratios west of this line (*Figure 10.5*) grade downwards to 0.702, indicating no influence from and hence the absence of underlying continental basement. East of this line the ratios grade to 0.707 and all the major batholiths lie just to the east of the 0.704 line. There is little doubt that this line represents the Mesozoic plate margin. The stratigraphy and sedimentary history of the Belt basin has been reviewed by Ross (1970) and strata-bound mineralization reviewed by Thompson, R.I. and Panteleyev (1976). The Belt Supergroup is known in Canada as the Purcell

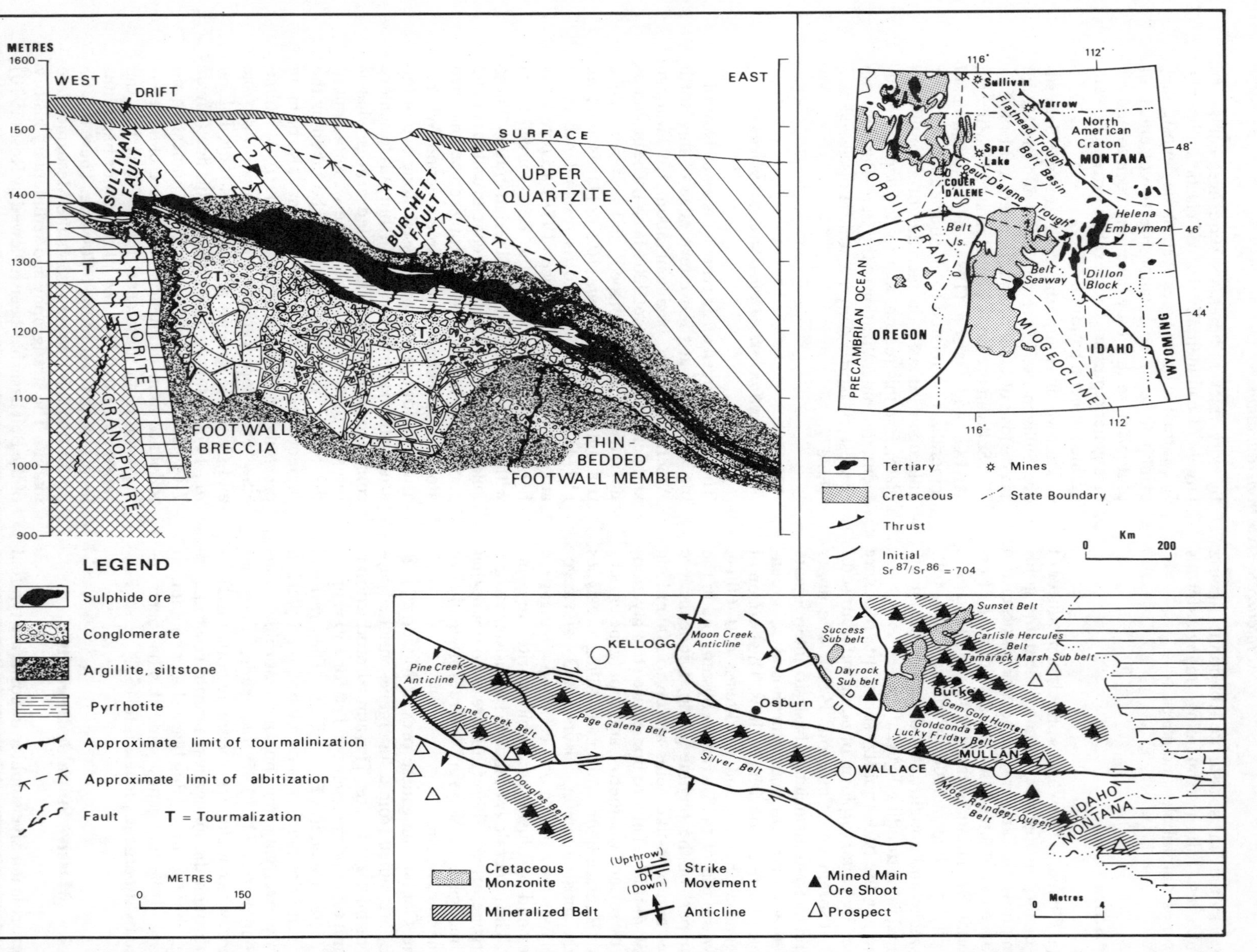

Figure 10.5. Mineralization in the Precambrian Belt aulacogen of western North America. Outline of the Belt basin in relation to the Precambrian craton margin (from Hutchison, 1981). Cross-section of the Sullivan ore body (after Thompson, R.I. and Panteleyev, 1976). Map of the Coeur d'Alene mineral belts of Idaho (after Hobbs, S.W. and Fryklund, 1968).

Supergroup. The thickness of the Belt–Purcell Supergroup is variable from about 1.2 km to 20 km (Harrison *et al.*, 1974). The rocks are unmetamorphosed in Canada and they progressively increase in regional metamorphic grade southwards, reaching sillimanite grade at the Idaho Batholith (Hutchison, 1981). The Belt–Purcell rocks are typically monotonous, composed of grey siltstone, mudstone and argillite with units of dolomite, limestone and sandstone. Abundant scapolite in some units may indicate they were evaporite-rich before metamorphism. Sedimentary structures indicate shallow-water deposition and there are some redbeds. Although igneous rocks, other than the Cretaceous–Tertiary batholiths, are rare, there is an important episode of Purcell lavas, dated at 1110 Ma and 1073 Ma. The earlier parts of the Belt Supergroup are represented by the Pritchard Formation in the USA and the Aldridge Formation in Canada. They are of interlayered green-to-red argillite, mudstone and siltstone. The middle part consists of a thick succession of turbidites that apparently developed as distal base of slope deposits (Thompson, R.I. and Panteleyev, 1976). The environment of deposition is likened to the aulacogen in West Africa—Benue Trough, which represents the failed arm of the Atlantic rift (*Figure 12.5*). The graben formation of the trough kept pace with the continentally derived sedimentation and a major delta prograded outwards towards the ocean (*see* page 286).

Copper in the Belt-Purcell Supergroup

Although there are many localities with anomalously high copper contents, there are two deposits of economic significance: the Spar Lake deposit (Montana) in the Revett Formation, a quartz sandstone unit of the lower part of the Belt Supergroup; and the Yarrow–Spionkop deposit (Alberta) in the Grinnel Formation, a redbed unit in the lower part of the Purcell Supergroup (Thompson, R.I. and Panteleyev, 1976).

Copper occurrences in the Spar Lake area comprise a zone 195 km long and 65 km wide called the Western Montana copper belt, extending from the Canadian border to Coeur d'Alene. Copper is concentrated in various parts of the Revett Formation. It occurs as fine disseminations, blebs and lenses, concentrated in the coarse quartzite beds, especially at the base of graded units, along bedding planes, in close association with sedimentary structures and as fissure fillings. The economic minerals are bornite, chalcopyrite, chalcocite, covellite, tetrahedrite and digenite. In places there is an upward zonation from bornite–chalcocite through a galena zone to an outer pyrite zone. Copper also occurs in the green (chloritic) argillite beds of the Spokane and Empire Formations.

Most of the copper occurrences are a product of remobilization and reconcentration of the disseminated mineralization into more permeable zones and beds. There has also been a remobilization on a regional scale into the Missoula Dome, where the ground-water solutions had been trapped beneath the overlying impermeable argillite of the St. Regis Formation.

Copper mineralization at the Yarrow–Spionkop deposit occurs in the quartzite and adjoining green argillite beds a few centimetres to some metres thick, mineralized for a strike length of over 750 m. The grade of mineralization has been enhanced by sulphides replacing argillite pebbles in some quartzite beds and in zones close to diorite and diabase sills. Isotope studies indicate that the sulphur is syngenetic in origin but has been remobilized and reconcentrated by ground-water convection cells driven by the hot sill and dyke intrusions.

Lead and Zinc Mineralization in the Belt–Purcell Supergroup

Unlike copper, which is widely distributed, stratabound lead–zinc deposits are few, though commercially very important. The Sullivan mine at Kimberley, British Columbia, is the largest ore body in the Belt–Purcell Supergroup basin and it ranks with the Broken Hill ore body of Australia (Both and Rutland, 1976) as one of the world's greatest lead–zinc deposits. Since 1920, over 6.8×10^6 t of lead, 5.4×10^6 t of zinc and 7087 t of silver have been extracted from more than 1.12×10^8 t of ore. The lead–zinc occurrences in the Aldridge Formation are mainly concordant but transgressive deposits are associated with the Precambrian Moyie intrusions. The lead is isotopically very uniform with a single stage age of 1200–1500 Ma (Zartman and Stacey, 1971).

The Sullivan ore body is a very large deposit of the stratiform class that exhibits some transgressive features (*Figure 10.5*). Hence a definitive statement of its mode of origin is not yet available. It occurs on the eastern flank of a broad northerly plunging anticline, and is contained within a 30–90-m interval of well-laminated argillite at the base of the middle Aldridge Formation (Freeze, 1966). The ore body lies above a conglomerate marker bed, separating the lower and middle members. The ore zone is truncated to the north by the major steeply dipping Kimberley Fault with over 3000 m of stratigraphic displacement. The ore body is bounded to the west by the Moyie intrusion (*Figure 10.5*) and mineralization passes gradually eastwards outward into an iron sulphide zone that persists laterally up to 3 km. In cross-section, the ore body can be divided into two contrasting zones (*Figure 10.5*). The upper zone is massive and somewhat discordant with intensely tourmalinized and albitized host rocks above and below. The lower zone comprises extremely regular and conformable sulphide layers. The pyrrhotite-rich argillite contains numerous folds, slumps and boudins indicative of penecontemporaneous soft sediment deformation. Well-laminated lead–zinc bands, centimetres to metres thick, are contained in laminated pyrrhotitic argillite. The well-laminated carbonaceous and pyrrhotitic argillite is transgressive on the conglomerate and may possibly represent a turbidite unit. The lead–zinc layers are laterally extensive regardless of their thickness. For example, one layer of 1 m thickness extends without change for more than 1 km. The contacts between all beds are sharp. The upper zone is massive, in contrast to the lower zone. The footwall rocks are in sharp contact with the ore body and are extensively tourmalinized. The hangingwall rocks are albitized and only slightly tourmalinized. Much of the footwall rocks are of megabreccia that extends downwards to a considerable but unknown depth. The breccia matrix is strongly mineralized with pyrrhotite in some areas, but sphalerite and galena are rare. An extensive cassiterite-bearing fracture, one of the numerous mineralized transgressive veins, cuts the sulphide zone and has been traced to depth. The metal distribution is distinctly zonal. An inner rectangular zone comprises pyrrhotite and pyrite with chlorite, which changes laterally to finely disseminated pyrite towards the southeast. In the overlying lead–zinc ore zone, lead/zinc ratios vary radially, with the highest ratios confined to an area around the pyrrhotite zone and decreasing gradually outwards. Cassiterite and arsenopyrite are confined predominantly to the outer margins of the pyrrhotite zone, whereas antimony is concentrated nearer the ore-body periphery. Silver is closely associated with the galena.

A rubidium–strontium age of 1265 Ma was obtained for the Aldridge Formation; potassium–argon dates are misleading because they have been reset by the East Kootenay orogeny.

Galenas from most of the large deposits and from a vein within the Moyie intrusion are isotopically similar and have single-stage ages of between 1200 Ma and 1400 Ma. It may be assumed that the lead was derived from the same general source terrain (namely the crystalline shield areas to the east) that supplied sediment to form the Aldridge Formation. Viewed on a regional context, Late Precambrian lead–zinc occurrences in the Aldridge Formation may represent the redistribution of metals, originally deposited in the sediments by circulating brine in which the sulphides were deposited at depth, as well as at the sediment–water interface during deposition. A driving mechanism for the circulating brine may have been the Moyie intrusions. This hydrothermal circulation mechanism therefore can be used to explain the stratiform nature combined with the zonal distribution of metals and the local cross-cutting ore deposits.

Coeur d'Alene Mining District

The Coeur d'Alene district of Idaho is one of the great mining districts of the world (*Figure 10.5*). First recorded production was in 1884, and since then more than 21 385 t of silver, 7×10^6 t of lead, 2.4×10^6 t of zinc and 1.25×10^5 t of copper have been extracted, in addition to substantial amounts of cadmium and antimony. The geology has been summarized by Hobbs, S.W. and Fryklund (1968). Mineralized veins occur and cut most of the Belt Formations, including (from oldest upwards): the Prichard Formation, the Burke Formation, the Revett Formation and the St. Regis Formation. Mineralized veins do extend into the overlying Wallace Formation but mineralization is limited. The stratigraphic thickness of the Belt Series in this region is estimated to be at least 6400 m and the base is not exposed. These formations have been intruded by several small Cretaceous monzonite stocks (Gem stocks) and by Tertiary diabase dykes and sills. Rhyodacite dykes are thought to be related to the monzonite. The diabase dykes are on the whole post-mineralization but they may contain minor local traces of mineralization.

The district occurs at the intersection of two major regional structures. One is a broad anticlinal arch that extends northwards from the Idaho Batholith towards Kimberley in Canada. The other is the Lewis and Clark (Coeur d'Alene Trough) Line, a prominent zone of faulting, shearing and complex folding trending eastsoutheast (*Figure 10.5*). The Osburn Fault is the major structure of the Lewis and Clark Line; it has a right-handed lateral displacement of 26 km. North of this fault the Belt strata are moderately deformed with fold axes trending northerly. South of the fault the folds are tightly compressed and the axes trend westerly. The major folding is Precambrian, and cannot be younger that uraninite dated at 1250 Ma. Some reverse faults are also Precambrian. Subsequent events include additional folding, complex faulting and mineralization.

Faults dominate the structure and control the mineralization. Permeable fault zones and fractures served as channelways for the ore solutions. Mineralization has been a long-lived process, as have the faulting and folding. The structural complexity has led to an apparent chaotic distribution of the ore and gangue minerals in the veins. Because of their complexity, they are difficult to classify. Superimposed multiple mineralization cannot be analysed in a simple manner.

The main-period deposits are steeply dipping tabular replacement veins of simple mineralogy. The major economic minerals are galena, sphalerite and tetrahedrite; but others are important locally. These include chalcopyrite, pyrrhotite, pyrite, arsenopyrite, magnetite, haematite, stibnite, uraninite, gold, gersdorffite, barite and siderite. The economic veins range from about 1 cm to 3 m in width. In a few places closely spaced stringers or narrow veins occur over a mineable width of 30 m or more. Individual ore shoots may extend up to a length of tens of metres to 1200 m. The mineral belts shown on *Figure 10.5* are defined by veins. The distribution of the veins within any mineral belt differs from place to place depending on the faulting and folding pattern.

The following features are evident north of the Osburn Fault. Magnetite is zoned concentrically around the Gem stocks, galena and sphalerite are the major ore minerals, and sphalerite predominates on the west and galena on the east of a mineral belt. The sulphide minerals are not related in their distribution to the Gem monzonite stocks. There is no evidence for any consistent change in mineral composition with depth in the veins.

The following features characterize the belts south of the Osburn Fault. The ore minerals are sphalerite, galena and tetrahedrite, while sphalerite predominates in the west, galena in the centre and tetrahedrite in the east. There is no concentric zoning around the monzonite intrusions. Generally there is no change in mineral content with depth.

Deposits associated with the Osburn Fault, and with faults that trend more northerly, contain Precambrian-type lead, characterized by $^{206}Pb/^{204}Pb$ ratios of 16.15–16.73, $^{207}Pb/^{204}Pb$ ratios of 15.37–15.45 and $^{208}Pb/^{204}Pb$ ratios of 35.88–36.46, with a model age of 1500–1200 Ma (Zartman and Stacey, 1971). Outside the Coeur d'Alene district, Mesozoic or Cenozoic lead has been identified, and indeed some minor mineralization occurs within the Gem stocks. It must therefore be concluded that a significant part of the structural activity and mineralization took place in the Precambrian. The exact age of the mineralization has long been controversial, and will continue to be as isotope data are reinterpreted. Harrison *et al.* (1974) suggest that the Coeur d'Alene ores were remobilized from Sullivan-type stratiform deposits that are buried in the lower parts of the Prichard Formation. Some time in the post-Belt, but pre-Cretaceous, these deposits were remobilized by hydrothermal activity and the ore solutions were redeposited in fault and fissure zones. The evidence is that the remobilization may have occurred several times. However the radiometric data on the lead support the same ultimate source for the stratiform Sullivan deposits and the remobilized Coeur d'Alene deposits. The more deformed and multiple folded and faulted tectonics of the Coeur d'Alene region has favoured remobilization. Clearly any genetic relationship to the Mesozoic and Cenozoic intrusions is ruled out by the data, but they may have acted as heat sources to partly redistribute some of the pre-existing mineral deposits.

The Great Australian Stratiform Ore Bodies

Australia is particularly well endowed with extremely large lead–zinc–silver stratiform deposits of Middle Proterozoic (Carpentarian) age. The most important of these are the unmetamorphosed zinc-rich McArthur deposit in the Northern Territory, the low-grade metamorphosed Mount Isa and Hilton deposits in northwest Queensland and the highly metamorphosed Broken Hill deposit in western New South Wales (*Figure 10.6*). The key to understanding these broadly similar deposits is likely to be found in the unmetamorphosed McArthur deposit, which has not been modified by post-lithification processes. With increasing modification, the genesis of the deposit becomes obscured and it becomes less amenable to interpretation. Thus controversy will continue to exist about the genesis of the highly metamorphic Broken Hill deposit.

The McArthur (or H.Y.C.) Deposit

The McArthur deposit occurs towards the eastern margin of the Batten Trough, which contains up to 5.5 km of unmetamorphosed Carpentarian (Middle Proterozoic) dolomites, siltstones, shales and tuffites. The thickness diminishes dramatically towards the sides of this fault-bounded north–south-trending trough, which like the

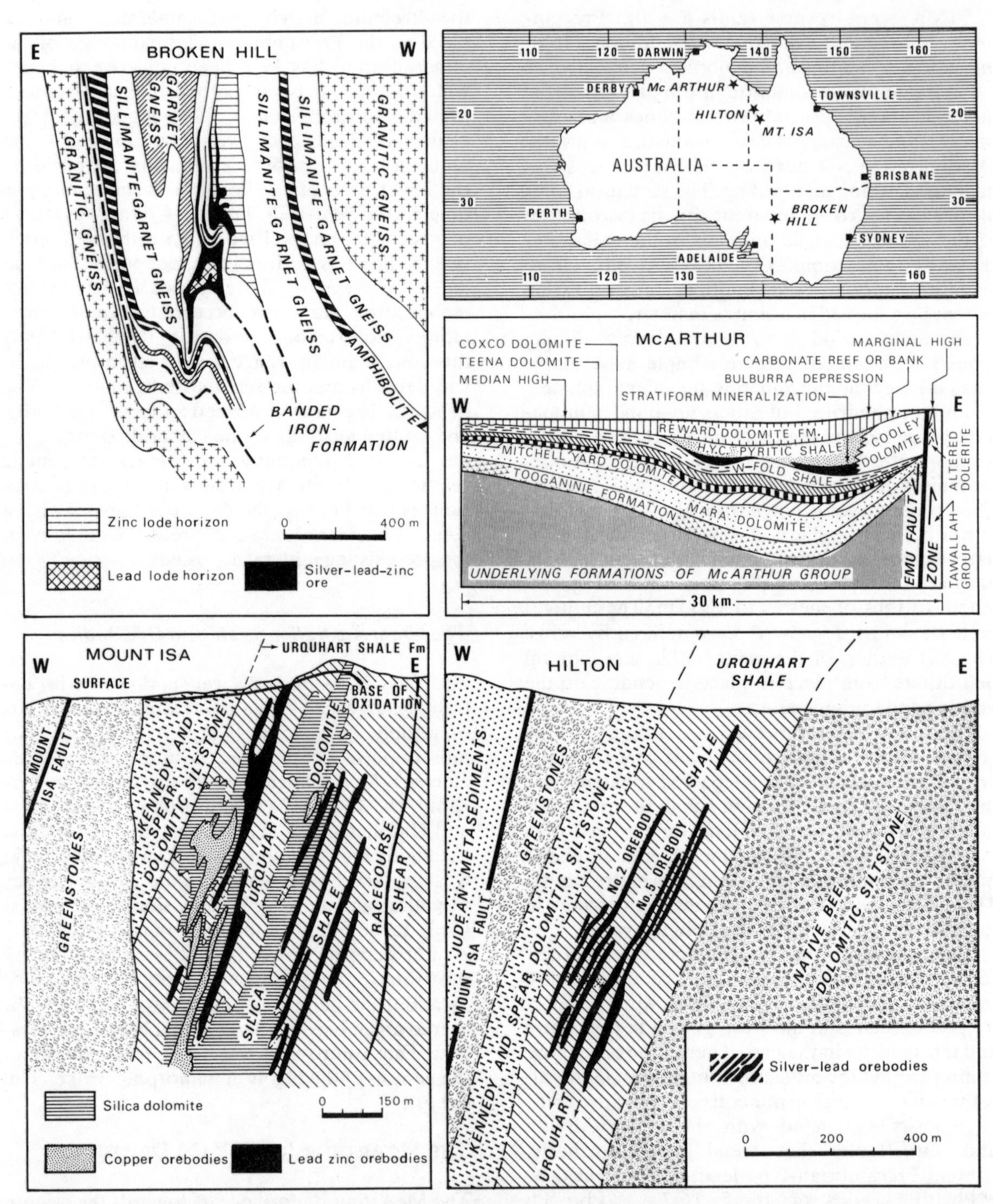

Figure 10.6. The great Australian Proterozoic stratiform ore bodies. Schematic cross-section through the McArthur area of the Northern Territory (after Lambert, 1976). Cross-section of the Hilton deposit of Queensland (after Mathias and Clark, 1975). Cross-section of the Mount Isa mine (after Bennett, 1965). Cross-section of the Broken Hill mine (after Both and Rutland, 1976).

Belt–Purcell trough of North America, may be likened to an aulacogen. Radiometric dating indicates deposition of the ore-bearing McArthur Group between 1400 Ma and 1600 Ma ago. The geology and mineralization have been summarized by Lambert (1976). The stratigraphic succession (*Figure 10.6*) indicates a major marine transgression beginning with subaerial siltstones, through intertidal and shallow marine carbonates, to deeper-water shaly carbonates, tuffaceous mudstones, and local graded arenite and breccia beds. Subsequent regression resulted in shallow water and intertidal carbonates. All the units thin towards the penecontemporaneous Emu Fault.

The H.Y.C. pyritic shale is not widely distributed. It contains the McArthur deposit and some smaller lead–zinc deposits and consists largely of black and dark-grey pyritic shales and siltstones, thin bedded and widely disturbed by preconsolidation slumping, sliding, scouring, load-casting and dewatering structures. This H.Y.C. shale was deposited only in the Bulburra Depression. It contains important tuff horizons, the fine-grained nature of which suggests that the volcanic source was distal and lay on the land or shallow water on the edge of the basin.

The geology of the ore deposit has been summarized by Murray (1975). It contains an estimated 200 million t of ore with approximately 10% zinc, 4% lead, 45 ppm silver and 0.2% copper. Highest-grade ore contains up to 24% zinc and 12% lead. The ore body extends over an area of 1.5 km^2 and is usually 55 m thick, increasing to 130 m on its eastern edge. It contains seven mineralized shale ore bodies separated by relatively mineral-poor dolomite-rich beds. The ore bodies consist of thin laminae of fine-grained sulphides and tuffaceous shales. Coarse lenses and rounded patches of pyrite are scattered throughout the ore. Occasionally there are small discordant sulphide–carbonate veins containing some barite.

The sulphide was obviously deposited prior to sediment consolidation, for the sulphide and shale bands are frequently broken up by early diagenetic dewatering structures. The main sulphide minerals are pyrite, sphalerite and galena, with minor chalcopyrite, arsenopyrite, marcasite, chalcocite and covellite.

Smith, J.W. and Croxford (1975) believe that biogenic sulphur was involved in pyrite formation throughout the H.Y.C. pyrite shale member, and that different sulphur-isotope values reflect changes in water depth, sulphate availability and degree of reduction. Richards, J.R. (1975) showed that the galena has a model age of 1600 Ma. It may be generally concluded that the McArthur deposit formed on the sea floor; it is syngenetic with the early diagenetic formation of the sulphide bands. It is possible that the metalliferous solutions were derived in some way as a result of igneous activity that gave rise to the tuffaceous debris in the enclosing sediments (Lambert, 1976). However it is also clear that this is a distal deposit, formed in a sedimentary basin far removed from contemporaneous igneous activity.

Mount Isa–Hilton

The Mount Isa and Hilton silver–lead–zinc deposits are situated about 20 km apart in the lower Proterozoic Urquhart Shale Formation, metamorphosed under lower greenschist facies (chlorite grade) (*Figure 10.6*). The Mount Isa Group and its syngenetic mineralization is likely to have been deposited 1500–1570 Ma ago (McClay, 1979). The geology of both mines has been described by Mathias and Clark (1975). There are two distinct ore types at Mount Isa, both of which occur in the pyritic Urquhart Shale Formation, considered to be the correlative of the H.Y.C. Pyrite Shale of the McArthur deposit to the northwest. The silver–lead–zinc ore bodies are well-bedded dolomitic siltstones and shales rich in sulphides, whereas the copper ore bodies are located in massive brecciated and recrystallized 'silica dolomites' (a local name for fractured and recrystallized siliceous dolomites, cherts and siltstones). The silver–lead–zinc mineralization occurs throughout the Urquhart Shale Formation, which is of well-bedded carbonaceous, dolomitic, quartzo–feldspathic siltstone, dolomites and tuffites. The mineralization occurs as distinct concordant bands of galena, sphalerite and pyrite. When there is a sufficient concentration of these bands, an ore body is constituted. Tuff beds occur in the Urquhart Shale and they are used as marker beds for stratigraphic analysis.

The main sulphide minerals are galena, sphalerite, pyrite and pyrrhotite, and there are minor amounts of chalcopyrite, arsenopyrite, marcasite and tetrahedrite. The pyrrhotite could have formed from the pyrite as a result of metamorphism.

The mineralization occurs discontinuously throughout a stratigraphic width of 1200 m and over a strike length of over 1600 m, and at least to a down-dip extent of 700 m. There are 16 groups of economic beds. Ore bodies 1–5 have a thickness of up to 50 m; they are fine grained and contain abundant carbonaceous material, whereas ore bodies 6–16 are narrower and coarser grained.

The mineralization is associated with framboidal pyrite and microfossils. An extensive assemblage of blue-green algae, of shallow-water origin, has been described. Folding in the silver–lead–zinc ore bodies, particularly 2–5, has produced some remobilization and redistribution of the sulphides. The total estimated size of the ore bodies is around 100 million t of average grade 7.8% lead, 6.0% zinc and 130 ppm silver.

The seven ore bodies at Hilton (Mathias and Clark, 1975) occupy a stratigraphic interval of 300 m and contain at least 35 million t of ore containing 7.7% lead, 9.6% zinc and 125 ppm silver. In both mines there is evidence of repetition of the stratigraphic sequence by isoclinal folding of amplitude up to 200 m (McClay, 1979).

A biogenic origin for all the sulphides is favoured, based on the isotope studies. The galena lead-isotope data indicate a model age of 1600 Ma (Ostic *et al.*, 1967).

The fine-grained sulphides in the silver–lead–zinc ore bodies are generally accepted as sedimentary or early diagenetic in origin. A volcanogenic source for the sulphur and metals has been suggested (Stanton, 1972). The discovery of pseudomorphed sulphate evaporites in the Urquhart Shale Formation suggests, however, that the silver–lead–zinc ores may have been formed from brines during evaporation of the basin (McClay and Carlile, 1978).

There are two schools of thought for the origin of the copper deposits at Mount Isa, which are not present at Hilton or McArthur. Bennett, E.M. (1965) and Stanton (1972) consider that the copper ore formed syngenetically in a near-shore algal reef or reef breccia environment, which was subsequently sheared and recrystallized. Smith, S.E. and Walker (1971) conclude that the copper was derived epigenetically from the underlying greenstones during diagenesis, metamorphism and/or hydrothermal activity. The absence of copper ore from the Hilton and McArthur bodies can be explained by the lack of a greenstone source rock in the immediate vicinity of the pyrite shales. Finlow-Bates and Stumpfl (1979) have argued that the copper ores are a product of subsurface deposition from the same hydrothermal solution, which, on entering the sea, precipitated the silver–lead–zinc ore bodies.

Broken Hill

The geology of this well-studied deposit has been summarized by Johnson and Klingner (1975). Its structural complexity and high-grade metamorphism have caused a lack of consensus regarding its origin. The problems and controversies have been summarized by Both and Rutland (1976).

It is the richest deposit for its size in the world, totalling 200 million t of ore with an average grade around 12% lead, 12% zinc and 115 ppm silver. The deposit is of Proterozoic age, associated with the Willyama Complex of granulite facies metamorphic rocks (*Figure 10.6*). The age of metamorphism has been dated at approximately 1700 Ma. The deposit was recrystallized by this high-grade metamorphism and therefore was deposited before the Carpentarian deposits of northern Australia.

Granulite facies metasediments are the most abundant rock–type in the Broken Hill sequence; they enclose the conformable sulphides. Stratigraphically beneath the sulphides are interlayered pelitic, semi-pelitic, psammitic and quartzo–feldspathic metasediments. A monotonous and massive sequence of pelitic metasediments overlies the sulphides. In their vicinity, the metasediments display rapid facies variation. Four major units of garnet–plagioclase gneiss (known locally as Potosi Gneiss) are recognized. They occur below, above, and also laterally equivalent to the sulphides. The sulphide rocks and the banded-iron formation in the south of the field occur around the margin of thicker masses of garnet–plagioclase gneiss. The sulphide rock and the banded-iron formation occur on the same stratigraphic plane, but are separated by and are on opposite sides of the thick unit of garnet–plagioclase gneiss. Amphibolite is a minor though common rock-type. The banded-iron formation is minor but prominent (*Figure 10.6*); it consists of discontinuous layers 1–3 m thick (Stanton, 1976). Eight banded-iron formations occur in the mine sequence; two are lateral equivalents of the sulphide ore. The principal occurrence lies stratigraphically above the sulphide ore bodies.

The deformational history of the Willyama Complex has not yet been completely resolved. Large isoclinal folds and small and large second-generation folds are the most abundant. The structural interpretations of, for example, Johnson and Klingner (1975) and Laing *et al.* (1978) are different. The latter are more likely to be correct because they have incorporated into their analysis sedimentological as well as structural features. They conclude that the main high-grade metamorphism coincided with the F_1 folding, which produced the major large-scale folds. The mine sequence occupies the single inverted limb of an F_1 fold. Three major tight F_2 folds occur in the mine. In the southern part of the field the ore bodies are in or near the hinge of an inferred F_2

fold. In the northern part the ore bodies are in the western limb of an inferred F_2 dragfold. Second-generation sillimanite, garnet and biotite were formed during the F_2 folding episode. An F_3 phase produced crenulation of the earlier planar structures. It is concluded that sulphides migrated into the F_2 and F_3 fold axes during metamorphism.

The sulphide and associated rocks have been described by Lawrence (1973). The Broken Hill field has five separate stratiform ore bodies, which are mineralogically, chemically and spatially distinct. Because the lead-rich ore bodies structurally overlie the zinc-rich ore bodies, some authors have suggested that the ore bodies could have been overturned, a hypothesis supported by structural studies (Laing *et al.*, 1978). The lode deposit is continuous, with a 7.3-km strike length, an 850-m vertical extent, and a 250-m horizontal extent. The stratigraphically lower zinc-rich ore bodies are enclosed and separated by slightly retrogressed quartz-rich rocks. The lead-rich ore bodies are characterized by a sheared marginal zone. Spessartite–garnet rocks are associated with the lead-rich ore bodies. The sulphide lenses have different lead/silver/zinc ratios. The stratigraphically lowest ore bodies, at the southern end of the field, are richer in zinc and copper and poorer in silver and lead than those of stratigraphically higher positions (*Table 10.1*). The vertical chemical changes are quite marked. The major ore minerals of the sulphide deposits are coarse-grained galena and sphalerite. Pyrrhotite is locally abundant, but overall is minor. Chalcopyrite is minor and there are traces of arsenopyrite, lollingite and tetrahedrite. Pyrite is minor and is probably a replacement of pyrrhotite. The main gangue minerals are quartz, calcite, manganiferous garnet, rhodonite, bustamite and fluorite. The lode horizons contain minor pyrrhotite for considerable distances away from the ore zones.

More than 300 km strike length of quartz–gahnite lode horizons are present in the Willyama Complex. These lode horizons are concordant with the banding of the metasediments and they occur in siliceous sillimanite gneiss and quartzite (Plimer, 1979). The quartz–gahnite horizon is 1–20 m thick. The lode horizon in the mine sequence is a rock composed of quartz and ferroan gahnite, with or without the following minor constituents—spessartite, plumbian orthoclase, fluorapatite and sulphides. In places they define the strike extensions of the massive sulphide ore.

Isotopic studies indicate that the sulphur of the sulphides has a juvenile origin. However Lambert (1976) believes that the sulphur could originally have been like that at McArthur and Mount Isa, and subsequently changed by the high-grade metamorphism. The lead has a model age of 1600 Ma.

The paucity of carbonates, and the older age of the deposit, distinguish it from the McArthur-type deposits. The other features can all be ascribed to complex folding and repeated high-grade metamorphism. Because of the complexity of the deposit, there is an understandable divergence of opinion, and no clear agreement on the interpretation of its main features (*see*, for example, Rutland

Table 10.1 Mineralogy and metal contents of principal ore bodies of Broken Hill[a]

Ore-body mineralogy	*Typical mining grade* Pb(%)	Ag (ppm)	Zn (%)
B lode			
A: quartz, sphalerite, galena. M: chalcopyrite, garnet, pyrrhotite, apatite, feldspar. T: rhodonite, calcite, manganhedenbergite, sillimanite, mica, staurolite, amphibole	5	40	17
A lode			
A: sphalerite, rhodonite, manganhedenbergite, quartz, galena. M: garnet, calcite, gahnite, lollingite, arsenopyrite, chalcopyrite, pyrrhotite, apatite, feldspars. T: sillimanite, amphibole, pyroxene, staurolite, mica, sulphides, sulpho-salts	4	40	10
No. 1 lenses			
A: sphalerite, galena, quartz, calcite. M: feldspars, bustamite, manganowollastonite, apatite, gahnite, fluorite, chalcopyrite, pyrrhotite. T: sillimanite, staurolite, mica, amphibole	8	50	20
No. 2 lenses			
A: galena, sphalerite, calcite, bustamite, manganohedenbergite, rhodonite. M: chalcopyrite, pyrrhotite, lollingite, arsenopyrite, garnet, quartz, manganese-containing olivines, feldspars, micas, amphiboles. T: sillimanite, staurolite, cubanite, ilvaite, vesuvianite, johannesenite, sulpho-salts	14	100	11
No. 3 lens (southern portion of field)			
A: sphalerite, galena, quartz, fluorite, rhodonite, garnet	11	200	15
No. 3 lens (northern portion of field)			
A: galena, sphalerite, rhodonite, fluorite, quartz, garnet. M: chalcopyrite, pyrrhotite, lollingite, arsenopyrite, gahnite, apatite, pyroxmangite, amphibole, calcite, feldspar. T: sillimanite, staurolite, chloritoid, sulpho-salts, bustamite, pyrosmalite	15	300	13

[a] A, abundant; M, minor; T, trace. After Plimer (1979).

et al., 1978). The most reasonable interpretation, based on the available data, is that the Broken Hill ore body is a stratiform deposit formed by exhalative-sedimentary processes. The relative importance of the roles of the volcanic and sedimentary processes will continue to be debated. The presence of the banded-iron formation clearly shows that the contiguous sulphide ore is a chemical sedimentary deposit. The stratigraphic sequence includes metamorphic rocks that could be interpreted as mafic tuffs and rhyolite horizons. Exhalations into the sedimentary basin could have come from volcanoes associated with these igneous rocks. There is evidence of hydrothermal alteration of the footwall rocks beneath the sulphide ore bodies (Plimer, 1979), so that the exhalative-sedimentary nature of the Broken Hill deposit may have been more proximal than the deposits of McArthur, which are clearly distal.

The Central African Copper Belt

The Central African Copper Belt of northern Zambia and southern Zaire is one of the great stratiform metallogenic provinces of the world (*Figure 10.7*). Its study forms a basis for understanding synsedimentary and syndiagenetic deposition of metals in extensive intracratonic basins of fluctuating water depth. The eight operating mines of the Zambian copper belt produced in 1974 a total of about 750 000 t of copper. The ore reserves of the producing mines range from 374 000 000 t at Nchanga containing an average of 4.11% copper to 8 000 000 t at Bwana Mkubwa containing an average of 3.40% copper. The geology and mineralization has been summarized by Fleischer *et al.* (1976) and Bowen and Gunatilaka (1977).

The copper belt is situated on the southern slope of the flat watershed between the Zaire and Zambesi drainage systems, about 1300 m above sea level. The stratiform ore bodies occur in the Roan Formation of the Proterozoic Katanga Supergroup. The basins occupy complex synclines lying between positive basement features. These Katanga Supergroup formations were metamorphosed about 550 Ma ago in the Lufilian Orogenic belt, which lies between the Archaean Congo Craton to the north and the Kalahari Craton to the south. The Zambian copper belt occupies 160 km at the southeast end of the 800-km long fold arc, which stretches from Angola, through Zaire, to Zambia.

Katanga Supergroup sediments overlie an extremely irregular unconformity, with hills and ridges of 1975 Ma old granite and quartzite rising up to 100 m above adjoining valleys. The first Lower Roan sediments to overlie the unconformity are terrestrial talus screes, valley boulder conglomerate and aeolian sands, followed by marine-sediment deposition by a transgressive sea of beach gravel and algal bioherms, passing seawards into sands and muds, which must have been rich in anaerobic bacteria that produced hydrogen sulphide. The copper and cobalt sulphide deposits are restricted to the near-shore sediments. Seawards and upwards in the succession, the Lower Roan gives way to dolomite and dolomitic argillite containing beds of anhydrite of the Upper Roan Formation. Transgression was followed by regression, and the three major ore horizons can be related to the three major transgressions (*Figure 10.8*).

Regional metamorphism accompanied folding of the Katanga sedimentary succession, with increasing grade from lower greenschist in the east to high greenschist over most of the mining area to lower amphibolite facies in the west. The metamorphism has increased the grain size of the ore minerals and hence has aided in ore beneficiation. In the Katanga deposits of Zaire, there is no metamorphism of the ore-bearing strata (Bartholome *et al.*, 1976).

The Lower Roan is a transgressive marine sequence. At Mufulira, three strata-bound ore bodies are contained in identical phases of the cyclic sedimentation. The host rock is a fine-grained carbonaceous quartzite (Van Eden, 1974).

Copper-belt ore bodies are tabular; they have extremely long strike lengths and are relatively thick, extending from near-surface to depth with a down-dip length of 2000 m and more. There are two main types of ore-bearing rock—shale and arenite–arkose; about 60% of the reserves are in the shale-type. A remarkable lineament of the shale-type ore bodies exists through Luanshya-Nkana–Chambishi–Chingula–Karila Bomwe–Mushoshi in the ore–shale alignment lying immediately west of the basement Kafue anticline (*Figure 10.7*). In each case carbonaceous shale with pyrite gives way eastwards to low carbon cupriferous shale. These deposits are considered to be aligned along a palaeo-shoreline, the pyritic facies representing deeper water. Algal reefs associated with the ore bodies suggest a relatively shallower environment at the time of ore deposition. A common feature of the ore shale is intense tight drag folding, in contrast to the large-amplitude open folding of the more competent arenite ore bodies, which characterize the belt lying east of the Kafue basement anticline

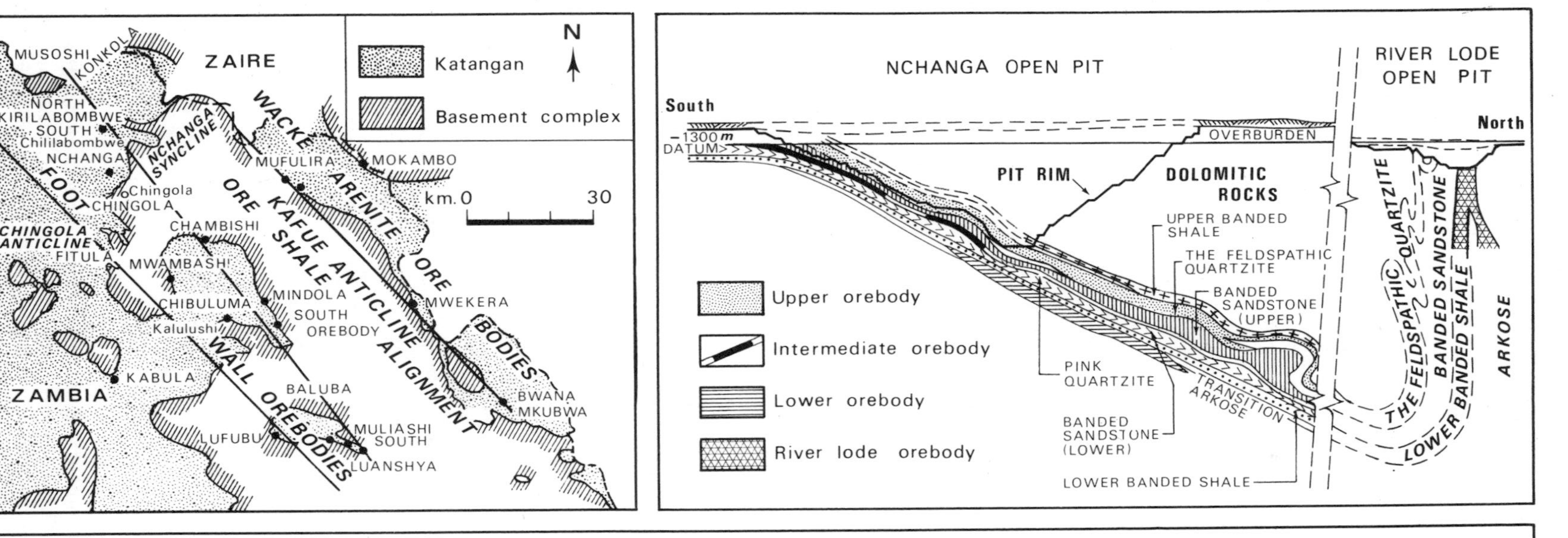

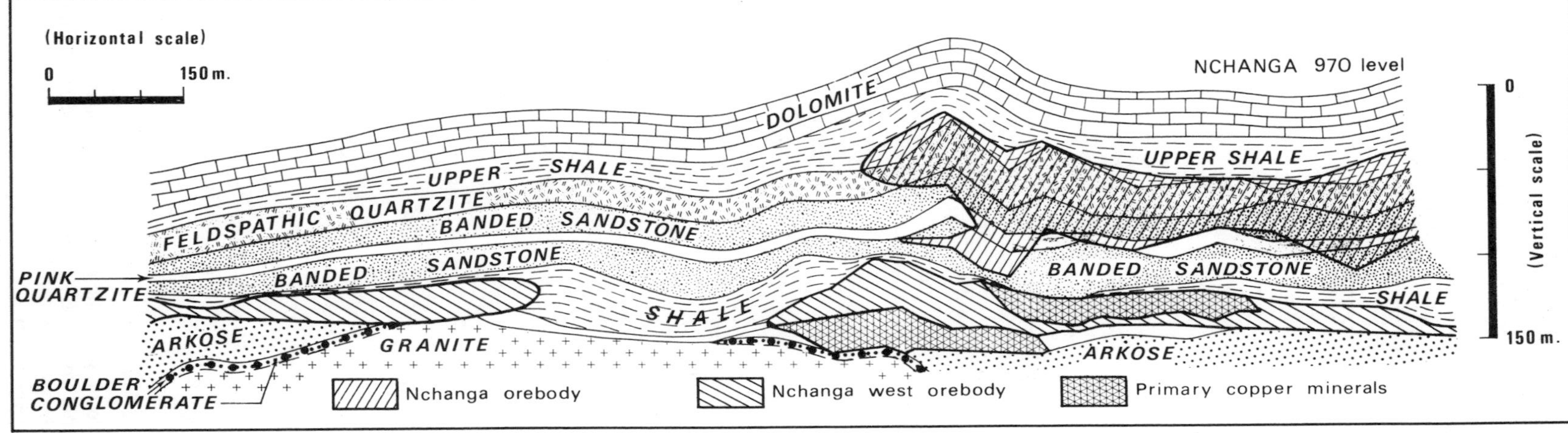

Figure 10.7. Distribution of copper-belt mines in Zambia. Structure and detail of the ore-bearing strata of the Nchanga mine (after Fleischer *et al.*, 1976; Bowen and Gunatilaka, 1977).

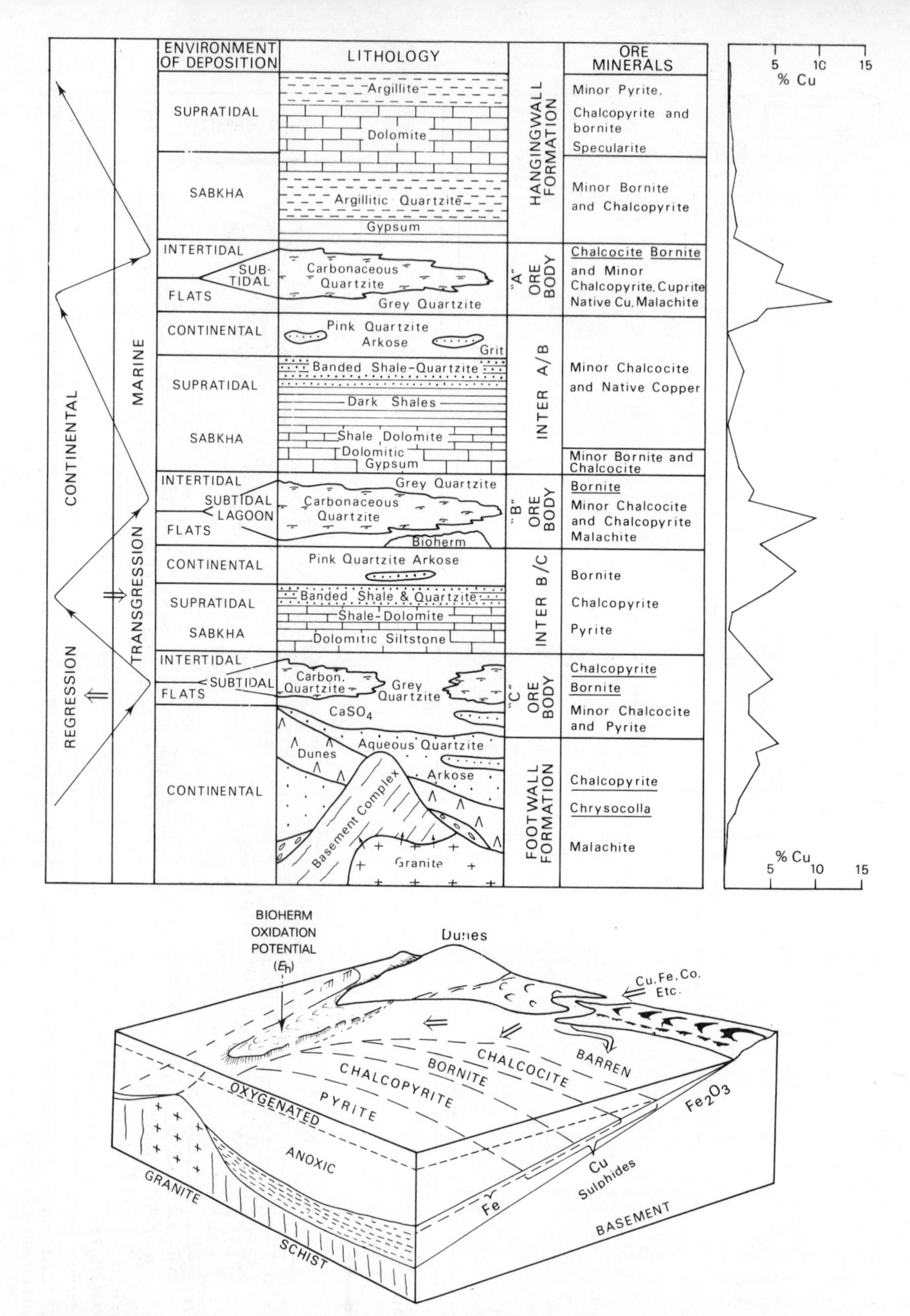

Figure 10.8. Cyclic sedimentation and its environment and control on copper deposition in the Zambia Copper Belt (after Renfro, 1974). Block diagram showing diffusion of copper, cobalt and iron in solution into the Roan Sea and the control of the palaeo-geography (after Garlick and Fleischer, 1972).

(*Figure 10.7*). The major sulphide minerals are chalcopyrite, bornite, chalcocite and pyrite (Notebaart and Vink, 1972). Carrolite and cobaltiferous pyrite occur where the deposits contain cobalt. There is a large number of other primary minerals but they are of minor importance. Where the deposits have been oxidized by deep weathering, they contain the copper carbonates malachite and azurite, native copper, cuprite and tenorite. The primary mineralization at each mine is sulphide with a cap of secondary oxidization due to weathering, usually to a depth of 30–70 m below surface depending on the permeability of the host rocks. Deep leaching down thin porous beds to a 500-m depth is common. The mineralization is disseminated in the sedimentary rocks. Cobalt occurs in economically extractable concentrations (up to 0.25% cobalt) in the shales and arenites in the mines situated west of the Kafue anticline (*Figure 10.7*). The grain size of the sulphides is similar to that of the enclosing sedimentary rocks. Lateral secretion veins occur abundantly in all ore bodies; they traverse both ore and country rock but the copper mineralization is confined within the limits of the stratiform ore body. Clearly these veins must not be considered as feeder veins for the ore is distinctly syngenetic with sedimentation.

At the start of each sulphide deposition, there is evidence of a change from prevailing oxidizing to reducing conditions as shown in the upward changes in an ore-body sequence from siderite to sulphide, barren brown shale to bluish shale with sulphide, and red limonitic arenite to grey arenite with sulphides (*Figure 10.8*). There is also a notable reduction of the anhydrite content within the ore body itself compared with the adjacent sediments. Sedimentary reworking took place in the basal parts of the arenaceous ore bodies.

The sulphide mineralization is regarded by most geologists as synsedimentary, and strongly controlled by the palaeo-topography. The depositional environment has been discussed by Garlick and Fleischer (1972). Prior to formation of the major ore deposits in marine waters, evaporation of playa lakes gave rise to a high concentration of sulphides and borates, and under anoxic conditions sulphide was precipitated. The drying out of the lakes gave mud flats rich in gypsum on which iron and cobalt sulphates formed (*Figure 10.8*). Sulphide in the sediments became sufficiently indurated to be eroded and redeposited as detrital grains, concentrated on foreset beds and truncation planes along with tourmaline, sphene, epidote and zircon grains.

The older basement rocks contain many occurrences of copper. For example, 1–3 m thick veins of quartz–carbonate with pyrite–chalcopyrite–bornite assemblages are exposed in the basement rocks in several mines. The metals could therefore have been derived by erosion of the basement highs; another source could have been distant volcanic-exhalative centres. Whatever the ultimate source, it is clear that they were eventually transported into the Lower Roan basins and concentrated by chemicobacterial processes in favourably confined basins at river mouths or gulfs under anaerobic conditions. Carbon occurs in all deposits and it is acknowledged that organic carbon provided some nutrients for the bacterial activity.

The origin of stratiform metalliferous deposits underlain by continental redbeds or other oxidized strata, and overlain by evaporites, which include the Kupferschieffer of Germany and the Roan of Zambia and Zaire, can be understood by a study of the coastal sabkhas (Arabic: uninhabitable evaporite flat bordering landlocked sea). Coastal sabkhas and their evaporite facies prograde seawards across adjacent algal-mat facies (Renfro, 1974). Upon burial, the algal-mat facies become saturated with hydrogen sulphide generated by anaerobic bacteria. Terrestrial formation water initially has low pH and high E_h values and thus can remobilize and transport considerable amounts of elements such as copper, lead, silver and zinc. As the terrestrial formation water passes through the hydrogen sulphide-charged algal mat, its load of solute material is reduced, and precipitated as sulphides. Deposits thus formed are conformable with the geometry of the strata. Such deposits are zoned from landward to seaward according to the relative solubilities in the presence of hydrogen sulphide (Renfro, 1974).

NCHANGA

The Nchanga group of deposits consists of the extensive Lower or Nchanga West ore body in shale and underlying arkose and a similar extensive Upper or Nchanga ore body in feldspathic quartzite. Both are on the south limb of the Nchanga syncline (*Figure 10.7*). The small River Lode is in shale on the steeply dipping north limb. There are at least twelve ore bodies along a strike of 24 km. The Katanga sediments are in the higher greenschist facies of regional metamorphism. Scapolite in the Lower and Upper Banded Shales indicates the original anhydritic nature of the sequence. The grade of the Lower (Nchanga West) ore body is in part due to surficial enrichment causing the deposit to contain as much copper oxide as sulphide. Many of the lower ore bodies are closely controlled by the surface eroded on basement granite, gneiss and schist during the first deposition of the Lower Roan. Ore in the Lower Roan arkose and in the Lower Banded Shale flanks both sides of the basement granite hill (*Figure 10.7*). As the arkose wedges out, the mineralization extends for 6 m into the basement granite, partly in fractures and partly in a metamorphosed palaeo-soil developed on the weathered granite. For a distance of 17 km, the ore is consistently close to the basement unconformity and it steps up gradually across the strata into the Lower Banded Shale. The Nchanga ore body may have formed in a local basin ascribed to compaction of the lower beds, while lesser compaction of the thinner sediments over the basement granite ridge caused shallower water above it and hence copper was not deposited.

Early Proterozoic Hydrothermal Uranium Deposits

Some of the largest and highest-grade uranium deposits occur in the Darwin region of northern Australia, and near Lake Athabasca in Canada. In these, the pitchblende was hydrothermally deposited in fracture-controlled structures within high-grade complexly folded metamorphic rocks, but the low-temperature mineralization continued to be remobilized. The human need for a classification system has led to unnecessary controversy. Childers and Bailey (1979) classify these deposits as structure-controlled and fracture-controlled, which is clearly correct, but tells little about their possible genesis. Genetic classifications do not have universal support. There is a widespread well-founded belief that the uranium was initially concentrated in the Late Archaean–early Proterozoic sediments, during the period 1900–2200 Ma ago, and further concentrated during the Hudsonian orogeny (1900–1800 Ma ago), which caused amphibolite facies metamorphism and anatexis of the rock sequences. There is a clear association of the deposits with envelopes of metasediments retrogressed to the greenschist facies. Thus Cornelius (1976) classified the deposits as metamorphic–hydrothermal. This is part truth, but does not tell the whole story. It has also been noted that major unconformities either beneath a redbed succession or overlying a deep lateritic regolith are always in close proximity to the ore concentrations, and some authors classify the deposits as unconformity-related and of supergene origin (Knipping, 1974; Langford, 1977). They hold that continental weathering and deep circulation of hydrothermal ground water was responsible for the ore concentration. Nothing is to be gained by trying to pursue a perfect classification further, for as shown below, these Proterozoic deposits have a long evolutionary history. It seems more than likely that they were initially synsedimentary, formed hydrothermally during regional metamorphism of the sedimentary basin, while some were predominantly concentrated by supergene processes related to an early Proterozoic regolith, and others were prevented from this type of concentration by a further thick cover of strata. Although they are primarily uranium deposits, some have significant economic contents of nickel and gold.

Whether or not the early Proterozoic unconformities were of genetic importance in their formation, such deposits are characteristically found in close proximity to such unconformities, which are therefore important for exploration purposes. Indeed the overlying rock strata have prevented the deposits from erosion.

All the deposits described in this section occur predominantly in crystalline basement rocks associated with, or overlain by, either redbed or clean sandstone units. If they are overlain by clean sandstones, then the upper part of the crystalline basement has been deeply lateritized. Rich *et al.* (1977) have summarized the formation of these hydrothermal deposits as follows. Pitchblende is probably deposited from oxidizing solutions in response to the reduction of U^{6+} to U^{4+}. The ore solutions may either be descending or deeply circulating surface waters that have equilibrated with atmospheric oxygen, or they may be deep waters of non-surface origin that have equilibrated with a haematitic aquifer (most deposits are associated with an alteration envelope that contains red haematite). The uranium that was deposited in veins, shear zones, breccia zones and folds was probably scavenged from uranium-rich rocks within the crystalline basement by these oxidizing fluids. This model suggests that the present or past regional association of redbeds with uranium-rich rocks (especially granites) may be a useful guide to exploration.

Alligator Rivers–Rum Jungle, Northern Australia

This area, extending 80 km south and 250 km east of Darwin, is one of the world's greatest uranium fields. With the exception of a few minor deposits, all known uranium deposits occur in Lower Proterozoic metasediments of the Pine Creek syncline (Dodson *et al.*, 1974). They are concentrated in the Rum Jungle and Alligator Rivers areas. The deposits are stratigraphically confined to the Lower Proterozoic Cahill Formation or its equivalents (South Alligator Group).

The region is occupied mostly by regionally metamorphosed Lower Proterozoic metasediments, some of which have been transformed by anatexis into migmatite complexes. In addition, granite stocks, dolerite and phonolite bodies intrude the sequence (Smart *et al.*, 1975). An undulating layered Oenpelli dolerite sheet (250 m thick in places) extends discontinuously through most of the region. The Lower Proterozoic rocks are unconformably overlain by the Kombolgie Formation, a generally flat-lying metasandstone and conglomerate.

Pink biotite granite, typically with a high uranium content, forms stocks up to 25 km in diameter. The Lower Proterozoic sediments are regionally metamorphosed, intensely folded and commonly faulted. The intensity increases with increasing metamorphic grade from southwest to northeast.

The individual deposits have been separately described: the south Alligator Valley deposits (Foy, 1975), the Nabarlek deposit (Anthony, 1975), Ranger One (Eupene *et al.*, 1975), Koongarra (Foy and Pederson, 1975), Jabiluka deposits (Rowntree and Mosher, 1975) and Rum Jungle (Spratt, 1965). A general review of the region is to be found in Hegge and Rowntree (1978).

Rum Jungle

This region lies 60 km southsoutheast of Darwin. Lower Proterozoic metasediments dip steeply away from a basement of the Late-Archaean Rum Jungle Complex. The contact was sheared during its domal uplift. The post-Archaean activity is also indicated by quartz–tourmaline veins radiating from the centres of the Complex into the overlying Lower Proterozoic rocks. The Complex is cut by the Giant's Reef Fault, which strikes northeast and has a horizontal displacement of about 6 km (Dodson *et al.*, 1974).

The Lower Proterozoic Batchelor Group sediments were deposited against and around the Rum Jungle Complex. They are in turn overlain by the Goodparla Group. Five small uranium deposits occur in the Lower Proterozoic schist, siltstone and shale of the Golden Dyke Formation of the Goodparla Group. The lodes are located in shears, faults and in a tight fold. The host rocks are carbonaceous sediments or chloritic slate, near the contact with the underlying Coomalie Dolomite. Pitchblende is the only important ore mineral, and quartz the most important gangue. Chalcopyrite and pyrite are common. Magnetite and fluorite occur as accessories in the metasediments. Granite has been dated at 1700–1760 Ma and the pitchblende at 650 Ma (Rich *et al.*, 1977).

Jabiluka

This is the world's largest uranium deposit, lying 220 km east of Darwin. The estimated ore reserves of Jabiluka One are 1 300 000 ore t containing, on average, 0.25% U_3O_8, and Jabiluka Two 52 000 000 t, containing 0.39% U_3O_8.

The Jabiluka One and Two deposits occur in the Cahill Formation within 300 m of each other. Jabiluka One coincides with an erosional window through the unconformably overlying sandstones of the Kombolgie Formation (Rowntree and Mosher, 1975). Jabiluka Two lies underneath 20–220 m of the Kombolgie Formation (*Figure 10.9*). There are four separate horizons, which acted as favourable hosts for mineralization, folded into open flexures dipping south and striking eastsoutheast. The hosts are mainly chlorite and/or graphite schists and their brecciated equivalents. Mineralization is of open-space fillings of amorphous pitchblende and lesser crystalline uraninite occurring as granular disseminations. Accompanying sulphides are pyrite with lesser chalcopyrite and galena. The main gangue is of chlorite, quartz and sericite. There is evidence of some silicification and either extensive magnesium metasomatism or depletion of alkalis as shown by the increase of magnesium oxide and decrease in potassium oxide and sodium oxide in the mineralized as compared with the unmineralized schists. There is extensive replacement of feldspar and iron–magnesium minerals by chlorite in the Mine Sequence and in the pegmatites and pre-existing garnet. A portion of Jabiluka Two contains economic native gold accompanied by traces of lead, bismuth and nickel tellurides, but there is no connection between gold and uranium grades. The uranium mineralization has been dated at 920 Ma (Hills and Richards, 1976).

Ranger One

These ore bodies occur approximately 15 km south of Jabiluka. The estimated ore reserves are 37 733 000 t with an average U_3O_8 content of 0.27%. The ore bodies occur just below the land surface and the depth of weathering fluctuates from a few metres to 35 m. The deposits occur close to the Lower Proterozoic–Middle Proterozoic unconformity. The main host is chloritized biotite–quartz–feldspar schist of the Upper Mine sequence (Eupene *et al.*, 1975). Some mineralization also occurs in chlorite veins con-

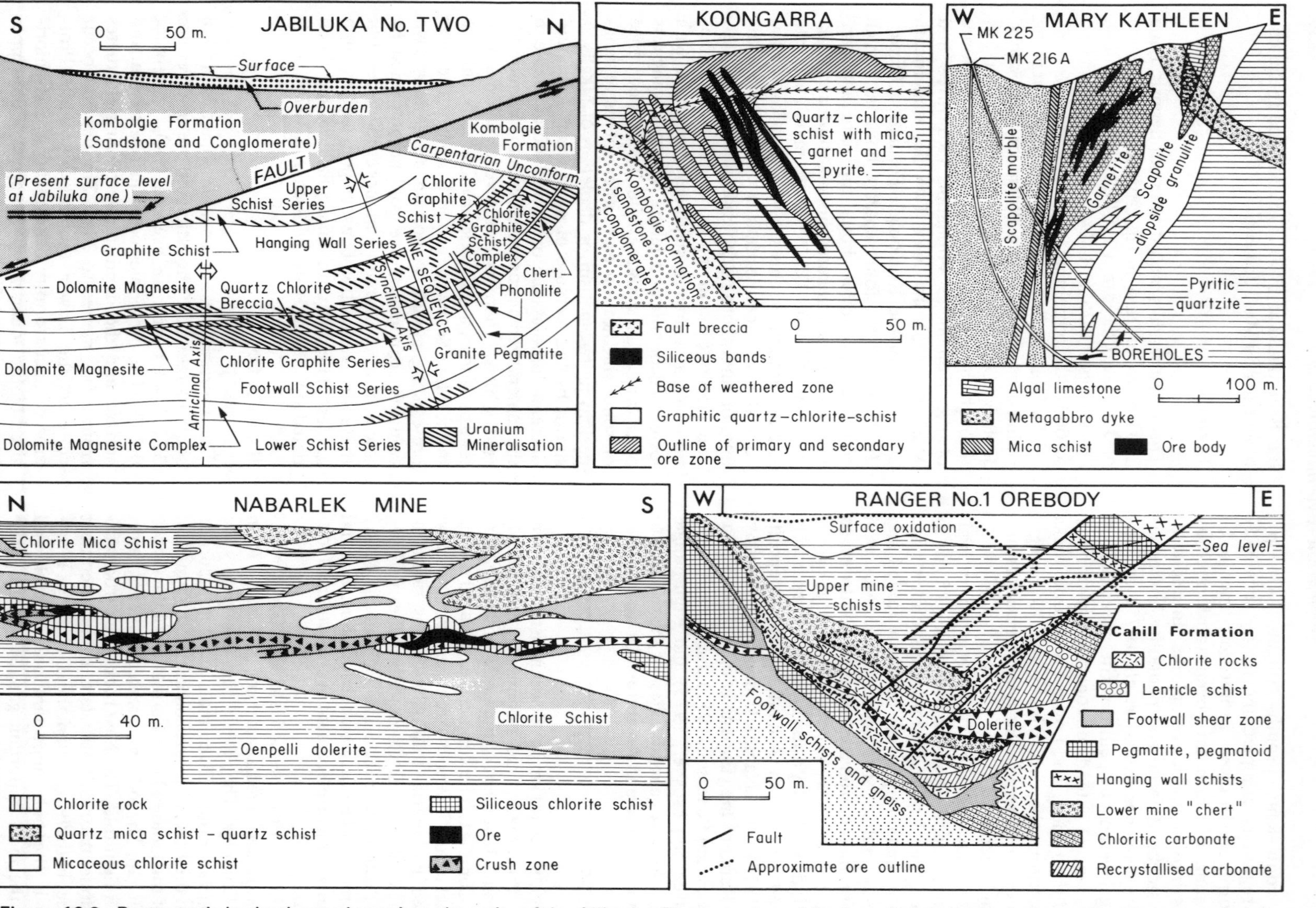

Figure 10.9. Proterozoic hydrothermal uranium deposits of the Alligator Rivers region of Arnhem Land, northern Australia. The ore occurs in shear, fracture and fold zones within regionally metamorphosed and retrogressed Lower Proterozoic Cahill Formation. After Rowntree and Mosher (1975), Foy and Pederson (1975), Eupene *et al.* (1975) and Anthony (1975). The cross-section of the Mary Kathleen deposit of Queensland is after Hawkins, B.W. (1975).

tained within silicified carbonate of the Lower Mine sequence (*Figure 10.9*).

The primary mineralization consists of vein-type and disseminated pitchblende. Near-surface secondary minerals include saleeite, skodowskite and torbernite. Chlorite, quartz and pyrite are common gangue minerals and there are traces of native gold. There is a strong structural control and most of the mineralization is in fractures and breccia zones. Just as at Jabiluka, there is a strong correlation between mineralization and magnesium metasomatism. The mineralization has been dated at 1700 Ma and possibly 900 Ma (Hills and Richards, 1976).

KOONGARRA

This region lies 20 km southeast of the Ranger deposits. The uranium mineralization occurs in Lower Proterozoic schist above a reverse-faulted contact with the younger Kombolgie Formation (*Figure 10.9*). The main deposit occurs in the Cahill Formation quartz–chlorite schist adjacent to the faulted contact (Foy and Pederson, 1975). Host rocks are variably fractured and brecciated because of the proximity of the reverse fault. Massive chlorite commonly fills the fractures. Mineralization is of uraninite in a crystalline or sooty amorphous form. Associated sulphides include pyrite, and traces of chalcopyrite and galena; gold occurs in trace amounts, while chlorite and quartz are the main gangue and haematite staining is prominent along the fault zones. There is quite a lot of secondary mineralization down-slope from the primary ore and along the fault zone. A number of secondary uranyl phosphates and silicates have been identified. Most of the mineralization occurs in fractures or breccia zones. The ore is spatially associated with chloritization.

Galena has been dated at 870 Ma and there are some lead dates as old as 1820 Ma (Hills and Richards, 1976).

NABARLEK

This is situated in Arnhem Land about 270 km due east from Darwin. Ore reserves are estimated at 402 000 t containing an average grade of 2.37% U_3O_8. The ore zone consists of very high-grade pods of pitchblende occurring at the intersection of a transcurrent crush zone and horizons of massive dark-green chlorite rock (*Figure 10.9*). The ore body has a tabular shape along the crush zone and is terminated at depth by the thick Oenpelli dolerite sill (Anthony, 1975).

The host rock is essentially of chlorite. The primary ore is pitchblende, either massive or disseminated. Galena, chalcopyrite and traces of pyrite are the main sulphides. Haematite occurs in association with sericite and coarse pitchblende. The mineralization is clearly related to the crush zone, and has been dated from galena at 920 Ma (Hills and Richards, 1976).

ORIGIN

The deposits all provide similar evidence for ore genesis. The stratigraphical control of the mineralization suggests that uranium was syngenetically enriched during the Lower Proterozoic Cahill Formation sedimentation. However it has been remobilized and the mineralization occupies faults, shear zones, breccias and fold hinge zones (Dodson *et al.*, 1974). The ore genesis is still a matter of contention; it seems reasonable to presume that the ultimate source was the Archaean basement, and that the uranium was deposited in the Proterozoic sediments of Pine Creek syncline. It may originally have been deposited in a reducing environment like the basal Proterozoic placers of South Africa and Canada. Smart *et al.* (1975) propose the following possible alternative processes for the final concentrations. (1) Uranium was mobilized by migmatization of the Lower Proterozoic sediments, and was deposited where a heterogeneous sequence of rocks provided suitable chemical and physical interfaces for precipitation. The ore was probably transported hydrothermally or by circulation of ground water and the presence of several phases of chlorite in the uranium deposits is considered to have resulted from the hydrothermal activity. (2) The anomalously radioactive pink granites and comagmatic acid volcanics exposed to the southwest and south of the region may have provided additional uranium.(3) Uranium was leached from the overlying Kombolgie Formation. The solution may have concentrated the uranium beneath its basal unconformity.

Dating of the mineralization (pitchblende and galena) has given ages ranging from 1800 Ma to 450 Ma (Hills and Richards, 1976). This suggests that the mineralization dates back to the metamorphic migmatization, and the ore has experienced subsequent remobilization since then, perhaps more than once.

The theory of genesis must take into account the following features that are common to the deposits: all lie within the transitional zone above migmatite complexes; all lie within or adjacent to post-migmatization structures such as brecciated fault zones, shear zones, or faulted collapse structures; and the host rocks show pronounced retrogressive metamorphism, such as amphibolite replaced by chlorite schist; quartz–chlorite schist, massive haematite–chlorite rock, and less commonly graphite schists are the major host rocks (Smart *et al.*, 1975).

The mineralization thus appears to be related to retrogressive metamorphism and remobilization of the uranium into structurally favourable traps. The sequence of events suggested by Hegge and Rowntree (1978) is as follows. (1) The Archaean granites, which contain, on average, 9.6 ppm uranium, were the original source. (2) The erosion of the Archaean highlands during Lower Proterozoic time resulted in miogeoclinal deposition of psammites, pelites and limestones. Sufficient oxygen was present to allow the uranium to be transferred as hydrated U^{6+}. The uranium was fixed in black carbonaceous shales under reducing conditions. Cahill Formation carbonaceous shales contain average values of 34 ppm uranium. (3) Regional metamorphism to amphibolite facies occurred at 1700–1800 Ma, accompanied by folding, faulting, anatexis and rejuvenation of the Archaean granites as gneissic migmatites. (4) Anatectic pegmatites were intruded. These events were perhaps accompanied by the initial stages of uranium remobilization. (5) Retrogressive metamorphism to greenschist facies was accompanied by magnesium metasomatism and chloritization, which resulted from the widespread circulation of uranium-scavenging hydrothermal solutions, probably rich in carbon dioxide and chlorine. The uranium was probably transported as a uranyl carbonate complex and precipitated by a decrease in pressure and subsequent escape of carbon dioxide. This could explain the introduction of pitchblende into open low-pressure spaces created structurally or by chemical means. (6) Uplift and erosion was followed by deposition of the Kombolgie Formation at 1600 Ma, and regional cratonization. (7) Remobilizations of uranium around 900 Ma and also at 500 Ma are related to the rejuvenation of the region and ground-water hydrothermal circulation.

The continued protective cover of the Kombolgie Formation preserved the deposits near the unconformity, and they have only recently been exposed by erosion, and in some cases subjected to near-surface secondary weathering.

Mary Kathleen Deposit, Queensland

The discovery of uranium in pegmatites in the highly metamorphosed Lower and Middle Proterozoic rocks near Mount Isa led to the discovery of the Mary Kathleen, which is Australia's largest uranium mine. Its reserves are estimated to be 9 483 000 t at a grade of 0.131% U_3O_8 (Hawkins, B.W., 1975). Davidite-bearing pegmatites are common throughout the area, but the Mary Kathleen mineralization does not occur in pegmatites. It has frequently erroneously been referred to as an igneous-related uranium deposit (for example, in Derry, 1978). This deposit is unique and several concepts of its origin have been proposed, but none appears to be generally accepted (Langford, 1978b).

The ore body consists of elongate lensoid shoots that are roughly parallel to the margins of a broader garnetite zone that encloses the ore. The ore zones dip at 30–50° west in the upper zone and become near-vertical with depth (*Figure 10.9*). The geology of the mine has been described by Hawkins, B.W. (1975). The envelope of the ore consists of garnetized and scapolitized calc-silicates, which are part of a highly metamorphosed Lower Proterozoic clastic-carbonate sequence. The garnet of the host is of andradite–grossularite, and the mineralization occurs within a network of zones controlled by a complex pattern of joints. The mine formations apparently lens out westwards, which gives an apparent synclinal structure to the area (*Figure 10.9*). The garnetite envelope is interfoliated with scapolite–diopside granulite, metamorphosed about 1670 Ma ago. The uraninite ore occurs as fine disseminations enclosed within allanite crystals, which usually replace the garnet. The uraninite forms oval grains, 0.01–0.1 mm across, surrounded by a thin silica shell. Stillwellite and fluorapatite also contain uraninite grains, and a few sulphide minerals are also present in minor quantities. The allanite that contains the uranium also is rich in rare earths, and these are extracted in the ore processing. For example, ore containing 0.05% U_3O_8 contains approximately 1.8% total rare earths. The rare earths are predominantly of ceric oxide (51.5%) and lanthanum oxide (33.5% of the total).

The source of the uranium and rare earths is uncertain. It has been suggested that they were introduced from a granite to the east, but there is no evidence for this; it is more likely that these elements were originally in the sedimentary sequence. Although highly metamorphosed, there are enough sedimentological features to indicate that the environment of deposition was a near-shore one within a restricted basin in which silty calcareous sediments were deposited behind sand bars. Minor algal reefs developed in places (Hawkins, B.W., 1975). The common occurrence of scapolite should not be regarded as metasomatic, but rather to have resulted from the metamorphism of the rock sequence that originally contained evaporite beds. It is imagined that the uranium and rare earths were remobilized and concentrated during the high-grade regional metamorphic event. The deposit is somewhat of an enigma, but it should not be further quoted as igneous-related.

Athabasca Basin Uranium Deposits

The uranium deposits of this region of northern Saskatchewan, Canada, are all located at the immediate vicinity of an unconformity between metamorphosed Archaean–early Proterozoic basement and overlying unmetamorphosed sandstone of the Helkinian Middle Proterozoic Athabasca Formation (*Figure 10.10*). They have been reviewed by several writers, including McMillan (1977a,b), Hoeve and Sibbald (1978a,b) and Tremblay (1978), and their mineralogy described and discussed by Rimsaite (1978). They have many similarities with the Alligator Rivers deposits of northern Australia, including their relationships with strongly deformed and metamorphosed Lower Proterozoic formations, a major unconformity, structural control of the hydrothermal mineralization, and their association with chlorite and graphite alteration of the country rocks. There are two main zones of ore occurrence—along the southeast and east margins of the Athabasca basin, within the Wollaston Foldbelt; and around the northern shore of Lake Athabasca in the Beaverlodge region. These two regions show certain differences in their style of mineralization, as demonstrated by Tremblay (1978), and summarized in *Table 10.2*.

Wollaston Foldbelt Deposits

The ore bodies occur within metasedimentary rocks of the Wollaston lithostratigraphic domain. The Wollaston Group comprises an Aphebian supracrustal succession of schists, meta-arkose and pegmatoids, of at least 3500 m thickness, which overlies the Archaean granites dated at 2470 Ma. The Wollaston Group is unconformably overlain by a succession of fluviatile quartz sands, conglomerates and greywackes of at least 1750 m. Beneath the unconformity, the Wollaston Group has been weathered to form a regolith, generally of 15–50 m depth, which has lateritic characteristics. The palaeo-latitude of deposition of the Athabasca Formation is calculated to be 20–25° N using an assumed 1350 Ma age of deposition, and this is consistent with the type of weathering of the pre-unconformity regolith.

At least four Hudsonian ductile deformation events are recorded in the region. Earlier deformation events were accompanied by low-pressure upper amphibolite facies metamorphism, and the later events by low-grade retrogressive metamorphism (Hoeve and Sibbald, 1978b), and at least two generations of faulting are recognized.

The Rabbit Lake deposit

This deposit has reserves estimated at about 32 000 t of U_3O_8 (McMillan, 1977b). The ore body has the shape of a laterally flattened, doubly plunging pipe tilted to the west. In the north, it is cut off at depth by the gently southerly dipping Rabbit Lake fault, and in the south it terminates above the fault at a depth of 200 m. A high-grade core is surrounded by a dispersed halo. The core occupies an ill-defined zone of repeated brecciation, which dips steeply.

Host-rock alteration pervades and envelopes the ore body (*Figure 10.10*). It is typified by chloritization, dolomitization, silicification and tourmalinization. The age of initial mineralization is difficult to determine because of frequent remobilization of the ore, but it is thought to predate considerably the oldest pitchblende radiometric date of 1075 Ma (Hoeve and Sibbald, 1978b).

The host rocks are interlayered calc-silicates and meta-arkoses with segregation pegmatites resulting from anatexis of the Wollaston Group. There is also a plagioclase-rich calc-silicate rock (called plagioclasite) and a biotite microgranite. Host-rock alteration includes an early dark-green chloritization, a red alteration caused by finely divided haematite in breccia zones, and a later pale-green alteration.

The mineralization is wholly contained within the altered-rock envelope, and comprises several generations of pitchblende (*see* p. 246) and coffinite, accompanied by a suite of secondary minerals. Three stages have been identified. (1) Fracture and breccia fillings in dark-green chloritized rock. The veins are predominantly of pitchblende and euhedral quartz and calcite, accompanied by adularia, chlorite, sulphides, coffinite and haematite. (2) Euhedral quartz veins that separate the episodes of red and pale-green alteration. (3) Impregnation in pale-green altered rocks of sooty

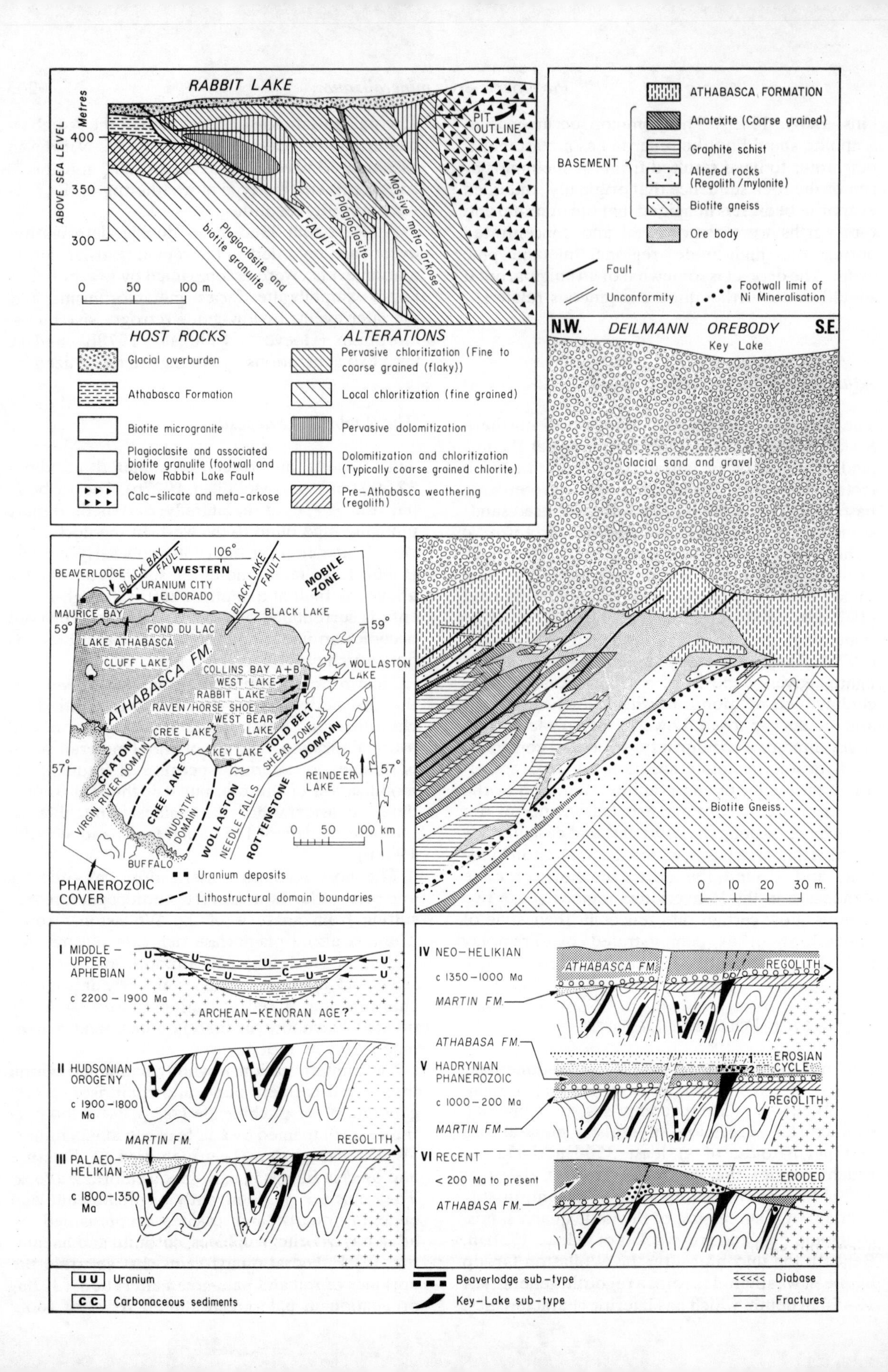

RABBIT LAKE
Metres
ABOVE SEA LEVEL
400
350
300
PIT OUTLINE
Massive meta-arkose
Plagioclasite
FAULT
Plagioclasite and biotite granulite
0 50 100 m.
HOST ROCKS
Glacial overburden
Athabasca Formation
Biotite microgranite
Plagioclasite and associated biotite granulite (footwall and below Rabbit Lake Fault)
Calc-silicate and meta-arkose
ALTERATIONS
Pervasive chloritization (Fine to coarse grained (flaky))
Local chloritization (fine grained)
Pervasive dolomitization
Dolomitization and chloritization (Typically coarse grained chlorite)
Pre-Athabasca weathering (regolith)
ATHABASCA FORMATION
BASEMENT
Anatexite (Coarse grained)
Graphite schist
Altered rocks (Regolith /mylonite)
Biotite gneiss
Ore body
Fault
Unconformity
Footwall limit of Ni Mineralisation
N.W. DEILMANN OREBODY S.E.
Key Lake
Glacial sand and gravel
Biotite Gneiss
0 10 20 30 m.
106°
BEAVERLODGE
BLACK BAY FAULT
WESTERN
URANIUM CITY
ELDORADO
BLACK LAKE FAULT
MOBILE ZONE
MAURICE BAY
59°
FOND DU LAC
BLACK LAKE
LAKE ATHABASCA
CLUFF LAKE
ATHABASCA FM.
COLLINS BAY A+B
WEST LAKE
WOLLASTON LAKE
RABBIT LAKE
RAVEN/HORSE SHOE
WEST BEAR LAKE
FOLD BELT
CREE LAKE
CRATON
KEY LAKE
SHEAR ZONE
DOMAIN
57°
VIRGIN RIVER DOMAIN
CREE LAKE
MUDJATIK DOMAIN
WOLLASTON
NEEDLE FALLS
ROTTENSTONE
REINDEER LAKE
0 50 100 km
BUFFALO
Uranium deposits
PHANEROZOIC COVER
Lithostructural domain boundaries
I MIDDLE – UPPER APHEBIAN
c 2200 – 1900 Ma
ARCHEAN–KENORAN AGE?
II HUDSONIAN OROGENY
c 1900 – 1800 Ma
MARTIN FM.
REGOLITH
III PALAEO-HELIKIAN
c 1800 – 1350 Ma
IV NEO-HELIKIAN
ATHABASCA FM.
REGOLITH
c 1350 – 1000 Ma
MARTIN FM.
ATHABASA FM.
EROSIAN CYCLE
V HADRYNIAN PHANEROZOIC
c 1000 – 200 Ma
REGOLITH
MARTIN FM.
VI RECENT
< 200 Ma to present
ERODED
ATHABASA FM.
U U Uranium
C C Carbonaceous sediments
Beaverlodge sub-type
Key-Lake sub-type
Diabase
Fractures

pitchblende and coffinite along fractures and joints.

There is a close relationship between brittle deformation, mineralization and alteration. The sequence suggested by Hoeve and Sibbald (1978b) is: 1350 Ma ago the basement rocks were weathered to form a regolith; the Athabasca Formation was unconformably deposited on the regolith; brecciation; dark-green chloritization; at 1100 Ma, stage-1 mineralization; brecciation, and development of the Rabbit Lake fault; red alteration, chloritization, tourmalinization, silicification and dolomitization; brecciation; stage-2 mineralization accompanying the euhedral quartz veins; pale-green alteration, and some silicification; stage-3 mineralization, and younger reworking of the deposits.

There were oscillations between oxidizing and reducing conditions, and the uranium mineralization followed each oxidizing episode.

The Key Lake deposit

This has reserves estimated to be more than 46 000 t of U_3O_8 (McMillan, 1977a). The deposit contains two major ore bodies that occur in a northeast-trending fault zone that dips northwest at 50–70° (Dahlkamp and Tan, 1977; Dahlkamp, 1978b). The Gaertner ore body exceeds 1400 m in length and the mineralization is normally 50–80 m below surface. It has grades up to 45% U_3O_8 and 45% nickel over a core length of 0.3 m. The Deilman ore body exceeds 1400 m in length, occurs 65–140 m below surface, and its highest grades are 59% U_3O_8 and/or 30% nickel over a core length of 0.3 m (Figure 10.10).

The main metallic elements are uranium and nickel. The nickel occurs as sulphide and sulpharsenide and the uranium as oxide and silicates. Lead, zinc and copper are present in accessory minerals. The pitchblende occurs in two forms —euhedral and radiating aggregates, and as sooty amorphous pitchblende. Coffinite forms rims around and patches within the sooty pitchblende. Gersdorffite is the most abundant nickel mineral; millerite and niccolite also occur. The gangue minerals are kaolinite, chlorite, quartz, sericite, calcite, sphene and epidote, but the massive ore is free of gangue. The mineralization occurs preferentially along or within shears that cut the rocks in large numbers. Mineralization has extended into the overlying Athabasca sandstone only where tectonic activity has produced cracks and fissures, and is confined to sooty pitchblende, gersdorffite and millerite (Dahlkamp, 1978b).

The ore body shows a vertical zoning. In the lower section, niccolite prevails, whereas in the upper part millerite predominates. Also massive pitchblende gives way upwards to sooty pitchblende.

Radiometric dating indicates the following sequence: pitchblende I was formed 1160–1228 Ma ago, which coincides with the age of the diabase dyke that cuts the Athabasca Formation. Pitchblende II was formed 370–960 Ma ago as dated by the coffinite. Sooty pitchblende has yielded radiometric ages of 107–250 Ma (Dahlkamp, 1978b).

The presence of tetragonal pitchblende and bravoite indicates that the ore formed at temperatures of around 135°C, and the ore texture conforms to a low temperature of formation (Dahlkamp and Tan, 1977).

The following genetic evolution of the deposit is offered by Dahlkamp (1978b), and illustrated in *Figure 10.10*. (I) During the middle-to-upper Aphebian time, uranium was transported into marine or lagoonal basins from erosion of the Archaean highlands, and deposited together with carbonaceous sediments. (II) The Hudsonian orogeny at 1800 Ma caused a concentration of the strata-bound uranium as the rocks underwent high-grade regional metamorphism. (III) In the Helikian–Middle Proterozoic time the area north of Lake Athabasca was covered with Martin Group sediments and protected from weathering (*see* description of Beaverlodge deposits, p. 246). In the Key Lake area there was no such cover and the uranium and other metals were remobilized by ground water as the regolith was formed. The mobile elements moved into structural traps. (IV) Subsequently the Athabasca Formation covered the area. Its possible thickness of several thousand

Figure 10.10 (opposite). Athabasca basin uranium deposits of northern Saskatchewan, Canada. Locality map of deposits situated around the Athabasca Formation outcrop (from Hoeve and Sibbald, 1978a). Cross-section of the Rabbit Lake ore body (after Hoeve and Sibbald, 1978b). Cross-section of the ore body and schematic sequence of events in the genesis of Athabasca-type ore bodies (after Dahlkamp, 1978a). (I) Transport of uranium into shelf or lagoonal basin; (II) mobilization and preconcentration during metamorphism; (III) sedimentation of Martin Formation in the north and deep regolith weathering in the south—ore bodies are leached where not protected and redeposited in structural traps; (IV) sedimentation of Athabasca Formation (1350 Ma) and diabase dyke emplacement—recrystallization of uranium; (V) periodic uplifts and erosion; (VI) continuing erosion and ore-body destruction.

metres could have further caused remobilization of the buried deposits at about 1160–1230 Ma and temperatures as high as 200°C could have been possible in the deeply circulating meteoric waters. (V) Successive periodic uplifting resulted in erosion of the cover and some redistribution of the deposits would have resulted, especially to give sooty pitchblende and coffinite. (VI) Continuing erosion and ore-body destruction.

An alternative hypothesis would be to disregard the whole weathering cycle and to suggest that the mineralization is entirely diagenetic and diagenetic–hydrothermal, as Hoeve and Sibbald (1978b) have done for the Rabbit Lake deposit. The feature that seems to favour the weathering element of the hypothesis is that pitchblende I, molybdenum and galena occur only within the crystalline basement and not in the overlying Athabasca Formation, which received only sooty pitchblende, coffinite, gersdorfite and millerite.

Mineralogical indications of continuing remobilization

Five types of pitchblende have been identified (Rimsaite, 1978). *Pitchblende I* is massive or colloform and occurs as bands (less than 1 mm thick) and crusts coating fractures along active faults and breccia zones. It appears to be the oldest and has highest PbO/UO_2 ratios of 1:2.8 to 1:8. *Pitchblende II* is lead-depleted and occurs along fractures and grain boundaries of pitchblende I. The PbO/UO_2 ratio varies from 1:8 to less than 1:100. *Pitchblende III* occurs as recrystallized euhedral cubes formed at the edges of fractured and massive pitchblende; crystal size is around 10 μm, and the cubes are associated with galena. This variety formed from uranium released during brecciation and alteration of pitchblende I and pitchblende II bands. The PbO/UO_2 ratio normally is 1:9.5 to 1:17. *Pitchblende IV* is a very fine aggregate of pitchblende, coffinite, chalcopyrite, bornite, covellite and minor carrolite and cobalt–nickel sulphide. It is low in lead with a PbO/UO_2 ratio around 1:700. *Pitchblende V* occurs as very thin veins and coatings a few micrometres thick on quartz, calcite, sulphides and hydrocarbons and as fracture coatings. The PbO/UO_2 ratio is very low, about 1:800. This variety crystallized after repeated fracturing and remobilization of uranium over a span of mineralization events. Any one ore sample will contain more than one type, inferring superposition of several remobilizations, and this is why radiometric dating is imprecise.

There is a distinct correlation between lead/uranium ratio and radiometric age of pitchblende: massive pitchblende of lead/uranium ratio 1:3.5 gives a $^{207}Pb/^{235}U$ apparent age of 1695 Ma and a $^{207}Pb/^{206}Pb$ age of 1795 Ma. Pitchblende containing gold and $NiAs_2$, with a lead/uranium ratio of 1:7.1 gives ages of 970 Ma and 1115 Ma. At the lower end of the scale, oxidized pitchblende with a lead/uranium ratio of 1:53 gives ages of 133 Ma and 360 Ma (Koeppel, 1968).

The mineralization complexity shows clearly that the primary mineralization was subsequently remobilized and each subsequent mobilization became depleted in lead. One of the important factors causing differences between individual uranium deposits of the region arises from their different alteration histories after the initial uranium mineralization. This clearly would be related to local geology differences, especially whether they were protectively buried by the Martin Formation during the prolonged Proterozoic weathering, or exposed to ground water–hydrothermal circulation at that time, and also the subsequent protective cover or access to hydrothermal circulation, even through to the present.

The Beaverlodge Mining Area

The uranium mineralization is contained within the Archaean-to-Lower Proterozoic Tazin Group composed of a thick (about 10 000 m) succession of interbedded greywacke, shale, feldspathic sandstone and basic tuffs metamorphosed 1930 Ma ago to amphibolite facies and associated with anatectic granites and gneisses (Tremblay, 1972). It has been deduced that the rocks were first metamorphosed to the granulite facies, and later to the amphibolite facies. In the areas of mineralization, most of the rocks have been retrogressed to the greenschist facies. The uranium mineralization is found mainly in the Fay Complex, but some also occurs at higher stratigraphic levels. Unlike the Wollaston Foldbelt deposits, in the Beaverlodge area the uranium-bearing rocks are overlain by a 14 000-m thick redbed succession called the Martin Formation, composed mainly of arkose, conglomerate and siltstone. It also includes basic flows and sills; the flows have been dated at 1630 Ma and the sills at 1410 Ma (Tremblay, 1978). It was deposited under subaerial conditions in a continental environment between 1780 Ma and 1930 Ma ago (Koeppel, 1968). Uranium mineralization occurs in the basal conglomerate above the Martin–Tazin unconformity, and also in the volcanic rocks. In both cases it occurs along and disseminated on either side of fracture systems.

The 1500-m thick Athabasca Formation is of

unmetamorphosed cross-bedded pure quartz sandstone with thin conglomerates and shales. Uranium mineralization occurs as disseminations and fracture fillings at or near its base (Tremblay, 1972).

The rocks are extensively deformed. The oldest deformation is represented by east–west trends in the Tazin Belt, related to the Kenoran orogeny. The Hudsonian orogeny imprinted northeast–southwest folds, and both of these dominate the area. Faults are pronounced in the Beaverlodge area and are represented by wide breccia and mylonite zones as well as by widely spaced fractures. At least four periods of faulting have been analysed; most strike east or southeast. The uranium mineralization is found within the faults or as disseminations on either side of the faults, or within folds.

Near the uranium deposits the wall rocks are characteristically altered. The alteration minerals are haematite, chlorite, carbonate, albite, quartz and epidote. They represent retrogressive assemblages imposed on the high-grade Tazin Group metasediments.

There are three distinct mineralization-types in the area (Tremblay, 1972, 1978). (1) Syngenetic deposits, represented by country rocks richer in uranium than most of the other country rocks. The deposits are not economic. The main minerals are uraninite, monazite, zircon and uranothorite occurring within basement pegmatites. This mineralization has been dated by Koeppel (1968) at two distinct periods, greater than 2200 Ma and around 1930 Ma. (2) The epigenetic deposits, which are the most common and important. Pitchblende is the main ore, and the deposits have no thorium or rare earths. Most of the pitchblende is colloform or massive and generally cryptocrystalline; it is rarely sooty or euhedral. The deposits also contain secondary uranium minerals when near the surface. Thucholite was noted in a few deposits. Haematite and pyrite are abundant locally, whereas chalcopyrite, galena, and sphalerite occur only as traces. There were several periods of ore deposition but the initial pitchblende mineralization occurred at 1780 Ma (Koeppel, 1968). The most common gangue is of haematite, calcite, quartz and chlorite. (3) The supergene deposits are characterized by uranophane, liebigite, kasolite and several other secondary minerals. They are small and seem to be of fairly recent origin; some contain admixed clay, and some have been worked.

The fluid-inclusion studies of Sassano *et al.* (1972) do not explicitly define the stages of pitchblende deposition, but represent the filling temperatures of the gangue minerals of the veins. Pre-uranium veins have been analysed to give temperatures of 145–195°C, and post-uranium veins to give temperatures of 60–410°C. The fluids have salinities in the range 26–28% sodium chloride equivalent by weight.

The uranium deposits of the Fay–Ace–Verna Mine and Bolger Open Pit occur in the footwall zone of the St. Louis fault. All bodies are within 100 m of the fault. The bodies are of breccia-type, stockwork-type, and vein-type (Tremblay, 1972). Sassano *et al.* (1972) have distinguished isotopically at least five generations of carbonate in the mine samples.

The Gunnar deposit is located near the intersection of two major faults. The pipe-like ore body consists of stockworks and breccia zones in carbonatized syenitic rock. The syenite appears to have been derived from granitic gneiss through albitization, carbonatization and desilication. Uranium in the Lorado mine occurs in a band of contorted graphitic and chloritic schists. The main ore control is structural within a syncline. Uranium occurs at the Cayzor mine in granitized chlorite schist and impure quartzite, and all ore occurs in fractures. At the Cinch Lake mine, ore occurs in a wedge-shaped fault block of Tazin quartzo–feldspathic paragneiss. The ore is both disseminated and fracture filling. Ore bodies at the Rix mine occur within brecciated and mylonitized rocks, both as stockwork-type and vein-type. Mining has largely been exhausted in the district. McMillan (1977a,b) records the following totals of U_3O_8 mined: Beaverlodge 22 675 t, Rex–Smitty 635 t, Cayzor 227 t, Cinch 680 t and Gunnar 8730 t.

Comparison of Beaverlodge with Wollaston Foldbelt Deposits

It is clear from *Table 10.2* that the deposits of the Lake Athabasca area (Beaverlodge) are not identical to those of the Wollaston Foldbelt to the east. Superficially they appear to be similar, in that they represent hydrothermal fillings of breccia zones, shear zones and occur as veins and disseminations in the neighbourhood of these structural traps. A major difference is the absence of the 1780 Ma and 1930 Ma old redbed Martin Formation from the Wollaston Foldbelt. The presence of this sequence in the west clearly prevented the formation of a deep regolith beneath the Athabasca Formation, as happened in the Wollaston Foldbelt. Most of the ore has been concentrated during the regolith formation, but clearly it was a remobilization of an earlier type of mineralization. It is the earlier type of mineralization that is seen at

Table 10.2 Comparison of the Beaverlodge-type metamorphic–hydrothermal and the Wollaston Foldbelt unconformity–hydrothermal uranium ore deposits of the Athabasca basin of Saskatchewan (after Tremblay, 1978)

Beaverlodge deposits	*Wollaston Foldbelt deposits*
Not stratigraphically confined to Martin Formation	Stratigraphically confined to Athabasca unconformity
Related to major faults or wide mylonite zones and to zones of closely spaced fractures superimposed on the mylonite zones	Related to faults in both basement and Athabasca sandstones; faults are old with late movements
Mineralization along fractures and disseminated in the wall rocks of the faults	Mineralization is along faults and disseminated along wall rocks. It is entirely either in basement or in Athabasca sandstone; locally it is in both
Ore is related to a specific lithological succession at a definite stratigraphic level in Tazin rocks	Not restricted stratigraphically but associated in basement with graphite, calc-silicate and amphibole-bearing rocks
Related to areas of intense granitization and widely retrograded to greenschist facies	Not related to granitization, but ore is confined to host rocks that have been retrograded to greenschist facies
Associated with areas intensely altered with red haematite, dark-green chlorite and white carbonate; some silicification, rare argillization	Associated with clay-white altered regolith areas in basement and with late green chloritic and sericitic areas in basement and sandstone
Mechanical detritus along unconformity is thin near ore zone and distribution is erratic	Maximum development of regolith at unconformity; widespread chemical alteration
Large horizontal and vertical ore extents	Large horizontal, but little vertical extent of ore
Pitchblende, minor secondary uranium minerals, rare brannerite	Pitchblende, coffinite and locally abundant secondary uranium minerals; sooty pitchblende common
Generally simple mineralogy, locally complex	Simple and complex mineralogy
Cu, V, Se and Ti: in variable amount and of erratic distribution; crude local zonation	Cu, Ni, Se, As, Au, Ag and/or Te: in variable amount and of erratic distribution; crude local zonation
Age of mineralization 1780 ± 20 Ma, 1140 ± 50 Ma and several younger remobilization periods	Age of mineralization 1200 Ma, with younger remobilization
Temperature of associated hydrothermal fluids 440–80°C, though the actual pitchblende temperature of mineralization has not been directly determined	Temperature of mineralization 260–60°C
Age of host rocks: mainly Archaean, locally Aphebian and Helikian	Age of host rocks: mainly Aphebian and Helikian, only locally Archaean

Beaverlodge, where it has not been significantly overprinted by hydrothermal remobilization associated with the 1200 Ma old basal Athabasca Formation regolith. However, on the face of it, the Beaverlodge deposits appear to be clearly older and less connected with the unconformity, whereas the Wollaston-Foldbelt deposits appear to be younger and unconformity-related. There can be no doubt that the initial mineralization in both cases was related to amphibolite metamorphism and anatexis of the host rocks, and that the main remobilization of the proto-ore was attended by greenschist facies downgrading of the country rocks. The role of the major unconformity and regolith formation appears not to be universal, and there is no justification for classifying all these deposits into an unconformity-related style. The differences in ages and mineralization styles have been resolved by Dahlkamp (1978a) as summarized in *Figure 10.10.*

11

Surficial deposits of continental areas

The long exposure of a low-relief continental land surface to weathering, and the inter-reaction of rainfall and ground water with the upper layers of its crust can give rise to a number of deposits of economic importance. Semi-arid climates are more conducive to the formation of deposits of this kind. Included in this chapter are bauxite deposits, the principal ore of aluminium. Laterite is formed by an analogous process, but is generally not of economic interest, except for nickel–laterite formed over ultramafic masses. Because nickel–laterite is associated with uplifted oceanic lithosphere in island arcs and continental margins, it is discussed in Chapter 2.

The inland drainage systems that develop on lateritized continental land surfaces may also concentrate uranium in the calcrete or caliche deposits associated with the alluvium-filled valleys. The process has analogies with the formation of evaporite deposits in intracontinental basins, but evaporites are not included in this chapter because their formation is not confined to surficial processes.

The deep weathering of continental crust, especially along its tectonically active up-faulted margins with oceanic plates, may give rise to important secondary mineral deposits by erosion and mechanical hydraulic transportation into fluviodeltaic or beach placers. Placer deposits are important for cassiterite, gold, diamond, ilmenite, rutile and zircon. Uraninite–pyrite placer deposits were restricted to the Basal Proterozoic, when the anoxygenic atmosphere permitted uraninite and pyrite to be transported detritally.

Bauxite

Bauxite is a residual or sedimentary formation enriched in aluminium hydroxide minerals and is the most important ore of aluminium. As much as 96% of the world's alumina comes from bauxite, only 4% comes from nephelinites and alunite. General reviews of bauxites are to be found in Bárdossy (1979), Kirpal and Tanyakov (1977) and Valeton (1972). The largest new discoveries have been in Australia, Jamaica, South America and India, but deposits are well known in North America, USSR, eastern Europe, Africa and Asia.

Classification

There may be significant differences between bauxites formed in upland areas and those formed on lowland peneplains (Grubb, 1973). The former are predominantly residual in origin, formed under free-leaching conditions. Lowland deposits depend on a fluctuating ground-water table for their formation, are therefore more complex, and may include residual, detrital and allochthonous material. However, subsequent chemical mobility frequently renders the various source materials indistinguishable, so that a simplified classification must be adopted.

Lateritic Crusts

These deposits are a result of *in situ* weathering resulting from wet tropical climatic conditions acting on a stable peneplained topography. They are underlain by and derived from aluminosilicate rocks. Fully 85% of the world's supply of bauxite is of this type. With increasing iron content, there is a continuous spectrum with uneconomic laterite.

The formation of lateritic bauxite is an *in situ* process. Minerals of the parent rock are directly weathered into alumina and iron hydroxide minerals by a very intense leaching of the alkalis and silica. An indirect formation of bauxite may occur by a process in which the aluminosilicate parent-rock minerals are first converted by weathering to clay minerals, which are then decomposed during a second stage of weathering. This indirect process is slower and less intense. The two processes, direct and indirect, may be concurrent or may occur in different geographic areas.

KARST-TYPE

These deposits overlie an uneven karstified surface of limestone or dolomite with a sharp disconformity. This category has been unnecessarily subdivided (Valeton, 1972). All occur as lensoid accumulations in solution hollows on the karst surface. About 14% of the world's deposits are of this type.

Karst bauxite formation always involves one or more reworkings of the primary weathered products. Therefore it is difficult in most cases to resolve the problem of the initial parent rock. There is no general agreement that the underlying carbonate rock is the parent material. For example, in the case of the karst deposits of Jamaica, wind-blown volcanic ash is held responsible for the bauxite parent material (*see* p. 251). It seems, however, likely that several parent materials contributed in different percentages to the accumulation of any karst bauxite deposit (Bárdossy, 1979).

SEDIMENTARY OR REDEPOSITED BAUXITES

These overlie an older weathered rock surface, with which they have no direct connection. They have been transported and sedimented as definite strata, which may be fossiliferous. These are best known from the Carboniferous and Mesozoic deposits of the USSR and China. Only about 1% of the world's supply is from this type.

The highly aluminous sedimentary deposits of China and the USSR must have been water-deposited in fresh-water lakes. In this case the drainage system must have been transporting alumina-rich material, so that the surrounding land must have been of a laterized blanket deposit.

Geographical Distribution

Cenozoic bauxites have formed only within the latitudes 30° N and 30° S. Palaeo-geographic reconstructions (Scotese *et al.*, 1979) for the Devonian bauxites show also that they similarly lay within the tropics during the time of their formation.

The lateritic bauxites occur on stable continental platforms, for example, on the Guyana Shield of South America, in West Africa, India and Australia. Others occur on younger cratonized areas, for example, on the Malay Peninsula and adjacent Indonesia, which cratonized only in the Triassic. The karstic bauxite belts are located to the north of the main laterite areas. They are concentrated in the Caribbean (Jamaica) and in the North Mediterranean bauxite belt (Greece, Hungary, Yugoslavia). Sedimentary types are most common on the Russian Platform, Ural Mountains and China. The laterite and sedimentary deposits are almost entirely confined to stable continental platforms; the karstic types to younger orogenic belts.

Distribution in Time

There is a definite periodicity of bauxite formation (Kirpal and Tanyakov, 1977): (most intense) Cretaceous to Recent, possibly subdivided into two submaxima-early Cretaceous to Palaeocene, and Miocene to Recent—most of the economic deposits are in this category; Devonian to Late Carboniferous; Uppermost Proterozoic to early Cambrian; and (least intense) Middle Ordovician. In addition, there are some earlier deposits of minor significance. The oldest known are about 3100 Ma. They are the low-grade metamorphic bauxitic laterites of Swaziland, Southern Africa.

Composition

Bauxites are characterized by a combined enrichment in aluminium, iron and titanium with a complementary depletion in potassium, sodium, calcium, magnesium and silicon, relative to their source rock. High-grade bauxites generally contain 40–60% aluminium oxide (Al_2O_3), 10–24% iron oxide (Fe_2O_3) and 2–5% titanium dioxide. There is also a low-iron bauxite that is white and contains less than 5% Fe_2O_3 with as much as 70% Al_2O_3, but this is rare and largely confined to the sedimentary (reworked) bauxites of Chinese-type. The titanium dioxide content can be as high as 10–20%. Trace elements such as beryllium, chromium, gallium, nickel, vanadium and zirconium may be enriched in bauxites.

More than 170 minerals have been identified in bauxites, but only ten occur in rock-forming quantities, namely gibbsite, boehmite, diaspore, corundum, goethite, haematite, kaolinite, halloysite, anatase and rutile (Bárdossy, 1979). Gibbsite is the main aluminium mineral, but boehmite becomes of increasing importance in older deposits.

Cenozoic deposits are predominantly of gibbsite. Mesozoic are predominantly of boehmite, and Palaeozoic of diaspore. Metamorphism changes the mineralogy from gibbsite to boehmite, to diaspore and eventually at higher grades to corundum (emery deposits). However, microcrystalline corundum occurs in some lateritic and karstic deposits that have never been buried or metamorphosed. This corundum is concentrated in oolites and pisolites, and was formed diagenetically under surface conditions (Bárdossy, 1979).

Origin of Bauxite

Bauxitization is a surface-weathering phenomenon. The main factors are detailed below.

CLIMATIC

Tropical monsoon climate with annual mean temperature around 26°C, annual rainfall 1200–4000 mm, number of rainy months 10 or 11.

VEGETATION

Generally between a tropical forest and savannah-type. Micro-organisms probably play an important role in decomposing the minerals of the parent rock and in buffering the pH of the percolating ground water.

GEOMORPHOLOGY

Gently undulating land surfaces where surface erosion and detrital sedimentation are insignificant.

HYDROLOGY

Weathering profile must be above the ground-water table or at least above its minimum level. Good drainage is absolutely necessary for bauxite formation; without it, leaching cannot proceed.

PHYSICOCHEMICAL CONDITIONS

In lateritic profiles the pH of the ground water is neutral to slightly acid and the water itself is strongly unionized. In the karst environment the ground water is slightly alkaline and contains dissolved alkalis. The higher the temperature of the ground water, the more intense the leaching. Oxidizing conditions are favourable for bauxitization; slightly reducing conditions result in a much slower formation rate.

PARENT-ROCK COMPOSITION

Under favourable conditions, practically all rocks undergo laterization, but the speed varies. High porosity and permeability of the parent rock are favourable, combined with a low surface stability of the minerals. High initial alumina and low silica are especially favourable for the formation of high-grade bauxite. Such rocks are feldspar-rich, and may be gneisses, granites, rhyolites, shales or arkoses.

TIME FACTOR

Laterization is considered to be a slow process, taking some 1–5 Ma. However, under favourable conditions, a complete bauxite mantle may have formed in a few hundreds of thousands of years.

Jamaica

Jamaican bauxite occurs as pocket and blanket deposits overlying almost exclusively relatively pure Middle Eocene to Lower Miocene karst limestone (*Figure 11.1*). Thick pockets occur, mainly filling sink holes in the Eocene and Oligocene beds; some deposits are more than 30 m thick. The blanket deposits occur in depressions in tilted fault blocks in Lower Miocene limestone (Comer, 1974). The high-grade bauxite has an average Al_2O_3 content of 50% and SiO_2 5% or less. Lower-grade bauxite has the same Al_2O_3 content with SiO_2 up to 10%. *Terra rossa* is a red soil containing less Al_2O_3 and SiO_2 greater than 10%. The bauxite and the terra rossa both contain 6–26% Fe_2O_3 and 1.7–3.5% TiO_2. The red colour is because of haematite. Formerly the bauxites were held to be residual because they exclusively overlie limestone. According to this theory, solution of the limestone left an argillaceous residue that subsequently altered to bauxite as a result of tropical weathering. However the Middle Eocene to Lower Miocene host limestone contains insufficient alumina to give rise to the bauxite by a residual process. Although Cretaceous and Lower Tertiary igneous, sedimentary and metamorphic rocks are thought to be ideal source rocks, they were protected from the climate by the pure limestone capping.

Miocene volcanic ash seems to be the best choice for bauxite parent material. Bentonitic clay occurs in the Middle and Upper Miocene limestones both as interbeds and as dispersed insoluble residue. Its high alumina and low silica show it to be chemically ideal for bauxite formation. It is believed that the volcanic ash alters to bauxite by desilication of the glass, plagioclase, biotite and ferromagnesian minerals through either a phyllosilicate phase or a gel stage. The volcanic ash is interpreted as falling on to the limestone karst topography during the Middle and Upper Miocene

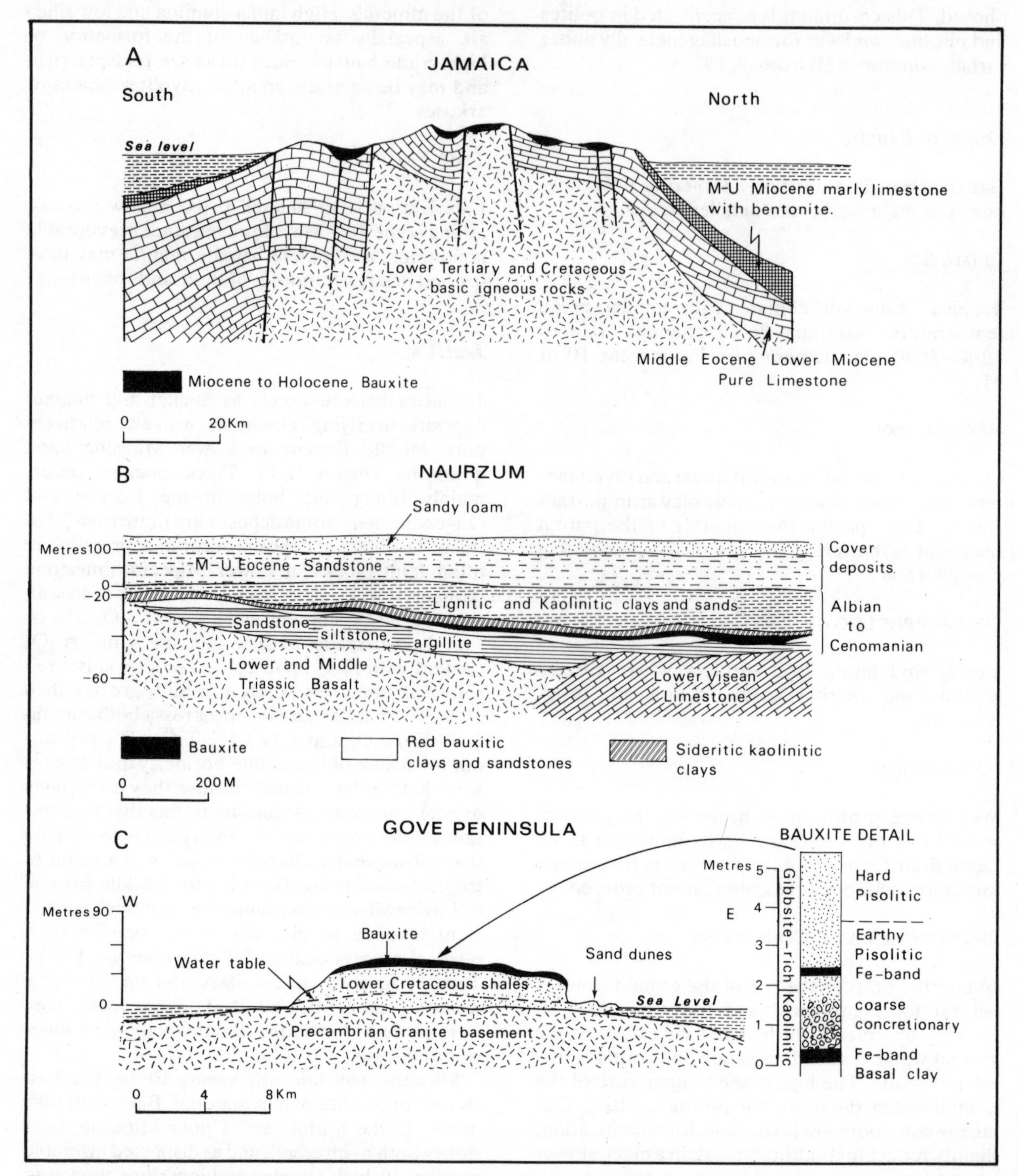

Figure 11.1. Cross-sections through three bauxite deposits. (A) Across the island of Jamaica showing the pocket occurrences of bauxite in karst depressions (after Comer, 1974). (B) The sedimentary Naurzum deposit of Tugai, northeast of the Aral Sea (after Kirpal and Tanyakov, 1977). (C) Gove Peninsula, northern Australia. Tertiary lateritic bauxite blanket deposits were formed *in situ* on horizontally bedded Cretaceous shales (after Grubb, 1970).

epochs and filling in solution hollows and depressions as well as forming a blanket cover over the flanks of the uplifted island core.

The most important control on the genesis and distribution of the bauxite is the subsurface drainage. Vertical drainage of vadose water through the faulted host limestone is extremely efficient in leaching silica from the overlying soils and concentrating the alumina.

Australia

Australia is now the world's major bauxite producer. The three major mining centres are: Weipa on the Cape York Peninsula of northern Queensland; the Darling Ranges inland from Perth, Western Australia; and the Gove Peninsula of the Northern Territory.

WEIPA

Laterite residual deposits, formed on Tertiary sedimentary rocks, cover an area of at least 1300 km^2. Proven reserves are 500×10^6 t with indications of an additional 1500×10^6 t (Evans, H.J., 1973, 1975). The bauxite is a surficial flat-lying blanket averaging 2.5 m thick with less than 1 m of overburden. The ore is strongly pisolitic (ranging in size from 1 mm to 20 mm) in a red-brown silty matrix. The ore minerals are a mixture of gibbsite and boehmite. Alumina distinctly decreases and silica increases regionally away from the coast. The grade therefore seems to be related to the drainage system within the laterite field. The bauxite has been developed by *in situ* weathering of arkose and sandy clay of Tertiary age. Rutile concentrations are formed by alteration of ilmenite during intense laterization, which is thought to be still going on at the present day.

GOVE PENINSULA

Laterite caps a continuous plateau system of more than 12 km^2 area, and elevation of 10–100 m above sea level (*Figure 11.1*). The plateau is underlain by flat-lying Cretaceous sandstone and shale, which in turn overlie the irregular Archaean granite basement (Somm, 1973, 1975).

The bauzite thickness ranges up to 10 m with a 3–4-m average. The upper layer is of loose pisolitic bauxite, with a middle layer of cemented pisolitic ore, and a bottom layer of tubular or vesicular ore. Non-bauxitic laterite overlies the commercial bauxite while gibbsite and haematite are the predominant minerals—only a little boehmite is present. Silica is low and increases rapidly at the bottom of the profile. The geomorphology of the peninsula suggests former eustatic sea-level changes. Part of the plateau was once below sea level and hence some of the bauxite may have been reworked. The age of bauxitization is presumed to be Tertiary (Grubb, 1970).

DARLING RANGES

The main deposits occur within a 50-km wide zone along the western margin of the continental plateau. The bauxites are considered to be the *in situ* fossil soil of Tertiary age developed on a partly dissected peneplain at elevations of 250–500 m (Baker, G.F.U., 1973, 1975).

The chief ore mineral is gibbsite. Ore bodies are discontinuous and are enclosed within ferruginous laterite. The average ore thickness is 4 m with 60 cm of overburden. The bauxite profile is of a hardcap, underlain by a friable zone, which passes abruptly into decomposed bedrock of Archaean granite, gneiss and metasediments. The laterites occur as ridges parallel to the structural features in the basement. Bedrock dolerite dykes often form boundary limits to the bauxite ore bodies. The bauxitization is thought to be related to the high rainfall regime during the Tertiary period. The deposits are directly derived from the Archaean granites and gneisses by *in situ* weathering and there are no signs of reworking.

Grubb (1979) has made a comparison of the Australian and the Amazon bauxites of South America. An outstanding difference is the general lack of true pisolitic material in the Amazon basin. Wherever pisolitic formations occur in Australia there is lack of a clay overburden. Hence the resulting good drainage ensures optimum leaching with only minimum weathering so that pisolites often show complex peripheral zoning.

Cenozoic lateritic bauxites have an almost exclusive association with stable low-lying platform areas. However most deposits show the effect of epeirogenic uplift of the platform, with lowering of the erosion base levels causing renewed incisement of rivers along existing drainage patterns followed by secondary bauxitization with or without some physical reworking of the deposits (Grubb, 1979).

Bedded Deposits of China

The two major bauxite mining regions of China are in the Kunghsien district of Honan and the Hsiuwen district of Kweichow. Both districts mine

Middle-to-Late Carboniferous highly aluminous bedded bauxite (Ikonnikov, 1975).

KUNGHSIEN

The Late Carboniferous horizontally stratified beds contain long lenses of bauxite that occur in extensive clay beds of even thickness. Diaspore is the major mineral; it is accompanied by muscovite, chlorite, kaolinite, rutile and quartz. The ore is light-grey or white, oolitic or pisolitic and contains little iron. The lower parts of the deposit turn gradually into aluminous grey-green shales. The ore composition is Al_2O_3 60–73%, SiO_2 6–17% and Fe_2O_3 3.4–11%. The associated shales are thick and contain haematite and siderite. The basin contains six layers. The fifth from the base contains the high-alumina bauxite. Late Carboniferous fossil plants occur in the third layer from the base. The lowest bed is iron-rich and unconformably overlies Middle Ordovician limestone. The uppermost part of the formation is covered by shales that contain a thin coal layer, and is then overlain by Upper Carboniferous limestone. The ore bodies are of large extent, tens of kilometres east–west and a few kilometres north–south. Gallium is an important trace element and it is chemically extracted.

HSIUWEN

Huge, thick, originally horizontally bedded, Middle Carboniferous bedded deposits have been folded and faulted during the Cretaceous period. Erosional remnants of these beds are worked by opencast methods. The major ore mineral is diaspore with lesser boehmite. The Al_2O_3 content averages 70%, SiO_2 11% and Fe_2O_3 2%. Kaolinite varies from 0% to 20%. Bauxite beds occur between two layers of massive aluminous shales, into which they may grade. The ore is earthy, oolitic, detrital, or massive. At the base, the shales contain iron and lie unconformably on Middle Ordovician or Late Cambrian limestone. The bauxite deposits are overlain by Upper Carboniferous shales and limestones.

The Naurzum Sedimentary Deposit

This deposit is located in the centre of the Turgai down-warp, about 500 km northeast of the Aral Sea (*Figure 11.1*). The pre-Cretaceous basement consists of Carboniferous limestone and Triassic basalts. The rocks of the Palaeozoic basement are overlain by a weathered crust down to a depth of 50 m in places. The weathered crust is overlain by Cretaceous continental bauxite-bearing strata (Kirpal and Tanyakov, 1977); their total thickness is 30–80 m. The bauxite-bearing strata are overlain by 10 m of marine Maastrichtian glauconitic sandstone, and then by Eocene beds. The bauxite-bearing sedimentary-rock sequence fills a gentle shallow depression in the pre-Cretaceous basement, termed the Naurzum basin. The upper-bauxite horizon consists of stony, friable and argillaceous bauxite. The beds that lie above the bauxite are lignite-bearing. The bauxite is iron-rich, and the main ore mineral is gibbsite. Average compositions are Al_2O_3 39.98%, Fe_2O_3 22.5% and TiO_2 4.8%. The source of the bauxite is presumed to be the weathered basalt, weathered *in situ*, then eroded and redeposited in the Naurzum depression. The high-iron content is to be expected from such an origin.

Uranium Deposits in Calcrete

Carnotite is a potassium uranyl vanadate that commonly occurs as a secondary accessory mineral around reduced uranium deposits. In recent years, however, carnotite has been recognized to be an important primary ore mineral in the calcreted drainage systems of the arid and semi-arid portions of the Archaean Yilgarn Block of Western Australia (Langford, 1974, 1978a). Testing of the Yeelirrie deposit has led to estimates of 46 000 t of ore, containing an average grade of 0.15% U_3O_8, half of the deposit being above 0.3%.

Calcrete is a surficial limestone formed by ground water, derived from the whole catchment area, rising towards the surface in the axis of valleys. Uranium leached from the underlying Archaean granite and gneiss occurs as carnotite in vugs and fractures within the calcrete. Since the calcretes are surficial, it follows that the carnotite deposits are also of surficial origin. Vanadium has been important in fixing uranium as insoluble carnotite and it appears to have been leached from the surrounding laterite of the peneplained countryside (Langford, 1974). The main Yeelirrie ore body is 6 km long, 0.5 km wide and 8 m thick (*Figure 11.2*) with depth to the base of the ore only 9 m.

The most common variety of calcrete is white, hard massive dolomite, with a porcellaneous lustre and conchoidal fracture. It is vuggy and cavernous. There is also an earthy fine-grained soft friable calcrete, which contains more calcite. Sometimes the dolomite variety occurs as nodules in the earthy variety. The calcrete contains lenses of

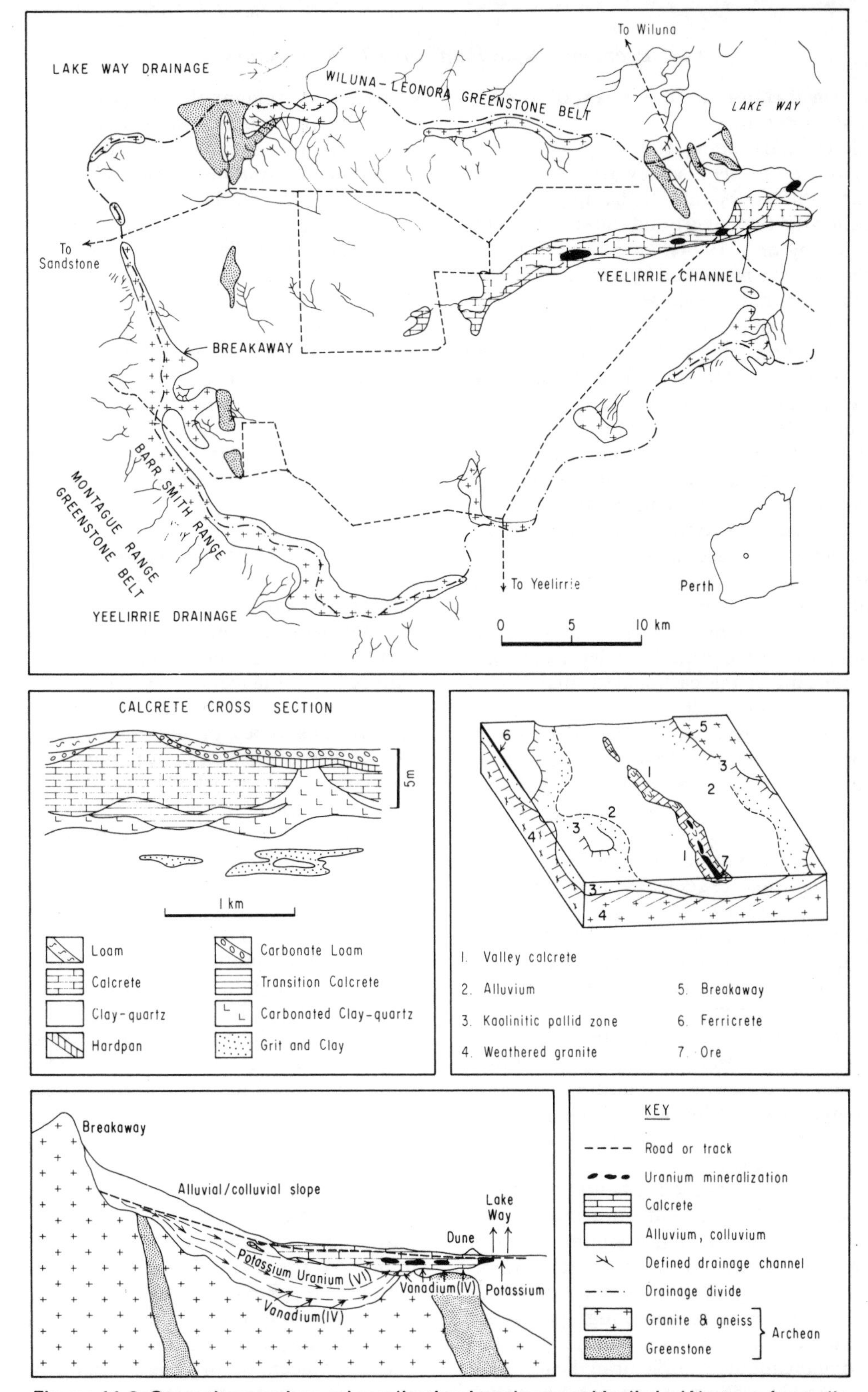

Figure 11.2. Carnotite uranium mineralization in calcrete at Yeelirrie, Western Australia. The carnotite-bearing calcrete is formed along the axis of the basin, filled with alluvium and colluvium, and underlain by weathered granite, gneiss and greenstone of Archaean age, which have acted as a source of the uranium, transported by subsurface ground water (after Langford, 1978a; Mann and Deutscher, 1978).

alluvium. Silicification in the form of opaline silica is quite common.

The calcrete starts about 15 km upstream from the main ore body. The valley narrows downstream, and this is thought to be important in concentrating the flow of ground water to form the ore body (Langford, 1974). In most places the base of the calcrete is marked by a transition zone in which the carbonate is calcite.

The major part of the valley is filled with brown alluvium consisting of quartz and clay eroded from the outcrops, locally called breakaways (*Figure 11.2*). The uranium is entirely of greenish carnotite and in places is covered by a thin shell of silica. It forms thin films on voids and fractures in the porcellaneous calcrete, and is disseminated in the earthy calcrete. It also occurs in the underlying calcareous alluvium as coatings on quartz grains. The carnotite is the last mineral to form, and, unless protected by a thin shell of silica, it is continually remobilized and reprecipitated (Langford, 1978a). The economic mineralization occurs throughout the calcrete profile to a depth of 9 m. It may be abundant in the carbonated quartz–clay zone, although the bulk is in the calcrete zone.

Valley calcretes are usually ascribed an age of Tertiary or more recent, but in some cases they are thought to be forming at present. They are confined to the lower parts of main drainages and to the upstream side of evaporite zones of salt lakes or playas, which act as drainage sumps.

Three mechanisms are seen as vital for carnotite formation in calcreted drainage channels—evaporation, decomplexing and reduction/oxidation of the ground water. Present-day ground waters provide the variation in salinity and E_h value required for carnotite formation. Remobilization is suggested by Mann and Deutscher (1978) as one of the important processes. The optimum conditions for carnotite deposition are a combination of the following three. (1) Constriction of a large slow-moving volume of ground water from an extensive granitoid catchment so that large amounts of leached uranium are available. That the granite basement is the source of the uranium is supported by high uranium values of 10–442 ppb in present-day ground water (Mann and Deutscher, 1978). (2) Partial evaporation of these waters in order to promote high potassium and uranyl ion concentrations in solution, the latter by decomplexing the soluble uranyl carbonate complexes. (3) Up-welling of waters and/or mixing of the waters with deeper-seated ground waters that have had access to a source of vanadium. The principle of moderately reducing ground water mobilizing and transporting uranium and vanadium to precipitate carnotite at a subhorizontal redox front and/or salinity boundary that complexes the uranium could be equally applicable to other aquifer systems in arid and semi-arid regions throughout the world.

It is possible that carnotite could also accumulate in surficial caliche deposits of sand and gravel. Caliche is a calcareous accumulation in soil that forms as a result of evaporation of soil moisture as the ground dries out after a rainy period. At present the only known giant uranium calcrete is at Yeelirrie, Western Australia, but Langer Heinrich in Namibia (Southwest Africa) may turn out to be as large (Cheney, 1981). There is also a similar deposit at Dusa Mareb in Somalia (Derry, 1980), but details of both these deposits are not available.

Placer Deposits

Placers are surficial sedimentary deposits formed by mechanical concentration of mineral grains eroded from exposed primary sources. An excellent review of placer characteristics and classification is given by Smirnov (1976) and a summary of the various types of deposits by Hails (1976). Regardless of the economic mineral, all placers are formed by and reworked by similar concentration processes. The dominant process is fluvial, but can also be marine, eolian, lacustrine or glacial. The economic material concentrated is usually a heavy mineral such as cassiterite, gold, or ilmenite. The cause of placer formation can be traced to hydraulic base-level changes, resulting from sea-level fluctuation or more spectacularly from tectonic uplift, resulting in rapid erosion (Henley and Adams, 1979). It is the tectonic uplift that is crucial to the formation of giant placers, which may be formed at any time when an uplifted land mass is exposed to long periods of deep weathering. Temperature seems to be less important than heavy seasonal rainfall, although tropical conditions leading to deep laterization seem to be conducive to the deep weathering of the primary deposits, and heavy rainfall to rapid erosion. The alluvial plain in which the placers are formed may drain towards the sea or inland towards a continental lake. Beach placers are formed by surf action during storms and preserved during beach accretion during fair weather.

Giant placers occur where older orogenic belts containing low-grade epigenetic ore veins (usually quartz-rich) have been uplifted and eroded. The vertical uplifts are characteristic of cordillera consuming plate margins, well documented in the circum-Pacific region, where important placer formation occurred throughout the Mesozoic and the Cenozoic eras.

The basal Proterozoic gold–uranium placers of Canada and South Africa show a remarkable similarity to the Quaternary cassiterite placers of Southeast Asia, although separated by about 2500 Ma of earth history, illustrating the fact that the processes of heavy-mineral concentration have not changed with time.

Tin Placers of Southeast Asia

Over 95% of the present tin output of Malaysia, Thailand and Indonesia is from Late Cenozoic alluvial placers. The major tin sources are in braided streams, piedmont fans and residual elutriation placers together with rich eluvial and colluvial placers. They are spatially restricted mainly to the flanks of geomorphologically positive granite features and their contact zones with the country rocks (Batchelor, 1979a).

The stratigraphic relationships of the Late Cenozoic alluvial deposits of Southeast Asia are summarized in *Figure 11.3*. The Old Alluvium (or Older Sedimentary Cover) is a complex of unconsolidated sediments that overlies the scattered occurrences of boulder beds and bedrock (Stauffer, 1973). The sequences are often complicated, slumped, extremely lenticular, and often difficult to decipher. It has been conclusively shown from sedimentological studies of the great alluvial tin field of the Kinta Valley (Sivam, 1969) that the Old Alluvium is of fluvial origin. Its extreme lateral variability (*Figure 11.4*) is a typical feature of fluvial sedimentation (Stauffer, 1973). The fluvial environment is also indicated by peat, peaty clay and occasional piles of partly lignitized branches and trunks of trees occurring as discontinuous beds and lenses. The existence of a deeply weathered

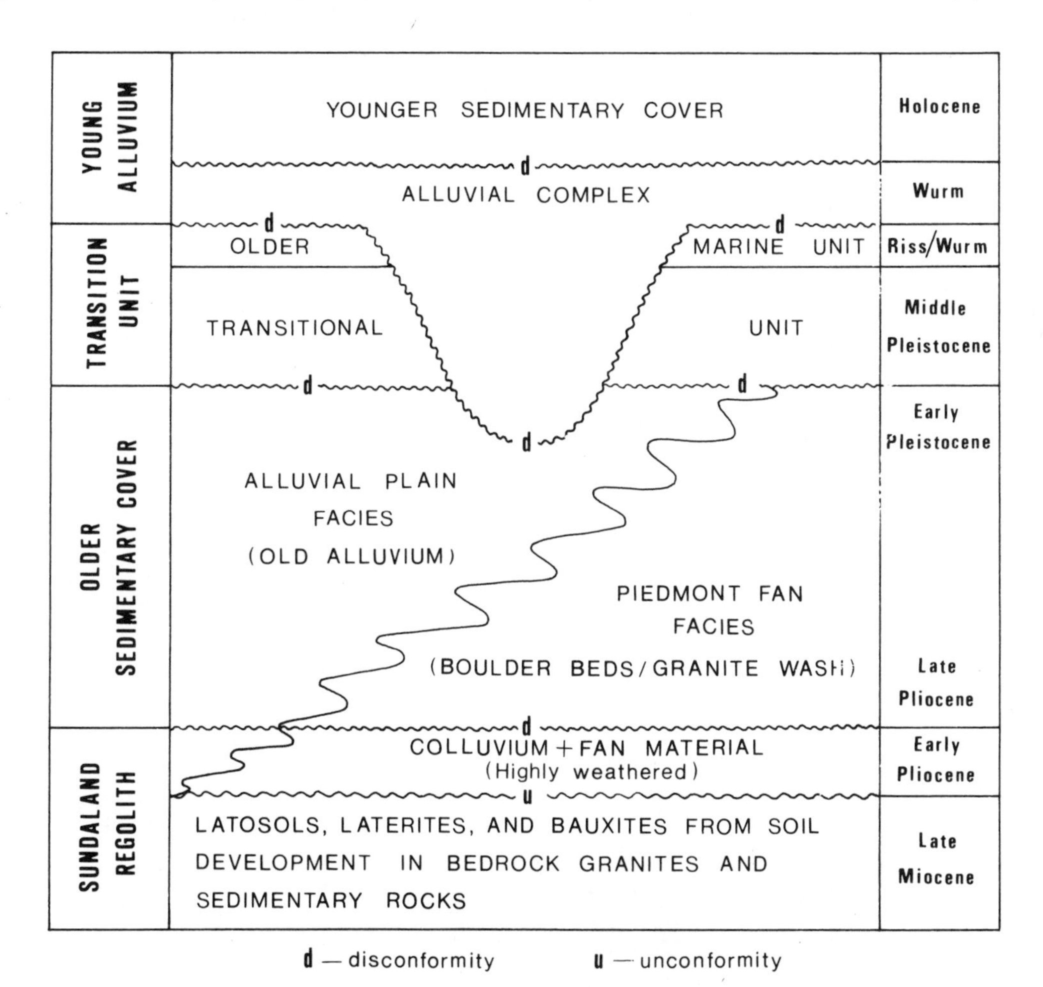

Figure 11.3. Regional Late Cenozoic stratigraphic relationships in Sundaland, Southeast Asia (after Batchelor, 1979a).

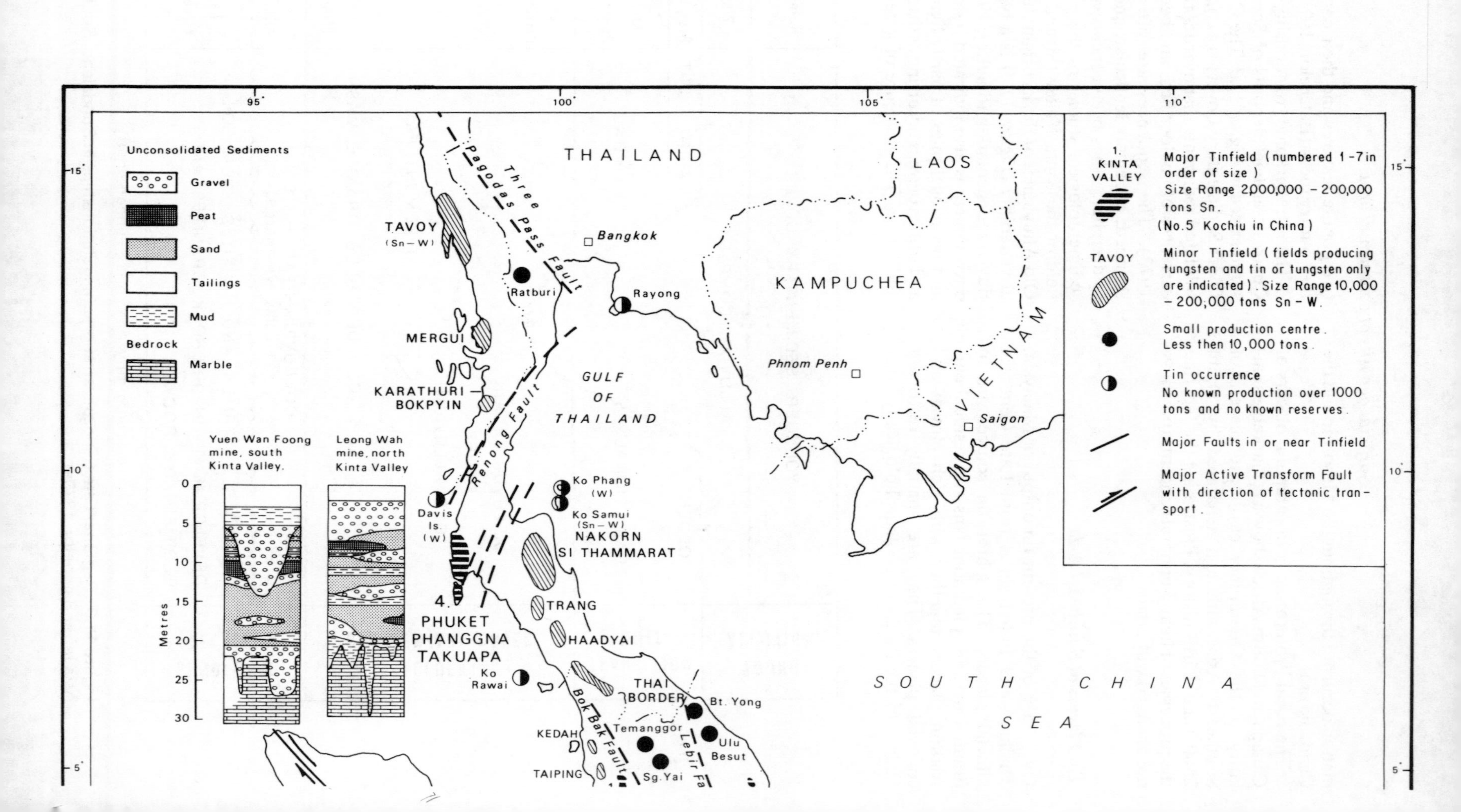

95°
100°
105°
110°
15°
10°
5°
Unconsolidated Sediments
Gravel
Peat
Sand
Tailings
Mud
Bedrock
Marble
THAILAND
LAOS
KAMPUCHEA
VIETNAM
Bangkok
Rayong
Ratburi
Phnom Penh
Saigon
Three Pagodas Pass Fault
Renong Fault
Bok Bak Fault
Lebir Fa
GULF OF THAILAND
SOUTH CHINA SEA
TAVOY (Sn–W)
MERGUI
KARATHURI – BOKPYIN
Davis Is. (W)
4. PHUKET PHANGGNA TAKUAPA
Ko Rawai
Ko Phang (W)
Ko Samui (Sn–W)
NAKORN SI THAMMARAT
TRANG
HAADYAI
THAI BORDER
Bt. Yong
Ulu Besut
Temanggor
Sg. Yai
KEDAH
TAIPING
Yuen Wan Foong mine, south Kinta Valley.
Leong Wah mine, north Kinta Valley
Metres
0
5
10
15
20
25
30
1. KINTA VALLEY
Major Tinfield (numbered 1-7 in order of size) Size Range 2000,000 – 200,000 tons Sn. (No.5 Kochiu in China)
TAVOY
Minor Tinfield (fields producing tungsten and tin or tungsten only are indicated). Size Range 10,000 – 200,000 tons Sn – W.
Small production centre. Less then 10,000 tons.
Tin occurrence No known production over 1000 tons and no known reserves.
Major Faults in or near Tinfield
Major Active Transform Fault with direction of tectonic transport.

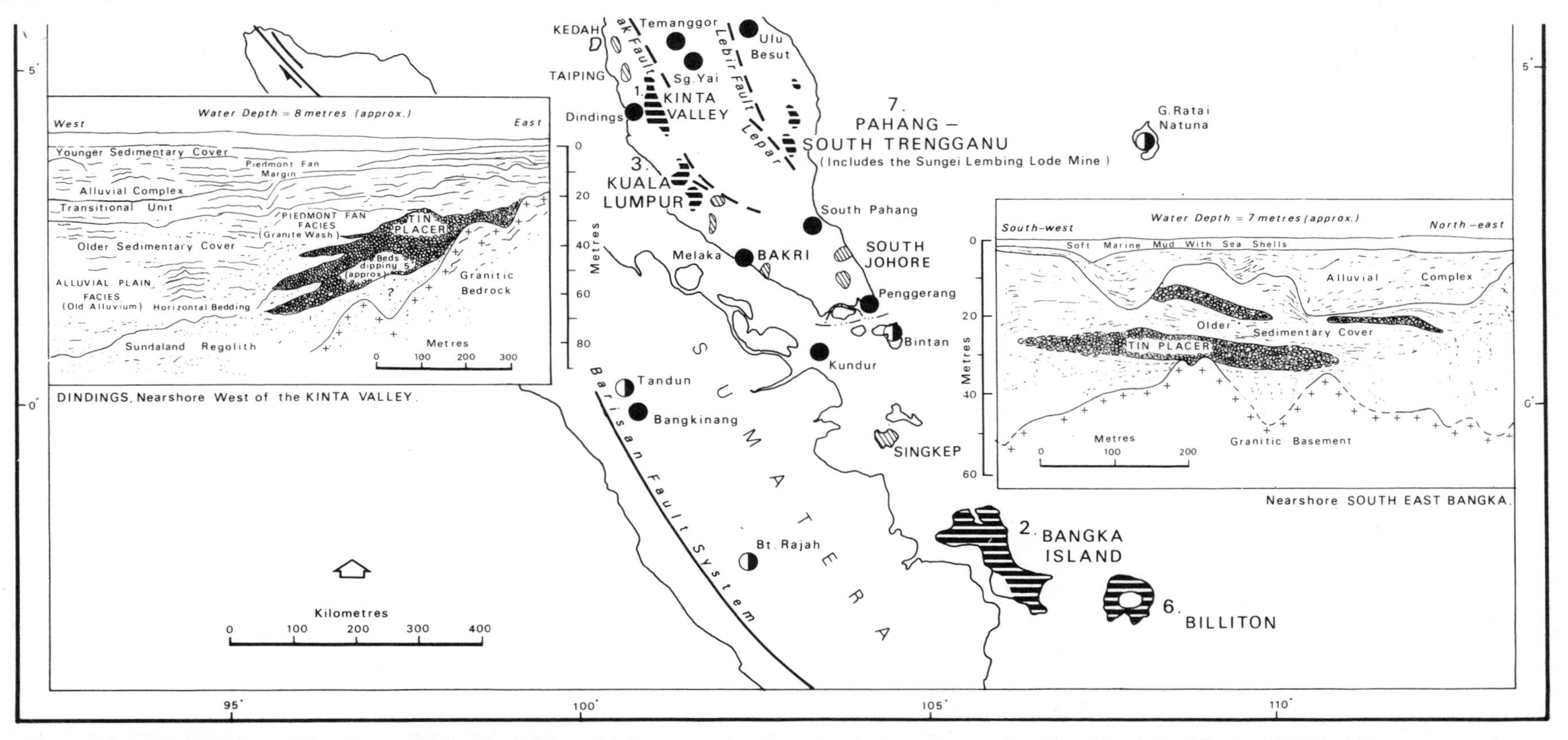

Figure 11.4. Distribution of Southeast Asian tin fields, which are predominantly of placer-type, after Hutchison and Taylor (1978), with cross-sections from Aleva (1973), Stauffer (1973) and Batchelor (1979b).

regolith bedrock is illustrated by the decomposed nature of the clasts in the Old Alluvium.

The Young Alluvium is unconformable on the Old. In contrast, its clasts are less weathered, it is less slumped and less consolidated than the Old Alluvium. It is also fluvial and its occurrences are limited to positions adjacent to the present-day river system (Stauffer, 1973).

The evolution of the alluvium and its associated placers can be summarized as follows, largely after Batchelor (1979a). The extremely low sea levels of the Late Miocene to Early Pliocene resulted in the extensive Sundaland continent (Batchelor, 1979b). Conditions were ideal for the formation of a deep laterized regolith over the land mass (*Figure 11.3*). Greisen-bordered vein-swarms in the border zones of the predominantly Triassic granite have been the main source of cassiterite to the placers of the western tin belt (*Figure 11.4*). The eroding hydrothermal veins contributed large amounts of vein quartz to the regolith and hence to the alluvium. Placer formation generally requires deep lateritic weathering of the primary sources, producing quartz and resistate grains such as cassiterite and ilmenite in a sectile kaolinite matrix. Sinking of the freed cassiterite grains in the soil to a position above the unweathered bedrock sometimes allowed sufficient concentration to form *eluvial* placers. Down-slope creep of the regolith, assisted by seasonal heavy rainfall led to the formation of *colluvial placers*, generally found as superficial cover on stream divides, at less than a few hundred metres from the primary source. Iron-cemented tin-bearing conglomerates are also formed by this process.

As sea level continued to rise and the size of the Sundaland continental land mass decreased, increased rainfall led to gullying, especially along more easily erodable fault and mineralized zones in the bedrock. Major piedmont fan placers were thus formed in an oxidizing environment. Scrivenor (1928) called these 'torrential deposits'. The poorly sorted 'granite wash' of Willbourn (1936) can also be included in this category. This granite wash contains some of the highest heavy-mineral contents (Newell, 1971). The piedmont fan placers form massive discontinuous wedges along valley sides, thickening and dipping away from the bedrock at angles from 5° to 30°. They have the greatest economic importance and include the Western Boulder Clays of the Kinta Valley, which lie adjacent to the granite contacts. They are extremely rich in cassiterite, but are of patchy occurrence and lie less than 1 km from their primary provenance. The cassiterite is coarse grained, as is also that in the eluvial and colluvial placers, and the crystals are angular with some faces preserved. There is a tendency for higher

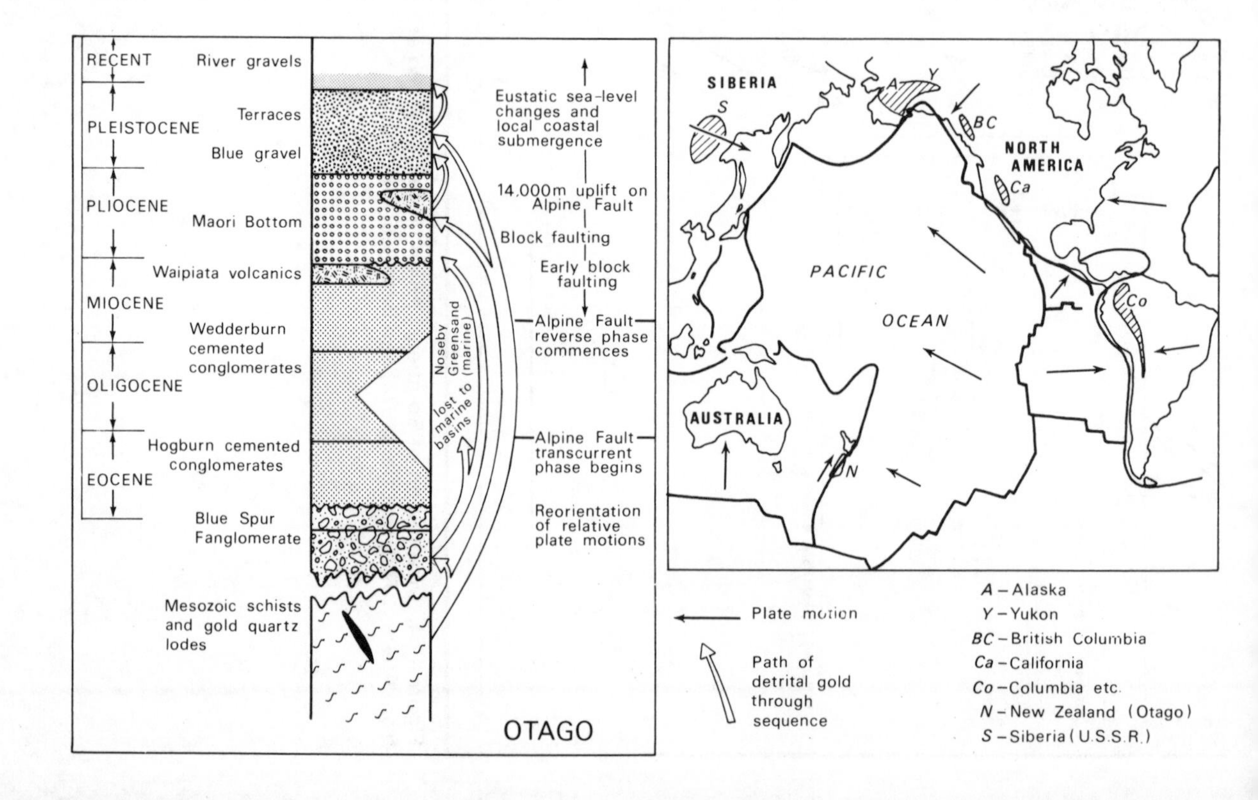

heavy-mineral contents to be associated with the sediments of coarser grain size (Newell, 1971).

Braided streams began to extend across the coalescing fans. This resulted in enrichment of the piedmont placers by removal of the clay and sand that accumulated down-slope in broad basement troughs. Increased precipitation near the end of the early Pleistocene resulted in the entrenchment of perennial braided streams. Their headwards erosion into the bedrock opened up new sources of primary deposits for erosion. During down-working of the streams, the stanniferous gravels remained *in situ* but were subjected to additional concentration by elutriation to form the exceedingly rich *kaksa* (Indonesian) or *karang* (Malaysian) placers, which lie as a thin layer on bedrock piedmont fans (Aleva, 1973). These also contain angular coarse-grained cassiterite. Gradients of 1 in 200 allowed cassiterite dispersion but gradients as low as 1 in 400 allowed concentration by selective fluviatile transportation to form *braided stream placers*, with transportation up to 5 km from the source, especially when there was a smooth clay bottom to the streams. These braided stream placers show a marked decrease of cassiterite grade and grain size downstream, but they have greater thicknesses and a wider and more uniform distribution than the piedmont fan deposits.

Following planation and laterization of the Old Alluvium surface, late Middle Pleistocene sea levels rose to form a great expanse of shallow seas across the Sunda Shelf. Increased precipitation gave rise to rivers with high discharge volumes, forming a second major phase of braided-stream development. The large *Padang Placers* (Osberger, 1968) are of gigantic width and extend up to 10 km from the primary source. Slumping of the old placers into solution hollows in limestone bedrock gave rise to some cave placers. Streams flowing through underground karst river systems filled some large caves with coarse current-bedded alluvium containing rich tin placers, exemplified by Kaki Bukit in northwest Malaysia (Stauffer, 1973).

Two major marine transgressions 120 000 years and 10 000 years ago resulted in elutriation and enrichment by wave action and tides, and longshore currents redistributed the coastal placers. Rich beach placers were formed along relatively stationary strand lines. The Thai deposits of Tongkah Harbour, Phuket, may be of this type.

Economic meander stream placers in the Young Alluvium are rare, except on Bangka (*Figure 11.4*). They formed as channel deposits, or point bar deposits on the inside of meanders, especially after the minimum sea level was attained 17 000 years ago.

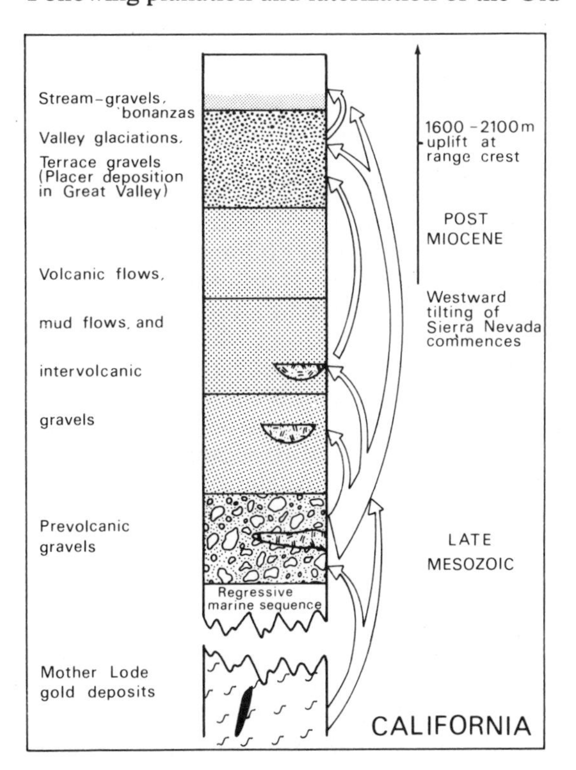

Gold Placers

With the outstanding exception of the Witwatersrand, which has produced 26 438 t of gold from an intermontaine fresh-water basin in Archaean–Proterozoic times, all the world's giant gold-placer fields (exceeding 125 t of gold, with placer production exceeding twice that of the associated bedrock) lie on the Pacific margins and were formed during the Cenozoic period in similar tectonic and sedimentary conditions. In each, the erosion and reworking of the alluvium and its placers is related to regional uplift and mountain formation, which can be related to changes of relative plate motions in the late Tertiary (Henley and Adams, 1979).

The circum-Pacific giant gold-placer fields are (largest first): California, Lena-Amur (USSR), Columbia (South America), Chile, Bolivia, Klondike (Yukon), Otago (New Zealand), Fairbanks (Alaska), British Columbia, Westplain (New Zealand) and Peru (*Figure 11.5*).

Figure 11.5 (opposite). The circum-Pacific giant gold-placer fields (after Henley and Adams, 1979).

CALIFORNIA

Placer formation occurred from the Late Mesozoic to the present (*Figure 11.5*). The fluviatile pre-volcanic gravels were derived from the Mother Lode bedrock deposits (*see* Chapter 8) to the east. The gold-bearing gravels were protected from re-working by widespread volcanic formations, but some were reworked in intravolcanic channel gravels. Post-Miocene tilting of the Sierra Nevada block with a total uplift at the crest of around 2000 m resulted in a new drainage pattern, which rapidly cut down through the volcanic formations and redistributed the detrital gold. Pleistocene valley glaciers also played a role as fluvioglacial transport. Post-glacial erosion and transport led to the spectacular 'bonanza' placers of the younger gravels.

OTAGO

The generalized stratigraphy is shown in *Figure 11.5.* The placer gold originated from a number of gold–quartz lodes occurring in greenschist facies rocks formed during the Cretaceous period. Rapid uplift and erosion resulted in a sequence of fluviatile sediments containing the important placer deposits. The detritus was obtained from the uprising Alpine Mountains. Hydraulic base-level variation and episodic regional uplift resulted in continual hydraulic redistribution of the detrital gold into younger sediments. The deposits can be closely related to activity along the Alpine Fault zone between the Australian and Pacific plates (Henley and Adams, 1979).

Diamond Placers

The placer diamonds of Namibia (Southwest Africa) are the best known of this type of deposit. The diamonds occur in relatively rich concentrations but nowhere today can they be found easily. They have to be won by large-scale mining methods involving the removal of large amounts of overburden. The only place where diamonds could formerly be freely picked from the ground was at Idatal, near Pomona. Workers picked stones by moving across the sand on their hands and knees, but this was a unique deposit. In Southwest Africa there are no primary deposits, all are in beach placers. The geological occurrence has been reviewed by Hallam (1964).

Raised beaches of Pliocene-to-Upper Pleistocene age are a spectacular feature of the west coast. They contain marine gravels and marine shells. They are best developed on the coast northwards from the major river mouths, as a result of the prevailing northwards longshore drift (*Figure 11.6*). Detrital diamonds occur in the raised beach gravels, in any portion of the raised beach, but the best concentrations are in the gravels at the top and bottom of a beach deposit. The best concentrations are on the northern side of a south-opening bay and on the south side of a headland where the shoreline turns from north to northwest. This is a direct result of the prevailing northwards marine current and the longshore drift.

The accepted theory is that the diamonds are transported down the Orange and other main rivers from the eroding kimberlite pipes, which are known to occur in the Brukaros Mountains, between Gibeon and Keetmanshoop. Many of the pipes are now covered over by desert sand. The diamonds are subsequently concentrated by wave action and northwards longshore currents. Some African kimberlites have a direct association with palaeo-placer deposits (Hawthorne *et al.*, 1979).

The main mining regions are shown on *Figure 11.6.* Each region extends northwards along the coast from the outlet of a major river. The northwards longshore current effect is clearly shown in each district by the northwards diminishing size of the detrital diamonds. Group 1 (Orange River to Dreimaster): stones diminish from 1 stone per carat (0.2 g) in the south to 12 stones per carat in the north. Group 2 (Buffels River): diminishing from 1 stone per carat in the south to 3 stones per carat in the north. Group 3 consists of an older 'Eocene' marine section. The stones progressively diminish from 2–5 per carat to 10–15 per carat in the north. The source is presumed to be the proto Orange River because the stones are similar to those of Field 1 (Oranjemund). Group 4 (Huab and Ugab River, northward to the Hoanib River): the stones diminish from 5 to 8 per carat. Group 5 (Olifants River): the stones diminish northwards from 8 to 20 per carat.

The diamonds are of a characteristically good quality, with good crystal shapes and curved faces of the octahedron and dodecahedron. Approximately 60% are completely colourless and 30% yellowish. Because they are detrital, they show percussion marks, cracks, iron staining and worn edges. The largest ever found was 246 carats (49.2 g) from a trench in the Main Terrace at Oranjemund. The associated detrital minerals are called 'bantams' and comprise ilmenite, garnet, staurolite, epidote, agate, chalcedony and jasper.

Many of the diamonds at Oranjemund are found in shelf gravels, in hollows and gullies in the bed-

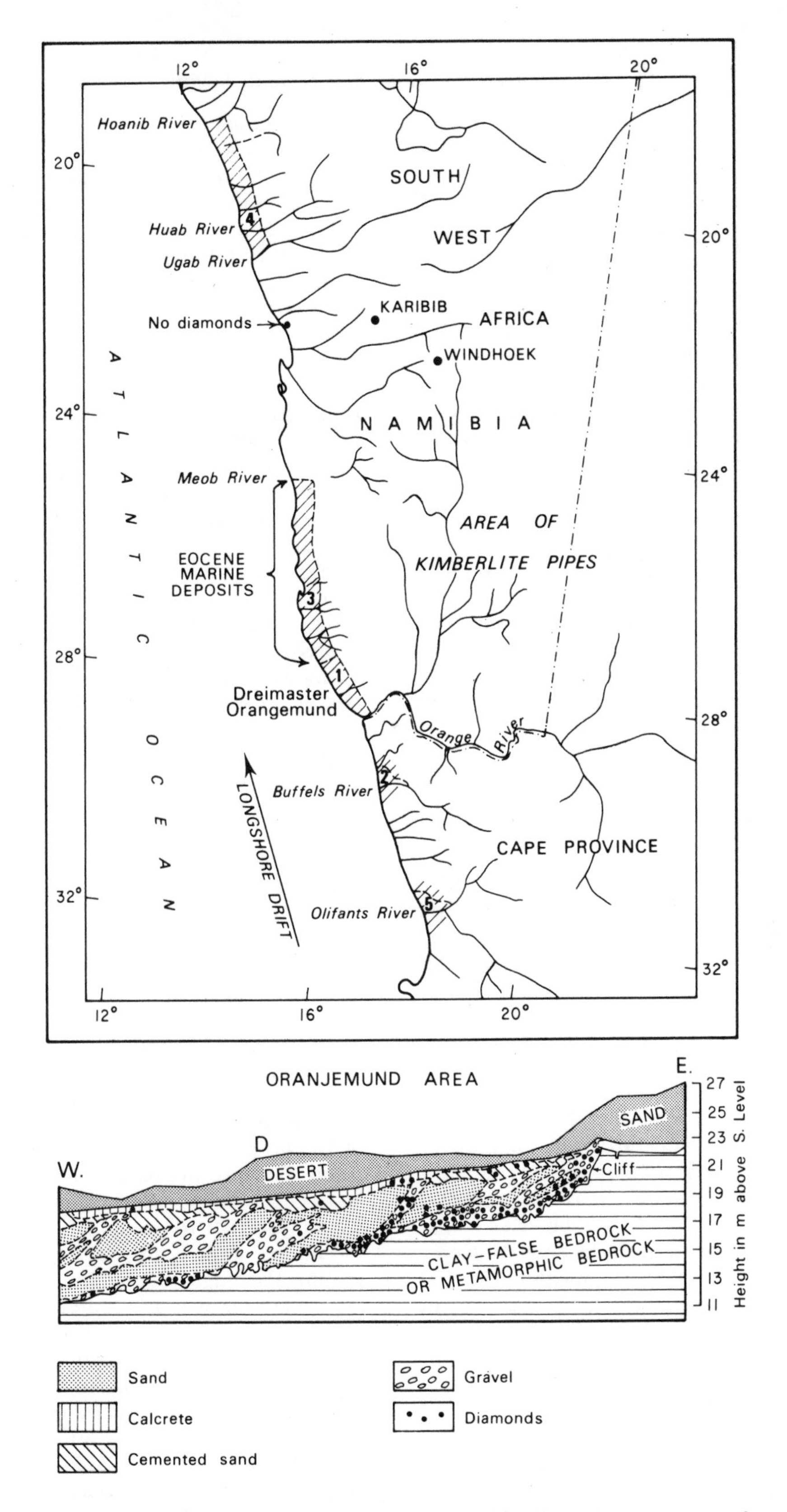

Figure 11.6. Distribution of the main diamond-bearing beach placers of Namibia (Southwest Africa) and a cross-section through the Oranjemund (Group 1) raised beaches (after Hallam, 1964).

rock. Destruction of a raised beach by a marine transgression will remove the sands but the diamonds and heavy minerals will stay in place and work downwards to concentrate on the bedrock. The bedrock may be a metamorphic or granitic basement, or a false bedrock of cemented sand or clay. In mining operations, bedrock sweeping and cleaning is the most important procedure.

Many of the raised beaches are covered by desert sand overburden. There are Upper Terraces at 14–20 m and Lower Terraces at 2–9 m height. Most of the gravels are of quartzite, vein quartz and lava clasts. The bedrock is of mica schist that has been eroded into gullies and potholes; the main gullies trend normal to the coast. Concentration by wave action on the beaches is the main method of diamond concentration. Longshore currents also moved the diamonds from south to north. The richest concentrations are on the north side of a bay or the south side of a headland. The Upper Terraces contain bigger diamonds than the Lower Terraces. The overall average size is 1 carat (0.2 g) per stone but some portions of the trenches on the Upper Terraces have up to 3 carats per stone. Coarse sand and gravel brought down by the Orange River, which is rich in detrital quartz and feldspar, is called 'Kies'. It is derived from weathered granite and is known to be free from diamonds.

Not all placer diamonds can be related to a kimberlite provenance. Diamonds are recovered from the tin dredges in the Phuket district of Thailand, and no kimberlites are known from this region (Garson *et al.*, 1975). The placers are in Quaternary alluvium and it is assumed that the diamonds have eroded from the pebbly mudstone of the Carboniferous Phuket Group. The stones range from 1.8 mm (0.04 carat) to about 4.7 mm (0.89 carat). The quality is generally poor.

Placer deposits are mined from the Quaternary river terraces and alluvial river gravels in southeast Borneo (Bemmelen, 1970b). It is deduced that they eroded from kimberlite-like breccias in the Meratus Ophiolite, or from Cretaceous conglomerate. The output has reached a maximum annual production of 3500 carats and the stones are cut and polished at Martapura.

Basal Proterozoic Uranium Placers

The uraniferous, sometimes gold-bearing, quartz–pebble conglomerate ores are restricted to an interval between 2.7 and 2.2×10^3 Ma. The basal Proterozoic rocks are typically grey-green or drab quartz-rich clastic strata that overlie the Archaean sequences of interbedded greywacke and volcanic rocks with a pronounced unconformity. The basal Proterozoic sequence is of restricted sedimentation distribution. Pyrite is always present and quartz–pebble conglomerates are always uraniferous (Robertson, D.S., 1974). The underlying Archaean rocks are normally contorted and metamorphosed to a greater degree than the uraniferous basal Proterozoic strata, and clean quartzites are limited or absent from the Archaean. Quartz–pebble conglomerates are not found in the Archaean. The Archaean sequence contains abundant granites and gneisses, whose erosion under an anoxygenic atmosphere resulted in detrital uraninite, pyrite and gold being transported and deposited as placers in the basal Proterozoic sediments, which filled depressions in the Archaean regolith.

In all four continents (North America, Australia, South Africa, South America) the basal Proterozoic sequence, which contains the uraniferous palaeo-placers, are conformably overlain by red-coloured rocks that contain haematite but no pyrite. The quartz–pebble rocks in the red sequence carry only monazite and no uraninite. Overlying the red-coloured clastic sequence are haematitic iron ores of widespread distribution, described in Chapter 10.

It seems obvious that the uraniferous conglomerate ores exist only in environments that reflect sedimentation throughout the period of the earth's evolution in which the atmosphere was non-oxidizing. The ores are found only stratigraphically below the oldest red sediments. They were deposited in relatively restricted basins prior to the development of the younger red Aphebian (earliest Proterozoic) cover rocks, that is in the pre-Lorraine Formation of Canada, the pre-Transvaal Supergroup of South Africa, the pre-Itabira group of Brazil, and the pre-Hamersley group of Australia. Because the basal Proterozoic uraniferous quartz–pebble conglomerates have been subjected to 2.2×10^9 years of earth history, they should be looked for as inlier remnants only around the edges of the younger red rocks that covered them over and protected them from erosion (Robertson, D.S., 1974). Thus in Canada they occur at the contact between the Huronian Supergroup and the underlying Archaean (*Figure 11.7*).

Pyritic uraniferous quartz–pebble conglomerates and their detrital minerals are of similar character. The host rocks are yellow-green to grey clastics, some clean and well sorted as in the Elliot Lake deposits of Canada, but some poorly sorted and dirty. They are, with one exception (Sakami Lake, Quebec), only very slightly metamor-

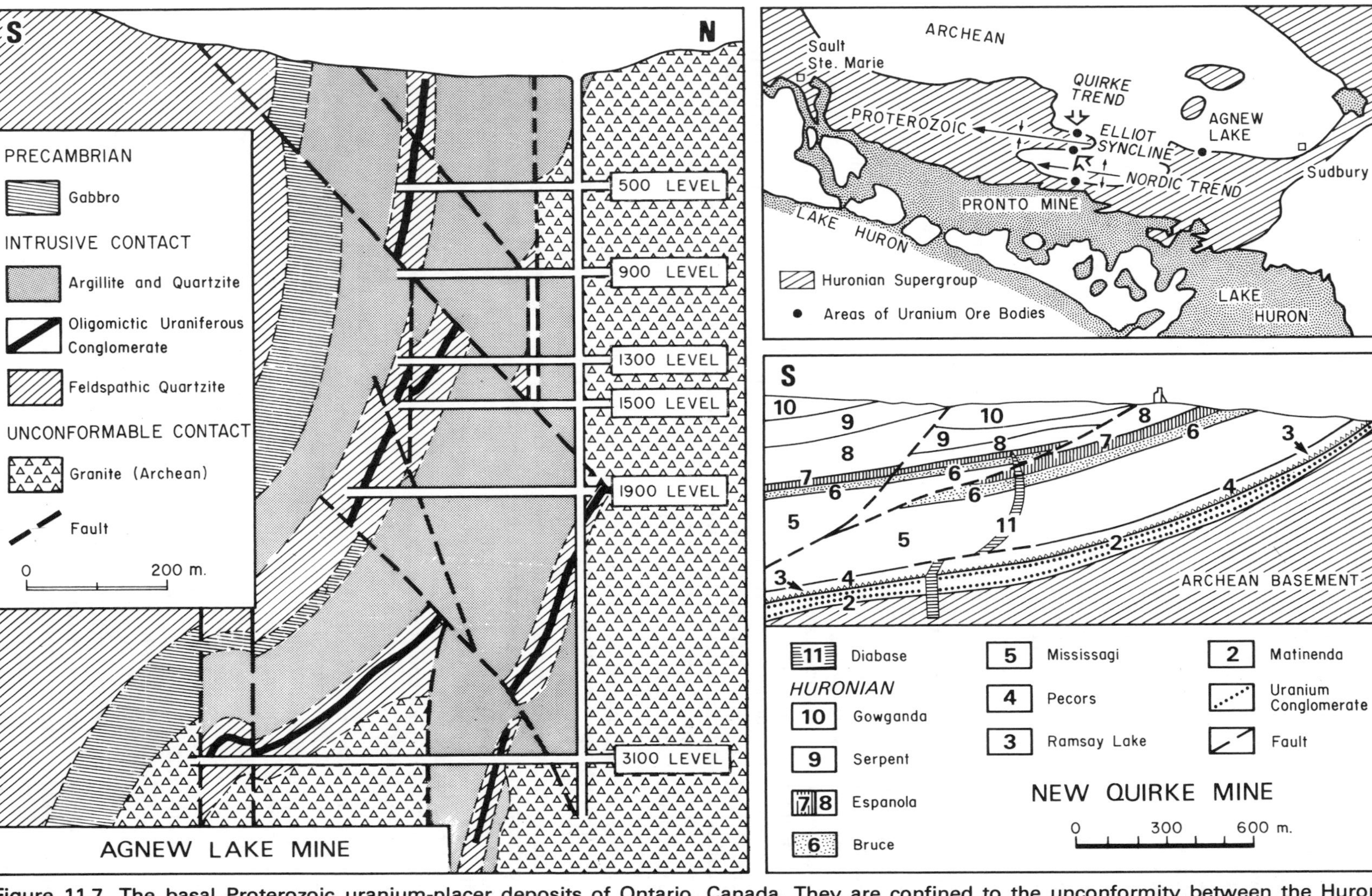

Figure 11.7. The basal Proterozoic uranium-placer deposits of Ontario, Canada. They are confined to the unconformity between the Huronian Supergroup and the underlying Archaean. The oligomictic uraniferous conglomerates are associated with feldspathic quartzite mainly on the north and south limbs of the Elliot Syncline and at Agnew Lake (after Robertson, D.S., 1974; Robertson, J.A., 1978).

phosed, and lie with pronounced unconformity on highly metamorphosed Archaean, and they underlie a redrock sequence that contains no pyrite. Robertson, D.S. (1974) holds that they are all of similar age ($2.2-2.8 \times 10^3$ Ma) and of the same fluviodeltaic origin, and that they are unique to this period of geological history, which preceded the formation of epigenetic uranium deposits. The atmosphere was anoxygenic and uraninite and pyrite eroded from pegmatites and granitic gneisses in the Archaean outcrops and these were carried detritally to be deposited as heavy minerals in the now palaeo-placers. At no subsequent time in earth history was it possible to transport uranium and pyrite like this again because of the oxygenated atmosphere. The uraninite is detrital for the following four reasons (Robertson, D.S., 1974). (1) It exists within a normal detrital assemblage and occurs in the same way as other detrital minerals. In particular, the distribution of ThO_2/U_3O_8 and rare-earth concentrations varies with pebble size in a manner explicable by the processes of density grading during sedimentation (Theis, 1978). (2) The minerals are compositionally of a kind that is accessory in granitic rocks. The uraninite, for example, with its high content of thorium and rare earths, is distinctly different from the pitchblende, which forms from ground-water transport in younger formations. (3) There is no primary connection between the uranium minerals and faults, veins or other epigenetic phenomena, although the uranium may remobilize later along such features. (4) At places where the ore units have been eroded by younger conglomerates, the radioactive debris from the former contaminates the latter, which are otherwise non-radioactive. The distribution of radioactivity is clearly related to the erosion process and implies the existence of uranium minerals in the older conglomerate beds prior to the development of the younger conglomerates.

Gold occurs in the matrix of the basal 2200–2700 Ma old Fortescue Group of the Hamersley Basin of northwest Western Australia, and has been mined at Nullagine since 1888. It is closely associated with pyritic horizons and its deposition related to the Lower Proterozoic unconformity (Blockley, 1975). The conglomerate is generally weakly radioactive, in the range 50–100 ppm U_3O_8, and one drill hole averaged 425 ppm U_3O_8 over 1.5 m (Robertson, D.S., 1974). The uranium is associated with thucholite pellets; thucholite is a complex mixture of hydrocarbons and uranium oxide. Uranium may constitute up to 53% of the ash and thorium up to 48%. Thucholite, named after Th, U, C, H, and O, is jet black and brittle; it occurs in pegmatite and quartz–pebble conglomerates. The associated heavy minerals are pyrite, monazite, zircon and anatase. Elsewhere in the Hamersley Basin, gold and uranium are found more abundantly in the overlying Wyloo Group.

Pyritic quartz–pebble conglomerates containing uranium and gold occur at several localities around the edge of the Archaean Sao Francisco Craton, Brazil—at Jacobina, Belo Horizonte, Pitangui and Cavalcante (Robertson, D.S., 1974). At Belo Horizonte, the conglomerates contain branerite, xenotime and zircon. Thucholite is locally prominent, and occurs as rounded balls carrying inclusions of uraninite and having a high thorium content. Some conglomerates also contain gold. Overlying the drab-coloured rocks, which contain pyrite and uranium minerals, is the red coloured iron-bearing Itabira Group. It is believed that prior to 2200 Ma ago the Sao Francisco Craton was shedding clastic debris off its flanks, and that the quartz–pebble conglomerates represent fluviodeltaic deposits in a non-oxidizing environment.

In Canada, small remnants of basal Proterozoic (basal Aphebian) sediments are preserved at three areas—the Elliot Lake area of Ontario, the Montgomery Lake–Padlei area of the Northwest Territories, and the Sakami Lake area in northwest Quebec (*Figure 11.7*). They all carry oligomictic quartz–pebble conglomerate with pyrite and uraninite, and are overlain by younger sediments that have their iron in red form and carry no pyrite. The overlying sediments may be radioactive but it is due to thorium in the monazite (Robertson, D.S., 1974). The conglomerates of the Montgomery Lake–Padlei area were laid down between 2500 Ma and 1800 Ma ago. Drilling has encountered values up to 260 ppm U_3O_8 and 350 ppm ThO_2 over 60 cm of conglomerate. The principal radioactive minerals are uranothorite, a titanium-bearing complex, and zircon. The conglomerates are not well sorted and assay results are disappointing. At Sakami Lake, Quebec, the conglomerates are intensely deformed and are in the amphibolite facies, with metamorphic ages of 2500 Ma. They carry more than 10% pyrite and contain uraninite and a thorium silicate.

The Elliot Lake Region of Ontario

The Elliot Lake district is the largest single uranium district in the world, with production of almost 91 000 t of U_3O_8 and a total uranium content of possibly 460 000 t (McMillan, 1977a). The average mined grade is 1000–1500 ppm U_3O_8. The uranium deposits occur in three zones or trends, named Quirke, Nordic and Pronto (*Figure 11.7*). The deposits are all similar and occur within 150 m of the base of the Matinenda Formation, which overlies the Archaean basement in the Quirke syncline. The arkose interbedded with the ore conglomerate is generally greenish in colour and is cross-bedded, normally of the festoon-type, indicative of a fluviatile environment (Roscoe, 1969). Arkose and pebbly arkose both show lensoid bedding and scour and fill structures. The conglomerate was apparently deposited in anastomosing or braided stream channels. Lateral migration of these channels coalesced to form sheets, or reefs. The cross-bedding and grain-size distribution clearly indicate that the currents flowed towards the southeast. The underlying Archaean basement consists of Keewatin-type greenstone, Algoma granite and minor mafic intrusive rocks. Uranium ore occurs only within the Matinenda Formation of the Huronian Group. The Huronian Supergroup has been folded into westwards-plunging folds, and the metamorphic grade is low. To the south and west, the folding is severe and the metamorphic grade may reach amphibolite facies (Robertson, J.A., 1978). The Elliot Lake ore bodies have been described by Robertson, D.S. (1974), Robertson, J.A. (1978) and McMillan (1977b).

The ore zones are primarily pyritic quartz–pebble conglomerates that occur within an envelope of poorly sorted coarse-grained subarkose, which has a greenish sericitic matrix. The radioactive conglomerates are composed of quartz and chert pebbles in a matrix of feldspathic quartzite containing a heavy mineral suite of pyrite, monazite, uraninite, zircon and titanium minerals. Uraninite and brannerite are the main ore minerals. Yttrium is recovered at one mine. Gold is generally not present in sufficient quantity for detection, and this is one of the major differences between this and the Witwatersrand palaeoplacers of South Africa (*Table 11.1*). Thucholite, a radioactive hydrocarbon, occurs in laminae within the ore and also as a secondary mineral in cross-cutting fractures. Ruzicka and Steacy (1976) suggested a biogenic origin for the laminated variety of thucholite, but the globular variety on open surfaces may be the result of radioactive polymerization of diesel fumes from the underground mining equipment.

The Quirke zone dips south and is about 2.4 km × 11 km in size (*Figure 11.7*). It consists of seven conglomerate horizons or reefs, which are found through a 52-m thick section. The reefs average 1.5–3.4 m thickness, and are separated by beds of subarkose. Locally the reefs pinch out where they lap on to basement highs or are truncated by the overlying Ramsay Lake Formation.

Theis (1978) has presented powerful sedimentological evidence from the Rio Algom–Denison ore reef of the Quirke Zone. Evidence is acquired from continuous mining over 5 km of its length. The following generalizations can be made. (1) The ore reef is wedge-shaped. The thin edge is towards the northwest where the ore reef pinches out against basement. It thickens to more than 9 m towards the southeast (representing the down-channel direction). (2) The percentage of well-packed conglomerate decreases and poorly packed conglomerate and quartzite increase towards the southeast, suggesting a loss of depositional energy down the channel of sedimentation. Cross-bedding clearly indicates that the currents flowed towards the southeast. (3) The pebble size decreases from northwest towards southeast. In addition the pebble size tends to decrease up a given stratigraphic section.

Pyrite, the uranium-bearing minerals, and other accessory minerals occur as discrete grains within the matrix of the ore conglomerates. The detrital nature of the pyrite grains is shown by a plot of pyrite grain size against pebble size, giving the following relationship: pyrite grain size (mm) = 0.038(pebble size in mm) with a correlation coefficient of +0.93. The range of pyrite grains is 0.7–3.0 mm, and of pebbles 21–70 mm. This clearly proves that the pyrite was detritally transported along with the pebbles. The U_3O_8 content generally increases with pebble size but the correlation coefficient is only +0.47. The U_3O_8 is contained mainly in uraninite and brannerite. Samples with larger pebbles tend to contain more uranium (56-mm pebbles are associated with approximately 6000 ppm U_3O_8; while 32-mm pebbles associate with 400 ppm U_3O_8). The ThO_2 content shows a moderate and negative correlation with pebble size of −0.63. The thorium is contained mainly in monazite, thorite and uranothorite, which transport farther than the uraninite. (Pebble size 32 mm is associated with 1500 ppm ThO_2; pebbles size 64 mm is associated with around 200 ppm ThO_2.) The U_3O_8/ThO_2 ratio increases with pebble size, showing clearly that the thorium is contained in the less-dense grains, associated with smaller pebbles farther down-current. The correlation coefficient is +0.79 (pebble size 16 mm is associated with a U_3O_8/ThO_2 ratio of 0.3; 54-mm pebbles with a ratio of 8.0). Cerium is the principal rare earth and is contained in the monazite. It correlates with pebble size with a correlation coefficient of −0.85 (pebble size 16 mm is associated with 10 000 ppm cerium; pebble size 64 mm with around 500 ppm cerium). The plots of zirconium and titanium dioxide are similar to cerium. The zircon content increases down-channel. The correlation coeffient is −0.87 (pebble size 64 mm is associated with 10 ppm zirconium; pebble size 20 mm with around 500 ppm zirconium).

Uraninite is the most abundant ore mineral. Samples with pebble sizes from 30 mm to 35 mm contain uraninite only rarely. Brannerite becomes more common as pebble size decreases. These distributions of minerals and their contained elements are clearly of detrital origin, related to grain size and density in a current waning in intensity from northwest towards the southeast. A mineralogical zonation, coincident with the chemical zonation, grades southeastwards from uraninite-rich, through brannerite-rich, to monazite–zircon-rich conglomerates.

The Nordic zone occurs on the south side of the Quirke syncline and dips north. It is about 1.6 km × 8 km in size and consists of three reefs in the lower 46-m of the Matinenda Formation. The three mineable reefs average in descending order; Upper or Pardee reef (1.5 m), Main reef (3 m) and Lower reef (3 m). They are truncated down-dip by conglomerates of the overlying McKim Formation, whose conglomerates also locally contain ore-grade uranium. The Nordic-zone conglomerates generally are lower in grade, less pyritic and have lower thorium/uranium ratios than those of the Quirke zone.

The Pronto zone consists of a single conglomerate reef that averages 2.3 m and dips 15–20° to the south. It is truncated at a depth of about 305 m at its southern end by the Pronto Thrust fault. In addition to the conglomerate ore, there is some fracture-related ore associated with carbonate, albite and chlorite, in which pyrite is absent and the uranium values are associated with haematite. This uranium has therefore been remobilized in an oxidizing environment, and is younger than the detrital ore deposition.

Radioactive quartz–pebble conglomerates occur at several stratigraphic levels within the Huronian Supergroup. Within the older Huronian units, such as the Matinenda or Mississagi, the conglomerates are pyritic and contain both uranium and thorium. This contrasts with quartz–pebble conglomerates in the overlying Lorrain formation, which are thorium-bearing and rich in haematite but deficient in uranium. The Lorrain Formation contains jasper pebbles and abundant redbeds and is believed to have been deposited after the atmosphere changed from anoxygenic to oxygenic (Robertson, D.S., 1974).

The Agnew Lake deposit (*Figure 11.7*) is located about 80 km east of Elliot Lake. The ore zone is of quartz–pebble conglomerate interbedded with arkose in metamorphosed Matinenda Formation. The deposit occurs in a steeply dipping faulted synclinal basin of Huronian rocks in a structurally complex environment. Pebbles are smaller and sparser, and pyrite less conspicuous than at Elliot Lake. The main economic mineral is uranothorite, which is associated with monazite and other heavy minerals. There is a high rare-earth content and the thorium/uranium ratio is approximately 3, higher than at Elliot Lake. Apparently much of the uranium has been remobilized along fractures.

Origin

After the Kenoran orogeny (2600 Ma ago), this area was deeply eroded and the topography was controlled by basement lithology and structure. The regolith was formed under a reducing anoxygenic atmosphere. The depositional basin extended east–west and deepened towards the southeast. The quartz–pebble conglomerates and associated sediments were derived from Archaean terrain to the north and northwest, transported by rapidly moving water, and deposited detritally in a near-shore fluviatile environment under cold, possibly frigid (association with tilloids), reducing conditions about 2500 Ma ago. The deposits were subjected to diagenesis and slight remobilization but not to either leaching or erosion (Robertson, J.A., 1978). The streams were episodic or seasonal and most of the uranium was transported as detrital uraninite into deltaic braided streams at or near sea level during periods of heavy runoff. During periods of relative quiescence, algal mats formed in a shallow marine environment. This organic material was instrumental in localizing rich concentrations of uranium (McMillan, 1977a).

The lack of gold in the uraniferous conglomerates is attributed to the general lack of good greenstone belts in the cratonic block north of the Huronian contact and to the low frequency of gold occurrences in the area believed to have been the source area for the basal Proterozoic placer deposits (Robertson, J.A., 1978).

Robertson, J.A. (1978) believes that the radioactive quartz monzonite bodies and related pegmatites of Keewatin age, north and northwest of the Elliot Lake area, acted as an erosional source area for the uranium minerals in the placers. A comparison of the Elliot Lake deposits with those of the Witwatersrand of South Africa is given in *Table 11.1*.

Table 11.1 Comparison of main features of the two main basal Proterozoic placer-uranium deposits, after Kimberley (1978a)

	Witwatersrand, South Africa	*Elliot Lake, Ontario, Canada*
Age of host-rock sequence	2800–2500 Ma	2450–2250 Ma
Lithology of host rocks	Sandstone, conglomerate, argillite and volcanics	Sandstone, conglomerate, argillite and volcanics
Sedimentary environment of host rocks	Fluviodeltaic	Fluviatile
Typical grade of metamorphism	Virtually unmetamorphosed	Slightly more altered than Witwatersrand
Lateral extent of mineralized beds	Tens of kilometres	Several kilometres
Lithology of mineralized rocks	Pyritic conglomerate, pyritic sandstone, upper surfaces of mudrocks, and thin carbonaceous beds above unconformable surfaces	Pyritic conglomerate, pyritic sandstone, which locally include hydrocarbon-bearing veins and carbonaceous laminations
Major mineralogy of ore beds	Quartz, sericite, rock fragments, chlorite and pyrite	Quartz, sericite, feldspar, rock fragments and pyrite
Minor mineralogy and chemistry of ore beds	Heavy minerals, for example, ilmenite, zircon, rutile, chromite and monazite, are more abundant than in typical Phanerozoic sandstone. Average uranium concentration (280 ppm) is about 90 times the mean crustal abundance. Average gold concentration (10 ppm) is about 2500 times the mean crustal abundance. Copper concentration is 212 ppm, about four times the crustal average, in the basal reef of the Free State Geduld mine	Average Zr, Ti and Cr contents are similar to those of world-average sandstone. Hence concentration of zircon, rutile and chromite is only moderate. The average concentration of uranium mined to date is about 600 times the mean for terrigenous sands and 500 times the crustal abundance. The gold content is usually below the limits of normal detection, but 1.1 ppm Au was reported in one sample of anomalously high uranium content (11 800 ppm). The average Cu content (360 ppm) and average Mo content (39 ppm) of 47 samples containing more than 340 ppm uranium are about 6 and 26 times the crustal average, respectively
Pyrite, uraninite, gold and hydrocarbon distribution and grain shapes	Close correlation exists between pyrite, uraninite, gold, silver and hydrocarbon abundances, although unmineralized pyrite is common. These elements are concentrated in conglomerate and sandstone above disconformable surfaces, especially where carbonaceous. Laterally extensive beds are more richly mineralized than lithologically similar beds of minor extent. Pyrite, and uraninite and silica, locally replace algal and fungal structures. Pyrite and uraninite also partially replace quartz clasts. Most pyrite and uraninite grains are rounded. Uraninite grains are commonly about 0.1 mm in diameter	Close correlation exists between pyrite, uraninite and, where present, hydrocarbon. Unmineralized pyrite is common. Uraninite is locally concentrated near the base of conglomerate beds and, at least in the high-grade ore, is commonly concentrated near quartz-clast surfaces at the rims of interstices. Partial replacement of quartz-clast edges and infilling of cracks by pyrite and uraninite is common. Uraninite grains are typically about 0.1 mm, but grade to 0.001 mm in diameter. Some pyrite and uraninite are intergrown. Uraninite grains have commonly coalesced and locally display serrated edges. Uraniferous conglomerate is relatively well packed, poorly sorted and coarse grained

THE WITWATERSRAND GOLD–URANIUM PALAEO-PLACERS

The intermontaine basin formed on the Kaapvaal Archaean craton (*Figure 9.1*). Its active northwest edge was fault-bounded and its southeast edge was a more passive gentle down-warp (Pretorius, 1975). The basin was 350 km long and 200 km wide, and was filled with 13 900 m of sediments and volcanics in the period between 2500 Ma and 2750 Ma ago. Six major gold fields and several smaller mineralized areas were discovered since 1886. From these, 28 722 t of gold and 76 012 t of uranium were recovered up to 1972 (Pretorius,

1976). The average gold tenor is 10 ppm and uranium 280 ppm.

The alluvial fans were all formed on the fault-bounded northwest margin of the basin, and no fans have been found on the more gentle southeast rim (*Figure 9.1*). A wet climate prevailed. The basin was occupied by a shallow-water lake or inland sea and no connection with the open sea has ever been demonstrated. The basin floor became more unstable with time and the various gold fields developed in the down-warps between the basement uplifts.

A high-energy transfer system took the form of short linear fluvial arrays. The rivers emerged from a canyon, flowed over a piedmont plain and dispersed into the basin via a braided-stream system (Pretorius, 1975). The characteristics of the cross-bedding point to a fluvial environment. Cross-bedding vectors are bimodal; one of the modes is produced by currents moving down-slope into the basin, the other is related to distribution by longshore currents within the basin itself. Sand waves of amplitude up to 1 m were produced by the longshore currents, which all moved clockwise within the basin (Pretorius, 1976).

A detailed study of the Kimberley Reef, East Rand gold field, shows it to be a fluviatile placer deposited in about 40 cm depth of water in a small-scale meandering stream pattern (Hirdes, 1978). The placers were formed on the insides of the channel curves (*Figure 11.8*) next to and alongside elongate conglomerate bodies that form point bars.

The conclusion is that the minerals of economic importance were eroded and transported from two different sets of mineralized Archaean outcrops. The provenance areas for the Witwatersrand placer gold was from the north and west (Anhaeusser, 1976b). Obviously the gold was eroded from the Archaean gold-rich greenstone belts (*see* Chapter 9) that lie south of the gold-poor high-grade metamorphic Limpopo Mobile Belt (*Figure 9.1*). The uraninite came from the younger granites that envelope the greenstones.

The placers

By far the greatest proportion of gold and uranium occurs in or immediately adjacent to bands of conglomerate (locally called *bankets*), which are preferentially developed at or near the base of each cycle of sedimentation. Within the Lower and Upper Witwatersrand divisions in the Johannesburg area (*Figure 11.8*), the conglomerates occupy about 8% of the total thickness. On the opposite southeast side of the basin they form only 1% of the stratigraphy. Only about 0.2% of the stratigraphy is economic on the northwest side of the basin, because not all conglomerates are economic (Pretorius, 1976). A typical average analysis of the banket, based on analyses of the Vaal and Ventersdorp Reefs, is given by Feather and Koen (1975): 44–50 ppm gold, 5–8 ppm silver, 290–870 ppm U_3O_8, 3–4.4% muscovite, 0.1–0.2% pyrophyllite, 0.8–4.9% chlorite, 88.3–88.9% quartz, 0.1% titanium, 0.1–0.2% zircon, 0.2% chromite and 3.2–6.6% pyrite.

The conglomerate reefs consist of well-rounded pebbles, forming 70% of the rock volume; they range from 20 mm to 50 mm in diameter, the larger are oval and the smaller subangular but the bulk are of vein quartz, accompanied by chert, jasper, quartzite, quartz porphyry and metamorphic rocks. The matrix is compact and of secondary quartz, phyllosilicates, sericite, pyrophyllite, muscovite, chlorite and chloritoid. Unaltered pyroxene, epidote and biotite may also occur (Feather and Koen, 1975).

The gold and uranium occur in the following places: in the matrix of conglomerates; in the strongly pyritic sands that usually fill erosion channels—the gold, uranium and pyrite particles lying on the foreset beds of cross-bedded sands; on sand along planes of unconformities that separate two cycles of sedimentation; on mud along the planes of unconformity that separate cycles of sedimentation; and in carbon seams that are developed on or immediately adjacent to planes of unconformity. Average values for the Vaal Reef are: 50 ppm gold, 8 ppm silver, 870 ppm U_3O_8; and for the Ventersdorp Reef, 44 ppm gold, 5 ppm silver and 290 ppm U_3O_8.

The gold particles are distributed according to five different patterns within the reefs (Hallbauer and Joughin, 1972), and none of the patterns is constant over large areas: (1) particles dispersed throughout the reef matrix, with individual particles spaced approximately evenly a few millimetres apart; (2) as discrete particles separated by distances up to 100 mm; (3) as isolated clusters, very rich in gold particles, inside a few cubic metres of matrix—200 grains of gold might be present in a patch measuring 0.4 mm × 0.7 mm; (4) as thin streaks on either the hangingwall or the footwall contact; and (5) as very rich isolated clusters occupying a few cubic centimetres and spaced 50 mm or more apart.

On a local scale, good correlation exists between gold and uranium concentrations, because both elements were concentrated hydraulically. However, gold/uranium ratios are not consistent with grade changes. Irregular distributions of gold and

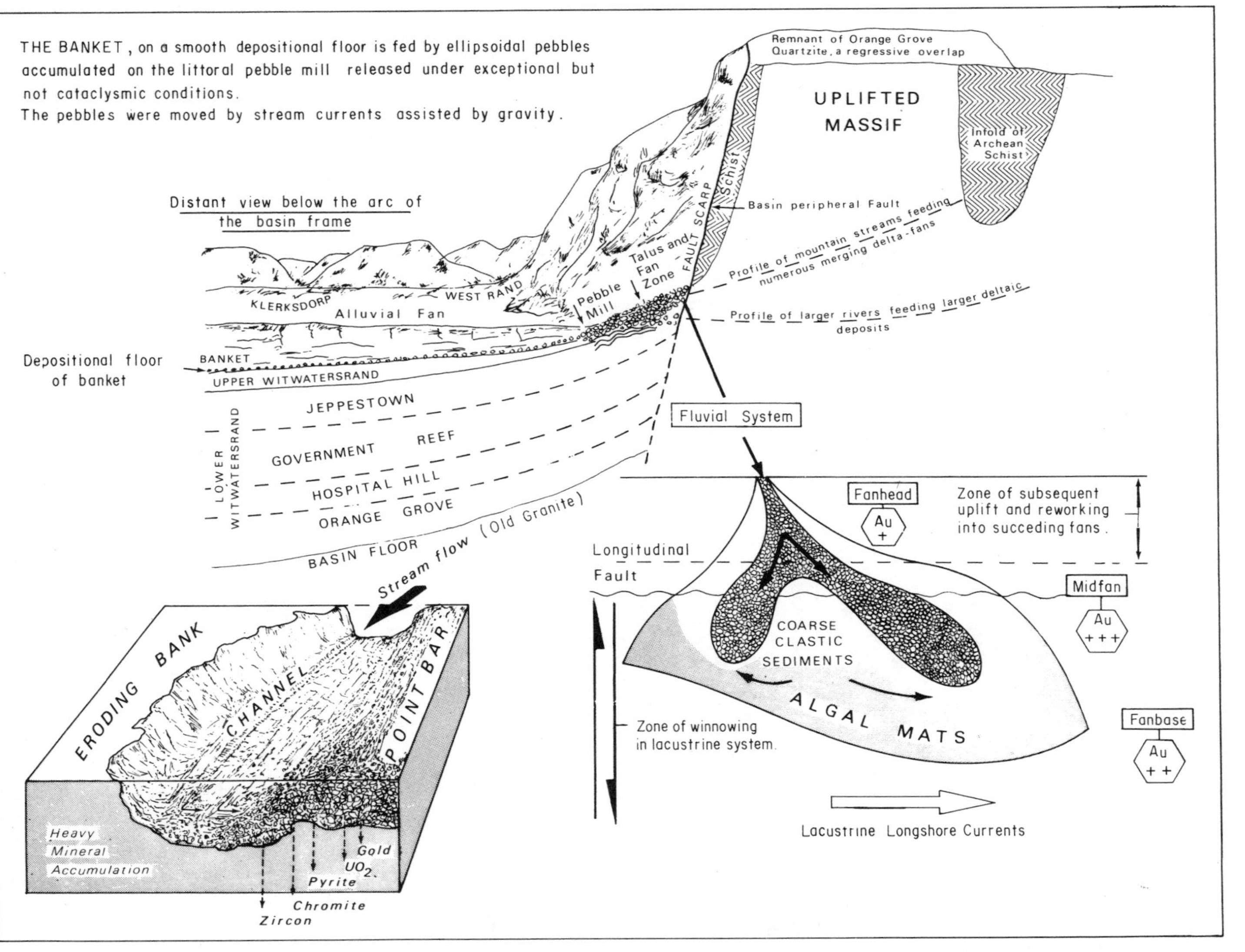

Figure 11.8. Details of the Witwatersrand alluvial gold field. Top: diagrammatic cross-section (after Pretorius, 1975). Bottom left: placer deposition in river channel (after Hirdes, 1978). Lower right: diagrammatic plan view of the alluvial fan system (after Pretorius, 1975). The gold was too fine to settle in the fanhead facies and its highest concentrations are in the mid-fan lobes. The peak uraninite deposition was a little farther downslope from the gold peak.

uranium are thought to be due partly to random erosion and temporary redeposition of bars and channel sediments (Smith, N.D. and Minter, 1980).

In the Klerksdorp gold field, the maximum concentrations of uranium are more towards the depositional axis of the basin than are the greater enrichments of gold (Minter, 1976). In any particular horizon within a fan, the reef tends to have a higher gold tenor relatively closer to the fanhead and greater uranium contents closer to the fanbase (*Figure 11.8*). This distribution is related to the differences in detrital densities. In the Vaal Reef, the palaeo-surface produced a parallel hydraulic-energy configuration that controlled heavy-mineral deposition and oriented pay shoots of gold. Uranium is less concentrated on the palaeo-surface and is spread throughout the reef, with the result that, in channelled areas, the total uranium content is proportional to reef thickness. Nodular pyrite is distributed similarly to uranium (Minter, 1976). The uranium/gold ratio is considerably enhanced where thicker reefs are developed, because gold has a tendency to be concentrated near the base of a conglomerate band or banket.

There were three stages of mineralization (Feather and Koen, 1975): detrital deposition of gold, pyrite, uraninite, ziron and rare diamonds—the placers were reworked in braided streams; formation of secondary pyrite and gold in the placers; and remobilization of gold and formation of secondary sulphides such as gersdorffite.

Mineralogy

Pyrite may make up more than 90% of the opaque minerals of the reefs; usually it occurs as rounded grains of detrital origin (pebble pyrite). Uraninite occurs as detritally rounded grains. It occurs in the lowest portions of the reefs. An age of approximately 2000 Ma has been obtained. This is younger than the age of the Witwatersrand System (2500–2700 Ma) and probably dates the last metamorphic recrystallization of the uraninite. The average size of grains of gold is 0.035 mm, of uraninite 0.065 mm and platinoid grains 0.055 mm (Pretorius, 1976). No large-sized or even medium-sized nuggets are ever found, and it can be concluded that the placer deposits of the Witwatersrand are not comparable to modern placers, where large nuggets can exist.

Carbon usually contains minute uraninite crystals. The carbon resembles a medium-volatile bituminous to semi-anthracite coal. Gold and uraninite are always associated with the carbon. It can be shown without doubt that the carbon represents the fossilized remains of Precambrian organisms, including bacteria, algae, fungi and lichen-like plants. The latter played a role in the concentrating of the gold (Hallbauer, 1975). The possibility exists that large amounts of gold were transported as organic-supported colloids, which became unstable when the braided streams entered the lake of the main basin. Gold is confined to the conglomerates and occurs in lesser amounts in the banded pyritic quartzites. The bulk forms a fine-grained matrix between the quartz pebbles. Gold veinlets may intersect the pebbles, and may also coat the pebbles. The bulk of the gold is no longer in its original detrital form; it has recrystallized as a result of hydrothermal conditions attained during metamorphism, but it has not migrated far from its original placer deposition sites.

Ilmenite, Rutile and Zircon Placers

Australia is the most outstanding producer of ilmenite, zircon and rutile. These minerals occur in placer concentrations on present and fossil sea shores. All except a few scattered occurrences are grouped into three provinces: the eastern Australian rutile province, which extends for about 1000 km from the Hawkesbury River, New South Wales, to Fraser Island, Queensland, described by McKellar (1975); the ilmenite province of the Capel area of Western Australia, south of Perth (described by Welch *et al.*, 1975); and the recently discovered Eneabba rutile–zircon–ilmenite deposit north of Perth (described by Lissiman and Oxenford, 1975).

THE EASTERN AUSTRALIAN RUTILE PROVINCE

This is a vast area of marine sands of recent origin that blanket the coastal plain and adjacent islands. They contain sporadic natural concentrations of heavy minerals. The coastal sand-mass top is predominantly less than 15 m above sea level, and its base is defined by bedrock at 25–37 m below sea level. There are remnants of high dunes in places. The sand is predominantly of quartz and heavy minerals average 0.05%; these are predominantly rutile, zircon and ilmenite. The sands are variously cemented by iron oxide beneath swampy areas or by carbonaceous materials. The hinterland peneplain is considered to be the source area for the sands, erosion of which was related to sea-level lowering during the Northern Hemisphere Ice Age. The retreat of the ice caused sea-level rises resulting in the drowned topography of the eastern coast. New erosional base lines were established, giving a complicated history of erosion.

Surf action is the most efficient mechanism for producing economic concentrations of heavy minerals. During cyclonic storms, large volumes of sand are removed from a beach and the heavy minerals remain as black seams on the upper limits of a severely eroded beach. These seams may extend along the beach or they may be concentrated in lenses. They mainly occupy the top 10 m of the beach and are up to 30 cm thick (McKellar, 1975). Grades as high as 70% (ilmenite + rutile + zircon) are common. The northwards directed longshore current concentrates the heavy minerals towards the northern ends of beaches. The concentrated seams are short lived; high tides erode them and only fair-weather buildup of the beaches preserves them by sand burial. Older buried concentrations on the coastal plain give the bulk of the high-grade rutile reserves. A typical beach shows a number of parallel dipping fossil beach deposits (*Figure 11.9*). The associated dune tops have been truncated. The series of beaches was formed at a common sea level, approximately the same as the present day. They represent phases in the buildup of the coastal plain when surf erosion, which produced the seams, alternates with periods of accretion, which buries and preserves them. Wind action in the coastal region may cause heavy mineral concentration by blowing the lighter materials landwards where bluffs reduce the wind velocity.

Generally from north (Fraser Island) to south (Sydney) the rutile and zircon contents increase at the expense of ilmenite. The fraction (rutile + zircon)/ilmenite increases from 0.23 to 8.04. In general the heavy-mineral grains become coarser towards the south. This is related to the northerly longshore current. The concentration of rutile, ilmenite and zircon in this region bears many similarities to the placer diamonds of Namibia, described on p. 262. These similarities have been amplified by Hails (1976) in a general description of beach placers.

THE ENEABBA RUTILE–ZIRCON–ILMENITE SANDS

These sands of Western Australia were deposited and concentrated in a favourable coastline environment during a series of high sea levels in later Tertiary or early Pleistocene times (Lissiman and Oxenford, 1975). The deposits are now 30 km inland from the Indian Ocean, and lie about 300 km north of Perth. The depositional environment was a north-facing bay fed by streams carrying dispersed heavy minerals.The bay area was exposed to strong prevailing southwest winds and wave action.

The mineral-bearing palaeo-shorelines now range from 82 m to 128 m above present sea level. The heavy minerals are presumed to have been eroded out of the Mesozoic rocks and from the crystalline metamorphic rocks of the Archaean shield farther to the east. There are several parallel higher-grade zones (about 20% heavy minerals) trending north–south and representing the positions of old strands, the bases of which occur on wave-cut platforms at successively lower elevations to the west (*Figure 11.9*). The higher-grade basal portion of each deposit represents bedded and well-sorted water-laid material in which the principal gangue is white quartz sand. The upper portion contains lesser heavy minerals and yellow quartz sand, which appears to be of dune character.

The bases of the strands all tilt slightly to the north. The principal heavy minerals in the strands are ilmenite, zircon and rutile, but minor kyanite and monazite are recovered. The 115-m strand is characterized by a high zircon content (61% of the heavy fraction). All other strands have lower zircon and higher rutile and ilmenite contents. Mineral proportions vary within strands. The difference between the 115-m strand and the lower ones is because the higher strand derived the bulk of its heavy minerals by direct erosion of the existing cliff faces. The 109-m and lower strands were formed by concentrations of heavy minerals brought into the bay.

The prevailing longshore drift has resulted in a sediment-size reduction southwards from average grain sizes of 0.207 mm in the north to 0.156 mm in the south. Some sandstones have been iron-cemented. The iron consists of cryptocrystalline goethite cement rimming the clastic quartz grains, and it is presumed to have been introduced soon after the deposition of the eastern strands (Lissiman and Oxenford, 1975).

THE CAPEL BEACH SANDS

These occupy a north-facing bay about 250 km south of Perth. The deposits have been described by Welch (1964) and Welch *et al.* (1975). They are located on or close to the Pleistocene and recent shore lines, now standing at levels of up to 76 m above present sea level. Each strand line is located in a notch cut into a wave-cut platform of Mesozoic sediments (*Figure 11.9*). The source rocks are presumed to be the high-grade metamorphic ter-

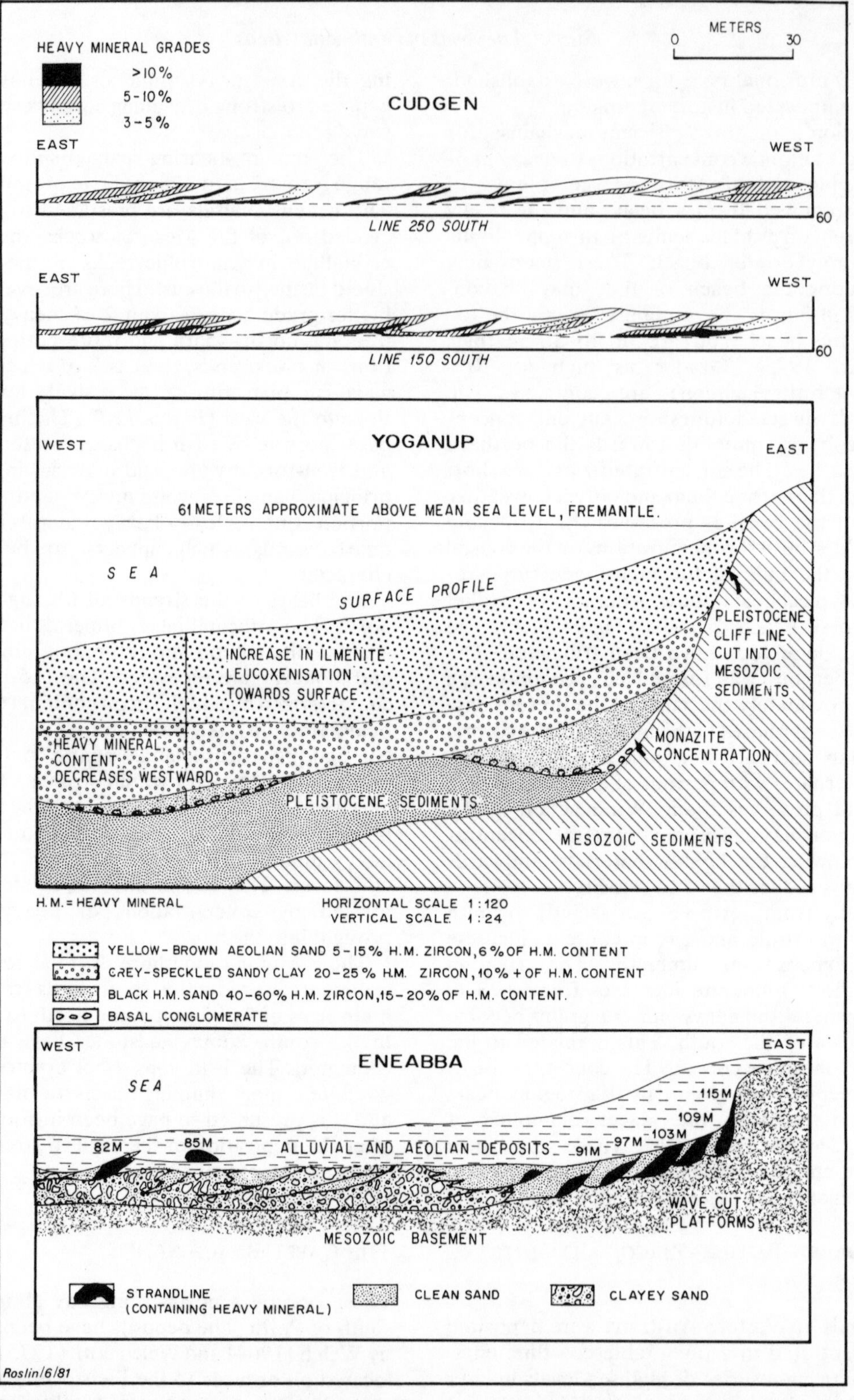

Figure 11.9. Recent rutile, ilmenite and zircon placer deposits in the coastal marine deposits of Australia. The Cudgen deposit lies on the coast about 120 km south of Brisbane and forms part of the extensive eastern Australian rutile province (after McKellar, 1975). The Yoganup deposit lies 18–50 km southeast of Banbury in Western Australia (after Welch *et al.*, 1975). The early Pleistocene deposit of Eneabba lies about 300 km north of Perth, Western Australia (after Lissiman and Oxenford, 1975).

rain to the east in addition to Mesozoic sediments closer at hand.

Ilmenite is the dominant heavy mineral. Zircon and rutile contents are highest in the oldest deposits. The gangue is mainly of quartz, calcareous shell fragments and kaolinite. The heavy-mineral contents from various strands can range from a few percent in the upper sandy sections to as much as 60% in the lower clayey sections. The ilmenite in the upper sandy sections tends to be altered to leucoxene (*Figure 11.9*).

12

Fuels

Petroleum and Natural Gas

The occurrence of fluid hydrocarbons in rocks depends on four critical properties of a sedimentary basin (Dickinson, 1975a): there must be organic-rich source beds within the sedimentary sequence; there must have been appropriate heat for sufficient time for thermal maturation of petroleum or gas; there must have been permeable migration paths for the fluid hydrocarbons to move from the source beds to the reservoirs; and there must be porous reservoir beds confined within some trapping configuration by impermeable, finer-grained or over-pressurized capping beds, which prevent the hydrocarbons from escaping.

Source Beds

The modern geochemical concept is that petroleum was formed from disseminated organic matter (kerogen) in sedimentary rocks. The quantity and variety of hydrocarbons formed after thermal maturation is related to the concentration and type of the kerogen in the source beds (Staplin, 1969).

KEROGEN CONCENTRATION

An arbitrary useful lower limit of about 0.5% organic matter (0.4% organic carbon by weight) is generally accepted as the minimum required before a rock can act as a source bed (Dow, 1977a). However, most acknowledged source beds contain 0.8–2.0% organic carbon by weight, and the best as much as 10%. However, in the deeper parts of a sedimentary basin, where the maturation of the organic matter has reached the dry-gas stage (*see* p. 279), a residual 0.5% organic carbon may witness an original 1% or more. Thus it may have been a potential source for gas (Tissot, 1977).

KEROGEN-TYPE

Elemental analysis of the major constituents, carbon, hydrogen and oxygen, has been used to define three main types of kerogen (Tissot, 1977) (*Figure 12.1*).

Type I

This type originally has a high-hydrogen and low-oxygen content. It is very rich in lipids and particularly in aliphatic chains, and contains only a small proportion of polyaromatic nuclei and oxygenated functional groups. It may be derived from the accumulation of algae, as in the case of oil shales, or may result from an intense reworking of various organic matters by microbes, leaving mostly the lipid fraction of the original material, and microbial lipids. It has a high potential for generating petroleum, and it also comprises the richest oil shales (Tissot, 1977). The Green River Shales of Colorado and Utah are a good example.

Type II

This type originally has a fairly high hydrogen content, but lower than type I. It contains abundant aliphatic chains of various length as well as saturated rings. It also comprises polyaromatic nuclei and hetero-atomic functional groups such as oxygen-containing carbonyl and carboxyl groups. The organic matter is usually derived from marine phytoplankton and zooplankton deposited in a confined environment. Many well-known oil fields are of this type—the Cretaceous of the Middle East and the Jurassic of western Europe are good examples.

Type III

This type was originally low in hydrogen and high in oxygen. Chemically it comprises mainly polyaromatic nuclei and oxygen-containing func-

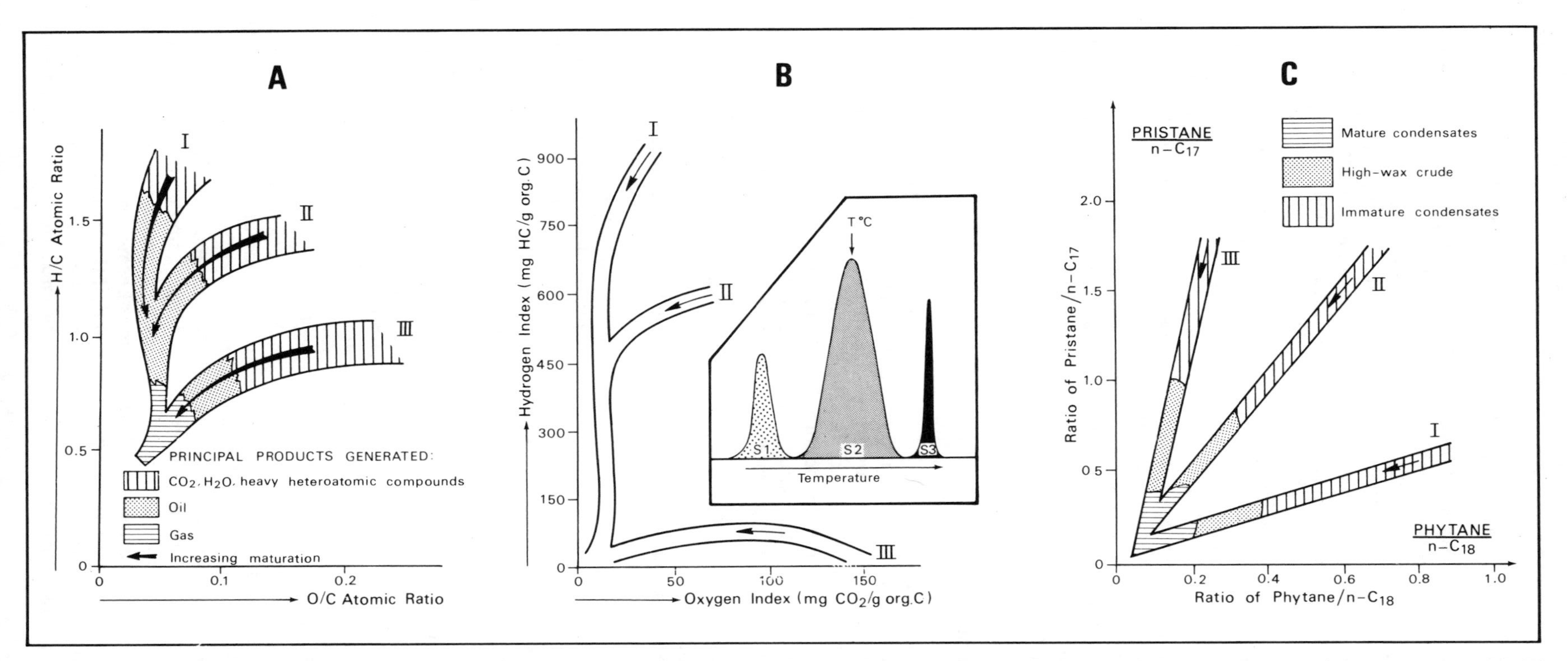

Figure 12.1. Kerogen-evolution path, or its change with maturation. (A) Based on atomic ratios of hydrogen, oxygen and carbon (after Tissot, 1977). (B) Based on pyrolysis analysis (from Espitalié *et al.*, 1977a,b). Inset is an example of a pyrolysis measurement. (C) Pristane to n-C_{17}, phytane to n-C_{18} ratios of crude oils and condensates (after Connan *et al.*, 1978).

tional groups, and in addition some aliphatic chains, including long straight chains from natural wax, attached to the polycyclic networks. The organic matter is mostly derived from the higher land plants. This terrestrial input may be transported by rivers and offshore currents and laid down in non-marine or in deltaic or continental margin environments. Compared with types I and II, type-III kerogen has a low potential for oil but can provide a good gas source at depth in a basin. Examples are the Miocene oil fields of Borneo or the Lower Mannville Shales of western Canada.

Intermediate types of organic matter commonly occur in basins, particularly between types II and III. This situation is found in a marine environment where an autochthonous marine input of kerogen, derived from plankton, is superimposed on a regional supply of terrigenous organic debris brought in by rivers and currents. The resulting petroleum potential is then intermediate in nature. An example is in the Malay Basin between Borneo, the Malay Peninsula and Indochina.

Marine organic matter is generally deposited under reducing conditions and has a high convertibility to oil and gas. The source rocks are frequently therefore rich in pyrite, often in the form of microcrystals coated with organic matter and assembled to form raspberry-like aggregates or framboids, as in the Miocene source rocks of Trinidad. On the other hand, terrestrial organic matter is commonly incorporated in the sediments under oxidizing conditions and therefore has a low convertibility and then only to gas.

Type IV

There is a fourth type of kerogen, which is recycled or oxidized material called inertinite, referred to as 'dead carbon' because it is particularly unaffected by heat and its convertibility to oil or gas is practically zero (Erdman, 1975).

Production and Accumulation of Organic Matter

At the present time, terrestrial plants and marine phytoplankton produce about equal amounts of organic carbon. The process results from photosynthesis on land or in the upper 200 m of the sea. Terrestrial organic matter is most abundant along continental margins especially in areas of major river runoff in giant deltas. Marine organic matter production is controlled by light, temperature and nutrients in the sea. Up-welling of deep ocean water can introduce nutrients such as phosphate and nitrate into the euphotic zone, resulting in regions of high organic production (Dow, 1977b). Some of the richest source beds were deposited along continental margins in the area of deep-water up-welling. It is estimated that less than 1% of the organic matter produced becomes incorporated into the sediments. The main destructive processes are chemical oxidation and consumption by grazing heterotrophic organisms.

Organic matter is deposited most readily where bottom conditions are not particularly oxidizing. Such conditions are attained in closed anoxic basins, in the oxygen minimum zone of the sea along certain upper continental slopes and in areas where organic productivity exceeds the ability of available free oxygen to convert organic matter to carbon dioxide. The average organic carbon content (in % by weight) in continental margin marine environments may be summarized as follows (Dow, 1977b):

Zone	*Water depth*	*Average organic carbon content*
Neritic		
inner	0–20 m	0.17
mid	20–100 m	0.31
outer	100–200 m	0.47
Bathyal		
upper	200–500 m	0.61
lower	500–2000 m	0.59
Abyssal	Exceeding 2000 m	0.63

Organic matter in a water environment is either dissolved or suspended as fine particles and is easily transported by water currents. Certain mineral particles, especially the clay minerals, absorb certain polar organic compounds and therefore convert dissolved organic matter to particulate form and reduce its residence time in the oxygenated water where it may have been oxidized. Because the clays have a greater absorption capacity than carbonates, they also contain more organic matter and comprise the bulk of the sediments that are rich in organic carbon. Mudstones and shales are therefore much more likely to become petroleum source rocks than any other sedimentary rock-type.

Muds accumulate in areas of quiet water where wave action and current activity are minimal, such as in deep water or in restricted basins like the Mediterranean or Black Sea.

High sedimentation rates, such as occur in deltas, may result in low organic carbon concentrations because of the dilution effect. On the other hand, if the sedimentation rate is too low, much of the organic matter reaching the bottom may be consumed by heterotrophic organisms before it can be buried. Intermediate sedimentation rates, which minimize the effects of both dilution and consumption, should result in the most organic-rich sediments. The upper portions of some continental slopes are the most favoured sites for deposition of organic-rich muds. Continental shelves and rises are often oxidizing, contain high-energy coarse clastic deposits (sands), or have high rate sedimentation values (turbidites). The organic carbon content in most shelf and rise deposits is consequently reduced (Dow, 1977b). The deposition of potential oil source beds requires special conditions, which are the exception rather than the rule. Almost any continental margin can satisfy such conditions, and none can unconditionally be ruled out. Statistically, however, continental slope sediments and restricted basins appear to be the most favourable for rich source-bed formation.

Maturation of Organic Matter

The instability of organic matter, as the containing sediments are subjected to increasing temperature resulting from burial, is the primary cause of coalification. Coal rank increases with burial. The thermal evolution of kerogen with increasing burial is called catagenesis, a process which is virtually homologous with coalification (Hood *et al.*, 1975). The process depends on temperature and duration of heating. It is generally accepted that organic matter will begin to generate liquid and gaseous hydrocarbons between temperatures of 70°C to 85°C with a transition to dry gas occurring between 150°C and 175°C, leading to the 'liquid window' concept of Pusey (1973). The depth at which the liquid window occurs will naturally vary with the prevailing geothermal gradient operating in the sedimentary basin, shallower in basins of high geothermal gradient and deeper in those with low gradient (*Figure 12.2*). Of course, the present geothermal gradient may not have been that operating during the maturation of the oil, but there is a high degree of stability in oil basins and the present gradient usually correlates approximately with the observed liquid-window depth.

Kerogen evolution paths

The total amount of hydrocarbons that can be generated and the relative importance of oil and gas depend on the composition of the original organic matter; some formations may generate mostly gas. The elemental composition, in terms of hydrogen, oxygen and carbon, changes with maturity according to the original composition. The change has been called 'evolution path' (Tissot *et al.*, 1974). All curves tend to the same final product approaching pure carbon (*Figure 12.1*). Pristane to n-C_{17} and phytane to n-C_{18} ratios steadily decrease through increasing maturity (Connan *et al.*, 1978).

Pyrolysis analysis

The degree of maturation of the kerogen in the source rocks may be conveniently measured by heating 50–100 mg of ground source rock in an inert atmosphere at a controlled rate up to 550°C (Barker, C., 1974; Claypool and Reed, 1976). The area of peak S_1 represents the free hydrocarbons present in the rock that are volatilized at a temperature below 300°C (*Figure 12.1*), expressed in milligrams of hydrocarbons (oil + gas) per gram of rock. The area of peak S_2 represents hydrocarbon-type compounds produced by cracking of the kerogen up to 550°C and equals the residual petroleum potential of the rock expressed in milligrams of hydrocarbons per gram of rock. The ratio S_2 divided by the total organic carbon of the rock is called the 'hydrogen index' (Espitalié *et al.*, 1977b).

The area of peak S_3 corresponds to the carbon dioxide produced by the pyrolysis of the organic matter expressed in milligrams per gram of rock. The ratio of S_3 divided by total organic carbon in the rock is called the 'oxygen index'. The peak temperature T°C (*Figure 12.1*) is characteristic of the evolution level of the organic matter. The 'production index' is given by the ratio $S_1/(S_1 + S_2)$. It characterizes the evolution level of the organic matter and makes possible the detection of oil shows.

The hydrogen index and oxygen index, when plotted against each other (*Figure 12.1*) show a good correlation with the atomic ratios H/C and O/C obtained from the elemental analysis of kerogen (Cooper, B.S., 1977). The techniques are identical to those used for coal petrology. Random rocks, and has the advantage of being carried out on the well site.

Vitrinite reflectance analysis

A high magnification photometric microscope with stabilized quartz halogen incident light source is used to measure the reflectivity in oil, R_o, of the polished surface of the vitrinite components of the kerogen (Cooper, B.S., 1977). The techniques are identical to those used for coal petrology. Random mean values are reported. In coal, most vitrinite is autochthonous, but kerogen samples, however, frequently contain allochthonous vitrinite from

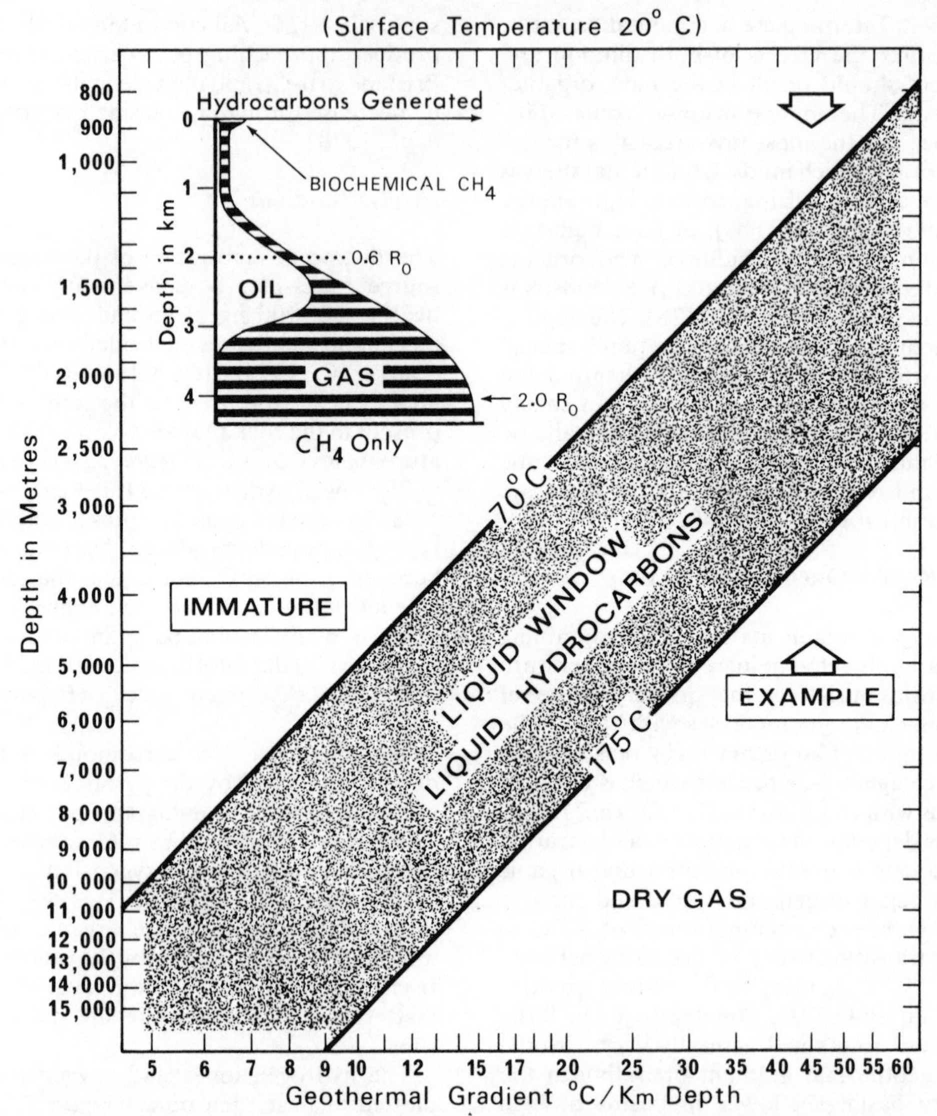

Figure 12.2. The 'liquid window' or depth at which liquid hydrocarbons will form under different geothermal gradients. Inset: scheme of hydrocarbon generation for the example of 45°C km^{-1} geothermal gradient (after Tissot *et al.*, 1974).

recycled organic matter as well as physically and chemically weathered vitrinite and caved samples from the drill hole. A large number (50–100) are therefore recorded and plotted on a histogram and the extraneous readings eliminated. The average R_o value reflects the thermal maturity of the source rock (*Figure 12.3*) and is a useful way to estimate the palaeo-temperature that pertained to the stratum (Dow, 1977a).

Spore colouration

The colouration of spores and pollen is a useful indication of the level of maturation (Cooper, B.S., 1977). With increasing maturity, spore colours change from pale-yellow, through golden-yellow, orange, brown to opaque, and twelve colour stages have been designated. There are some anomalies in spore colour. In redbeds, the palynomorphs have been bleached during deposition and retain their pale colour throughout maturation. Strongly coloured palynomorphs occur in bituminous shales, owing probably to their uptake of heavy metals.

All the maturity indicators are shown together with the equivalent coal rank scale in *Figure 12.3* (Dow, 1977b). The 'liquid window' approximates to depths of high volatile bituminous coal.

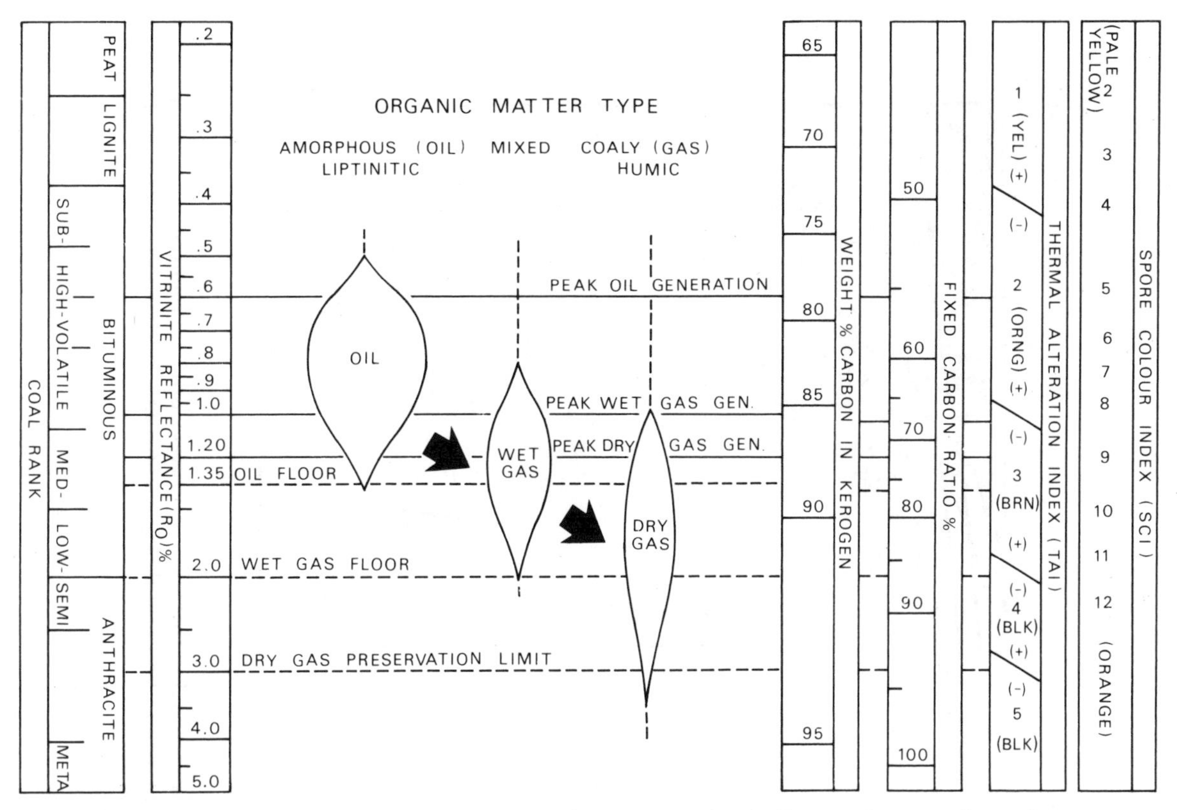

Figure 12.3. Correlation of coal-rank scale with various maturation indices and zones of petroleum generation and destruction (after Dow, 1977b and Cooper, B.S. 1977). The relative importance of each petroleum-generation zone depends on the composition of the original kerogen.

The time factor in maturation

The reactions start early as the sediments increase in temperature because of burial within the prevailing geothermal gradient. However, change is very slow and the rate of conversion of kerogen to hydrocarbons only increases exponentially with temperature increase. Hence ultimately there results a period of rapid oil generation that may be termed 'peak generation' or the 'principal phase of oil formation'. The depth of this occurrence depends on the geothermal gradient and the burial history. A good illustration of this is given by Dow (1978) for the Louisiana Gulf Coast (*Figure 12.4*). The wells selected have very similar geothermal gradients of about 25.5°C km^{-1} and a surface temperature of 21°C. The variation in maturation–gradient slopes is due primarily to temperature exposure time or average sediment age. The depths and present temperatures at the top of the oil-generation zone ($R_o = 0.6$) and the floor ($R_o = 1.35$) are different for each trend (*Figure 12.4*). For example, the 0.6R_o level occurs at 5578 m and 164°C in the Plio–Pleistocene trend, but only at 2469 m and 84°C in the Cretaceous trend. The difference is due to the temperature exposure time because kerogen takes time to change to hydrocarbons.

It is evident that the depth of burial and temperature required to reach the principal oil-generating zone (0.6R_o–1.35R_o) vary according to both exposure time and geothermal gradient. The values for the Louisiana Gulf Coast will not apply elsewhere because of differences in geological history. Changing sedimentation rates, periods of erosion or non-deposition and varying geothermal gradients can all affect the actual maturities in any given basin (Dow, 1977b).

Primary Migration

Hydrocarbons generated in fine-grained sedimentary rocks are probably disseminated at first, but eventually they move into more permeable and porous sediments to form commercial accumulations. The movement from source rocks to reservoir rocks is known as primary migration. **There is**

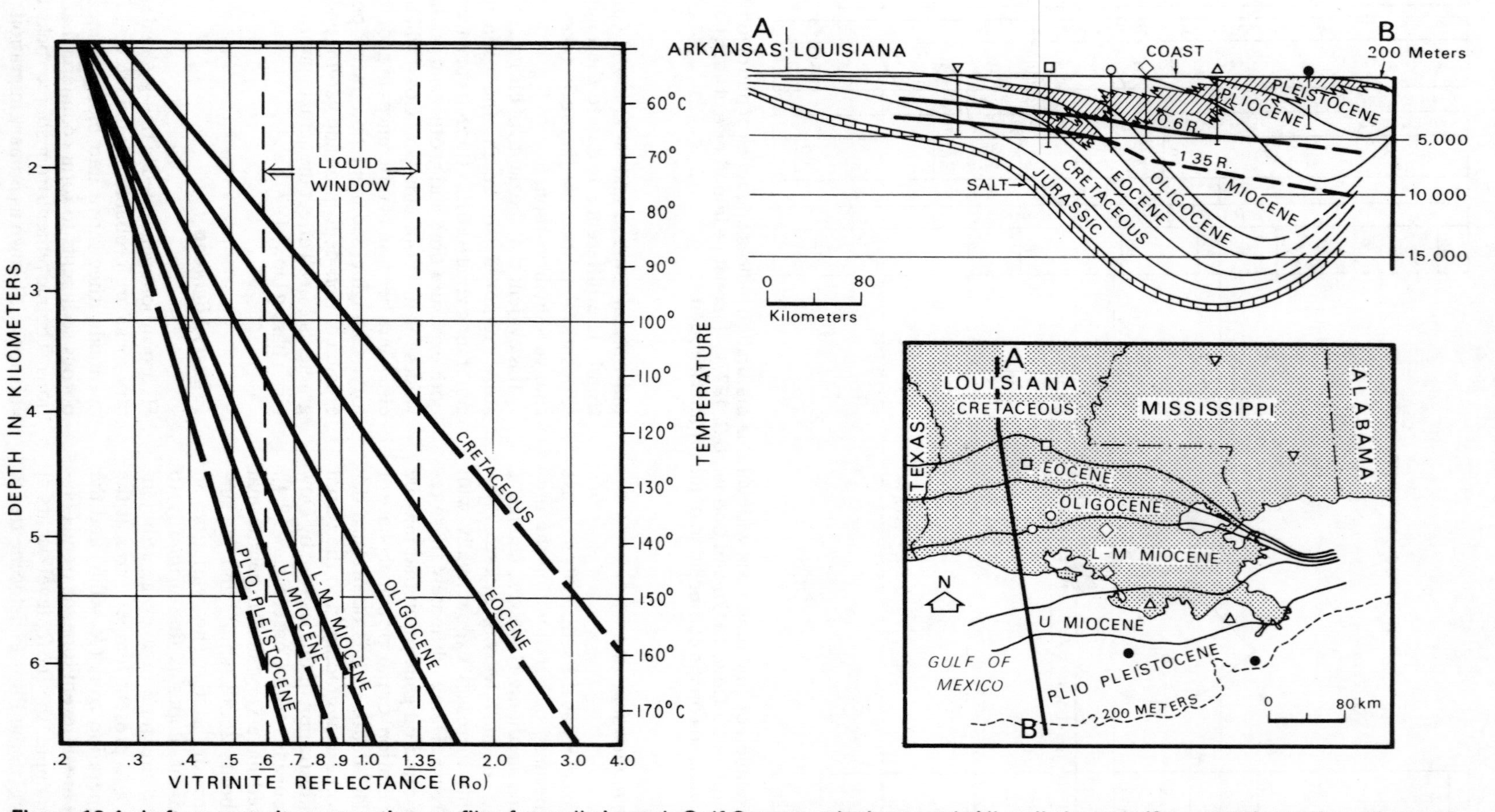

Figure 12.4. Left: composite maturation profiles for wells in each Gulf Coast producing trend. All wells have uniform geothermal gradients of 25.5°C km^{-1} and surface temperatures of 21°C. Top right: cross-section showing liquid window ($0.6R_o$–$1.35R_o$) and shaded reservoir facies. Lower right: line of cross-section and age of formation outcrops (from Dow, 1978).

no consensus on the method by which hydrocarbons migrate, as the heated discussions in the *Bulletin of the American Association of Petroleum Geologists* bear witness. All agree that it must be difficult for hydrocarbons to migrate through the fine-grained source rocks, just as all agree that migration has definitely occurred.

There can be no doubt that the petroleum in reservoirs is allochthonous. The example of *Figure 12.4* can be taken to illustrate the general principle. Most of the Louisiana Gulf Coast production is from sandstones associated with thermally immature shales. The oil must therefore have migrated vertically from the mature source beds into the overlying shelf deposits of sand–shale magnafacies. Calculations by Young, A. *et al.* (1977) show that the ages of the offshore Gulf Coast oils are, on average, 8.7 Ma older than the reservoirs in which they occur and that this age difference indicates an average vertical migration of 3350 m. The Gulf Coast is especially productive because mature shelf-slope shale-source rocks are overlain by prograding delta sands that are connected by faults and fracture systems to the source rocks, which have provided channelways for vertical migration. The porous reservoir sands and relatively impermeable cap-rock seals have combined in excellent harmony to provide a perfect oil field.

Possible Mechanisms

Gaseous solution

Compacted shales are very susceptible to microfracturing and faulting during tectonic events. These faults and microfractures provide an easy vertical escape route for gas that has been generated in the deeper parts of a basin (below R_o 1.35). Natural gas is continuously generated in the deep hot parts of a basin provided that the shales still contain hydrogen.

A solubilization process of liquid hydrocarbon in the highly pressurized gas as it rises up through the 'liquid window' is thought to be the main mechanism for the extraction of oil from the source rocks (Neglia, 1979). The gas percolates up through microfractures in the compacted shales.

The little-compacted shales at higher levels in the sedimentary basin are plastic and can hold gas and oil provided that the displacement threshold is not reached. Retrograde concentration occurs in the shallower horizons where the temperature and pressure of the migrating gas decrease. Differences in the displacement pressure of the oil and gas will separate the gas. The primary migration terminates when the oil and gas reach an adjacent permeable bed. An accumulation takes place when there is a suitable stratigraphic or structural trap overlain by a sealing cap rock. Most cap rocks are not truly water-impermeable, though they deter the upward migration of liquid hydrocarbon and they allow the gas to pass upwards, thus concentrating oil. The only truly impermeable cap rock are evaporite and permafrost.

As the gas phase percolates up through the cap rock, the oil will remain below it. Bacterial degradation processes generally occur in the hydropressurized layers, converting the condensate into heavy oil (Neglia, 1979).

Aqueous solution

As deeper levels are encountered, the solubility of oil in water increases from less than 22 ppm at 100°C to as high as 500 ppm at 175°C (Price, 1976). However, a decrease in oil solubility attends the increasing salinity of the water. Benzene at 25°C has a solubility in salt-free water of 1500 ppm but at 200 000 ppm sodium chloride, the water solubility of benzene drops to as low as 500 ppm.

The upward migration of the water is facilitated by microfractures and growth faults in the compacted shale and by three-dimensional kerogen fragment networks wedged between the clay platelets keeping the pores interconnected. As the water rises through the stratigraphic column, a decrease of temperature causes exsolution of the hydrocarbons from the aqueous phase since solubility falls off rapidly with temperature. As deep-basin compaction proceeds to completion, the hot waters carrying large amounts of hydrocarbons to the higher levels, focused along faults and microfractures, move towards higher levels until permeable reservoir rocks are encountered and further upward migration is impeded by impermeable cap rocks (Price, 1976). Temperatures of at least 225°C and possibly as high as 400°C are required for initial aqueous solution in the Gulf Coast Basin. However, Barker, C. (1977) has strongly objected to water solubility as a carrying mechanism because of the low solubility of saline water, and the high temperatures needed; temperatures that would exceed the major thermal breakdown of the oil. The large quantity of water needed for this mechanism far exceeds the amount available. However Price (1977) disagrees that oil has predominantly broken down by 150°C and says there is evidence for significant oil still existing in oil wells at temperatures up to 200–300°C.

Micellar solution

Hydrocarbon solubility in water is substantially increased if the water contains micelles formed by soaps of organic acids (Baker, E.G., 1962). A naphthenic acid soap concentration of at least 500 ppm is necessary at 25°C and 8300 ppm at 90°C for critical micelle solutions. Surfactant concentrations in oil-filled waters are commonly only 2–3 ppm and rarely over 10 ppm. These values are too low for critical micelle concentration (Price, 1976).

Oil phase

The difficulties of the above mechanisms have led Magara (1977) to conclude that the larger amount of liquid hydrocarbons must migrate as a separate phase, although some may migrate in solution. The amount of hydrocarbons in the fine-grained source rocks is small compared with the water content and obviously the movement of large quantities of water must have an influence and may have controlled the direction and effectiveness of the oil migration. Dickey (1975) has argued that the conclusion cannot be avoided that oil must have migrated as an oil phase. The possibility that much of the water in compacted source rocks may have behaved as part of the solid matrix, bounded to the clay minerals, would imply that oil forms the larger portion of the pore fluid and would totally wet the rock surfaces bounding the pores, so that the saturation at which oil could flow as a continuous phase would be much less than 10% of the pore fluid. As compaction of the source rock proceeds, more and more water is bound up in the clay structures so that oil would be preferentially expelled (Dickey, 1975). The migration of oil as a separate phase may be facilitated by continuous three-dimensional networks of kerogen fused together during the source-rock compaction (McAuliffe, 1979).

The general conclusion must be drawn that oil does migrate out of the source rocks, but the mechanism is not understood.

Secondary Migration

Hydrocarbons, on their upward or lateral passage from the source rocks, may encounter coarser-grained rock bodies that act as reservoirs in which commercial accumulation of oil and gas are trapped. The concentration and accumulation in reservoirs is known as *secondary migration.*

The coarser reservoir rocks have been subjected to physical conditions that did not exist in the source rocks. These are: larger pore spaces, fewer capillary restrictions, less semi-solid water (water held in the clay minerals) and less fluid pressure. These new conditions help to enlarge and connect the hydrocarbon globules. The vertical connection of the globules will produce a significant buoyant force and the oil will migrate to a higher structural position in the reservoir. The mechanics of secondary hydrocarbon migration and entrapment have been discussed by Schowalter (1979).

The main driving force is buoyancy, which depends on the relative densities of the hydrocarbon phase and the water, which vary significantly. Subsurface oil densities range from 0.5 g ml^{-1} to 1.0 g ml^{-1} and water from 1.0 g ml^{-1} to 1.2 g ml^{-1}. The buoyancy forces are different between hydrodynamic and hydrostatic subsurface conditions. The hydrodynamic effect is more complicated, but can be quantified if the potentiometric gradient and dip of the strata are known.

The main resistant force to secondary migration is capillary pressure, which prevents the further migration of oil and allows it to accumulate beneath a cap rock or stratum of smaller pore spaces than the reservoir. The equation can be used to calculate the height (Z_c) of an oil column in a reservoir that can remain in equilibrium without escaping upwards out of the reservoir (Berg, 1975):

$$Z_c = \left\{2\gamma\left(\frac{1}{r_t} - \frac{1}{r_p}\right)\right\} \Big/ g(P_w - P_o)$$

where γ is the interfacial tension between oil and water, r_t the radius of the pores in the barrier rock overlying the reservoir, r_p the reservoir pore radius, g the acceleration due to gravity, P_w the density of water and P_o the density of oil. In another form, the equation may be expressed as follows (Schowalter, 1979):

$$P_d = (2\gamma \cos\theta)/R$$

where P_d is the hydrocarbon–water displacement pressure or the resistant force to migration, γ the interfacial oil–water tension, $\cos\theta$ the wetability term and R the radius of the largest connected pore throats (interconnected channelways between the pores). The value of R can be measured by capillary techniques on cores or drill cuttings. The subsurface hydrocarbon–water interfacial tension ranges from 5 dyne cm^{-1} to 35 dyne cm^{-1} for oil–water systems and from 30 dyne cm^{-1} to 70 dyne cm^{-1} for gas–water systems. The slugs of oil are thought to always pass through water-wet rocks. Therefore the contact angle, θ, of hydrocarbons and water against the solid rock surface, as measured through a water phase, is assumed to be 0°, and the wetability term cos θ is therefore assumed to be 1.

The capillary-pressure mechanism for petroleum migration and trapping is a valid hypothesis. Calculations show that the upward migration of oil in reservoirs will be impeded by a change in lithology, not necessarily to an impermeable clay horizon, but simply to a moderate fining of grain size. For example, the upward fining in graded bedding can form an effective upward barrier or seal to migration. A barrier facies may be porous and permeable and yet trap a significant oil column simply by virtue of its finer grain size (Berg, 1975).

A large amount of oil is trapped in simple stratigraphic sand lenses, overlain and surrounded by silt or clay, but such reservoirs are difficult to find because there is no structural evidence for their existence and they cannot be seen on seismic profiles.

Whereas capillary seals have been known for many years, pressure seals (Evans, C.R. *et al.*, 1975) are relatively new. The authors described a 'mixed compaction facies' in the Tertiary sequence of the Beaufort Basin. The slightly undercompacted shales are overpressurized and the normally compacted shales and sandstones, with which they are interbedded, are normally pressurized. The slightly overpressurized shales are considered to have restricted the vertical escape of fluids in sandstones that underlie them, and they are referred to as *pressure seals*. About 90% of the accumulated hydrocarbon in the Beaufort Basin (MacKenzie Delta of the Arctic Coastal Plain), here predominantly gas, are found to be in the mixed compaction facies and overpressurized shales have trapped the gas accumulations.

A comparison of the capillary and pressure seals is given in *Table 12.1* (after Evans, C.R. *et al.*, 1975). Pressure seals direct the migrating fluids along bedding planes, thus focusing the migration of hydrocarbons towards structurally high areas, where they are trapped by buoyancy, or may escape upwards along fault zones. A structural high may be composed of a dipping sandstone bed on one side and a steep fault on the other, which offsets it against an impermeable shale. In this case the sealing capacity of the trap will be determined by the sealing capacity of the mineral along the fault plane. Some faults make good seals while others provide good leakaway channels for the upward migration of oil and gas. The fault seal appears to result from the presence of boundary fault-zone material emplaced along the fault plane by mechanical or chemical processes (Smith, D.A., 1980).

Secondary migration under hydrostatic conditions will lead to the separation of a gas phase overlying an oil, which in turn overlies water, in order of their buoyancy. However, the geometry of separation may be different under hydrodynamic conditions. If the direction of fluid flow along a permeable bed is different from the vertical flow caused by buoyancy, hydrocarbon accumulations may be forced to move to locations other than the structural tops, or even may be completely carried out of the sedimentary basin (Hubbert, 1953). Some accumulations may be moved down-dip by the direction of the fluid flow through a basin owing to the hydrodynamic force directed in response to the potentiometric surface.

Some Examples of Oil Fields

The kinds of oil fields are too numerous to detail adequately in this volume, but the few chosen examples serve to illustrate some important features of the stratigraphic and structural trapping mechanisms.

Table 12.1 Comparison of seals which impede upward migration of hydrocarbons

	Pressure seal	*Capillary seal*
1.	Seal for any form of hydrocarbons whether they are in solution in water, or as a separate phase	Seal for hydrocarbons only in the hydrogen phase
2.	This seal is developed during the early stages of the compaction history of the shales when limited fluid expulsion has occurred	This type of seal may become important during the later stages of compaction because it becomes more effective as shales become more compacted
3.	May be more important for gas than for oil because gas is more soluble in water than oil	More important for oil than gas because of the higher solubility of gas in water
4.	Prevents the vertical migration of water and delays its expulsion from the compacting shales	Does not prevent or significantly delay the vertical migration of water

Niger Delta

The Niger Delta formed in the Benue Trough graben or aulacogen, which is taken to represent the failed arm of a triple spreading junction related to the opening of the Atlantic as South America separated from Africa in the Cretaceous period (*Figure 12.5*).

Rapid sand deposition along the delta edge, on top of undercompacted clay, has resulted in the development of a large number of synsedimentary gravitational faults (*Figure 12.6*). These so-called

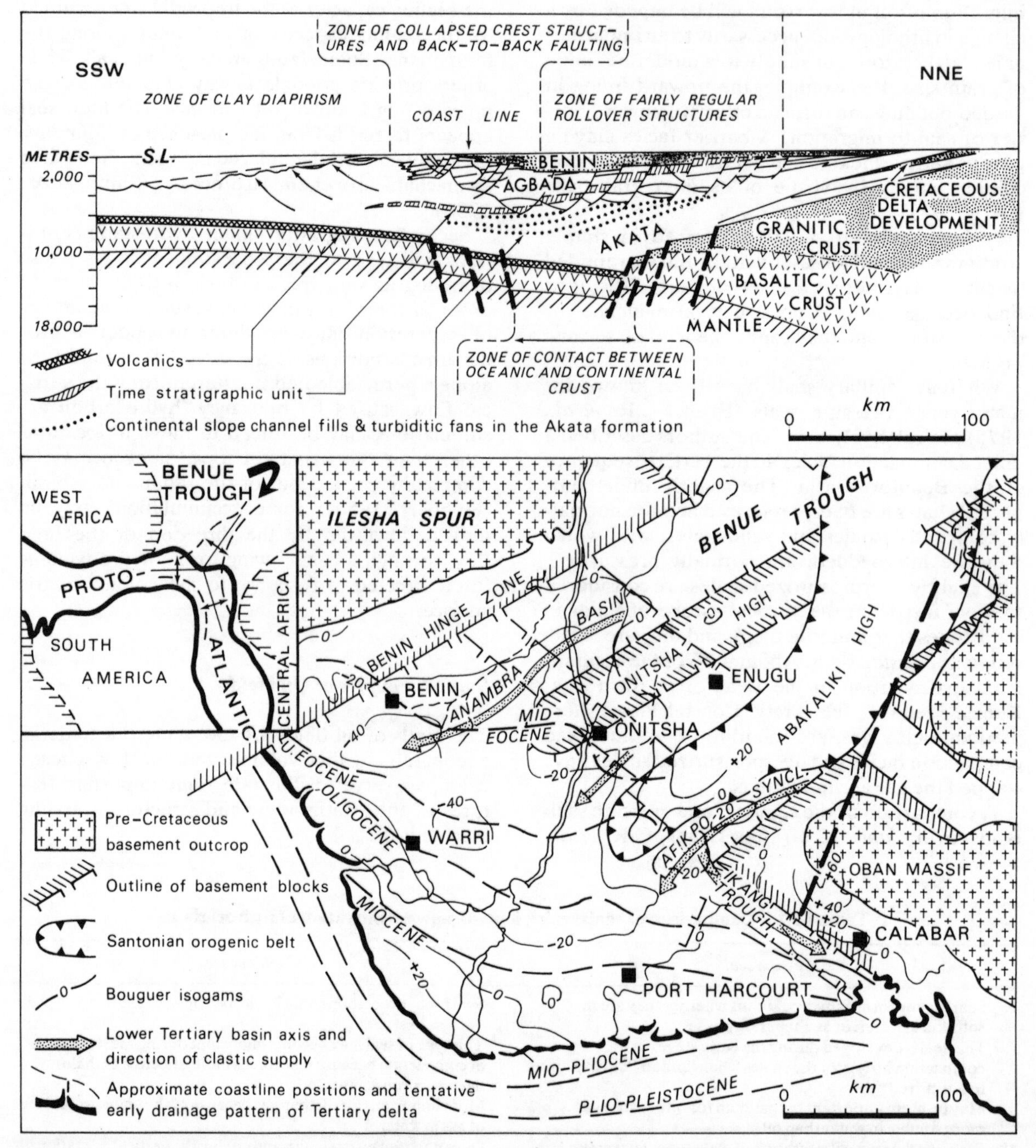

Figure 12.5. Bottom: megatectonic framework of the Niger Delta in relation to the Benue Trough and the opening of the Atlantic. Top: schematic dip section of the Niger delta (after Weber, K.J. and Daukoru, 1975).

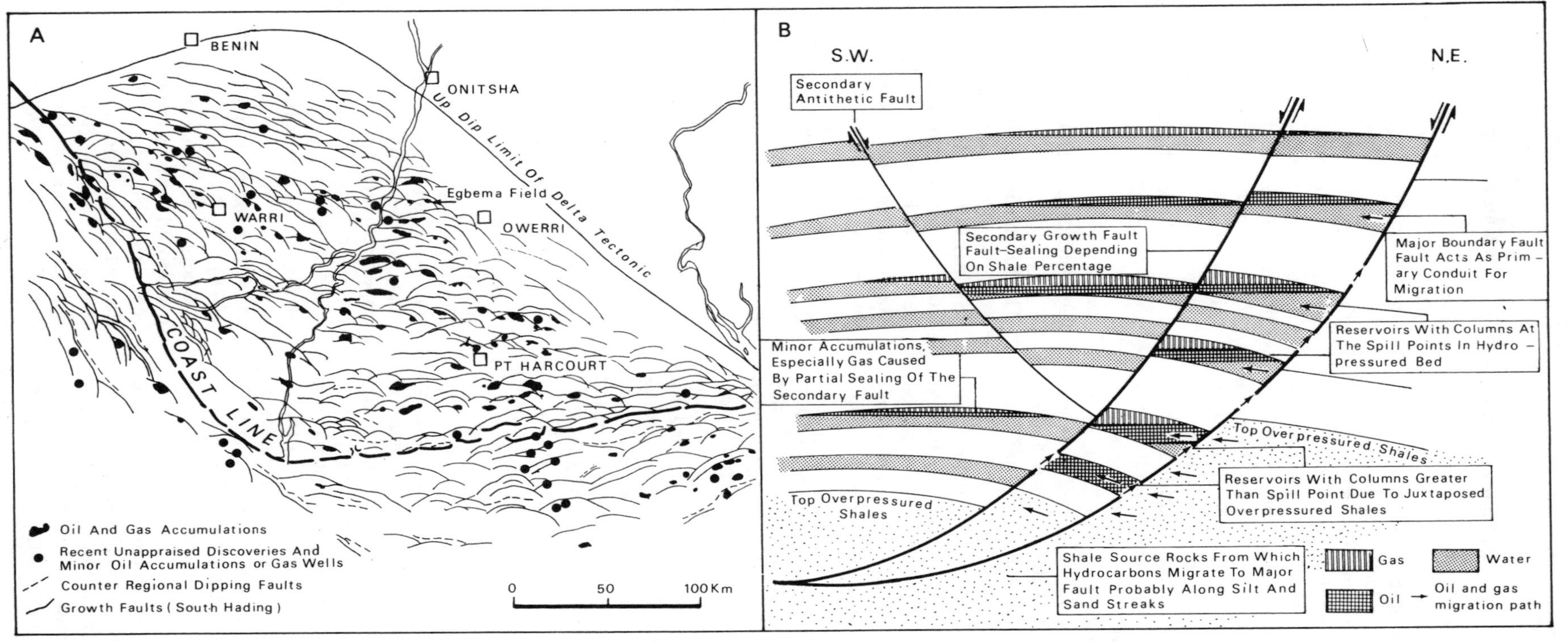

Figure 12.6. Listric growth faults and hydrocarbon accumulations in the Niger Delta. (A) Outcrop pattern; (B) schematic cross-section to show the principal features that cause the oil trapping (after Weber, K.J. and Daukoru, 1975).

'growth faults' are also well known from the Gulf Coast of the USA. Their origin and listric shape can be explained on the basis of the theory of soil plasticity (Weber, K.J. and Daukoru, 1975). After their formation, the faults remained active, allowing faster sedimentation on the down-thrown relative to the up-thrown block. The ratio of the thickness of a stratigraphic unit on the down-thrown block, relative to the up-thrown block, is called the 'growth index', which in Nigeria can be as high as 2.5. The enhanced sedimentation rates along the growth fault causes a rotational movement that tilts the beds towards the fault. In this way anticlinal so-called 'roll-over' structures are formed along the fault. Some 25 oil fields in the Niger Delta are basically unfaulted roll-over anticlines. More commonly the intersection of a south-hading fault with a structure forms 70 oil fields. About 10 fields are on simple anticlines with one or more antithetic faults. About 20 fields in the coastal region are much more intensely faulted. Their fault pattern is of the collapsed crest-type, with a series of closely spaced growth faults and a series of antithetic faults intersecting an anticlinal structure (*Figure 12.6*).

Shale ridges or shale diapirs can be of three kinds—those at zones behind major growth faults; those at bulges in front of the growth faults, which act as positive features causing collapsed crest structures and unconformities; and those along the continental slope, shale bodies being extruded in a seaward direction as a result of differential loading over the plastic marine shale (Weber, K.J. and Daukoru, 1975). The shale diapirs become buried during sedimentation but continue to grow after burial.

Oil migration and trapping

The main source rock is the Eocene-to-early Miocene Akata Formation shale, ranging in thickness from 600 m to 6000 m, deposited under anoxic conditions on the continental slope in front of the delta where the nutrient supply for planktonic organisms was plentiful. Migration of hydrocarbons was through the numerous fault zones from the overpressurized shale. Apart from along fault migration, another likely path was from a seaward facies up-dip into the south flank of roll-over structures.

Clay streaks in fault zones hamper or prevent flow across a fault plane, whilst sand streaks can give the fault zone a permeability along its length. In Nigeria the best evidence for vertical conductivity of the major boundary faults is that in most cases the fault intersection with the upper bedding plane of the reservoir functions as the spill point of the accumulations (*Figure 12.6*). It is thought likely that these spill points are also the entry points of the hydrocarbons from the fault zone into the reservoir (Weber, K.J. and Daukoru, 1975). The major role played by faults in oil migration is illustrated by the fact that only 5% of the oil columns exceed 50 m and hydrocarbon columns in excess of 150 m are very rare indeed. Most trapping is in a combination of structural and fault traps, and these have been well documented from the Niger Delta.

OFFSHORE LOUISIANA–GULF OF MEXICO

The Grand Isle Block 16 Field represents a good example of oil trapping from salt-dome tectonics (Steiner, 1976). The salt domes are diapiric intrusions or spines on an ancestral salt ridge. The ridge extends 45 km and covers an area of 362 km^2. The Bay Marchand salt ridge is paralleled to the north by a salt-withdrawal syncline called the Terrebonne Trough (*Figure 12.7*). A large, buried growth fault, down-faulted to the north, influenced sedimentation at the Grand Isle Block 16 Field. Deposition contemporaneous with salt tectonics not only caused thickening towards the trough but also caused several sedimentary units to become thinner towards the uprising salt.

Strata that are pierced by the salt dome generally dip 20–30° but near the salt the dip increases to as much as 55°. Large radially oriented faults divide the field into several peripheral segments (*Figure 12.7*). All the oil is produced from the peripheral fault segments.

Ninety-five percent of the hydrocarbons occur in the Upper Miocene strata, particularly in the *Bigenerina floridana, Cristellaria K.* and *Bigenerina Discorbis* biozones. The sequence of sands and shales in these zones was deposited in neritic-to-upper bathyal conditions. The oil is found in commercial concentrations in 26 Upper Miocene sandstone members. Up-dip reservoir seals are located at the contact with the salt, or with abnormally pressurized shales, or in stratigraphic traps where sandstone pinches out to shale up-dip (Steiner, 1976).

Many fault traps exist where the reservoirs are juxtaposed across a fault by shale. Some faults are sealing and others non-sealing to lateral oil migration where sandstones are juxtaposed with sandstones across a fault. Impermeable material deposited along the fault plane may form a perfect seal (Smith, D.A., 1980).

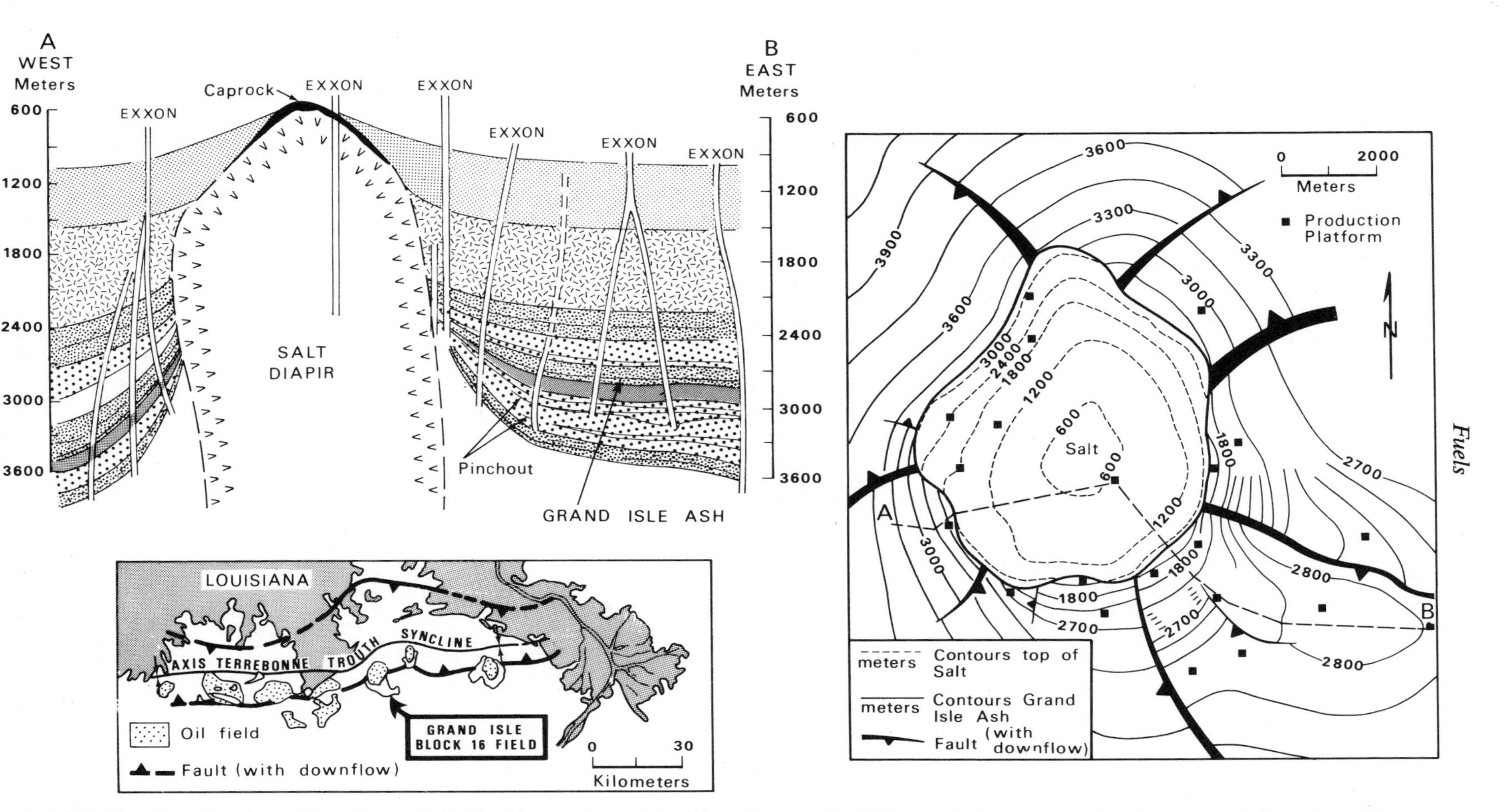

Figure 12.7. Details of the Grand Isle Block 16 Field, offshore Louisiana (after Steiner, 1976). Upper left: cross-section showing relation of the Miocene strata to the salt diapir. Lower left: locality map showing the salt-ridge trend and the Terrebonne syncline caused by salt withdrawal. Right: structural contour map of the top of the salt diapir and of the Grand Isle Ash. Radial faults are shown by thickened lines.

THE NORTH SEA VIKING–CENTRAL GRABEN

The North Sea Basin was developed on cratonic Europe. The Jurassic taphrogenetic or rift development of the Viking–Central Graben has been described by Ziegler, P.A. (1975). There were several stages of down-faulting and differential subsidence of the graben floor. In the northern North Sea there is clear evidence for continuous differential subsidence of the graben floor since the Triassic and throughout the Jurassic period. The Upper Jurassic is represented by deeper-water shales that locally contain turbidite sands, indicating that the rift system had developed into a submarine trough. Upper Jurassic shales are frequently very rich in organic matter and constitute an important oil-source rock. Further major rifting continued throughout the graben at the Jurassic–Cretaceous boundary. Volcanism, related to rift tectonics, is found widely throughout the graben in its development stages.

The Rhaetic-to-Lower Jurassic Statfjord and Dogger Brent sands form the main reservoirs for accumulation in the Viking Graben. However, their equivalents are of less importance in the Central Graben, where Upper Jurassic shallow-water sands constitute the major Mesozoic oil reservoir.

During the Albian–Aptian, much of Europe was covered by shallow sea in which the Chalk was deposited (Ziegler, W.H., 1975). In the Viking Graben the Chalk grades northwards into shales (*Figure 12.8*). The Chalk forms important reservoirs in some parts of the graben.

The taphrogenic stage of the North Sea came to a close in the Late Palaeocene. During the Eocene epoch to Recent time, the North Sea was dominated by regional subsidence resulting in the formation of a symmetrical saucer-shaped intracratonic basin, the axis of which coincides with the now inactive Central Graben (Ziegler, P.A., 1975). During the Palaeocene and Eocene times, sands and shales poured into the Viking–Central Graben trough from a predominantly northwestern source, forming deep-water turbidite tongues into the deep-basin parts, becoming the reservoirs for the Forties and other fields (*Figure 12.8*).

The Forties Field

Oil occurs in a broadly domed sandstone reservoir of Palaeocene age below a thick Cenozoic section consisting primarily of mudstone (Walmsley, 1975). The Palaeocene sandstone and mudstone section lies above the Danian and Maastrichtian micritic limestones, which in turn overlie presumed Jurassic alkaline volcanic rocks, associated with the rifting tectonics of the graben.

The trapping structure is a broad low-relief anticlinal feature with a closed area of 90 km^2 and an oil column of 155 m. Because the underlying rocks are predominantly micritic limestones and lava flows, the question of a source rock must suggest that the oil may have migrated downwards from overpressurized overlying mudstones, an explanation given for the Ekofisk Field, which has a similar tectonic setting, but different reservoir character.

The Ekofisk Field

The Danian–Upper Cretaceous is represented by chalky limestone consisting of a very fine-grained calcite–silt matrix, in which is set varying amounts of coarser-grained constituents (Byrd, 1975). The environment of chalk deposition must have been one of very quiet water, as shown by the well-preserved fragile coccospheres. The oil is contained in the Chalk. Its porosity is the original intergranular space between the loosely packed coccolith grains; the average value is 28%, but porosity occasionally exceeds 40%. Pressure solution stylolite formation has locally reduced porosity. The reservoir rock is highly fractured, resulting from tectonic stresses induced by the Miocene growth of salt structures beneath the Chalk. This tectonic fracturing is the most important characteristic and has allowed the Chalk reservoir to produce oil at prolific rates.

The overlying Tertiary shales are considered to be the source rocks; they are extremely rich in organic matter and are abnormally pressurized. Such overpressurizing has prevented the vertical migration of hydrocarbons up out of the Danian reservoirs and has facilitated the downward migration into the less-pressurized Danian reservoir (Byrd, 1975). This field is an excellent example of undoubted source rock overlying a perfect reservoir and vertical downwards migration assisted by the overpressurization of the overlying shales, which provide a perfect seal for the reservoir.

The Dan Field

Near the axis of the Central Graben, a deep trough was filled with a thick sequence of Permian-to-Cretaceous sediments (*Figure 12.8*). Upper Cretaceous–Danian Chalk at the top of this sequence provides the reservoir. Geochemical studies indicate that the deeper Upper Jurassic marine shales are the probable source rocks (Childs and Reed, 1975). The Dan Field is a halokinematically induced domal anticlinal struc-

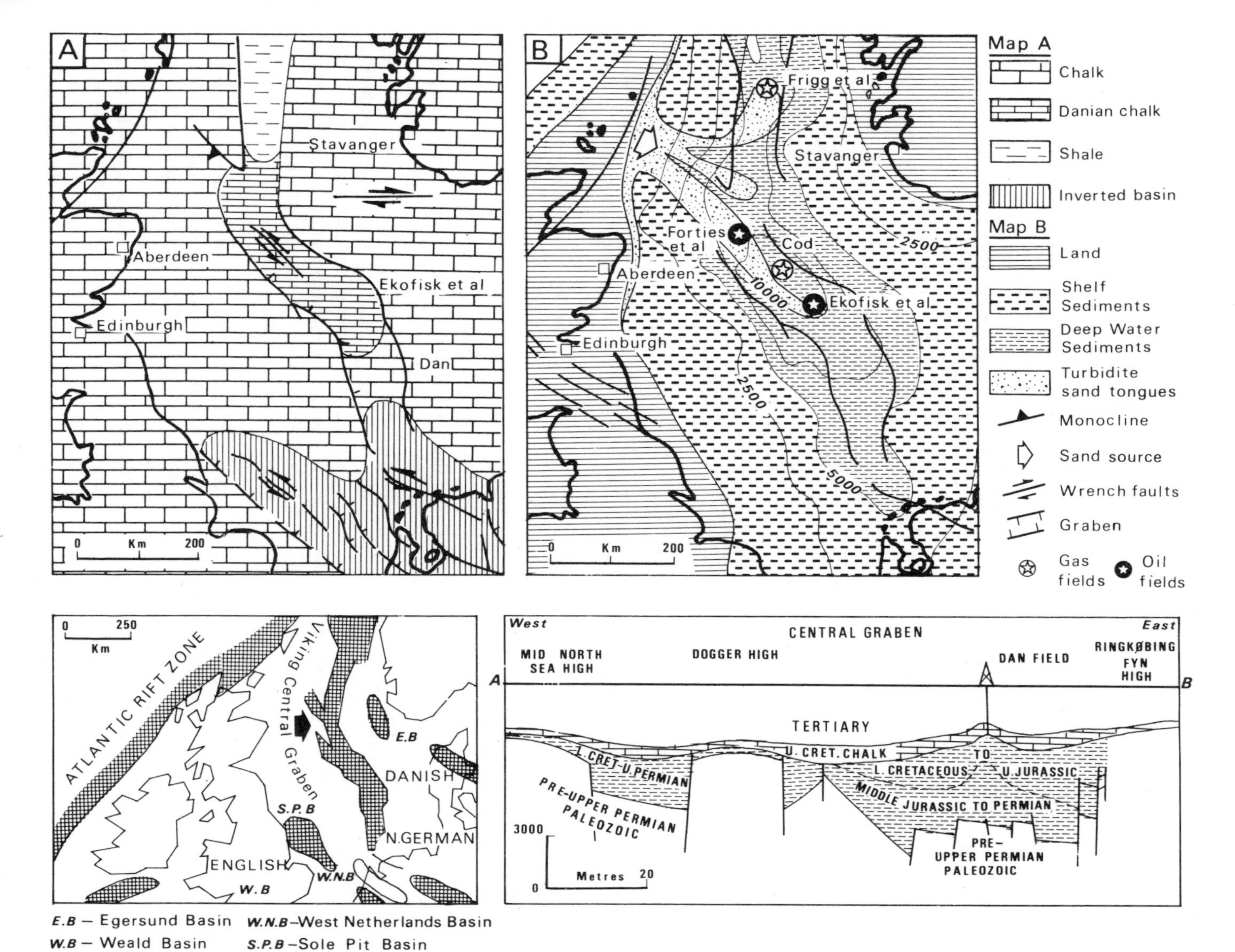

Figure 12.8. Tectonic setting of the North Sea oil and gas fields. (A) Cretaceous palaeo-geography; (B) Tertiary palaeo-geography. Lower left: major geotectonic features of the North Sea basins. Lower right: cross-section across the Dan Field (based on Childs and Reed, 1975; Ziegler, P.A., 1975).

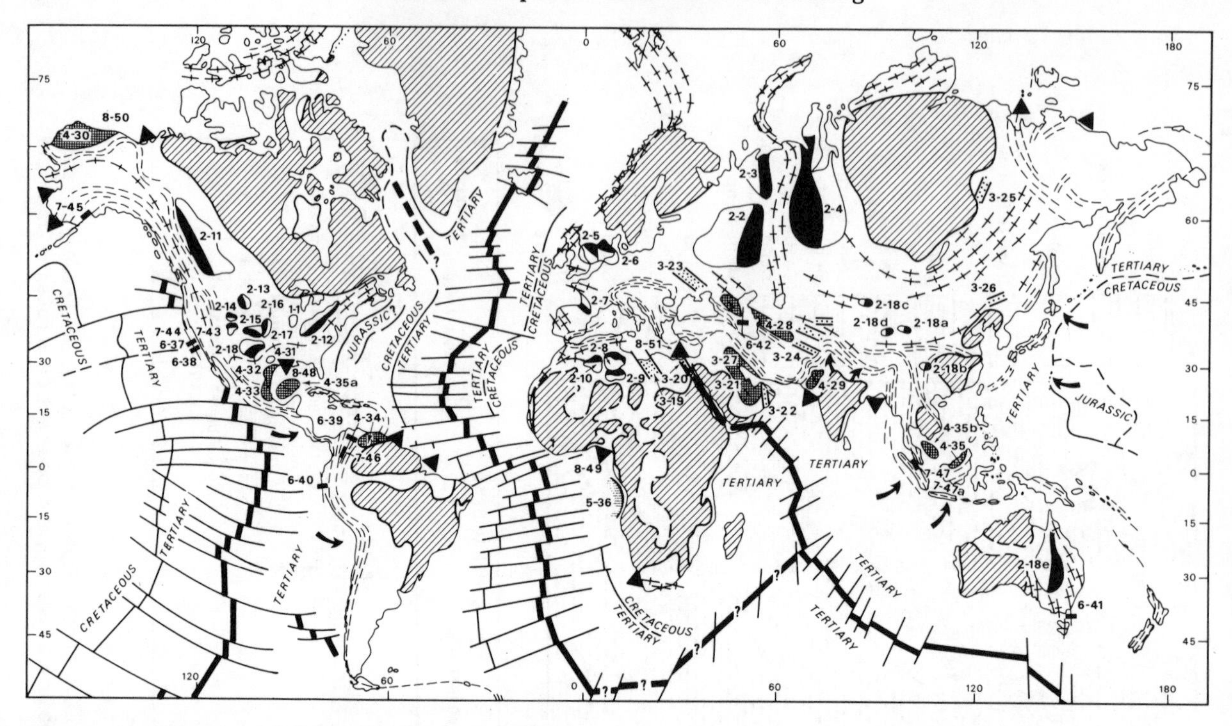

ture with an average porosity of 28% but extremely low permeability, offering slow production rates. The field can be divided into a gas cap of 80 m thickness, an oil zone of 96 m and an underlying oil–water transition zone of 68 m thickness.

Classification of Oil Traps

Hydrocarbon traps have been divided into three types: structural; stratigraphic; and a combination of the two (Levorsen and Berry, 1967). Most geologists, including the Russians, follow this classification and more recently (Klemme, 1973, 1975) has crystallized the classification scheme as detailed below.

Structural traps

- I. Anticlinal
  - Ia. Anticlinal fold, regional arch, block uplift or horst (cratonic)
  - Ib. Compressional fold (younger orogenic-type)
  - Ir. Depositional fold (roll-over)
- II. Fault
- III. Flowage
- IV. *Combination traps*
  - IVa. Combination of types I and Va
  - IVb. Combination of types I and Vb
  - IVc. Combination of type II or III with type Va or Vb
- V. *Stratigraphic traps*
  - Va. Unconformity—related to unconformity, truncation traps
  - Vb. Lithological—facies-related, and caused by pinch-outs, wedges, reefs or diagenetic pinch-outs

The anticlinal traps are subdivided on the basis of the genesis of the structure. Group Ia comprises folds, regional anticlines and block uplifts or horsts, over which a trap forms by reversal of the dip of the strata. Usually these traps result from basement movement in the cratonic crust.

Younger orogenic movements and the character of the sediments have exerted greater influence on group-Ib compressional folds, and faulted compressional folds. Group-Ia anticlines occur most commonly in cratonic basins and Ib more often in marginal basins at the continental edge. Group Ir are located on the edge of a continent and appear to be caused by tension resulting in depositional folds or 'roll-over' traps related to growth faulting (contemporaneous faulting). Compressional anticlines Ib are most often found in basins located along plate margins where oceanic crust is converging against continental crust (subduction zones of the Pacific and Tethys regions), including the mountain belts inland from these zones. Depositional folds Ir are most often associated with a plate margin where oceanic and continental crust

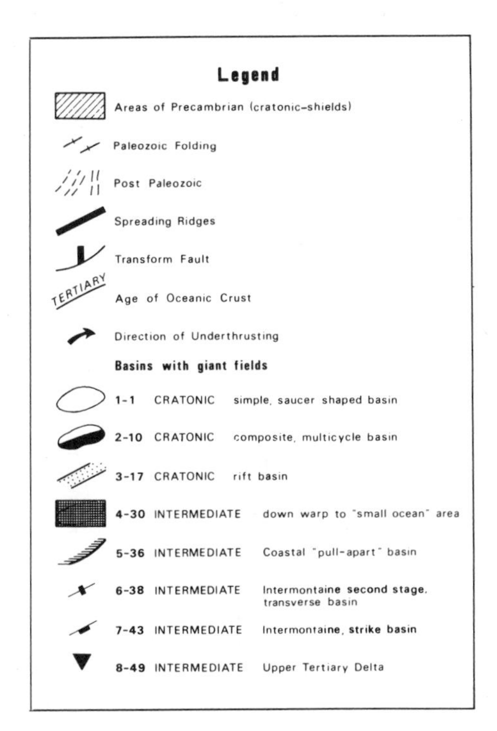

Figure 12.9 (opposite). Worldwide distribution of the giant and supergiant oil fields (after Klemme, 1971). For details of the classification, *see Figure 12.10.* Giant fields have more than 80×10^9 litres of reserves. Supergiant fields have more than 636×10^9 litres of oil reserves. For this purpose gas is converted to an oil equivalent with 1 litre oil ≡ 1069 litres of gas. *Type-1 field:* 1.1 = Illinois. *Type-2 fields:* 2.2 = Volga–Ural, 2.3 = Pechora, 2.4 = West Siberia, 2.5 = North Sea, 2.6 = Netherlands–German, 2.7 = Aquitaine, 2.8 = Erg-Oriental, 2.9 = Erg-Occidental, 2.10 = Fort Polignac, 2.11 = Alberta, 2.12 = Appalachian, 2.13 = Powder River, 2.14 = Uinta, 2.15 = San Juan, 2.16 = Anadarko–Amarillo, 2.17 = Oklahoma–Mid Continent, 2.18 = Permian, 2.18a = Kansu, 2.18b = Szechuan, 2.18c = Dzungaria, 2.18d = Tsaidam and 2.18e = Coopers Creek. *Type-3 fields:* 3.19 = Sirte, 3.20 = Suez, 3.21 = Red Sea, 3.22 = Oman, 3.23 = Dneiper–Donetz, 3.24 = Bukhara–Tadzik, 3.25 = Ver Khayano–Vilyuy and 3.26 = Heilungkiang. *Type-4 fields:* 4.27 = Arabian–Iranian (4a), 4.28 = Pre-Caucusus–Mangyshiak–Turkmen (4a), 4.29 = Indus (4b), 4.30 = North Slope (4c), 4.31 = East Texas (4c), 4.32 = Gulf Coast (4c), 4.33 = Tampico (4c), 4.34 = East Venezuela (4a), 4.35 = North Borneo (4b), 4.35a = Gulf of Mexico (4a) and 4.35b = Malay Basin (4b). *Type-5 field:* 5.36 = Cabinda. *Type-6 fields:* 6.37 = Ventura, 6.38 = Los Angeles, 6.39 = Maracaibo, 6.40 = Pivura, 6.41 = Gippsland and 6.42 = Baku. *Type-7 fields:* 7.43 = San Joaquin, 7.44 = Sacramento, 7.45 = Cook Inlet, 7.46 = Magdalena, 7.47 = Central Sumatra and 7.47a = Java Sea. *Type-8 fields:* 8.48 = Mississippi, 8.49 = Niger, 8.50 = Mackenzie and 8.51 = Nile.

are being separated by pull-apart or foundering (tension).

Fault traps II and flowage traps III are in areas of closure where the basic structural trap is due to either faulting or flowage (salt and shale diapirism predominating).

About 95% of the giant reserves and more than 85% of the oil fields are located in visible structure-related traps. Three-quarters of these are in some kind of anticlinal uplift (Klemme, 1975). Structurally visible combination traps have recently increased the world's giant reserves in, for example, the North Sea, Libya, Middle East, West Siberia and Australia. Anticlines or structural combination traps dominate the compressional sides of continents in basin types 2, 4a, 4b, 6 and 7 (*Figure 12.9*), while subsidence-related anticlines, combination traps and stratigraphic traps predominate in basin-types 3, 4c, 5 and 8.

Although anticlinal-types predominate in continental and continental margin basins, in the offshore areas a greater number of oil fields are of the combination structural–stratigraphic trap-type.

Classification of Oil Basins

Basin classification is an art rather than a science, for whereas there are some clearly defined types of oil-field tectonic setting, undoubtedly a whole spectrum exists and a rigid classification would ignore the complexities of nature. Nevertheless the spectrum of kinds of basins can be described. For this, I have drawn on the work of Klemme (1971, 1977), Bally (1975), Dickinson (1975a) and Murphy, R.W. (1975); only those tectonic settings that give rise to oil fields are included. The basin classification of Klemme (1975) is illustrated in *Figure 12.10* and the world's giant ($>80 \times 10^9$ litres of reserves) and supergiant fields ($>636 \times 10^9$ litres of reserves) are shown in *Figure 12.9.* Some plate tectonic interpretations of oil basins are given in *Figures 12.11* and *12.12* after Dickinson (1975a).

Basins Located on Continental Crust

Type 1—cratonic interior

These are saucer-shaped moderately large, flat, single-cycle basins, often with Palaeozoic platform

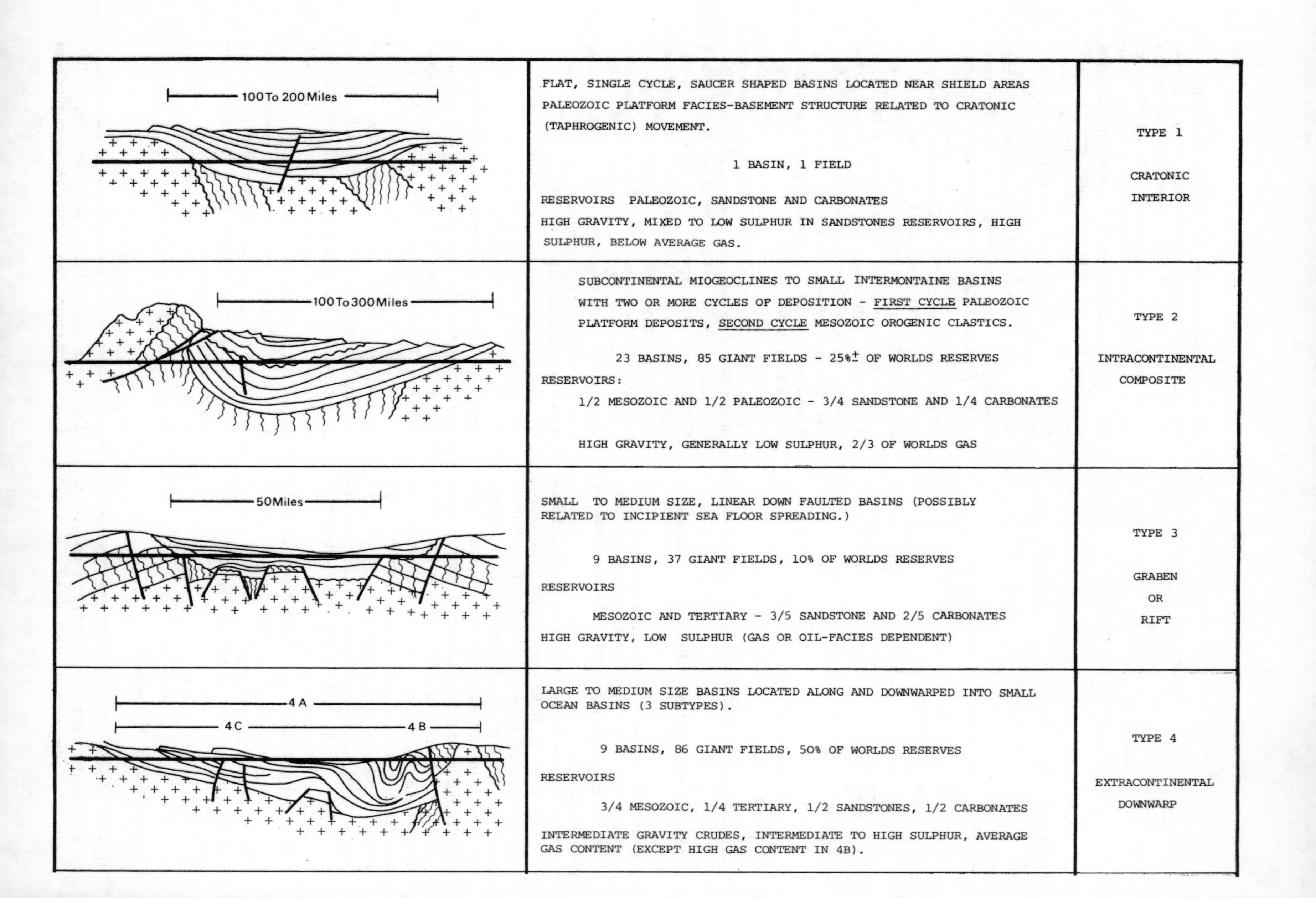

Cross-section	Description	Type
100 To 200 Miles	FLAT, SINGLE CYCLE, SAUCER SHAPED BASINS LOCATED NEAR SHIELD AREAS PALEOZOIC PLATFORM FACIES-BASEMENT STRUCTURE RELATED TO CRATONIC (TAPHROGENIC) MOVEMENT. 1 BASIN, 1 FIELD RESERVOIRS PALEOZOIC, SANDSTONE AND CARBONATES HIGH GRAVITY, MIXED TO LOW SULPHUR IN SANDSTONES RESERVOIRS, HIGH SULPHUR, BELOW AVERAGE GAS.	TYPE 1 CRATONIC INTERIOR
100 To 300 Miles	SUBCONTINENTAL MIOGEOCLINES TO SMALL INTERMONTAINE BASINS WITH TWO OR MORE CYCLES OF DEPOSITION - FIRST CYCLE PALEOZOIC PLATFORM DEPOSITS, SECOND CYCLE MESOZOIC OROGENIC CLASTICS. 23 BASINS, 85 GIANT FIELDS - 25%± OF WORLDS RESERVES RESERVOIRS: 1/2 MESOZOIC AND 1/2 PALEOZOIC - 3/4 SANDSTONE AND 1/4 CARBONATES HIGH GRAVITY, GENERALLY LOW SULPHUR, 2/3 OF WORLDS GAS	TYPE 2 INTRACONTINENTAL COMPOSITE
50 Miles	SMALL TO MEDIUM SIZE, LINEAR DOWN FAULTED BASINS (POSSIBLY RELATED TO INCIPIENT SEA FLOOR SPREADING.) 9 BASINS, 37 GIANT FIELDS, 10% OF WORLDS RESERVES RESERVOIRS MESOZOIC AND TERTIARY - 3/5 SANDSTONE AND 2/5 CARBONATES HIGH GRAVITY, LOW SULPHUR (GAS OR OIL-FACIES DEPENDENT)	TYPE 3 GRABEN OR RIFT
4 A; 4 C; 4 B	LARGE TO MEDIUM SIZE BASINS LOCATED ALONG AND DOWNWARPED INTO SMALL OCEAN BASINS (3 SUBTYPES). 9 BASINS, 86 GIANT FIELDS, 50% OF WORLDS RESERVES RESERVOIRS 3/4 MESOZOIC, 1/4 TERTIARY, 1/2 SANDSTONES, 1/2 CARBONATES INTERMEDIATE GRAVITY CRUDES, INTERMEDIATE TO HIGH SULPHUR, AVERAGE GAS CONTENT (EXCEPT HIGH GAS CONTENT IN 4B).	TYPE 4 EXTRACONTINENTAL DOWNWARP

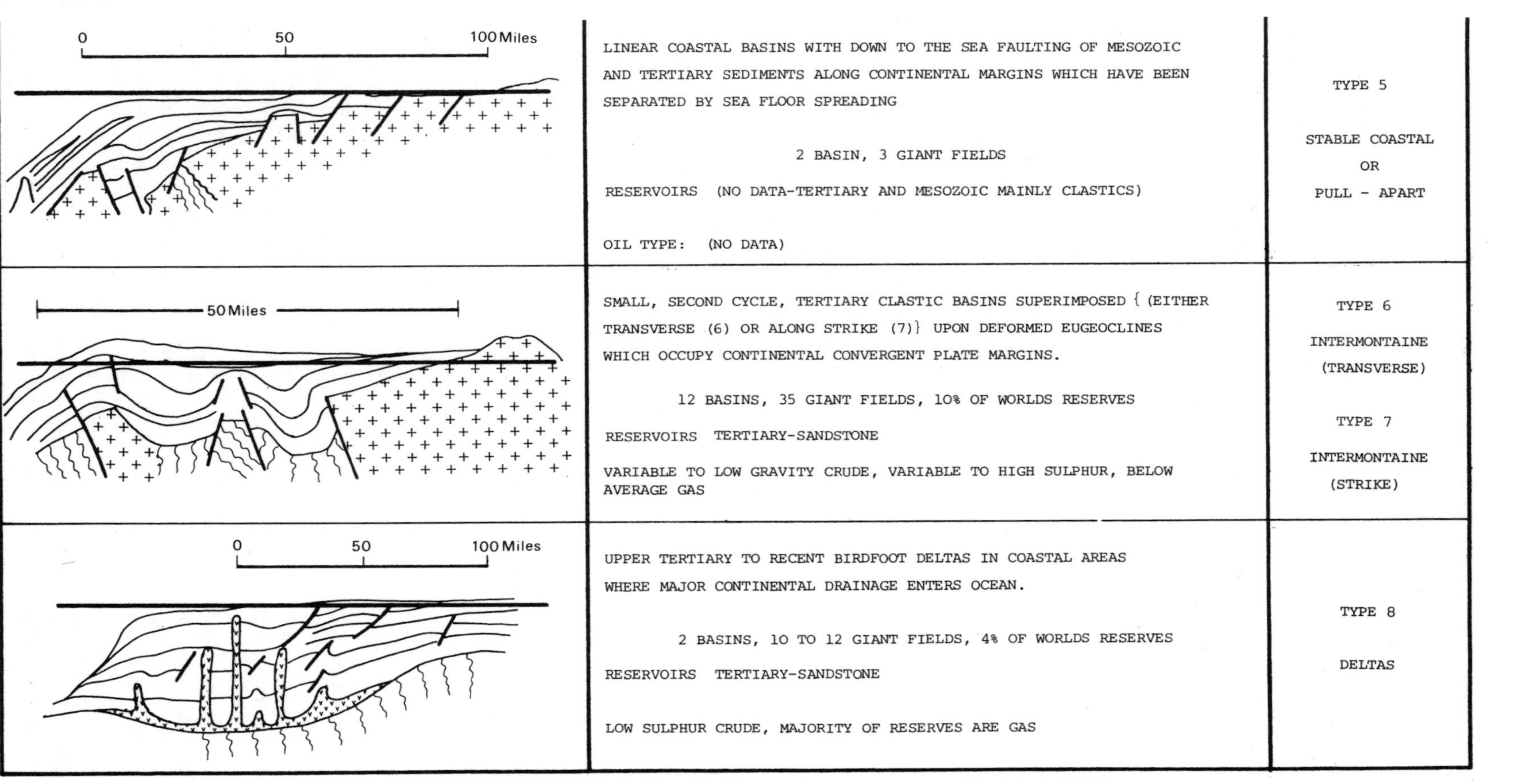

Figure 12.10. Oil-field basin classification system (based on Klemme, 1975).

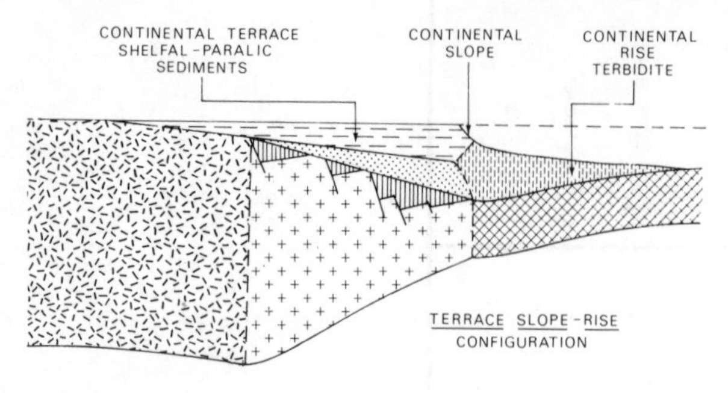

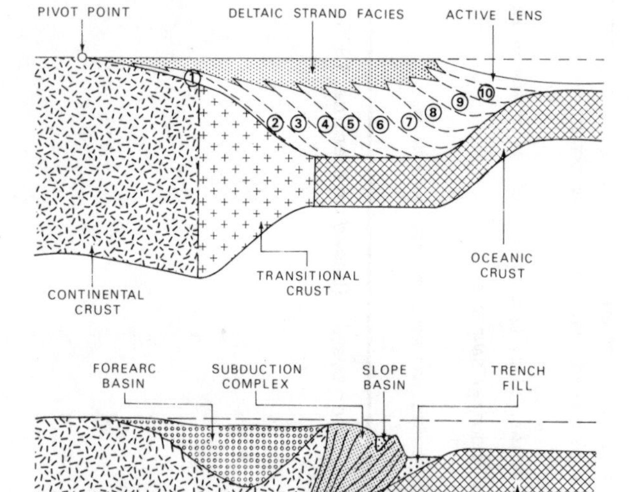

Figure 12.11. Schematic cross-sections to show the plate tectonic settings of oil basins. Top: Atlantic-type miogeoclinal margin; middle: growth of a continental embankment in a miogeoclinal setting; bottom: the basins in a cordilleran-type consuming plate margin (after Dickinson, 1975a).

or embayment facies with basement uplifts and sedimentary structures; they are located on interior portions of a craton. The plate tectonic setting of these basins is obscure; they occur entirely on cratonic crust, and their subsidence, shape and fill are related to deformation along a converging plate margin along the edge of the craton, causing warping of the whole continent far inland from the plate margin. The giant-field example is the Illinois Basin (*Figure 12.9*). Smaller examples are the Paris Basin, Michigan Basin and North Sahara. Largely unexplored examples may be the Hudson Bay and the Gulf of Carpentaria of Australia.

Type 2—cratonic intracontinental

This category includes composite foreland shelf and remote interior-to-intermontaine basins. They are large subcontinental miogeoclinal to small intermontaine basins with two or more cycles of deposition. The first cycle is often of platform or shelf sediments and the second cycle of orogenic clastic rocks. They are located around the margins of cratons. Partial continental separation of a cratonic area occurred prior to subsidence. The best example is the North Sea Basin, in which the crust is thinner than normal beneath the major graben trends (*Figure 12.8*). Other examples are the East African Rift system and the Lake Baikal Rift. The importance of thermal gas in the North Sea and West Siberian basins may reflect a corresponding high heat flow early in the basin development. Many basins on the cratonic areas are related to back-arc underthrusting of the craton towards the active arc (*Figure 12.12*). This kind of underthrust was termed A-subduction, as opposed to B-subduction along the trench (Bally, 1975). Some basins of this category relate to the closing of marine basins that were floored by oceanic crust (*Figure 12.12*), but in others no oceanic crust was involved.

Type 2 (a)—peripheral basins

The foreland basin is formed on the continental block as it tilts downwards towards the A-subduction zone. It represents the extinction of a basin floored by oceanic crust, after the oceanic crust has been consumed by subduction. The Arkoma and Fort Worth basin adjacent to the Ouachita orogenic belt are examples. Where a crustal collision occurs, the marginal sediment prism is drawn first into the subduction zone and the strata are tilted. This allows oil to migrate up-dip away from the flank of the orogenic belt. The tectonic and sedimentary loading both increase down-dip, hence there is a strong tendency for oil to migrate upwards. The Persian Gulf peripheral foreland may reflect such conditions (*Figure 12.12*). An immense rifted margin sediment prism of Mesozoic age along the edge of the Arabian Platform has been drawn down against the Zagros Suture Belt and covered by a Tertiary fore-arc basin. For climatic reasons and because of insufficient uplift in the adjacent foreland, some foredeeps are dominated by carbonate sedimentation. The Tertiary of the Middle East furnishes the most striking example.

Type 2 (b)—retro-arc basins

These basins are formed entirely on continental crust behind the continental margin magmatic arcs

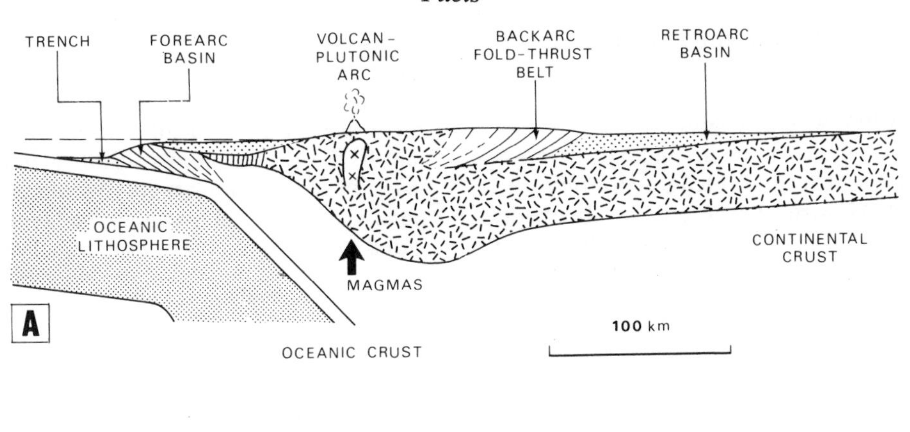

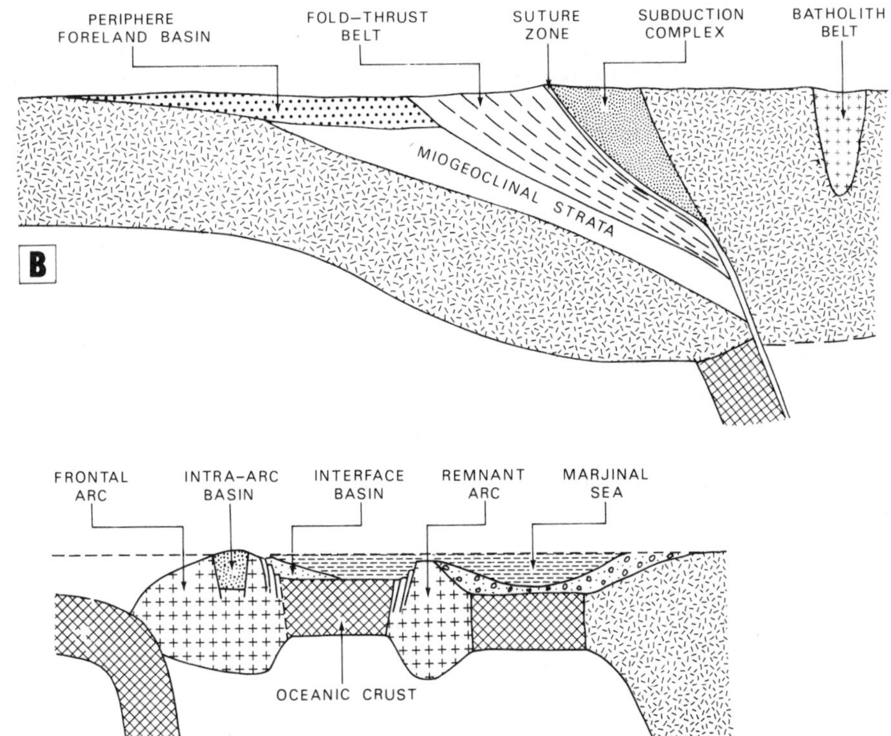

Figure 12.12. Schematic cross-sections to show the plate tectonic settings of oil basins. (A) Retro-arc basin formed as a result of continental underthrusting (A-type subduction) in the back-arc region; (B) schematic collision orogen where the leading edge of a continent with its miogeoclinal shelf has collided and partly subducted with a continent; (C) intra-arc basin, marginal sea and remnant arcs (after Dickinson, 1975a).

(*Figure 12.12*). Their obvious relationship to the active arc–trench system is shown by their parallelism to the trend of the volcanic arc. The Cenozoic Sub-Andean basins and the marine Cretaceous of the Rocky Mountains east of the Mesozoic batholiths are good examples of retro-arc foreland basins. A limited amount of subduction occurs here as the continental crust is underthrust at a low angle towards the arc–trench system (A-type subduction of Bally, 1975). The tilting of the older miogeoclinal prisms beneath the retro-arc basins have a strong influence on hydrocarbon concentration just as in the peripheral basin-type. Smaller basins develop owing to breakup by block faulting, for example, Bighorn and Green River.

Type 2 (c)—Chinese-type basins

In China there appears to be no underthrusting of the foreland towards the volcano–plutonic arc. Instead the Palaeozoic basement has been deformed with a characteristic block-faulting style, often with associated folding. The basins of Central Asia are block-faulted uplifts. Examples are the Dzungaria and Pre-Nan Shan basins.

Type 3—craton graben, half graben, or rift

The basins are small-to-medium sized and are graben or rift-faulted. In some cases they may be caused by incipient or aborted sea-floor spreading, so that the floors of the grabens may be of crust intermediate between truly continental and oceanic. Most of these basins occur on continental crust and may extend offshore.

This category of Klemme (1975) clearly shows the difficulty of basin classification, for it can also include what other authors refer to as marginal aulacogens, which are aborted oceans or failed arms of a branching rift system whose other members continued to evolve into fully fledged oceanic basins. The structure is intermediate between an infracratonic basin and an oceanic basin. A good example is the Benue Trough of Nigeria. The oil field is related to the development of a major delta along the aulacogen graben structure (*Figure 12.5*). Aulacogens are ideal settings for the development of major deltas, and hence the Niger Delta is classified under type 8 on page 299.

Some basins in this category may be best referred to as *Proto-oceanic rifts.* These are deep rift valleys that formed during the early stages of continental separation. Good examples are the Red Sea and the Gulf of California. True oceanic crust forms at a spreading axis along the length of the rift, and sedimentary sequences are usually rich in redbeds, lavas and evaporates, which form as a result of the high heat flow.

BASINS LOCATED ON CRUST INTERMEDIATE BETWEEN CONTINENTAL AND OCEANIC

Type 4—extracontinental down-warp to small ocean basin

Such basins are large-to-medium sized, located along the margin of small ocean basins. They also developed along the Tethys Ocean as type 4a (platform and trough-types) and type 4b (trough only), and along small ocean basins. Type 4c is a down-warp into a Mediterranean-like sea. The most prolific of these basins is the unique Arabian/Iranian basin of the Middle East, which at the present time is locally submerged by an inland sea. In the terminology of Bally (1975) and Dickinson (1975a), this category would be classified as passive continental margin-type (*Figure 12.11*) divided into the subtypes detailed below.

(a) Miogeoclinal prisms (Atlantic-type passive margins)

The classic miogeoclinal sequences of past geosynclinal theory are now regarded as miogeoclinal sediment prisms deposited along formerly rifted continental margins in settings open to the ocean (*Figure 12.11*). The margin between the continent and the ocean is now tectonically passive and there is no plate boundary between continent and ocean. Modern examples are the Mesozoic and Cenozoic sequences along the eastern edge of the Americas. Palaeozoic analogues include the Appalachians and cordilleran miogeoclines as developed prior to the Taconic and Antler orogenies, respectively. Abundant carbonate platforms may be characteristic with carbonate buildups near the shelf break. Miogeoclinal prisms may also border small ocean basins, or marginal seas, for example, the Sarawak Basin of north Borneo.

(b) Continental embankments

This is a modification of the miogeoclinal prism where very active sedimentary progradation of the edge of the continent has occurred (*Figure 12.11*). The shelf break advances from the original continent–ocean interface until it reaches a position above oceanic basement. The Gulf Coast Tertiary and the Niger Delta are outstanding examples, although both may also be classified as major deltas.

(c) Marginal basins

These occur in the Western Pacific and are the oceanic or near-oceanic counterparts of shelf or backdeep basins. They are associated with extensional tectonics in the back-arc region and have a high heat flow regime. Examples are the Andaman Sea of Burma, the Sea of Japan and the South China and Sulu Seas (*Figure 12.12*).

(d) Archipelagic or intra-arc basins

Relatively small basins may occur on island arcs that have been formed by the consolidation of arc–trench elements and probably began as complex grabens along a volcanic chain (*Figure 12.12*). Examples are in the Philippine islands. If one is to include the Molucca Sea basin, then such arcs may result from the complex merging and collision of opposed polarity arc–trench systems.

Type 5—stable coastal or pull-apart

These are linear coastal basins with down-faulted Mesozoic and Tertiary strata, located along cratonic margins. They may be separated from the craton by active sea-floor spreading. There are two subtypes: parallel and transform pull-aparts. Production has been established in both types—Gabon and northeast Brazil. These basins may be subdivided on the basis of whether they are underlain by continental or oceanic basement.

Type 6 and type 7—intermontaine, second-cycle coastal

These are small second-cycle Tertiary clastic basins lying transversely (type 6) or parallel (type 7) on the deformed eugeoclinal troughs that were developed parallel to certain coast lines, possibly as a result of subduction of oceanic plates beneath the continental margins. The parallel types (type 7) have greater linear extent than type 6. These basins have been subdivided into fore-arc, back-arc and median-zone foundered basins (Klemme, 1977). Most production is limited to those basins in a back-arc setting. Many of the basins have been recipients of shallow semi-deltaic to deep-water turbidite sedimentation. Most of the basins have adequate shale source rocks, but they suffer from poor sand reservoirs of suitable extent.

All these basins are related to converging plate margins. In plate tectonic terminology, they may be subdivided as detailed below.

(a) Fore-arc basins

These basins occur within the arc–trench gap. They occur outside the active subduction zone and do not undergo intense deformation (*Figure 12.11*). A good example is the Late Mesozoic Great Valley sequence, deposited between the coeval Franciscan subduction complex and the Sierra Nevada batholith belt. However very few basins in this class have established oil production and little is known about the hydrocarbon habitat. Extensive drilling in Southeast Asia (the continental margin basins of Murphy, R.W., 1975) has resulted in failure for the oil companies. The Anchorage Basin is an example of one that produces oil.

(b) Shelf or backdeep basins

These are located on continental or intermediate crust formed during an earlier deformation phase, and the active arc–trench systems characteristically form island arcs peripheral to the basins. For example, the basins of Southeast Asia (Murphy, R.W., 1975), which include the Sumatra Basins, the Malay Basin and the Java Sea Basin, formed on crust of continental-to-intermediate character that cratonized only in the Late Triassic. Extreme normal faulting caused uplift and graben formation owing to thermal expansion of the crust behind the volcanic arc. These basins are characterized by high heat flow regimes, which have matured the oil in Miocene strata.

Type 8—Upper Tertiary deltas

These are Upper Miocene-to-Recent birdfoot deltas in coastal areas where major continental drainage areas exist and flow out into the ocean. The deltas may begin on continental crust and flow out over the edge of the continent on to oceanic crust. Reservoirs are entirely in sandstone, often at the interface with high-pressure shale zones. Trap-types are predominantly roll-over related to growth faults. Because the deltas are related to underlying tectonic features, these basins can also be alternatively classified, as discussed above.

It is obvious that there can be no perfect and generally agreed classification and, depending on the criteria used, equally appropriate alternatives may exist.

Worldwide Oil Generation and Preservation

Oil generation is a function of planktonic productivity, which in turn is a function of water-body geometry, climate, and availability of nutrients—all these factors being ultimately controlled by plate tectonics. The trapping of oil in reservoirs is largely a function of tectonism. The preservation of oil in oil fields is a function of tectonism and little else (Irvine *et al.*, 1974).

The worldwide distribution of oil throughout the Phanerozoic cannot be simply related to climate, however there is a strong tendency for oil basins to occur in regions that had low latitudes at the time of their formation, hence warm seas obviously played an important role. The distribution of oil with time reveals that 72% of all known reserves were probably formed in the late Mesozoic, and mostly (60%) in the Mid Cretaceous. These statistics are biased by the supergiant fields of the Middle East and Middle Americas.

Why was so much oil formed near the present Persian Gulf, and to a lesser extent along the African coast of the Mediterranean and Middle

Americas (*Table 12.2*) during the brief time span of the Late Mesozoic–early Cenozoic? All the controlling factors—climate (especially temperature), mineral nutrients in the sea, tectonic factors controlling the initial basin development and tectonic factors controlling the trapping and preservation of oil—were optimized in those places during the Mesozoic-to-early Cenozoic eras. Two large marine embayments opened astride the equator in the Late Mesozoic, and they may have been connected through the Mediterranean Sea (the Tethys–Atlantic Seaway). One embayment contained the Persian Gulf and the other the Gulf of Mexico. The renewal of mantle convection about 100 Ma ago activated these embayments, abruptly increased the rate of sea-floor spreading and enlarged the mid-ocean ridges, causing maximum development of warm shallow seas, and releasing through igneous activity greatly increased quantities of mineral nutrients, as in the Red Sea at the present time (Irvine *et al.*, 1974). Hence prolific oil source rocks were formed along the Tethys–Atlantic equatorial Seaway.

The geometry of subsequent plate activity was such that the Persian Gulf was protected from severe tectonic activity by the rapid northwards movement of the Indian Plate, and India absorbed most of the impact with the Eurasian Plate. The gulf of Mexico was protected from severe tectonic activity by the northeastwards movement of the Antillean Arc (Irvine *et al.*, 1974). Hence these rich oil deposits have been protected from any severe tectonic disturbance that could have destroyed them.

Coal

Coal is a readily combustible sedimentary rock containing more than 50% by weight and more than 70% by volume of carbonaceous material including inherent moisture, formed from compaction and induration of variously altered plant remains similar to those in peat. Coal is classified by type (differences in the kinds of plant material it contains), by rank (differences in the degree of metamorphism it has undergone) and by grade (the range of impurity it contains). The normal coal-type is known as bituminous coal. Coals form an integral part of a characteristic sedimentary rhythm known as cyclothem, which is essentially a post-Devonian phenomenon. Though Devonian coals do occur, they are rare and insignificant.

Origin

It has long been recognized that bituminous coals owe their origin to accumulations of peat that formed in ancient swamps and marshes. They originated in a variety of geographical settings, including deltas, lagoons and estuaries (Wanless *et al.*, 1969). There is little doubt that some ancient coals developed from peats that accumulated in delta-plain environments. Many comparisons can be made between repetitive sedimentary cycles (cyclothems) of coal measures and repetitive sequences of Recent deltaic deposits beneath the coastal plain of, for example, Louisiana, USA (Frazier and Osanik, 1969). Swamps and marshes are the principal sites of present-day peat accumulation (*Figure 12.13*). Significant peat deposits are derived from vegetation of fresh-to-brackish water marshes and levee–flank swamps located in coastal interdistributary basins and from vegetation of cypress–gum swamps in broad inland flood basins of the Mississippi deltaic plain (Frazier and Osanik, 1969). The cypress–gum swamp peats are the thickest because the environment of their accumulation remains more stable. Except for a few

Table 12.2 List of the Supergiant fields of the Tethys–Atlantic Seaway

Field	*Age of oil*	*Estimated reserves (barrels)*[a]
Persian Gulf (Arabia, Kuwait, Iraq)	Jurassic–Cretaceous	500×10^9
Sirte, Libya	Cretaceous–Palaeocene	33×10^9
Maracaibo, Venezuela, Columbia	Cretaceous–Tertiary	38×10^9
Persian Gulf (Iran, Iraq)	Jurassic–Cretaceous–Palaeocene, Miocene	90×10^9
Gulf Coast (Louisiana, Texas)	Palaeocene–Miocene	30×10^9
Gulf of Mexico (Mexico)	Palaeocene–Miocene	Not yet fully assessed

[a]1 barrel = 158.98 litres.

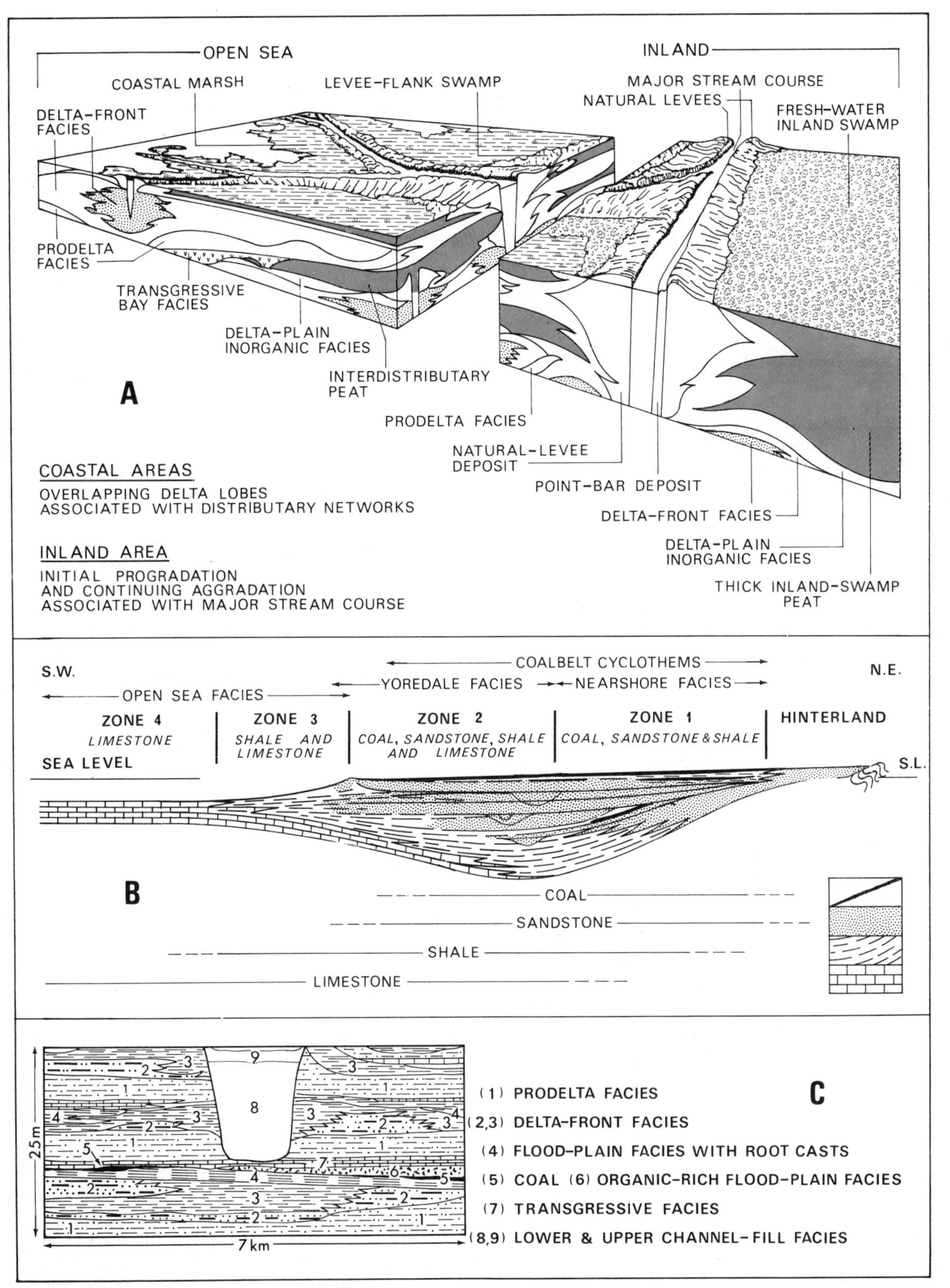

Figure 12.13. (A) Facies relationships between coastal repetitive sequences and the inland sequence of a typical delta complex (after Frazier and Osanik, 1969). (B) Schematic Yoredale-type cyclothem succession of the English Carboniferous coal measures (after Westoll, 1968). (C) Repetitive delta sequences in the Carboniferous Yoredale Series of northern England (after Frazier and Osanik, 1969).

local occurrences of detrital peat debris, all peat accumulations are autochthonous. Peat derived from the floating-marsh environment, however, could appear as an allochthonous variety, since it may overlie a rootless underclay.

Interdistributary peats occur in delta-plain facies of the delta sequences (*Figure 12.13*). Each sequence consists of basal prodelta facies, overlain by delta-front and delta-plain facies, which themselves are unconformably overlain in the seaward portions by transgressive deposits. Complete repetitive sequences are found beneath the coastal region of the deltaic plain. These sequences, or cyclothems, provide the clue to some of the depositional environments and facies relationships within coal sequences of the geological past.

The general stratigraphy of Carboniferous Coal Measure cyclothems can also be recognized in a modern succession of sedimentary units underlying the coastal mangrove swamps of southwestern Florida (Scholl, 1969). Cyclothems are thought to have formed in part in a deltaic environment, yet southwest Florida has no deltas and most of the paralic sediments are derived from coastal sources. Hence the Florida cyclothems must be brought about by a relative rise in the sea level across a low coastal platform that supports peat-depositing paralic and fresh-water swamps and forests. It may therefore be concluded that some cyclothems formed in a swampy environment penecontemporaneously with a relative sea-level rise. The coal members, therefore, are in part a transgressive unit.

Studies of distribution and character of Pennsylvanian coals in the eastern and central USA (Wanless *et al.*, 1969) have shown that their accumulation resulted from several environmental patterns: building up of a widespread delta system, for example, Illinois; unfilled channels, either of alluvial streams (for example, western Illinois) or unfilled delta distributary channels (for example, Illinois)—thinner coal seams occur away from the unfilled channels; accumulation in an estuary prior to drowning (for example, Illinois); deposition along a narrow coastal strip comparable with the Atlantic coastal marshes of today (for example, Missouri and Kansas); local accumulation in a lagoon behind an offshore barrier (for example, southern Ohio); local accumulation in a cut-off meander of a stream, with clastic partings and no underclay (examples occur in western Illinois); a plain exposed following an abrupt marine regression (for example, Missouri and Iowa); and a level depositional plain formed by burial of pre-Pennsylvanian topographic irregularities (for example, Illinois).

Some coals, which formed on delta plains or flood plains, are separated from channels by natural levees, an example of which is found in southeastern Illinois (Wainless *et al.*, 1969). Where a fluviatile or deltaic channel was filled, because of greater compaction of shales away from the channel, a coal may thin or wedge out over the channel sandstone.

Pennsylvanian coals of the USA generally are underlain by strata in which root impressions, some coalified, are more or less abundant. These are generally referred to as the *underclay. Seatearth* implies an underclay in which the plants were directly rooted. There is no relationship between the thickness of the coals and their associated seatearths and furthermore it is unreasonable to suppose that a seatearth provided rootage for the entire succession of changing plant communities that constituted the flora of the thick overlying peat mass (Moore, L.R., 1968b). The term *fireclay* implies an underclay that is rich in kaolinite and can be used commercially as a refractory clay.

The underclay sediments are usually argillaceous but examples of sandstones or limestones are known. When root impressions are not found in the underclay, it is concluded that the vegetation was probably locally transported to form the stratum of coal directly overlying it.

Underclays have the following distinctive features (Wanless *et al.*, 1969): a lack of shaly lamination; root impressions; slickensides; a non-calcareous zone in the upper part overlying a calcareous clay below; a dark zone at the top resembling humus in a soil; and a high kaolinite or illite content. These characteristics suggest that the underclays served as soil for the plants to grow in. However, the detailed clay mineralogy, and the biological and microbiological contents of underclays, when taken together, indicate an accumulation of sediment at or near base-level conditions (Moore, L.R., 1968b). Underclays may not therefore represent soils. The chemical characteristics are largely original and not a result of post-depositional weathering. The origin of the underclays has been summarized by Wanless *et al.* (1969) in the following alternatives: deposited in an alluvial environment or in the proximal portion of a delta, in which case they are rich in kaolinite; or deposited in the distal portion of a delta, lagoon, estuary or shallow sea preceding the formation of the regressive coals—these underclays are likely to be rich in illite. The underclay certainly served as a soil whenever plants were rooted in it. However, this was generally under wet swampy conditions; they did not represent upland soils. The underclay was never sufficiently above groundwater level to favour appreciable chemical leaching.

The Yoredale Series of the Upper Viséan and Lower Namurian of northern England has been recognized as the typical cyclothem (Westoll, 1968). It is characterized by a repetition of an upward sequence of limestone, shale, sandstone, seatearth and coal (*Figure 12.13*). The cyclothem loses its clastic components southwards towards the open sea and loses its limestone component northwards towards the existing land mass at the time of deposition. The coal-belt cyclothems represent rather rapid marine transgressions over deltaic areas; from the new shoreline and prodelta, the delta-front and delta-top regimes migrate away from the land until the next transgression (Westoll, 1968).

Tectonic Setting of Coal Basins

From a general review of the tectonic setting of major deltas (Audley-Charles *et al.*, 1977), it would appear that deltas formed on stable passive rifted continental margins are the most likely to give rise to extensive coal deposits of *paralic-type*. The present-day Niger, Amazon and Mississippi River deltas are excellent examples. The Carboniferous coals of northwest Europe were formed on shallow coastal marine areas and are excellent examples of paralic coal basins (Mackowsky, 1968). Most of the North American Pennsylvanian coals were likewise deposited in paralic settings (Wanless *et al.*, 1969). The rivers responsible for paralic coal basins flowed off stable cratons towards the sea and the transition from continental to oceanic crust is not marked by tectonic activity but by miogeoclinal sedimentation. Similar examples also border smaller marginal seas in Southeast Asia, for example, in Sumatra and Borneo (Bemmelen, 1970a).

Intracratonic settings are also favoured sites for river deltas (Audley-Charles *et al.*, 1977); a present-day example is the Rhine. These rivers form entirely on cratonic continental crust and the course of the river is controlled by faulting and graben formation. Fresh-water lakes may also form in the grabens within intracratonic settings. Such basins are likely to be preserved in a relatively undeformed condition because their framework is preserved by a stable cratonic infrastructure. Rivers draining into fresh-water lakes formed on cratonic crust give rise to *limnic basins.* Many young examples of coal-bearing limnic basins occur. For example, Stauffer (1973) has described the Tertiary lignitic coal basins of the Malay Peninsula. These fresh-water basins appear to be located on the now cratonized continental crust by fault-controlling mechanisms. Indeed large inland swamp lakes have formed at the present time in the Malay Peninsula in response to fault control. All South African Permian coals are of the fresh-water lake (limnic) origin (Falcon, 1977, 1978). They can be directly compared with those in the Moscow Coal Field, which is similarly regarded as a continental basin formed in a cratonic setting (Mackowsky, 1968). Permian coal formation in Gondwanaland began in limnic basins formed on the cratonic crust of South Africa, Australia, India, Antarctica and Southern America before the breakup of the major continent (Hall, T.F.C., 1950).

Climate and Coal Formation

A tropical climate is favourable, though not essential, to coal formation (*Figure 12.14*). The Carboniferous paralic coal basins were developed adjacent to the Mid-European Ocean, which probably connected with the Tethys. During the times of coal formation in North America and Europe, the latitudes ranged from the equator to 20° S (Keppie, 1977).

The Miocene brown coals of Germany formed initially under near-tropical conditions, which became more temperate with time. Palaeo-latitudes increased from 38°N to 45°N during the time of coal formation (Smith, A.G. and Briden, 1977). The North Sea reached far into Germany in the Upper Oligocene and Lower Miocene (Teichmüller and Teichmüller, 1968a). Coastal swamps were protected from inundation by the sea by a long sand bar. Peat accumulation kept pace with subsidence, thus preventing the sea entering the coal basin.

In contrast, the Permian coal formation in Gondwanaland followed the Carboniferous glaciation (*Figure 12.14*); with the disappearance of the ice cap the coals formed in limnic subpolar peat bogs (Hall, T.F.C., 1950). The palaeo-latitudes varied from 50° S to 70° S during the period of coal formation (Keppie, 1977).

Sapropelic Coals

Cannel and boghead coals constitute members of the sapropelic coals that are formed basically from the biological and physical degradation products of contemporaneous peat swamps (Moore, L.R.,

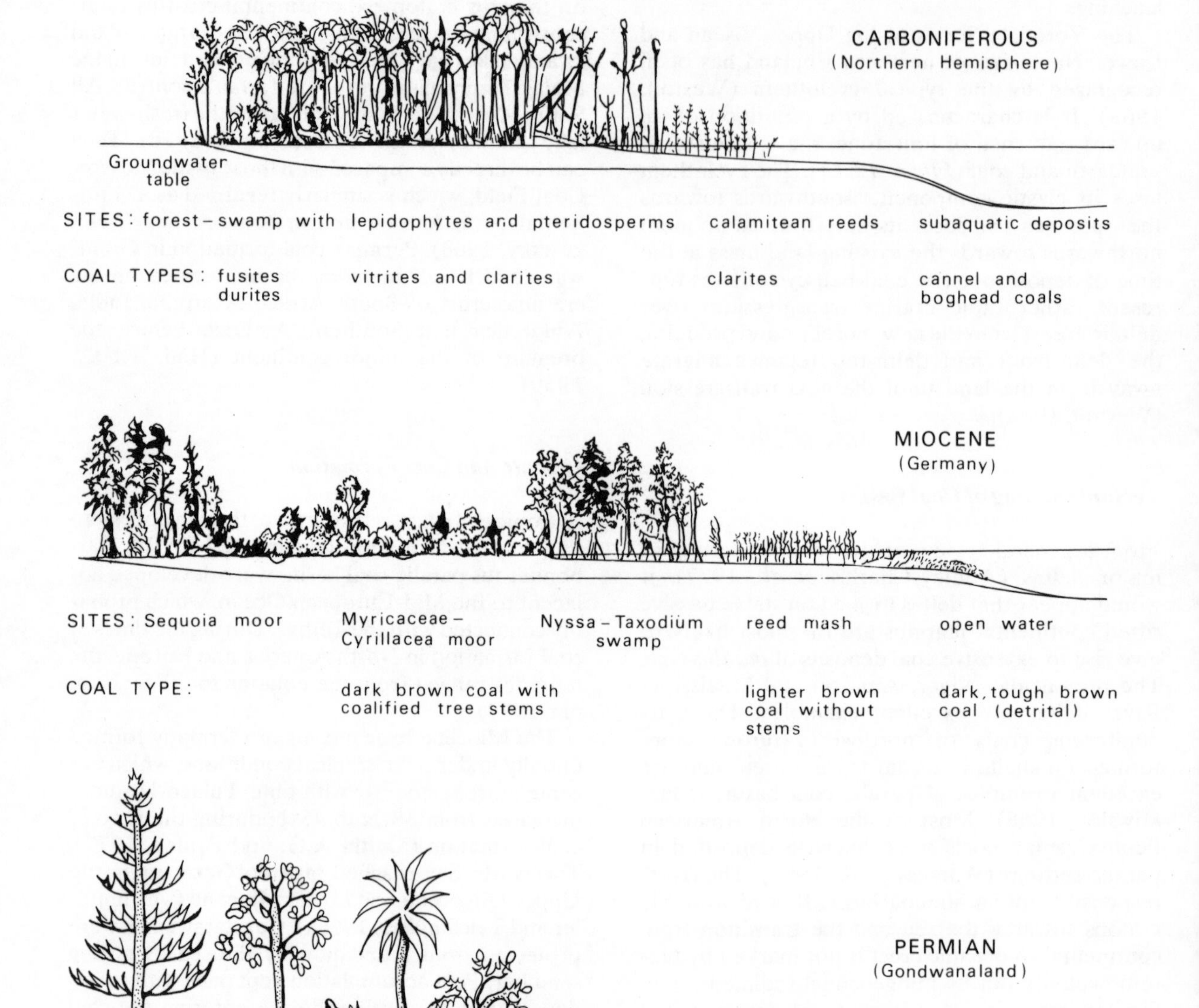

Figure 12.14. Schematic representation of the plant varieties in relation to the basin of sedimentation for Carboniferous, Permian and Tertiary coals (after Teichmüller and Teichmüller, 1968b; Falcon, 1978).

1968a). This allochthonous mass is associated with autochthonous material commonly in the form of surface-living algae, with the addition of wind-borne or water-borne spores. The constituents may be present in variable quantities that give the rocks their character, for example, algal cannels and spore cannels. Much of the organic material suggests attack by fungus. The rocks so produced are related to the durain content of bituminous coals in their chemical composition, mode of occurrence and constituents.

The typical cannel coal consists of the degradation products of an existing coal peat denuded to accumulate as an organic ooze on the bottom of a shallow lake (Moore, L.R., 1968a). The environment of formation is that of a peat substrate upon which broad shallow lakes have formed and are surrounded by vegetation.

Boghead coals are algal coals. The whole mass of the coal originated from algal material. They may have formed either in the centres of larger basins into which transport of organic material was restricted, or in small shallow basins with little drainage intake (Moore, L.R., 1968a).

A well-known example of a boghead coal is the Scottish torbanite, named from its occurrence at Torban Hill, Linlithgow. The Scottish oil-shale industry was based on this occurrence.

Oil Shales

Oil shales of the basin type, such as occur in the Midland Valley of Scotland, are related to cannel coal and boghead coals. They incorporate similar allochthonous and autochthonous constituents that have been subjected to a high degree of microbiological degradation. The resultant selection of more resistant organic material and its transportation across or through peat swamps culminates in the accumulation of an organic-rich ooze on the bottom of shallow-water, barred estuaries, or lagoons under euxinic conditions. These deposits include a higher proportion of detrital inorganic matter and are commonly varved. The rich organic bands are referred to as oil-shale seams and they form part of a sequence of shales, sandstones, thin limestones and coals, which follow a modified cyclothem pattern.

Important kerogen or oil-shale deposits are found in the following countries (Irving *et al.*, 1974): Kolm, Sweden (Cambrian); Kukkersite, Estonia (Ordovician); Midlothian, Scotland (Mississippian); Albert, New Brunswick (Mississippian); Irati, Brazil (Permian); oil-shale series of Svalbard (Triassic); Fushun, China (Jurassic); Kashpirian, USSR (Jurassic); Neuquen, Argentina (Jurassic); Frontier, Wyoming (Cretaceous); Green River, Utah, Colorado and Wyoming (Palaeogene); and Coal Measures of New Zealand (Palaeogene). The Green River Formation of Utah, Wyoming and Colorado has been well studied.

The Eocene Green River Formation of south-western Wyoming was deposited in a lake that evolved from a saline alkaline lake to a fresh-water lake (Surdam and Stanley, 1979). There is an important laminated carbonate lithofacies. Kerogeneous laminated carbonate (oil shale) characterizes sequences in the geographic centre of the Eocene lake, but dolomicrite and oolitic or stromatolitic limestones increase towards the margins. Some of the oil-shale laminae are mud-cracked and some are varve-like in character. Fish fossils are common in the laminated carbonate. The kerogen content of the oil shale varies widely but is generally as high as 70 litres t^{-1}. The kerogen-rich carbonate is usually calcitic, whereas the kerogen-poor carbonate is a mixture of calcite and dolomite (Surdam and Stanley, 1979). The oil-shale beds contain such normally perishable things as delicate larvae and the remains of green algae. The oil shales must have originated from an algal ooze that formed at the bottom of a very shallow spring-fed lake (Bradley, 1973). The blue-green algae grew on and in the flocculated ooze. At intervals the lake levels fell enough to expose the algal ooze to the air to cause partial drying and mud-cracking.

Kerogen usually amounts to about 80% of the organic fraction of the Green River Formation rocks. The kerogen defines an evolution path for type I, rich in lipid material (Tissot *et al.*, 1978). The influence of microbial degradation during deposition of the organic matter is intense in the lower stratigraphic horizons and decreases upwards, where fossil molecules derived from higher plants, algae and other planktonic organisms and bacteria become more abundant. The great depth of the oil-generation stage (4000–5600 m) is explained by a relatively low geothermal gradient.

Petrology of Coal

Two major types of coals can be recognized (Falcon, 1978): dull, non-banded compact coals, usually considered to be allochthonous (drifted) or sapropelic (deep water); and banded with alternating fine layers of varying brightness. They are usually considered to be autochthonous humic coals. The layers represent different lithotypes.

The term *maceral* denotes the basic organic

constituents of coal that can be recognized under the microscope. There are three groups of macerals (Stach, 1968): vitrinite, exinite and inertinite. They are examined under oil in polished section under reflected light microscopy. Vitrinite varies from dark to light-grey, frequently showing signs of botanical structures. Exinite appears much darker and is composed of spores, cuticles and resin bodies. Inertinite is the highest reflecting maceral and is frequently characterized by cellular structure. It shows high relief because of its harder nature compared to other macerals. Vitrinite is the most important in amount and forms the familiar, brilliant black bands in coal. When considering the economic use of coal, the exinite macerals are the most important. In the coking and low-temperature carbonization processes, exinites are the most valuable component of tar (Stach, 1968). Of the three maceral groups, vitrinite is relatively rich in oxygen, exinite in hydrogen, and inertinite in carbon.

The microlayers of coal are made up of macerals and are known as lithotypes or microlithotypes, depending on the scale. Lithotypes may be named from the hand specimen. They are: vitrain, which occurs as shiny black bands; clarain, which occurs as laminated shiny and dull bands; fusain, which is usually massive and dull; and durain, which occurs as charcoal-like fragments.

Coal contains both syngenetic as well as epigenetic mineral impurities. The former were deposited in the basin of sedimentation or crystallized in the basin where the peat was forming. They include clay minerals, quartz, siderite, calcite, pyrite, marcasite, limonite and haematite. Epigenetic minerals form later along joints and fissures in the coal. They include calcite, pyrite, galena and chalcedony.

Coalification

Coalification is a process of diagenesis and metamorphism by which fuels are formed from peat by way of brown coals and the different stages of bituminous coals, eventually into anthracite (*Table 12.3*). The rank of a coal refers to the degree of coalification or coal metamorphism reached. The principal change with increased coalification is an increase in the carbon content relative to volatiles and moisture. The subject has been reviewed by Teichmüller and Teichmüller (1968a) and Falcon (1978). The process of coalification increases with depth of burial, with concomitant increase of temperature, and it may be accelerated adjacent to igneous intrusions.

At the first stages, which may be called peatification, biological–bacterial processes play an important role. At higher ranks the changes are predominantly chemical and physical. The degree of coalification, or rank, is commonly measured by chemical parameters, as shown in *Table 12.3.* The conventional diagrams for plotting these changes is by means of a Seyler chart, of rectangular coordinates, of mass percent (dry-ash free) hydrogen versus carbon (Battaerd and Evans, 1979). As oxygen is the only other major component present in appreciable quantity, an oxygen grid may be superimposed over the carbon/hydrogen diagram (*Figure 12.15*). The trend of coalification from brown coal to anthracite is characterized by an increase in carbon content from 70% to 95%, a decrease in hydrogen from 5% to 3% and a decrease in oxygen from 20% to about 3% (*Table 12.3*). An alternative method of plotting the chemical evolution paths is by use of a ternary bond–equivalent carbon–hydrogen–oxygen diagram (Battaerd and Evans, 1979).

The coal rank may be determined by the reflectivity R_o of vitrinite separated from the coal, polished and studied under oil immersion in a reflecting microscope. The scale is given in *Table 12.3* and is identical to that used in assessing petroleum source rocks (*see* p. 279). The vitrinite is chosen for defining coal rank because it is the most characteristic maceral and is the most homogeneous. However, it is not completely homogeneous and studies of R_o as a percentage have to be statistically based on a large number of measurements (Teichmüller and Teichmüller, 1968a).

Factors Affecting Coal Rank

'Hilt's Law' of 1873 established that rank increases with depth, since temperature increases with depth. However, higher temperatures may be attained at shallow depth by proximity to igneous intrusions. Therefore anthracitization may occur in their proximity at abnormally shallow depths.

As with oil maturation, the geothermal gradient as well as the age are important factors in coalification. Coal rank increases more or less strongly and steadily with depth and is independent of the composition of the coals. This is such a reliable characteristic that any abnormality in the rank found in a coal field must be interpreted as being due to tectonic or magmatic influences.

With increasing rank, the woody cells found in brown coals become homogenized and compacted, leading to the formation of homogeneous vitrinite of the bituminous coals. This process of

Table 12.3 The equivalence of different rank parameters during the coalification process[a]

Rank stages (scientific)	ASTM classification of coals (commercial)	Carbon (% daf) of vitrinite	Volatile matter of vitrinite (% daf)	Moisture in situ (%)	Calorific value of vitrinite (kJ g^{-1}) (af)	Reflectance of vitrinite (%)	Important microscopic characteristics
Peat							Free cellulose; details of initial plant material often recognizable; large pores
		60		75			
Soft brown coal	Brown coal						No free cellulose; plant structures still recognizable; cell cavities frequently empty
			c.53	35	17	c.0.3	
Hard brown coal Dull	Lignite						Marked gelification and compaction takes place; vitrinite is formed
		c.71	c.49	25	23		
Bright	Sub-bituminous coal						
		c.77	c.42	8.1	29	c.0.5	
Bituminous hard coal Low rank	High-volatile bituminous coal						Low-reflecting exinite; exinite becomes markedly higher in reflection (coalification jump)
		87	29		36	1.1	
High rank	Medium-volatile bituminous coal Low-volatile bituminous coal Semi-anthracite						Exinites no longer distinguishable from vitrinites
		91	8		36	2.5	
Anthracite	Anthracite Meta-anthracite						Pronounced reflection anisotropy

[a] daf = dry-ash free; af = ash free; kJ g^{-1} = joules × 10^3 per gram; *c.* = *circa*.

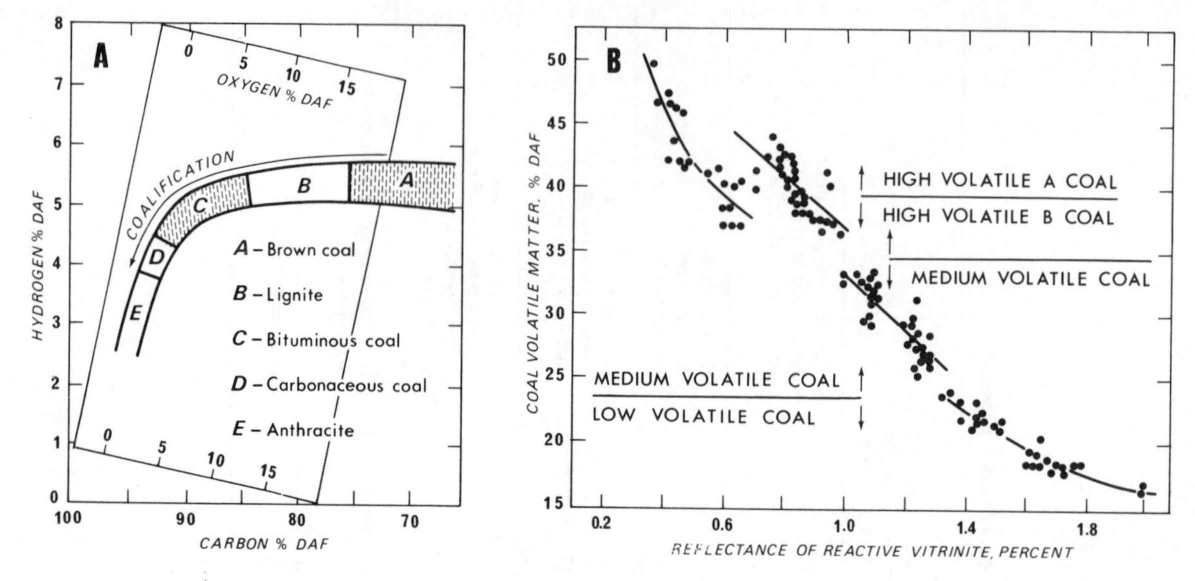

Figure 12.15. (A) Seyer's coal chart showing coalification trends from brown coal to anthracite (after Battaerd and Evans, 1979). (B) Discontinuous relationship between reflectance of vitrinite and coal volatile matter content (after Thompson, R.R. and Benedict, 1974).

homogenization is referred to as *gelification*. By it, the soft earthy and dull brown coals become black, hard bright sub-bituminous and eventually a bright true bituminous coal.

Rising overburden pressure forces water out of the coal, so that one of the striking features of increase in rank is a decrease in moisture content during the early stages of coalification. This coincides with the transition from lignite to sub-bituminous coal.

Whereas the transition between lignite and bituminous coal takes place gradually, there is a distinct boundary between high-volatile and medium-volatile bituminous coal at approximately 87% carbon and 29% volatile matter (dry-ash free).

The discontinuities in the coal metamorphism series are shown by plotting average vitrinite reflectivity R_o in percent against coal volatile matter (dry-ash free) (*Figure 12.15*). The lines are separated by distinct discontinuities that occur around R_o 0.7%, 1.0% and 1.35% reflectance levels; they coincide with the boundaries defined by the American Society for Testing Materials (ASTM) between different volatile levels of the coals.

The anthracite stage is marked by a graphite-like structure as determined by X-ray diffraction. At this stage the vitrinite reflection becomes strongly anisotropic and hence difficult to measure.

When coals reach a high rank through deep burial, strong tectonic deformation, or heating by igneous intrusions, they give off considerable amounts of gas. There is little doubt that the gas found in the Rotliegendes (Lower–Mid Permian) reservoirs of the southern North Sea Basin has its source in the underlying Carboniferous coals (Eames, 1975).

The Carboniferous Coal Field of the Rhine–Ruhr, Germany

The European coal belt stretches from Britain, under the North Sea eastwards and through Germany. In the Late Carboniferous, large paralic coal deposits formed along the northern border of the Variscan Mountains. Somewhat later the limnic coals of the Saar, Lorraine and central France developed in troughs between the mountains. The Ruhr coal field has been extremely well studied (Mackowsky, 1968; Teichmüller and Teichmüller, 1968b). Vitrinite is the most important maceral group in northwest Europe coal fields, forming on average more than 60% by volume of the coals. The Carboniferous rock sequence is complexly folded and tilted, so that the depth of the Upper Carboniferous strata increases from north to south. There are six anticlines and five synclines striking southwest to eastnortheast (*Figure 12.16*). Across the strike, the strata are dissected by a series of faults. North of Essen, Bochum and Dortmund, the coal-bearing strata are overlain by Cretaceous sedimentary rocks, indicating that the re-

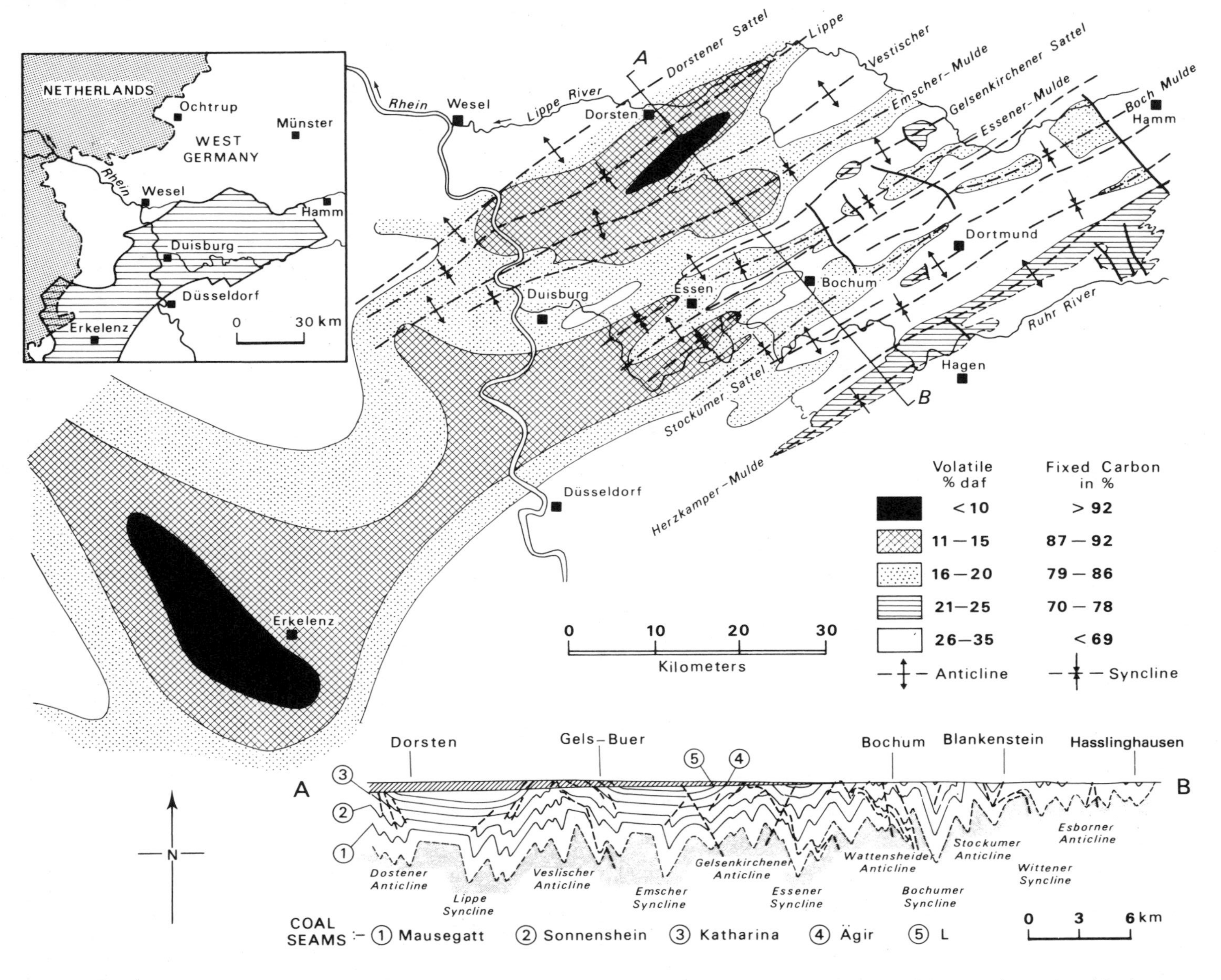

Figure 12.16. Coalification pattern and structural cross-section of the Rhine–Ruhr coal field, West Germany (based on Mackowsky, 1968; Teichmüller and Teichmüller, 1968a).

gion had emerged and been eroded between the Upper Carboniferous and the Cretaceous periods. Uplift, erosion and block faulting has made correlation of the coal seams difficult.

The Ruhr Basin is a classic example of the preservation of the pre-orogenic coalification pattern during the intense folding phase (Damberger, 1974). The Merlenbach anticline and the adjacent Marienau syncline of the Saar–Lorraine coal district are good examples of continued coalification during the folding phase. The amount of volatile matter in the vitrinite is highest where the tectonic pressure was most severe, namely at the southern margin of the coal basin (Teichmüller and Teichmüller, 1968a). As a result of differential subsidence of the seams before the orogenesis, the coals in the megasynclines are of slightly higher rank than those in the mega-anticlines. In the Bochum megasyncline the contours of equal rank run parallel to the folded beds, which means that the coalification must be older than the folding and that it cannot be a result of tectonic pressure. Since the folding pressures must have been more severe than the overburden pressure, neither folding nor overburden pressures have been responsible entirely for the increase in coal rank.

The coalification map of the Sonnenschein coal seam of the Ruhr–Rhine district (*Figure 12.16*) depicts a general increase in rank towards Krefeld, west of the river Rhine. This increase cannot be attributed to greater depth of burial, because the formations thin in this direction. It has been proposed by several workers that the mineralization, igneous intrusions and high rank of the coals that are present in this area are related to the same cause—an extensive magmatic up-welling at depth in this western portion of the Rhine–Ruhr coal field (Teichmüller and Teichmüller, 1968a). This process of coalification acceleration by increased geothermal gradient has been called *telemagmatic coalification.* An even more conspicuous geothermal impact on the coalification pattern is associated with a large basic intrusion near Erkelenz in the westernmost part of the district (*Figure 12.16*). The presence of the igneous body was demonstrated by strong magnetic anomalies.

Pennsylvanian Coal Basins of the Eastern USA

Coalification patterns reflect: depth of burial during later Pennsylvanian and Permian times, when the main coalification took place; and regional thermal disturbances that post-date the main coalification (Damberger, 1974). The general decrease in rank towards the Canadian Shield corresponds well with the general northwards thinning of the sedimentary cover over the basement.

Superimposed upon this broad-scale rather simple coalification pattern that reflects burial, there are several areas of anomalously high rank that must be interpreted as being a result of unusually high heat flow sometime after the main coalification phase had terminated, probably connected with deep-seated igneous activity. These areas include the Rhode Island meta-anthracite and the Pennsylvania anthracite regions.

Sydney Basin, New South Wales, Australia

The Permian Gondwana coal-measure sequences of the Sydney Basin are characterized by coal-seam splitting. In the Singleton Coal Measure the coal seams frequently develop zig-zag splits in a process of multidirectional splitting, indicating an environment where sediments were deposited from river channels switching back and forth across the Permian peat swamps (Britten *et al.*, 1975). The Illawarra Coal Measure does not show such splitting, indicating slower and more stable subsidence of the basin.

Coalification in the Sydney Basin has been expressed as R_o–depth factors (reflectance increase per 100 m thickness of overburden) (*Figure 12.17*). Since coalification is a function of both temperature and time, the R_o–depth factors differ with geothermal gradients and with diagenetic maturity reached by the coals (Diessel, 1975).

In low-rank coals the increase in rank with increasing depth of burial is small compared with the higher coalification gradients observed in high-rank coals. The R_o–depth factors range from $0.028R_o$/100 m in the Upper Hunter Valley to $0.073R_o$/100 m to the south of Sydney (*Figure 12.17*). It is thought that a higher heat flow, possibly as a result of deep igneous activity, has been responsible for these high values south of Sydney.

Uranium Resources

Uranium ores have been described under suitable headings in various chapters throughout the book, but it is appropriate here to give a general review of their age distribution and general classification. The worldwide distribution of the major deposits is given in *Figure 12.18*, from which the USSR and eastern Europe are excluded because of insufficient data, but a description of Russian uranium

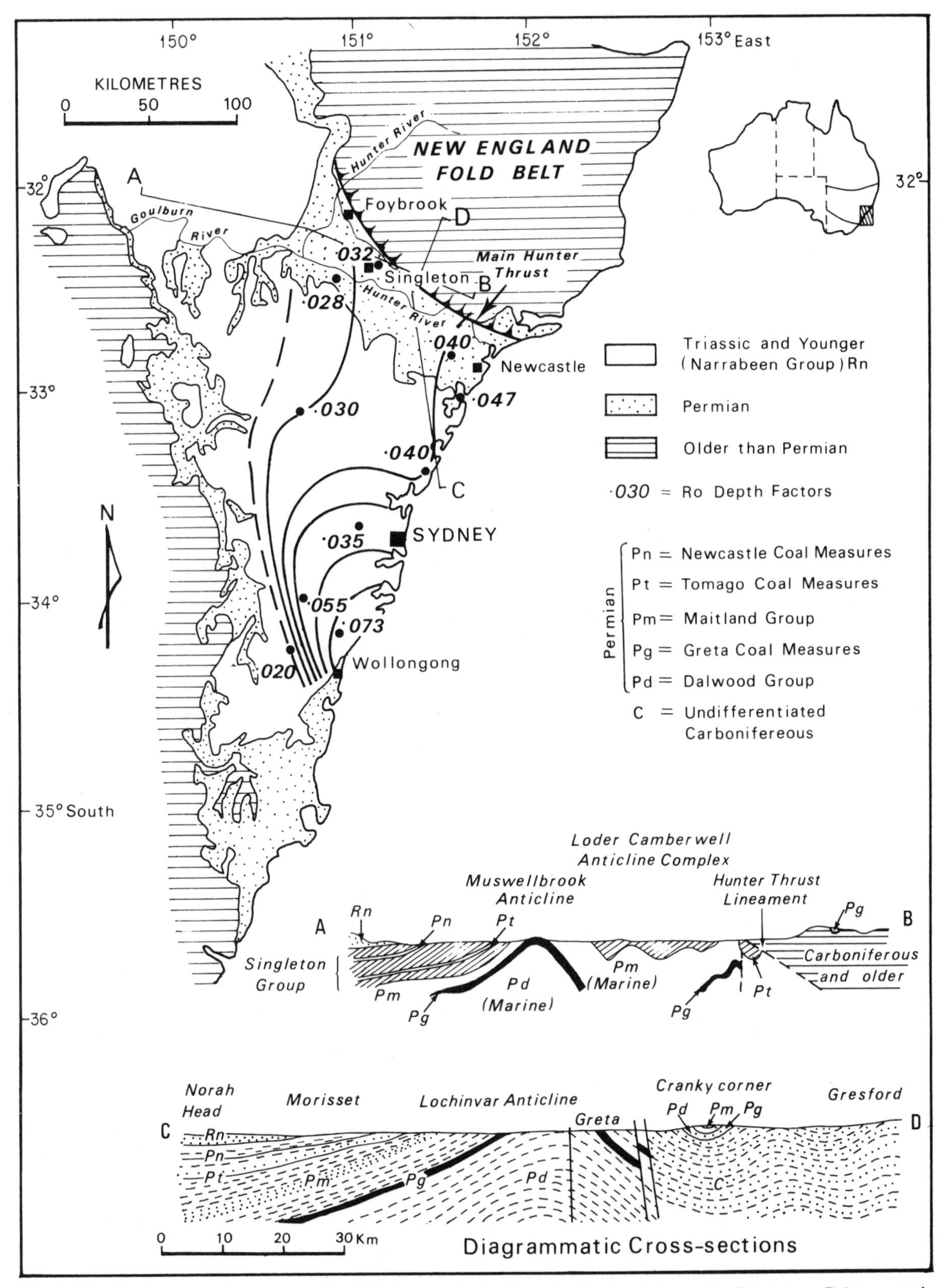

Figure 12.17. Permian Gondwana Coal Measures in the Sydney Basin of Australia (based on Britten *et al.*, 1975; Diessel, 1975) showing the regional variation in R_o–depth factors (coalification intensity).

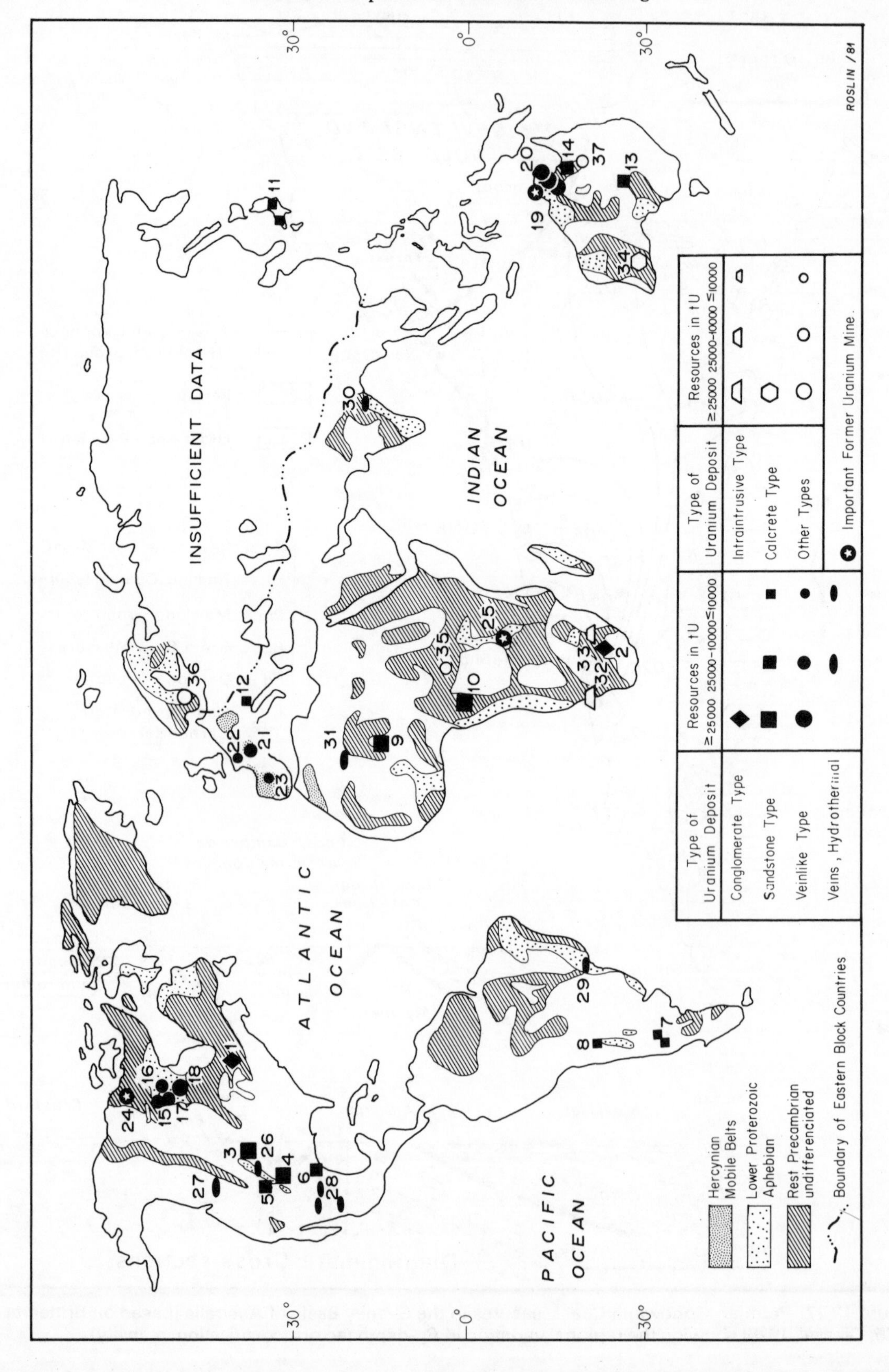
PACIFIC OCEAN
ATLANTIC OCEAN
INDIAN OCEAN
INSUFFICIENT DATA
30°
0°
30°
ROSLIN /81
Hercynian Mobile Belts
Lower Proterozoic Aphebian
Rest Precambrian undifferenciated
Boundary of Eastern Block Countries
Type of Uranium Deposit
Resources in tU
≥25000 25000–10000 ≤10000
Conglomerate Type
Sandstone Type
Veinlike Type
Veins, Hydrothermal
Intraintrusive Type
Calcrete Type
Other Types
Important Former Uranium Mine.

deposits may be found in Kazansky and Laverov (1977).

The time-bound character of uranium deposits has been noted in particular by Robertson, D.S. *et al.*, (1978), Derry (1980) and Cheney (1981). About 90% of the deposits, as known at present, occurs in environments in which uranium appears to have been deposited from either surficial or ground waters. Typical quartz–pebble conglomeratic ores are not present in the stratigraphic record after 2200 Ma ago, and they appear to form an important time-bound zone from 2200 Ma to 2800 Ma ago (*Figure 12.19*). This was the only known period when uranium was transported in detritus. Such placer deposits cannot have developed after 2200 Ma ago because of the inhibiting quality of the developing atmosphere. The atmosphere was essentially anoxygenic during the Archaean and the early Proterozoic, and uranium minerals were therefore insoluble (Robertson, D.S. *et al.*, 1978). They remained as detrital minerals and accumulated in conglomerates as heavy minerals. These palaeo-placer uranium deposits are described in Chapter 11. All uranium younger than 2000 Ma has been transported in solution.

The classification of uranium ore deposits has been the subject of increasing concern as the search for resources has recently accelerated. The scheme adopted is based on recent papers by Bailey, R.V. and Childers (1977), Dahlkamp (1978a), McMillan (1978), Childers and Bailey (1979) and Cheney (1981).

Figure 12.18 (opposite). Distribution of known uranium deposits in the western world, after Dahlkamp (1978a). Conglomerate-type: 1 = Elliot Lake district, Canada; 2 = Witwatersrand, South Africa. Sandstone-type: 3 = Wyoming district, USA; 4 = New Mexico district, USA; 5 = Colorado Plateau, USA; 6 = Gulf Coast, USA; 7 = Malargue, Sierra Pintata, Argentina; 8 = Salta, Argentina; 9 = Arlit, Niger; 10 = Mounana, Oklo, Gabon; 11 = Ningyo toge, Tono, Japan; 12 = Zirowski Vhr, Yugoslavia; 13 = Lake Frome Basin, Australia; 14 = Westmoreland district, Australia. Vein-like-type: 15 = Beaverlodge District, Canada; 16 = Rabbit Lake, Canada: 17 = Cluff Lake, Canada; 18 = Key Lake, Canada; 19 = Rum Jungle, Australia; 20 = Alligator Rivers district, Jabiluka, Koongarra, Ranger, Nabarlek, Australia; 21 = Central Massif, France; 22 = Vendée, France; 23 = Iberian Meseta, Spain/Portugal. Hydrothermal veins: 24 = Port Radium, Canada; 25 = Shinkolobwe, Zaire; 26 = Schwartzwalder Mine, Colorado, USA; 27 = Spokane district, Washington, Sunshine Mine, Midnite Mine, USA; 28 = Sonora, Durango, Chihuahua province, Mexico; 29 = Pocos de Caldas, Agostinho, Cercado, Brazil; 30 = Singhbum district, India; 31 = Hoggar, Algeria. Intraintrusive-type: 32 = Rössing, Namibia; 33 = Phalaborwa, South Africa. Calcrete-type: 34 = Yeelirrie, Australia. Other types: 35 = Bakouma (uranium phosphate), Central African Republic; 36 = Ranstad (black shales), Sweden; 37 = Mary Kathleen (pyrometasomatic), Australia.

Strata-controlled Deposits

This category contains the largest proportion of known uranium reserves and it includes both oxidized and reduced lithofacies. It may be further subdivided.

Sandstone–Conglomerate Host Rocks

This category may be further subdivided.

Trend deposits (p. 78)

In these, the ore bodies are distributed along mineralized belts or trends. The ore bodies are generally tabular and subparallel to the host-rock stratification. This category may be further subdivided.

The mineralized trends are parallel to that of the host rocks (palaeo-drainage control)

Examples are: the Westwater Canyon deposits of the Ambrosia Lake trend, New Mexico; and the Chinle deposits in Lisbon Valley, White Canyon and Monument Valley trends, Utah and Arizona.

The mineralized trends generally cross the palaeodrainage pattern

An example is the Salt Wash deposits in the Uravan mineral belt, Colorado and Utah.

Roll-front deposits (p.79)

The ore bodies are distributed along lateral and distal margins of altered complexes in sandstones and conglomerates. The ore bodies occur in permeable hosts generally discordant to the stratification. This category may be further subdivided.

Biofacies roll-fronts

These occur in formations with both oxidized and reduced facies. Examples are: the uranium occurrences in the Wind River, Wasatch; the Battle

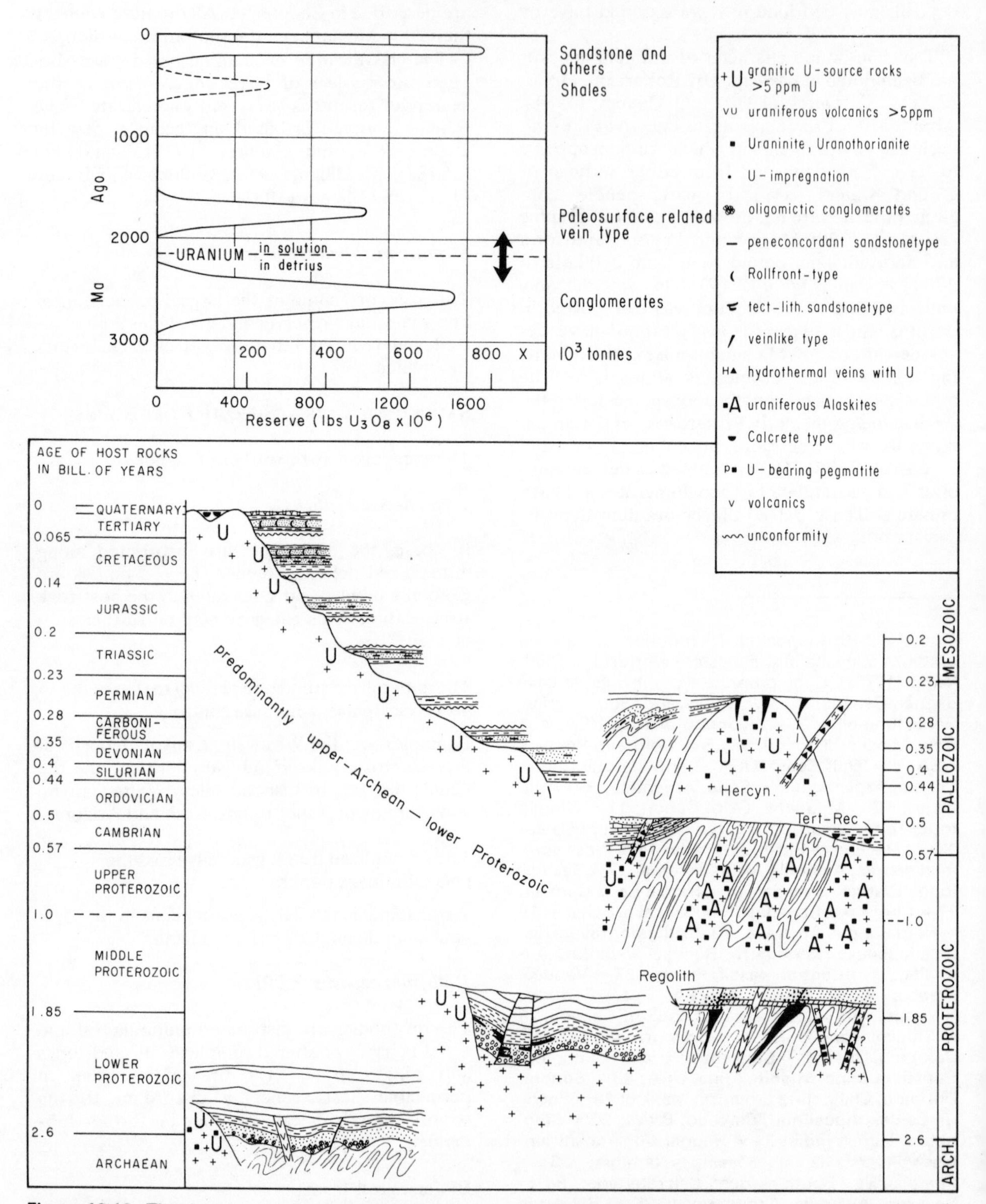

Figure 12.19. The time-bound character of uranium deposits (after Robertson, D.S. *et al.*, 1978). The change in character of uranium deposits with time, as shown schematically by Dahlkamp (1978a).

Spring Formations of Wyoming; and the Goliad, Oakville and Catahoula deposits in south Texas.

Monofacies roll-fronts

These occur in formations that are uniformly reducing. Examples are: the Inyan Kara, Fox Hills, Lance and Fort Union deposits of Wyoming and nearby areas; and the Jacksonville deposits of south Texas.

Stack deposits

The ore bodies are of irregular shape, associated with trend ore, and frequently controlled by structures. An example is the West Water Canyon ore occurrences in Grants uranium region, New Mexico, generally regarded as redistributed ore.

Precambrian heavy-mineral or placer deposits (p. 264)

Uraninite and other heavy minerals are concentrated detritally along stratification in sandstones and pebbly conglomerates. They appear to represent placer concentrations in ancient reducing atmosphere. Examples are: the Witwatersrand of South Africa; and the Elliot Lake area of Canada.

CARBONATE HOST ROCKS

This category may be further subdivided.

Palaeo-karst deposits

The ore bodies occur in solution cavities. An example is the Pryor and Bighorn Mountains of Montana and Wyoming.

Reef-trend deposits

The ore bodies occur along reef fronts. An example is the Todilto deposits in Grants uranium region, New Mexico.

Calcrete deposits (p. 254)

These are of irregular distribution of carnotite in calcareous material (caliche or calcrete) and are distributed in roughly horizontal tabular bodies along major drainageways in arid to semi-arid regions. An example is the Yeelirrie deposit of Western Australia.

LIGNITE, BLACK SHALE AND PHOSPHATIC HOST ROCKS

This category may be subdivided.

Uraniferous lignites

The ore occurs as irregular dissemination in lignite and associated carbonaceous shale and siltstone. The category may be further subdivided.

Low-grade wide distribution in lignites marginal to oxidizing environments

An example is the Wasatch deposits of the Great Divide Basin, Wyoming.

Deposits (including high-grade) in lignites associated with permeable sandstones below a regional unconformity

Examples are the Fort Union and Hell Creek deposits of the southwest Williston Basin in Montana, South Dakota and North Dakota.

Uraniferous black shales (p. 95)

The deposits occur in reducing environments marginal to oxidizing environments. Examples are: the Chattanooga Shale in central Tennessee, Kentucky and Alabama; the Phosphoria Formation in eastern Utah and Idaho; the Lacustrine beds of Eocene age in Wyoming and Utah; the Lodeve area of southern France; and the black shales of Sweden.

Structure or Fracture-controlled (Vein-type and Similar) Deposits (p. 238)

These are fractured and brecciated host rocks in which pitchblende and other uranium minerals fill voids and coat fractures, and partially replace the host rocks. They are typically what have been called the *hydrothermal uranium deposits.* Examples are at: Rabbit Lake, Key Lake, and Cluff Lake, Saskatchewan, Canada; the Pine Creek Geosyncline Province of the Northern Territory of Australia; the Front Range District of Colorado; and the Midnite Mine of Washington.

Intrusion Controlled (p. 160)

The ore minerals occur as disseminations in igneous bodies. Finely disseminated uranium minerals occur in alaskite, nepheline syenite and other felsic igneous rocks. Examples are: the Ilimaussaq area of Greenland (where uranium occurs in nepheline syenite); the Rössing area of Namibia, Southwest Africa (where uraninite is disseminated in syntectonic alaskite in a migmatite); and the Colorado

Front Range (where uranium is disseminated in dykes of bostonite).

The above classification by Childers and Bailey (1979) largely coincides with that of Dahlkamp (1978a), used in *Figure 12.18.* The conglomerate-type coincides with the 2200–2800 Ma old peak and represents the Late Archaean–early Proterozoic peak of *Figure 12.19.* Unfortunately data on deposits in communist countries are lacking. The conglomerate placer-types account for reasonably assured reserves of 450×10^3 t of uranium, with an additional estimated 280×10^3 t. The major occurrences are in Canada and South Africa. The sandstone-type deposits have 500×10^3 t of reasonably assured, with an additional possible further 800×10^3 t in the western world. The major occurrences are in the USA, Niger and Gabon. Vein-like deposits account for 320×10^3 t with a further possible 260×10^3 t of reserves. Their major occurrences are in Australia, Canada and France. Hydrothermal vein deposits are of minor significance with only 50×10^3 t of estimated reserves. Their main occurrences are in Zaire, USA and Canada. The intraintrusive-type occurs within igneous rocks of felsic-type, and accounts for known reserves of 110×10^3 t and a further possible reserve of 500×10^3 t. The main occurrence is in Namibia. The calcrete-type occurs in Australia and accounts for reserves of 40×10^3 t. Black shales and phosphorite deposits occur in Sweden, USA, Brazil and Angola (Dahlkamp, 1978a). The time-bound characteristic of uranium deposits by type is clearly shown in *Figure 12.19.*

The peak shown between 2000 Ma and 1500 Ma is created by the uranium contained in palaeo-surface-controlled, vein-like deposits (Robertson, D.S. *et al.*, 1978). The time-bounding limits indicated are based on age determinations on the basement rocks of the weathered and eroded palaeo-surface, with which most of the vein-like deposits are associated, and on the ages of the overlying cover rocks. An unconformity is therefore considered important in the concentration process of such deposits.

Derry (1980) has estimated the relative importance of the various types of deposits in terms of annual production in the western world. In 1977, the production from magmatic deposits was 7.6%, from unconformity/vein-like deposits 17%, from palaeo-placer conglomerates 23.6% and from sandstone-types 51.7%. The trend is for the sandstone-type to drop in importance in favour of an increase in the unconformity-type.

The time-bound character of uranium can be overstressed. Although the uranium may have been concentrated by processes related to an unconformity, primarily by weathering, erosion and ground-water action, the deposits associated with the unconformity may be remobilized several times in the same area. Uranium has an extreme susceptibility to be remobilized and reconcentrated by supergene or epigene agents. Thus the Key Lake deposit of Saskatchewan was initially concentrated before 1800 Ma ago. The unconformity of Athabasca was dated at 1350 Ma. The uranium was remobilized by diabase intrusion at 1230 Ma and there was a general remobilization of the same deposit at 900 Ma ago (Dahlkamp, 1978b). Likewise the East Alligator River deposit, Northern Territory, Australia, was originally concentrated in the Cahill Formation 2500–1700 Ma ago, the Kombolgie unconformity was dated at 1600 Ma ago, but the uranium was remobilized 900 Ma and 500 Ma ago (Hegge and Rowntree, 1978). There is also a case to be made for uranium metallogenic provinces, or regions that are enriched in uranium, and therefore may contain more than one style of deposit. Thus in Gabon some deposits are of the unconformity-type, and others of the sandstone-type.

The Proterozoic deposits are spatially related to Archaean granites, which are more uraniferous than average, and presumably were the ultimate source. It is indeed possible that uranium-rich provinces, no matter what age the mineral concentration took place, owe their existence to an original enrichment in uranium in the Archaean crust, and especially in acid igneous rocks that formed at the early stages of the crustal evolution. In this, one may see a similarity with tin deposits that are regionally restricted, for example, to deposits in Bolivia and the Malay Peninsula. However, in the latter, the economic deposits are not restricted to any one age of granite activity, and may be Triassic or Cretaceous. The model of Hutchison and Chakraborty (1979) suggests that the regional infrastructure was originally enriched in tin, and that the tin is remobilized by anatectic events, of whatever age, that can tap the continental infrastructure. Uranium may be similar in that it was originally concentrated in Archaean granitoids, but younger granitoids may equally well be enriched in uranium. However uranium differs from tin in that it is extremely susceptible to meteoric hydrothermal remobilization, and especially to low-temperature systems related to regolith formation.

Appendix
The Economic Commodities

This section provides a commodity cross-reference to the geological descriptions of the main text. The page numbers given after most section headings (main entries are enumerated in *italic* type) are intended to direct the reader to the pages in which the geological descriptions are found. The brief entries also give some idea of the world's annual production, and the five major countries that share in this. Commodity production figures are based on those of the Bureau of Mines (1980a,b) and United Nations (1979). Other important references are Brobst and Pratt (1973) and LeFond (1975).

There is no unified quantity of measure. Commodities may be traditionally quoted in metric tonnes (denoted by t) (=2205 pounds), long tons (=2240 pounds) and short tons (2000 pounds). Precious metals are usually quoted in troy ounces (=31.103 grams), diamonds in carats (= 0.2 gram) and mercury in flasks (= 76 pounds). The standard unit used throughout this book is the metric tonne (tonne), which is equal to 1 000 000 grams (1 Mg).

This Appendix is not intended to be read by students using the book as a course textbook. It is a cross-reference chapter for professional geologists.

ANTIMONY

(145, *186–188*, *210–212*)

World 1979 production 71 998 t. Share of production: Bolivia 21%, China 17%, South Africa 16%, USSR 11%, Thailand 7%. Antimonian lead is used in starting–lighting–ignition batteries, but the newer maintenance-free batteries do not use antimony. Antimonial–lead alloys are used in solder, ammunition, chemical pump pipes, roofing sheets, communication equipment, anti-friction bearings and pewter alloys. Antimony oxide is used in plastics as a stabilizer and flame-retardant and in textiles. Antimony sulphides are used in fireworks and ammunition. Because of its military uses, antimony is a strategic metal. The main ore minerals are stibnite Sb_2S_3, tetrahedrite $(Cu,Fe)_8Sb_2S_7$ and jamesonite $Pb_2Sb_2S_5$ (Miller, 1973; Rowland and Cammarota, 1980).

ARSENIC

(*209–210*)

World 1979 production of arsenic trioxide 32 824 t. Share of production: USSR 23%, France 22%, Sweden 21%, Mexico 18%, Namibia 8%. These figures exclude the USA, whose production figures are withheld. Arsenic compounds are used in herbicides and plant desiccants, wood preservatives, mineral flotation agents, glassware and pharmaceuticals. Arsenic is produced always as a by-product of some metal, and the principal ore minerals are arsenopyrite FeAsS, löllingite $FeAs_2$, smaltite $CoAs_2$, niccolite NiAs, tennantite $(Cu,Fe)_8As_2S_7$, enargite Cu_3AsS_4, proustite Ag_3AsS_3 and relagar AsS (Gualtieri, 1973).

ASBESTOS

World 1979 production 5 277 591 t. Share of production: USSR 46%, Canada 29%, South Africa 7%, Zimbabwe 4%, China 4%. The main uses are in asbestos–cement pipes and asbestos–cement sheets. It is used for flooring, roofing, insulation, friction products, coatings and textiles, which are fire-resistant. Other uses include packing gaskets for bearings. Legislation now prevents the use of asbestos in brake linings because of the health hazard. Main mineral is chrysolite (serpentine) $Mg_3(Si_2O_5)(OH)_4$ (Shride, 1973; Winson, 1975; Clifton, 1980a).

BARITE

(*62–67*)

World 1979 production 7 039 227 t. Share of production: USA 25%, USSR 8%, Ireland 6%, China 6%, India 6%. Barite is primarily used for increasing the specific gravity of oil-well and gas-well drilling muds. Other uses include as a filler in paint, plastics, paper and rubber. It is used as a flux, oxidizer and decolourizer in glass manufacture and heavy concrete aggregates. Witherite is more valuable in the chemical industry. The major minerals are barite $BaSO_4$ and witherite $BaCO_3$ (Brobst, 1973, 1975; Fulkerson, 1980).

BAUXITE AND ALUMINIUM

(249–254)

Bauxite world 1979 production 86 814 000 t. Share of production: Australia 32%, Jamaica 13%, Guinea 14%, Surinam 6%, USSR 5%. Aluminium metal is made by extracting alumina from bauxite by the Bayer process and by reducing the oxide electrolytically to metal. Because of its resistance to oxidation, efficient electrical conductivity and low price, it is widely used in the manufacturing industry—mainly for aircraft (low specific gravity of 2.7), motor vehicles, electrical equipment, machinery, drink cans, containers and a variety of home-consumer products. Although most bauxite ends up as metal, it is also used to make alumina abrasives. The conversion from bauxite to aluminium metal is carried out only where electricity is plentiful and cheap (such as that from hydroelectric power). The principal metals of bauxite are gibbsite $Al(OH)_3$, boehmite $AlO(OH)$ and diaspore $AlO(OH)$ (Patterson and Dyni, 1973; Shaffer, 1975; Baumgardner and Hough, 1980).

BENTONITE

World 1979 production 5 847 419 t. Share of production: USA 62%, Greece 8%, Japan 7%, Italy 5%, Hungary 2%. The major uses are as bleaching clay (oil refining, filtering, clarifying and decolourizing), drilling mud, iron-ore pellitizing and foundry-sand bonding. In addition it is used for filtering and clarifying wine and waste water, and in pharmaceuticals. The main component of bentonite is montmorillonite $(0.5Ca,Na)_{0.7}$-$(Al,Mg,Fe)_4(Si,Al)_8O_{20}(OH)_4 \cdot nH_2O$ (Hosterman, 1973; Patterson and Murray, 1975; Ampian, 1980).

BERYLLIUM

(129)

World 1979 production (excluding USA) 2795 t. Share of production (excluding USA): USSR 70%, Brazil 18%, Argentina 4%, Zimbabwe 3%, Rwanda 2%. Its uses reflect its excellent thermal properties and high stiffness-to-weight ratio. The main consumption is in beryllium–copper alloys, business machines and appliances, transportation and the communications industries. It is used in electronic systems to form connectors, sockets, switches and temperature–pressure sensors. Beryllium oxide is used in lasers, microwave tubes and semi-conductors. Beryllium metal is used in satellites, navigation instruments, space optics, nuclear devices and military aircraft brakes, and X-ray tube windows. The principal mineral is beryl $Be_3Al_2(Si_6O_{18})$ (Griffiths, 1973; Petkof, 1980a).

BISMUTH

(192)

World 1979 production 4273 t. Share of production: Australia 22%, Mexico 16%, Japan 15%, Peru 15%, Bolivia 11%. The principal uses are in low-melting alloys, in metallurgical additives for aluminium, carbon steel and malleable iron, in pearlescent cosmetic pigments and in medicines. The largest use is in the pharmaceutical industry. Bismuth–cadmium alloy is used in selenium rectifiers. It is also used in bismuth–tellurium electronic devices and as an accelerating agent in rubber vulcanizing. There is no specific bismuth ore. It is usually recovered in refining processes from ores of lead and copper, molybdenum, tin, tungsten and zinc. The main mineral is bismuthinite Bi_2S_3, which occurs widely as an accessory in ores of the above metals (Hasler *et al.*, 1973; Carlin and Bascle, 1980).

BITUMEN

Asphalts refined from crude petroleum have almost completely replaced native-types. Asphalt cements are used as hot mixes to pavement roads, etc. Cut-back asphalts are used in road mixes. Oxidized asphalts are used in undersealing and roofing. Slow-curing asphalts, or road oils, are used for dust control of unpaved roads. Emulsified asphalts are used in macadam road pavements. Native asphalts occur as lakes or rock impregnations. The largest is at La Brea, Trinidad; it is used for paving roads and waterproofing roofs. Rock asphalt occurs as bituminous sandstone or limestone and is used for paving. The Athabaska tar sands in Canada are being mined and processed to produce crude oil. Tar sands in Albania, Romania and the USSR are also being exploited (Cashion, 1973; Bostwick, 1975).

BORON MINERALS

World 1979 production 2 632 330 t. Share of production: USA 55%, Turkey 34%, USSR 8%, Argentina 2%, Chile 1%. Boron in the form of borates is used in insulation products and glass fibre-reinforced plastics. Glass wool is the main product, followed by textile-grade glass fibres, and special borosilicate glasses. Borates are also used by the chemical industries. The major minerals are: borax (tincal) $Na_2B_4O_7 \cdot 10H_2O$, which is the primary world source, colemanite $Ca_2B_6O_{11} \cdot 5H_2O$, produced in the USA and Turkey, and Szaibelyite $MgBO_2(OH)$, the principal source in the USSR (Kistler and Smith, 1975; Absalom, 1980a).

BROMINE

World 1979 production 345 570 t. Share of production: USA 66%, Israel 13%, United Kingdom 7%, France 5%, USSR 4%. The largest use is in the manufacture of ethylene dibromide, which, together with tetraethyl lead, is used as an anti-knock additive in gasolene. Disinfectants and pharmaceuticals use the next largest amount of bromine; then come fire-retardants and fire-extinguishing agents. Insecticides and soil fumigants use bromine. Calcium bromide is used in the drilling industry for high-density, solids-free completion of wells. Bromine occurs in nature in oil-well brines, which provide the major source in the USA. Israel obtains bromine from the Dead Sea, and the United Kingdom from sea water. France obtains bromine from potash evaporate fields (Jensen, J.H., 1975; Absalom, 1980b).

CADMIUM

(228)

World 1979 smelter production 18 280 t. Share of production: USSR 16%, Japan 14%, USA 9%, Canada 8%, Belgium 8%. Cadmium is used in the electroplating of fabricated steel products. It is also used to make batteries, and as pigments, plastics and other alloys. All cadmium is recovered as a by-product of zinc mining and smelting. None of the cadmium minerals is abundant enough to form an ore, although trace greenockite (CdS) in slightly oxidized sphalerite will enhance the cadmium content. The cadmium content in the zinc ores is within the sphalerite itself. Mississippi Valley sphalerites may contain as much as 5000 ppm cadmium, while the Kupferschiefer ore of Germany contains, on average, 500 ppm of cadmium (Wedow, 1973; Lucas, 1980).

COAL

(*300–310*)

World 1979 production 2 763 187 000 t. Share of production: USA 22%, USSR 19%, China 20%, Poland 7%, West Germany 5%. This major fuel is used primarily for electricity generation and in steel-smelting plants. Coal occurs in a variety of ranks, related to the geological environment (Averitt, 1973; United Nations, 1979).

COBALT

(*209–210*, 237)

World 1979 mine production in metal 28 513 t. Share of production: Zaire 53%, Zambia 11%, Cuba 6%, USSR 6%, Canada 5%, Australia 5%. Cobalt is used in high-temperature alloys, magnets and tool steels. The high-temperature alloys find their use in turbine engines for aircraft and power plants. The largest market is for use in a cobalt–molybdenum–alumina catalyst for desulphurization of light petroleum distillates. Cobalt is also used in paint driers and oil-based paints. It is usually a by-rodut of copper and nickel ore. he main minerals are siegenite $(Co,Ni)_3S_4$, carrollite $(Co_2Cu)S_4$, cobaltite (Co,Fe)AsS, skutterudite $(Co,Fe)As_3$, erythrite $(Co,Ni)_3(AsO_4)_2.8H_2O$ and gersdorffite (Ni,Co)AsS. Cobalt may also be contained in pyrrhotite, pentlandite, pyrite, sphalerite, arsenopyrite and manganese oxides (Vhay *et al.*, 1973; Sibley, 1980).

COPPER

(31–42, 46, 92–93, 102, 104, *168–175*, 183–185, 202–204, 207–210, 227, *234–238*)

World 1979 production from mines 7 606 800 t. Share of production: USA 18%, Chile 14%, USSR 11%, Canada 10%, Zambia 6%. The high electrical and thermal conductivities of copper, its good resistance to corrosion, its good ductility and high strength mean that copper finds innumerable industrial applications. Copper and its alloys can be joined by welding, brazing and odering. The four alloys brass, bronze, nickel–silver and cupronickel contain no less than 40% copper. About 53% of copper is used in electrical applications, 16% in construction, 12% in industrial machinery and the remainder in numerous other uses. The main minerals are chalcopyrite $CuFeS_2$, bornite Cu_5FeS_4, chalcocite Cu_2S, covellite CuS, enargite Cu_3AsS_4, cuprite Cu_2O, malachite $Cu_2(OH)_2(CO_3)$ and native copper Cu. There are significant copper contents in the following minerals: tetrahedrite $(Cu,Fe)_8Sb_2S_7$, tennantite $(Cu,Fe)_8As_2S_7$, famatinite Cu_3SbS_4, stannite Cu_2FeSnS_4 and several others. Many ore deposits are mined primarily for copper, but yield significant amounts of by-products (Cox, D.P. *et al.*, 1973; Jolly, 1980b).

CORUNDUM

World 1979 production 9830 t. Share of production: USSR 81%, India 13%, Uruguay 4%, South Africa 1%. Corundum (Al_2O_3) is the hardest natural substance except for diamond. It is used exclusively as crushed sieved and sized grit and powder for finishing optical parts and polishing metals. The main attraction as an abrasive is that a transverse crystal parting presents to the work a sharp chisel-like edge, which scrapes rather than scratches the polished surface. When the grains wear, a flake spalls off to expose a new cutting edge. This property ensures a continuing use for corundum as an abrasive. Several varieties of corundum are of gemstone quality; they are ruby, sapphire, star-ruby and star-sapphire (Thaden, 1973a,b; Hight, 1975; Jahns, 1975; Baskin, 1980a,b).

CHROMIUM

(*19–23, 104–112*)

World 1979 production 9 521 686 t. Share of production: South Africa 34%, USSR 22%, Albania 9%, Turkey 7%, Philippines 6%. Most chromium consumption goes to form an alloyingelement in stainless and higher-alloyed heat-resistant and corrosin-resitant alloys, and full-alloy steels. Chromite is used to form refractory bricks and casting items for furnaces. Chromium is used in the chemical industry and in the leather industry for tanning. Chromite represents a composition range of the spinel series $(Mg,Fe^{2+})(Cr,Al,Fe^{3+})_2O$. Chromites of economic interest have Cr_2O_3 contents ranging from 25% to 65% by weight (Thayer, 1973; Mikami, 1975; Matthews and Morning, 1980).

DIAMONDS (NATURAL)

(*119–123*, 144, *262–264*)

World 1979 production 7.94 t (39 698 000 carats) (industrial 73%, gem 27%). Share of production: South Africa 20%, Zaire 28%, USSR 26%, Botswana 7%, Ghana 6%. Diamond (C) is the hardest natural substance. Only 27% of the production is of gemstone quality. The majority of the non-gem material is used in industry. A small proportion goes into hard-wearing uses such as record-player needles, instrument bearings and wire-drawing dies. Most is used as abrasive. Pencil-like tools for cutting glass, or trimming abrasive wheels, have a whole diamond set in a tool face. Rock drills have diamond crystals bonded into the cutting end. Crushed and sieved diamond is made into a paste for polishing a variety of hard materials. To a large extent artificially produced diamonds have taken the place of the natural form, but the rich deposits at Kimberley, Western Australia, may change this (Thaden 1973a,b; Hight, 1975; Jahns, 1975; Reckling *et al.*, 1975; Baskin, 1980a,b).

DIATOMITE

World 1979 production 1 587 302 t. Share of production: USA 33%, USSR 24%, France 12%, Romania 2%, West Germany 1%. The widest use of processed diatomite is as a filter aid for the separation of suspended solids from fluids. It is also used s a filler in paint and paper, and as a carrier in rug cleansers, pesticides and catalysts. The highly refined diatomite is used as a mild abrasive in metal polishes. Diatomite is composed of fossilized skeletons of diatoms composed of silica $SO_2.nH_2O$ (Durham, 1973; Kadley, 1975; Meisinger, 1980a).

EMERY

The world's supply comes from Greece, which produced 6530 t in 1979 and from Turkey, which produced 65 509 t in 1977. Emery is a black granular material consisting of corundum (Al_2O_3) and magnetite ($(Fe,Mg)Fe_2O_4$) with admixed spinel, haematite, garnet and other minerals. The use of natural emery has largely been replaced by synthetic materials as an abrasive (Thaden, 1973a; Baskin, 1980a).

FELDSPAR

(115, 119)

World 1979 production 3 092 870 t. Share of production: USA 24%, West Germany 16%, USSR 11%, Italy 8%, France 7%. More than half the feldspar mined is consumed in glass-making; the second major use is in pottery. The remainder is used in glazes, enamels, soaps, abrasives, sanitary ware, rubber products and electrical insulators. The main ore minerals are orthoclase $KAlSi_3O_8$ and albite $NaAlSi_3O_8$. Most mined feldspars will contain a small amount of the calcium end-member, anorthite $CaAl_2Si_2O_8$ (LeSure, 1973a; Rogers, C.P. and Neal, 1975; Potter, 1980a). Nepheline syenite is commercially mined for its feldspar and nepheline content. It is processed to remove the iron-bearing minerals. Its uses are in glass manufacture and in ceramics. The main producing countries are Canada and Norway (Minnes, 1975). Aplite, when low in iron minerals, is used in the manufacture of glass, especially container glass. Japan is the world's foremost producer, with an annual production of 450 000 t (Potter, 1980a).

FLUORITE

(53–55, *59–62, 107–108*)

World 1979 production 4 865 165 t. Share of production: Mexico 21%, USSR 11%, Spain 9%, South Africa 8%, France 8%. The major use of fluorite is in the production of hydrofluoric acid, from which is made synthetic cryolite, used as an electrolyte in the manufacture of aluminium metal. Fluorite (CaF_2) is used as a flux in steel-making furnaces; it promotes fluidity of the slag and passage of sulphur and phosphorus into the slag. It is used to make flint glass, white or coloured glass and enamels (Worl *et al.*, 1973; Grogan and Montgomery, 1975; Kostick and DeFilippo, 1980). Natural cryolite (Na_3AlF_6) has been found in commercial quantities only at Ivigtut, Greenland, within a pegmatite, but the deposit is now exhausted. Cryolite is now produced synthetically. It is needed for the electrolytic production of aluminium.

FULLERS EARTH

World 1979 production 1 675 381 t. Share of production: USA 81%, United Kingdom 10%, Mexico 4%, Algeria 2%, Morocco 1%; excluding communist countries for which data are not available. The palygorskite (attapulgite) clay is used as a component in drilling mud, and as a carrier for insecticides and fungicides. It is used as an absorbent for greases, oil, water and other undesirable substances on the floors of factories and other installations. Some grades are used for pharmaceuticals designed to absorb toxins and bacteria. It is used as a thickener in paints, cements and plastics. The mineralogy of fullers earth is palygorskite $(OH_2)_4(OH)_2Mg_5Si_8O_{20}.4H_2O$, but substitution of Al^{3+} for either Mg^{2+} or Si^{4+} takes place (Hosterman, 1973; Patterson and Murray, 1975; Ampian, 1980).

GALLIUM, GERMANIUM, INDIUM

(254)

World production of gallium is not available, but the USA imported 6.4 t of metal in 1979, and Switzerland supplied 86% of this total. Germanium world production in 1979 is estimated to be 109 t, of which France, Italy and Austria are the main producers. Production figures for indium are not available, but the USA imported 9.144 t in 1979, 44% from Belgium and Luxembourg, 32% from Peru and 13% from Canada. The principal uses for these three metals are in the manufacture of transistors, diodes and rectifiers. Gallium, germanium and indium do not form their own ore minerals and they are recovered as a by-product from zinc ore. Sphalerite is the main ore mineral that contains all three elements; other sulphides that contain them are chalcopyrite, bornite, enargite and tennantite. They may also be concentrated in wood tin. Germanium and gallium may be recovered from coal ash (Weeks, 1973; Petkof, 1980b).

GEM STONES

With the exception of diamonds (listed above) production figures are not available. Some of the most important and the main producers are: agate (Australia, Brazil, Mexico, China), amber (Italy, Sicily), amethyst (USSR, Sri Lanka, India), emerald (Egypt, USSR), jadeite jade (Burma, China), nephrite jade (Taiwan), opal (Australia, Mexico), ruby (Burma, Indochina, Afghanistan), sapphire (Burma, Thailand), topaz (Sri Lanka, India), tourmaline (USSR, Burma, India) and turquois (USSR, Australia, Arizona). For a complete discussion, *see* Thanden (1973a,b), Jahns (1975) and Baskin (1980a,b).

GOLD

(168–175, *180–194*, *204–207*, *261–262*, *269–272*)

World 1979 production 1220.4 t (39 238 343 troy ounces). Share of production: South Africa 58%, USSR 20%, Canada 4%, USA 3%, Zimbabwe 2%. Gold is used primarily as bullion, stored in banks. The next major use is in jewellery and art works. Its soft, easily malleable nature and resistance to corrosion makes it perfect for dental repairs. It is used in electronics equipment, but increasing cost has led to its replacement by palladium, tin and other metals wherever possible. Gold occurs mainly as native gold Au. Other important minerals are tellurides, calaverite $AuTe_2$ and sylvanite $(Au,Ag)Te_2$. Gold may occur as small contents in sulphide minerals, extractable during processing of the ore (Simons and Prinz, 1973; Butterman, 1980).

GRAPHITE

World 1979 production (excluding the USA, for which data are withheld) 523 049 t. Share of production: USSR 21%, North Korea 17%, Mexico 13%, China 11%, Austria 8%. The largest single use is in foundry facings. It is also used in steel-making, and in crucibles. Other uses include batteries, bearings, brake linings, carbon electrical brushes, lubricants, pencils, and in graphite–oil packings. Graphite is crystalline carbon C (Weis, 1973; Graffin, 1975; Taylor, H.A., 1980).

GYPSUM

World 1979 production 74 332 278 t. Share of production: USA 19%, Canada 11%, Iran 10%, France 9%, USSR 8%. The largest use of gypsum is based on its property of readily giving up or taking in water of crystallization. When calcined, gypsum is converted to plaster of paris. This material is manufactured into plasters and wallboard. Uncalcined gypsum is used as a retardant in portland cement, as soil conditioner and as mineral filler. It is also used to produce sulphuric acid. Alabaster is a compact fine-grained variety of rock gypsum, prized by sculptors because of its workability. Gypsum $CaSO_4.2H_2O$ is usually closely associated with anhydrite $CaSO_4$ and bassanite $CaSO_4.0.5\ H_2O$ (Smith, G.I. *et al.*, 1973; Appleyard, 1975; Pressler, 1980a).

HALITE

(*89–94*)

World 1979 production 167 756 910 t. Share of production: USA 23%, China 18%, USSR 9%, West Germany 7%, United Kingdom 5%. There are so many uses for salt that only a few can be mentioned: as a nutrient or flavour in food, as a preservative, as a food-processing material, in chemical manufacturing, as a freezing-point depressant, largely for highway de-icing, in metallurgical processing, dye processing, herbicides, in the chemical industry to form soda ash, in the manufacture of sodium sulphate, hydrogen, hydrochloric acid, chlorine and caustic soda. Altogether there are over 14 000 uses for salt. Halite or rock salt is NaCl, but it may be associated with other evaporite minerals (Smith, G.I. *et al.*, 1973; LeFond and Jacoby, 1975; Foster, 1980).

ILMENITE

(*272–275*)

World 1979 production 3 490 784 t. Share of production: Australia 32%, Norway 25%, USA 17%, USSR 12%, Malaysia 5%. Titanium metal is made from ilmenite and rutile by the Kroll process. The metal has a high strength-to-weight ratio, so is used in the aerospace industry, military aircraft and spaceships and rockets. It is also used in the manufacture of aircraft engines and gas turbines. Pigments are made by converting ilmenite to water-soluble sulphates. Ilmenite has the composition $FeTiO_3$ (Klemic *et al.*, 1973c; Lynd and LeFond, 1975; Lynd and Hough, 1980).

IRON ORE

(14, 15, *26–30*, *37–42*, 46, *68–75*, 102, *112*, *156–160*, *198–202*, 207–210, *214–219*)

The 1979 iron metal production from ore was 510 679 370 t. Share of production: USSR 28%, Australia 12%, Canada 7%, USA 7%, China 7%. Iron is the principal metal employed in modern industry. Iron is alloyed with other metals to form a variety of steels used in manufacturing, equipment, tools, containers, building, etc. The main minerals are magnetite Fe_3O_4, haematite Fe_2O_3, goethite $Fe_2O_3.H_2O$, siderite $FeCO_3$, pyrite FeS_2 and chamosite $(Mg,Fe,Al)_6(Si,Al)_4O_{14}(OH)_8$ (Klemic *et al.*, 1973b;

Peterson, E.C. and Collins, 1980). Natural iron oxides are also used as pigments for colouring glass, plastics, paper and ceramics. The main producers of these are the USA and India (Collins, 1980).

KAOLINITE

World 1979 production 20 365 200 t. Share of production: USA 34%, United Kingdom 23%, USSR 13%, France 5%, Czechoslovakia 3%. The uses result from the chemically inert nature of the clay. The major use is in paper manufacture. Glossy paper of magazines uses as much as 30% by weight of kaolinite. It is also used as a filler in natural and synthetic rubber, and as an extender in paints. An important use is as a filler in plastics. Historically its main use was in ceramics, and this is still a major use for the clay, especially in porcelain. Because of its chemical inertness and whiteness it has a wide range of other uses. Kaolinite has the composition $Si_4Al_4O_{10}(OH)_8$ (Hosterman, 1973; Patterson and Murray, 1975; Ampian, 1980).

KYANITE, SILLIMANITE, ANDALUSITE

(113)

World 1979 production, based on incomplete data, 239 758 t. Share of production: South Africa 63%, India 28%, France 5%, Spain 3%. The major use is in the manufacture of refractory mortars, cements, castables and plastic ramming mixes. A major use is in refractory linings for furnaces. Kyanite, sillimanite and andalusite are polymorphs with the formula Al_2SiO_5 (Espenshade, 1973; Bennett, P.J. and Castle, 1975; Potter, 1980b).

LEAD

(39, *47–59*, 92–93, 175–176, 183–185, *227–238*)

World 1979 production 3 512 700 t. Share of production: USA 16%, USSR 15%, Australia 13%, Canada 8%, Peru 5%. Lead is used mainly in storage batteries, leaded gasolene and the construction industry. Some of its main uses are in ammunition, as bearing metals, in brass and bronze, in cable coverings, as a casting metal and as a caulking material. It is also used as a pigment. The main ore mineral is galena PbS. Galena ore bodies are commonly near-surface altered to cerussite $PbCO_3$, anglesite $PbSO_4$ and pyromorphite $Pb_4(PbCl)(PO_4)_3$. Most ores of lead also contain iron, copper and zinc (Morris *et al.*, 1973; Rathjen and Rowland, 1980).

LIMESTONE

Carbonate rocks are perhaps the world's most widely used materials. Huge quantities are used in construction, in agriculture, and in the chemical and metallurgical industries. High-purity calcite limestone is used in the production of portland cement, made by fusing a finely ground mixture of limestone and aluminous and siliceous materials such as shale and silica sand. The world 1979 production of hydraulic cement was 868 716 440 t. Share of production: USSR 14%, Japan 10%, USA 9%, China 9%, West Germany 4% (Dikeou, 1980). Apart from cement manufacture, limestone is used in construction, as a flux in open-hearth furnaces, to manufacture lime for smelting ores and for the chemical industry, as refractory dolomite, as agricultural limestone and several other uses such as filler and whiting. The main ore minerals are calcite $CaCO_3$ and dolomite $CaMg(CO_3)_2$ (Hubbard and Ericksen, 1973; Ames, 1975; Boynton and Gutschick, 1975; Carr and Rooney, 1975; Pressler, 1980b). The world 1979 production of lime (quicklime, hydrated lime, dead-burned dolomite) was 112 972 290 t. Share of production: USSR 21%, USA 16%, West Germany 9%, Poland 7%, Japan 8%.

LITHIUM MINERALS

(*127–129*, 143)

World 1979 production (data from USA withheld) 74 410 t of metal. Share of production: USSR 65%, China 13%, Zimbabwe 12%, Brazil 4%, Namibia 3%. Spodumene $LiAlSi_2O_6$ is used in ceramics, glass, enamels, glass–ceramics, and manufacture of lithium salts. Petalite $LiAlSi_4O_{10}$ is used in ceramics, glasses, glazes and enamels. Lepidolite $K_2(Li,Al)_{5-6}(Si_{6-7}Al_{2-1}O_{20})(OH,F)_4$ is used in the manufacture of glasses. Amblygonite $LiAl(PO_4)(F,OH)$ is used for the manufacture of lithium chemicals; the other silicate is eucryptite $LiAlSiO_4$. In the metal form, lithium is used in deoxidation, degasification, and the manufacture of alloys, organic compounds, synthetic rubber, vitamins and batteries, as well as in the manufacture of a wide variety of chemicals and in the nuclear-energy industry (Norton, J.J., 1973; Kunasz, 1975; Singleton and Searles, 1980a). Caesium is produced as pollucite $(Cs,Na)AlSi_2O_6$ from pegmatite at Bikita, Zimbabwe.

MAGNESITE

(113)

World 1979 production 10 054 866 t. Share of production: USSR 21%, North Korea 17%, Greece 12%, China 11%, Austria 11%. Most refractory magnesia is used by the steel industry. The magnesia is converted to calcined products or chemicals. Dead-burned magnesite is used as a refractory material, used in the cement, glass, steel and copper industries. The advantage of magnesia as a refractory material is its great ability to resist basic slags at high temperatures. Caustic-calcined magnesia is used in cements for fire-proofing buildings. Other major uses are in the paper, fertilizer and chemical industries. Processed olivine (Mg_2SiO_4) is used in limited quantities for refractories and in fertilizer. Magnesite has the composition $MgCO_3$ (Bodenlos and Thayer, 1973; Wicken and Duncan, 1975; Petkof, 1980c).

MANGANESE

(14, *15*, *26–30*, 46, 74–75, *222–225*)

World 1979 production 24 455 397 t. Share of production: USSR 39%, South Africa 23%, Gabon 8%, India 8%, Australia 6%. Manganese is used in two ways in the manufacture of steel: as a scavenger in molten steel—it combines with sulphur and oxygen and removes them in the slag; and as an alloying material. It makes steel more resistant to shock or abrasion. It is also used in the manufacture of various chemicals. Pyrolusite is used in the manufacture of batteries. Other uses of manganese include the manufacture of paints and fertilizers and in solution mining. Manganese is contained in a large number of minerals. The main ore minerals are hausmannite Mn_3O_4, polianite MnO_2, pyrolusite MnO_2, cryptomelane $K(Mn^{2+},Mn^{4+})_8O_{16}$, psilomelane $BaMn^{2+}Mn^{4+}_8O_{18}.2H_2O$ (Dorr *et al.*, 1973; Jacoby, 1975; De Huff, 1980).

MERCURY

(*96–101*)

World 1979 production 6646 t (192 845 flasks). Share of production: USSR 29%, Spain 19%, USA 14%, Algeria 13%, China 10%. It is used in the manufacture of thermometers and switches. The biggest use is by chemical and allied industries for processes such as chlorine and caustic preparation, and in the manufacture of pigments, catalysts, pharmaceuticals and agricultural chemicals. It is also used in dental preparations and batteries. The major mineral of mercury is cinnabar HgS (Bailey, E.H. *et al.*, 1973; Drake, 1980a).

MICA

(127–129)

World 1979 production 239 352 t. Share of production: USA 60%, USSR 17%, India 13%, South Korea 4%. Sheet mica is hand-trimmed and stamped into shape for industrial use; this use is because of its excellent electrical and thermal insulating qualities. Major uses are as electrical insulations in capacitors and other electrical applications. Scrap and flake mica is reconstituted into mica paper or ground up. It is used in oil-drilling fluids, decorative finishes, and protective coatings for electrodes and lubricants. The main ore minerals are muscovite $K_2Al_4(Si_6Al_2O_{20})$-$(OH,F)_4$, phlogopite $K_2(Mg,Fe)_6(Si_6Al_2O_{20})(OH,F)_4$, biotite $K_2(Mg,Fe)_{6-4}(Fe,Al,Ti)_{0-2}(Si_{6-5}Al_{2-3}O_{20})O_{0-2}$-$(OH,F)_4$ and lepidolite $K_2(Li,Al)_{5-6}(Si_{6-7}Al_{2-1}O_{20})$-$(OH,F)_4$ (LeSure, 1973b; Petkof, 1975; Zlobik, 1980).

MOLYBDENUM

(168–175, *177–179*)

World 1979 production 102 992 t. Share of production: USA 59%, Canada 18%, Chile 12%, USSR 10%, China 2%. The principal uses are in the iron and steel industries, mainly to form high-speed, tool, stainless and low-alloy construction steels. It is also used in special alloys and castings. Other uses are in the manufacture of pigments, catalysts, agriculture, chemicals and lubricants. The main minerals are molybdenite MoS_2, powellite $Ca(Mo,W)O_4$, wulfenite $PbMoO_4$, ferrimolybdite $Fe_2(MoO_4)_3.8H_2O$ and jordisite, which is amorphous MoS_2 (King *et al.*, 1973; Kummer, 1980).

MONAZITE

(257–261, 272–275)

World 1979 production 23 686 t (data from the USA withheld). Share of production: Australia 56%, India 19%, Malaysia 13%, Brazil 12%. Monazite is the world's foremost source of thorium. Thorium is used to make nuclear fuel cells, incandescent gas-light mantles and in alloys. Mixed with yttrium, thorium makes incandescent lamps. Monazite has the composition $(Ce,La,Nd,Th)PO_4$. Other thorium ores are thorite $ThSiO_4$ and brannerite (a multiple oxide of Ti, U, Ca, Fe, Th and Y) (Staatz and Olson, 1973; Shannon, 1975; Kirk, 1980).

NATURAL GAS

(*276–300*)

World 1979 production $1\,677\,559 \times 10^6$ m^3. Natural gas is measured at a pressure of 14.73 psi ($= 1.13$ kg cm^{-2}) at 15°C under saturated water vapour conditions. Share of production: USA 35%, USSR 21%, Holland 6%, Canada 6%, Iran 3%, Saudi Arabia 3%. Natural gas is used for home heating and cooking, as an industrial and engine fuel, in electricity generation, in fuel cells, for oil-field use to increase oil production, in protein synthesis, as natural gas liquids and liquefied petroleum gas and in the manufacture of chemical products. Natural gas is formed of methane, ethane, propane, butane, pentane and nitrogen in various proportions, but methane is the major component (McCulloh, 1973; Tiratsoo, 1979).

NICKEL

(15, *23–26*, *102–104*, *202–204*, *209–210*, 245)

World 1979 mine production 704 323 t. Share of production: Canada 30%, New Caledonia 14%, USSR 19%, Australia 11%, Cuba 5%. The main uses of nickel are for making stainless steels and super alloys, in electroplating, and in the manufacture of high-nickel heat-resistant and corrosion-resistant alloys. It is also used in making alloys for permanent magnets and in the chemical industry. The two major nickel minerals are pentlandite $(Fe,Ni)_9S_8$ and garnierite $(Ni,Mg)_3Si_2O_5(OH)_4$. It is not a true mineral, but rather a mixture of nickel serpentine, nickel talc and possibly other silicates. Other minor nickel minerals are bravoite $(Fe,Ni)S_2$, millerite NiS, siegenite $(Ni,Co)_3S_4$, linnaeite $(Co,Fe,Ni)_3S_4$, gersdorffite NiAsS and niccolite NiAs (Cornwall, 1973; Matthews, 1980).

NIOBIUM AND TANTALUM

(*118, 123–125, 127–129*)

World 1979 production 25 093 t. Share of production: Brazil 76%, Canada 20%, Nigeria 3%, Australia 1%. Columbium is used as ferrocolumbium by the steel industry to give a high-strength low-alloy steel and in the automotive industry to replace carbon steel and save weight. Columbium is also used in pipeline and tubular steels, and in super-conducting alloys. Tantalum is used in the electronics industry for the manufacture of capacitors. The main mineral is pyrochlore $(Na,Ca)_2(Nb,Ta)_2O_6(O,OH,F)$, in which the amount of niobium is greater than the tantalum; it is isomorphous with microlite $(Na,Ca)_2(Ta,Nb)_2O_6(O,OH,F)$, in which the amount of tantalum is greater than the niobium. Another important mineral is columbite $(Fe,Mn)(Nb,Ta)_2O_6$, which is isomorphous with tantalite $(Fe,Mn)(Ta,Nb)_2O_6$ (Parker and Adams, 1973; Jones, T.S., 1980).

OIL SHALE

(*305*)

This fine-grained sedimentary rock contains insoluble organic matter and can yield substantial amounts of oil by destructive distillation. It occurs in large volumes in many parts of the world, and constitutes an enormous reserve of oil, if mining and distilling problems can be overcome (Culbertson and Pitman, 1973).

PEAT

(*302–303, 306*)

World 1979 production 201 773 000 t. Share of production: USSR 94%, Ireland 3%, West Germany 1%. Peat is used for agriculture and horticulture. In some countries it is used as a fuel. Peat consists of partly decayed vegetable matter, inorganic minerals and water in varying proportions. It occurs in peat bogs, and almost 80% of the world's peat lands are in the USSR (Cameron, C.C., 1973; Singleton and Searls, 1980b).

PERLITE

World 1979 production 1 424 897 t. Share of production: USA 42%, USSR 25%, Greece 10%, Hungary 7%, Italy 6%. Expanded perlite is used because of its low density, low thermal conductivity and high resistance to fire. It is primarily used as an aggregate in insulation boards, acoustical material in buildings, and cavity-fill insulation. Perlite is a glassy volcanic rock that will, on rapid controlled heating, expand into frothy material. Most deposits are as volcanic flows and domes of Cenozoic age (Chesterman, 1975; Meisinger, 1980b).

PETROLEUM (CRUDE)

(*276–300*)

World 1977 production 2 994 275 000 t. Share of production: USSR 18%, Saudi Arabia 15%, USA 14%, Iran 9%, Venezuela 4%. Petroleum is the major energy source of the western world, upon which it is so dependent that no satisfactory alternative has yet been developed. Crude petroleum is refined into the following products: gasolene, kerosene and jet fuel, liquefied gases, ethane and ethylene, distillate fuel oils, residual fuel oil, asphalt and road oil. Other products include petrochemical feedstocks, special naphthas, lubricants, waxes, petroleum coke, still gas, unfinished and finished oils. The finished oils form the basis of the petrochemical industry (McCulloh, 1973; Carleton *et al.*, 1975; United Nations, 1979).

PHOSPHATE ROCK

(*82–89, 114–119*)

World 1979 production 127 842 000 t. Share of production: USA 41%, USSR 21%, Morocco 15%, China 4%, Tunisia 3%. Fertilizers consume 90% of the world's rock production. Phosphorus is also used in the chemical industry and gypsum is produced as a by-product. The phosphate in both igneous and sedimentary deposits is contained in minerals of the apatite family, $Ca_5(PO_4)_3(OH,F,Cl)$ (Cathcart and Gulbrandsen, 1973; Emigh, 1975; Stowasser, 1980).

PLATINUM GROUP

(*105–110*)

World 1979 production 207.13 t (6 659 520 troy ounces). Share of production: South Africa 48%, USSR 48%, Canada 3%, Japan 0.5%, Colombia 0.2%. The uses are because of the catalytic properties, electrical conductivity, and resistance to chemical corrosion, heat and oxidation. The main uses are as electrical contacts in high-precision instruments and insoluble anodes for metallurgy, in the chemical industry as catalytic gauze for oxidation of ammonia, in petroleum refining as a catalyst to produce high-octane gasolene, in dental and medical devices, in decorative arts and jewellery, in spinnerettes used in the manufacture of glass and in laboratory crucibles. The metals of the group are platinum, palladium, iridium, osmium, rhodium and ruthenium. The first two are the most abundant. The platinum-group minerals occur in a variety of forms: ferroplatinum (Pt, Fe, Ir, Os, Ru, Rh, Pd, Cu, Ni), iridium–ruthenium–osmium, osmium, osmium–iridium, braggite (Pt, Pd, Ni)S, laurite $(Ru, Ir, Os)S_2$, erlichmanite OsS_2, cooperite PtS, mertieite $Pt_5(Sb,As)_2$ and sperrylite $PtAs_2$. In addition the platinum-group metals are contained in small amounts within pyrrhotite, pentlandite, pyrite and chalcopyrite. The metals may therefore be recovered during smelting of ores whose contents are extremely low (Page *et al.*, 1973; Jolly, 1980a).

POTASH

(89–91)

World 1979 production 26 345 000 t of K_2O equivalent. Share of production: USSR 32%, Canada 24%, East Germany 13%, West Germany 9%, USA 9%. Potash is primarily used as a fertilizer and in the chemical industry. The main minerals are sylvite KCl, carnallite $KCl.MgCl_2.6H_2O$, kainite $4KCl.4MgSO_4.11H_2O$, polyhalite $K_2SO_4.MgSO_4.2CaSO_4.2H_2O$, langbeinite $K_2SO_4.2MgSO_4$ and leonite $K_2SO_4.MgSO_4.4H_2O$. There are many minor minerals. Most ores contain admixed halite NaCl. For example, sylvinite is KCl + NaCl (Smith, G.I. *et al.*, 1973; Adams, S.S., 1975; Singleton and Searls, 1980c).

PUMICE AND VOLCANIC CINDERS

World 1979 production 17 686 000 t. Share of production: Italy 43%, USA 18%, West Germany 12%, Greece 12%, France 5%. The main use is in the construction industry in light-weight structural concrete, plaster aggregate and road-surfacing material. It is also used as an abrasive for polishing glass, metal, leather and stone. Pumice and cinders are products of explosive volcanic activity. The rock is generally basaltic and highly vesicular (Peterson, N.V. and Mason, 1975; Meisinger, 1980c).

RARE-EARTH ELEMENTS

(*118–119*, 127, 129, 242)

Rare earths have numerous uses, for example, in the manufacture of petroleum-cracking catalysts, glass and ceramics, colourizers for television tubes, phosphors, optical fibres, low-alloy steels and permanent magnets. Monazite $(Ce,La,Th,Y)PO_4$ and bastnaesite $CeFCO_3$ are the most abundant minerals. Xenotime YPO_4 is also mined. Apatite may incorporate re-earth elements (Adams, J.W. and Staatz, 1973; Moore, C.M., 1980).

RHENIUM

Rhenium is used as a petroleum-refining catalyst, in high-temperature thermocouples, and in the manufacture of X-ray tubes, electrical contacts and alloys. Porphyry-copper deposits are the major source of rhenium. One of its major occurrences is in molybdenite, which contains 100–2000 ppm rhenium (King, 1973; Alverson, 1980).

RUTILE

(272–275)

World 1979 production 361 143 t. Share of production: Australia 77%, South Africa 12%, Sri Lanka 4%, Sierra Leone 3%, India 2%. The uses of rutile (TiO_2) are similar to those of ilmenite (Klemic *et al.*, 1973c; Lynd and LeFond, 1975; Lynd and Hough, 1980).

SAND AND GRAVEL

The mining and use of sand and gravel represents one of the largest industries. It is used in the building industry as a portland cement concrete aggregate, and as asphaltic aggregate for paving roads. The main materials used are stream deposits, alluvial fans, glacial deposits, dredge tailings, and older geological formations and beach deposits. The sand and gravel may be monomineralic, or of lithic fragments (Yeend, 1973; Goldman and Reining, 1975; Tepordei, 1980).

SILICA SAND

High-purity silica sand is the main raw material for glass-making. Minor uses include for sand-blasting, in railroad traction and as blast-furnace linings. High-purity silica sands are the result of often-repeated cycles of weathering and transportation. The most extensive and purest are found in shallow seas, desert dunes and the flood plains of streams (Ketner, 1973; Murphy, T.D., 1975).

SILVER

(34–36, 168–175, *180–194*, 206–207, *228–238*)

World 1979 production 10 714 t (344 457 000 troy ounces). Share of production: Mexico 14%, USSR 14%, Canada 13%, USA 12%, Peru 9%. The main uses of silver are in photographic materials, in electrical and electronic products, in sterling ware, in electroplated ware, in brazing wares, and for other sundry uses such as in dentistry, medicine, as catalysts, as a bearing and for jewellery. Of the world's total production of silver, about 75% is a by-product of base-metal ores. In these deposits, silver is found within galena, chalcopyrite and sphalerite. The galena is commonly highly argentiferous (up to 0.2% content). The following are some silver minerals: silver Ag, electrum (Ag,Au), andorite $PbAgSb_3S_6$, acanthite Ag_2S, argentite Ag_2S, pearceite $Ag_{16}As_2S_{11}$, proustite Ag_3AsS_3, polybasite $Ag_{16}Sb_2S_{11}$, pyrargyrite Ag_3SbS_3, sylvanite $Ag_2Au_2Te_8$, argentiferous tennantite $(Cu,Fe,Ag)_{12}As_4S_{13}$, argentiferous tetrahedrite $(Cu,Fe,Ag)_{12}Sb_4S_{13}$ and argentian plumbojarosite $(Pb,Ag)Fe_6(SO_4)_4(OH)_{12}$. Silver is an important by-product of both porphyry-copper and massive sulphide and other base-metal deposits, but it also occurs in epithermal vein deposits (Heyl *et al.*, 1973; Drake, 1980b).

SODIUM CARBONATE

World 1979 production 28 433 543 t. Share of production: USA 26%, USSR 19%, China 6%, United Kingdom 6%, Canada 5%. However the production is both natural and synthetic. The world 1977 production of natural sodium carbonate was 5 780 000 t, of which the USA produced 98% and Kenya 2%. Soda ash is a white crystalline hygroscopic powder of composition Na_2CO_3. The market is dominated by the glass industry. Inorganic chemicals use most of the remainder. Other uses are in detergents, in the manufacture of caustic soda, and in metallurgy. The main minerals are predominantly trona $Na_2CO_3.NaHCO_3.2H_2O$, natron $Na_2CO_3.10H_2O$ and thermonatrite $Na_2CO_3.H_2O$ (Smith, G.I. *et al.*, 1973; Mannion, 1975; Kostick, 1980).

SODIUM SULPHATE

Natural 1979 production 1 746 882 t. Share of production: USA 31%, Canada 23%, USSR 17%, Mexico 12%, Spain 8%. The principal use is in the manufacture of kraft paper pulp. Other important uses are in the glass-manufacturing industry, and in detergents. The two important minerals are mirabilite $Na_2SO_4.10H_2O$ and thenardite Na_2SO_4 (Smith, G.I. *et al.*, 1973; Weisman and Tandy, 1975; Kostick, 1980).

STONE

Stone is used in the construction industry in two forms, crushed aggregate and dimension stone, either polished or unpolished and cut into slabs for decorative use in buildings and memorials. A large variety of rocks are suitable, but the most commonly used are limestone, marble, granite, gabbro, basalt, sandstone and slate (Laurence, 1973; Schenck and Torries, 1975; Singleton, 1980).

STRONTIUM

World 1979 production 78 900 t of strontium minerals. Share of production: Mexico 42%, Turkey 23%, Iran 10%, Spain 9%, Algeria 7%. The main uses are in glass manufacture, in electronics, in pyrotechnics and in metal refining. The major mineral is celestite $SrSO_4$ and strontianite is of second importance $SrCO_3$ (Fulton, 1975).

SULPHUR

(11–14, *37–39*, 46–50, 53–58, *94–95*)

World 1979 production 54 834 000 t (26% of total by Frasch process; 7% native; 18% from pyrite). Share of production: USA 21%, USSR 19%, Poland 10%, Canada 14%, Japan 5%. Most sulphur is made into sulphuric acid, which is mainly used to make fertilizers, explosives, lubricants, rayon, iron and steel, paints and enamels, and in various chemicals and pharmaceuticals. Sulphur is used directly in paper making, cellophane, rubber-processing chemicals, insecticides and in bleaching chemicals. Elemental sulphur is produced from cap rocks of salt domes, evaporite basins and volcanic deposits. Much is mined by the Frasch process in which hot water and air are pumped down a well to drive sulphur and air upwards. Elemental sulphur is also a by-product of the petroleum-refining industry. Sulphur is also recovered from pyrite FeS_2, marcasite FeS_2 and pyrrhotite Fe_nS_{n+1} from massive sulphide deposits (Bodenlos, 1973; Gittinger, 1975; Shelton, 1980).

TALC AND PYROPHYLLITE

World 1979 production 6 212 839 t. Share of production: Japan 24%, USA 19%, South Korea 11%, USSR 8%, France 5%. Although not related, pyrophyllite and talc have similar physical properties, so share a common use. The properties that are important are the flaky nature and softness. Ground talc is used in the manufacture of ceramics, paint, plastics, paper, cosmetics and rubber. Pyrophyllite is predominantly used in refractories, but can be used for the same end-products as talc. The main ore minerals are talc $Mg_6(Si_8O_{20})(OH)_4$ and pyrophyllite $Al_4(Si_8O_{20})(OH)_4$ (Brown, C.E., 1973; Roe, 1975; Clifton, 1980b).

TIN

(4, *112*, *123–125*, 127–129, *134–139*, *141–145*, *148–150*, *155–156*, 178, *188–193*, *257–261*)

World 1979 production 256 002 t. Share of production: Malaysia 25%, USSR 14%, Bolivia 14%, Thailand 10%, Indonesia 10%. The major uses of tin are in solder and tinplate. It is also used in the manufacture of bearing alloys, bronze, chemicals and coatings. Cassiterite SnO_2 is the major ore mineral. Small production may come from stannite Cu_2FeSnS_4, cylindrite $Pb_2Sn_4Sb_2S_{14}$ and teallite $PbSnS_2$ (Sainsbury and Reed, 1973; Carlin and Harris, 1980).

TUNGSTEN

(123–125, 135, *141*, *144–147*, *150–155*)

World 1979 production 46 013 t. Share of production: China 21%, USSR 19%, Bolivia 7%, South Korea 6%, Thailand 6%. The major use of tungsten is in cutting and wear-resistant materials, primarily tungsten carbide. It is also used in the manufacture of mill products, hard-facing steels, super alloys and chemicals. The main occurrence is as wolframite, which forms an isomorphous series between huebnerite $MnWO_4$ and ferberite $FeWO_4$. Scheelite $CaWO_4$ is the other major ore mineral (Hobbs, S.W. and Elliot, 1973; Stafford, 1980).

URANIUM

(*75–81*, 95–96, 119, *160–167*, *238–248*, *254–256*, *264–272*, *310–316*)

World 1977 production, excluding the USSR, eastern Europe and China, 28 615 t. Share of production: USA 39%, Canada 21%, South Africa 23%, France 8%, Niger 6%. Uranium is an important energy resource. It is mainly used as the fuel for nuclear-powered electricity generators and reactors. It has important military uses. It is likely to be a major fuel of the future if the disposal problems are overcome. The major ore minerals are uraninite UO_2, pitchblende (the same material that occurs as rounded botryoidal masses), coffinite $U(SiO_4)_{1-x}(OH)_{4x}$, another important primary mineral, brannerite $(U,Ca,Ce)(Ti,Fe)_2O_6$ and davidite

$A_6B_{15}(O,OH)_{36}$, where A = U, Fe, rare earths, Ca, Zr, Th, and B = Ti, Fe, V and Cr. There is a large number of brightly coloured secondary minerals such as carnotite $K_2(UO_2)_2(VO_4)_2$, tyuyamunite $Ca(UO_2)_2(VO_4)_2$, autunite $(H,Na,K)_2(UO_2)_2(PO_4)_2$, torbernite Cu $(UO_2)_2(PO_4)_2$ and uranophane $Ca(UO_2)_2(SiO_3OH)_2$ (Finch *et al.*, 1973; United Nations, 1979).

VANADIUM

(92–93, *105–112*)

World 1979 production 37 568 t. Share of production: South Africa 37%, USSR 30%, USA 20%, Finland 6%. Vanadium is usually added to steel to toughen and strengthen it and to control its grain size. It is also used in cast irons and super alloys. Vanadium rarely forms ore minerals. It is concentrated in magmatic magnetite deposits, especially those that are titaniferous, commonly ranging from 1000 ppm to 5000 ppm in hese deposits. Carnotite $K_2(UO_2)_2(VO_4)_2.3H_2O$ is a secondary mineral formed in near-surface deposits of vanadium and uranium. Vanadinite $Pb_5(VO_4)_3Cl$ is a secondary mineral formed in the oxidized parts of base-metal deposits. Trace amounts of vanadium accumulate in crude petroleum and bauxite (Fischer, 1973; Morgan, 1980).

VERMICULITE

(110)

World 1979 production 550 549 t. Share of production: USA 57%, South Africa 35%, Brazil 4%, Japan 3%, Argentina 1%. Heated and expanded vermiculite is mainly used in the construction industry for insulation and acoustical walls. Fertilizer and agricultural chemicals are often carried on vermiculite. The composition of vermiculite is $(H_2O)(Mg,Ca,K)(Al,Fe,Mg)(SiAl, Fe)_4O_{10}(OH)_2$ (Strand, 1975; Meisinger, 1980d).

ZEOLITES

Natural zeolites in the form of zeolitic tuff are used in the manufacture of pozzolanic cements and concrete and as a light-weight aggregate. High-brightness zeolite ore may be used as a paper filler. Important uses are in large-scale ion exchange processes, for example, to separate radioactive strontium and caesium from waste streams of nuclear facilities and also for sewage treatment. Zeolites are used for producing high-purity oxygen for smelting operations. Zeolites may be used as a dietary supplement for farm animals, as a fertilizer additive to the soil, and as a molecular sieve in the purification of gaseous hydrocarbons and the preparation of catalysts in the petroleum-refining industry. Zeolites occur concentrated in tuffaceous sedimentary rocks. There are at least 20 different species reported; for example, analcime $NaAlSi_2O_6.H_2O$, chabazite $(Ca,Na_2)Al_2Si_4O_{12}.6H_2O$, harmotone $(Ba,Na_2)_2Al_4Si_{12}O_{32}.12H_2O$ and stilbite $(Ca,Na_2)_4Al_8Si_{28}O_{72}.28H_2O$. The zeolites in sedimentary rocks are fine grained but voluminous (Mumpton, 1975; Olson, R.H., 1975; Sheppard, R.A., 1973, 1975).

ZINC

(*34–40*, 46, 47–49, *50–53*, *56–59*, 92–93, 175–176, 183–185, 207–210, *227–238*)

World 1979 production from mines 5 977 800 t. Share of production: Canada 21%, USSR 12%, Australia 8%, Peru 7%, USA 7%. Most commercial zinc is produced as metallic slab zinc. Zinc is utilized chiefly in the automobile, household-appliance and hardware industries. The three major uses are for zinc-based alloy die castings, for galvanizing iron and steel and in the manufacture of copper-based alloys and brass. Zinc oxide is used in the manufacture of chemicals, in the photocopying, fabrics and plastics industries, and generally as a pigment. There are several zinc ore minerals, the main ones being sphalerite ZnS, wurtzite ZnS, zincite ZnO, franklinite $(Fe,Zn,Mn)(Fe,Mn)_2O_4$, smithsonite $ZnCO_3$, hydrozincite $Zn_5(OH)_6(CO_3)_2$, willemite Zn_2SiO_4 and hemimorphite $Zn_4(OH)_2Si_2O_7.H_2O$ (Wedow *et al.*, 1973; Cammarota *et al.*, 1980).

ZIRCONIUM

(*272–275*)

World 1979 production of zirconium concentrate 543 865 t. Share of production: Australia 94%, South Africa 3%, India 2%, Brazil 1%, Sri Lanka 1%. Because of their refractory nature, zirconium sands are used as mould material in foundries. Zircon sands are also used in refractory bricks. Zirconium dioxide is produced by reacting zircon sand with dolomite at high temperature; the zirconia is then converted to metal zirconium. Most of the metal is used in water-cooled nuclear reactors for fuel cladding and pressure tubes. It is also used for corrosion-resistant applications in the chemical industry and in the manufacture of photographic flashbulbs. The major minerals in zirconium sands are zircon $ZrSiO_4$, baddeleyite ZrO_2 and eudialyte–eucolite $(Ca,Na)_5Zr_2Si_6(O,OH,Cl)_{20}$(Klemic *et al.*, 1973a; Klemic, 1975; Lynd, 1980).

References

Abel, M.K., Buchan, R., Coats, C.J.A. and Penstone, M.E. (1979). 'Copper mineralization in the Footwall Complex, Strathcona Mine, Sudbury, Ontario', *Can. Mineral.* Vol. 17, 275–285

Absalom, S.T. (1980a). 'Boron', *Minerals Yearbook 1978–79, Vol. 1. Metals and Minerals*, pp. 119–129. Bureau of Mines. Washington DC; US Government Printing Office

Absalom, S.T. (1980b). 'Bromine', *Minerals Yearbook 1978–79, Vol. 1. Metals and Minerals*, pp. 131–137. Bureau of Mines. Washington DC; US Government Printing Office

Adam, J.W.H. (1960). 'On the geology of the primary tin-ore deposits in the sedimentary formations of Billiton', *Geol. Mijnbouw.* Vol. 22, 405–426

Adams, J.W. and Staatz, M.H. (1973). 'Rare-earth elements', *United States Mineral Resources, US Geol. Surv. Prof. Paper 820*, pp. 547–556. Ed. D.A. Brobst and W.P. Pratt. Washington DC; US Geological Survey

Adams, S.S. (1975). 'Potash', *Industrial Minerals and Rocks (Nonmetallics Other Than Fuels)*, pp. 963–990. Ed. S.J. LeFond. New York; Amer. Inst. Min. Metall. Petrol. Engineers

Ahfeld, F. (1967). 'Metallogenic epochs and provinces of Bolivia', *Mineral. Deposita.* Vol. 2, 291–311

Albers, J.P. and Kleinhampl, F.J. (1970). *Spatial relation of mineral deposits to Tertiary volcanic centers in Nevada. US Geol. Surv. Prof. Paper 700-C*, pp. C1–C10. Washington DC; US Geological Survey

Aleva, G.J.J. (1973). 'Aspects of the historical and physical geology of the Sunda Shelf essential to the exploration of submarine tin placers', *Geol. Mijnbouw.* Vol. 52, 79–91

Alexander, J.B. and Flinter, B.H. (1965). 'A note on varlamoffite and associated minerals from the Batang Padang district, Perak, Malaya, Malaysia', *Mineral. Mag.* Vol. 35, 622–627

Alexandrov, E.A. (1973). 'The Precambrian banded iron-formations of the Soviet Union', *Econ. Geol.* Vol. 68, 1035–1062

Alexiev, E.T. (1970). 'Rare earth elements in younger granites of northern Nigeria and the Camerouns and their genetic significance', *Geochem. Int.* Vol. 7, 127–132

Alverson, L.J. (1980). 'Rhenium', *Minerals Yearbook 1978–79, Vol. 1, Metals and Minerals*, pp. 743–749. Bureau of Mines. Washington DC; US Government Printing Office

Ames, J.A. (1975). 'Construction materials. Cement and cement raw materials', *Industrial Minerals and Rocks (Nonmetallics Other Than Fuels)*, pp. 129–155. Ed. S. LeFond. New York; Amer. Inst. Min. Metall. Petrol. Engineers

Ampian, S.G. (1980). 'Clays', *Minerals Yearbook 1978–79, Vol. 1. Metals and Minerals*, pp. 207–247. Bureau of Mines. Washington DC; US Government Printing Office

Amstutz, G.C. and Bubenicek, L. (1967). 'Diagenesis in sedimentary mineral deposits', *Diagenesis in Sediments*, pp. 417–475. Ed. G. Larsen and G.V. Chilingar. Amsterdam, Elsevier

Amstutz, G.C. and Park, W.C. (1967). 'Stylolites of diagenetic age and their role in the interpretation of the southern Illinois fluorspar deposits', *Mineral. Deposita.* Vol. 2, 44–53

Anderson, G.M. (1978). 'Basinal brines and Mississippi Valley-type ore deposits', *Episodes.* No. 2, 15–19

Anderson, O.L. (1979). 'The role of fracture dynamics in kimberlite pipe formation', *Kimberlites, Diatremes and Diamonds: Their Geology, Petrology and Geochemistry. Proc. 2nd International Kimberlite Conference, Vol. 1*, pp. 344–353. Ed. F.R. Boyd and H.O.A. Meyer. Washington DC; American Geophysical Union

Anderson, R.N. and Halunen, A.J. Jr. (1974). 'Implications of heat flow for metallogenesis in the Bauer Deep', *Nature.* Vol. 251, 473–475

Anderson, S.B. and Swinehart, R.P. (1979). 'Potash salts in the Williston Basin, USA', *Econ. Geol.* Vol. 74, 458–476

Anger, G., Nielsen, H., Puchelt, H. and Ricke, W. (1966). 'Sulfur isotopes in the Rammelsberg ore deposit (Germany)', *Econ. Geol.* Vol. 61, 511–536

Anhaeusser, C.R. (1976a). 'Archean metallogeny in southern Africa', *Econ. Geol.* Vol. 71, 16–43

Anhaeusser, C.R. (1976b). 'The nature and distribution of Archean gold mineralization in southern Africa', *Mineral. Sci. Eng.* Vol. 8, 46–84

Anthony, P.J. (1975). 'Nabarlek uranium deposit', *Economic Geology of Australia and Papua New Guinea. Vol. 1, Metals*, pp. 304–308. Ed. C.L. Knight. Parkville, Victoria; Australasian Inst. Min. Metall.

Appleyard, F.C. (1975). 'Gypsum and anhydrite', *Industrial Minerals and Rocks (Nonmetallics Other Than Fuels)*, pp. 707–723. Ed. S. LeFond. New York; Amer. Inst. Min. Metall. Petrol. Engineers

Armands, G. (1972a). 'Geochemical studies of uranium, molybdenum and vanadium in a Swedish alum shale', *Stokh. Contrib. Geol.* Vol. 27, 1–148

Armands, G. (1972b). 'Caledonian geology and uranium-bearing strata in the Tåsjö Lake area, Sweden', *Geol. För. Stockh. Forh.* Vol. 94, 321–345

Armstrong, F.C. (1974). 'Uranium resources of the future—'porphyry' uranium deposits', *Formation of Uranium Ore Deposits. Proc. Symposium, Athens, 1974*, pp. 625–635. Vienna; International Atomic Energy Agency

Armstrong, R.L., Taubeneck, W.H. and Hales, P.O. (1977). 'Rb–Sr and K–Ar geochronometry of Mesozoic granite rocks and their Sr isotopic composition, Oregon, Washington and Idaho', *Geol. Soc. Amer. Bull.* Vol. 88, 397–411

Arndt, N.T., Francis, D. and Hynes, A.J. (1979). 'The field characteristics and petrology of Archean and Proterozoic komatiites', *Can. Mineral.* Vol. 17, 147–163

Arrykul, S., Ishihara, S., Bungbrakearti, N., Busaracome, P. and Sawata, H. (1978). 'Some characteristics of primary cassiterite and wolframite deposits in southern Thailand', *Proc. 3rd Regional Conference on Geology and Mineral Re-*

sources of Southeast Asia, pp. 415–417. Ed. Prinya Nutalaya. Bangkok; Asian Institute of Technology

Arth, J.C. (1976). 'A model for the origin of the early Precambrian greenstone–granite complex of north-eastern Minnesota', *The Early History of the Earth*, pp. 299–302. Ed. B.F. Windley. New York; John Wiley

Ashley, R.P. (1974). 'Goldfield mining district, Nevada', *Rep. Nevada Bur. Mines Geol.* No. 19, 49–66

Asnachinda, P. and Pitragool, S. (1978). 'Review of non-metallic deposits of Thailand', *Proc. 3rd Regional Conference on Geology and Mineral Resources of Southeast Asia*, pp. 795–804. Ed. Prinya Nutalaya. Bangkok; Asian Institute of Technology

Audley-Charles, M.G., Curray, J.R. and Evans, G. (1977). 'Location of major deltas', *Geology*. Vol. 5, 341–344

Averitt, P. (1973). 'Coal', *United States Mineral Resources, US Geol. Surv. Prof. Paper 820*, pp. 133–142. Ed. D.A. Brobst and W.P. Pratt. Washington DC., US Geological Survey

Badham, J.P.N. (1976). 'Orogenesis and metallogenesis with reference to the silver–nickel, cobalt arsenide ore association', *Metallogeny and Plate Tectonics, Geol. Assoc. Canada Special Paper 14*, pp. 559–571. Ed. D.F. Strong. Toronto; Geol. Assoc. Canada

Badham, J.P.N. (1978). 'Slumped sulphide deposits at Avoca, Ireland, and their significance', *Inst. Min. Metall. Trans.* Vol. 87, B21–B26

Bailey, E.H. and Everhart, D.L. (1964). *Geology and Quicksilver Deposits of the New Almaden District, Santa Clara County, California. US Geol. Surv. Prof. Paper 360.* Washington DC, US Geological Survey

Bailey, E.H., Clark, A.L. and Smith, R.M. (1973). 'Mercury', *United States Mineral Resources, US Geol. Surv. Prof. Paper 820*, pp. 401–414. Ed. D.A. Brobst and W.P. Pratt. Washington DC; US Geological Survey

Bailey, R.V. and Childers, M.O. (1977). *Applied Mineral Exploration with Special Reference to Uranium*, pp. 25–96. Boulder, Colorado; Westview Press

Baker, E.G. (1962). 'Distribution of hydrocarbons in petroleum', *Amer. Assoc. Petrol. Geol. Bull.* Vol. 46, 76–84

Baker, G.F.U. (1973). 'Bauxite deposits of the Darling Range, Western Australia', *Metallogenic Provinces and Mineral Deposits in the Southwestern Pacific, Bur. Min. Res. Geol. Geophys. Bull.* No. 141, pp. 223, 224. Ed. N.H. Fisher. Canberra; Bur. Min. Res. Geol. Geophys.

Baker, G.F.U. (1975). 'Darling Range bauxite deposits, W.A.', *Economic Geology of Australia and Papua New Guinea. Vol. 1. Metals*, pp. 980–986. Ed. C.L. Knight. Parkville, Victoria; Australasian Inst. Min. Metall.

Baldwin, J.T., Swain, H.D. and Clark, G.H. (1978). 'Geology and grade distribution of the Panguna Porphyry Copper Deposit, Bougainville, Papua New Guinea', *Econ. Geol.*, Vol. 73, 690–702

Bally, A.W. (1975). 'A geodynamic scenario for hydrocarbon occurrences', *Ninth World Petroleum Congress Proc. Vol. 2. Geology*, pp. 33–44. London; Applied Science Publishers

Bárdossy, G. (1979). 'Growing significance of bauxites', *Episodes*. Vol. 2, 22–25

Barker, C. (1974). 'Pyrolysis techniques for source rock evaluation', *Amer. Assoc. Petrol. Geol. Bull.* Vol. 58, 2349–2361.

Barker, C. (1977). 'Aqueous solubility of petroleum as applied to its origin and primary migration: a discussion', *Amer. Assoc. Petrol. Geol.* Vol. 61, 2146–2149

Barker, J.M., Cochran, D.E. and Semrad, R. (1979). 'Economic geology of the Mishraq native sulfur deposit, Northern Iraq', *Econ. Geol.* Vol. 74, 484–495

Barnes, H.L. (ed.) (1979). *Geochemistry of HydroRhermal Ore Deposits*, 2nd edn. New York; Wiley–Interscience

Barrett, F.M., Binns, R.A., Groves, D.I., Marston, R.J. and McQueen, K.G. (1977). 'Structural history and metamorphic modification of Archean volcanic-type nickel deposits, Yilgarn block, Western Australia', *Econ. Geol.* Vol. 72, 1195–1223

Bartholome, P., Evrard, P., Katekesha, F., Lopez-Ruiz, J. and Ngongo, M. (1976). 'Diagenetic ore-forming processes at Kamoto, Katanga, Republic of the Congo', *Ore in Sediments*, pp. 21–41. Ed. G.C. Amstutz and A.J. Bernard. Berlin; Springer-Verlag

Baskin, G.D. (1980a). 'Abrasive materials', *Minerals Yearbook 1978–79, Vol. 1. Metals and Minerals*, pp. 27–41. Bureau of Mines. Washington DC; US Government Printing Office

Baskin, G.D. (1980b). 'Gem stones', *Minerals Yearbook 1978–79, Vol. 1. Metals and Minerals*, pp. 363–375. Bureau of Mines. Washington DC; US Government Printing Office

Batchelor, B.C. (1979a). 'Geological characteristics of certain coastal and offshore placers as essential guides for tin exploration in Sundaland, Southeast Asia', *Geol. Soc. Malaysia Bull.* Vol. 11, 283–313

Batchelor, B.C. (1979b). Discontinuously rising Late Cenozoic eustatic sea-levels, with special reference to Sundaland, Southeast Asia', *Geol. Mijnbouw*. Vol. 58, 1–20

Bateman, A.M. (1950). *Economic Mineral Deposits*. New York; John Wiley

Bates, R.L. and Jackson, J.A. (1980). *Glossary of Geology*, 2nd edn. Falls Church, Virginia; Amer. Geol. Institute

Battaerd, H.A.J. and Evans, D.G. (1979). 'An alternative representation of coal composition data', *Fuel*. Vol. 58, 105–108

Baumgardner, L.H. and Hough, R.A. (1980). 'Bauxite and alumina', *Minerals Yearbook 1978–79, Vol. 1. Metals and Minerals*, pp. 95–110. Bureau of Mines. Washington DC; US Government Printing Office

Bayley, R.W. and James, H.L. (1973). 'Precambrian iron-formations of the United States', *Econ. Geol.* Vol. 68, 934–959

Beales, F.W. and Jackson, S.A. (1968). 'Pine Point—a stratigraphic approach', *Can. Inst. Min. Metall. Trans.* Vol. 61, No. 675, 867–878

Beales, F.W. and Onasick, E.P. (1970). 'Stratigraphic habitat of Mississippi-Valley type ore bodies', *Trans. Inst. Min. Metall.* Vol. 79, B145–B154

Bean, J.H. (1969). *The Iron Ore Deposits of West Malaysia. Geol. Surv. West Malaysia Econ. Bull. 2*, Ipoh, West Malaysia; Geological Survey of West Malaysia

Bell, R.T. (1978). 'Uranium in black shales—a review', *Uranium Deposits, Their Mineralogy and Origin. Short Course Handbook 3*, pp. 307–329. Ed. M.M. Kimberley. Ottawa; Miner. Assoc. Canada

Bemmelen, R.W. Van (1970a). *The Geology of Indonesia. Vol. 1A. General Geology of Indonesia*. The Hague; Martinus Nijhoff

Bemmelen, R.W. Van (1970b). *The Geology of Indonesia. Vol. 2. Economic Geology*, 2nd edn. The Hague; Martinus Nijhoff

Bennett, E.M. (1965). 'Lead–zinc–silver and copper deposit of Mt. Isa', *Geology of Australian Ore Deposits*, Eighth Commonwealth Min. and Met. Congr. Vol. 1, pp. 233–246. Ed. J. McAndrew. 2nd edn. Melbourne; Australasian Inst. Min. Metal

Bennett, P.J. and Castle, J.E. (1975). 'Kyanite and related minerals', *Industrial Minerals and Rocks (Nonmetallics Other Than Fuels)*, pp. 729–736. New York; Amer. Inst. Min. Metall. Petrol. Engineers

Berg, R.R. (1975). 'Capillary pressures in stratigraphic traps', *Amer. Assoc. Petrol. Geol. Bull.* Vol. 59, 939–956

Berning, J., Cooke, R., Hiemstra, S.A. and Hoffman, U. (1976). 'The Rössing uranium deposit, South West Africa',

Econ. Geol. Vol. 71, 351–368

Beukes, N.J. (1973). 'Precambrian iron-formations of southern Africa', *Econ. Geol.* Vol. 68, 960–1004

Bichan, R. (1969). 'Origin of chromite seams in the Hartley Complex of the Great Dyke, Rhodesia', *Econ. Geol. Monograph 4*, pp. 95–113. Ed. H.D.B. Wilson. Lancaster, Penn.; Economic Geology Publishing Company

Bichan, R. (1970). 'The evolution and structural setting of the Great Dyke, Rhodesia', *African Magmatism and Tectonics*, pp. 51–71. Ed. T.N. Clifford and I.G. Gass. Edinburgh; Oliver and Boyd

Bignell, R.D., Cronan, D.S. and Tooms, J.S. (1976). 'Red Sea metalliferous brine precipitates', *Metallogeny and Plate Tectonics, Geol. Assoc. Can. Special Paper 14*, pp. 147–179. Ed. D.F. Strong. Toronto; Geol. Assoc. Canada

Binns, R.A., Gunthorpe, R.J. and Groves, D.I. (1976). 'Metamorphic patterns and development of greenstone belts in the eastern Yilgarn block, Western Australia', *The Early History of the Earth*, pp. 303–313. Ed. B.F. Windley. London; John Wiley

Biq, C. (1974). 'Metallogeny in Taiwan: a plate tectonic approach', *Bull. Geol. Surv. Taiwan*. Vol. 24 (Oct.), 139–155

Bird, J.M. (ed.) (1980). *Plate Tectonics: Selected Papers from Publications of the American Geophysical Union*, 2nd edn. Washington DC; American Geophysical Union

Bird, J.M. and Isacks, B. (eds.) (1972). *Plate Tectonics: Selected Papers from the Journal of Geophysical Research*. Washington DC; American Geophysical Union

Bischoff, J.L. (1969). 'Red Sea geothermal brine deposits: their mineralogy, chemistry, and genesis', *Hot Brines and Recent Heavy Metal Deposits in the Red Sea*, pp. 368–401. Ed. E.T. Degens and D.A. Ross. New York; Springer-Verlag

Blockley, J.G. (1975). 'Hamersley Basin—mineralization', *Economic Geology of Australia and Papua New Guinea. Vol. 1. Metals*, pp. 413–415. Ed. C.L. Knight. Melbourne; Australasian Inst. Min. Metall.

Bloomer, A.G. and Nixon, P.H. (1973). 'The geology of the Letseng-la-terae kimberlite pipes', *Lesotho Kimberlites*, pp. 20–38. Ed. P.H. Nixon. Maseru, Lesotho; Lesotho National Development Corporation

Bodenlos, A.J. (1973). 'Sulfur', *United States Mineral Resources, US Geol. Surv. Prof. Paper 820*, pp. 605–618. Ed. D.A. Brobst and W.P. Pratt. Washington DC; US Geological Survey

Bodenlos, A.J. and Thayer, T.P. (1973). 'Magnesian refractories', *United States Mineral Resources, US Geol. Surv. Prof. Paper 820*, pp. 379–384. Ed. D.A. Brobst and W.P. Pratt. Washington DC; US Geological Survey

Borchert, H. (1960). 'Genesis of marine sedimentary ores', *Trans. Inst. Min. Metall.* Vol. 69, B261–B279

Borodaevskaya, M.B. and Rozhkov, I.S. (1977). 'Deposits of gold', *Ore Deposits of the USSR. Vol. III*, pp. 3–81. Ed. V.I. Smirnov. Translated by D.A. Brown. London; Pitman

Bostrom, R.C. (1978). 'Motion of the Pacific Plate and formation of marginal basins: asymmetric flow induction', *J. Phys. Earth*. Vol. 26 Supplement, S103–S122

Bostwick, J.M. (1975). 'Bituminous materials', *Industrial Minerals and Rocks (Nonmetallics Other Than Fuels)*, 4th edn, pp. 463–471. Ed. S.J. LeFond. New York; Amer. Inst. Min. Metall. Petrol. Engineers

Both, R.A. and Rutland, R.W.R. (1976). 'The problem of identifying and interpreting stratiform ore bodies in highly metamorphosed terrains: the Broken Hill example', *Handbook of Strata-bound and Stratiform Ore Deposits. Vol. 4*, pp. 260–325. Ed. K.H. Wolf. Amsterdam; Elsevier

Bott, M.H.P. and Masson-Smith, D. (1957). 'The geological interpretation of a gravity survey of the Alston Block and the Durham coalfield', *Quart. J. Geol. Soc. London.* Vol. 113, 93–117

Bowden, P. (1970). 'Origin of the younger granites of northern Nigeria', *Contrib. Mineral. Petrol.* Vol. 25, 153–162

Bowden, P. and Turner, D.C. (1974). 'Peralkaline and associated ring-complexes in the Nigeria–Niger province, west Africa', *Alkaline Rocks*, pp. 330–351. Ed. J. Sørensen. New York; John Wiley

Bowden, P., Van Breemen, O.J., Hutchinson, J. and Turner, D.C. (1976). 'Palaeozoic and Mesozoic age trends for some ring complexes in Niger and Nigeria', *Nature*. Vol. 259, 297–299

Bowen, R. and Gunatilaka, A. (1977). *Copper: its Geology and Economics.* London; Applied Science Publishers

Boyle, R.W. and Dass, A.S. (1971). 'The silver-arsenide deposits of the Cobalt–Gowganda region. The origin of the native silver veins at Cobalt, Ontario', *Can. Mineral.* Vol. 11, No. 1, 414–417

Boyle, R.W., Wanless, R.K. and Stevens, R.D. (1976). 'Sulfur isotope investigations of the barite, manganese, and lead–zinc–copper–silver deposits of the Walton–Cheverie area, Nova Scotia, Canada', *Econ. Geol.* Vol. 71, 749–762

Boynton, R.S. and Gutschick, K.A. (1975). 'Lime', *Industrial Minerals and Rocks (Nonmetallics Other Than Fuels)*, pp. 737–756. Ed. S. LeFond. New York; Amer. Inst. Min. Metall. Petrol. Engineers

Bradley, W.H. (1973). 'Oil shale formed in desert environment: Green River Formation, Wyoming', *Geol. Soc. Amer. Bull.* Vol. 84, 1121–1124

Brinkmann, R. (1976). *Geology of Turkey.* Amsterdam; Elsevier

Britten, R.A., Smyth, M., Bennett, A.J.R. and Shibaoka, M. (1975). 'Environmental interpretations of Gondwana Coal Measure sequences in the Sydney Basin of New South Wales', *Gondwana Geology*, pp. 233–247. Ed. K.S.W. Campbell. Canberra; Australian National University Press

Brobst, D.A. (1958). *Barite Resources of the United States. US Geol. Surv. Bull. 1072B.* Washington DC; US Geological Survey

Brobst, D.A. (1973). 'Barite', *United States Mineral Resources, US Geol. Surv. Prof. Paper 820*, pp. 75–84. Ed. D.A. Brobst and W.P. Pratt. Washington DC; US Geological Survey

Brobst, D.A. (1975). 'Barium minerals', *Industrial Minerals and Rocks (Nonmetallics Other Than Fuels)*, 4th edn, pp. 427–442. Ed. S.J. LeFond. New York; Amer. Inst. Min. Metall. Petrol. Engineers

Brobst, D.A. and Pratt, W.P. (eds.) (1973). *United States Mineral Resources. US Geol. Surv. Prof. Paper 820.* Washington DC; US Geological Survey

Bromley, A.V. (1975). 'Tin mineralization of western Europe: is it related to crustal subduction?', *Trans. Inst. Min. Metall.* Vol. 84, B28–B30

Brothers, R.N. and Blake, M.C. Jr. (1973). 'Tertiary plate tectonics and high pressure metamorphism in New Caledonia', *Tectonophysics*. Vol. 17, 337–358

Brown, C.E. (1973). 'Talc', *United States Mineral Resources. US Geol. Surv. Prof. Paper 820*, pp. 619–626. Ed. D.A. Brobst and W.P. Pratt. Washington DC; US Geological Survey

Brown, J.S. (1967). 'Isotopic zoning of lead and sulfur in southeast Missouri', *Genesis of Stratiform Lead–Zinc–Barite–Fluorite Deposits (Mississippi Valley-Type Deposits)—A Symposium. Econ. Geol. Monograph 3*, pp. 410–425. Ed. J.S. Brown. Lancaster, Penn.; Economic Geology Publishing Company

Brown, W.H. and Weinberg, E.C. (1968). 'Geology of the Austinville–Ivanhoe district, Virginia', *Ore Deposits of the United States 1933–1967. The Graton–Sales Volume, Vol. 1*, pp. 168–186. Ed. J.D. Ridge. New York; Amer. Inst.

Min. Metall. Petrol. Engineers

Brunstrom, R.G.W. and Walmsley, P.J. (1969). 'Permian evaporites in North Sea Basin', *Amer. Assoc. Petrol. Geol. Bull.* Vol. 53, 870–883

Bryner, L. (1969). 'Ore deposits of the Philippines—an introduction to their geology', *Econ. Geol.* Vol. 64, 644–666

Bubenicek, L. (1964). 'Étude sedimentologique du mineral de fer oolithique de Lorraine', *Sedimentology and Ore Genesis. Vol. 2. Developments in Sedimentology*, pp. 113–122. Ed. C.G. Amstutz. Amsterdam; Elsevier

Buchanan, M.S., MacLeod, W.N., Turner, D.C., Berridge, N.G. and Black, R. (1971). *The Geology of the Jos Plateau. Vol. 2. Younger Granite Complexes. Geol. Surv. of Nigeria Bull. 32.* Kaduna South; Geological Survey of Nigeria

Burbank, W.S. and Luedke, R.G. (1968). 'Geology and ore deposits of the Western San Juan Mountains, Colorado', *Ore Deposits of the United States, 1933–1967*, pp. 714–733. Ed. J.D. Ridge. New York; Am. Inst. Min. Metall. Petrol. Engineers

Bureau of Mines (1980a). *Minerals Yearbook for 1977, Vol. 1. Metals and Minerals.* Washington DC; Department of the Interior, Bureau of Mines

Bureau of Mines (1980b). *Minerals Yearbook 1978–79, Vol. 1. Metals and Minerals.* Washington DC; US Government Printing Office

Burger, P.A. (1979). 'The Greenvale nickel laterite orebody', *International Laterite Symposium*, pp. 24–37. Ed. D.J.I. Evans, R.S. Shoemaker and H. Veltman. New York; Soc. Mining Eng., Amer. Inst. Min. Metall. Petrol. Engineers

Burke, K., Dewey, J.F. and Kidd, W.S.F. (1976). 'Dominance of horizontal movements, arc and microcontinental collisions during the later Permobile regime', *The Early History of the Earth*, pp. 113–129. Ed. B.F. Windley. New York; John Wiley

Burke, K.C. and Wilson, J.T. (1976). 'Hot spots on the earth's surface', *Scientific American*, Vol. 235, No. 2; 46–57

Burnham, C.W. and Ohmoto, H. (1980). 'Late-stage processes of felsic magmatism', *Granitic Magmatism and Related Mineralization, Mining Geology Special Issue No. 8*, pp. 1–11. Ed. S. Ishihara and S. Takenouchi. Tokyo; Society of Mining Geologists of Japan

Butterman, W.C. (1980). 'Gold', *Minerals Yearbook 1978–79, Vol. 1. Metals and Minerals*, pp. 377–399. Bureau of Mines. Washington DC; US Government Printing Office

Button, A. (1976). 'Transvaal and Hamersley basins—review of basin development and mineral deposits', *Mineral. Sci. Eng.* Vol. 8, 262–293

Byrd, W.D. (1975). 'Geology of the Ekofisk Field, offshore Norway', *Petroleum and the Continental Shelf of Northwest Europe. Vol. 1. Geology*, pp. 439–445. Ed. A.W. Woodland. London; Applied Science Publishers

Callahan, W.H. (1968). 'Geology of the Friedensville Zinc Mine, Lehigh County, Pennsylvania', *Ore deposits of the United States 1933–1967. The Graton-Sales Volume. Vol. 1*, pp. 95–107. Ed. J.D. Ridge. New York; Amer. Inst. Min. Metall. Petrol. Engineers

Callender, E. and Bowser, C.J. (1976). 'Freshwater ferromanganese deposits', *Handbook of Strata-bound and Stratiform Ore Deposits, Vol. 7*, pp. 341–394. Ed. K.H. Wolf. Amsterdam; Elsevier

Callow, K.J. (1967). 'The geology of the Thanksgiving mine, Baguio district, Mountain province, Philippines', *Econ. Geol.* Vol. 62, 472–481

Calvo, F.A. and Guilemany, J.M. (1975). 'Structure and origin of mercury ore from Almaden, Spain', *Trans. Inst. Min. Metall. London.* Vol. 84, B146–B149

Cameron, C.C. (1973). 'Peat', *United States Mineral Resources. US Geol. Surv. Prof. Paper 820*, pp. 505–513. Ed. D.A. Brobst and W.P. Pratt. Washington DC; US Geological Survey

Cameron, E.N. and Desborough, G.A. (1969). Occurrence and characteristics of chromite deposits–eastern Bushveld Complex', *Magmatic Ore Deposits. Econ. Geol. Monograph 4*, pp. 23–40. Ed. H.D.B. Wilson. Lancaster, Penn.; Economic Geology Publishing Company

Cameron, E.N., Jahns, R.H., McNair, A.H. and Page, L.R. (1949). *Internal Structure of Granitic Pegmatites. Econ. Geol. Monograph 2*, Lancaster, Penn.; Economic Geology Publishing Company

Cammarota, V.A. Jr., Lucas, J.M. and Gorby, B.M. (1980). 'Zinc', *Minerals Yearbook 1978–79, Vol. 1. Metals and Minerals*, pp. 981–1019. Bureau of Mines. Washington DC; US Government Printing Office

Carleton, D.A., Harper, W.B., Michalski, B. and Moore, B.M. (1975). 'Crude petroleum and petroleum products', *Minerals Yearbook 1973, Vol. 1. Metals, Minerals and Fuels*, pp. 903–1018. Bureau of Mines. Washington DC; US Government Printing Office

Carlin, J.F. Jr. and Bascle, R.J. (1980). 'Bismuth', *Minerals Yearbook 1978–79, Vol. 1. Metals and Minerals*, pp. 115–118. Bureau of Mines. Washington DC; US Government Printing Office

Carlin, J.F. Jr. and Harris, K.L. (1980). 'Tin', *Minerals Yearbook 1978–79, Vol. 1. Metals and Minerals*, pp. 915–932. Bureau of Mines. Washington DC; US Government Printing Office

Carr, D.D. and Rooney, L.F. (1975). 'Limestone and dolomite', *Industrial Minerals and Rocks (Nonmetallics Other Than Fuels)*, pp. 757–789. Ed. S. LeFond. New York; Amer. Inst. Min. Metall. Petrol. Engineers

Casadevall, T. and Rye, R.O. (1980). 'The Tungsten Queen deposit, Hamme district, Vance county, North Carolina: a stable isotope study of a metamorphosed quartz–huebnerite vein', *Econ. Geol.* Vol. 75, 523–537

Cashion, W.B. (1973). 'Bitumen-bearing rocks', *United States Mineral Resources. US Geol. Surv. Prof. Paper 820*, pp. 99–103. Ed. D.A. Brobst and W.P. Pratt. Washington DC; US Geological Survey

Cathcart, J.B. and Gulbrandsen, R.A. (1973). 'Phosphate deposits', *United States Mineral Resources. US Geol. Surv. Prof. Paper 820*, pp. 515–525. Ed. D.A. Brobst and W.P. Pratt. Washington DC; US Geological Survey

CCOP (1980). *Studies in East Asian tectonics and resources (SEATAR). United Nations ESCAP, CCOP Technical Publication No. 7.* Bangkok; United Nations

Chappell, B.W. (1978). 'Granitoids from the Moonbi district, New England Batholith, eastern Australia', *Geol. Soc. Australia Bull.* Vol. 25, 267–283

Chappell, B.W. and White, A.J.R. (1974). 'Two contrasting granite types', *Pacific Geol.* Vol. 8, 173–174

Charoy, B. and Weisbrod, A. (1975). 'Charactéristiques de la phase fluide associée a la genèse des gisements d'étain d'Abbaretz et de la Villeder (Bretagne Meridionale)', *Mineral. Deposita.* Vol. 10, 89–99

Cheney, E.S. (1981). 'The hunt for giant uranium deposits', *American Scientist.* Vol. 69, No. 1, 37–48

Cheney, T.M., McClellan, G.H. and Montgomery, E.S. (1979). 'Sachura phosphate deposits, their stratigraphy, origin and composition, *Econ. Geol.* Vol. 74, 232–259

Chesterman, C.W. (1975). 'Perlite', *Industrial Minerals and Rocks (Nonmetallics Other Than Fuels)*, pp. 927–934. Ed. S.J. LeFond. New York; Amer. Inst. Min. Metall. Petrol. Engineers

Childers, M.O. and Bailey, R.V. (1979). 'Classification of uranium deposits', *Contrib. Geology, Univ. Wyoming.* Vol. 17, No. 2, 187–199

Childs, F.B. and Reed, P.E.C. (1975). 'Geology of the Dan Field and the Danish North Sea', *Petroleum and the Continental Shelf of Northwest Europe. Vol. 1. Geology*, pp. 429–438. Ed. A.W. Woodland, London; Applied Science Publishers

Clark, A.H., Farrar, E., Caelles, J.C., Haynes, S.J., Lortie, R.B., McBride, S.L., Quirt, G.S., Robertson, R.C.R. and Zentilli, M. (1976). 'Longitudinal variations in the metallogenic evolution of the central Andes: a progress report', *Metallogeny and Plate Tectonics. Geol. Assoc. Canada Special Paper 14*, pp. 23–58. Ed. D.F. Strong. Toronto; Geol. Assoc. Canada

Clark, W.B. (1970). *Gold Districts of California. California Div. Mines Geol. Bull. 193*. San Francisco; California Division of Mines and Geology

Claypool, G.E. and Reed, P.R. (1976). 'Thermal analysis technique for source rock evaluation: quantitative estimate of organic richness and effects of lithologic variation', *Amer. Assoc. Petrol. Geol. Bull.* Vol. 60, 608–626

Clifton, R.A. (1980a). 'Asbestos', *Minerals Yearbook 1978–79, Vol. 1. Metals and Minerals*, pp. 71–84. Bureau of Mines. Washington DC; US Government Printing Office

Clifton, R.A. (1980b). 'Talc and pyrophyllite', *Minerals Yearbook 1978–79, Vol. 1. Metals and Minerals*, pp. 899–905. Bureau of Mines. Washington DC; US Government Printing Office

Coleman, R.G. (1971). 'Plate tectonic emplacement of upper mantle peridotites along continental edges', *J. Geophys. Res.* Vol. 76, 1212–1222

Coleman, R.G. (1977). *Ophiolites*. Berlin; Springer-Verlag

Coleman, R.G. and Irwin, W.P. (1974). 'Ophiolites and ancient continental margins', *The Geology of Continental Margins*, pp. 921–931. Ed. C.A. Burk and C.L. Drake. Berlin; Springer-Verlag

Collerson, K.D. and Fryer, B.J. (1978). 'The role of fluids in the formation and subsequent development of early continental crust', *Contrib. Mineral. Petrol.* Vol. 67, 151–167

Colley, H. (1976). 'Classification and exploration guide for Kuroko-type deposits based on occurrences in Fiji', *Trans. Inst. Min. Metall.* Vol. 85, B190–B199

Collins, C.T. (1980). 'Iron oxide pigments', *Minerals Yearbook 1978–79, Vol. 1. Metals and Minerals*, pp. 457–464. Bureau of Mines. Washington DC; US Government Printing Office

Comer, J.B. (1974). 'Genesis of Jamaican bauxites', *Econ. Geol.* Vol. 69, 1251–1264

Connan, J., Cassou, A.M. and Van Der Weide, B. (1978). 'Kerogen maturity and gas/oil properties in shale–sandstone sequences', *Proceedings of the Seventh Indonesian Petroleum Association Annual Convention*, Jakarta, 1978, 179–195

Constantinou, G. (1976). 'Genesis of the conglomerate structure, porosity and collomorphic textures of the massive sulphide ores of Cyprus', *Metallogeny and Plate Tectonics. Geol. Assoc. Canada Special Paper 14*, pp. 187–210. Ed. D.F. Strong. Toronto; Geol. Assoc. Canada

Constantinou, G. and Govett, G.J.S. (1972). 'Genesis of sulphide deposits, ochre and umber of Cyprus', *Inst. Min. Metall. Trans.* Vol. 81, B34–B46

Constantinou, G. and Govett, G.J.S. (1973). 'Geology, geochemistry, and genesis of Cyprus sulfide deposits', *Econ. Geol.* Vol. 68, 843–858

Cook, P.J. (1976). 'Sedimentary phosphate deposits', *Handbook of Strata-bound and Stratiform Ore Deposits*, Vol. 7, pp. 505–535. Ed. K.H. Wolf. Amsterdam; Elsevier

Cook, P.J. and McElhinny, M.W. (1979). 'A reevaluation of the spatial and temporal distribution of sedimentary phosphate deposits in the light of plate tectonics', *Econ. Geol.* Vol. 74, 315–330

Cooper, B.S. (1977). 'Estimation of the maximum temperatures attained in sedimentary rocks', *Developments in Petroleum Geology*. Vol. 1, pp. 127–146. Ed. G.D. Hobson. London; Applied Science Publishers

Cooper, D.G. (1964). 'The geology of the Bikita pegmatite', *The Geology of Some Ore Deposits of Southen Africa*, pp. 441–461. Ed. S.H. Houghton. Johannesburg; Geol. Soc. South Africa

Cornelius, K.D. (1976). 'Preliminary rock type and genetic classification of uranium deposits', *Econ. Geol.* Vol. 71, 941–942

Cornwall, H.R. (1973). 'Nickel', *United States Mineral Resources. US Geol. Surv. Prof. Paper 820*, pp. 437–442. Ed. D.A. Brobst and W.P. Pratt. Washington DC; US Geological Survey

Cousins, C.A. (1969). 'The Merensky Reef of the Bushveld Igneous complex, *Magmatic Ore Deposits. Econ. Geol. Monograph 4*, pp. 239–251. Ed. H.D.B. Wilson. Lancaster, Penn.; Economic Geology Publishing Company

Cousins, C.A. and Vermaak, C.F. (1976). 'The contribution of southern African ore deposits to the geochemistry of the platinum group metals', *Econ. Geol.* Vol. 71, 287–305

Cox, D.P., Schmidt, R.G., Vine, J.D., Kirkemo, H., Tourtelot, E.B. and Fleischer, M. (1973). 'Copper', *United States Mineral Resources. US Geol. Surv. Prof. Paper 820*, pp. 163–190. Ed. D.A. Brobst and W.P. Pratt. Washington DC; US Geological Survey

Cox, K.G., Bell, J.D. and Pankhurst, R.J. (1979). *The Interpretation of Igneous Rocks*. London; George Allen and Unwin

Cox, R. and Glasson, K.R. (1970). 'The exploration and evaluation of a sulphide–cassiterite deposit at Cleveland mine, Tasmania', *A Second Technical Conference on Tin: Bangkok 1969, Vol. 1*, pp. 343–354. Ed. W. Fox. London; International Tin Council

Crawford, J. and Hoagland, A.D. (1968). 'The Mascot–Jefferson City zinc district, Tennessee', *Ore Deposits of the United States 1933–1967. The Graton–Sales Volume, Vol. 1*, pp. 242–256. Ed. J.D. Ridge. New York; Amer. Inst. Min. Metall. Petrol. Engineers

Creasey, S.C. (1977). 'Intrusives associated with porphyry copper deposits', *Geol. Soc. Malaysia Bull.* Vol. 9, 51–66

Crelling, J.C. and Dutcher, R.R. (1980). *Principles and Applications of Coal Petrology. Short Course Notes.* Indiana University at Bloomington; Society of Economic Paleontologists and Mineralogists

Cressman, E.R. and Swanson, R.W. (1964). *Stratigraphy and Petrology of the Permian Rocks of Southeastern Montana. US Geol. Surv. Prof. Paper 313-C*, pp. 275–569. Washington DC; US Geological Survey

Cronan, D.S. (1976). 'Basal metalliferous sediments from the eastern Pacific', *Bull. Geol. Soc. Amer.* Vol. 87, 928–934

Cronan, D.S. (1977). 'Deep sea nodules: distribution and geochemistry', *Marine Manganese Deposits*, pp. 11–44. Ed. G.P. Glasby. Amsterdam; Elsevier

Crough, S.T. (1979). 'Hotspot epeirogeny', *Tectonophysics.* Vol. 61, 321–333

Culbertson, W.C. and Pitman, J.K. (1973). 'Oil shale', *United States Mineral Resources. US Geol. Surv. Prof. Paper 820*, pp. 497–503. Ed. D.A. Brobst and W.P. Pratt. Washington DC; US Geological Survey

Cumming, G.L. and Robertson, D.K. (1969). 'Isotopic composition of lead from the Pine Point Deposit', *Econ. Geol.* Vol. 64, 731–732

Cuney, M. (1978). 'Geologic environment, mineralogy, and fluid inclusions of the Bois Noirs–Limouzat uranium vein, Forez, France', *Econ. Geol.* Vol. 73, 1567–1610

Curray, J.R., Emmel, F.J., Moore, D.G. and Raitt, R.W. (1982). 'Structure, tectonics and geological history of the northeastern Indian Ocean', *The Ocean Basins and Margins.*

6, *The Indian Ocean*. Ed. A.E.M. Nairn and F.G. Stehli. New York; Plenum Press

Curtis, C.D. and Spears, D.A. (1968). 'The formation of sedimentary iron minerals', *Econ. Geol.*, Vol. 63, 257–270

Dahl, A.R. and Hagmaier, J.L. (1974). 'Genesis and characteristics of the southern Powder River basin uranium deposits, Wyoming, USA', *Formation of Uranium Ore Deposits. Proc. Symposium, Athens, 1974*, pp. 201–218. Vienna; International Atomic Agency

Dahlkamp, F.J. (1978a). 'Classification of uranium deposits', *Mineral. Deposita*. Vol. 13, 83–104

Dahlkamp, F.J. (1978b). 'Geologic appraisal of the Key Lake U–Ni deposits, northern Saskatchewan', *Econ. Geol.* Vol. 73, 1430–1449

Dahlkamp, F.J. and Tan, B. (1977). 'Geology and mineralogy of the Key Lake U–Ni deposits, northern Saskatchewan, Canada', *Geology, Mining, and Extractive Processing of Uranium*, pp. 145–157. Ed. M.J. Jones. London; Inst. Min. Metall.

Damberger, H.H. (1974). 'Coalification patterns of Pennsylvanian coal basins of the eastern United States', *Carbonaceous Materials as Indicators of Metamorphism. Geol. Soc. Amer. Special Paper 153*, pp. 53–74. Ed. R.R. Dutcher, P.A. Hacquebard, J.M. Schopf and J.A. Simon. Boulder, Colorado; Geol. Soc. Amer.

Daniel, M.E. and Hood, W.C. (1975). 'Alteration of shale adjacent to the Knight orebody, Rosiclare, Illinois', *Econ. Geol.* Vol. 70, 1062–1069

Danielson, M.J. (1975). 'King island scheelite deposits', *Economic Geology of Australia and Papua New Guinea. Vol. 1. Metals*, pp. 592–598. Ed. C.L. Knight. Parkville, Victoria; Australasian Inst. Min. Metall.

Davidson, C.F. (1967). 'The kimberlites of the USSR', *Ultramafic and Related Rocks*, pp. 251–261. Ed. P.J. Wyllie. New York; John Wiley

Davies, D.N. (1964). 'The tin deposits of Swaziland', *The Geology of Some Ore Deposits in Southern Africa*, Vol. 2, pp. 59–75. Ed. S.H. Haughton. Geol. Soc. South Africa

Davies, H.L. (1971). *Peridotite–Gabbro–Basalt Complex in Eastern Papua, Bull. 128*, Canberra; Bur. Min. Res. Geol. Geophys. Australia

Davies, H.L. (1981). 'The major ophiolite complex in southeastern Papua–New Guinea: A review'. *The Geology and Tectonics of Eastern Indonesia, Spec. Pub. No. 2*, pp. 391–408. Ed. A.J. Barber and S. Wiryosujono. Bandung and London; Geol. Res. Dev. Centre and Pergamon Press

Davies, J.F. and Luhta, L.E. (1978). An Archean 'porphyry-type' disseminated copper deposit, Timmins, Ontario, *Econ. Geol.* Vol. 73, 383–396

Davis, J.B. and Kirkland, D.W. (1973). 'Native sulphur deposition in the Castile Formation, Culberson Country, Texas', *Econ. Geol.* Vol. 65, 107–121

Davis, J.B. and Kirkland, D.W. (1979). 'Bio-epigenetic sulfur deposits', *Econ. Geol.* Vol. 74, 462–468

Davis, S.G. (1964). 'Mineralogy and genesis of the wolframite ore deposit at Needle Hill mine, New Territories, Hong Kong', *Economic Geology of Hong Kong*, pp. 24–34. Ed. S.G. Davis. Hong Kong; Hong Kong University Press

Dawson, J.B. (1967). 'A review of the geology of kimberlites', *Ultramafic and Related Rocks*, pp. 241–251. Ed. P.J. Wyllie. New York; John Wiley

Dawson, J.B. (1971). 'Advances in kimberlite geology', *Earth Sci. Rev.* Vol. 7, 187–214

Dawson, J.B., Delaney, J.S. and Smith, J.V. (1978). 'Aspects of the mineralogy of alnöitic breccias, Malaita, Solomon Islands; comparison with continental kimberlites', *Contrib. Mineral. Petrol.* Vol. 67, 187–193

Deans, T. (1966). 'Economic mineralogy of African carbonatites', *Carbonatites*, pp. 385–413. Ed. O.F. Tuttle and J. Gittins. New York; Wiley–Interscience

Degens, E.T. and Ross, D.A. (eds.) (1974). *The Black Sea—Geology, Chemistry and Biology. Amer. Assoc. Petrol. Geol. Mem., Vol. 20*. Tulsa, Oklahoma; Amer. Assoc. Petrol. Geol.

Degens, E.T., Khoo, F. and Michaelis, W. (1977). 'Uranium anomaly in Black Sea sediments', *Nature*, No. 269 (13 Oct.), 566–569

De Huff, G.L. (1980). 'Manganese', *Minerals Yearbook 1978–79, Vol. 1. Metals and Minerals*, pp. 577–591. Bureau of Mines. Washington DC; US Government Printing Office

De Kun, N. (1965). '*The Mineral Resources of Africa*. Amsterdam; Elsevier

Dence, M.I.R. (1972). 'Meteorite impact craters and the structure of the Sudbury Basin', *New Developments in Sudbury Geology. Geol. Assoc. Canada Special Paper 10*, pp. 7–18. Ed. J.V. Guy-Bray. Toronto; Geol. Assoc. Canada

Denholm, L.J. (1967). 'Lode structures and ore shoots at Vatukoula, Fiji', *Proc. Australasian Inst. Min. Metall.* Vol. 222, 73–83

Dennis, J.G. (1972). *Structural Geology*. New York; Ronald Press

Derry, D.R. (1980). 'Uranium deposits through time', *The Continental Crust and its Mineral Deposits. Geol. Assoc. Canada Special Paper 20*, pp. 625–632. Ed. D.W. Strangway. Toronto; Geol. Assoc. Canada

De Ruiter, P.A.C. (1979). 'The Gabon and Congo Basins salt deposits', *Econ. Geol.* Vol. 74, 419–431

DeVoto, R.H. (1978). 'Uranium in Phanerozoic sandstone and volcanic rocks', *Uranium Deposits: Their Mineralogy and Origin. Short Course Handbook 3*, pp. 293–306. Ed. M.M. Kimberley. Ottawa; Mineral. Assoc. Canada

Dickey, P.A. (1968). 'Contemporary nonmarine sedimentation in Soviet central Asia', *Amer. Assoc. Petrol. Geol. Bull.* Vol. 52, 2396–2421

Dickey, P.A. (1975). 'Possible primary migration of oil from source rock in oil phase', *Amer. Assoc. Petrol. Geol. Bull.* Vol. 59, 337–345

Dickinson, W.R. (1966). 'Table Mountain serpentinite extrusion in California Coast Ranges', *Geol. Soc. Amer. Bull.* Vol. 77, 451–472

Dickinson, W.R. (1975a). 'Plate tectonic evolution of sedimentary basins', *A.A.P.G. Continuing Education Course Note Series 1 "Plate Tectonics and Hydrocarbon Accumulation"*, pp. 1–56. Tulsa, Oklahoma; Amer. Assoc. Petrol. Geol.

Dickinson, W.R. (1975b). 'Potash–depth (K–h) relations in continental margins and intra-oceanic magmatic arcs', *Geology*. Vol. 3, 53–56

Dickinson, W.R. (1978). 'Plate tectonic evolution of north Pacific rim', *J. Phys. Earth*. Vol. 26 Supplement, S1–N2S19

Dickson, F.W. and Tunnel, G. (1968). 'Mercury and antimony deposits associated with active hot springs in the western United States', *Ore Deposits of the United States 1933–1967. The Graton–Sales Volume*, Vol. 2, pp. 1673–1701. Ed. J.D. Ridge. New York; Amer. Inst. Min. Metall. Petrol. Engineers

Diessel, C.F.K. (1975). 'Coalification trends in the Sydney Basin, New South Wales', *Gondwana Geology*, pp. 295–309. Ed. K.S.W. Campbell. Canberra; Australian National University Press

Dietz, R.S. (1972). 'Sudbury astrobleme, splash emplaced sub-layer and possible cosmogenic ores', *New Developments in Sudbury Geology. Geol. Assoc. Canada Special Paper 10*, pp. 29–40. Ed. J.V. Gay-Bray. Toronto; Geol. Soc. Canada

Dikeou, J.T. (1980). 'Cement', *Minerals Yearbook 1978–79, Vol. 1. Metals and Minerals*, pp. 153–191. Bureau of Mines. Washington DC; US Government Printing Office

Dimroth, E. (1975). 'Paleo-environment of iron-rich sedimentary rocks', *Geol. Rundschau.* Vol. 64, 751–767

Dines, H.G. (1956). *The Metalliferous Mining Region of South-west England. Mem. Geol. Surv. Gt. Britain*, 2 vols. London; HMSO

Dixon, C.J. (1979). *Atlas of Economic Mineral Deposits.* Ithaca, New York; Cornell University Press

Dodge, F.C.W. and Bateman, P.C. (1977). 'The Sierra Nevada Batholith, California, USA, and spatially related mineral deposits', *Geol. Soc. Malaysia Bull.* Vol. 9, 17–29

Dodson, R.G., Needham, R.S., Wilkes, P.G., Page, R.W., Smart, P.G. and Watchman, A.L. (1974). 'Uranium mineralization in the Rum Jungle–Alligator Rivers province, Northern Territory, Australia', *Formation of Uranium Ore Deposits. Proc Symposium, Athens 1974*, pp. 551–568. Vienna; International Atomic Energy Agency

Doe, B.R. and Delevaux, H. (1972), 'Source of lead in southeast Missouri galena ores', *Econ. Geol.* Vol. 67, 409–425

Doe, B.R. and Stacey, J.S. (1974). 'The application of lead isotopes to the problems of ore genesis and ore prospect evaluation', *Econ. Geol.* Vol. 69, 757–776

Doe, B.R., Hedge, C.E. and White, D.E. (1966). 'Preliminary investigation of the source of lead in deep geothermal brines underlying the Salton Sea geothermal area', *Econ. Geol.* Vol. 61, 462–483

Dorr, J. Van N. II (1973a). 'Iron formations in South America', *Econ. Geol.* Vol. 68, 1005–1022

Dorr, J. Van N. II (1973b). 'Iron-formation and associated manganese in Brazil', *Genesis of Precambrian Iron and Manganese Deposits*, pp. 105–113. Paris; UNESCO

Dorr, J. Van N. II, Crittenden, M.D. Jr. and Worl, R.G. (1973). 'Manganese', *United States Mineral Resources. US Geol. Surv. Prof. Paper 820*, pp. 385–399. Ed. D.A. Brobst and W.P. Pratt. Washington D.C.; US Geological Survey

Dow, W.G. (1977a). 'Kerogen studies and geological interpretations', *J. Geochem. Exploration.* Vol. 7, 79–99

Dow, W.G. (1977b). 'Petroleum source beds on continental slopes and rises', *A.A.P.G. Continuing Education Course Notes 5 "Geology of Continental Margins"*, pp. D-1–D-37. Tulsa, Oklahoma; Amer. Assoc. Petrol. Geol.

Dow, W.G. (1978). 'Petroleum source beds on continental slopes and rises', *Amer. Assoc. Petrol. Geol. Bull.* Vol. 62, 1584–1606

Dozy, J.J. (1970). 'A geological model for the genesis of the lead–zinc ores of the Mississippi Valley, USA', *Trans. Inst. Min. Metall.* Vol. 79, B163–B170

Drake, H.J. (1980a) 'Mercury', *Minerals Yearbook 1978–79, Vol. 1. Metals and Minerals*, pp. 593–600. Bureau of Mines. Washington DC; US Government Printing Office

Drake, H.J. (1980b) 'Silver', *Minerals Yearbook 1978–79, Vol. 1. Metals and Minerals*, pp. 801–820. Bureau of Mines. Washington DC; US Government Printing Office

Dunbar, W.R. (1948), 'Structural relations of the Porcupine ore deposits', *Structural Geology of Canadian Ore Deposits*, pp. 442–456. Toronto; Can. Inst. Min. Metall.

Dunham, A.C. and Hanor, J.S. (1967). 'Controls on barite mineralization in the western United States', *Econ. Geol.* Vol. 62, 82–94

Dunham, K., Beer, K.E., Ellis, R.A., Gallagher, M.J., Nutt, M.J.C. and Webb, B.C. (1978). 'United Kingdom', *Mineral Deposits of Europe. Vol. 1. Northwest Europe*, pp. 263–317. Ed. S.H.U. Bowie, A. Kvalheim and H.W. Haslam. London; Inst. Min. Metall.

Dunham, K.C., Dunham, A.C., Hodge, B.L. and Johnson, G.A.L. (1965). 'Granite beneath Viséan sediments with mineralization at Rookhope, northern Pennines', *Quart. J. Geol. Soc. London.* Vol. 121, 383–417

Durham, D.L. (1973). 'Diatomite', *United States Mineral Resources. US Geol. Surv. Prof. Paper 820*, pp. 191–195. Ed. D.A. Brobst and W.P. Pratt. Washington DC, US Geological Survey

Durisova, J. (1978). 'Geothermometry in the minerals from the tin deposits of the eastern Krusnehory Mts. (Czechoslovakia)', *Metallization Associated with Acid Magmatism, Vol. 3*, pp. 325–335. Ed. M. Stemprok, L. Burnol and G. Tischendorf. Prague; Geological Survey.

Duthou, J.L. (1977). 'Chronologie Rb–Sr et géochimie des granitoids d'un segment de la chaine Varisque. Relations avec métamorphism: le Nord Limousin (Mjssif Central française)', *Ann. Fac. Sci. de l'Univ. Clermont-Ferrand.* Vol. 63, 1–294.

Eames, T.D. (1975). 'Coal rank and gas source relationships—Rotliegendes reservoirs', *Petroleum and the Continental Shelf of North-West Europe*, pp. 191–203. Ed. A.W. Woodland. London; Applied Science Publishers

Eastoe, C.J. (1978). 'A fluid inclusion study of the Panguna porphyry copper deposit, Bougainville, Papua New Guinea', *Econ. Geol.* Vol. 73, 721–748.

Edmond, J.M., Measures, C., Mangum, B., Grant, B., Sclater, F.R., Collier, R., Hudson, A., Gordon, L.I. and Corliss, J.B. (1979). 'On the formation of metal-rich deposits at ridge crests', *Earth Planet. Sci. Lett.* Vol. 46, 19–30.

Eichler, J. (1976). 'Origin of the Precambrian banded iron-formations', *Handbook of Strata-bound and Stratiform Ore Deposits, Vol. 7*, pp. 157–202. Ed. K.H. Wolf. Amsterdam; Elsevier

Eichmann, R., Saupé, F. and Schidlowski, M. (1977). 'Carbon and oxygen isotope studies in rocks in the vicinity of the Almaden Mercury deposit (Province of Cuidad Real, Spain)', *Time- and Strata-bound Ore Deposits*, pp. 396–405. Ed. D.D. Klemm and H.J. Schneider. Berlin; Springer-Verlag

Elthon, D. and Ridley, W.I. (1979). 'The oxide and silicate mineral chemistry of a kimberlite from the Premier Mine: implications for the evolution of kimberlitic magmas', *Kimberlites, Diatremes and Diamonds: Their Geology, Petrology and Geochemistry. Proc. 2nd International Kimberlite Conference, Vol. 1*, pp. 206–216. Ed. F.R. Boyd and H.O.A. Meyer. Washington DC, American Geophysical Union

Emery, J.A. (1968). 'Geology of the Pea Ridge iron ore body', *Ore Deposits of the United States 1933–1967. The Graton–Sales Volume, Vol. 1*, pp. 359–369. Ed. J.D. Ridge. New York; Amer. Inst. Min. Metall. Petrol. Engineers.

Emigh, G.D. (1975). 'Phosphate rock', *Industrial Minerals and Rocks (Nonmetallics Other Than Fuels)*, pp. 935–962. Ed. S.J. LeFond. New York; Amer. Inst. Min. Metall. Petrol. Engineers

Emmons, J. (1933). 'On the mechanism of the deposition of certain metalliferous lode systems associated with granitic batholiths', *Ore Deposits of the Western United States. Lindgren Volume*, pp. 327–439. New York; Amer. Inst. Min. Metall. Petrol. Engineers

Erdman, J.G. (1975). 'Time and temperature relations affecting the origin, expulsion and preservation of oil and gas', *Proc. 9th World Petroleum Congress, Vol. 2*, pp. 139–148. London; Applied Science Publishers

Erdosh, G. (1979). 'The Ontario carbonatite province and its phosphate potential', *Econ. Geol.* Vol. 74, 331–338

Espenshade, G.H. (1973). 'Kyanite and related minerals', *United States Mineral Resources. US Geol. Surv. Prof. Paper 820*, pp. 307–312. Ed. D.A. Brobst and W.P. Pratt. Washington DC; US Geological Survey

Espitalié, J., Laporte, J.L., Madec, M., Marquis, F., LePlat, P., Paulet, J. and Boutefeu, A. (1977a). 'Méthode rapide de caractérization des roches meres de leur potential pétrolier et de leur degré d'évolution', *Instit. Français Petrole Rev.*

Vol. 32, 23–42

Espitalié, J., Madec, M., Tissot, B., Mennig, J.J. and Leplat, P. (1977b). 'Source rock characterization method for petroleum exploration', *9th Annual Offshore Technology Conference, Houston, May 1977*, pp. 439–444

Eugster, H.P. and Chou, I-Ming (1979). 'A model for the deposition of Cornwall-type magnetite deposits', *Econ. Geol.* Vol. 74, 763–774

Eupene, G.S., Fee, P.H. and Colville, R.G. (1975). 'Ranger one uranium deposits', *Economic Geology of Australia and Papua New Guinea. Vol. 1. Metals*, pp. 308–317. Ed. C.L. Knight. Parkville, Victoria; Australasian Inst. Min. Metall.

Evans, A.M. (1976). 'Genesis of Irish base-metal deposits', *Handbook of Strata-bound and Stratiform Ore Deposits. Vol. 5*, pp. 231–256. Ed. K.H. Wolf. Amsterdam; Elsevier

Evans, A.M. (1980). *An Introduction to Ore Geology. Geoscience Texts Vol. 2.* Oxford; Blackwell Scientific Publications

Evans, C.R., McIvor, D.K. and Magara, K. (1975). 'Organic matter, compaction history and hydrocarbon occurrence—Mackenzie Delta, Canada', *Ninth World Petroleum Congress Proceedings, Vol. 2, Geology*, pp. 149–157. London; Applied Science Publishers

Evans, H.J. (1973). 'Weipa bauxite deposit, Queensland, Australia', *Metallogenic Provinces and Mineral Deposits in the Southwestern Pacific. Bur. Min. Res. Geol. Geophys. Bull. 141*, p. 222. Ed. N.H. Fisher, Canberra; Bur. Min. Res. Geol. Geophys.

Evans, H.J. (1975). 'Weipa bauxite deposit, Queensland', *Economic Geology of Australia and Papua New Guinea. Vol. 1. Metals*, pp. 959–964. Parkville, Victoria; Australasian Inst. Min. Metall.

Evernden, J.F., Kriz, S.J. and Cherroni, C. (1977). 'Potassium–argon ages of some Bolivian rocks', *Econ. Geol.* Vol. 72, 1042–1061

Falcon, R.M.S. (1977). 'Coal in South Africa, Part 1. The quality of South African coal in relation to its uses and world energy resources', *Mineral. Sci. Eng.* Vol. 9, 198–217

Falcon, R.M.S. (1978). 'Coal in South Africa, Part 2. The application of petrography to the characterization of coal', *Mineral. Sci. Eng.* Vol. 10, 28–52

Farrar, E., Clark, A.H. and Kim, O.J. (1978). 'Age of the Sangdong tungsten deposit, Republic of Korea and its bearing on the metallogeny of the Southern Korean Peninsula', *Econ. Geol.* Vol. 73, 547–552

Faure, G. (1977). *Principles of Isotope Geology.* New York; John Wiley

Faust, G.T. (1966). 'The hydrous nickel–magnesium silicates: the garnierite group', *Amer. Mineral.* Vol. 51, 279–298

Feather, C.E. and Koen, G.M. (1975). 'The mineralogy of the Witwatersrand Reefs', *Mineral. Sci. Eng.* Vol. 7, 189–224

Fehn, U., Cathles, L.M. and Holland, H.D. (1978). 'Hydrothermal convection and uranium deposits in abnormally radioactive plutons', *Econ. Geol.* Vol. 73, 1556–1566

Feiss, P.G. (1978). 'Magmatic sources of copper in porphyry copper deposits', *Econ. Geol.* Vol. 73, 397–404

Fernandez, H.E. (1968). 'The geology of cinnabar deposits of central Palawan', *Philipp. Geol.* Vol. 22, 91–105

Fernandez, N.S. (1960). 'Notes on the geology and chromite deposits of the Zambales range', *Philipp. Geol.* Vol. 14, 1–8

Fick, L.J. (1960). 'The geology of the tin pegmatites of Kamativi, southern Rhodesia', *Geol. Mijnbouw.* Vol. 39, 472–491

Finch, W.I. (1967). *Geology of Epigenetic Uranium Deposits in Sandstone in the United States. US Geol. Surv. Prof. Paper 538.* Washington DC; US Geological Survey

Finch, W.I., Butler, A.P. Jr., Armstrong, F.C. and Weissenborn, A.E. (1973). 'Uranium', *United States Mineral Resources. US Geol. Surv. Prof. Paper 820*, pp. 456–468. Ed. D.A. Brobst and W.P. Pratt. Washington DC; US Geological Survey

Finlow-Bates, T. and Stumpfl, E.F. (1979). The copper and lead–zinc–silver orebodies of Mt. Isa mine, Queensland: products of one hydrothermal system', *Ann. Soc. Geol. de Belgique.* Vol. 102, 497–517

Fischer, R.P. (1968). 'The uranium and vanadium deposits of the Colorado Plateau', *Ore Deposits of the United States 1933–1967. The Graton–Sales Volume, Vol. 1*, pp. 735–746. Ed. J.D. Ridge. New York; Amer. Inst. Min. Metall. Petrol. Engineers

Fischer, R.P. (1973). 'Vanadium', *United States Mineral Resources. US Geol. Surv. Prof. Paper 820*, pp. 679–688. Ed. D.A. Brobst and W.P. Pratt. Washington DC; US Geological Survey

Fitch, F.H. (1952). *The Geology and Mineral Resources of the Neighbourhood of Kuantan, Pahang. Geol. Surv. Dept. Federation Malaya Mem. 6* (new series). Ipoh, Malaysia; Malaysian Geological Survey

Fleet, A.J. and Robertson, A.H.F. (1980). 'Ocean-ridge metalliferous and pelagic sediments of the Semail Nappe, Oman', *J. Geol. Soc. London.* Vol. 137, 403–422

Fleischer, V.D., Garlick, W.G. and Haldane, R. (1976). 'Geology of the Zambian Copperbelt', *Handbook of Strata-bound and Stratiform Ore Deposits. Vol. 6.* pp. 223–352. Ed. K.H. Wolf. Amsterdam; Elsevier

Flint, D.E., Albear, J.F. de and Guild, P.W. (1948). 'Geology and chromite deposits of Camaguey district, Camaguey province, Cuba', *Bull. US Geol. Surv.* Vol. 954–B, 39–63

Flinter, B.H. (1960). 'The effects of heat, hydrochloric acid and lead chloride on some Malayan mineral grains', *Overseas Geol. Mineral Resources.* Vol. 8, 54–56

Floran, R.J. and Papike, J.J. (1975). 'Petrology of the low-grade rocks of the Gunflint iron-formation, Ontario–Minnesota', *Geol. Soc. Amer. Bull.* Vol. 86, 1169–1190

Folinsbee, R.E., Kirkland, K., Nekolaichuk, A. and Smejkal, V. (1972). 'Chinkuashih—a gold–pyrite–enargite–barite hydrothermal deposit in Taiwan', *Mem. Geol. Soc. Amer.* Vol. 135, 323–335

Foo, B.N. (1977). 'Mineral paragenesis, fluid inclusion studies and geochemistry of the Sungei Lembing tin lodes, West Malaysia', *Trans. Inst. Min. Metall.* Vol. 86, B163

Foose, M.P., Slack, J.F. and Casadevall, T. (1980). 'Textural and structural evidence for a predeformation hydrothermal origin of the Tungsten Queen deposit, Hamme District, North Carolina', *Econ. Geol.* Vol. 75, 515–522

Ford, J.H. (1978). 'A chemical study of alteration at the Panguna porphyry copper deposit, Bougainville, Papua New Guinea', *Econ. Geol.* Vol. 73, 703–720

Ford, T.D. (1976). 'The Ores of the south Pennines and Mendip Hills, England—a comparative study', *Handbook of Strata-bound and Stratiform Ore Deposits. Vol. 5*, pp. 161–195. Ed. K.H. Wolf. Amsterdam; Elsevier

Foshag, W.F. and Fries, C. Jr. (1942). 'Tin deposits of the Republic of Mexico', *Bull. US Geol. Surv.* Vol. 935–C, 99–176

Foster, R.J. (1980). 'Salt', *Minerals Yearbook 1978–79, Vol. 1. Metals and Minerals*, pp. 751–762. Bureau of Mines. Washington DC; US Government Printing Office

Foy, M.F. (1975). 'South Alligator Valley uranium deposits', *Economic Geology of Australia and Papua New Guinea. Vol. 1. Metals*, pp. 301–303. Ed. C.L. Knight. Parkville, Victoria; Australasian Inst. Min. Metall.

Foy, M.F. and Pederson, C.P. (1975). 'Koongarra uranium deposit', *Economic Geology of Australia and Papua New Guinea. Vol. 1. Metals*, pp. 317–321. Ed. C.L. Knight. Parkville, Victoria; Australasian Inst. Min. Metall.

Franklin, J.M., Kasarda, J. and Poulsen, K.H. (1975). 'Petrology and chemistry of the alteration zone of the Mattabi massive sulfide deposit', *Econ. Geol.* Vol. 70, 63–79

Frazier, D.E. and Osanik, A. (1969). 'Recent peat deposits—Louisiana coastal plain', *Environments of Coal Deposition. Geol. Soc. Amer. Special Paper 114*, pp. 63–85. Ed. E.C. Dapples and M.E. Hopkins. Boulder, Colorado; Geol. Soc. Amer.

Freeze, A.C. (1966). 'On the origin of the Sullivan orebody, Kimberley, BC, *Can. Inst. Min. Metall.* Spec. Vol. 8, 263–294

French, B.M. (1972). 'Shock metamorphism features in the Sudbury structure, Ontario: a review', *New Developments in Sudbury Geology. Geol. Assoc. Canada Special Paper, 10*, pp. 19–28. Ed. J.V. Guy-Bray. Toronto; Geol. Assoc. Canada

French, B.M (1973). 'Mineral assemblages in diagenetic and low-grade metamorphic iron-formation', *Econ. Geol.* Vol. 68, 1063–1074

Friedman, G.M. and Sanders, J.E. (1978). *Principles of Sedimentology.* New York; John Wiley

Frietsch, R. (1978). 'On the magmatic origin of iron ores of the Kiruna type', *Econ. Geol.* Vol. 73, 478–485

Fripp, R.E.P. (1976a). 'Gold metallogeny in the Archean of Rhodesia', *The Early History of the Earth*, pp. 455–466. London; John Wiley

Fripp, R.E.P. (1976b). 'Stratabound gold deposits in Archean banded iron-formations of Rhodesia', *Econ. Geol.* Vol. 71, 58–75

Fritz, P. (1969). 'The oxygen and carbon isotopic composition of carbonates from the Pine Point lead–zinc ore deposits', *Econ. Geol.* Vol. 64, 733–742

Fritz, P. (1976). 'Oxygen and carbon isotopes in ore deposits in sedimentary rocks', *Handbook of Strata-bound and Stratiform Ore Deposits, Vol. 2*, pp. 191–217. Ed. K.H. Wolf. Amsterdam; Elsevier

Fulkerson, F.B. (1980). 'Barite', *Minerals Yearbook 1978–79, Vol. 1. Metals and Minerals*, pp. 181–187. Bureau of Mines. Washington DC; US Government Printing Office

Fulton, R.B. III (1975). 'Strontium', *Industrial Minerals and Rocks (Nonmetallics Other Than Fuels)*, pp. 1099–1102. Ed. S.J. LeFond. New York; Amer. Inst. Min. Metall. Petrol Engineers

Galloway, W.E. (1978). 'Uranium mineralization in a coastal-plain fluvial aquifer system: Catahoula Formation, Texas', *Econ. Geol.* Vol. 73, 1655–1676

Garlick, W.G. and Fleischer, V.D. (1972). 'Sedimentary environment of Zambian copper deposition', *Geol. Mijnbouw.* Vol. 51, 277–298

Garson, M.S. and Mitchell, A.H.G. (1977). 'Mineralization at destructive plate boundaries: a brief review', *Volcanic Processes in Ore Genesis. Geol. Soc. London, Special Publication No. 7*, pp. 81–97. Ed. I. Gass. London; The Geological Society

Garson, M.S., Bradshaw, N. and Rattawong, S. (1970). 'Lepidolite pegmatites in the Phangnga area of Peninsular Thailand', *A Second Technical Conference on Tin: Bangkok, 1969*, pp. 327–339. Ed W. Fox. London; International Tin Council

Garson, M.S., Young, B., Mitchell, A.H.G. and Tait, B.A.R. (1975). *The Geology of the Tin Belt in Peninsular Thailand around Phuket, Phangnga and Takua Pa. Inst. Geol. Sci. London, Overseas Mem. 1.* London; Institute of Geological Sciences

Gass, I.G. (1977). 'Origin and emplacement of ophiolites', *Volcanic Processes in Ore Genesis. Geol. Soc. London Special Publication 7*, pp. 72–76. London; The Geological Society

Gass, I.G. and Masson-Smith, D. (1963). 'The geology and gravity anomalies of the Troodos massif, Cyprus', *Phil. Trans. Roy. Soc. London.* Vol. A255, 417–467

Geijer, P. and Ödman, O.H. (1974). 'The emplacement of the Kiruna iron ores and related deposits', *Sveriges Geol. Undersökning*, Ser. C, No. 700, 1–48

Gerasimovsky, V.I., Volkov, V.P., Kogarko, L.N. and Polyakov, A.I. (1974). 'Kola Peninsula', *The Alkaline Rocks*, pp. 206–221. Ed. H. Sørenson. New York; John Wiley

Gerdemann, P.E. and Myers, H.E. (1972). 'Relationships of carbonate facies patterns to ore distribution and to ore genesis in the Southeast Missouri lead district', *Econ. Geol.* Vol. 67, 426–433

Gervasio, F.C. (1973). 'Ore deposits of the Philippine mobile belt', *Metallogenic Provinces and Mineral Deposits in the Southwestern Pacific. Bur. Min. Resources Geol. Geophys., Canberra, Bull. 141*, pp. 191–207. Ed. N.H. Fisher. Canberra; Bur. Min. Resources Geol. Geophys.

Gibbs, A.K. (1980). 'The Archean–Proterozoic transition: perspective from the Guiana Shield', *Geol. Soc. Amer. Abstracts with Programs.* Vol. 12, No. 7, 433

Gittinger, L.B. Jr. (1975). 'Sulfur', *Industrial Minerals and Rocks (Nonmetallics Other Than Fuels)*, pp. 1103–1125. Ed. S.J. LeFond. Amer. Inst. Min. Metall. Petrol. Engineers

Glasby, G.P. (1976). 'Manganese nodules in the south Pacific: a review', *New Zealand J. Geol. Geophys.* Vol. 19, 707–736

Glasby, G.P. and Read, A.J. (1976). 'Deep-sea manganese nodules', *Handbook of Strata-bound and Stratiform Ore Deposits. Vol. 7*, pp. 295–340. Ed. K.H. Wolf. Amsterdam; Elsevier

Glikson, A.Y. (1976a). 'Archean to early Proterozoic shield elements: relevance of plate tectonics', *Metallogeny and Plate Tectonics. Geol. Assoc. Can. Special Paper 14*, pp. 489–516. Ed. D.F. Strong, Toronto; Geol. Assoc. Canada

Glikson, A.Y. (1976b). 'Stratigraphy and evolution of primary and secondary greenstones: significance of data from shields of the Southern Hemisphere', *The Early History of the Earth*, pp. 257–277. Ed. B.F. Windley. New York; John Wiley

Gold, D.P. (1967). 'Alkaline ultrabasic rocks in the Montreal Area, Quebec', *Ultramafic and Related Rocks*, pp. 288–302. Ed. P.J. Wyllie. New York; John Wiley

Gold, D.P., Vallee, M. and Charette, J.P. (1967). 'Economic geology and geophysics of the Oka alkaline complex, Quebec', *Can. Inst. Min. Metal.* Vol. 60, 1131–1144

Goldhaber, M.B., Reynolds, R.L. and Rye, R.O. (1978). 'Origin of a south Texas roll-type uranium deposit. II Sulfide petrology and sulfur isotope studies', *Econ. Geol.* Vol. 73, 1690–1705

Goldich, S.S. (1973). 'Ages of Precambrian banded iron formations', *Econ. Geol.* Vol. 68, 1126–1134

Goldman, H.B. and Reining, D. (1975). 'Sand and gravel', *Industrial Minerals and Rocks (Nonmetallics Other Than Fuels)*, pp. 1027–1042. Ed. S.J. LeFond. New York; Amer. Inst. Min. Metall. Petrol. Engineers

Goldsmith, L.H. (1969). 'Concentration of potash salts in saline basins', *Amer. Assoc. Petrol. Geol. Bull.* Vol. 53, 790–797

Golightly, J.P. (1979a). 'Nickeliferous laterites: a general description', *International Laterite Symposium*, pp. 3–23. Ed. D.J.I. Evans, R.S. Shoemaker and H. Veltman. New York; Amer. Inst. Min. Metall. Petrol. Engineers

Golightly, J.P. (1979b). 'Geology of Soroako nickeliferous laterite deposits', *International Laterite Symposium*, pp. 38–56. Ed. D.J.I. Evans, R.S. Shoemaker and H. Veltman. New York; Amer. Inst. Min. Metall. Petrol. Engineers

Gonzales, A.G. (1956). 'Geology of the Lepanto Copper mine, Mankayan, Mountain Province', *Bur. Min. Philip. Spec. Proj.*

Ser. No. 16, 17–50

Goodwin, A.M. (1956). 'Facies relations in the Gunflint iron formation', *Econ. Geol.* Vol. 51, 565–595

Goodwin, A.M. (1965). 'Mineralized volcanic complexes in the Porcupine–Kirkland Lake–Noranda region, Canada', *Econ. Geol.* Vol. 60, 955–971

Goodwin, A.M. (1973a). 'Archaean volcanogenic iron-formations of the Canadian Shield', *Genesis of Precambrian Iron and Manganese Deposits*, pp. 23–34. Paris; UNESCO

Goodwin, A.M. (1973b). 'Archean iron-formations and basins', *Econ. Geol.* Vol. 68, 915–930

Goodwin, A.M. and Ridler, R.H. (1970). 'The Abitibi orogenic belt', *Basins and Geosynclines of the Canadian Shield. Ottawa Geol. Surv. Canada Paper 70–40*, pp. 1–24. Ed. A.J. Baer. Ottawa; Geological Survey of Canada

Goossens, P.J. (1978). 'The metallogenic provinces of Burma: their definitions, geologic relationships and extensions into China, India and Thailand', *Geology and Mineral Resources of Southeast Asia, Proc. 3rd Regional Conference, Bangkok*, pp. 431–492. Ed. Prinya Nutalaya. Bangkok; Asian Institute of Technology

Graffin, G.D. (1975). 'Graphite', *Industrial Minerals and Rocks (Nonmetallics Other Than Fuels)*, 4th edn., pp. 691–705. Ed. S.J. LeFond. New York; Amer. Inst. Min. Metall. Petrol. Engineers

Grant, J.N., Halls, C., Avila, W. and Avila, G. (1977). 'Igneous geology and the evolution of hydrothermal systems in some sub-volcanic tin deposits of Bolivia', *Volcanic Processes in Ore Genesis. Geol. Soc. London, Special Publication 7*, pp. 117–126. London; The Geological Society

Grant, J.N., Halls, C., Avila, W. and Snelling, N.J. (1979). 'Potassium–argon ages of igneous rocks and mineralization in part of the Bolivian tin belt', *Econ. Geol.* Vol. 74, 838–851

Grant, J.N., Halls, C., Sheppard, M.F. and Avila, W. (1980). 'Evolution of the porphyry tin deposits of Bolivia', *Granitic Magmatism and Related Mineralization. Special Issue No. 8, Mining Geology*, pp. 151–173. Ed. S. Ishihara and S. Takenouchi. Tokyo; The Society of Mining Geologists of Japan

Greenbaum, D. (1977). 'The chromitiferous rocks of the Troodos ophiolite complex, Cyprus', *Econ. Geol.* Vol. 72, 1175–1194

Greenwood, D.A. (1977). 'Fluorspar mining in the northern Pennines', *Trans. Inst. Min. Metall. London.* pp. B181–B190

Griffiths, W.R. (1973). 'Beryllium', *United States Mineral Resources. US Geol. Surv. Prof. Paper 820*, pp. 85–93. Ed. D.A. Brobst and W.P. Pratt. Washington DC; US Geological Survey

Griffiths, W.R., Albers, J.P. and Oner, O. (1972). 'Massive sulfide copper deposits of the Ergani–Maden area, southeastern Turkey', *Econ. Geol.* Vol. 67, 701–716

Grip, E. (1978). 'Sweden', *Mineral Deposits of Europe. Vol. 1. Northwest Europe*, pp. 93–198. Ed. S.H.U. Bowie, A. Kvalheim and H.W. Haslam. London; Inst. Min. Metall.

Grogan, R.M. and Bradbury, J.C. (1968). 'Fluorite–zinc–lead deposits of the Illinois–Kentucky mining district', *Ore Deposits of the United States 1933–1967. The Graton–Sales Volume, Vol. 1*, pp. 370–399. Ed. J.D. Ridge. New York; Amer. Inst. Min. Metall. Petrol. Engineers

Grogan, R.M. and Montgomery, G. (1975). 'Fluorspar', *Industrial Minerals and Rocks (Nonmetallics Other Than Fuels)*, 4th edn., pp. 653–677. Ed. S. J. LeFond. New York; Amer. Inst. Min. Metall. Petrol. Engineers

Grogan, R.M., Cunningham-Dunlop, P.K., Bartlett, H.F. and Czel, L.J. (1974). 'The environments of deposition of fluorspar', *A Symposium on the Geology of Fluorspar. Kentucky Geol. Surv. Series X, Special Publication 22*, pp. 4–9. Ed. D.W. Hutcheson. Lexington, Kentucky; Kentucky Geological Survey

Gross, G.A. (1972). 'Primary features in cherty iron-formations', *Sed. Geol.* Vol. 7, 241–261

Gross, G.A. (1973). 'The depositional environment of principal types of Precambrian iron-formations', *Genesis of Precambrian Iron and Manganese Deposits*, pp. 15–21. Paris; UNESCO

Gross, W.H. (1975). 'New ore discovery and source of silver–gold veins, Guanajuato, Mexico', *Econ. Geol.* Vol. 69, 1078–1085

Groves, D.I., Barrett, F.M. and McQueen, K.G. (1979). 'The relative roles of magmatic segregation, volcanic exhalation and regional metamorphism in the generation of volcanic associated nickel ores of Western Australia', *Can. Mineral.* Vol. 17, 319–336

Grubb, P.L.C. (1970). 'Mineralogy, geochemistry, and genesis of the bauxite deposits on the Gove and Mitchell Plateaux, Northern Australia', *Mineral. Deposita.* Vol. 5, 248–272

Grubb, P.L.C. (1973). 'High-level and low-level bauxitization: a criterion for classification', *Mineral. Sci. Eng.* Vol. 5, 219–231

Grubb, P.L.C. (1979). 'Genesis of bauxite deposits in the lower Amazon basin and Guianas coastal plain', *Econ. Geol.* Vol. 74, 735–750

Grubb, P.L.C. and Hannaford, P. (1966). 'Magnetism in cassiterite', *Mineral. Deposita.* Vol. 2, 148–171

Gruss, H. (1973). 'Itabirite iron ores of the Liberia and Guyana shields', *Genesis of Precambrian Iron and Manganese Deposits*, pp. 335–359. Paris; UNESCO

Gualtieri, J.L. (1973). 'Arsenic', *United States Mineral Resources. US Geol. Surv. Prof. Paper 820*, pp. 51–61. Ed. D.A. Brobst and W.P. Pratt. Washington DC; US Geological Survey

Guild, P.W. (1947). 'Petrology and structure of the Moa district, Oriente province, Cuba', *Geophys. Union Trans.* Vol. 28, 218–246

Guild, P.W. (1978). 'Metallogenesis in the western United States', *J. Geol. Soc. London.* Vol. 135, 355–376

Guillon, J.H. (1974). 'New Caledonia', *Mesozoic–Cenozoic Orogenic Belts. Geol. Soc. London Special Publication 4*, pp. 445–452. Ed. A.M. Spencer. London; The Geological Society

Guillon, J.H. and Lawrence, K.J. (1973). 'The opaque minerals of the ultramafic rocks on New Caledonia', *Mineral. Deposita.* Vol. 8, 115–126

Gulbrandsen, R.A. and Robertson, C.E. (1973). 'Inorganic phosphorus in sea water', *Environmental Phosphorus Handbook*, pp. 117–140. Ed. E.J. Griffith, A. Beeton, J.M. Spencer and D.T. Mitchell. New York; John Wiley

Gurney, J.T., Harris, J.W. and Rickard, R.S. (1979). 'Silicate and oxide inclusions in diamonds from the Finsch kimberlite pipe', *Kimberlites, Diatremes and Diamonds: Their Geology, Petrology and Geochemistry. Proc. 2nd International Kimberlite Conference*, pp. 1–15. Ed. F.R. Boyd and H.O.A. Meyer. Washington DC; American Geophysical Union

Gustafson, L.B. (1978). 'Some major factors of porphyry copper genesis', *Econ. Geol.* Vol. 73, 600–607

Haapala, I. and Kinnunen, K. (1979). 'Fluid inclusions in cassiterite and beryl in greisen veins in the Eurajoki stock, southwestern Finland', *Econ. Geol.* Vol. 74, 1231–1238

Hackett, J.P. and Bischoff, J.L. (1973). 'New data on the stratigraphy, extent and geologic history of the Red Sea geothermal deposits', *Econ. Geol.* Vol. 68, 553–564

Hails, J.R. (1976). 'Placer deposits', *Handbook of Strata-bound and Stratiform Ore Deposits. Vol. 3*, pp. 213–244. Ed. K.H. Wolf. Amsterdam; Elsevier

Haldeman, E.G., Buchan, R., Blowes, J.H. and Chandler, T. (1979). 'Geology of lateritic nickel deposits Dominican Republic', *International Laterite Symposium*, pp. 57–84. Ed. D.J.I. Evans, R.S. Shoemaker and H. Veltman. New York; Amer. Inst. Min. Metall. Petrol. Engineers

Hall, T.C.F. (1950). 'The coal of Gondwanaland', *Min. Mag.* Vol. 82, 201–210

Hall, W.E. and Friedman, I. (1963). 'Composition of fluid inclusions, Cave-in-Rock fluorite district, Illinois, and Upper Mississippi Valley zinc–lead district', *Econ. Geol.* Vol. 58, 886–911

Hall, W.E. and Friedman, I. (1969). 'Oxygen and carbon isotopic composition of ore and host rocks of selected Mississippi Valley deposits', *US Geol. Surv. Prof. Paper 650-C*, pp. C140–C148. Washington DC; US Geological Survey

Hall, W.E., Friedman, I. and Nash, J.T. (1974). 'Fluid inclusion and light stable isotope study of the Climax molybdenum deposits, Colorado', *Econ. Geol.* Vol. 69, 884–901

Hallam, C.D. (1964). 'The geology of the coastal diamond deposits of southern Africa', *The Geology of Some Ore Deposits in Southern Africa, Vol. 2*, pp. 671–728. Ed. S.H. Haughton. Johannesburg; Geol. Soc. South Africa

Hallbauer, D.K. (1975). 'The plant origin of the Witwatersrand "carbon"', *Mineral. Sci. Eng.* Vol. 7, 111–131

Hallbauer, D.K. and Joughin, N.C. (1972). 'Distribution and size of gold particles in the Witwatersrand reefs and their affects on sampling procedures', *Inst. Min. Metall. Bull.* Vol. 788, A133–A142

Halliday, A.N. (1980). 'The timing of early and main stage ore mineralization in southwest Cornwall', *Econ. Geol.* Vol. 75, 752–759

Hamilton, W. (1970). *Bushveld Complex—product of impacts? Geol. Soc. South Africa Special Publication 1*, pp. 367–374. Johannesburg; Geol. Soc. South Africa

Hamilton, W. (1979). *Tectonics of the Indonesian Region. US Geol. Surv. Prof. Paper 1078*. Washington DC; US Geological Survey

Harrison, J.E., Griggs, A.B. and Wells, J.D. (1974). *Tectonic Features of the Precambrian Belt Basin and Their Influence on Post-belt Structures. US Geol. Surv. Prof. Paper 886*. Washington DC; US Geological Survey

Harshman, E.N. (1968). 'Uranium deposits of the Shirley Basin, Wyoming', *Ore Deposits of the United States 1933–1967. The Graton–Sales Volume, Vol. 1*, pp. 849–856. Ed. J.D. Ridge. New York; Amer. Inst. Min. Metall. Petrol. Engineers

Harshman, E.N. (1972). *Geology and Uranium Deposits, Shirley Basin Area, Wyoming. US Geol. Surv. Prof. Paper 745*. Washington DC; US Geological Survey

Harshman, E.N. (1974). 'Distribution of elements in some roll-type uranium deposits', *Formation of Uranium Ore Deposits. Proc. Symposium, Athens 1974*, pp. 169–183. Vienna; International Atomic Energy Agency

Hasler, J.W., Miller, M.H. and Chapman, R.M. (1973). 'Bismuth', *United States Mineral Resources. US Geol. Surv. Prof. Paper 820*, pp. 95–98. Ed. D.A. Brobst and W.P. Pratt. Washington DC; US Geological Survey

Hawkins, B.W. (1975). 'Mary Kathleen uranium deposit', *Economic Geology of Australia and Papua New Guinea. Vol. 1. Metals*, pp. 398–403. Ed. C.L. Knight. Parkville, Victoria; Australasian Inst. Min. Metall.

Hawkins, J.W. and Batiza, R. (1977). 'Petrology and geochemistry of an ophiolite complex: Zambales Range, Luzon, Republic of Philippines', *EOS, Trans. Amer. Geophys. Union*, Vol. 58, Dec., 244

Hawthorne, J.B., Carrington, A.J., Clement, C.R. and Skinner, E.M.W. (1979). 'Geology of the Dokolwayo kimberlite and associated palaeo-alluvial diamond deposits', *Kimberlites, Diatremes and Diamonds: Their Geology, Petrology and Geochemistry. Proc. 2nd International Kimberlite Conference, Vol. 1*, pp. 59–70. Ed. F.R. Boyd and H.O.A. Meyer. Washington DC; American Geophysical Union

Heaton, T.H.E. and Sheppard, S.M.F. (1977). 'Hydrogen and oxygen isotope evidence for sea-water–hydrothermal alteration and ore deposition, Troodos Complex, Cyprus', *Volcanic Processes in Ore Genesis. Geol. Soc. London, Special Publication 7*, pp. 42–57. London; The Geological Society

Hedge, C.E. (1974). 'Strontium isotopes in economic geology', *Econ. Geol.* Vol. 69, 823–826

Hegge, M.R. and Rowntree, J.C. (1978). 'Geologic setting and concepts of origin of uranium deposits in the East Alligator River region, N.T., Australia', *Econ. Geol.* Vol. 73, 1420–1429

Heinrich, E.W. (1966). *The Geology of Carbonatites*. Chicago; Rand McNally

Helke, A. (1962). 'The metallogeny of the chromite deposits of the Guleman district, Turkey', *Econ. Geol.* Vol. 57, 954–962

Hemingway, J.E. (1974a). 'Jurassic', *The Geology and Mineral Resources of Yorkshire*, pp. 161–223. Ed. D.H. Rayner and J.E. Hemingway. Leeds; Yorkshire Geological Society

Hemingway, J.E. (1974b). 'Ironstone', *The Geology and Mineral Resources of Yorkshire*, pp. 329–335. Ed. D.H. Rayner and J.E. Hemingway. Leeds; Yorkshire Geological Society

Henley, R.W. and Adams, J. (1979). 'On the evolution of giant gold placers', *Inst. Min. Metall. Trans.* Vol. 88, B41–B50

Henley, R.W. and Thornley, P. (1979). 'Some geochemical aspects of polymetallic massive sulfide formation', *Econ. Geol.* Vol. 74, 1600–1612

Hesp, W.R. and Varlamoff, N. (1977). 'Temporal and spatial relations between the formation of acid magmatic rocks and deposits', *Metallization Associated with Acid Magmatism, Vol. 2*, pp. 23–37. Ed. M. Stemprok, L. Burnol and G. Tischendorf. Prague; Geological Survey

Hewett, D.F. (1966). 'Stratified deposits of the oxides and carbonates of manganese', *Econ. Geol.* Vol. 61, 431–461

Heyl, A.V., Delevaux, M.H., Zartman, R.E. and Brock, M.R. (1966). 'Isotopic study of galenas from the Upper Mississippi Valley, the Illinois–Kentucky, and some Appalachian Valley mineral districts', *Econ. Geol.* Vol. 61, 933–961

Heyl, A.V., Hall, W.E., Weissenborn, A.E., Stager, H.K., Puffett, W.P. and Reed, B.L. (1973). 'Silver', *United States Mineral Resources. US Geol. Surv. Prof. Paper 820*, pp. 581–603. Ed. D.A. Brobst and W.P. Pratt. Washington DC; US Geological Survey

Heyl, A.V., Landis, G.P. and Zartman, R.E. (1974). 'Isotopic evidence for the origin of Mississippi Valley-type mineral deposits: a review', *Econ. Geol.* Vol. 69, 992–1006

Hiemstra, S.A. (1979). 'The role of collectors in the formation of the platinum deposits in the Bushveld complex', *Can. Mineral.* Vol. 17, 469–482

Hight, R.P. (1975). 'Abrasives', *Industrial Minerals and Rocks (Nonmetallics Other Than Fuels)*, 4th edn., pp. 11–31. Ed. S.J. LeFond. New York; Amer. Inst. Min. Metall. Petrol. Engineers

Hills, J.H. and Richards, J.R. (1976). 'Pitchblende and galena ages in the Alligator Rivers region, Northern Territory, Australia', *Mineral. Deposita.* Vol. 11, 133–154

Hirdes, W. (1978). 'Small-scale gold distribution patterns in the Precambrian Kimberley Reef Placer: a case study at Marievale GMC, East Rand Goldfield, South Africa', *Mineral. Deposita.* Vol. 13, 313–328

Hoagland, A.D. (1976). 'Appalachian zinc–lead deposits', *Handbook of Strata-bound and Stratiform Ore Deposits. Vol. 6*, pp. 495–534. Amsterdam; Elsevier

Hobbs, B.E., Means, W.D. and Williams, P.F. (1976). *An Outline of Structural Geology*. New York; John Wiley

Hobbs, S.W. and Elliot, J.E. (1973). 'Tungsten', *United States Mineral Resources. US Geol. Surv. Prof. Paper 820*, pp. 667–678. Ed. D.A. Brobst and W.P. Pratt. Washington DC; US Geological Survey

Hobbs, S.W. and Fryklund, V.C. Jr. (1968). 'The Coeur d'Alene district, Idaho', *Ore Deposits of the United States 1933–1967. Vol. 2*, pp. 1417–1435. Ed. J.D. Ridge. New York; Amer. Inst. Min. Metall. Petrol. Engineers

Hobson, G.D. and Tiratsoo, E.N. (1981). *Introduction to Petroleum Geology*, 2nd edn. Beaconsfield, England; Scientific Press

Hoeve, J. and Sibbald, T.I.I. (1978a). 'Mineralogy and geologic setting of unconformity-type uranium deposits in northern Saskatchewan', *Uranium Deposits: Their Mineralogy and Origin. Short Course Handbook 3*, pp. 457–474. Ed. M.M. Kimberley. Ottawa; Mineral. Assoc. Canada

Hoeve, J. and Sibbald, T.I.I. (1978b). 'On the genesis of Rabbit Lake and other unconformity-type uranium deposits in northern Saskatchewan, Canada', *Econ. Geol.* Vol. 73, 1450–1473

Höll, R. (1977). 'Early Paleozoic ore deposits of the Sb–W–Hg formation in the eastern Alps and their genetic interpretation', *Time- and Strata-bound Ore Deposits*, pp. 169–198. Ed. D.D. Klemme and H.J. Schneider. Berlin; Springer-Verlag

Höll, R. and Maucher, A. (1976). 'The strata-bound ore deposits in the eastern Alps', *Handbook of Strata-bound and Stratiform Ore Deposits. Vol. 5*, pp. 1–36. Ed. K.H. Wolf. Amsterdam; Elsevier

Höll, R., Maucher, A. and Westenberger, M. (1972). 'Synsedimentary–diagenetic ore fabrics in the strata- and time-bound scheelite deposits of Kleinartal and Felbertal in the eastern Alps', *Mineral. Deposita.* Vol. 7, 217–226

Hollister, V.F. (1975). 'An appraisal of the nature and source of porphyry copper deposits', *Mineral. Sci. Eng.* Vol. 7, 225–233

Hollister, V.F. (1978). *Geology of the Porphyry Copper Deposits of the Western Hemisphere.* New York; Amer. Inst. Min. Metall. Petrol. Engineers

Hood, A., Gutjahr, C.C.M. and Heacock, R.L. (1975). 'Organic metamorphism and the generation of petroleum', *Amer. Assoc. Petrol. Geologists Bull. 59*, p. 986. Tulsa, Oklahoma; Amer. Assoc. Petrol. Geol.

Hook, J.W. (1974). 'Structure of the fault systems in the Ilinois–Kentucky fluorspar district', *A Symposium on the Geology of Fluorspar. Kentucky Geol. Surv. Series X. Special Publication 22*, pp. 77–86. Ed. D.W. Hutcheson. Lexington, Kentucky; Kentucky Geological Survey

Horon, O. (1977). 'Les gisements de fer de la France', *The Iron Ore Deposits of Europe and Adjacent Areas. Vol. 1*, pp. 143–159. Ed. A. Zitzmann. Hannover; Bundesanstalt für Geowissenschaften und Rohstoffe

Hosking, K.F.G. (1964). 'Permo-Carboniferous and later primary mineralization of Cornwall and southwest Devon', *Present Views of Some Aspects of the Geology of Cornwall and Devon*, pp. 201–245. Ed. K.F.G. Hosking and G.J. Shrimpton. Camborne, Cornwall; Royal Geographical Society of Cornwall

Hosking, K.F.G. (1970a). 'The nature of the primary tin ores of the south-west of England', *Second Technical Conference on Tin, Bangkok 1969, Vol. 1*, pp. 1157–1244. Ed. W. Fox. London; International Tin Council

Hosking, K.F.G. (1970b). 'Aspects of the geology of the tin fields of south-east Asia', *Second Technical Conference on Tin, Bangkok 1969, Vol. 1*, pp. 41–80. Ed. W. Fox. London; International Tin Council

Hosking, K.F.G. (1973a). 'Primary mineral deposits', *Geology of the Malay Peninsula*, pp. 335–390. Ed. D.J. Gobbett and C.S. Hutchison. New York; Wiley–Interscience

Hosking, K.F.G. (1973b). The search for tungsten deposits. *Geol. Soc. Malaysia Bull.* Vol. 5, 1–70

Hosking, K.F.G. (1977). 'Known relationships between the 'hard-rock' tin deposits and the granites of Southeast Asia', *Geol. Soc. Malaysia Bull.* Vol. 9, 141–157

Hosking, K.F.G. (1979). 'Tin distribution patterns', *Geol. Soc. Malaysia Bull.* Vol. 11, 1–70

Hosterman, J.W. (1973). 'Clays', *United States Mineral Resources. US Geol. Surv. Prof. Paper 820*, pp. 123–131. Ed. D.A. Brobst and W.P. Pratt. Washington DC; US Geological Survey

Howard, P.F. and Hough, M.J. (1979). 'On the geochemistry and origin of the D tree, Wonorah, and Sherrin Creek phosphorite deposits of the Geogina Basin, Northern Australia', *Econ. Geol.* Vol. 74, 260–284

Hsu, L.C. (1976). 'The stability relations of the wolframite series', *Amer. Mineral.* Vol. 61, 944–955

Huang, C.K. (1955). 'Gold–copper deposits of the Chinkuashih Mine, Taiwan, with special reference to the mineralogy', *Acta Geol. Taiwan.* Vol. 7, 1–20

Hubbard, H.A. and Ericksen, G.E. (1973). 'Limestone and dolomite', *United States Mineral Resources. US Geol. Surv. Prof. Paper 820*, pp. 357–364. Ed. D.A. Brobst and W.P. Pratt. Washington DC; US Geological Survey

Hubbert, M.K. (1953). 'Entrapment of petroleum under hydrodynamic conditions', *Amer. Assoc. Petrol. Geol. Bull.* Vol. 37, 1954–2026

Hunter, D.R. (1973). 'The localization of tin mineralization with reference to southern Africa', *Mineral. Sci. Eng.* Vol. 5, 53–77

Hunter, D.R. (1976). 'Some enigmas of the Bushveld complex', *Econ. Geol.* Vol. 71, 229–248

Hunter, R.E. (1970). 'Facies of iron sedimentation in the Clinton Group', *Studies of Appalachian Geology: Central and Southern*, pp. 101–121. Ed. G.W. Fisher, F.J. Pettijohn, J.C. Reed Jr. and K.N. Weaver. New York; Wiley–Interscience

Hutchinson, R.W. (1973). 'Volcanogenic sulfide deposits and their metallogenic significance', *Econ. Geol.* Vol. 68, 1223–1246

Hutchinson, R.W. (1979). 'Evidence of exhalative origin for Tasmanian tin deposits', *Trans. Can. Inst. Min. Metall.* Vol. 82, 90–104

Hutchinson. R.W. (1980). 'Massive base metal sulphide deposits as guides to tectonic evolution', *The Continental Crust and its Mineral Deposits. Geol. Assoc. Canada Special Paper 20*, pp. 659–684. Ed. D.W. Strangway. Toronto; Geol. Assoc. Canada

Hutchinson, R.W., Ridler, R.H. and Suffel, G.G. (1971). 'Metallogenic relationships in the Abitibi belt Canada: a model for Archean metallogeny', *Trans. Can. Inst. Min. Metall.* Vol. 74, 106–115

Hutchison, C.S. (1972). 'Alpine-type chromite in North Borneo, with special reference to Darvel Bay', *Amer. Mineral.* Vol. 57, 835–856

Hutchison, C.S. (1975). 'Ophiolite in Southeast Asia', *Geol. Soc. Amer. Bull.* Vol. 86, 797–806

Hutchison, C.S. (1977). 'Granite emplacement and tectonic subdivision of Peninsular Malaysia', *Geol. Soc. Malaysia Bull.* Vol. 9, 187–207

Hutchison, C.S. (1978a). 'The distribution and origin of Southeast Asian ore deposits', *Malaysian Geographers.* Vol. 1, 13–36

Hutchison, C.S. (1978b). 'Ophiolite metamorphism in northeast Borneo', *Lithos.* Vol. 11, 195–208

Hutchison, C.S. (1978c). 'Southeast Asian tin granitoids of contrasting tectonic setting', *J. Phys. Earth*, Vol. 26 Supplement, S221–S232

Hutchison, C.S. (1981). *Belt Orogenesis along Northern Margin of Idaho Batholith. Idaho Bureau of Mines and Geology, Open File Report. 81–2.* Moscow, Idaho; Idaho Bureau of Mines and Geology

Hutchison, C.S. (1982). 'Southeast Asia', *The Ocean Basins and Their Margins: The Indian Ocean,* pp. 451–512. Ed. A.E.M. Nairn and F.G. Shehli. New York; Plenum Press

Hutchison, C.S. and Chakraborty, K.R. (1979). 'Tin: a mantle or crustal source?', *Geol. Soc. Malaysia Bull.* Vol. 11, 71–79

Hutchison, C.S. and Snelling, N.J. (1971). 'Age determination on the Bukit Paloh adamellite', *Geol. Soc. Malaysia Bull.* Vol. 4, 97–100

Hutchison, C.S. and Taylor, D. (1978). 'Metallogenesis in S.E. Asia', *J. Geol. Soc. London.* Vol. 135, 407–428

Ikonnikov, A.B. (1975). *Mineral Resources of China. Geol. Soc. Amer., Microfilm 2.* Boulder, Colorado; Geol. Soc. Amer.

Ineson, P.R. (1976). 'Ores of the northern Pennines, the Lake District, and north Wales', *Handbook of Strata-bound and Stratiform Ore Deposits. Vol. 5,* pp. 197–230. Ed. K.H. Wolf. Amsterdam; Elsevier

Ingham, F.T. and Bradford, E.F. (1960). *The Geology and Mineral Resources of the Kinta Valley, Perak. Federation of Malaya Geol. Surv. District Mem. 9.* Ipoh, Malaysia; Geological Survey of Malaysia

Irving, E., North, F.K. and Couillard, R. (1974). 'Oil, climate and tectonics', *Can. J. Earth. Sci.* Vol. 11, 1–17

Isacks, B., Oliver, J. and Sykes, L.R. (1968). 'Seismology and new global tectonics', *J. Geophys. Res.* Vol. 73, 5855–5899

Ishihara, S. (ed.) (1974). *Geology of Kuroko Deposits, Mining Geology Special Issue 6.* Tokyo; Soc. Min. Geol. Japan

Ishihara, S. (1977). 'The magnetite-series and ilmenite-series granitic rocks', *Min. Geol.* Vol. 27, 293–305

Ishihara, S. (1978). 'Metallogenesis in the Japanese island arc system', *J. Geol. Soc. London.* Vol. 135, 389–406

Ishihara, S. (1980). 'Significance of the magnetite-series and ilmenite-series of granitoids in mineral exploration', *Proc. Fifth Quadrennial IAGOD Symposium.* pp. 309–312. Stuttgart; Schweizerbart'sche

Ishihara, S. and Mochizuki, T. (1980). 'Uranium and thorium contents of Mesozoic granites from Peninsular Thailand', *Bull. Geol. Surv. Japan.* Vol. 31, No. 8, 369–376

Ishihara, S. and Sasaki, A. (1978). 'Sulfur of kuroko deposits—a deep seated origin?', *Min. Geol.* Vol. 28, 361–367

Ishihara, S., Sawata, H., Arpornsuwan, S., Busaracome, P. and Bungbrakearti, N. (1978). 'Granitic rocks in southern Thailand', *Geology and Mineral Resources of Southeast Asia,* pp. 265–267. Ed. Prinya Nutalaya. Bangkok; Asian Institute of Technology

Ishihara, S., Sawata, H., Arpornsuwan, S., Busaracome, P. and Bungbrakearti, N. (1979). 'The magnetite-series and ilmenite-series granitoids and their bearing on tin mineralization, particularly of the Malay Peninsula region', *Geol. Soc. Malaysia Bull.* Vol. 11, 103–110

Ishihara, S., Sawata, H., Shibata, K., Terashima, S., Arrykul, S. and Sato, K. (1980). 'Granites and Sn–W deposits of Peninsular Thailand', *Granitic Magmatism and Related Mineralization. Mining Geology, Japan, Special Issue No. 8,* pp. 223–241. Ed. S. Ishihara and S. Takenouchi. Tokyo; Geological Survey of Japan

Ivanova, G.F. (1978). 'Geochemical and physico-chemical conditions of tungsten migration and deposition', *Metallization Associated with Acid Magmatism. Vol. 3,* pp. 337–341. Ed. M. Stemprok, L. Burnol and G. Tischendorf. Prague; Geological Survey

Jackson, S.A. and Folinsbee, R.E. (1969). 'The Pine Point lead–zinc deposits, N.W.T., Canada, introduction and paleoecology of the Presqu'ile reef', *Econ. Geol.* Vol. 64, 711–717

Jackson, N.J. (1979). 'Geology of the Cornubian tin field—a review', *Geol. Soc. Malaysia Bull.* Vol. 11, 209–238

Jacobson, R.R.E. and Webb, J.S. (1946). *The pegmatites of central Nigeria. Geol. Surv. Nigeria Bull.* Vol. 17. Kaduna South; Geological Survey of Nigeria

Jacoby, C.H. (1975). 'Manganese', *Industrial Minerals and Rocks (Nonmetallics Other Than Fuels),* pp. 821–831. Ed. S. LeFond. New York; Amer. Inst. Min. Metall. Petrol. Engineers

Jahns, R.H. (1975). 'Gem materials', *Industrial Minerals and Rocks (Nonmetallics Other Than Fuels),* 4th edn., pp. 271–326. Ed. S.J. LeFond. New York; Amer. Inst. Min. Metall. Petrol. Engineers

Jahns, R.H. and Burnham, C.W. (1969). 'Experimental studies of pegmatite genesis I. A model for the derivation and crystallization of granitic pegmatites', *Econ. Geol.* Vol. 64, 843–864

Jambor, J.L. (1971). 'The silver-arsenide deposits of the Cobalt–Gowganda region, Ontario. General geology', *Can. Mineral.* Vol. 11, No. 1, 12–33

James, D.E. (1971). 'Plate tectonic model for the evolution of the Central Andes', *Geol. Soc. Amer. Bull.* Vol. 82, 3325–3346

James, L.P. (1976). 'Zoned alteration in limestone at porphyry copper deposits, Ely, Nevada', *Econ. Geol.* Vol. 71, 488–512

Jenkyns, H.C. (1977). 'Fossil nodules'. *Marine Manganese Deposits,* pp. 87–108. Ed. G.P. Glasby. Amsterdam; Elsevier

Jensen, J.H. (1975). 'Bromine', *Industrial Minerals and Rocks (Nonmetallics Other Than Fuels),* 4th edn., pp. 497–500. Ed. S.J. LeFond. New York; Amer. Inst. Min. Metall. Petrol. Engineers

Jensen, M.L. and Bateman, A.M. (1979). *Economic Mineral Deposits,* 3rd edn. New York; John Wiley

John, J.U. (1963). 'Geology and mineral deposits of east-central Balabac island, Palawan Province, Philippines', *Econ. Geol.* Vol. 58, 107–130

John, Y.W. (1963). 'Geology and origin of Sangdong tungsten mine, republic of Korea', *Econ. Geol.* Vol. 58, 1285–1300

Johnson, I.R. and Klingner, G.D. (1975). 'Broken Hill ore deposit and its environment', *Economic Geology of Australia and Papua New Guinea. Vol. 1. Metals,* pp. 476–491. Parkville, Victoria; Australasian Inst. Min. Metall.

Jolly, J. (1980a). 'Platinum-group metals', *Minerals Yearbook 1978–79, Vol. 1. Metals and Minerals,* pp. 699–712. Bureau of Mines. Washington DC; US Government Printing Office

Jolly, J.H. (1980b). 'Copper', *Minerals Yearbook 1978–79, Vol. 1. Metals and Minerals,* pp. 271–311. Bureau of Mines. Washington DC; US Government Printing Office

Jones, M.T., Reed, B.L., Doe, B.R. and Lanphere, M.A. (1977). 'Age of tin mineralization and plumbotectonics, Belitung, Indonesia', *Econ. Geol.* Vol. 72, 745–752

Jones, T.S. (1980). 'Columbium and tantalum', *Minerals Yearbook 1978–79, Vol. 1. Metals and Minerals,* pp. 259–270. Bureau of Mines. Washington DC; US Government Printing Office

Jones, W.A. (1948). 'Hollinger Mine', *Structural Geology of Canadian Ore Deposits,* pp. 464–481. Montreal; Can. Inst. Min. Metall.

Juan, V.C. (1946). 'Mineral Resources of China', *Econ. Geol.* Vol. 41, 399–474

Jung, W. and Knitzschke, C. (1976). 'Kupferschiefer in the German Democratic Republic (GDA) with special reference to the kupferschiefer deposit in the southeast Harz foreland', *Handbook of Strata-bound and Stratiform Ore Deposits. Vol. 6, Cu, Zn, Pb, and Ag Deposits,* pp. 353–406. Ed. K.H.

Wolf. Amsterdam; Elsevier

Kadley, F.L. Jr. (1975). 'Diatomite', *Industrial Minerals and Rocks (Nonmetallics Other Than Fuels)*, pp. 605–635. Ed. S.J. LeFond. New York; Amer. Inst. Min. Metall. Petrol. Engineers

Kajiwara, Y. (1970). 'Syngenetic features of the Kuroko ore from the Shakanai Mine', *Volcanism and Ore Genesis*, pp. 197–206. Ed. T. Tatsumi. Tokyo; University of Tokyo Press

Kanehira, K. (1970). 'Conformable copper-pyrite deposits in the Iimori mining district', *Volcanism and Ore Genesis*, pp. 93–104. Ed. T. Tatsumi. Tokyo. University of Tokyo Press

Kanehira, K. and Tatsumi, T. (1970). 'Bedded cupriferous iron sulphide deposits in Japan, a review', *Volcanism and Ore Genesis*, pp. 51–76. Ed. T. Tatsumi. Tokyo; University of Tokyo Press

Karig, D.E. (1972). 'Remnant arcs', *Geol. Soc. Amer. Bull.* Vol. 83, 1057–1068

Katz, M.B. and Tuckwell, K.D. (1979). 'Controls of tin-bearing pegmatites and granites in the Precambrian of Broken Hill, Australia', *Geol. Soc. Malaysia Bull.* Vol. 11, 253–265

Kazansky, V.I. and Laverov, N.P. (1977). 'Deposits of uranium', *Ore Deposits of the USSR. Vol. II*, pp. 349–424. Ed. V.I. Smirnov. Translated into English by D.A. Brown. London; Pitman Publishing Co.

Kelley, V.C., Kittel, D.F. and Melancon, P.E. (1968). 'Uranium deposits of the Grants Region', *Ore Deposits of the United States 1933–1967. The Graton–Sales Volume, Vol. 1*, pp. 747–769. Ed. J.D. Ridge. New York; Amer. Inst. Min. Metall. Petrol. Engineers

Kelly, W.C. and Rye, R.O. (1979). 'Geologic, fluid inclusion, and stable isotope studies of the tin–tungsten deposits of Panasqueira, Portugal', *Econ. Geol.* Vol. 74, 1721–1822

Kelly, W.C. and Turneaure, F.S. (1970). *Mineralogy, Paragenesis and Geochemistry of the Tin and Tungsten Deposits of the Eastern Andes, Bolivia. Econ. Geol.* Vol. 65, 609–680

Keppie, J.D. (1977). *Plate Tectonic Interpretations of Palaeozoic World Maps. Nova Scotia Department of Mines Paper 77-3.* Halifax; Nova Scotia Department of Mines

Kesler, S.E. (1977). 'Geochemistry of Manto fluorite deposits, northern Coahuila, Mexico', *Econ. Geol.* Vol. 72, 204–218

Kesler, S.E. (1978). 'Metallogenesis of the Caribbean region', *J. Geol. Soc. London.* Vol. 135, 429–441

Kesler, S.E., Jones, L.M. and Walker, R.L. (1975). 'Intrusive rocks associated with porphyry copper mineralization in island arc areas', *Econ. Geol.* Vol. 70, 515–526

Kesler, T.L. (1950). *Geology and Mineral Deposits of the Carterville District, Georgia. US Geol. Surv. Prof. Paper 224.* Washington DC; US Geological Survey

Ketner, K.B. (1965). 'Barite deposits', *Tectonic and Igneous Geology of the Northern Shoshone Range, Nevada. US Geol. Surv. Prof. Paper 465*, pp. 134–144. Ed. J. Gilluly and O. Gates. Washington DC; US Geological Survey

Ketner, K.B. (1973). 'Silica sand', *United States Mineral Resources. US Geol. Surv. Prof. Paper 820*, pp. 577–580. Ed. D.A. Brobst and W.P. Pratt. Washington DC; US Geological Survey

Ketner, K.B. (1975). 'Replacement barite deposit, southern Independence Mountain, Nevada', *US Geol. Surv. J. Res.* Vol. 3, No. 5, 547–551

Kimberley, M.M. (1978a). 'Origin of stratiform uranium deposits in sandstone, conglomerate, and pyroclastic rock', *Uranium Deposits, Their Mineralogy and Origin, Short Course Handbook 3*, pp. 339–381. Ed. M.M. Kimberley. Ottawa; Mineral. Assoc. Canada

Kimberley, M.M. (1978b). 'Paleo-environmental classification of iron-formations', *Econ. Geol.* Vol. 73, 215–229

Kimura, E.T., Bysouth, G.D. and Drummond, A.D. (1976). 'Endako', *Porphyry Deposits of the Canadian Cordillera. Can. Inst. Min. Metall. Special Volume 15*, pp. 444–454. Ed. A. Sutherland Brown. Montreal; Can. Inst. Min. Metall.

King, R.U. (1973). 'Rhenium', *United States Mineral Resources. US Geol. Surv. Prof. Paper 820*, pp. 557–559. Ed. D.A. Brobst and W.P. Pratt. Washington DC; US Geological Survey

King, R.U., Shawe, D.R. and McKevett, E.M. Jr. (1973). 'Molybdenum', *United States Mineral Resources. US Geol. Surv. Prof. Paper 820*, pp. 425–435. Ed. D.A. Brobst and W.P. Pratt. Washington DC; US Geological Survey

Kirk, W.S. (1980). 'Thorium', *Minerals Yearbook 1978–79, Vol. 1. Metals and Minerals*, pp. 907–913. Bureau of Mines. Washington DC; US Government Printing Office

Kirpal, G.R. and Tanyakov, V.A. (1977). 'Deposits of aluminium', *Ore Deposits of the USSR, Vol. 1*, pp. 273–352. Ed. V.I. Smirnov. Translated into English by D.A. Brown. London; Pitman Publishing Co.

Kistler, R.B. and Smith, W.C. (1975). 'Boron and borates', *Industrial Minerals and Rocks (Nonmetallics Other Than Fuels)*, 4th edn., pp. 473–496. Ed. S.J. LeFond. New York; Amer. Inst. Min. Metall. Petrol. Engineers

Kisvarsanyi, G. (1977). 'The role of the Precambrian igneous basement in the formation of the stratabound lead–zinc–copper deposits in southeast Missouri', *Econ. Geol.* Vol. 72, 435–442

Klemic, H. (1975). 'Zirconium and hafnium minerals', *Industrial Minerals and Rocks (Nonmetallics Other Than Fuels)*, pp. 1275–1283. Ed. S.J. LeFond. New York; Amer. Inst. Min. Metall. Petrol. Engineers

Klemic, H., Gottfried, D., Cooper, M. and Marsh, S.P. (1973a). 'Zirconium and hafnium', *United States Mineral Resources. US Geol. Surv. Prof. Paper 820*, pp. 713–722. Ed. D.A. Brobst and W.P. Pratt. Washington DC; US Geological Survey

Klemic, H., James, H.L. and Eberlein, G.D. (1973b). 'Iron', *United States Mineral Resources. US Geol. Surv. Prof. Paper 820*, pp. 291–306. Ed. D.A. Brobst and W.P. Pratt. Washington DC; US Geological Survey

Klemic, H., Marsh, S.P. and Cooper, M. (1973c). 'Titanium', *United States Mineral Resources. US Geol. Surv. Prof. Paper 820*, pp. 653–665. Ed. D.A. Brobst and W.P. Pratt. Washington DC; US Geological Survey

Klemme, H.D. (1971). 'To find a giant, find the right basin', *Oil and Gas Journal*, March 8th

Klemme, H.D. (1973). 'Structure-related traps expected to dominate world reserve statistics', *Oil and Gas Journal*, Dec. 31st

Klemme, H.D. (1975). 'Giant oil fields related to their geologic setting: a possible guide to their exploration', *Bull. Can. Petrol. Geol.* Vol. 23, 30–66

Klemme, H.D. (1977). 'One-fifth of reserves lies offshore', *Oil and Gas Journal*, Petroleum 2000 seventy-fifth anniversary issue, Aug.

Klingspor, A.M. (1969). 'Middle Devonian Muskeg evaporites of western Canada', *Amer. Assoc. Petrol. Geol. Bull.* Vol. 53, 927–948

Knight, C.L. (1975). 'Mount Bischoff tin orebody', *Economic Geology of Australia and Papua New Guinea. Vol. 1. Metals*, pp. 591, 592. Ed. C.L. Knight. Parkville, Victoria; Australasian Inst. Min. Metall.

Knipping, H.D. (1974). 'The concepts of supergene versus hypogene emplacement of uranium at Rabbit Lake, Saskatchewan, Canada', *Formation of Uranium Ore Deposits*, pp. 531–549. Vienna; International Atomic Energy Agency

Kobelski, B.J., Gold, D.P. and Deines, P. (1979). 'Variations in stable isotope compositions for carbon and oxygen in some South African and Lesothan kimberlites', *Kimberlites, Diatremes and Diamonds: Their Geology, Petrology and Geochemistry. Proc. 2nd International Kimberlite Conference, Vol. 1*, pp. 252–271. Ed. F.R. Boyd and H.O.A. Meyer. Washington DC; American Geophysical Union

Koeppel, V. (1968). *Age and History of the Uranium Mineralization of the Beaverlodge Area, Saskatchewan. Can. Geol. Surv. Paper 67-31.* Ottawa; Canadian Geological Survey

Köppel. V. and Saager, R. (1976). 'Uranium-, thorium- and lead-isotope studies of stratabound ores', *Handbook of Strata-bound and Stratiform Ores. Vol. 2*, pp. 267–316. Ed. K.H. Wolf. Amsterdam; Elsevier

Koo, J. and Mossman, D.J. (1975). 'Origin and metamorphism of the Flin Flon stratabound Cu–Zn sulfide deposit, Saskatchewan and Manitoba', *Econ. Geol.* Vol. 70, 48–62

Kostick, D.S. (1980). 'Sodium and sodium compounds', *Minerals Yearbook 1978–79, Vol. 1. Metals and Minerals*, pp. 835–842. Bureau of Mines. Washington DC; US Government Printing Office

Kostick, D.S. and DeFilippo, R.J. (1980). 'Fluorspar', *Minerals Yearbook 1978–79, Vol. 1. Metals and Minerals*, pp. 341–357. Bureau of Mines. Washington DC; US Goverment Printing Office

Kraume, E. (1960). 'Stratigraphie und Tektonik der Rammelsberger Erzlager unter besonderer Berücksichtigung des neuen Lagers unter der 10 Sohle', *Zeit. Erzberg. Metallhuttenw.* Vol. 13, 7–12

Krauskopf, K.B. (1979). *Introduction to Geochemistry*, 2nd edn. New York; McGraw-Hill

Krishnan, M.S. (1973). 'Occurrence and origin of the iron ores of India', *Genesis of Precambrian Iron and Manganese Deposits*, pp. 69–76. Paris; UNESCO

Kummer, J.T. (1980). 'Molybdenum', *Minerals Yearbook 1978–79, Vol. 1. Metals and Minerals*, pp. 615–628. Bureau of Mines. Washington DC; US Government Printing Office

Kunasz, I.A. (1975). 'Lithium raw materials', *Industrial Minerals and Rocks (Nonmetallics Other Than Fuels)*, pp. 791–803. Ed. S.J. LeFond. New York; Amer. Inst. Min. Metall. Petrol. Engineers

Laing, W.P., Marjoribanks, R.W. and Rutland, R.W.R. (1978). 'Structure of the Broken Hill mine area and its significance for the generation of the orebodies', *Econ. Geol.* Vol. 73, 1112–1136

Lambert, I.B. (1976). 'The McArthur zinc–lead–silver deposit; features, metallogenesis and comparisons with some other stratiform ores', *Handbook of Strata-bound and Stratiform Ore Deposits. Vol. 6*, pp. 535–585. Ed. K.H. Wolf. Amsterdam; Elsevier

Lambert, I.B. and Sato, T. (1974). 'The kuroko and associated ore deposits of Japan: a review of their features and metallogenesis', *Econ. Geol.* Vol. 69, 1215–1236

Lameyre, J. (1980). 'Les magmas granitiques: leurs compartements, leurs associations et leurs sources'. *Mem. h. ser. Soc. geol. de France, No. 10, Livre Jubilaire de la Soc. geol. de France, 1830–1980*, pp. 51–62

Landis, G.P. and Rye, R.O. (1974). 'Geologic fluid inclusion, and stable isotope studies of the Pasto Bueno tungsten-base metal ore deposit, northern Peru', *Econ. Geol.* Vol. 69, 1025–1059

Langford, F.F. (1974). 'A supergene origin for vein-type uranium ores in the light of the Western Australia calcrete–carnotite deposits', *Econ. Geol.* Vol. 69, 516–526

Langford, F.F. (1977). 'Surficial origin of North American pitchblende and related uranium deposits', *Amer. Assoc. Petrol. Geol. Bull.* Vol. 61, 28–42

Langford, F.F. (1978a). 'Mobility and concentration of uranium in arid surficial environments', *Uranium Deposits: Their Mineralogy and Origin, Short Course Handbook 3*, pp. 383–394. Ed. M.M. Kimberley. Ottawa; Mineral. Assoc. Canada

Langford, F.F. (1978b). 'Uranium deposits in Australia', *Uranium Deposits, Their Mineralogy and Origin. Short Course Handbook 3*, pp. 205–216. Ed. M.M. Kimberley. Ottawa; Mineral. Assoc. Canada

Lapham, D.M. (1968). 'Triassic magnetite and diabase at Cornwall, Pennsylvania', *Ore Deposits of the United States, 1933–1967. The Graton–Sales Volume, Vol. 1*, pp. 72–94. Ed. J.D. Ridge. New York; Amer. Inst. Min. Metall. Petrol. Engineers

Lapham, D.M. and Gray, C. (1973). *Geology and Origin of the Triassic Magnetite Deposits and Diabase at Cornwall, Pennsylvania. Geol. Survey Mineral Resources Rept. M56*, Washington DC; US Geological Survey

Larsen, K.G. (1977). 'Sedimentology of the Bonneterre Formation, Southeast Missouri', *Econ. Geol.* Vol. 72, 408–419

Lau, J.W.E. (1971). 'Mineralogical study of the arsenical gold ore from the Bau mining district, Sarawak, Malaysia', *Discussion Meeting Geol. Soc. Malaysia*, Dec. 1970, Paper 7

Laurence, R.A. (1973). 'Construction stone', *United States Mineral Resources. US Geol. Surv. Prof. Paper 820*, pp. 157–162. Ed. D.A. Brobst and W.P. Pratt. Washington DC; US Geological Survey

Lawrence, L.J. (1973). 'Polymetamorphism of the sulphide ores at Broken Hill', *Mineral. Deposita.* Vol. 8, 211–236

Laznicka, P. (1973). 'Development of non-ferrous metal deposits in geological time', *Can. J. Earth Sci.* Vol. 10, 18–25

LeBas, M.J. (1977). *Carbonatite–Nephelinite Volcanism.* New York; Wiley–Interscience

LeFond, S.J. (ed.) (1975). *Industrial Minerals and Rocks (Nonmetallics Other Than Fuels)*, 4th edn. New York; Amer. Inst. Min. Metall. Petrol. Engineers

LeFond, S.J. and Jacoby, C.H. (1975). 'Salt', *Industrial Minerals and Rocks (Nonmetallics Other Than Fuels)*, pp. 995–1025. Ed. S.J. LeFond. New York; Amer. Inst. Min. Metall. Petrol. Engineers

Leow, J.H., Hosking, K.F.G. and Teh, G.H. (1969). 'Aspects of the mineralogy of the N.E.–S.W. trending veins at Tekka, Perak', *Geol. Soc. Malaysia Newsletter.* Vol. 21, 2–8

LeRoy, J. (1978). 'The Margnac and Fanay uranium deposits of the La Crouzille district (Western Massif Central, France): geologic and fluid inclusion studies', *Econ. Geol.* Vol. 73, 1611–1634

LeSure, F.G. (1973a). 'Feldspar', *United States Mineral Resources. US Geol. Surv. Prof. Paper 820*, pp. 217–222. Ed. D.A. Brobst and W.P. Pratt. Washington DC; US Geological Survey

LeSure, F.G. (1973b). 'Mica', *United States Mineral Resources. US Geol. Surv. Prof. Paper 820*, pp. 415–423. Ed. D.A. Brobst and W.P. Pratt. Washington DC; US Geological Survey

Levinson, A.A. (1974). *Introduction to Exploration Geochemistry.* Calgary, Canada; Applied Publishing Ltd.

Levorsen, A.I. and Berry, F.A.F. (1967). *Geology of Petroleum*, 2nd edn. San Francisco; W.H. Freeman

Lillie, A.R. and Brothers, R.N. (1970). 'The geology of New Caledonia', *New Zealand J. Geol. Geophys.* Vol. 13, 145–183

Lindgren, W. (1933). *Mineral Deposits*, 4th edn. New York; McGraw-Hill

Lipman, P.W., Prostka, H.J. and Christiansen, R.L. (1971). 'Evolving subduction zones in the western United States as interpreted from igneous rocks', *Science.* Vol. 74, 821–825

Lipman, P.W., Fisher, F.S., Mehnert, H.H., Naeser, C.W., Luedke, R.G. and Steven, T.A. (1976). 'Multiple stages of Mid Tertiary mineralization and alteration in the western San Juan Mountains, Colorado', *Econ. Geol.* Vol. 71, 571–588

Lissiman, J.C. and Oxenford, R.J. (1975). 'Eneabba rutile–zircon–ilmenite sand deposit', *Economic Geology of Australia and Papua New Guinea. Vol. 1. Metals*, pp. 1062–1070. Ed. C.L. Knight. Melbourne; Australasian Inst. Min. Metall.

Livingston, D.E. (1973). 'A plate tectonic hypothesis for the genesis of porphyry copper deposits of the Southern Basin and Range Province', *Earth Planet. Sci. Lett.* Vol. 20, 171–179

Lowell, G.R. (1976). 'Tin mineralization and mantle hot spot activity in south-eastern Missouri', *Nature*. Vol. 261, 482–483

Lowell, J.D. and Guilbert, J. (1970). 'Lateral and vertical alteration mineralization zoning in porphyry ore deposits', *Econ. Geol.* Vol. 65, 373–408

Lucas, J.M. (1980). 'Cadmium', *Minerals Yearbook 1978–79, Vol. 1. Metals and Minerals*, pp. 139–145. Bureau of Mines. Washington DC; US Government Printing Office

Lufkin, J.L. (1976). 'Oxide minerals in miarolitic rhyolite, Black Range, New Mexico', *Amer. Mineral.* Vol. 61, 425–430

Lynd, L.E. (1980). 'Zirconium and hafnium', *Minerals Yearbook 1978–79, Vol. 1. Metals and Minerals*, pp. 1021–1031. Bureau of Mines. Washington DC; US Government Printing Office

Lynd, L.E. and Hough, R.A. (1980). 'Titanium', *Minerals Yearbook 1978–79, Vol. 1. Metals and Minerals*, pp. 993–947. Bureau of Mines. Washington DC; US Government Printing Office

Lynd, L.E. and LeFond, S.J. (1975). 'Titanium minerals', *Industrial Minerals and Rocks (Nonmetallics Other Than Fuels)*, pp. 1149–1208. Ed. S.J. LeFond. New York; Amer. Inst. Min. Metall. Petrol. Engineers

MacKevett, E.A. (1963). *Geology and Ore Deposits of the Bokan Mountain Uranium–Thorium Area, Southeastern Alaska. US Geol. Surv. Bull. 1154.* Washington DC; US Geological Survey

Mackowsky, M.T. (1968). 'European Carboniferous coalfields and Permian Gondwana coalfields', *Coal and Coal-bearing Strata*, pp. 325–345. Ed. D. Murchison and T.S. Westoll. Edinburgh; Oliver and Boyd

MacLeod, W.N., Turner, D.C. and Wright, E.P. (1971). *The Geology of the Joss Plateau. Vol. 1. General Geology. Geol. Surv. Nigeria Bull. 32.* Kaduna South; Geological Survey of Nigeria

Magara, K. (1977). 'Petroleum migration and accumulation', *Developments in Petroleum Geology—1*, pp. 83–126. Ed. G.D. Hobson. London; Applied Science Publishers

Maher, S.W. (1970). *Barite Resources of Tennessee. Report of Investigations 28.* Nashville; Tennessee Division of Geology

Mamen, C. (1971). 'Almaden—quicksilver centre of the world', *Can. Min. J.* Vol. 92, No. 1, 40–43

Mann, A.W. and Deutscher, R.L. (1978). 'Genesis principles for the precipitation of carnotite in calcrete drainages in Western Australia', *Econ. Geol.* Vol. 73, 1724–1737

Mannheim, B.R.S. (1979). 'A new conception of the formation of the Kola Peninsular apatite deposit: the coprogenic impact theory—CIT', *J. Irreproducible Results.* Vol. 25, 6, 7

Mannion, L.E. (1975). 'Sodium carbonate deposits', *Industrial Minerals and Rocks (Nonmetallics Other Than Fuels)*, pp. 1061–1079. Ed. S.J. LeFond. New York; Amer. Inst. Min. Metall. Petrol. Engineers

Martin, H.J. (1964). 'The Bikita tinfield', *Rhodesia Geol. Surv. Bull.* Vol. 58, 114–132

Martini, J.E.J. (1976). 'The fluorite deposits in the Dolomite Series of the Marico District, Transvaal, South Africa', *Econ. Geol.* Vol. 71, 625–635

Mason, J.E. (1974). 'Geology of the Derbyshire fluorspar deposits, United Kingdom', *A Symposium on the Geology of Fluorspar. Kentucky Geol. Surv. Series X, Special Publication 22*, pp. 10–22. Ed. D.W. Hutcheson. Lexington, Kentucky; Kentucky Geological Survey

Mason, R. (1978). *Petrology of the Metamorphic Rocks.* London; George Allen and Unwin

Matheis, G. (1979). 'Geochemical exploration around the pegmatitic Sn–Nb–Ta mineralization of southwest Nigeria', *Geol. Soc. Malaysia Bull.* Vol. 11. 333–351

Mathias, B.V. and Clark, G.J. (1975). 'Mount Isa copper and silver–lead–zinc orebodies—Isa and Hilton mines', *Economic Geology of Australia and Papua New Guinea. Vol. 1. Metals*, pp. 351–372. Ed. C.L. Knight. Parkville, Victoria; Australasian Inst. Min. Metal.

Matthews, N.A. (1980). 'Nickel', *Minerals Yearbook 1978–79. Vol. 1. Metals and Minerals*, pp. 629–641. Bureau of Mines. Washington DC; US Government Printing Office

Matthews, N.A. and Morning, J.L. (1980). 'Chromium', *Minerals Yearbook 1978–79. Vol. 1. Metals and Minerals*, pp. 193–205. Bureau of Mines. Washington DC; US Government Printing Office

Matsuhisa, Y. (1978). 'Some new data on low $_{18}$O igneous rocks and minerals from the Western San Juan Mountains, Colorado', *Bull. Geol. Surv, Japan.* Vol. 29, No. 8, 45–50

Matsukuma, T. and Horikoshi, E. (1970). 'Kuroko deposits in Japan, a review', *Volcanism and Ore Genesis*, pp. 153–179. Ed. T. Tatsumi. Tokyo; University of Tokyo Press

Maucher, A. (1972). 'Time- and strata-bound ore deposits and the evolution of the earth', *Int. Geol. Congr. 24th, Montreal*, Sec. 4, pp. 83–87

Maucher, A. (1976). 'The strata-bound cinnabar–stibnite–scheelite deposits (discussed with examples from the Mediterranean region)', *Handbook of Strata-bound and Stratiform Ore Deposits. Vol. 7*, pp. 477–503. Ed. K.H. Wolf. Amsterdam; Elsevier

McAuliffe, C.D. (1979). 'Oil and gas migration—chemical and physical constraints', *Amer. Assoc. Petrol. Geol. Bull.* Vol. 63, 761–781

McClay, K.R. (1979). 'Folding in silver–lead–zinc orebodies, Mount Isa, Australia', *Inst. Min. Metall. Trans.* Vol. 88, B5–B14

McClay, K.R. and Carlile, D. (1978). 'Mid Proterozoic sulphate evaporites at Mount Isa mine, Queensland, Australia', *Nature.* Vol. 274, 240–241

McCulloh, T.H. (1973). 'Oil and gas', *United States Mineral Resources. US Geol. Surv. Prof. Paper 820*, pp. 477–496. Ed. D.A. Brobst and W.P. Pratt. Washington DC; US Geological Survey

McKellar, J.B. (1975). 'The eastern Australian rutile province', *Economic Geology of Australia and Papua New Guinea. Vol. 1. Metals*, pp. 1055–1062. Ed. C.L. Knight. Parkville, Victoria; Australasian Inst. Min. Metall.

McKelvey, V.E. (1967). *Phosphate Deposits. US Geol. Surv. Bull. 1252-D.* Washington DC; US Geological Survey

McKelvey, V.E., Williams, J.S., Sheldon, R.P., Cressman, E.R., Cheney, T.M. and Swanson, R.W. (1959). 'The Phosphoria, Park City and Shedhorn formations in the western phosphate field', *US Geol. Surv. Prof. Paper 313-A*, pp. 1–47. Washington DC; US Geological Survey

McLimans, R.K., Barnes, H.L. and Ohmoto, H. (1980). 'Sphalerite stratigraphy of the Upper Mississippi Valley zinc–lead district, Southeast Wisconsin, *Econ. Geol.* Vol. 75, 351–361

McMillan, R.H. (1977a). 'Uranium in Canada', *Bull. Can. Petrol. Geol.* Vol. 25, 1222–1249

McMillan, R.H. (1977b). 'Metallogenesis of Canadian uranium deposits: a review', *Geology, Mining and Extractive Processing of Uranium*, pp. 43–55. Ed. M.J. Jones. London; Institute of Mining and Metallurgy

McMillan, R.H. (1978). 'Genetic aspects and classification of important Canadian uranium deposits', *Uranium Deposits, Their Mineralogy and Origin, Short Course Handbook 3*, pp. 187–204. Ed. M.M. Kimberley. Ottawa; Mineral. Assoc. Canada

Meisinger, A.C. (1980a). 'Diatomite', *Minerals Yearbook 1978–79. Vol. 1. Metals and Minerals*, pp. 313–315. Bureau of Mines. Washington DC; US Government Printing Office

Meisinger, A.C. (1980b). 'Perlite', *Minerals Yearbook 1978–79. Vol. 1, Metals and Minerals*, pp. 671–675. Bureau of Mines. Washington DC; US Government Printing Office

Meisinger, A.C. (1980c). 'Pumice and volcanic cinder', *Minerals Yearbook 1978–79. Vol. 1. Metals and Minerals*, pp. 729–733. Bureau of Mines. Washington DC; US Government Printing Office

Meisinger, A.C. (1980d). 'Vermiculite', *Minerals Yearbook 1978–79. Vol. 1. Metals and Minerals*, pp. 977–980. Bureau of Mines. Washington DC; US Government Printing Office

Melcher, G.C. (1966). 'The carbonatites of Jacupiranga, Sao Paulo, Brazil', *Carbonatites*, pp. 169–181. Ed. O.F. Tuttle and J. Gittins. New York; Wiley–Interscience

Mengel, J.T. (1976). 'Physical sedimentation in Precambrian cherty iron formations of the Lake Superior type', *Ores in Sediments*, pp. 179–193. Ed. G.C. Amstutz and A.J. Bernard. Berlin; Springer-Verlag

Mero, J.L. (1977). 'Economic aspects of nodule mining', *Marine Manganese Deposits*, pp. 327–355. Ed. G.P. Glasby. Amsterdam; Elsevier

Meyer, H.O.A. (1976). 'Kimberlites of the continental United States: a review', *J. Geology*. Vol. 84, 377–403

Mikami, H.M. (1975). 'Chromite', *Industrial Minerals and Rocks (Nonmetallics Other Than Fuels)*, 4th edn., pp. 501–517. Ed. S.J. LeFond. New York; Amer. Inst. Min. Metall. Petrol. Engineers

Miller, M.H. (1973). 'Antimony', *United States Mineral Resources. US Geol. Surv. Prof. Paper 820*, pp. 45–50. Ed. D.A. Brobst and W.P. Pratt. Washington DC; US Geological Survey

Minnes, D.G. (1975). 'Nepheline syenite', *Industrial Minerals and Rocks (Nonmetallics Other Than Fuels)*, 4th edn., pp. 861–894. Ed. S.J. LeFond. New York; Amer. Inst. Min. Metall. Petrol. Engineers

Minter, W.E.L. (1976). 'Detrital gold, uranium and pyrite concentrations related to sedimentology in the Precambrian Vaal Reef placer, Witwatersrand, South Africa', *Econ. Geol.* Vol. 71, 157–176

Mitchell, A.H.G. (1974). 'Southwest England granites; magmatism and tin mineralization in a post-collision tectonic setting', *Trans. Inst. Min. Metall.* Vol. 83, B95–B97

Mitchell, A.H.G. (1977). 'Tectonic settings for emplacement of Southeast Asian tin granites', *Geol. Soc. Malaysia Bull.* Vol. 9, 123–140

Mitchell, A.H.G. (1979). 'Rift-, subduction- and collision-related tin belts', *Geol. Soc. Malaysia Bull.* Vol. 11, 81–102

Mitchell, A.H.G. and Garson, M.S. (1976). 'Mineralization at plate boundaries', *Mineral. Sci. Eng.* Vol. 8, 129–169

Mitchell, R.H. (1973). 'Composition of olivine, silica activity and oxygen fugacity in kimberlite', *Lithos*. Vol. 6, 65–81

Mitsuchi, T. (1952). 'Iron ore deposits in Japan', *Symposium sur les gisements de fer du monde. 19th International Geol. Congress, Algiers*, pp. 537–560

Miyashiro, A. (1973). *Metamorphism and Metamorphic Belts*. London; George Allen and Unwin

Mlakar, I. (1964). 'The role of post-mineralization tectonics in the search for new mineralized zones in the Idria area', *Rudarsko–Metalurski, Zbornik*. Vol. 1, 19–25

Mlakar, I. and Drovenik, M. (1971). 'Structural and genetic particularities of the Idrija mercury ore deposit', *Geol. Razprave Poroćila, Ljubljana*. Vol. 14, 67–126

Moh, G.H. and Hutchison, C.S. (1977). 'Hydrocassiterites', *Neues Jahrbuch Miner. Abh.* Vol. 131, No. 1, 13–17

Moore, A.C. (1973). 'Carbonatites and kimberlites in Australia: a review of the evidence', *Min. Sci. Eng.* Vol. 5, 81–91

Moore, C.M. (1980). 'Rare-earth minerals and metals', *Minerals Yearbook 1978–79. Vol. 1. Metals and Minerals*, pp. 735–742. Bureau of Mines. Washington DC; US Government Printing Office

Moore, F. and Moore, D.J. (1979). 'Fluid inclusion study of mineralization at St. Michael's Mount, Cornwall', *Trans. Inst. Min. Metall.* Vol. 88, B57–B60

Moore, L.R. (1968a). 'Cannel coal, bogheads and oil shales', *Coal and Coal-bearing Strata*, pp. 19–29. Ed. D. Murchison and T.S. Westoll. Edinburgh; Oliver and Boyd

Moore, L.R. (1968b). 'Some sediments closely associated with coal seams', *Coal and Coal-bearing Strata*, pp. 105–123. Ed. D. Murchison and T.S. Westoll. Edinburgh; Oliver and Boyd

Moores, E.M. (1969). *Petrology and Structure of the Vourinos Ophiolitic Complex of Northern Greece. Special Paper Geol. Soc. Amer. 118*. Boulder, Colorado; Geol. Soc. Amer.

Moores, E.M. and Vine, F.J. (1971). 'The Troodos massif, Cyprus and other ophiolites as oceanic crust: evaluation and implications', *Phil. Trans. Roy. Soc. London*. Vol. A268, 443–466

Moreau, M. (1977). 'L'Uranium et les granitoids: essai d'interprétation', *Geology, Mining and Extractive Processing of Uranium*, pp. 83–102. Ed. M.J. Jones. London; Inst. Min. Metall.

Morey, G.B. (1973). 'Mesabi, Gunflint and Cuyuna Ranges, Minnesota', *Genesis of Precambrian Iron and Manganese Deposits*, pp. 193–208. Paris; UNESCO

Morgan, G.A. (1980). 'Vanadium', *Minerals Yearbook 1978–79. Vol. 1. Metals and Minerals*, pp. 969–975. Bureau of Mines. Washington DC; US Government Printing Office

Morris, H.T., Heyl, A.V. and Hall, R.B. (1973). 'Lead', *United States Mineral Resources. US Geol. Surv. Prof. Paper 820*, pp. 313–332. Ed. D.A. Brobst and W.P. Pratt. Washington DC; US Geological Survey

Morton, P.K. (1977). *Geology and Mineral Resources of Imperial County, California. County Report 7.* San Francisco; California Div. of Mines and Geology

Mossoulos, L. (1964). 'Les gisements lateritiques de fer chromonickelifere de Grece', *Methods of Prospection for Chromite*, pp. 61–77. Ed. R. Woodtli. Paris; Organization for Economic Co-operation and Development

Motica, J.E. (1968). 'Geology and uranium–vanadium deposits in the Uravan mineral belt, Southwestern Colorado', *Ore Deposits of the United States 1933–1967. The Graton–Sales Volume, Vol. 1*, pp. 805–813. Ed. J.D. Ridge. New York; Amer. Inst. Min. Metall. Petrol. Engineers

Muff, R. (1978). *The Antimony Deposits in the Murchison Range of the Northeastern Transvaal, Republic of South Africa. Monograph Series on Mineral Deposits, 16.* Berlin; Gebrüder Borntraeger

Muir, J.D. and Comba, C.D.A. (1979). 'The Dundonald deposit: an example of volcanic-type nickel-sulfide mineralization', *Can. Mineral.* Vol. 17, 351–359

Mukaiyama, H. (1970). 'Volcanic sulphur deposits in Japan', *Volcanism and Ore Genesis*, pp. 285–294. Ed. T. Tatsumi. Tokyo; University of Tokyo Press

Mumpton, F.A. (1975). 'Commercial utilization of natural zeolites', *Industrial Minerals and Rocks (Nonmetallics Other Than Fuels)*, pp. 1262–1274. Ed. S.J. LeFond. New York; Amer. Inst. Min. Metall. Petrol. Engineers

Murphy, T.D. (1975). 'Silica and silicon', *Industrial Minerals and Rocks (Nonmetallics Other Than Fuels)*, pp. 1043–1060. Ed. S.J. LeFond. New York; Amer. Inst. Min. Metall. Petrol. Engineers

Murphy, R.W. (1975). 'Tertiary basins of southeast Asia', *Southeast Asia Petroleum Exploration Society Proc.* Vol. 2, 1–36

Murray, W.J. (1975). 'McArthur River H.Y.C. lead–zinc and related deposits', *Economic Geology of Australia and Papua New Guinea. Vol. 1. Metals.* Ed. C.L. Knight. Parkville, Victoria; Australasian Inst. Min. Metall.

Muskatt, H.S. (1972). 'The Clinton Group of east-central New York', *New York Geol. Assoc. Guidebook 44th Annual Meeting*, pp. A1–A37

Nakamura, T. (ed.) (1976). *Genesis of Vein-type Deposits in Japan. Mining Geology, Japan, Special Issue No. 7*, pp. 1–131. Tokyo; Soc. Min. Geol. Japan

Naldrett, A.J. (1973). 'Nickel sulfide deposits—their classification and genesis, with special emphasis on deposits of volcanic association', *Can. Inst. Min. Metall. Trans.* Vol. 76, 183–201

Naldrett, A.J. and Cabri, L.J. (1976). 'Ultramafic and related mafic rocks: their classification and genesis with special reference to the concentration of nickel sulfides and platinum group elements', *Econ. Geol.* Vol. 71, 1131–1158

Naldrett, A.J. and Kullerud, G. (1967). 'A study of the Strathcona mine, and its bearing on the origin of the nickel–copper ores of the Sudbury district, Ontario', *J. Petrology.* Vol. 8, 453–531

Naldrett, A.J., Greenman, L. and Hewins, R.H. (1972). 'The main irruptive and the sub-layer at Sudbury, Ontario', *Int. Geol. Congr. 24th.* Vol. 4, 206–214

Nash, J.T. (1975). 'Fluid inclusion studies of vein, pipe, and replacement deposits, northwestern San Juan Mountains, Colorado', *Econ. Geol.* Vol. 70, 1448–1462

Nash, J.T. and Lehrman, N.J. (1975). *Geology of the Midnite Uranium Mine, Stevens County, Washington—A Preliminary Report. US Geol. Surv. Open File Report 75–402*, pp. 5–32. Washington DC; US Geological Survey

Neglia, S. (1979). 'Migration of fluids in sedimentary basins', *Amer. Assoc. Petrol. Geol. Bull.* Vol. 63, 573–597

Nesbitt, R.W., Sun, Shen-Su, and Purvis, A.C. (1979). 'Komatiites: geochemistry and genesis', *Can. Mineral.* Vol. 17, 165–186

Neumann-Redlin, C., Walther, H.W. and Zitzman, A. (1977). 'The iron ore deposits of the Federal Republic of Germany', *The Iron Ore Deposits of Europe and Adjacent Areas. Vol. 1*, pp. 165–186. Ed. A. Zitzmann. Hannover; Bundesanstalt für Geowissenschaften und Rohstoffe

Newell, R.A. (1971). 'Characteristics of the stanniferous alluvium in the southern Kinta Valley, West Malaysia', *Geol. Soc. Malaysia Bull.* Vol. 4, 15–37

Newnham, L.A. (1975). 'Renison Bell tinfield', *Economic Geology of Australia and Papua New Guinea. Vol. 1. Metals*, pp. 581–583. Ed. C.L. Knight. Parkville, Victoria; Australasian Inst. Min. Metall.

Nielsen, R.L. (1976). 'Recent developments in the study of porphyry copper geology—a review', *Porphyry Deposits of the Canadian Cordillera. Can. Inst. Min. Metall. Special Vol. 15*, pp. 487–500. Ed. A. Sutherland Brown. Montreal; Can. Inst. Min. Metall.

Nishiwaki, C., Matsukuma, T. and Urashima, Y. (1971). 'Neogene gold–silver ores in Japan', *Proc. IMA–IAGOD Meeting 1970: IAGOD Volume, Soc. Min. Geol. Japan, Special Issue 3*, pp. 409–417. Ed. Y. Takeuchi. Tokyo; Soc. Min. Geol. Japan

Nolan, T.B. (1935). 'The underground geology of the Tonopah mining district, Nevada', *Bull. Nevada Univ.* Vol. 29, No. 5, 1–49

Norton, D.L. and Cathles, L.M. (1973). 'Breccia pipes: products of exsolved vapour from magmas', *Econ. Geol.* Vol. 68, 540–547

Norton, J.J. (1973). 'Lithium, cesium, and rubidium—the rare earth alkali elements', *United States Mineral Resources, US Geol. Surv. Prof. Paper 820*, pp. 365–378. Ed. D.A. Brobst and W.P. Pratt. Washington DC; US Geological Survey

Notebaart, C.W. and Vink, B.W. (1972). 'Ore minerals of the Zambian copperbelt', *Geol. Mijnbouw.* Vol. 51, 337–345

Notholt, A.J.G. (1979). 'The economic geology and development of igneous phosphate deposits in Europe and the USSR', *Econ. Geol.* Vol. 74, 339–350

Notholt, A.J.G. (1980). 'Economic phosphatic sediments: mode of occurrence and stratigraphical distribution', *J. Geol. Soc. London.* Vol. 137, 793–805

Ohle, E.L. (1980). 'Some considerations in determining the origin of ore deposits of the Mississippi Valley-type—Part II', *Econ. Geol.* Vol. 75, 161–172

Ohmoto, H. (1978). 'Submarine calderas: a key to the formation of volcanogenic massive sulfide deposits?', *Mining Geology.* Vol. 28, 219–231

Ohmoto, H. and Rye, R.O. (1974). 'Hydrogen and oxygen isotopic compositions of fluid inclusions in the Kuroko deposits of Japan', *Econ. Geol.* Vol. 69, 947–953

Ohtagaki, T., Tsukada, Y., Hirayama, H., Fujioka, H. and Miyoshi, T. (1974). 'Geology of the Shakanai Mine, Akita Prefecture', *Geology of Kuroko Deposits. Mining Geology, Special Issue 6*, pp. 131–139. Ed. S. Ishihara. Tokyo; Soc. Min. Geol. Japan

Olson, J.C. and Pray, L.C. (1954). 'The Mountain Pass rare earth deposits', *Geology of Southern California. Bull. 170*, pp. 23–29. Ed. R.H. Jahns. San Francisco; Division of Mines, State of California

Olson, J.C., Shawe, D.R., Pray, L.C. and Sharp, W.N. (1954). *Rare Earth Mineral Deposits of the Mountain Pass District, Bernardino County, California. Prof. Paper US Geol. Surv. 261.* pp. 1–75. Washington DC; US Geological Survey

Olson, R.H. (1975). 'Zeolites: introduction', *Industrial Minerals and Rocks (Nonmetallics Other Than Fuels)*, pp. 1235–1243. Ed. S.J. LeFond. New York; Inst. Min. Metall. Petrol. Engineers

O'Neil, J.R. and Silberman, M.L. (1974). 'Stable isotope relations in epithermal Au–Ag deposits', *Econ. Geol.* Vol. 69, 902–909.

Osberger, R. (1968). 'Billiton tin placers: types, occurrences, and how they were formed', *World Mining.* June, 34–40

Ostic, R.G., Russel, R.D. and Stanton, R.L. (1967). 'Additional measurements of the isotopic composition of lead from stratiform deposits', *Can. J. Earth Sci.* Vol. 4, 245–267

Page, N.J., Clark, A.L., Desborough, G.A. and Parker, R.L. (1973). 'Platinum-group metals', *United States Mineral Resources, US Geol. Surv. Prof. Paper 820*, pp. 537–545. Ed. D.A. Brobst and W.P. Pratt. Washington DC; US Geological Survey

Pan, Y.-S. and Ypma, P.J.M. (1974). 'Heating and recrystallization of wood-tin: an indication of low temperature genesis of cassiterite and wood-tin veins in volcanics', *Abstr. Progr. Geol. Soc. Amer.* Vol. 6, 62, 63

Panayiotou, A. (1978). 'The mineralogy and chemistry of the podiform chromite deposits in the serpentinites of the Limassol Forest, Cyprus', *Mineral. Deposita.* Vol. 13, 259–274

Parák, T. (1975). 'Kiruna iron ores are not "intrusive-magmatic ores of the Kiruna type"', *Econ. Geol.* Vol. 70, 1242–1258

Parasnis, D.S. (1973). *Mining Geophysics.* Amsterdam; Elsevier

Pardo, G. (1975). 'Geology of Cuba', *The Ocean Basins and Margins: 3, The Gulf of Mexico and the Caribbean*, pp. 553–615. Ed. A.E.M. Nairn and F.G. Stehli. New York;

Plenum Press

Park, C.F. Jr. (1961). 'A magnetite "flow" in northern Chile', *Econ. Geol.* Vol. 56, 431–436

Park, C.F. Jr. (1972). 'The iron ore deposits of the Pacific Basin', *Econ. Geol.* Vol. 67, 339–349

Park, C.F. Jr. and MacDiarmid, R.A. (1970). *Ore Deposits*, 2nd edn. San Francisco; W.H. Freeman

Parker, R.L. and Adams, J.W. (1973). 'Niobium (columbium) and tantalum', *United States Mineral Resources. US Geol. Surv. Prof. Paper 820*, pp. 443–454. Ed. D.A. Brobst and W.P. Pratt. Washington DC; US Geological Survey

Patterson, S.H. and Dyni, J.R. (1973). 'Aluminum and bauxite', *United States Mineral Resources. US Geol. Surv. Prof. Paper 820*, pp. 35–43. Ed. D.A. Brobst and W.P. Pratt. Washington DC; US Geological Survey

Patterson, S.H. and Murray, H.H. (1975). 'Clays', *Industrial Minerals and Rocks (Nonmetallics Other Than Fuels)*, pp. 519–585. Ed. S.J. LeFond. New York; Amer. Inst. Min. Metall. Petrol. Engineers

Pattison, E.F. (1979). 'The Sudbury Sublayer', *Can. Mineral.* Vol. 17, 257–274

Pawlowski, S., Pawlowska, K. and Kubica, B. (1979). 'Geology and genesis of the Polish sulfur deposits', *Econ. Geol.* Vol. 74, 475–483

Pearce, J.A. and Gale, G.H. (1977). 'Identification of ore-deposition environment from trace element geochemistry of associated igneous host rocks', *Volcanic Processes in Ore Genesis, Special Publication 7*, pp. 14–24. Ed. I.G. Gass. London; The Geological Society

Peredery, W.V. and Naldrett, A.J. (1975). 'Petrology of the Upper Irruptive rocks, Sudbury, Ontario', *Econ. Geol.* Vol. 70, 164–175

Peters, W.C. (1978). *Exploration and Mining Geology.* New York; John Wiley

Petersen, U. (1965). 'Regional geology and major ore deposits of central Peru', *Econ. Geol.* Vol. 60, 407–476

Petersen, U. (1970). 'Metallogenic provinces in South America', *Geol. Rundschau.* Vol. 59, 834–897

Petersen, U. (1972). 'Geochemical and tectonic implications of South American metallogenic provinces', *Annals of the New York Acad. of Sciences.* Vol. 196, 1–38

Peterson, E.C. and Collins, C.T. (1980). 'Iron ore', *Minerals Yearbook 1978–79. Vol. 1. Metals and Minerals*, pp. 433–456. Bureau of Mines. Washington DC; US Government Printing Office

Peterson, N.V. and Mason, R.S. (1975). 'Pumice, pumicite, and volcanic cinders', *Industrial Minerals and Rocks (Nonmetallics Other Than Fuels)*, pp. 991–994. Ed. S.J. LeFond. New York; Amer. Inst. Min. Metall. Petrol. Engineers

Petkof, B. (1975). 'Mica', *Industrial Minerals and Rocks (Nonmetallics Other Than Fuels)*, pp. 837–850. Ed. S.J. LeFond. New York; Amer. Inst. Min. Metall. Petrol. Engineers

Petkof, B. (1980a). 'Beryllium', *Minerals Yearbook 1978–79. Vol. 1. Metals and Minerals*, pp. 111–114. Bureau of Mines. Washington DC; US Government Printing Office

Petkof, B. (1980b). 'Gallium', *Minerals Yearbook 1978–79. Vol. 1. Metals and Minerals*, pp. 359–361. Bureau of Mines. Washington DC; US Government Printing Office

Petkof, B. (1980c). 'Magnesium compounds', *Minerals Yearbook 1978–79. Vol. 1. Metals and Minerals*, pp. 567–575. Bureau of Mines. Washington DC; US Government Printing Office

Petruk, W. (1971a). 'The silver-arsenide deposits of the Cobalt–Gowganda region, Ontario. Depositional history of the ore minerals'. *Can. Mineral.* Vol. 11, No. 1, 396–401

Petruk, W. (1971b). 'The silver-arsenide deposits of the Cobalt–Gowganda region, Ontario. Geochemistry of the ores'. *Can. Mineral.* Vol. 11, No. 1, 140–149

Pickard, G.W. (1974). 'Geology of Mexican fluorspar deposits', *A Symposium on the Geology of Fluorspar. Kentucky Geol. Surv. Series X, Special Publication 22*, pp. 23–30. Ed. D.W. Hutcheson. Lexington, Kentucky; Kentucky Geological Survey

Pinar-Erdem, N. and Ilhan, E. (1977). 'Outlines of the stratigraphy and tectonics of Turkey, with notes on the geology of Cyprus', *The Ocean Basins and Margins: 4A, The Eastern Mediterranean*, pp. 277–318. Ed. A.E.M. Nairn, W.H. Kanes and F.G. Stehli. New York; Plenum Press

Pitcher, W.S. (1979). 'The nature, ascent and emplacement of granitic magmas', *J. Geol. Soc. London.* Vol. 136, 627–662

Platt, J.W. (1977). 'Volcanogenic mineralization at Avoca, Co. Wicklow, Ireland, and its regional implications', *Volcanogenic Processes in Ore Genesis. Geol. Soc. London, Special Publication 7*, pp. 163–170. London; The Geological Society

Plimer, I.R. (1979). 'Sulphide rock zonation and hydrothermal alteration at Broken Hill, Australia', *Inst. Min. Metall. Trans.* Vol. 88, B161–B176

Plummer, L.N. (1971). 'Barite deposition in central Kentucky', *Econ. Geol.* Vol. 66, 252–258

Potter, M.J. (1980a). 'Feldspar, nepheline syenite, and aplite', *Minerals Yearbook 1978–79. Vol. 1. Metals and Minerals*, pp. 317–325. Bureau of Mines. Washington DC; US Government Printing Office

Potter, M.J. (1980b). 'Kyanite and related materials', *Minerals Yearbook 1978–79. Vol. 1. Metals and Minerals*, pp. 503–506. Bureau of Mines. Washington DC; US Government Printing Office

Poty, B.P., LeRoy, J. and Cuney, M. (1974). 'Les inclusions fluides dans les minerals des gisements d'uranium intra granitiques du Limousin et du Forez (Massif Central, France)', *Formation of Uranium Ore Deposits. Proc. Symposium, Athens 1974*, pp. 569–582. Vienna; International Atomic Energy Agency

Powell, J.L. and Bell, K. (1974). 'Isotopic composition of strontium in alkalic rocks', *The Alkaline Rocks*, pp. 412–421. Ed. H. Sørensen. London; Wiley–Interscience

Powell, T.G., Cook, P.J. and McKirdy, D.M. (1975). 'Organic geochemistry of phosphorites: relevance to petroleum genesis', *Amer. Assoc. Petrol. Geol. Bull.* Vol. 59, 618–632

Pressler, J.W. (1980a). 'Gypsum', *Minerals Yearbook 1978–79. Vol. 1. Metals and Minerals*, pp. 411–422. Bureau of Mines. Washington DC; US Government Printing Office

Pressler, J.W. (1980b). 'Lime', *Minerals Yearbook 1978–79. Vol. 1. Metals and Minerals*, pp. 539–549. Bureau of Mines. Washington DC; US Government Printing Office

Pretorius, D.A. (1975). 'The depositional environment of the Witwatersrand gold fields: a chronological review of speculations and observations', *Mineral. Sci. Eng.* Vol. 7, 18–47

Pretorius, D.A. (1976). 'The nature of the Witwatersrand gold–uranium deposits', *Handbook of Strata-bound and Stratiform Ore Deposits. Vol. 7*, pp. 29–88. Ed. K.H. Wolf. Amsterdam; Elsevier

Price, L.C. (1976). 'Aqueous solubility of petroleum as applied to its origin and primary migration', *Amer. Assoc. Petrol. Geol. Bull.* Vol. 60, 213–244

Price, L.C. (1977). 'Aqueous solubility of petroleum as applied to its origin and primary migration: reply', *Amer. Assoc. Petrol. Geol. Bull.* Vol. 61, 2149–2156

Priem, H.N.A., Verschure, R.H., Bon, E.H., Boelrijk, N.A.I.M., Hebeda, E.H. and Verdurmen, E.A.T. (1971). 'Granitic complexes and associated tin mineralization of "Grenville" age in Rondonia, western Brazil', *Bull. Geol. Soc. Amer.* Vol. 82, 1095–1102

Pusey, W.C. III (1973). 'The ESR–Kerogen method. How to

evaluate potential gas and oil source rocks', *World Oil*. Vol. 176, 71
Quade, H. (1976). 'Genetic problems and environmental features of volcano sedimentary iron-ore deposits of the Lahn–Dill type', *Handbook of Strata-bound and Stratiform Ore Deposits. Vol. 7*, pp. 255–294. Ed. K.H. Wolf. Amsterdam; Elsevier
Rackley, R.I. (1976). 'Origin of western-States type uranium mineralization', *Handbook of Strata-bound and Stratiform Ore Deposits. Vol. 7*, pp. 98–156. Ed. K.H. Wolf. Amsterdam; Elsevier
Raguin, E. (1961). *Geologie des gites mineraux*, 3rd edn. Paris; Masson
Ramdohr, P. (1980). *The Ore Minerals and Their Intergrowths*, 2nd edn., Vols. 1 and 2. Oxford; Pergamon
Ramsay, W.R.H. and Kobe, H.W. (1974). 'Great Barrier Island silver–gold deposits, Hauraki Province, New Zealand', *Mineral. Deposita*. Vol. 9, 143–153
Ransom, D.M. and Hunt, F.L. (1975). 'Cleveland tin mine', *Economic Geology of Australia and Papua New Guinea. Vol. 1. Metals*, pp. 584–591. Parkland, Victoria; Australasian Inst. Min. Metall.
Rathjen, J.A. and Rowland, T.J. (1980). 'Lead', *Minerals Yearbook 1978–79. Vol. 1. Metals and Minerals*, pp. 507–538. Bureau of Mines. Washington DC; US Government Printing Office
Reading, H.G. (ed.) (1978). *Sedimentary Environments and Facies*. New York; Elsevier
Reckling, K., Hoy, R.B. and LeFond S.J. (1975). 'Diamonds', *Industrial Minerals and Rocks (Nonmetallics Other Than Fuels)*, 4th edn., pp. 587–604. Ed. S.J. LeFond. New York; Amer. Inst. Min. Metall. Petrol. Engineers
Reedman, J.H. (1979). *Techniques in Mineral Exploration*. London; Applied Science Publishers
Renfro, A.R. (1974). 'Genesis of evaporite associated stratiform metalliferous deposits—a sabkha process', *Econ. Geol.* Vol. 69, 33–45
Reynolds, C.D., Havryluk, I., Bastaman, S. and Atmowidjojo, S. (1973). 'The exploration of the nickel laterite deposits in Irian Barat, Indonesia', *Geol. Soc. Malaysia Bull.* Vol. 6, 309–323
Reynolds, R.L. and Goldhaber, M.B. (1978). 'Origin of a south Texas roll-type uranium deposit. I. Alteration of iron–titanium oxide minerals', *Econ. Geol.* Vol. 73, 1677–1689
Rich, R.A., Holland, H.D. and Petersen, U. (1977). *Hydrothermal Uranium Deposits*. Amsterdam; Elsevier
Richards, F.A. (1965). 'Anoxic basins and fjords', *Chemical Oceanography, Vol. 1*, pp. 611–644. Ed. J.P. Riley and G. Skirrow. London; Academic Press
Richards, J.R. (1975). 'Lead isotope data on three Australian galena localities', *Mineral. Deposita*. Vol. 10, 287–301
Richardson, J.A. (1939). *The Geology and Mineral Resources of the Neighbourhood of Raub, Pahang, Federated Malay States, with an Account of the Geology of the Raub Australian Gold Mine, Geol. Surv. Dept. Fed. Malay States. Mem. 3 (New Series)*. Ipoh, Malaysia; Geological Survey of Malaysia
Rife, D.L. (1971). 'Barite fluid inclusion geothermometry, Cartersville mining district, Northwest Georgia', *Econ. Geol.* Vol. 66, 1164–1167
Riggs, S.R. (1979a). 'Petrology of the Tertiary Phosphorite System of Florida', *Econ. Geol.* Vol. 74, 195–220
Riggs, S.R. (1979b). 'Phosphorite sedimentation in Florida—a model phosphogenic system', *Econ. Geol.* Vol. 74, 285–314
Riggs, S.R. (1980). 'Intraclast and pellet phosphorite sedimentation in the Miocene of Florida', *J. Geol. Soc. London*. Vol. 137, 741–748
Riley, G. (1967). 'The cassiterite–stannite occurrence at Tekka Mines, Kinta', *Geol. Soc. Malaysia Newsletter*. Vol. 7, 10–12
Rimsaite, J. (1978). 'Application of mineralogy to the study of multi-stage uranium mineralization in remobilized uranium deposits, Saskatchewan', *Uranium Deposits: Their Mineralogy and Origin. Short Course Handbook 3*, pp. 403–430. Ed. M.M. Kimberley. Ottawa; Mineral. Assoc. Canada
Rivas, S. (1979). 'Geology of the principal tin deposits of Bolivia', *Geol. Soc. Malaysia Bull.* Vol. 11, 161–180
Robbins, D.A. (1978). 'Applied geology in the discovery of the Spokane Mountain uranium deposit, Washington', *Econ. Geol.* Vol. 73, 1523–1538
Roberts, R.G. (1975). 'The geological setting of the Mattagami Lake mine, Quebec: a volcanogenic massive sulfide deposit', *Econ. Geol.* Vol. 70, 115–129
Robertson, D.S. (1974). 'Basal Proterozoic units as fossil time markers and their use in uranium prospection', *Formation of Uranium Deposits, Proc. Symposium, Athens 1974*, pp. 495–512. Vienna; International Atomic Energy Agency
Robertson, D.S., Tilsley, J.E. and Hogg, G.M. (1978). 'The time-bound character of uranium deposits', *Econ. Geol.* Vol. 73, 1409–1419
Robertson, J.A. (1978). 'Uranium deposits in Ontario', *Uranium Deposits, Their Mineralogy and Origin. Short Course Handbook 3*, pp. 229–280. Ed. M.M. Kimberley. Ottawa; Mineral. Assoc. Canada
Robinson, B.W. (1974). 'The origin of mineralization at the Tui mine, Te Aroha, New Zealand, in the light of stable isotope studies', *Econ. Geol.* Vol. 69, 910–925
Robinson, D.N. (1978). 'The characteristics of natural diamond and their interpretation', *Mineral. Sci. Eng.* Vol. 10, 55–72
Rodriguez, R.E., Cabre, R.S.J. and Mercado, A. (1976). 'Geometry of the Nazca Plate and its geodynamic implications', *The Geophysics of the Pacific Ocean Basin and its Margins. Geophysics Monograph 19*, pp. 87–103. Ed. G.H. Sutton, M.H. Manghnani, R. Moberly and E.U. McAfee. Washington DC; American Geophysical Union
Roe, L.A. (1975). 'Talc and pyrophyllite', *Industrial Minerals and Rocks (Nonmetallics Other Than Fuels)*, pp. 1127–1147. Ed. S.J. LeFond. New York; Amer. Inst. Min. Metall. Petrol. Engineers
Roedder, E.H. (1976). 'Fluid inclusion evidence in the genesis of ores in sedimentary and volcanic rocks', *Handbook of Strata-bound and Stratiform Ore Deposits, Vol. 2*, pp. 67–110. Ed. K.H. Wolf. Amsterdam; Elsevier
Roedder, E.H. (1977). 'Fluid inclusion studies of ore deposits in the Viburnum Trend, southeast Missouri', *Econ. Geol.* Vol. 72, 474–479
Roedder, E. (1979). 'Fluid inclusions as samples of ore fluids', *Geochemistry of Hydrothermal Ore Deposits*, 2nd edn., pp. 684–737. Ed. H.L. Barnes. New York; Wiley-Interscience
Roering, C. and Gevers, T.W. (1964). 'Lithium- and beryllium-bearing pegmatites in the Karibib district, Southwestern Africa', *The Geology of Some Ore Deposits in Southern Africa, Vol. 2*, pp. 463–495. Ed. S.H. Haughton. Johannesburg; Geol. Soc. South Africa
Rogers, C.P. Jr. and Neal, J.P. (1975). 'Feldspar and aplite', *Industrial Minerals and Rocks (Nonmetallics Other Than Fuels)*, 4th edn., pp. 637–651. Ed. S.J. LeFond. New York; Amer. Inst. Min. Metall. Petrol. Engineers
Rogers, J.J.W., Ragland, P.C., Nishimori, R.K., Greenberg, J.K. and Hauck, S.A. (1978). 'Varieties of granitic uranium deposits and favourable exploration areas in the eastern United States', *Econ. Geol.* Vol. 73, 1539–1555
Rogers, P.J. (1977). 'Fluid inclusion studies in fluorite from the Derbyshire orefield', *Trans. Inst. Min. Metall. London*.

Vol. B128–B132

Rolfe, D.G. (1973). 'The geology of the Kao kimberlite pipes', *Lesotho Kimberlites*, pp. 101–106. Ed. P.H. Nixon. Maseru, Lesotho; Lesotho National Development Corporation

Rona, P.A. (1980). 'Tag Hydrothermal field: Mid-Atlantic ridge crest at latitude 26° N', *J. Geol. Soc. London.* Vol. 137, 385–402

Roscoe, S.M. (1969). *Huronian Rocks and Uraniferous Conglomerates in the Canadian Shield. Geol. Surv. Canada Paper 68–40.* Ottawa; Geol. Surv. Canada

Ross, C.P. (1970). 'The Precambrian of the United States of America: Northwestern United States—the Belt Series', *Precambrian, Vol. 4*, pp. 145–251. Ed. K. Rankama. New York; Wiley–Interscience

Rossmann, R., Callender, E. and Bowser, C.J. (1972). 'Interelement geochemistry of Lake Michigan ferromanganese nodules', *Proc. 24th Intl. Geol. Congr., Montreal*, Sect. 10, pp. 336–341

Routhier, P. (1963). *Les gisements metallifères, Vols. 1 and 2.* Paris; Masson

Rowe, R.B. (1958). 'Niobium (columbium) deposits of Canada', *Geol. Surv. Can. Econ. Geol. Series.* Vol. 18, 1–108

Rowland, T.J. and Cammarota, V.A. Jr. (1980). 'Antimony', *Minerals Yearbook 1978–79. Vol. 1. Metals and Minerals*, pp. 63–70. Bureau of Mines. Washington DC; US Government Printing Office

Rowntree, J.C. and Mosher, D.V. (1975). 'Jabiluka uranium deposits', *Economic Geology of Australia and Papua New Guinea. Vol. 1. Metals*, pp. 321–326. Ed. C.L. Knight. Parkville, Victoria; Australasian Inst. Min. Metall.

Roy, S. (1973). 'Genetic studies of the Precambrian manganese formations of India with particular reference to the effects of metamorphism', *Genesis of Precambrian Iron and Manganese Deposits*, pp. 229–242. Paris; UNESCO

Roy, S. (1976). 'Ancient manganese deposits', *Handbook of Strata-bound and Stratiform Ore Deposits, Vol. 7*, pp. 395–476. Ed. K.H. Wolf. Amsterdam; Elsevier

Ruckmick, J.C. (1963). 'The iron ores of Cerro Bolivar, Venezuela', *Econ. Geol.* Vol. 58, 218–236

Ruckmick, J.C., Wimberly, B.H. and Edwards, A.F. (1979). 'Classification and genesis of biogenic sulfur deposits', *Econ. Geol.* Vol. 74, 469–474

Ruiz, J., Kesler, E., Jones, L.M. and Sutter, J.F. (1980). 'Geology and geochemistry of the Las Cuevas fluorite deposit, San Luis Potosi, Mexico', *Econ. Geol.* Vol. 75, 1200–1209

Rutland, R.W.R., Marjoribanks, R.W., Laing, W.P., Glen, R.A. and Stanton, R.L. (1978). 'Tectonic deformations at Broken Hill, New South Wales, Australia, and their significance for interpretations of ore environment', *Inst. Min. Metall. Trans.* Vol. 87, B172–B180

Ruzicka, V. (1971). *Geological Comparisons between East European and Canadian Uranium Deposits. Can. Geol. Surv. Paper 70-48.* Ottawa; Geol. Survey Canada

Ruzicka, V. (1978). 'Phanerozoic uranium deposits and occurrences in Europe and eastern North America', *Uranium Deposits: Their Mineralogy and Origin. Short Course Handbook 3*, pp. 217–228. Ed. M.M. Kimberley. Ottawa; Mineral. Assoc. Canada

Ruzicka, V. and Steacy, H.R. (1976). 'Some sedimentary features of conglomeratic uranium ore from Elliot Lake, Ontario', *Geol. Surv. Can. Paper 76-1A*, pp. 343–346. Ottawa; Geol. Survey Canada

Rye, R.O. and Ohmoto, H. (1974). 'Sulphur and carbon isotopes and ore genesis: a review', *Econ. Geol.* Vol. 69, 826–842

Rye, R.O. and Sawkins, F.J. (1974). 'Fluid inclusion and stable isotope studies on the Casapalca Ag–Pb–Zn–Cu deposit, Central Andes, Peru', *Econ. Geol.* Vol. 69, 181–205

Saegart, W.E. and Lewis, D.E. (1976). *Characteristics of Philippine Porphyry Copper Deposits and Summary of Current Production and Reserves. Preprint 76-1-79.* Salt Lake City; Soc. of Mining Engineers of AIME

Sainsbury, C.L. and Reed, B.L. (1973). 'Tin', *United States Mineral Resources. US Geol. Surv. Prof. Paper 820*, pp. 637–651. Ed. D.A. Brobst and W.P. Pratt. Washington DC; US Geological Survey

Sangster, D.F. (1969). *The Contact Metasomatic Magnetite Deposits of Southwestern British Columbia. Geol. Survey Can. Bull. 172.* Ottawa; Geol. Survey Canada

Sangster, D.F. (1976). 'Sulphur and lead isotopes in stratabound deposits', *Handbook of Strata-bound and Stratiform Ore Deposits, Vol. 2*, pp. 219–266. Ed. K.H. Wolf. Amsterdam; Elsevier

Sasaki, A. and Krouse, H.R. (1969). 'Sulfur isotopes and the Pine Point lead–zinc mineralization', *Econ. Geol.* Vol. 64, 718–730

Sassano, G.P., Fritz, P. and Morton, R.D. (1972). 'Paragenesis and isotopic composition of some gangue minerals from the uranium deposits of Eldorado, Saskatchewan', *Can. J. Earth Sci.* Vol. 9, 141–157

Sato, T. (1974). 'Distribution and geological setting of the kuroko deposits', *Geology of Kuroko deposits. Mining Geology Special Issue 6*, pp. 1–9. Ed. S. Ishihara. Tokyo; Soc. Min. Geol. Japan

Sato, T. (1977). 'Kuroko deposits: their geology, geochemistry and origin', *Volcanic Processes in Ore Genesis. Geol. Soc. London Special Publication 7*, pp. 153–161. London; The Geological Society

Saupé, F. (1967). 'Note preliminaire concernant la genese du gîsement de mercure d'Almaden', *Mineral. Deposita.* Vol. 2, 26–33

Saupé, F. (1973). 'La géologie du gîsement de mercure d'Almaden', *Sciences de la terre*, Mem. 29, Nancy; University of Nancy

Sawkins, F.J. (1966). 'Ore genesis in the north Pennine ore field in the light of fluid inclusion studies', *Econ. Geol.* Vol. 61, 385–401

Sawkins, F.J. (1969). 'Chemical brecciation, an unrecognized mechanism for breccia formation?', *Econ. Geol.* Vol. 64, 613–617

Sawkins, F.J. (1976). 'Massive sulphide deposits in relation to geotectonics', *Metallogeny and Plate Tectonics. Geol. Assoc. Can. Special Paper 14*, pp. 221–240. Ed. D.F. Strong. Toronto; Geol. Assoc. Canada

Sayles, F.L. and Bischoff, J.L. (1973). 'Ferromangoan sediments in the equatorial east Pacific', *Earth. Planet. Sci. Lett.* Vol. 19, 330–336

Scheibner, E. and Markham, N.L. (1976). 'Tectonic setting of some stratabound massive sulphide deposits in New South Wales, Australia', *Handbook of Strata-bound and Stratiform Ore Deposits, Vol. 6*, pp. 55–77. Ed. K.H. Wolf. Amsterdam; Elsevier

Schenck, G.H.K. and Torries, T.F. (1975). 'Constructional materials: aggregates–crushed stone', *Industrial Minerals and Rocks (Nonmetallics Other Than Fuels)*, pp. 66–84. Ed. S.J. LeFond. New York; Amer. Inst. Min. Metall. Petrol. Engineers

Schmalz, R.F. (1969). 'Deep-water evaporite deposition: a genetic model', *Amer. Assoc. Petrol. Geol. Bull.* Vol. 53, 798–823

Schneider, H.J. and Lehmann, B. (1977). 'Contribution to a new genetical concept of the Bolivian tin province', *Time- and Strata-bound Ore Deposits*, pp. 153–168. Ed. D.D. Klemm and H.J. Schneider. Berlin; Springer-Verlag

Scholl, D.W. (1969). 'Modern coastal mangrove swamp stratigraphy and the ideal cyclothem', *Environments of Coal Deposition. Geol. Soc. Amer. Special Paper 114*, pp. 37–61. Ed. E.C. Dapples and M.E. Hopkins. Boulder, Colorado; Geol. Soc. Amer.

Schowalter, T.T. (1979). 'Mechanics of secondary hydrocarbon migration and entrapment', *Amer. Assoc. Petrol. Geol. Bull.* Vol. 63, 723–760

Schuiling, R.D. (1967a). 'Tin belts on the continents around the Atlantic Ocean', *Econ. Geol.* Vol. 62, 540–550

Schuiling, R.D. (1967b). 'Tin belts around the Atlantic Ocean: some aspects of the geochemistry of tin', *Technical Conference on Tin*, pp. 1–15. London; International Tin Council

Schweller, W.J. and Karig, D.E. (1979). 'Constraints on the origin and emplacement of the Zambales ophiolite, Luzon, Philippines', *Geol. Soc. Amer. Abstr. with Programs.* Vol. 11, 512–513

Scotese, C.R., Bambach, R.K., Barton, C., Van Der Voo, R. and Ziegler, A.M. (1979). 'Palaeozoic base maps', *J. Geology.* Vol. 87, 217–277

Scott, R.B., Malpas, J., Rona, P.A. and Udintsev, G. (1976). 'Duration of hydrothermal activity at an oceanic spreading centre mid-Atlantic Ridge (at 26°N)', *Geology.* Vol. 4, 233–236

Scrivenor, J.B. (1928). *The Geology of the Malayan Ore-deposits.* London; MacMillan

Sears, J.W. and Price, R.A. (1978). 'The Siberian connection: a case for Precambrian separation of the North American and Siberian cratons', *Geology.* Vol. 6, 267–270

Semenov, E.I. (1974). 'Economic mineralogy of alkaline rocks', *The Alkaline Rocks*, pp. 543–552. Ed. H. Sørensen. London; Wiley–Interscience

Shaffer, J.W. (1975). 'Bauxitic raw materials', *Industrial Minerals and Rocks (Nonmetallics Other Than Fuels)*, pp. 443–462. Ed. S.J. LeFond. New York; Amer. Inst. Min. Metall. Petrol. Engineers

Shannon, S.S. Jr. (1975). 'Monazite and related minerals', *Industrial Minerals and Rocks (Nonmetallics Other Than Fuels)*, pp. 851–859. Ed. S. LeFond. New York; Amer. Inst. Min. Metall. Petrol. Engineers

Shatskiy, N.S. (1964). 'On manganiferous formations and the metallogeny of manganese, Paper 1—volcanic–sedimentary manganese formations', *Int. Geol. Rev.* Vol. 6, 1030–1056

Shawe, D.R., Poole, F.G. and Brobst, D.A. (1969). 'Newly discovered bedded barite deposits in eastern Northumberland Canyon, Nye County, Nevada', *Econ. Geol.* Vol. 64, 245–254

Shcherba, G.N. (1970). 'Greisens', *Int. Geol. Rev.* Vol. 12. 114–150, 239–255

Sheldon, R.P. (1963). *Physical Stratigraphy and Mineral Resources of Permian Rocks in Western Wyoming. US Geol. Surv. Prof. Paper 313-B*, pp. 49–273. Washington DC; US Geological Survey

Shelton, J.E. (1980). 'Sulfur and pyrites', *Minerals Yearbook 1978–79. Vol. 1. Metals and Minerals*, pp. 877–897. Bureau of Mines. Washington DC; US Government Printing Office

Sheppard, R.A. (1973). 'Zeolites in sedimentary rocks', *United States Mineral Resources. US Geol. Surv. Prof. Paper 820*, pp. 689–695. Ed. D.A. Brobst and W.P. Pratt. Washington DC; US Geological Survey

Sheppard, R.A. (1975). 'Zeolites in sedimentary rocks', *Industrial Minerals and Rocks (Nonmetallics Other Than Fuels)*, pp. 1257–1262. Ed. S.J. LeFond. New York; Amer. Inst. Min. Metall. Petrol. Engineers

Sheppard, S.M.F. (1977a). 'Identification of the origin of ore-forming solutions by the use of stable isotopes', *Volcanic Processes in Ore Genesis. Special Publication 7*, pp. 25–41. London; The Geological Society

Sheppard, S.M.F. (1977b). 'The Cornubian batholith, S.W. England: D/H and $^{18}O/^{16}O$ studies on kaolinite and other alteration minerals', *J. Geol. Soc. London.* Vol. 133, 573–591

Sheppard, S.M.F., Nielsen, R.L. and Taylor, H.P. Jr. (1971). 'Hydrogen and oxygen isotope ratios in minerals from porphyry copper deposits', *Econ. Geol.* Vol. 66, 515–542

Shride, A.F. (1973). 'Asbestos', *United States Mineral Resources. US Geol. Surv. Prof. Paper 820*, pp. 63–73. Ed. D.A. Brobst and W.P. Pratt. Washington DC; US Geological Survey

Sibley, S.F. Jr. (1980). 'Cobalt', *Minerals Yearbook 1978–79. Vol. 1. Metals and Minerals*, pp. 249–258. Bureau of Mines. Washington DC; US Government Printing Office

Silberman, M.L. and McKee, E.H. (1974). 'Ages of Tertiary volcanic rocks and hydrothermal precious-metal deposits in central and western Nevada', *Rep. Nevada Bur. Mines Geol.* No. 19, 67–72

Sillitoe, R.H. (1974). 'Tin mineralization above mantle hot spots', *Nature.* Vol. 248, 497–499

Sillitoe, R.H. (1976). 'Andean mineralization: a model for the metallogeny of convergent plate margins', *Metallogeny and Plate Tectonics. Geol. Assoc. Can. Special Paper 14*, pp. 59–100. Ed. D. Strong. Toronto; Geol. Assoc. Canada

Sillitoe, R.H. (1977). 'Metallic mineralization affiliated to subaerial volcanism: a review', *Volcanic Processes in Ore Genesis. Geol. Soc. London Special Publication 7*, pp. 99–116. London; The Geological Society

Sillitoe, R.H., Halls, C. and Grant, J.N. (1975). 'Porphyry tin deposits in Bolivia', *Econ. Geol.* Vol. 70, 913–927

Silver, E.A., Yoko Joyodiwiryo and McCaffrey, R. (1978). 'Gravity results and emplacement geometry of the Sulawesi ultramafic belt, Indonesia', *Geology.* Vol. 6, 527–531

Simons, F.S. and Prinz, W.C. (1973). 'Gold', *United States Mineral Resources. US Geol. Surv. Prof. Paper 820*, pp. 263–275. Ed. D.A. Brobst and W.P. Pratt. Washington DC; US Geological Survey

Simons, F.S. and Straczek, J.A. (1958). *Geology of the Manganese Deposits of Cuba, US Geol. Surv. Bull. 1057.* Washington DC; US Geological Survey

Simpson, P.R., Plant, J. and Cope, M.J. (1977). 'Uranium abundance and distribution in some granites from northern Scotland and southwest England as indicators of uranium provinces', *Geology, Mining, and Extractive Processing of Uranium*, pp. 126–139. Ed. M.J. Jones. London; Inst. Min. Metall.

Singleton, R.H. (1980). 'Stone', *Minerals Yearbook 1978–79. Vol. 1. Metals and Minerals*, pp. 843–876. Bureau of Mines. Washington DC; US Government Printing Office

Singleton, R.H. and Searls, J.P. (1980a). 'Lithium', *Minerals Yearbook 1978–79. Vol. 1. Metals and Minerals*, pp. 551–557. Bureau of Mines. Washington DC; US Government Printing Office

Singleton, R.H. and Searls, J.P. (1980b). 'Peat', *Minerals Yearbook 1978–79. Vol. 1. Metals and Minerals*, pp. 655–669. Bureau of Mines. Washington DC; US Government Printing Office

Singleton, R.H. and Searls, J.P. (1980c). 'Potash', *Minerals Yearbook 1978–79. Vol. 1. Metals and Minerals*, pp. 713–727. Bureau of Mines. Washington DC; US Government Printing Office

Sittig, M. (1980). *Geophysical and Geochemical Techniques for Exploration of Hydrocarbons and Minerals.* Park Ridge, New Jersey; Noyes Data Corporation

Sivam, S.P. (1969). 'Quaternary alluvial deposits in the north Kinta Valley, Perak', M.Sc. thesis (unpublished), University of Malaya, Kuala Lumpur

Skall, H. (1975). 'The paleoenvironment of the Pine Point lead–zinc district', *Econ. Geol.* Vol. 70, 22–47

Skinner, B.J. (1979). 'The many origins of hydrothermal mineral deposits', *Geochemistry of Hydrothermal Ore Deposits*, 2nd edn., pp. 1–21. Ed. H.L. Barnes. New York; Wiley–Interscience

Skinner, E.M.W. and Clement, S.C.R. (1979). 'Mineralogical classification of southern African kimberlites', *Kimberlites, Diatremes and Diamonds: Their Geology, Petrology and Geochemistry. Proc 2nd International Kimberlite Conference, Vol. 1*, pp. 129–139. Ed. F.R. Boyd and H.O.A. Meyer. Washington DC; Amer. Geophysical Union

Slater, D. and Highley, D.E. (1977). 'The iron ore deposits in the United Kingdom of Great Britain and Northern Ireland', *The Iron Ore Deposits of Europe and Adjacent Areas. Vol. 1*, pp. 393–409. Ed. A. Zitzmann. Hannover. Bundesanstalt für Geowissenschaften und Rohstoffe

Sloss, L.L. (1969). 'Evaporite deposition from layered solutions', *Amer. Assoc. Petrol. Geol. Bull.* Vol. 53, 776–789

Smart, P.G., Wilkes, P.G., Needham, R.S. and Watchman, A.L. (1975). 'Geology and geophysics of the Alligator Rivers region', *Economic Geology of Australia and Papua New Guinea. Vol. 1. Metals*, pp. 285–301. Ed. C.L. Knight. Parkville, Victoria; Australasian Inst. Min. Metall.

Smirnov, V.I. (1976). *Geology of Mineral Deposits* (English version). Moscow; MIR publishers

Smith, A.G. and Briden, J.C. (1977). *Mesozoic and Cenozoic Paleocontinental Maps.* Cambridge; Cambridge University Press

Smith, D.A. (1980). 'Sealing and non-sealing faults in Louisiana Gulf Coast Salt Basin', *Amer. Assoc. Petrol. Geol. Bull.* Vol. 64, 145–172

Smith, D.B. and Crosby, A. (1979). 'The regional and stratigraphical context of Zechstein 3 and 4 Potash deposits in the British sector of the southern North Sea and adjoining land areas', *Econ. Geol.* Vol. 74, 397–408

Smith, G.I., Jones, C.L., Culbertson, W.C., Ericksen, G.E. and Dyni, J.R. (1973). 'Evaporites and brines', *United States Mineral Resources. US Geol. Surv. Prof. Paper 820*, pp. 197–216. Ed. D.A. Brobst and W.P. Pratt. Washington DC; US Geological Survey

Smith, J.W. and Croxford, N.J.W. (1975). 'An isotopic investigation of the environment of deposition of the McArthur mineralization', *Mineral. Deposita.* Vol. 10, 269–276

Smith, N.D. and Minter, W.E.L. (1980). 'Sedimentological controls of gold and uranium in two Witwatersrand paleoplacers', *Econ. Geol.* Vol. 75, 1–14

Smith, S.E. and Walker, K.R. (1971). 'Primary element dispersion associated with mineralization at Mount Isa', *Bur. Min. Res. Australia Bull.* Vol. 131, 1–80

Snethlage, R. and Klemm, D.D. (1978). 'Intrinsic oxygen fugacity measurements on chromites from the Bushveld Complex and their petrogenetic significance', *Contrib. Mineral. Petrol.* Vol. 67, 127–138

Snyder, F.G. (1968). 'Geology and mineral deposits, Midcontinent United States', *Ore Deposits of the United States 1933–1967. The Graton–Sales Volume*, pp. 257–286. Ed. J.D. Ridge. New York; Amer. Inst. Min. Metall. Petrol. Engineers

Snyder, F.G. (1969). 'Precambrian iron deposits in Missouri', *Magmatic Ore Deposits. A Symposium. Econ. Geol. Monograph 4*, pp. 231–238. Ed. H.D.B. Wilson. Lancaster, Penn.; Economic Geology Publishing Company

Snyder, F.G. and Gerdemann, P.E. (1968). 'Geology of the southeast Missouri lead district', *Ore Deposits of the United States 1933–1967. The Graton–Sales Volume*, pp. 326–358. Ed. J.D. Ridge. New York; Amer. Inst. Min. Metall. Petrol. Engineers

Solomon, M., Rafter, T.A. and Dunham, K.C. (1971). 'Sulphur and oxygen isotope studies in the northern Pennines in relation to ore genesis', *Trans. Inst. Min. Metall.* Vol. 80, B259–B275

Solomon, M., Walshe, J.L. and Palomero, F.G. (1980). 'Formation of massive sulphide deposits at Rio Tinto, Spain', *Inst. Min. Metall. Trans.* Vol. 89, B16–B24

Somm, A. (1973). 'Bauxite deposits of Gove Peninsula, Northern Territory, Australia', *Metallogenic Provinces and Mineral Deposits in the Southwestern Pacific. Bur. Min. Res. Geol. Geophys. Bull. 141*, p. 223. Ed. N.H. Fisher. Canberra; Bur. Min. Res. Geol. Geophys.

Somm, A.F. (1975). 'Gove Bauxite deposits, N.T.', *Economic Geology of Australia and Papua New Guinea. Vol. 1. Metals*, pp. 964–968. Ed. C.L. Knight. Parkville, Victoria; Australasian Inst. Min. Metall.

Soregaroli, A.E. and Sutherland Brown, A. (1976). 'Characteristics of Canadian Cordilleran molybdenum deposits', *Porphyry Deposits of the Canadian Cordillera. Can. Inst. Min. Metall. Special Volume 15*, pp. 417–431. Ed. A. Sutherland Brown. Montreal; Can. Inst. Min. Metall.

Sorem, R.K. (1956). 'Geology of three major manganese districts of the Philippines', *Proc. 8th Pacific Sci. Congr. 1953.* Vol. 2a, 661–668. Quezon City; National Research Council

Sorem, R.K. and Cameron, E.N. (1960). 'Manganese oxides and associated minerals of the Nsuta manganese deposits, Ghana, West Africa', *Econ. Geol.* Vol. 55, 278–310

Sørensen, H. (ed.) (1974). *The Alkaline Rocks.* London; Wiley–Interscience

Sørensen, H., Nielsen, B.L. and Jacobsen, F.L. (1978). 'Denmark and Greenland', *Mineral Deposits of Europe. Vol. 1. Northwest Europe*, pp. 251–261. Ed. S.H.U. Bowie, A. Kvalheim and H.W. Haslam. London; Inst. Min. Metall.

Souch, B.E., Podolsky, T. and Geological Staff (1969). 'The sulfide ores of Sudbury: their particular relationship to a distinctive inclusion-bearing facies of the Nickel Irruptive', *Magmatic Ore Deposits. Econ. Geol. Monograph 4*, pp. 252–261. Ed. H.D.B. Wilson. Lancaster, Penn.; Economic Geology Publishing Company

Spence, C.D. (1975). 'Volcanogenic features of the Vauze sulfide deposit, Noranda, Quebec', *Econ. Geol.* Vol. 70, 102–114

Spence, C.D. and De Rosen-Spence, A.F. (1975). 'The place of sulfide mineralization in the volcanic sequence at Noranda, Quebec', *Econ. Geol.* Vol. 70, 90–101

Spencer, A.C. (1908). *Magnetite Deposits of the Cornwall Type in Pennsylvania. US Geol. Surv. Bull. 359.* Washington DC; US Geological Survey

Spencer, E.W. (1977). *Introduction to the Structure of the Earth*, 2nd edn. New York; McGraw-Hill

Spooner, E.T.C. and Fyfe, W.S. (1973). 'Sub sea-floor metamorphism, heat and mass transfer', *Contrib. Mineral. Petrol.* Vol. 42, 287–304

Spratt, R.N. (1965). 'Uranium ore deposits of Rum Jungle', *Geology of Australian Ore Deposits. 8th Commonwealth Min. Metall. Congress, Melbourne*, pp. 201–206. Ed. J. McAndrew. Melbourne; Aust. Inst. Min. Metall.

Staatz, M.H. and Olson, J.C. (1973). 'Thorium', *United States Mineral Resources. US Geol. Surv. Prof. Paper 820*, pp. 468–476. Ed. D.A. Brobst and W.P. Pratt. Washington DC; US Geological Survey

Stach, E. (1968). 'Basic principles of coal petrology: macerals, microlithotypes and some effects of coalification', *Coal and Coal-bearing Strata*, pp. 3–17. Ed. D. Murchison and T.S. Westoll. Edinburgh; Oliver and Boyd

Stach, E., Mackowsky, M.T., Teichmüller, M., Taylor, G.H., Chandra, D. and Teichmüller, R. (eds.) (1975). *Stach's Textbook of Coal Petrology*, 2nd revised edn. Berlin; Gebrüder Borntraeger

Stafford, P.T. (1980). 'Tungsten', *Minerals Yearbook*

1978–79. Vol. 1. Metals and Minerals, pp. 949–963. Bureau of Mines. Washington DC; US Government Printing Office

Stanton, R.L. (1972). *Ore Petrology*. New York; McGraw-Hill

Stanton, R.L. (1976). 'Petrochemical studies of the ore environment at Broken Hill, New South Wales: 1—constitution of the banded iron-formations', *Trans. Inst. Min. Metall.* Vol. 85, B33–B46

Staplin, F.L. (1969). 'Sedimentary organic matter, organic metamorphism, and oil and gas occurrence', *Bull. Can. Petrol. Geol.* Vol. 17, 47

Stauffer, P.H. (1973). 'Cenozoic', *Geology of the Malay Peninsula: West Malaysia and Singapore*, pp. 143–176. Ed. D.J. Gobbett and C.S. Hutchison. New York; Wiley–Interscience

Steiner, R.J. (1976). 'Grand Isle Block 16 Field, offshore Louisiana', *North American Oil and Gas Fields. Mem. 24, Amer. Assoc. Petrol. Geol.* pp. 229–238. Ed. J. Braunstein. Tulsa, Oklahoma; Amer. Assoc. Petrol. Geol.

Steven, T.A. (1975). 'Middle Tertiary volcanic field in the southern Rocky Mountains', *Mem. Geol. Soc. Amer.* Vol. 144, 75–94

Steven, T.A. and Lipman, P.W. (1976). *Calderas of the San Juan volcanic field, southwest Colorado. US Geol. Surv. Prof. Paper 958*. Washington DC; US Geological Survey

Stoll, W.C. (1962). 'Geology and petrology of the Masinloc chromite deposit, Zambales, Luzon, Philippine islands', *Philippine Geologist*. Vol. 16, 1–41

Stowasser, W.F. (1980). 'Phosphate rock', *Minerals Yearbook 1978–79. Vol. 1. Metals and Minerals*, pp. 677–697. Bureau of Mines. Washington DC; US Government Printing Office

Stracke, K.J., Ferguson, J. and Black, L.P. (1979). 'Structural setting of kimberlites in south-eastern Australia', *Kimberlites, Diatremes and Diamonds: Their Geology, Petrology and Geochemistry. Proc. 2nd International Kimberlite Conference, Vol. 1*, pp. 71–91. Ed. F.R. Boyd and H.O.A. Meyer. Washington DC, American Geophysical Union

Strand, P.R. (1975). 'Vermiculite', *Industrial Minerals and Rocks (Nonmetallics Other Than Fuels)*, pp. 1219–1226. Ed. S.J. LeFond. New York; Amer. Inst. Min. Metall. Petrol. Engineers.

Strauss, G.K. and Madel, J. (1974). 'Geology of massive sulphide deposits in the Spanish–Portuguese pyrite belt', *Geol. Rundschau*. Vol. 63, 191–211

Streckeisen, A. (1976). 'To each plutonic rock its proper name', *Earth Sci. Rev.* Vol. 12, 1–33

Strong, D.F. (ed.) (1976). *Metallogeny and Plate Tectonics. Geol. Assoc. Can. Special Paper 14*. Toronto; Geol. Assoc. Canada

Stuckless, J.S. (1979). 'Uranium and thorium concentrations in Precambrian granites as indicators of a uranium province in central Wyoming', *Contrib. Geol., Univ. Wyoming.* Vol. 17, No. 2, 173–178

Stumpfl, E.F. (1977). 'Mineralogical aspects of ores. Sediments, ores and metamorphism: new aspects', *Phil. Trans. Royal Soc. London.* Vol. A286, 507–525

Stumpfl, E.F. (1979a). 'Manganese haloes surrounding metamorphic stratabound base metal deposits', *Mineral. Deposita.* Vol. 14, 207–217

Stumpfl, E.F. (1979b). 'Precambrian stratabound base metal deposits in Namaqualand, S. Africa', *Schriftenreihe der GDMB Gesellschaft Deutscher Metallhütten und Bergleute.* Vol. 33, 101–114

Surdam, R.C. and Stanley, K.O. (1979). 'Lacustrine sedimentation during the culminating phase of Eocene Lake Gosiute, Wyoming (Green River Formation)', *Geol. Soc. Amer. Bull.* Vol. 90, 93–110

Sutherland Brown, A. (1976). 'Morphology and classification', *Porphyry Deposits of the Canadian Cordillera. Can. Inst. Min. Metall. Special Volume 15*, pp. 44–51. Ed. A. Sutherland Brown. Montreal; Can. Inst. Min. Metall.

Sutherland Brown, A. and Cathro, R.J. (1976). 'A perspective of porphyry deposits', *Porphyry Deposits of the Canadian Cordillera. Can. Inst. Min. Metall. Special Volume 15*, pp. 7–16. Ed. A. Sutherland Brown. Montreal; Can. Inst. Min. Metall.

Suzuki, M. and Kubota, Y. (1980). 'The Shimokawa diabase and wall rock alteration of the Shimokawa ore deposit, Hokkaido', *Mining Geology*. Vol. 30, 1–18

Swanson, R.W. (1970). *Mineral Resources in Permian Rocks of Southwest Montana. US Geol. Surv. Prof. Paper 313-E*, pp. 661–777. Washington DC; US Geological Survey

Swanson, R.W. (1973). *Geology and Phosphate Deposits of the Permian Rocks in Central Western Montana. US Geol. Surv. Prof. Paper 313-F*, pp. 779–883. Washington DC; US Geological Survey

Swanson, V.E. (1961). *Geology and Geochemistry of Uranium in Marine Black Shales: A Review. US Geol. Surv. Prof. Paper 356-C*, pp. 67–112. Washington DC; US Geological Survey

Takahashi, M., Aramaki, S. and Ishihara, S. (1980). 'Magnetite-series/ilmenite-series vs. I-type/S-type granitoids', *Granitic Magmatism and Related Mineralization. Mining Geol. Japan, Special Issue No. 8*, pp. 13–28. Ed. S. Ishihara and S. Takenouchi. Tokyo; Soc. Min. Geol. Japan

Takeda, H. (1970). 'Structural study on the stratified pyritic deposits in the Shirataki and Sazare mining district, Shikoku', *Volcanism and Ore Genesis*, pp. 77–91. Ed. T. Tatsumi. Tokyo; University of Tokyo Press

Talbot, M.R. (1974). 'Ironstones in the Upper Oxfordian of southern England', *Sedimentology*. Vol. 71, 433–450

Tan, L.P. (1976). 'Taiwan Copper mineralization in relation to plate tectonics', *Bull. Geol. Surv. Taiwan.* Vol. 25, 1–17

Tantisukrit, C. (1975). 'Review of the metallic deposits', *Proc. Conference on Geology of Thailand. Special Publication 1*, pp. 185–209. Ed. R.B. Stokes, C. Tantisukrit and K.V. Campbell. Chiengmai, Thailand; Department of Geological Science, Chiengmai University

Tarling, D.H. (ed.) (1981). *Economic Geology and Geotectonics*. Oxford; Blackwell Scientific

Tarney, J., Dalziel, I.W.D. and De Wit, M.J. (1976). 'Marginal basin "Rocas Verdes" complex from S. Chile: a model for Archaean Greenstone Belt formation', *The Early History of the Earth*, pp. 131–146. Ed. B.F. Windley. New York; John Wiley

Taylor, D. (1971). 'An outline of the geology of the Bukit Ibam orebody, Rompin, Pahang', *Geol. Soc. Malaysia Bull.* Vol. 4, 71–89

Taylor, D. (1973). 'The liberation of minor elements from rocks during plutonic igneous cycles and their subsequent concentration to form workable ores, with particular reference to copper and tin', *Geol. Soc. Malaysia Bull.* Vol. 7, 1–16

Taylor, D. and Hutchison, C.S. (1979). 'Patterns of mineralization in Southeast Asia, their relationship to broad-scale geological features and the relevance of plate-tectonic concepts to their understanding', *Proc. 11th Comm. Mining and Metall. Congr., Hong Kong, 1978*, pp. 93–107. Ed. M.J. Jones. London; Inst. Min. Metall.

Taylor, H.A. Jr. (1980). 'Graphite', *Minerals Yearbook 1978–79. Vol. 1. Metals and Minerals*, pp. 401–409. Bureau of Mines. Washington DC; US Government Printing Office

Taylor, H.P. Jr. (1973). 'O^{18}/O^{16} evidence for meteoric–hydrothermal alteration and ore deposition in the Tonopah, Comstock Lode, and Goldfield mining districts, Nevada', *Econ. Geol.* Vol. 68, 747–764

Taylor, H.P. Jr. (1974). 'The application of oxygen and hydrogen isotope studies to problems of hydrothermal alteration and ore deposition', *Econ. Geol.* Vol. 69, 843–883

Taylor, H.P. Jr. (1979). 'Oxygen and hydrogen isotope relationships in hydrothermal mineral deposits', *Geochemistry of Hydrothermal Ore Deposits*, 2nd edn., pp. 236–277. Ed. H.L. Barnes. New York; Wiley–Interscience

Taylor, J.C.M. and Colter, V.S. (1975). 'Zechstein of the English sector of the Southern North Sea Basin', *Petroleum and the Continental Shelf of Northwest Europe. 1, Geology*, pp. 249–263. Ed. A.W. Woodland. London; Applied Science Publishers

Taylor, R.G. (1979a). 'Some observations upon the tin deposits of Australia', *Geol. Soc. Malaysia Bull.* Vol. 11, 181–207

Taylor, R.G. (1979b). *Geology of Tin Deposits.* Amsterdam; Elsevier

Taylor, S. and Andrew, C.J. (1978). 'Silvermines orebodies, Tipperary, Ireland', *Trans. Inst. Min. Metall.* Vol. 87, B111–B124

Teh, G.H. (1978). 'The Tekka tin deposits of Perak, Peninsular Malaysia', *Geol. Soc. Malaysia Newsletter 4*, No. 2 (annex), 52, 53

Teichmüller, M. and Teichmüller, R. (1968a). 'Geological aspects of coal metamorphism,' *Coal and Coal-bearing Strata*, pp. 233–267. Ed. D. Murchison and T.S. Westoll. Edinburgh; Oliver and Boyd

Teichmüller, M. and Teichmüller, R. (1968b), 'Cainozoic and Mesozoic Coal deposits of Germany', *Coal and Coal-bearing Strata*, pp. 347–379. Ed. D. Murchison and T.S. Westoll. Edinburgh; Oliver and Boyd

Tepordei, V.V. (1980). 'Sand and gravel', *Minerals Yearbook 1978–79. Vol. 1. Metals and Minerals*, pp. 763–791. Bureau of Mines. Washington DC; US Government Printing Office

Thacker, J.L. and Anderson, K.H. (1977). 'The geologic setting of the southeast Missouri lead district—regional geologic history, structure and stratigraphy', *Econ. Geol.* Vol. 72, 339–348

Thaden, R.E. (1973a). 'Abrasives', *United States Mineral Resources. US Geol. Surv. Prof. Paper 820*, pp. 27–33. Ed. D.A. Brobst and W.P. Pratt. Washington DC; US Geological Survey

Thaden, R.E. (1973b). 'Gemstones', *United States Mineral Resources. US Geol. Surv. Prof. Paper 820*, pp. 247–250. Ed. D.A. Brobst and W.P. Pratt. Washington DC; US Geological Survey

Thayer, T.P. (1942). 'Chrome resources of Cuba', *Bull. US Geol. Surv.* Vol. 935A, 1–47

Thayer, T.P. (1964). 'Principal features and origin of podiform chromite deposits, and some observations on the Guleman–Soridag district, Turkey', *Econ. Geol.* Vol. 59, 1497–1524

Thayer, T.P. (1966). 'Chromite', *Mineral Resources of California. Bull. 191*, pp. 120–126. San Francisco, Calif.; Div. Mines and Geology

Thayer, T.P. (1970). *Chromite Segregations as Petrogenetic Indicators. Geol. Soc. South Africa Special Publication 1* (Symposium on the Bushveld Igneous Complex and other layered intrusions), pp. 380–390. Johannesburg; Geol. Soc. South Africa

Thayer, T.P. (1973). 'Chromium', *United States Mineral Resources. US Geol. Surv. Prof. Paper 820*, pp. 111–121. Ed. D.A. Brobst and W.P. Pratt. Washington DC; US Geological Survey

Theis, N.J. (1978). 'Mineralogy and setting of Elliot Lake deposits', *Uranium Deposits, Their Mineralogy and Origin. Short Course Handbook 3*, pp. 331–338. Ed. M.M. Kimberley. Ottawa; Miner. Assoc. Canada

Theodore, T.G. (1977). 'Selected copper-bearing skarns and epizonal granitic intrusions in the southwestern United States', *Geol. Soc. Malaysia Bull.* Vol. 9, 31–50

Thompson, R.I. and Panteleyev, A. (1976). 'Stratabound mineral deposits of the Canadian Cordillera', *Handbook of Strata-bound and Stratiform Ore Deposits. Vol. 5*, pp. 37–108. Ed. K.H. Wolf. Amsterdam; Elsevier

Thompson, R.R. and Benedict, L.G. (1974). 'Vitrinite reflectance as an indicator of coal metamorphism for coke making', *Carbonaceous Materials as Indicators of Metamorphism. Geol. Soc. Amer. Special Paper 153*, pp. 95–108. Ed. R.R. Dutcher, P.A. Hacquebard, J.M. Schopf and J.A. Simon. Boulder, Colorado; Geol. Soc. Amer.

Thrush, P.W. (ed.) (1968). *A Dictionary of Mining, Mineral and Related Terms.* Washington DC; Bureau of Mines. US Department of the Interior

Thurlow, J.G., Swanson, E.A. and Strong, D.F. (1975). 'Geology and lithogeochemistry of the Buchans polymetallic sulfide deposits, Newfoundland', *Econ. Geol.* Vol. 70, 130–144

Tiratsoo, E.N. (1979). *Natural Gas.* Beaconsfield, England; Scientific Press

Tissot, B. (1977). 'The application of the results of organic geochemical studies in oil and gas exploration', *Developments in Petroleum—1*, pp. 53–82. Ed. G.D. Hobson. London; Applied Science Publishers

Tissot, B., Durand, B., Espitalié, J. and Combaz, A. (1974). 'Influence of nature and diagenesis of organic matter in formation of petroleum', *Amer. Assoc. Petrol. Geol. Bull.* Vol. 58, 499–506

Tissot, B., Deroo, G. and Hood, A. (1978). 'Geochemical study of the Uinta Basin: formation of petroleum from the Green River Formation', *Geochim. Cosmochim. Acta.* Vol. 42, 1469–1485

Tissot, B.P. and Welte, D.H. (1978). *Petroleum Formation and Occurrence: A New Approach to Oil and Gas Exploration.* Berlin; Springer-Verlag

Tomich, S.A. (1974). 'A new look at Kalgoorlie Golden Mile geology', *Proc. Australasian Inst. Min. Metall.* No. 251 (Sept.), 27–35

Tomich, S.A. (1976). 'Further thoughts on the application of the volcanogenic theory to the Golden Mile ores at Kalgoorlie', *Proc. Australasian Inst. Min. Metall.* No. 258 (June), 19–29

Tooms, J.S., Summerhayes, C.P. and Cronan, D.S.L. (1969). 'Geochemistry of marine phosphate and manganese deposits, *Oceanogr. Mar. Biol. Ann. Rev.* Vol. 7, 49–100

Trace, R.D. (1974). 'Illinois–Kentucky fluorspar district', *A Symposium on the Geology of Fluorspar. Kentucky Geol. Surv. Series X, Special Publication 22*, pp. 58–76. Ed. D.W. Hutcheson. Lexington, Kentucky; Kentucky Geological Survey.

Tremblay, L.P. (1972). *Geology of the Beaverlodge Mining Area, Saskatchewan. Can. Geol. Surv. Memoir 367.* Ottawa; Can. Geol. Surv.

Tremblay, L.P. (1978). 'Geologic setting of the Beaverlodge-type of vein-uranium deposit and its comparison to that of the unconformity type', *Uranium Deposits: Their Mineralogy and Origin. Short Course Handbook 3*, pp. 431–456. Ed. M.M. Kimberley. Ottawa; Min. Assoc. Canada

Trendall, A.F. (1973). 'Precambrian iron-formations of Australia', *Econ. Geol.* Vol. 68, 1023–1034

Trescases, J.J. (1973). 'Weathering and geochemical behaviour of the elements of ultramafic rocks in New Caledonia', *Metallogenic Provinces and Mineral Deposits in the Southwestern Pacific. Bur. Min. Resources Geol. Geophys., Canberra, Bull. 141*, pp. 149–161. Ed. N.H. Fisher. Canberra; Bureau of Mineral Resources, Geology and Geophysics

Troly, G., Esterle, M., Pelletier, B. and Reibell, W. (1979). 'Nickel deposits in New Caledonia. Some factors influencing their formation', *International Laterite Symposium*, pp. 85–119. Ed. D.J.I. Evans, R.S. Shoemaker and H. Veltman. New York; Amer. Inst. Min. Metall. Petrol. Engineers

Turneaure, F.S. (1971). 'The Bolivian tin-silver province', *Econ. Geol.* Vol. 66, 215–225

Tuttle, O.F. and Gittins, J. (1966). *Carbonatites.* New York; Wiley-Interscience

United Nations (1979). *World Energy Supplies 1973–1978. Statistical Papers J-22.* New York; United Nations

Upadhyay, H.D. and Strong, D.F. (1973). 'Geological setting of the Betts Cove copper deposits: an example of ophiolite sulphide mineralization', *Econ. Geol.* Vol. 68, 161–167

Usselman, T.M., Hodge, D.S., Naldrett, A.J. and Campbull, I.H. (1979). 'Physical constraints of the characteristics of nickel-sulfide ore in ultramafic lavas', *Can. Mineral.* Vol. 17, 361–372

Uytenbogaart, W. and Burke, E.A.J. (1971). *Tables for Microscopic Identification of Ore Minerals*, 2nd edn. Amsterdam; Elsevier

Vaché, R. (1966). 'Zur geologie der Varisziden und ihrer Lagerstätten im Südanatolischen Taurus', *Mineral. Deposita.* Vol. 1, 30–42

Valeton, I. (1972). *Bauxites.* Amsterdam; Elsevier

Van Eden, J.G. (1974). 'Depositional and diagenetic environment related to sulphide mineralization, Mufulira, Zambia', *Econ. Geol.* Vol. 69, 59–79

Varlamoff, N. (1972). 'Central and west African rare metal granitic pegmatites, related aplites, quartz veins and mineral deposits', *Mineral. Deposita.* Vol. 7, 202–216

Varlamoff, N. (1978). 'Classification and spatio-temporal distribution of tin and associated mineral deposits', *Metallization Associated with Acid Magmatism, Vol. 3*, pp. 139–159. Ed. M. Stemprok, L. Burnol and G. Tischendorf. Prague; Geological Survey

Varne, R. and Rubenach, M.J. (1973). 'Geology of Macquarie island in relation to tectonic environment', *The Western Pacific: Island Arcs, Marginal Seas, Geochemistry*, pp. 535–541. Ed. P.J. Coleman. Perth; University of Western Australia Press

Vartiainen, H. and Paarma, H. (1979). 'Geological characteristics of the Sokli carbonatite complex, Finland', *Econ. Geol.* Vol. 74, 1296–1306

Veeh, H.H., Burnett, W.C. and Soutar, A. (1973). 'Contemporary phosphorites on the continental margin of Peru', *Science.* Vol. 181, 844–845

Vermaak, C.F. (1976). 'The Merensky Reef—thoughts on its environment and genesis', *Econ. Geol.* Vol. 71, 1270–1298

Vermaak, C.F. and Hendriks, C.P. (1976). 'A review of the mineralogy of the Merensky Reef, with special reference to new data on the precious metal mineralogy', *Econ. Geol.* Vol. 71, 1244–1269

Vhay, J.S., Brobst, D.A. and Heyl, A.V. (1973). 'Cobalt', *United States Mineral Resources. US Geol. Surv. Prof. Paper 820*, pp. 143–155. Ed. D.A. Brobst and W.P. Pratt. Washington DC; US Geological Survey

Vialette, Y. (1973). 'Age des granites du Massif Central', *Soc. Géol. France Bull.* Vol. 15, 260–270

Vokes, F.M. (1976). 'Caledonian massive sulphide deposits in Scandinavia: a comparative view', *Handbook of Strata-bound and Stratiform Ore Deposits. Vol. 6*, pp. 79–127. Ed. K.H. Wolf. Amsterdam; Elsevier

Von Backström, J.W. (1974). 'Other uranium ore deposits', *Formation of Uranium Ore Deposits. Proc. Symposium, Athens 1974*, pp. 605–624. Vienna; International Atomic Energy Agency

Von Gruenewaldt, G. (1977). 'The mineral resources of the Bushveld Complex', *Mineral. Sci. Eng.* Vol. 9, 83–95

Von Gruenewaldt, G. (1979). 'A review of some recent concepts of the Bushveld Complex, with particular reference to sulfide mineralization', *Can. Mineral.* Vol. 17, 233–256

Walker, R.R., Matulich, A., Amos, A.C., Watkins, J.J. and Mannard, G.W. (1975). 'The geology of Kidd Creek Mine', *Econ. Geol.* Vol. 70, 80–89

Wallace, S.R., Muncaster, N.K., Jonson, D.C., Mackenzie, W.B., Bookstrom, A.A. and Surface, V.E. (1968). 'Multiple intrusion and mineralization at Climax, Colorado', *Ore Deposits of the United States, 1933–1967. Vol. 1*, pp. 605–640. Ed. J.D. Ridge. New York; Amer. Inst. Min. Metall. Petrol. Engineers

Wallace, S.R., Mackenzie, W.B., Blair, R. and Muncaster, N.K. (1978). 'Geology of the Urad and Henderson molybdenite deposits, Clear Creek County, Colorado, with section on a comparison of their deposits with those of Climax, Colorado', *Econ. Geol.* Vol. 73, 325–368

Walmsley, P.J. (1975). 'The Forties Field', *Petroleum and the Continental Shelf of Northwest Europe. Vol. 1. Geology*, pp. 477–485. Ed. A.W. Woodland. London; Applied Science Publishers

Walter, M.R. (1972). 'A hot spring analog for the depositional environment of Precambrian iron formations of the Lake Superior region', *Econ. Geol.* Vol. 67, 965–980

Wang, L.K., Zhao, B., Zhu, W.F., Cai, Y.J. and Li, T.J. (1980). 'Characteristics and melting experiments of granites in southern China', *Granitic Magmatism and Related Mineralization. Mining Geology, Japan, Special Issue No. 8*, pp. 29–38. Ed. S. Ishihara and S. Takenouchi. Tokyo; Soc. Min. Geol. Japan

Wanless, H.R., Baroffio, J.R. and Trescott, P.C. (1969). 'Conditions of deposition of Pennsylvanian coal beds', *Environments of Coal Deposition. Geol. Soc. Amer. Special Paper 114*, pp. 105–142. Ed. E.C. Dapples and M.E. Hopkins. Boulder, Colorado; Geol. Soc. Amer.

Warren, R.G. (1973). 'A brief review of metallogenesis in the Precambrian part of the Australian craton', *Metallogenic Provinces and Mineral Deposits in the Southwestern Pacific. Bur. Min. Res. Geol. Geophys. Bull. Canberra, 141*, pp. 69–79. Ed. N.H. Fisher. Canberra; Bur. Min. Res. Geol. Geophys.

Watanabe, T., Yui, S. and Kato, A. (1970a). 'Bedded manganese deposits in Japan, a review', *Volcanism and Ore Genesis*, pp. 119–142. Ed. T. Tatsumi. Tokyo; University of Tokyo Press

Watanabe, T., Yui, S. and Kato, A. (1970b). 'Metamorphosed bedded manganese deposits of the Noda-Tamagawa Mine', *Volcanism and Ore Genesis*, pp. 143–152. Ed. T. Tatsumi. Tokyo; University of Tokyo Press

Watanabe, T., Iwao, S., Tatsumi, T. and Kanehira, K.n(1970c). 'Folded ore bodies of the Okuki mine', *Volcanism and Ore Genesis*, pp. 105–117. Ed. T. Tatsumi. Tokyo; University of Tokyo Press

Watson, J. (1976). 'Mineralization in Archaean provinces', *The Early History of the Earth*, pp. 443–453. Ed. B.F. Windley. New York; John Wiley

Weber, F. (1973). 'Genesis and supergene evolution of the Precambrian sedimentary manganese deposit at Moanda (Gabon)', *Genesis of Precambrian Iron and Manganese Deposits*, pp. 307–322. Paris; UNESCO

Weber, K.J. and Daukoru, E. (1975). 'Petroleum geology of the Niger Delta', *Ninth World Petroleum Congress Proceedings, Vol. 2*, pp. 209–221. London; Applied Science Publishers

Wedow, H. Jr. (1973). 'Cadmium', *United States Mineral Resources. US Geol. Surv. Prof. Paper 820*, pp. 105–109. Ed. D.A. Brobst and W.P. Pratt. Washington DC; US Geological Survey

Wedow, H. Jr., Kiilsgaard, T.H., Heyl, A.V. and Hall, R.B. (1973). 'Zinc', *United States Mineral Resources. US Geol. Surv. Prof. Paper 820*, pp. 697–711. Ed. D.A. Brobst and

W.P. Pratt. Washington DC; US Geological Survey

Weeks, R.A. (1973). 'Gallium, germanium, and indium', *United States Mineral Resources. US Geol. Surv. Prof. Paper 820*, pp. 237–246. Ed. D.A. Brobst and W.P. Pratt. Washington DC; US Geological Survey

Weis, P.L. (1973). 'Graphite', *United States Mineral Resources. US Geol. Surv. Prof. Paper 820*, pp. 277–283. Ed. D.A. Brobst and W.P. Pratt. Washington DC; US Geological Survey

Weisman, W.I. and Tandy, C.W. (1975). 'Sodium sulfate deposits', *Industrial Minerals and Rocks (Nonmetallics Other Than Fuels)*, pp. 1081–1093. Ed. S.J. LeFond. New York; Amer. Inst. Min. Metall. Petrol. Engineers

Welch, B.K. (1964). 'The ilmenite deposits of Geographe Bay', *Proc. Australasian Inst. Min. Metall.* Vol. 211, 25–48

Welch, B.K., Sofoulis, J. and Fitzgerald, A.C.F. (1975). 'Mineral sand deposits of the Capel area W.A.', *Economic Geology of Australia and Papua New Guinea. Vol. 1. Metals*, pp. 1070–1088. Ed. C.L. Knight. Parkville, Victoria; Australasian Inst. Min. Metall.

Westoll, T.S. (1968). 'Sedimentary rhythms in coal-bearing strata', *Coal and Coal-bearing Strata*, pp. 71–103. Ed. D. Murchison and T.S. Westoll. Edinburgh; Oliver and Boyd

White, A.J.R. and Chappell, B.W. (1977). 'Ultrametamorphism and granitoid genesis', *Tectonophysics.* Vol. 43, 7–22

White, D.E. (1968). 'Environment of generation of some base-metal deposits', *Econ. Geol.* Vol. 63, 301–335

White, D.E. (1974). 'Diverse origins of hydrothermal ore fluids', *Econ. Geol.* Vol. 69, 954–973

White, W.C. and Warin, O.N. (1964). 'A survey of phosphate deposits in the South-west Pacific and Australian Waters', *Bur. Min. Res. Geophys., Bull. 69.* Canberra; Comm. of Australia

Whitford, D.J., Nicholls, I.A. and Taylor, S.R. (1979). 'Spatial variations in the geochemistry of Quaternary lavas across the Sunda arc in Java and Bali', *Contrib. Mineral. Petrol.* Vol. 70, 341–356

Wicken, O.M. and Duncan, L.R. (1975). 'Magnesite and related minerals', *Industrial Minerals and Rocks (Nonmetallics Other Than Fuels)*, pp. 805–820. Ed. S. LeFond. New York; Amer. Inst. Min. Metall. Petrol. Engineers

Wilford, G.E. (1955). *The Geology and Mineral Resources of the Kuching–Lundu Area, West Sarawak, Including the Bau Mining District, Geol. Surv. Dept. British Territories in Borneo, Mem. 3*, Kuching, Sarawak; Geol. Surv. Malaysia

Willbourn, E.S. (1932). 'The Beatrice Mine, Selibin, F.M.S.', *Min. Mag.* Vol. 46, 20–24

Willbourn, E.S. (1936). 'A short account of the geology of those tin deposits of Kinta that are mined by alluvial methods', *J. Eng. Assoc. Malaya.* Vol. 4, 255–264

Willbourn, E.S. and Ingham, F.T. (1933). 'Geology of the scheelite mine, Kramat Pulai, F.M.S.', *Quart. J. Geol. Soc. London.* Vol. 89, 449–479

Willemse, J. (1969a). 'The geology of the Bushveld Igneous Complex, the largest respository of magmatic ore deposits in the world', *Magmatic Ore Deposits: A Symposium.* Monograph 4, pp. 1–22. Ed. H.D.B. Wilson. Lancaster, Penn.; Economic Geology Publishers

Willemse, J. (1969b). 'The vanadiferous magnetite iron ore of the Bushveld Igneous Complex', *Magmatic Ore Deposits: A Symposium.* Monograph 4, pp. 187–208. Ed. H.D.B. Wilson. Lancaster, Penn.; Economic Geology Publishers

Williams, C.E. and McArdle, P. (1978). 'Ireland', *Mineral Deposits of Europe. Vol. 1. Northwest Europe*, pp. 319–345. Ed. S.H.U. Bowie, A. Kvalheim and H.W. Haslam. London; Inst. Min. Metall.

Williams, D. (1962). 'Further reflections on the origin of the porphyries and ores of Rio Tinto, Spain', *Trans. Inst. Min. Metall.* Vol. 71, B265–B266

Williams, D., Stanton, R.L. and Rambaud, F. (1975). 'The Planes–San Antonio pyritic deposit of Rio Tinto, Spain: its nature, environment and genesis', *Trans. Inst. Min. Metall.* Vol. 84, B73–B82

Williams, G.J. (1965). 'Tertiary propylitic mineralization', *Economic Geology of New Zealand*, pp. 87–126. Melbourne; Australasian Inst. Min. Metall.

Wilson, J.G. (1979). 'The major controls of tin mineralization in the Bushveld Igneous complex, South Africa', *Geol. Soc. Malaysia Bull.* Vol. 11, 239–251

Winson, R.W. (1975). 'Asbestos', *Industrial Minerals and Rocks (Nonmetallics Other Than Fuels)*, pp. 379–425. Ed. S.J. LeFond. New York; Amer. Inst. Min. Metall. Petrol. Engineers

Wisser, E. (1966). 'The epithermal precious-metal provinces of north-west Mexico', *Rep. Nevada Bur. Mines*, No. 13, Part C, 63–92

Wolf, K.H. (1976a). 'Ore genesis influenced by compaction', *Compaction of Coarse Grained Sediments II. Developments in Sedimentology. Vol. 18B*, pp. 475–675. Ed. G.V. Chilingarian and K.H. Wolf. Amsterdam; Elsevier

Wolf, K.H. (ed.) (1976b). *Handbook of Strata-bound and Stratiform Ore Deposits.* Vol. 1. Classifications and historical studies. Vol. 2. Geochemical studies. Vol. 3. Supergene and surficial ore deposits; textures and fabrics. Vol. 4. Tectonics and metamorphism. Index to volumes 1–4. Vol. 5. Regional studies. Vol. 6. Cu, Zn, Pb, and Ag deposits. Vol. 7. Au, U, Fe, Mn, Hg, Sb, W, and P deposits. Index to volumes 5–7. Amsterdam; Elsevier

Wong, D.H.W. (1979). 'Geology and mineralization of the Bukit Young gold mine area, Bau, West Sarawak, Malaysia', B.Sc. thesis (unpublished). Geological Department, University of Malaya; Kuala Lumpur

Wood, H.B. (1968). 'Geology and exploitation of uranium deposits in the Lisbon Valley area, Utah', *Ore Deposits of the United States 1933–1967. The Graton–Sales Volume. Vol. 1*, pp. 770–789. Ed. J.D. Ridge. New York; Amer. Inst. Min. Metall. Petrol. Engineers

Woods, P.J.E. (1979). 'The geology of Boulby Mine', *Econ. Geol.* Vol. 74, 409–418

Wopfner, H. and Schwarzbach, M. (1976). 'Ore deposits in the light of paleoclimatology', *Handbook of Strata-bound and Stratiform Ore Deposits, Vol. 3*, pp. 43–92. Ed. K.H. Wolf. Amsterdam; Elsevier

Worl, R.G., Van Alstine, R.E. and Shawe, D.R. (1973). 'Fluorine', *United States Mineral Resources. US Geol. Surv. Prof. Paper 820*, pp. 223–235. Ed. D.A. Brobst and W.P. Pratt. Washington DC; US Geological Survey

Worsley, N. and Fuzesy, A. (1979). 'The potash-bearing members of the Devonian Prairie Evaporite of southeastern Saskatchewan, south of the mining area', *Econ. Geol.* Vol. 74, 377–388

Worst, B.G. (1964). 'Chromite in the Great Dyke of Southern Rhodesia. The geology of some ore deposits in Southern Africa', *Geol. Soc. South Africa.* Vol. 2, 209–224

Worzel, J.L. (1974). 'Standard oceanic and continental structure', *The Geology of Continental Margins*, pp. 59–66. Ed. C.A. Burk and C.L. Drake. Berlin; Springer-Verlag

Wright, J.B. (1970). 'Controls of mineralization in the older and younger tin fields of Nigeria', *Econ. Geol.* Vol. 65, 945–951

Wyllie, P.J. (1976). *The Way the Earth Works: An Introduction to the New Global Geology and its Revolutionary Development.* New York; John Wiley

Wyllie, P.J. (1979). 'Kimberlite magmas from the system peridotite–CO_2–H_2O', *Kimberlites, Diatremes and Diamonds: Their Geology, Petrology and Geochemistry. Proc.*

2nd International Kimberlite Conference, Vol. 1, pp. 319–329. Ed. F.R. Boyd and H.O.A. Meyer. Washington DC; Amer. Geophysical Union

Xu Keqin, Guo Lingzhi, Hu Shouxi, Ji Shouyuan, Shi Yangshen, Wang Zhengran, Sun Minggzhi, Mou Weixi, Lin Chengyi and Ye Jun (1980). 'Investigation on the time and spatial distribution of the granitic rocks of Southeastern China, their petrographic evolution, petrogenetic types, and metallogenetic relations', *J. Nanjing Univ. Spec. Issues on Geological Science*, pp. 1–56

Yan, M.Z., Wu, Y.L. and Li, C.Y. (1980). 'Metallogenetic systems of tungsten in southern China and their mineralization characteristics', *Granitic Magmatism and Related Mineralization. Mining Geology, Japan, Special Issue No. 8*, pp. 215–221. Ed. S. Ishihara and S. Takenouchi. Tokyo; Soc. Min. Geol. Japan

Yeap, E.B. (1978a). 'Hydrothermal tin-bearing breccias of the Yap Peng mine, Sungai Besi, Selangor, Peninsular Malaysia', *Proc. 3rd Regional Conference on the Geology and Mineral Resources of Southeast Asia*, pp. 367–375. Ed. Prinya Nutalaya. Bangkok; Asian Institute of Technology

Yeap, E.B. (1978b). 'The carbona type replacement ore bodies of the No. 2 opencast Sungai Besi mines', *Geol. Soc. Malaysia Newsletter 4*, No. 2 annex, 23, 24

Yeend, W. (1973). 'Sand and gravel', *United States Mineral Resources. US Geol. Surv. Prof. Paper 820*, pp. 561–565. Ed. D.A. Brobst and W.P. Pratt. Washington DC; US Geological Survey

Yoder, H.S. (ed.) (1979). *The Evolution of the Igneous Rocks: Fiftieth Anniversary Perspectives.* Princeton, New Jersey; Princeton University Press

Young, A., Monaghan, P.H. and Schweisberger, R.T. (1977). 'Calculation of ages of hydrocarbons in oils—physical chemistry applied to petroleum geochemistry', *Amer. Assoc. Petrol. Geol. Bull.* Vol. 61, 573–600

Young, E.J. (1979). 'Genesis of the Schwartzwalder uranium deposit, Jefferson County, Colorado', *Contrib. Geol., Univ. Wyoming.* Vol. 17, No. 2, 179–186

Zachos, K. (1964). 'Mineralization of the Greek serpentinites', *Methods of Prospection for Chromite*, pp. 49–54. Ed. R. Woodtli. Paris; Organization for Economic Co-operation and Development

Zartman, R.E. and Stacey, J.D. (1971). 'The use of lead isotopes to distinguish between Precambrian and Mesozoic–Cenozoic mineralization in Belt Supergroup rocks, northwestern Montana and northern Idaho', *Econ. Geol.* Vol. 66, 849–860

Zeissink, E.H. (1969). 'The mineralogy and geochemistry of a nickeliferous lateritic profile (Greenvale, Queensland, Australia)', *Mineral Deposita.* Vol. 4, 132–152

Ziegler, P.A. (1975). 'North Sea Basin History in the tectonic framework of Northwest Europe', *Petroleum and the Continental Shelf of Northwest Europe. Vol. 1. Geology*, pp. 131–149. Ed. A.W. Woodland. London; Applied Science Publishers

Ziegler, W.H. (1975). 'Outline of the geological history of the North Sea', *Petroleum and the Continental Shelf of Northwest Europe . Vol. 1. Geology*, pp. 165–190. Ed. A.W. Woodland. London; Applied Science Publishers

Zimmerman, R.A. (1969). 'Stratabound barite deposits in Nevada', *Mineral. Deposita.* Vol. 4, 401–409

Zimmerman, R.A. and Amstutz, G.C. (1964). 'Small scale sedimentary features in the Arkansas barite district', *Sedimentology and Ore Genesis. Developments in Sedimentology, Vol. 2*, pp. 157–163. Ed. C.G. Amstutz. Amsterdam; Elsevier

Zitzmann, A. and Neumann-Redlin, C. (1977). 'The genetic types of iron ore deposits in Europe and adjacent areas', *The Iron Ore Deposits of Europe and Adjacent Areas. Vol. 1*, pp. 13–35. Ed. A. Zitzmann. Hannover; Bundesanstalt für Geowissenschaften und Rohstoffe

Zlobik, A.B. (1980). 'Mica', *Minerals Yearbook 1978–79. Vol. 1. Metals and Minerals*, pp. 601–613. Bureau of Mines. Washington DC; US Government Printing Office

Zwart, H.J. (1967). 'The duality of orogenic belts', *Geol. Mijnbouw.* Vol. 46, 283–309

Author and Subject Index*

Abitibi greenstone belt, Canada 202, 203, 206, *208–209*
  Lake, Canada 204
abyssal zone, organic matter 278
Acoje, Philippines 20
actualistic model 4
ADAM, J.W.H. 137
ADAMS, J. 262
African Copper Belt *234–238*
Agnew Lake deposit 268
AHFELD, F. 188, 190
albitite 115
ALEVA, G.J.J. 259
algal ore, fluorite 108
  structures, ironstone 215
Algoman-type ironstone 199–200
alkali granite 130
  pyroxenite 115
  syenite 115
alkaline igneous rocks *114–123*
Alligator River, Australia *238–242*
alluvial deposit 302
alluvium, Southeast Asia 257
Almaden mercury deposit, Spain *97–99*
alnöite 115
Alpinotype orogenies 131–132
Alps, tungsten 154
Alston Block, England 53–54
alteration tongues 80
Alum Shale, Sweden 96
American coal 310
AMSTUTZ, G.C. 62–63
anatexis 9, 194
andalusite, Bushveld 113
ANDERSON, G.M. 48, 50
Andes mineralization 189, 191, *194–195*
Andinotype orogenies 131–132
ANHAEUSSER, C.R. 197, 205, 206, 212, 213
anthracite 306
anticlinal traps 292
antimony, Bolivia 188
  , Burma 188
  , China 145, 188
  –gold, Malaysia 187
  –gold mineralization *186–188*
  , greenstone belts *210–212*
Antimony Line, South Africa 211
antimony, Peru 188
  , South Africa *210–212*
antithetic faults 288
apatite, Brazil 117
  deposits, igneous *114–119*
  , iron ore 219
  –nepheline ore 115–116
Appalachian zinc–lead deposits *58–59*
aqueous solution 283
Aral Sea 68
Archaean mineralization *196–213*
  , model for 212–213
  tectonics *196–198*
Arkansas barite *62–63*, 66
ARTH, J.C. 198
Ashio Mountains, Japan 28–29
Askrigg Block, England 53–55
astrobleme 8
  , Sudbury 104
Athabasca Formation 243, 245
  Trough 90
  uranium *243–247*
Atlantis Deep 45–46
AUDLEY-CHARLES, M.G. 303
aulacogen 44, 225, 286
Australian bauxite *253*
  coal *310–311*
  gold *204*
  nickel *202–204*
  pegmatites 129
  phosphorites 88
  placers *272–275*
  stratiform ore bodies *229–233*
  sulphide ores *229–234*
  uranium *238–242*
Avoca sulphide, Ireland 40
Azov Sea 96

back-arc spreading 12
back-arc thrusting 14
backdeep basins 299
BAILEY, R.V. 78, 79, 100, 238, 313, 315
BAKER, E.G. 284
BALDWIN, J.T. 173, 175
BALLY, A.W. 293, 297
banded-iron formations 201, 214
banded spar fluorite 108
Bangka, Indonesia 259
banket 270, 271
BÁRDOSSY, G. 249, 250, 251
barite bedded deposits *62–63*
  deposits *62–67*
  , Pennines 53
  replacement deposits *63–64*
  residual deposits *66–67*
  vein and cavity deposits *64–66*
BARKER, C. 279
BARNES, H.L. 5
BARRETT, F.M. 202
basal zone, Bushveld complex 108
basalt 17
basin classification *293–299*
BATCHELOR, B.C. 257, 258
BATEMAN, P.C. 183
batholith geology *130–133*
bathyal zone, organic matter 278
Bau, gold and antimony 186
  , mercury 97
  , stibnite 186
bauxite *249–254*
  , age 250
  , classification 249–250
  , composition 250
  , geographical distribution 250
  , origin 251
BEAN, J.H. 136, 137, 159
Beatrice Pipe, Malaysia 139, 140
Beaverlodge mining area *246–248*
bedded bauxite 253–254
  manganese deposits *26–30, 222–225*
BELL, R.T. 95
Belt–Purcell aulacogen *225–227*
  copper 227
  lead–zinc *227–228*
  palaeo-geography 225
Benioff Zones 12–13, 17, 131
Benue Trough, Africa 44, 286
BERNING, J. 167
Besshi deposits *40–42*
BEUKES, N.J. 223, 224

* Main entries are enumerated in *italic* type. Author entries are in capitals.

BICHAN, R. 113
Big Indian Ore Belt, Utah 78–79
BIGNELL, R.D. 45
Bikita pegmatite 127–129
Billiton, Indonesia 137–139
bioepigenetic sulphur deposits 94
biosyngenetic sulphur deposits 95
BISHOFF, J.L. 44
Biwabik Iron Formation *214–217*
Black Sea 68, 93, 95
Black shales, uraniferous *95–96*
blackspar fluorite 108
bog iron ore 75
boghead coal 305
Bolivia tin *189–193*
Bonneterre Formation, Missouri 51–53
Borneo 186
BOTT, M.H.P. 55
Bougainville, porphyry copper *173–175*
Boulby potash, England 91–92
BOWDEN, P. 124
BOWEN, R. 234, 235
BOYLE, R.W. 64
BRADFORD, E.F. 139, 141
braided stream placers 261
Brazil apatite 115, 117
  tin 124
breccia ore 141–142
  spar fluorite 108
brine 4, 14, 45, 46, 50, 62
  circulation 33, 45
BRITTEN, R.A. 310, 311
BROBST, D.A. 62
Broken Hill, Australia 230, *232–233*
  mineralogy 233
brown coal 306
BRYNER, L. 21, 157, 185
Buchans sulphide, Newfoundland 40
Bukit Besi, Malaysia 136–137
Bukit Ibam, Malaysia 157–158
buoyancy 284
BURBANK, W.S. 183, 184
Burma, antimony 188
  , barite 66
Bushveld Granite 107, 110
  igneous complex *104–113,* 197
  igneous stratigraphy 109
  tectonic setting 105–106

calc-alkaline series 130
calcrete, uraniferous *254–256*
caliche 256
California, chromite 23
  , gold 262
  , mercury 100
CALLOW, K.J. 157
Camborne, England 149
CAMERON, E.N. 111, 129, 225
Canada, gold *204–205*
  , uranium *243–247*
  , uranium placers *264–265*
cannel coal 303
cap rock 95, 289
Capel Beach sands 273–274
capillary pressure 284–285
  seal 285
carbon isotopes 57
  , Witwatersrand 272
carbonas 135, 143, 149, 150
carbonatite 115, *116–119*
Carboniferous coals 304, *308–310*
carnotite 78, 255–256
Cartersville, Georgia, barite 65
cassiterite, Malaysia 136
  , pleochroic 143
  , temperature of deposition 134
Catahoula Formation, Texas 80
catazonal emplacement 134
cauldron subsidence 36, 180, 184
Cave-in-Rock, Illinois 60
Central Graben, North Sea 290–291
Cerro Boliva, Venezuela 217–218
Cerro Chorolque, Bolivia 189
Cerro Colorado, Spain 37–39
CHAKRABORTY, K.R. 194
CHAPPELL, B.W. 130
Chattanooga Shale, USA 96
chemical brecciation 141
CHENEY, E.S. 167, 256, 313
CHENEY, T.M. 89
cherty iron formation 201, 215
CHILDERS, M.O. 238, 313, 315
chill zone, Bushveld complex 108
China, bauxite *253–254*
  , iron ore *159*
  , tungsten *144–147*
Chinese-type basins 298
Chinkuashih, Taiwan 185
Chinle Formation 79
Christmas Island 86
chromite *19–23*
  , Bushveld *111–112*
  , Great Dyke *113–114*
  , podiform *19–23*
circum-Pacific mercury 97
CLARK, A.H. 190, 194
classification *4–7*
clay alteration zones 36–37, 172–173
Cleveland ironstone, England 72
Cleveland, Tasmania 156
Climax porphyry mine, Colorado *178–180*
Clinton Group ironstones *69–71*
  palaeo-geography 71
coal, climate control 303
  deposits *300–310*
  , origin *300–303*
  measures 302
  petrology *305–306*
  ranks 281, *306–307*
  , tectonic setting 303
coalification 279, *306–308,* 310
Cobalt, Canada 210
Cobalt–Gowanda ore body, Canada 209–210
Coeur d'Alene 226, *228–229*
coffinite 77
COLEMAN, R.G. 15, 17
COLLEY, H. 32, 43
collision orogen 297
collophane 84
colluvial placers 260
columbite, Nigeria 125
combination traps 292
COMER, J.B. 251
Comstock Lode 182
conglomerate ore, sulphide 34
  , uraniferous *265–267,* 270, 313
Congo pegmatite 128
CONNAN, J. 279
connate water 5
CONSTANTINOU, G. 26, 32
consuming plate margin 32
Consuzo stock, Peru 151
continental embankment 298
  margin basin 298
COOK, P.J. 82, 83, 88
COOPER, B.S. 280
copper skarn *175–176*
Copper Belt, Africa *235–238*
cordierite hornfels 143
cordillera 12–13
Cordillera Real, Bolivia *190–192*
cordilleran margin basin 296
  porphyry copper *172*
Cornish-type lodes 149
Cornwall, England *148–150*
Cornwall, Pennsylvania 158, *159–160*
Coto, Philippines 20–21
cratonic basins *293–294*
  igneous intrusions *102–129*
  interior basins 89
Creighton Embayment, Sudbury 103–104
Cretaceous ironstones, England 72
critical zone, Bushveld 108
CRONAN, D.S. 14–15
cryptic layering, Bushveld 105
cryptovolcano 8
Cuba, chromite 23
  , manganese 28
Cudgen placers 274
CUNEY, M. 163
cupriferous sulphide 27
cyclothem 301–302
Cyprus 16, 18, 22, 26, 32
Cyprus-type sulphides *31–34,* 40

DAHLKAMP, F.J. 244, 245, 248, 312, 313, 314, 316
Dalcoath mine, England 149, 150
DAMBERGER, H.H. 310
Dan Field, North Sea *290–291*
Darling Ranges, bauxite *253*
Dartmoor granite 148
DAUKORU, E. 286, 287
DAWSON, J.B. 119, 120, 121, 122
DEGENS, E.T. 95
delta deposits 299–302
deposit 1
Derbyshire Dome, England 54–55
DEUTSCHER, R.L. 256
diamond *119–123*
formation 119
placers *262–264*
, Thailand 144, 264
diatomite 89
DICKEY, P.A. 284
DICKINSON, W.R. 12, 293, 296, 297
DIESSEL, C.F.K. 310
DIETZ, R.S. 104
DIMROTH, E. 68
distal deposits 43, 214
DODSON, R.G. 238, 239, 241
DOE, B.R. 53
Dogger ironstones, Germany 73
DOW, W.G. 276, 278, 280, 282
DOZY, J.J. 50
Duluth gabbro 215
Dundonald, Canada 203
DUNHAM, A.C. 63
DUNHAM, K.C. 53, 54, 55, 72, 148, 149

East Pacific Rise 14
EASTOE, C.J. 175
economic deposit 2
Ekofisk, North Sea *290,* 291
El Laco, magnetite 219
Elliot Lake, Ontario *266, 267,* 269
Elliot syncline 265
eluvial placers 260
elvans 148
Ely, Nevada 175, 176
EMERY, J.A. 221, 222
EMMONS, J. 133, 135
emplacement depth 133, 134
Endako, British Columbia *177–178*
Eneabba placers *273–274*
English ironstones *72–73*
epicontinental basin deposits *82–101,* 214
epicrustal rocks, Bushveld 110
epigenetic deposits *168–195*
uranium 75
epithermal vein deposits *180–194*
epizonal pluton emplacement 133, 134, 168

EUGSTER, H.P. 160
EVANS, A.M. 53, 56, 66
EVANS, C.R. 285
EVANS, D.G. 306
EVANS, H.J. 253
evaporite *89–94*
, cycles in North Sea 91, 92
, salina model 93
, water depth 93
exhalative processes *9–10,* 212–213

FALCON, R.M.S. 303, 304, 305
Fanay mine, France 162
fault control, migration 288
seals 288
Felbertal, Alps 154
fenite 115
ferromanganese deposits *26–30*
, fresh water *74–75*
nodules *74–75*
Fiji, gold 185
, Kuroko deposits 43
FINCH, W.I. 76, 77, 84
fireclay 302
fissure veins 59
, Cornwall 149
FITCH, F.H. 136
flats 53
FLEISCHER, V.D. 235
Flin Flon ore body, Canada 209
FLINTER, B.H. 139
flood plains 302
FLORAN, R.J. 215
Florida phosphorite 85
fluid inclusions 5, *47,* 60, 62, 134, 152, 175, 181, 184, 247
fluorite, Bushveld complex *107–108*
deposits *59–62*
, Mexico *59–60*
, Pennines 54
FOLINSBEE, R.E. 56
FORD, T.D. 54, 56
fore-arc basins 296, 299
foreland basins 296
Forez uranium deposit *161–163*
Forties Field *290,* 291
foyaite 115
Franciscan melange 100
FRAZIER, D.E. 301
FREEZE, A.C. 227
French ironstones *74*
FRIETSCH, R. 219, 220
FRIPP, R.E.P. 207, 213
FRYKLUND, V.C. Jr. 226
fumarole 208, 212, 213

Gabon manganese 225
galena, Missouri 51
, Pennines 53
, Pine Point 56

Gällivare, Sweden 220
gangue 2
GARLICK, W.G. 236, 237
garnierite 23
GARSON, M.S. 18, 32, 43, 143, 144
gaseous solution 283
GASS, I.G. 15, 16, 18
Georgia barite 65, 67
Georgina Basin phosphorite *88*
geothermal gradient 280, 282
GERASIMOVSKY, V.I. 115
German coal *308–310*
ironstones *74*
GEVERS, T.W. 128, 129
Ghana manganese 225
GLASBY, G.P. 15
GLIKSON, A.Y. 196, 199
goethite 23
Golalan, Turkey 22
gold–antimony mineralization *186–188*
, Malaysia *186–188*
gold deposits *182–188, 204–207, 261–262, 269–272*
, USA 182, 183
, greenstone belts *204–207*
lode deposits 207
placers *261–262, 269–272*
, South Africa *205–207, 269–272*
, stratiform 206–207
–uranium placers *270–272*
Zimbabwe 205
GOLDICH, S.S. 214
GOLIGHTLY, J.P. 25
gondite 223, 225
Gondwanaland 217, 218, 303
coal 303, 304, 310, 311
GOODWIN, A.M. 199, 200, 201, 205, 212, 215
GOOSSENS, P.J. 66
gossan 9
Gove Peninsula bauxite 252, 253
GOVETT, G.J.S. 26, 32
graben basins 298
grade, cut-off 2–3
, ore 2–3
Grand Isle Block Oil Field 288–289
granite series 130, 131
GRANT, J.N. 189, 190, 192, 193
Gravelotte mine 211
Gravelotte–Mica pegmatite 129
Great Dyke, Zimbabwe 111, *113–114,* 197, 206
Greece, nickel laterite *26*
Green River Shale 276, *305*
green tuff, Japan 35, 37, 181
GREENBAUM, D. 22
greenstone belts *196–199*
, antimony *210–212*
, evolution 198–199
, gold 205, 207

greenstone belts (contd)
mineralization *198–213*
, South Africa 197, *205–207*
, stratigraphy 203
greisen 135, 143, 145
-bordered veins 135, 148
GRIP, E. 96, 220
GROGAN, R.M. 61
GROSS, G.A. 199
ground water, uranium-bearing 80, 256
GROVES, D.I. 202, 203
growth faults *286–287*
GRUBB, P.L.C. 137, 249, 252, 253
GRUSS, H. 217, 218
Guanajuato, Mexico 182, 183
guano *85–86*
GUILBERT, J. 171
GUILD, P.W. 23, 182, 183
GUILLON, J.H. 24, 25
GULBRANDSEN, R.A. 88
Guleman–Soridag, Turkey 22
Gulf Coast oil fields 282, *288–289*
Gulf of Mexico sulphur 94
GUNATILAKA, A. 234, 235
Gunflint Iron Formation *214–217*
Range 216
GUSTAFSON, L.B. 168, 171, 173
Guyana iron formation 218

Hainan 159
HALL, T.C.F. 303
HALL, W.E. 178, 180
HALLAM, C.D. 262
Hamme tungsten, North Carolina *154*
HARRISON, J.E. 227
HARSHMAN, E.N. 80
Hauraki, New Zealand 185
HAWKINS, B.W. 242
HEGGE, M.R. 242
Henderson mine *177*
HENLEY, R.W. 261, 262
Hercynian granite *161–165*
uranium *161–165*
Hercyno-type orogenies *131–133*
HEYL, A.V. 48, 49, 59, 62
Hicks Dome, Illinois 61
Highland area, Powder River 76
Hilton deposit, Australia *230–232*
Hilt's Law 306
HIRDES, W. 270, 271
HOAGLAND, A.D. 58
HOBBS, S.W. 226, 228
HOEVE, J. 243
Hokuroko, Japan 36
HÖLL, R. 155, 210
Hollinger, Canada *205–206*
HOLLISTER, V.F. 168
Hong Kong tungsten 147
HOSKING, K.F.G. 66, 133, 135, 136, 137, 138, 140, 141, 144, 148, 149, 150, 155, 159
hot-spot 12–13
tracks 13, 123
Hsiuwen bauxite 254
Huallatani, Bolivia 191, 192
Hunan, China 96, 188
HUNTER, D.R. 105, 106, 107
HUNTER, R.E. 69, 70
Hunter Valley coal 310
HUTCHINSON, R.W. 155, 207, 212, 225
HUTCHISON, C.S. 22, 97, 131–133, 135, 136, 150, 188, 194, 259
H.Y.C. pyritic shale 230–231
hydrocarbon generation 278, 280
hydrogen isotopes 5
hydrothermal alteration *171–173, 175,* 177, *185*
exhalations 62
pipes 139, 140
processes *8*
solutions 2, *5–7,* 19, 133, 152
temperatures *134–135*
uranium *238–248*

I-granite 130, 131, 133
Iberian pyrite belt 37
Idaho batholith 227
phosphorite 87
Springs Formation 81, 178
Idria mine, Yugoslavia *98–100*
igneous layering, Bushveld *105,* 109
ijolite 115
IKONNIKOV, A.B. 96, 145
Ilimuassaq, Greenland 119
Illinois, fluorite *60–62*
, fluorite–zinc–lead 61
–Kentucky fluorite 49
ilmenite placers *272–275*
-series granite 131, 144
Imperial Valley, California 44
Indian manganese deposits 223
Indonesia tin mineralization 138–139
inertinite 278
INESON, P.R. 53
INGHAM, F.T. 139–141
intermontaine basins 299
intra-arc basin 297–298
intracratonic basins *44–81,* 89
Ireland barite 64, 66
Kuroko deposits 40
Mississippi Valley deposits 56
iron deposits, China *159*
, contact *156–160*
, Japan 159
, Malaysia 136–137, *157–158*
, Philippines 157
formations *198–202, 214–219*
Iron Mountain, Missouri 222
iron skarns 156
–tin mineralization, Malaysia 136–137
ironstones, reworked 69, 73
, sedimentary *68–74,* 214
ISHIHARA, S. 131, 135, 142, 144
island arc 12–13, 43
porphyry copper 172
itabirite 217–218

Jabiluka deposit *239*
JACKSON, S.A. 56
Jacupiranga, Brazil *115–117*
jacupirangite 115
Jamaica bauxite *251–252*
JAMBOR, J.L. 209
Japan, Hokuroko district 35
, iron deposits 159
, Kuroko deposits *34–37*
, manganese deposits *28–30*
JOHNSON, I.R. 232
JONES, M.T. 139
JONES, W.A. 205, 206
Jos–Bukuru, Nigeria 125–126
JUNG, W. 93
Jurassic ironstones, Europe *72–74*

Kafue anticline *234–235,* 237
kaksa 261
Kalahari manganese *223–224*
Kalgoorlie, Australia 204
Kalimantan 186
Kamativi pegmatite *129*
Kambalda, Australia *202–204*
KANEHIRA, K. 42
Kao kimberlite, Lesotho *120, 123*
Karamken, Siberia 185
karang 261
karst mineralization *58–59, 250*
Katanga Supergroup 234
Keewatin flows 209
KELLY, W.C. 152, 153
Kentucky barite *66*
fluorite *60–62*
fluorite–zinc–lead 61
keratophyre, magnetite ore 220
kerogen concentration 276
evolution paths 277, *279*
in shale 305
types *276–278*
KESLER, S.E. 60, 171, 182, 183
KESLER, T.L. 65, 67
Key Lake deposit *245*
Khakandzha, Siberia 185
Khao Soon, Thailand 142
Khibina, Kola Peninsula *114–116*

khibinite 115
Kiangsi, China 145
Kidd Creek mine, Canada *208–209*
Kiirunavaara 220
killas 148
KIMBERLEY, M.M. 30, 68, 199, 201, 214, 269
kimberlite 114, 115, *119–123*
  age 122
  classification 119
  diatremes 119, 120
  distribution 122
  dykes 119, 120
King Island, Australia *155*
Kinta Valley, Malaysia *139–140, 257,* 258–259
KIRPAL, G.R. 254
Kiruna 219, 221
  iron ore *219–221*
Klamath Mountains, California 23
Klappa Kampit, Indonesia 138–139, 155
Kleinarltal, Alps 154
KLEMME, H.D. 292–295, 298
Klerksdorp gold field 272
KNIGHT, C.L. 156
KNITZSCHKE, C. 92
Kochiu, China 148
KOEPPEL, V. 246
Kola Peninsula, USSR *114–116*
komatiites 196, 202
Koongarra, Australia *241*
Korea tungsten *154–155*
Kosaka, Japan 36
Kramat Pulai scheelite–fluorite *140–141*
Kuching, Sarawak 186
Kunghsien bauxite 254
Kupferschiefer 91, *92–93*
Kuroko deposit 32, *34–40*
  ore 35
  –porphyry comparisons *42–43*
Kwantung, China 145

Lahat pipe, Malaysia 141
Lahn–Dill iron deposits 29, *30*
Lake Michigan ferromanganese 75
LAMBERT, I.B. 35, 185, 230, 231, 233
Lamotte Sandstone, Missouri 53
lamprophyre 115
LANDIS, G.P. 151
LANGFORD, F.F. 75, 77, 254–256
LAPHAM, D.M. 158–159
Larap, Philippines 157
LARSEN, K.G. 51
Las Cuevas fluorite 60
lateral secretion 8, 47
laterite *23–26,* 249
LAVEROV, N.P. 313
LAWRENCE, L.J. 232
layered sequence, Bushveld *108–110*
LAZNICKA, P. 204
lead isotopes 47, 48, 56, 59, 229, 246
LEOW, J.H. 143
Lepanto, Philippines 185
lepidolite 134, 144
LEROY, J. 162, 164
Lesotho kimberlite 120, *122–123*
Letseng-la-terae, Lesotho 120, *122–123*
Levack embayment, Sudbury 103, *104*
LEWIS, D.E. 171
Liberia iron formation *217–219*
Liesegang rings 79
lignite, uraniferous 315
Limassol Forest, Cyprus 22
limestone mineralization 50–51
limnic basins 303
Limousin, France *164*
LINDGREN, W. 4
LIPMAN, P.W. 183, 184
liquid immiscibility *8,* 219
  window *279–280*
Lisbon Valley anticline 78
listric growth faults *286–287*
Llallagua mine, Bolivia 193
lode 2
  gold deposits *207*
  tin deposits, Malaysia *136,* 138
longshore drift *262–263,* 273
lopolith, Sudbury 102
Lorraine ironstones, France 73–74
Louisiana Gulf Coast oil fields 281, 282, *288–289*
Lousal, Portugal 38–39
Loussavaara 220
LOWELL, J.D. 171
LUEDKE, R.G. 183, 184

maceral 305
Macquarie Island 17
Madhya Pradesh iron *223–225*
magmatic cumulates 7–8, 19
  disseminations 7
  processes *7–8*
  water 5
magnesite, Bushveld 113
magnetite, Bushveld 109, *112*
  , Canada 159
  , Cornwall, Pennsylvania *158–160*
  , Kiruna-type *219–222*
  lava flows 219
  series 131
Maharashtra manganese *223*
Main Chromite seam, Bushveld 109
Main Magnetite seam, Bushveld 109
Main Range, Malaysia 135, *139–143*
Main Zone, Bushveld 109, 110
Malay Peninsula tectonics 132, 133, *135–137*
malayaite 139
Malaysia barite 66
  iron *157–158*
  tin mineralization *135–137*
manganese, bedded ore *26–30, 222–225*
  nodules *15,* 74–75
mangrove swamps 302
MANN, A.W. 255, 256
manto deposits 59–60
marginal basins 12–13, 17, 297–298
Margnac mine, France 162
Mary Kathleen uranium deposit 240, *242–243*
Masinloc, Philippines 20
Massif Central 161–162
massive pyritite 39
  sulphide deposits 14, *31–43, 207–210, 225–238*
MATHIAS, B.V. 231
Mattabi mine, Canada 208, 209
Mattagami Lake deposit, Canada 209
maturation, organic matter 277, *279–280,* 281
  profile 282
  , time factor *281*
MAUCHER, A. 99, 210
McArthur deposit, Australia *229–231*
McELHINNY, M.W. 82, 88
McKELLAR, J.B. 273, 274
McMILLAN, R.H. 243, 266, 268
Meade Peak phosphorite 86, 87
mechanical accumulation 9
Mediterranean Sea mercury 97
MELCHER, G.C. 115, 117
melteigite 115
Menglembu, Malaysia 142
Meratus Mountains 264
mercury deposits *96–101*
  mobilization 97
  ore genesis 99
Merensky Reef, Bushveld 108, 109, *110–111*
Mesabi Range 216, 217
mesozonal emplacement 133, 134
metal-rich sediments 46
metamorphic processes *8–9*
  water 5
meteoric water 5
Mexico fluorite *59–60*
  tin *190*
Michipicoten Basin *200–202*
mid-Atlantic Ridge 14
Midnite mine, Washington *165–166*
migration of oil *281–285,* 288
  , primary *281–284*
  , secondary *284–285*

minette iron ore 74
Miocene coals 303, 304
miogeoclinal basin 89, 296, *298*
miscellar solution 284
Mishrag sulphur 94
Mississippi Delta 282
Embayment 49–50
Valley deposits 6, *47–59*
, mineralogy 51–52
, lead isotopes 48
, ore genesis 49–50
Missouri iron ore *221–222*
, Mississippi Valley deposits *50–53*
MITCHELL, A.H.G. 18, 32, 43, 123, 133, 150
MLAKER, I. 98, 99
molybdenum porphyry *177–179*
Montana phosphorite 87
MOORE, L.R. 305
MOREAU, M. 160, 161, 165
MOREY, G.B. 215, 216
Mother Lode, California 182, 183
Mount Bishoff, Tasmania *156*
Mount Isa, Australia *230–232*
Mountain Pass, California *117–119*
MUFF, R. 210, 211, 212
Murchison Range, South Africa *210–212*
Muskeg–Prairie evaporite 56–57, *89–91*

Nabarlek, Australia *241*
NALDRETT, A.J. 102, 202
Namaqualand tungsten 154
Namibia diamonds *262–264*
pegmatites 128, *129*
uranium *166–167*
natural gas *276–300*
Naurzum bauxite 252, *254*
Nchanga copper mine 235, *237–238*
NEGLIA, S. 283
nepheline syenite 115
nephelinite 115
neritic zone, organic matter 278
NEUMANN-REDLIN, C. 73, 74
Nevada barite *63–64*
New Almaden mine, California 98, *100*
New Caledonia 19, *24–25*
New Zealand gold *261–262*
Newcastle, Australia 311
Newfoundland, ophiolite 33
, sulphides 34
NEWNHAM, L.A. 155
nickel, Australia *202–204*
Nickel Eruptive, Sudbury 102
nickel laterite *23–26*
, allochthonous 26
, Sudbury *102–104*
sulphide deposits *202–204*
Niger Delta oil field *286–288*
Nigeria granite 126
tin *124–126*
Nimba iron formation *217–219*
niobium deposits 117, *118, 123–125*
, Canada 117–118
, cratonic settings *123–125*
NIXON, P.H. 120, 122
Noda–Tamagawa, Japan 28–29
Noranda region, Canada 205, *208*
norite, Sudbury 102
North Carolina tungsten *154*
North Sea evaporites *91–92*
oil fields *290–291*
palaeo-geography 291
stratigraphy 92, 291
Norway tungsten 154
NOTHOLT, A.J.G. 115
Nova Scotia barite *64–65*
Nsuta, Ghana 225

obduction 17–18
Oberfalz ironstone, Germany 74
ocean-floor mineralization *14–30*
oceanic lithosphere *11–30*
unwellings 88
ochre 26–27
OHMOTO, H. 37, 141, 171, 181
oil basin classification *293–299*
tectonics 296, 297
oil generation 299
-phase migration 284
preservation *299–300*
shale *305*
-trap classification *292–293*
oil-fields classification *293–299*
distribution 293, *299–300*
occurrence 292, *299–300*
, supergiants 300
Oka alkaline complex, Canada 117, *118*
okaite 115
Okuki mine, Japan 42
Old Lead Belt, Missouri 52
olistostrome 19
Oman ophiolite *26–27*
Ongeluk lava 223, 224
Ontario placers, uranium 264–266
oolitic ironstone 68, 215
ophiolite *15–23*
tectonics 18
ore 2
grade 2–3
minerals 2
reserves 2
shoot 2–3
organic-matter maturation 277, *279–280*
production *278*
orogenic belts *131–132*
OSANIK, A. 300, 301
OSBERGER, R. 261
Osburn Fault 229
Otago gold fields *261–262*
Oxfordian ironstone 72–73
oxidation of iron 79
oxidative sulphur 95
oxygen isotopes 5, *49*, 57, 58, 151, 152, 173, 185, 215

Pachuca, Mexico 182, 183
padang placers 261
Pahang Consolidated Co. mine, Malaysia *136*, 138
paired granite belts 132, 133
orogenic belts 132, 135
Palabora, South Africa 119
Palawan mercury 100
Panasqueira, Portugal *152–153*
Panguna, Bougainville *173–175*
PAPIKE, J.J. 214, 215
Papua New Guinea 18
PARÁK, T. 219, 220, 221
paralic coals 303
PARK, C.F. Jr. 156, 157, 219
Pasto Buena, Peru *150–152*
PATTISON, E.F. 104
Pea Ridge, Missouri 219, *221–222*
peat deposits 300, 302, 306
pegmatite 8, *126–129*, 134, 144, 148
, Bushveld 110
classification *127–128*
mineralization *126–129*, 144
, Nigeria 124, 126
Peine detrital iron ore 73–74
Pelepah Kanan, Malaysia *137*
Pennine Hills, England *53–56*
fluorite 53
, ore genesis 55
Pennsylvania coals 302, 304, *310*
Per Geijer 220
peridotite 17, 115
Permian coals 303, 304, *310–311*
Peru, tungsten *150–152*
PETERSEN, U. 188, 190, 191, 194
petroleum *276–300*
, relation to Mississippi Valley deposits 48, 50
PETRUK, W. 210
Philippines, Barlo mine 34
, chromitite *20–21*
, iron *157*
, Lorraine ore *33–34*
, manganese 28–29
, mercury 100–101
phoscrete 88
phosphate, igneous *114–119*
phosphatic guano 86
sediments 82
source rocks 85

Phosphoria Formation, western USA 84, *86–87*
phosphorites *82–89*
classification *83–85*
distribution 82, 83
petrology *83, 84*
, platform deposits 85
, residual 85
, west coast-type 85
phosphorogenesis 82, *88, 89*
Phuket, Thailand *143–144,* 264
phyllic alteration 172–173
Pilbara Block, Australia 204
Pilot Knob 222
Pine Point zinc–lead deposit, Canada *56–58*
pipe deposits *139–140,* 150
pitchblende 81, 164, 238, 246
PITCHER, W.S. 131, 132
placer deposits 4, *256–276*
diamonds *262–264*
, gold *261–262, 269–272*
, gold–uranium *270–272*
ilmenite, rutile, zircon *272–275*
tin *257–261*
Planes–San Antonio, Spain *38–39*
platinum, Bushveld 105, *110–111*
Platreef, Bushveld *111*
PLIMER, I.R. 233
Porcupine, Canada 204–206
porphyry-copper deposits 5, 32, *168–175*
, classification *168–169*
, distribution 169–170
porphyry-molybdenum deposits 169, *177–179*
porphyry stocks 171
porphyry-tin deposits 169, 188, 193
porphyry-tungsten deposits 147
Portugal, massive sulphide deposits *37–40*
, tungsten *152–153*
potash salts 89
potassic alteration 172, 175
Potosi, Bolivia 189, 192
POTY, B.P. 163
Powder River Basin, Wyoming 80
Prairie Formation *89–91*
Presqu'ile Barrier Reef 56–57
pressure seal *285*
PRETORIUS, D.A. 269–272
Příbram, Czechoslovakia 165
PRICE, L.C. 283
primary migration *281–284*
propylitic alteration 173, 175, 180
Proterozoic mineralization *214–248,* 264–272
uranium placers *264–272*
protore 2
proximal deposits 43
pull-apart basins 299
PUSEY, W.C. 279
pyrochlore deposits 118
pyrolysis 277, *279*
pyrometasomatic deposits 148, 156
pyroxenite 115

QUADE, H. 30, 201
Quirke zone 267

Rabbit Lake deposit *243–244*
radiogenic lead *48*
rakes 53, 56
Rammelsberg, Germany 43
Ranger One, Australia *239–241*
rare-earth deposits *118–119*
, California *117–119*
Raub Australian gold mine *186–188*
recycled tin 193–195
Red Sea 4, 6, *44–47*
redeposited bauxite *250,* 253–254
reef mineralization 47, 51
reflectance of coal 307–308
remnant arcs 14
RENFRO, A.R. 236, 237
Renison–Bell, Tasmania 153, *155–156*
replacement ore bodies *143*
residual barite *66–67*
phosphorite *85*
processes 9
Retort member phosphorite 86
retro-arc basins *297*
Rhine–Ruhr coal field *308–310*
RICHARDSON J.A. 186, 187
rift-related mineralization 123
rift valleys 298
RIGGS, S.R. 83–85, 88
RILEY, G. 143
ring complex 115, 124
, Nigeria 124, 126
Rio Tinto, Spain 37–39
rischorrite 115
RIVAS, S. 193
Roan Formation 234
ROBBINS, D.A. 165
ROBERTSON, D.S. 264–266, 313, 314
ROBERTSON, J.A. 265, 268
ROEDDER, E.H. 5, 47, 49, 58
ROGERS, J.J. 161
roll-front uranium deposits *79–81*
roll-over structures 288, *292*
Rondonia, Brazil *124*
Rössing, Namibia *166–167*
Rotliegendes evaporite 91
ROWNTREE, J.C. 242
ROY, S. 223
Rubicon pegmatite 128–129
RUCKMICK, J.C. 94, 95, 217
Ruhr-Basin coal *308–310*
Rum Jungle, Australia *238–242*
rutile placers *272–275*
Rwanda pegmatites 128
RYE, R.O. 151–153, 171, 181

S-granite 130–133, 160
Sabah, Borneo 20, 22
SAEGART, W.E. 171
salina deposits 93
salt deposits *89–94*
diapirs 288–289
withdrawal 288–289
Salton Sea 4, 44
San Juan, Colorado *183–185*
sandstone uranium deposits 75, 76, 77, 79
Sangdong, Korea *153–155*
SANGSTER, D.F. 47, 59
saprolite 25
sapropelitic coals *303–305*
Sarawak 186
Saskatchewan potash deposits 90, 91
uranium 244
SASSANO, G.P. 247
SATO, T. 35, 185
SAUPÉ, F. 97, 98
SAWKINS, F.J. 42, 181, 225
scheelite, China 145
–fluorite skarns *141*
, King Island *155*
, Malaysia *141*
SCHNEIDER, H.J. 191, 192, 195
SCHOWALTER, T.T. 284
SCHUILING, R.D. 191, 194
Schwartzwalder deposit, Colorado 81
scrins 53, 56
SCRIVENOR, J.B. 260
sea-floor spreading 12–13
sealing of faults 288
seatearth 302
secondary migration *284–285*
sedimentary bauxite *250,* 254
ironstone *68–74*
precipitation 9
Semail nappe, Oman 27
SEMENOV, E.I. 116
serpentinite diapirs 19
melange 100
Seyler chart 308
Shakanai, Japan 36
shale, uraniferous *95–96,* 315
SHCHERBA, G.N. 137
shelf basins *299*
SHEPPARD, S.M.F. 33, 135, 148, 150, 173, 180
Shihlu, China 159
Shikoku, Japan 41–42
Shimokawa, Japan 41

Shirataki, Japan 41
Shirley basin, Wyoming 80
shonkinite 115
Shor-Su sulphur 94
SIBBALD, T.I.I. 243
silica–carbonate rock 100
SILLITOE, R.H. 124, 180–183, 191, 193
silver deposits *182–186*
, USA 182
Silverton cauldron 184
skarn 136–137, 139, *145,* 156, 169, *175–176*
, China *145*
, copper *175–176*
, iron 156
, Malaysia 136–137
, tin 139
SKINNER, E.M.W. 119
SLOSS, L.L. 89
SMART, P.G. 239, 241
SMIRNOV, V.I. 256
SNYDER, F.G. 47, 51, 52, 222
Sokli carbonatite, Lapland 118
SOMM, A.F. 253
Sonnenschein coal seam 309, 310
SØRENSEN, H. 119
SOUCH, B.E. 103, 104
source beds *276–279*
South Africa andalusite 113
antimony *210–212*
chromite *111*
fluorite *107*
gold *205–207*
magnetite *112*
pegmatites 129
platinoids *110*
tin 107, 112
South China Sea 20
Southeast Asia Cenozoic stratigraphy 257
placers *257–261*
sövite 115
Spain massive sulphides *37–40*
mercury *97–99*
Spar Lake copper 227
SPENCE, C.D. 209
SPENCER, A.C. 159, 160
sphalerite, Appalachians 58
, Pennsylvania 58
, Pine Point 56
, Tennessee 59
, Virginia 58
spinifex texture 203
Spokane Mountain *165–166*
spore colouration *280–281*
spreading axis 6, 12, 13
St. Francois Mountains, Missouri 50–51, 53, 124
STACH, E. 305, 306
STANTON, R.L. 232
STAUFFER, P.H. 257, 261, 303
STEINER, R.J. 288
stockwork deposits 148
strata-bound 5
sulphides 207, *229–238*
tin *154–155, 191–192*
tungsten *154–155*
Strathcona mine, Sudbury 103, *104*
stratiform gold *206–207*
ore bodies 5, *229–238*
stratigraphic traps 292
strontium isotopes 114, 124, 130, 171
structural traps 292
STUMPFL, E.F. 154
subduction 296
zone, mercury 100
submarine caldera 36, 37
hydrothermal activity 14, 35
Sudbury Basin, Canada *102–104*
Nickel Eruptive 102
platinoids 111
sublayer 104
Sulawesi, Indonesia *24–25*
Sullivan ore body, British Columbia *226–228*
Sulphide Queen carbonatite 118–119
sulphur, bacteriogenic origin 50
deposits *94–95*
isotopes *49,* 56, 64, 171
Sungei Besi, Malaysia 142
supergene enrichment 2, *9, 217–219*
supergiant oil fields 300
Superior iron formation 214, 216
surface processes *9–10*
surficial deposits *249–275*
SUTHERLAND-BROWN, A. 169, 177
swamp deposits 301, 302
SWANSON, R.W. 86, 87
Swaziland pegmatites *127*
Swedish black shales *96*
magnetite *219–222*
Sydney Basin, Australia *310,* 311
coal field *311*
syenite 115
syndiagenetic 5
syngenetic deposition 238
synvolcanic nickel deposits *202–204*

taconite 215
TAKAHASHI, M. 144
TALBOT, M.R. 72, 73
Tamiao, China 159
tantalite, cratonic settings *123–125*
, Nigeria 125
TARNEY, J. 198, 199
Tarnpbrzeg sulphur 94
Tasmania tin fields 153, *155–156*
Tayeh, China 159
TAYLOR, D. 97, 135, 157, 158, 188, 259
TAYLOR, H.P. 135, 152, 185
Taylor Creek, New Mexico 190
Teesdale Dome, England 54
TEH, G.H. 142
TEICHMÜLLER, M. 304, 306, 308–310
Tekka mines, Malaysia 142–143
telescoped ore deposit 143
Tennessee barite *67*
TENYAKOV, V.A. 254
terra rossa 251
Terrebonne syncline 289
Tertiary coals 304
Tethys Sea 22, 299–300
Thailand barite 66
Thanksgiving mine, Philippines 157
THAYER, T.P. 19, 22, 23
THEIS, N.J. 267
THEODORE, T.G. 176
thermogenic sulphur 95
THOMPSON, R.I. 226, 227
thucolite 247, 266, 267
time in maturation *281*
tin breccia 142
, Bushveld complex 107, *112*
, cratonic settings *123–125*
deposit classification 134
, Indonesia 138, 139
–iron mineralization, Malaysia *136–137*
, Malaysia *136–138, 139–143*
, Nigeria *124–126*
pegmatites 127, 129, 143
placers *257–261*
skarns, Malaysia *139–141*
, strata-bound 191, 192, 194
TISSOT, B. 276, 277, 279, 280, 305
TOOMS, J.S. 84
trachytoid texture 115
Transvaal fluorite 107, 108
System *107*
trapping of oil *288, 292–293*
TREMBLAY, L.P. 243, 246, 247, 248
trenches 12, 13
trend uranium deposits *78–79*
Troodos, Cyprus 16, 17, 21, *22*
tungsten breccia 142
, China *144–147*
, cratonic settings 123–125
, Korea *153–155*
, Peru *150–152*
, Portugal *152–153*
skarn *141, 155*
Turkey, Ana Yatak sulphide 34
, ophiolite *22*
TURNEAURE, F.S. 190

umber 26, 27
unconformity, uranium *243–248*
underclay 302
upper zone, Bushveld 109, 110
Urad mine *177*
uraninite 77, 78
uranium, black shales *95–96*
, calcrete *254–256*
deposits, ages 314, *316*
, classification *312–316*
, Sweden *96*
deposition, climate control 77
, detrital 267–268
, geographical distribution 312
–gold placers *270–272*
, granitoids *160–167*
, Hercynian *161–165*
, hydrothermal *238–248*
, penecontemporaneous deposits 77
, peralkaline plutons *119*
, phosphorite 84
placers *264–272*
mineralogy 269
, remobilization 244, 246, 316
, resources *310–316*
, roll-front deposits *79–81*
, sandstones 76
, strata-controlled epigenetic *75–81*
, trend deposits *78–79*
Urquhart Shale 232
urtite 115

Vaal Reef 272
VALETON, I. 249
vanadium, Bushveld 112
, uranium deposits 77
VARLAMOFF, N. 127, 128, 133, 134
Vauze, massive sulphide *208, 209*
vein swarms *142,* 180–194
Venezuela iron formations *217–218*
Viburnum Trend, Missouri *50–53*
Viking, North Sea *290–291*
Viloco, Bolivia 191, 192
vitrinite reflectance *279–281, 306–308*
volatiles of coal *307–308*
volcanic dome 208–209
exhalations *9–10*
sublimates 97
sulphur 95
volcanogenic sulphides *207–210*
tin *188–193*
VON GRUENEWALDT, G. 105, 106

WALLACE, S.R. 178, 179
Walton–Cheverie barite, Nova Scotia 65
WANLESS, H.R. 300, 302
WATANABE, T. 28
Weardale granite, England 55
WEBER, F. 225
WEBER, K.J. 286, 287
Weipa bauxite 253
WESTOLL, T.S. 301, 303
Westphalian ironstones, England 72–73
WHITE, A.J.R. 130
WILLBOURN, E.S. 139–140, 260
WILLEMSE, J. 105, 109, 112
Williston basin 89
Witwatersrand gold field 197, *269–272*
wolframite, China *145–146*
, Malaysia 136
, Nigeria 125
Wollaston Foldbelt deposits *243–247*
wood tin 150, 190
WORSLEY, N. 91
WYLLIE, P.J. 120
Wyoming phosphorite 87

xenoliths 130
xenothermal ore deposit 143
XU, K. 144, 147

YAN, M.Z. 145–147
Yarrow–Spionkop copper 227
YEAP, E.B. 142, 143
Yeelirrie, Australia *254, 255*
Yilgarn Block, Australia 204
Yoganup placers 274
Yoredale series 301, 303
Yugoslavia mercury *99–100*
Yunnan, China 148

Zaire copper 234, 235
Zambales, Philippines *20–21*
Zambia copper *234–238*
Zechstein salt *91–92*
Zechstein Sea 92, 93
ZIEGLER, W.H. 91, 92, 290, 291
Zimbabwe, gold *205–206*
, Great Dyke *113–114*
, greenstone 206
, pegmatites *127–129*
, tungsten 154
ZIMMERMAN, R.A. 63
zircon placers *272–275*
zoning, vein deposits 146–147